AF731589

A TEXTBOOK OF ENGINEERING MATHEMATICS

A TEXTBOOK OF

ENGINEERING MATHEMATICS

For

B.TECH. I YEAR, *SEMESTER I*

According to the New revised Syllabus BAS-103 of
Dr. A.P.J. ABDUL KALAM TECHNICAL UNIVERSITY, LUCKNOW

By

N.P. BALI
Former Principal
S.B. College, Gurugram
Haryana

Dr. MANISH GOYAL
M.Sc. (Mathematics), Ph.D., CSIR-NET
Professor and Associate Head
Department of Mathematics
Institute of Applied Sciences & Humanities
G.L.A. University, Mathura
U.P.

UNIVERSITY SCIENCE PRESS
(An Imprint of Laxmi Publications Pvt. Ltd.)
An ISO 9001:2015 Company
BENGALURU • CHENNAI • GUWAHATI • HYDERABAD
KOCHI • KOLKATA • LUCKNOW • RANCHI
NEW DELHI

A TEXTBOOK OF ENGINEERING MATHEMATICS Sem–I

Printed and bound in India

Typeset at : Excellent Graphics, Delhi

First Edition : 2002, *Second Edition* : 2003, *Third Edition* : 2006, *Fourth Edition* : 2008

Fifth Edition : 2009, *Sixth Edition* : 2012, *Seventh Edition* : 2013, *Eight Edition* : 2014, *Ninth Edition* : 2016 *Tenth Edition* : 2017, *Eleventh Edition* : 2019, 2021, *Twelfth Edition* : 2022

Thirteenth Revised Edition : 2023, 2024, 2025, *Edition* : 2026

ISBN : 978-93-83828-63-0

Branches

©	Bengaluru	080-26 75 69 30
©	Chennai	044-24 34 47 26
©	Guwahati	0361-254 36 69
©	Hyderabad	040-27 55 53 83
©	Kochi	0484-405 13 03
©	Kolkata	033-40 04 77 79
©	Lucknow	0522-430 36 13
©	Ranchi	0651-224 24 64

Published in India by

UNIVERSITY SCIENCE PRESS

(An Imprint of Laxmi Publications Pvt. Ltd.)

An ISO 9001:2015 Company

113, GOLDEN HOUSE, GURUDWARA ROAD, DARYAGANJ,

NEW DELHI - 110002, INDIA

Corporate Office : B-33, Sector-5, Noida,

Gautambuddha Nagar (U.P.)-201 301, INDIA

Telephone : +91-120-44 46 900

www.laxmipublications.com

info@laxmipublications.com

C—00000/025/08

Printed at : Ajit Printing Press, Delhi

To
My Sweet Little Son
Yash Goyal

— Dr. Manish Goyal

CONTENTS

PREFACE TO THE THIRTEENTH EDITION

We are glad to present the **Thirteenth** edition of **"A Textbook of Engineering Mathematics"** for the engineering students of B.Tech. I year I semester of Dr. A.P.J. Abdul Kalam Technical University, Lucknow.

The speciality of this textbook is its lucid presentation and addition or submission of all the questions asked in the examination papers till date.

The University has implemented new revised syllabus of Engineering Mathematics-I (BAS 103) for B. Tech. I semester students. In view of this, the textbook has been revised strictly as per it. Separate assignments containing **short answer type questions** for Section-A have been given at the end of each unit to familiarize the students with the recent trend of questions.

We have no words to convey our very special gratitude to **Dr. Hari Kishan**, a renowned author of Mathematics and father of Dr. Manish Goyal, one of the authors, for providing necessary instructions and help round the clock during the entire revision.

The authors are also thankful to all those teachers who gave necessary suggestions and helped us in making this book **'THE FIRST CHOICE'** of readers all over India.

We are also obliged to all those authors, whom we consulted while giving final shape to the matter. We are indebted to **GOD** for shower of blessings.

The suggestions for the welfare of dear students regarding the book are always welcome.

—AUTHORS

B.Tech. First Year, Semester-I

(All Branches except Agriculture Engineering and Biotechnology)

3-WEEKS STUDENT INDUCTION PROGRAMME in the beginning of the session														
					Evaluation Scheme									
SN	Subject Code	Subject Name	Type	Category	Period			Sessional Component		Sessional (SW) (TS/PS)	End Semester Examination (ESE)	Total	Credit
					L	T	P	CT	TA	CT+TA	TE/PE	SW+ESE	Cr
1.	BAS101/ BAS102	Engineering Physics / Engineering Chemistry	T	BS	3	1	0	20	10	30	70	100	4
2.	BAS103	Engineering Mathematics-I	T	BS	3	1	0	20	10	30	70	100	4
3.	BEE101/ BEC101	Fundamentals of Electrical Engineering / Fundamentals of Electronics Engineering	T	ES	2	1	0	20	10	30	70	100	3
4.	BCS101/ BME101	Programming for Problem Solving / Fundamentals of Mechanical Engineering	T	ES	2	1	0	20	10	30	70	100	3
5.	BAS104/ BAS105	Environment and Ecology / Soft Skill	T	BS / HS	3	0	0	20	10	30	70	100	3
6.	BAS151/ BAS152	Engineering Physics Lab / Engineering Chemistry Lab	P	BS	0	0	3		50	50	50	100	1
7.	BEE151/ BEC151	Basic Electrical Engineering Lab / Basic Electronics Engineering Lab	P	ES	0	0	3		50	50	50	100	1
8.	BCS151/ BAS155	Programming for Problem Solving Lab / English Language Lab	P	ES / HS	0	0	3		50	50	50	100	1
9.	BCE151/ BWS151	Engineering Graphics & Design Lab / Workshop Practice Lab	P	ES	0	1	3		50	50	50	100	2
					13	5	12			350	550	900	22

Abbreviation Used:

BS: Basic Science Course
ES: Engineering Science Course
HS: Humanities and Social Science Course
VA: Value Added Course

SYLLABUS

BAS103 : ENGINEERING MATHEMATICS-I

Course Objectives:

The objective of this course is to familiarize the graduate engineers with techniques in matrix, calculus, multivariate analysis and vector calculus. It aims to equip the students with standard concepts and tools from intermediate to advanced level that will enable them to tackle more advanced level of mathematics and applications that they would find useful in their disciplines. The students will learn:

- The essential tools of matrices, Eigen values and its application in a Comprehensive manner.
- To apply the knowledge of differential calculus in the field of engineering.
- To deal with functions of several variables that is essential in optimizing the results of real life problems.
- To apply integral calculus in various field of engineering and have a basic understanding of Beta and Gamma functions and application of Dirichlet's integral.
- To deal with vector calculus that is required in different branches of Engineering to graduate engineer.

Content	Contact Hours
Unit-1: Matrices	8
Elementary transformations, Inverse of a matrix, Rank of matrix, Solution of system of linear equations, Characteristic equation, Cayley-Hamilton Theorem and its application, Linear Dependence and Independence of vectors, Eigen values and Eigen vectors, Complex Matrices, Hermitian, Skew-Hermitian and Unitary Matrices, Applications to Engineering problems.	
Unit-2: Differential Calculus-I	8
Successive Differentiation (nth order derivatives), Leibnitz theorem, Curve tracing, Partial derivatives, Euler's Theorem for homogeneous functions, Total derivative.	
Unit-3: Differential Calculus-II	8
Expansion of functions by Taylor's and Maclaurin's theorems for functions of one and two variables, Maxima and Minima of functions of several variables, Lagrange's method of multipliers, Jacobians, Approximation of errors.	
Unit-4: Multiple Integration	8
Double integral, Triple integral, Change of order of integration, Change of variables, Beta and Gama function and their properties, Dirichlet's integral and its applications to area and volume, Liouville's extensions of Dirichlet's integral.	

Unit-5: Vector Calculus	8
Vector Differentiation: Gradient, Curl and Divergence and their Physical interpretation, Directional derivatives. **Vector Integration:** Line integral, Surface integral, Volume integral, Gauss's Divergence theorem, Green's theorem and Stoke's theorem (without proof) and their applications.	

Course Outcomes:

	Course Outcome (CO)	Bloom's Level
At the end of this course, the students will be able to:		
CO 1	Understand the concept of complex matrices, Eigen values, Eigen vectors and apply the concept of rank to evaluate linear simultaneous equations	K2 & K5
CO 2	Remember the concept of differentiation to find successive differentiation, Leibnitz Theorem, and create curve tracing, and find partial and total derivatives	K1, K6 & K5
CO 3	Applying the concept of partial differentiation to evaluate extrema, series expansion, error approximation of functions and Jacobians	K3 & K5
CO 4	Remember the concept of Beta and Gamma function; analyze area and volume and Dirichlet's theorem in multiple integral	K1 & K4
CO 5	Apply the concept of Vector Calculus to analyze and evaluate directional derivative, line, surface and volume integrals.	K3, K4 & K5

Kl – Remember, K2 – Understand, K3 – Apply, K4 – Analyze, K5 – Evaluate, K6 – Create

STANDARD RESULTS

1. $\frac{d}{dx}(x^n) = nx^{n-1}$

2. $\frac{d}{dx}(a^x) = a^x \log_e a$

3. $\frac{d}{dx}(e^x) = e^x$

4. $\frac{d}{dx}(\log_e x) = \frac{1}{x}$

5. $\frac{d}{dx}(\log_{10} x) = \frac{1}{x} \log_{10} e$

6. $\frac{d}{dx}(\sin x) = \cos x$

7. $\frac{d}{dx}(\cos x) = -\sin x$

8. $\frac{d}{dx}(\tan x) = \sec^2 x$

9. $\frac{d}{dx}(\text{cosec } x) = -\text{cosec } x \cot x$

10. $\frac{d}{dx}(\cot x) = -\text{cosec}^2 x$

11. $\frac{d}{dx}(\sec x) = \sec x \tan x$

12. $\frac{d}{dx}(\sin^{-1} x) = \frac{1}{\sqrt{1-x^2}}$

13. $\frac{d}{dx}(\cos^{-1} x) = \frac{-1}{\sqrt{1-x^2}}$

14. $\frac{d}{dx}(\tan^{-1} x) = \frac{1}{1+x^2}$

15. $\frac{d}{dx}(\sec^{-1} x) = \frac{1}{x\sqrt{x^2-1}}$

16. $\frac{d}{dx}(\cot^{-1} x) = \frac{-1}{1+x^2}$

17. $\frac{d}{dx}(\text{cosec}^{-1} x) = -\frac{1}{x\sqrt{x^2-1}}$

18. $\sinh x = \frac{e^x - e^{-x}}{2}$

19. $\cosh x = \frac{e^x + e^{-x}}{2}$

20. $\tanh x = \frac{e^x - e^{-x}}{e^x + e^{-x}}$

21. $\cosh^2 x - \sinh^2 x = 1$, $\text{sech}^2 x + \tanh^2 x = 1$, $\coth^2 x = 1 + \text{cosech}^2 x$

22. $\cosh^2 x + \sinh^2 x = \cosh 2x$

23. $\sinh^{-1} x = \log(x + \sqrt{x^2+1})$, $\cosh^{-1} x = \log(x + \sqrt{x^2-1})$

24. $\frac{d}{dx}(\sinh x) = \cosh x$

25. $\frac{d}{dx}(\cosh x) = \sinh x$

26. $\frac{d}{dx}(\tanh x) = \text{sech}^2 x$

27. $\frac{d}{dx}(\coth x) = -\text{cosech}^2 x$

28. $\frac{d}{dx}(\text{sech } x) = -\text{sech } x \tanh x$

29. $\frac{d}{dx}(\text{cosech } x) = -\text{cosech } x \coth x$

30. Product rule: $\frac{d}{dx}(uv) = u\frac{dv}{dx} + v\frac{du}{dx}$

31. Quotient rule: $\frac{d}{dx}\left(\frac{u}{v}\right) = \frac{v\frac{du}{dx} - u\frac{dv}{dx}}{v^2}$

32. $\frac{dy}{dx} = \frac{dy}{dt} \cdot \frac{dt}{dx}$ if $y = f_1(t)$ and $x = f_2(t)$

33. $\sin^{-1} x + \cos^{-1} x = \frac{\pi}{2}$, $\tan^{-1} x + \cot^{-1} x = \frac{\pi}{2}$, $\sec^{-1} x + \text{cosec}^{-1} x = \frac{\pi}{2}$

34. $\tan^{-1}\left(\frac{a-b}{1+ab}\right) = \tan^{-1} a - \tan^{-1} b$, $\tan^{-1}\left(\frac{a+b}{1-ab}\right) = \tan^{-1} a + \tan^{-1} b$

35. $\tan^{-1}\left(\frac{2x}{1-x^2}\right) = \sin^{-1}\left(\frac{2x}{1+x^2}\right) = 2\tan^{-1} x$

36. $\sin 3x = 3\sin x - 4\sin^3 x$, $\cos 3x = 4\cos^3 x - 3\cos x$, $\tan 3x = \frac{3\tan x - \tan^3 x}{1 - 3\tan^2 x}$

$\sin 2x = 2\sin x \cos x$, $\tan 2x = \frac{2\tan x}{1-\tan^2 x}$,

$\cos 2x = 2\cos^2 x - 1 = 1 - 2\sin^2 x = \cos^2 x - \sin^2 x = \frac{1-\tan^2 x}{1+\tan^2 x}$

37. $\sin x = x - \frac{x^3}{3!} + \frac{x^5}{5!} - \frac{x^7}{7!} + \ldots$, $\qquad \cos x = 1 - \frac{x^2}{2!} + \frac{x^4}{4!} - \frac{x^6}{6!} + \ldots$

$e^x = 1 + x + \frac{x^2}{2!} + \frac{x^3}{3!} + \ldots$

$(1-x)^{-1} = 1 + x + x^2 + x^3 + \ldots; \ |x| < 1 \qquad (1-x)^{-2} = 1 + 2x + 3x^2 + 4x^3 + \ldots$

$(1+x)^{-1} = 1 - x + x^2 - x^3 + \ldots \qquad (1+x)^{-2} = 1 - 2x + 3x^2 - 4x^3 + \ldots$

38. $\sin C + \sin D = 2\sin\frac{C+D}{2}\cos\frac{C-D}{2}$, $\sin C - \sin D = 2\cos\frac{C+D}{2}\sin\frac{C-D}{2}$

$\cos C + \cos D = 2\cos\frac{C+D}{2}\cos\frac{C-D}{2}$, $\cos C - \cos D = 2\sin\frac{C+D}{2}\sin\frac{D-C}{2}$

39. $2\cos A\cos B = \cos(A+B) + \cos(A-B)$, $2\sin A\sin B = \cos(A-B) - \cos(A+B)$

$2\sin A\cos B = \sin(A+B) + \sin(A-B)$, $2\cos A\sin B = \sin(A+B) - \sin(A-B)$

40. $\sin(A+B) = \sin A\cos B + \cos A\sin B$, $\sin(A-B) = \sin A\cos B - \cos A\sin B$

$\cos(A+B) = \cos A\cos B - \sin A\sin B$, $\cos(A-B) = \cos A\cos B + \sin A\sin B$

41. $\frac{d}{dx}(\sinh^{-1} x) = \frac{1}{\sqrt{1+x^2}}$, $\frac{d}{dx}(\cosh^{-1} x) = \frac{1}{\sqrt{x^2-1}}$

$\frac{d}{dx}(\tanh^{-1} x) = \frac{1}{1-x^2}$, where $|x| < 1$, $\qquad \frac{d}{dx}(\coth^{-1} x) = \frac{1}{x^2-1}$, where $|x| > 1$

$\frac{d}{dx}(\text{sech}^{-1} x) = -\frac{1}{x\sqrt{1-x^2}}$, $\frac{d}{dx}(\text{cosech}^{-1} x) = -\frac{1}{x\sqrt{x^2+1}}$

42. $(\cos\theta + i\sin\theta)^n = \cos n\theta + i\sin n\theta$, $(\cos\theta + i\sin\theta)^{-n} = \cos n\theta - i\sin n\theta$

43. $\sin^2\theta + \cos^2\theta = 1$, $\sec^2\theta - \tan^2\theta = 1$, $1 + \cot^2\theta = \text{cosec}^2\theta$

44.

θ	0°	30°	45°	60°	90°	180°	270°	360°
sin θ	0	1/2	$1/\sqrt{2}$	$\sqrt{3}/2$	1	0	– 1	0
cos θ	1	$\sqrt{3}/2$	$1/\sqrt{2}$	1/2	0	– 1	0	1
tan θ	0	$1/\sqrt{3}$	1	$\sqrt{3}$	∞	0	∞	0

45.

θ	90° – θ	90° + θ	$\pi - \theta$	$\pi + \theta$
sin θ	cos θ	cos θ	sin θ	– sin θ
cos θ	sin θ	– sin θ	– cos θ	– cos θ
tan θ	cot θ	– cot θ	– tan θ	tan θ

46. sine formula: $\dfrac{a}{\sin A} = \dfrac{b}{\sin B} = \dfrac{c}{\sin C}$; cosine formula: $\cos A = \dfrac{b^2 + c^2 - a^2}{2bc}$

47. Area of triangle $\Delta = \sqrt{s(s-a)(s-b)(s-c)}$, where $s = \dfrac{a+b+c}{2}$

48. ${}^nC_r = \dfrac{n!}{r!\,n-r!}$

49. $\int x^n\,dx = \dfrac{x^{n+1}}{n+1} + c;\ n \neq -1$

$\int \dfrac{1}{x}\,dx = \log_e x + c;\ \int e^x\,dx = e^x + c;\ \int a^x\,dx = \dfrac{a^x}{\log_e a} + c$

$\int \sin x\,dx = -\cos x + c;\ \int \cos x\,dx = \sin x + c$

$\int \tan x\,dx = \log \sec x + c;\ \int \cot x\,dx = \log \sin x + c$

$\int \sec x\,dx = \log(\sec x + \tan x) + c = \log \tan\left(\dfrac{\pi}{4} + \dfrac{x}{2}\right) + c$

$\int \text{cosec}\, x\,dx = \log(\text{cosec}\, x - \cot x) + c = \log \tan \dfrac{x}{2} + c$

$\int \sec x \tan x\,dx = \sec x + c;\ \int \text{cosec}\, x \cot x\,dx = -\text{cosec}\, x + c$

50. $\int \dfrac{dx}{\sqrt{a^2 - x^2}} = \sin^{-1}\left(\dfrac{x}{a}\right) + c;\ \int \dfrac{-dx}{\sqrt{a^2 - x^2}} = \cos^{-1}\left(\dfrac{x}{a}\right) + c$

$\int \dfrac{dx}{a^2 + x^2} = \dfrac{1}{a}\tan^{-1}\left(\dfrac{x}{a}\right) + c;\ \int \dfrac{-dx}{a^2 + x^2} = \dfrac{1}{a}\cot^{-1}\left(\dfrac{x}{a}\right) + c$

$\int \dfrac{dx}{a^2 - x^2} = \dfrac{1}{2a}\log\left(\dfrac{a+x}{a-x}\right) + c;\ \int \dfrac{dx}{x^2 - a^2} = \dfrac{1}{2a}\log\left(\dfrac{x-a}{x+a}\right) + c$

$\int \dfrac{dx}{x\sqrt{x^2 - a^2}} = \dfrac{1}{a}\sec^{-1}\left(\dfrac{x}{a}\right) + c;\ \int \dfrac{-dx}{x\sqrt{x^2 - a^2}} = \dfrac{1}{a}\text{cosec}^{-1}\left(\dfrac{x}{a}\right) + c$

51. $\int \text{sech}^2 x\, dx = \tanh x + c, \int \text{cosech}^2 x\, dx = -\coth x + c$

$\int \sinh x\, dx = \cosh x + c, \int \cosh x\, dx = \sinh x + c$

$\int \text{sech}\, x \tanh x\, dx = -\text{sech}\, x + c, \int \text{cosech}\, x \coth x\, dx = -\text{cosech}\, x + c$

52. $\int \sqrt{a^2 - x^2}\, dx = \frac{1}{2} x \sqrt{a^2 - x^2} + \frac{1}{2} a^2 \sin^{-1} \frac{x}{a} + c$

$\int \sqrt{a^2 + x^2}\, dx = \frac{1}{2} x \sqrt{a^2 + x^2} + \frac{1}{2} a^2 \log (x + \sqrt{a^2 + x^2}) + c$

$\int \sqrt{x^2 - a^2}\, dx = \frac{1}{2} x \sqrt{x^2 - a^2} - \frac{1}{2} a^2 \log (x + \sqrt{x^2 - a^2}) + c$

$\int \frac{dx}{\sqrt{a^2 + x^2}} = \sinh^{-1} \left(\frac{x}{a}\right) + c; \int \frac{dx}{\sqrt{x^2 - a^2}} = \cosh^{-1} \left(\frac{x}{a}\right) + c$

53. $\int_a^b f(x)\, dx = \int_a^b f(y)\, dy; \int_a^b f(x)\, dx = -\int_b^a f(x)\, dx; \int_0^a f(x)\, dx = \int_0^a f(a - x)\, dx$

$$\int_{-a}^{a} f(x)\, dx = \begin{cases} 2\int_0^a f(x)\, dx &, \text{ if } f(x) \text{ is even function} \\ 0 &, \text{ if } f(x) \text{ is odd function} \end{cases}$$

$$\int_0^{2a} f(x)\, dx = \begin{cases} 2\int_0^a f(x)\, dx &, \text{ if } f(2a - x) = f(x) \\ 0 &, \text{ if } f(2a - x) = -f(x) \end{cases}$$

54. Leibnitz rule for differentiation under the integral sign

$$\frac{d}{dx} \int_{\phi(\alpha)}^{\psi(\alpha)} f(x, \alpha)\, dx = \int_{\phi(\alpha)}^{\psi(\alpha)} \frac{\partial}{\partial \alpha}\{f(x, \alpha)\}\, dx + f\{\psi(\alpha), \alpha\} \frac{d\psi(\alpha)}{d\alpha} - f\{\phi(\alpha), \alpha\} \frac{d\phi(\alpha)}{d\alpha}$$

55. If $\vec{r} = x\hat{i} + y\hat{j} + z\hat{k}$ then $|\vec{r}| = \sqrt{x^2 + y^2 + z^2}$ and $\hat{r} = \frac{\vec{r}}{|\vec{r}|} = \frac{x\hat{i} + y\hat{j} + z\hat{k}}{\sqrt{x^2 + y^2 + z^2}}$

56. $\overrightarrow{AB}$ = position vector of B-position vector of A = $\overrightarrow{OB} - \overrightarrow{OA}$

57. $\vec{a} \cdot \vec{b} = |\vec{a}|\,|\vec{b}| \cos \theta$; work done = $\int_c \vec{F} \cdot \vec{dr}$

58. $\vec{a} \times \vec{b} = |\vec{a}|\,|\vec{b}| \sin \theta\, \hat{n}$

59. Area of parallelogram = $\vec{a} \times \vec{b}$, Moment of force = $\vec{r} \times \vec{F}$

60. $\vec{a}\cdot(\vec{b}\times\vec{c}) = [\vec{a}\ \vec{b}\ \vec{c}] = \begin{vmatrix} a_1 & a_2 & a_3 \\ b_1 & b_2 & b_3 \\ c_1 & c_2 & c_3 \end{vmatrix} = (\vec{a}\times\vec{b})\cdot\vec{c}$

where $\vec{a} = \Sigma a_1\hat{i}$, $\vec{b} = \Sigma b_1\hat{i}$ and $\vec{c} = \Sigma c_1\hat{i}$

If $\vec{a}\cdot(\vec{b}\times\vec{c}) = 0$, then $\vec{a}, \vec{b}, \vec{c}$ are coplanar.

61. $\vec{a}\times(\vec{b}\times\vec{c}) = (\vec{a}\cdot\vec{c})\vec{b} - (\vec{a}\cdot\vec{b})\vec{c}$

62. $(\vec{a}\times\vec{b})\cdot(\vec{c}\times\vec{d}) = \begin{vmatrix} \vec{a}\cdot\vec{c} & \vec{a}\cdot\vec{d} \\ \vec{b}\cdot\vec{c} & \vec{b}\cdot\vec{d} \end{vmatrix}$

63. $(\vec{a}\times\vec{b})\times(\vec{c}\times\vec{d}) = [\vec{a}\ \vec{b}\ \vec{d}]\,\vec{c} - [\vec{a}\ \vec{b}\ \vec{c}]\,\vec{d}$

64. A (Adj. A) = | A | I

65. $AA^{-1} = I = A^{-1}A$

66. AI = A = IA

67. (ABC)′ = C′B′A′

68. (AB)C = A(BC); A(B + C) = AB + AC

69. A + B = B + A; A + (B + C) = (A + B) + C

70. $(AB)^{-1} = B^{-1}A^{-1}$

71. **Walli's formula**

$$\int_0^{\pi/2} \sin^n\theta\, d\theta = \int_0^{\pi/2} \cos^n\theta\, d\theta = \begin{cases} \dfrac{n-1}{n}\cdot\dfrac{n-3}{n-2}\cdot\dfrac{n-5}{n-4}\ldots\ldots\dfrac{3}{4}\cdot\dfrac{1}{2}\cdot\dfrac{\pi}{2} & \text{if } n \text{ is even} \\ \dfrac{n-1}{n}\cdot\dfrac{n-3}{n-2}\cdot\dfrac{n-5}{n-4}\ldots\ldots\dfrac{4}{5}\cdot\dfrac{2}{3} & \text{if } n \text{ is odd} \end{cases}$$

72. $\int e^{ax}\sin bx\, dx = \dfrac{e^{ax}}{a^2+b^2}(a\sin bx - b\cos bx) + c$

$\int e^{ax}\cos bx\, dx = \dfrac{e^{ax}}{a^2+b^2}(a\cos bx + b\sin bx) + c$

73. $\Gamma(1/2) = \sqrt{\pi}$, $\Gamma(-1/2) = -2\sqrt{\pi}$

74. $\log(1+x) = x - \dfrac{x^2}{2} + \dfrac{x^3}{3} - \dfrac{x^4}{4} + \dfrac{x^5}{5} - \dfrac{x^6}{6} + \ldots\ldots$

$\log(1-x) = -x - \dfrac{x^2}{2} - \dfrac{x^3}{3} - \dfrac{x^4}{4} - \dfrac{x^5}{5} - \dfrac{x^6}{6} - \ldots\ldots$

75. $\sin n\pi = 0$; $\cos n\pi = (-1)^n$, $\sin\left(n+\dfrac{1}{2}\right)\pi = (-1)^n$; $\cos\left(n+\dfrac{1}{2}\right)\pi = 0$, where $n \in I$.

SYMBOLS

Greek Alphabets

A	α	Alpha	I	ι	Iota	P	ρ	Rho
B	β	Beta	K	κ	Kappa	Σ	σ	Sigma
Γ	γ	Gamma	Λ	λ	Lambda	T	τ	Tau
D	δ	Delta	M	μ	Mu	Y	υ	Upsilon
E	ε	Epsilon	N	ν	Nu	Φ	φ	Phi
Z	ζ	Zeta	Ξ	ξ	Xi	X	χ	Chi
H	η	Eta	O	ο	Omicron	Ψ	ψ	Psi
Θ	θ	Theta	Π	π	Pi	Ω	ω	Omega
	∃	there exists		∀	for all			

Metric Weights and Measures

LENGTH

10 millimetres	= 1 centimetre
10 centimetres	= 1 decimetre
10 decimetres	= 1 metre
10 metres	= 1 decametre
10 decametres	= 1 hectometre
10 hectometres	= 1 kilometre

CAPACITY

10 millilitres	= 1 centilitre
10 centilitres	= 1 decilitre
10 decilitres	= 1 litre
10 litres	= 1 decalitre
10 decalitres	= 1 hectolitre
10 hectolitres	= 1 kilolitre

VOLUME

1000 cubic centimetres	= 1 centigram
1000 cubic decimetres	= 1 cubic metre

AREA

100 square metres	= 1 are
100 ares	= 1 hectare
100 hectares	= 1 square kilometre

WEIGHT

10 milligrams	= 1 centigram
10 centigrams	= 1 decigram
10 decigrams	= 1 gram
10 grams	= 1 decagram
10 decagrams	= 1 hectogram
10 hectograms	= 1 kilogram
100 kilograms	= 1 quintal
10 quintals	= 1 metric ton (tonne)

ABBREVIATIONS

kilometre	km	tonne	t
metre	m	quintal	q
centimetre	cm	kilogram	kg
millimetre	mm	gram	g
kilolitre	kl	are	a
litre	l	hectare	ha
millilitre	ml	centiare	ca

MODULE 1

Matrices

1.1. INTRODUCTION

A matrix is defined as a rectangular array (or arrangement in rows or columns) of scalars subject to certain rules of operations.

If mn numbers (real or complex) or functions are arranged in the form of a rectangular array A having m rows (horizontal lines) and n columns (vertical lines) then A is called an $m \times n$ matrix. Each of the mn numbers is called an element of the matrix.

There are different notations of enclosing the elements constituting a matrix *viz.* [ ], ( ). || || but square bracket is generally used. An $m \times n$ matrix is also called a matrix of order $m \times n$.

For example, $\begin{bmatrix} 2 & -1 & 5 \\ 3 & 0 & 4 \end{bmatrix}$ is a 2 × 3 matrix or matrix of order 2 × 3. It has two rows and three columns. The numbers 2, – 1, 5, 3, 0, 4 are its elements.

An $m \times n$ matrix is usually written as $\begin{bmatrix} a_{11} & a_{12} & a_{13} & \dots a_{1n} \\ a_{21} & a_{22} & a_{23} & \dots a_{2n} \\ a_{31} & a_{32} & a_{33} & \dots a_{3n} \\ \dots & \dots & \dots & \dots \\ a_{m1} & a_{m2} & a_{m3} & \dots a_{mn} \end{bmatrix}$

Here each element has two suffixes. The first suffix indicates the row and the second suffix indicates the column in which the element lies. Thus, a_{23} is the element lying in the second row and third column, a_{ij} is the element lying in the ith row and j^{th} column.

For brevity, a matrix is usually denoted by a single capital letter A, B or C etc.

Thus, an $m \times n$ matrix A may be written as

$$A = [a_{ij}]_{m \times n} \quad \text{or} \quad A = [a_{ij}], \quad \text{where } i = 1, 2, 3, ..., m \text{ ; } j = 1, 2, 3, ..., n$$

1.2. APPLICATIONS OF MATRICES

In Algebra, the matrices have their largest application in the theory of simultaneous equations and linear transformations. *e.g.*, the set of simultaneous equations,

$$a_{11}x_1 + a_{12}x_2 + a_{13}x_3 = b_1$$
$$a_{21}x_1 + a_{22}x_2 + a_{23}x_3 = b_2$$
$$a_{31}x_1 + a_{32}x_2 + a_{33}x_3 = b_3$$

may be symbolically represented by the equation,

$$AX = B$$

where $A = \begin{bmatrix} a_{11} & a_{12} & a_{13} \\ a_{21} & a_{22} & a_{23} \\ a_{31} & a_{32} & a_{33} \end{bmatrix}, X = \begin{bmatrix} x_1 \\ x_2 \\ x_3 \end{bmatrix}, B = \begin{bmatrix} b_1 \\ b_2 \\ b_3 \end{bmatrix}$

The theory of matrices has been found of great utility in many branches of higher mathematics such as algebraic and differential equations, astronomy, mechanics, theory of electrical circuits, quantum mechanics, nuclear physics and aerodynamics.

Note. There is a similarity in the way of writing a determinant and a matrix but the two are entirely different. A determinant has a numerical value but a matrix has no numerical value. Matrix is only a convenient way of storing information.

1.3. TYPES OF MATRICES

(1) **Real Matrix.** A matrix is said to be real if all its elements are real numbers.

e.g., $\begin{bmatrix} \sqrt{5} & -3 & 1 \\ 0 & -\sqrt{2} & 7 \end{bmatrix}$ is a real matrix.

(2) **Square Matrix.** A matrix in which the number of rows is equal to the number of columns is called a **square matrix,** otherwise, it is said to be a **rectangular matrix.**

Thus, a matrix $A = [a_{ij}]_{m \times n}$ is a square matrix if $m = n$ and a rectangular matrix if $m \neq n$.

A square matrix having n rows and n columns is called "a square matrix of order n" or "an n-rowed square matrix",

e.g., $\begin{bmatrix} a_{11} & a_{12} & a_{13} \\ a_{21} & a_{22} & a_{23} \\ a_{31} & a_{32} & a_{33} \end{bmatrix}$ is a 3-rowed square matrix.

The elements a_{11}, a_{22}, a_{33} of a square matrix are called its **diagonal elements** and the diagonal along which these elements lie is called the **principal** or **leading diagonal.**

In a square matrix $A = [a_{ij}]$,

(*i*) for elements along the principal diagonal, $i = j$

(*ii*) for elements above the principal diagonal, $i < j$

(*iii*) for elements below the principal diagonal, $i > j$

(*iv*) for non-diagonal elements, $i \neq j$

The sum of the diagonal elements of a square matrix is called its **trace** or **spur.** Thus, trace of the n rowed square matrix $A = [a_{ij}]$ is $a_{11} + a_{22} + a_{33} + \ldots + a_{nn} = \sum_{i=1}^{n} a_{ij}$

(3) **Row Matrix.** A matrix having only one row and any number of columns *i.e.,* a matrix of order $1 \times n$ is called a **row matrix.** *e.g.* $[2 \;\; 5 \;\; -3 \;\; 0]$ is a row matrix.

(4) **Column Matrix.** A matrix having only one column and any number of rows *i.e.,* a matrix of order $m \times 1$ is called a **column matrix.** *e.g.,* $\begin{bmatrix} \sqrt{2} \\ 0 \\ -1 \end{bmatrix}$ is a column matrix.

(5) **Null Matrix.** A matrix in which each element is zero is called a **null matrix** or **void matrix** or a **zero matrix.** A null matrix of order $m \times n$ is denoted by $O_{m \times n}$.

$$\begin{bmatrix} 0 & 0 \\ 0 & 0 \\ 0 & 0 \end{bmatrix} = O_{3 \times 2}, \begin{bmatrix} 0 & 0 & 0 & 0 \\ 0 & 0 & 0 & 0 \end{bmatrix} = O_{2 \times 4}$$

(6) **Sub-matrix.** A matrix obtained from a given matrix A by deleting some of its rows or columns or both is called a **sub-matrix of A.**

Thus, $B = \begin{bmatrix} 3 & 0 \\ 1 & 4 \end{bmatrix}$ is a sub-matrix of $A = \begin{bmatrix} 0 & -1 & 2 & 5 \\ 3 & 5 & 0 & 7 \\ 1 & 6 & 4 & -2 \end{bmatrix}$

obtained by deleting the first row, second and fourth columns of A.

(7) **Diagonal Matrix. A square matrix** in which all non-diagonal elements are zero is called a **diagonal matrix.**

Thus, $A = [a_{ij}]_{n \times n}$ is a diagonal matrix if $a_{ij} = 0$ for $i \neq j$.

For example, $A = \begin{bmatrix} 2 & 0 & 0 \\ 0 & -1 & 0 \\ 0 & 0 & 0 \end{bmatrix}$ is a diagonal matrix.

An n-rowed diagonal matrix is briefly written as diag. $[d_1, d_2, ..., d_n]$, where $d_1, d_2, ..., d_n$ are the diagonal elements. Thus, the above diagonal matrix A can be written as diag. $[2, -1, 0]$.

(8) **Scalar Matrix.** A diagonal matrix in which all the diagonal elements are equal to a scalar, say k, is called a **scalar matrix.**

Thus a scalar matrix is a square matrix in which all non-diagonal elements are zero and all diagonal elements are equal to a scalar, say k.

i.e., $A = [a_{ij}]_{n \times n}$ is a scalar matrix if $a_{ij} = \begin{cases} 0 & \text{when } i \neq j \\ k & \text{when } i = j \end{cases}$

For example, $\begin{bmatrix} 2 & 0 & 0 \\ 0 & 2 & 0 \\ 0 & 0 & 2 \end{bmatrix} \begin{bmatrix} -5 & 0 & 0 & 0 \\ 0 & -5 & 0 & 0 \\ 0 & 0 & -5 & 0 \\ 0 & 0 & 0 & -5 \end{bmatrix}$ are scalar matrices.

(9) **Unit Matrix or Identity Matrix.** A scalar matrix in which each diagonal element is unity (*i.e.*, 1) is called a **unit matrix or identity matrix.**

Thus, a unit matrix is a square matrix in which all non-diagonal elements are zero and all diagonal elements are equal to 1.

i.e., $A = [a_{ij}]_{n \times n}$ is a unit matrix if $a_{ij} = \begin{cases} 0 & \text{when } i \neq j \\ 1 & \text{when } i = j \end{cases}$

A unit matrix of order n is denoted by I_n. If the order is evident, it may be simply denoted by I.

Thus, $\begin{bmatrix} 1 & 0 \\ 0 & 1 \end{bmatrix} = I_2, \begin{bmatrix} 1 & 0 & 0 \\ 0 & 1 & 0 \\ 0 & 0 & 1 \end{bmatrix} = I_3$

(10) **Upper Triangular Matrix.** A square matrix in which all the elements **below** the principal diagonal are zero is called an **upper triangular matrix.**

Thus, $A = [a_{ij}]_{n \times n}$ is an upper triangular matrix if $a_{ij} = 0$ for $i > j$.

For example, $\begin{bmatrix} 2 & 3 & 4 \\ 0 & -1 & 5 \\ 0 & 0 & 3 \end{bmatrix}$ is an upper triangular matrix.

(11) **Lower Triangular Matrix.** A square matrix in which all the elements **above** the principal diagonal are zero is called a **lower triangular matrix.**

Thus $A = [a_{ij}]_{n \times n}$ is a lower triangular matrix if $a_{ij} = 0$ for $i < j$.

For example, $\begin{bmatrix} -1 & 0 & 0 \\ 5 & 6 & 0 \\ 3 & 2 & 0 \end{bmatrix}$ is a lower triangular matrix.

(12) **Triangular Matrix.** A square matrix in which all the elements either below or above the principal diagonal are zero is called a triangular matrix. Thus, a triangular matrix is either upper triangular or lower triangular.

(13) **Equal Matrices.** Two matrices A and B are said to be equal (written as A = B) if and only if they have the same order and their corresponding elements are equal.

Thus, if $A = [a_{ij}]_{m \times n}$ and $B = [b_{ij}]_{p \times q}$, then A = B if and only if

(*i*) $m = p$ and $n = q$ (*ii*) $a_{ij} = b_{ij}$ for all *i* and *j*.

(14) **Single Element Matrix.** A matrix having only one element is called a single element matrix. Thus any matrix [3] is a single element matrix.

(15) **Singular and Non-singular Matrices.** A square matrix A is said to be singular if $|A| = 0$ and non-singular if $|A| \neq 0$.

For example, $A = \begin{bmatrix} 2 & -3 & 4 \\ 2 & -3 & 4 \\ 0 & -1 & 1 \end{bmatrix}$ is a singular matrix since $|A| = 0$.

(16) **Tridiagonal Matrix.** Tridiagonal matrix is the matrix having non-zero entries only in the leading diagonal, sub-diagonal and super diagonal. In other words, we can define it as:

A real matrix $A = [a_{ij}]$ is said to be tridiagonal if $a_{ij} = 0$ for $|i - j| > 1$

For example, $A = \begin{bmatrix} a_{11} & a_{12} & 0 & 0 \\ a_{21} & a_{22} & a_{23} & 0 \\ 0 & a_{32} & a_{33} & a_{34} \\ 0 & 0 & a_{43} & a_{44} \end{bmatrix}$ is a tridiagonal matrix.

(17) **Stochastic Matrix.** A square matrix A is called a stochastic matrix if all its elements are non-negative and the sum of the elements in each row is 1. Thus, the matrix

$A = [a_{ij}]_{n \times n}$ will be a stochastic matrix if $a_{ij} \geq 0$ for $0 \leq i \leq n$, $0 \leq j \leq n$ and $\sum_{j=1}^{n} a_{ij} = 1$ for $i = 1, 2,$ ..., n. *e.g.*, $A = \begin{bmatrix} 2/3 & 1/3 \\ 1/4 & 3/4 \end{bmatrix}$ is a stochastic matrix.

1.4. ADDITION *AND* SUBTRACTION OF MATRICES

Two matrices are said to be conformable for addition and subtraction if they have the same order.

If A and B are two matrices of the same order, then their sum A + B is a matrix each element of which is obtained by adding the corresponding elements of A and B.

In general, if $A = [a_{ij}]_{m \times n}$ and $B = [b_{ij}]_{m \times n}$, then $A + B = C = [c_{ij}]_{m \times n}$, where $c_{ij} = a_{ij} + b_{ij}$

Similarly, if A and B are two matrices of the same order, then their difference A – B is a matrix whose elements are obtained by subtracting the elements of B from the corresponding elements of A.

In general, if $A = [a_{ij}]_{m \times n}$ and $B = [b_{ij}]_{m \times n}$, then $A - B = C = [c_{ij}]_{m \times n}$, where $c_{ij} = a_{ij} - b_{ij}$.

For example, if $A = \begin{bmatrix} 2 & 5 & -1 \\ 3 & 0 & 4 \end{bmatrix}$ and $B = \begin{bmatrix} 1 & -6 & 2 \\ -2 & 5 & 7 \end{bmatrix}$

then $$A + B = \begin{bmatrix} 2+1 & 5-6 & -1+2 \\ 3-2 & 0+5 & 4+7 \end{bmatrix} = \begin{bmatrix} 3 & -1 & 1 \\ 1 & 5 & 11 \end{bmatrix}$$

and $$A - B = \begin{bmatrix} 2-1 & 5-(-6) & -1-2 \\ 3-(-2) & 0-5 & 4-7 \end{bmatrix} = \begin{bmatrix} 1 & 11 & -3 \\ 5 & -5 & -3 \end{bmatrix}.$$

1.5. MULTIPLICATION OF A MATRIX BY A SCALAR

The product of a matrix $A = [a_{ij}]$ by a scalar k is denoted by kA and is obtained by multiplying every element of A by k.

Thus, $$kA = [ka_{ij}]$$

If $$A = \begin{bmatrix} a_1 & a_2 & a_3 \\ b_1 & b_2 & b_3 \end{bmatrix}, \text{ then } kA = \begin{bmatrix} ka_1 & ka_2 & ka_3 \\ kb_1 & kb_2 & kb_3 \end{bmatrix}$$

In particular, $$-A = (-1)\,A = \begin{bmatrix} -a_1 & -a_2 & -a_3 \\ -b_1 & -b_2 & -b_3 \end{bmatrix}.$$

1.6. PROPERTIES OF MATRIX ADDITION

(*i*) **Matrix addition is commutative** *i.e.*, A + B = B + A

(*ii*) **Matrix addition is associative** *i.e.*, (A + B) + C = A + (B + C)

(*iii*) For any matrix A, there exists a null matrix O of the same order as A such that

$$A + O = O + A = A.$$

(*iv*) For any matrix A, there exists a matrix – A of the same order as A such that

$$A + (-A) = (-A) + A = O.$$

ILLUSTRATIVE EXAMPLES

Example 1. *Find x, y, z and w if* $3\begin{bmatrix} x & y \\ z & w \end{bmatrix} = \begin{bmatrix} x & 6 \\ -1 & 2w \end{bmatrix} + \begin{bmatrix} 4 & x+y \\ z+w & 3 \end{bmatrix}$.

Sol. The given equation is $\begin{bmatrix} 3x & 3y \\ 3z & 3w \end{bmatrix} = \begin{bmatrix} x+4 & 6+x+y \\ -1+z+w & 2w+3 \end{bmatrix}$

Equating the corresponding elements on the two sides

$$3x = x + 4, \quad 3y = 6 + x + y, \quad 3z = -1 + z + w, \quad 3w = 2w + 3$$

$$\Rightarrow \quad 2x = 4, \quad 2y = 6 + x, \quad 2z = -1 + w, \quad w = 3$$

$$\Rightarrow \quad x = 2, \quad y = 4, \quad z = 1, \quad w = 3.$$

Example 2. *Express* $\begin{bmatrix} 2 & 5 & -7 \\ -9 & 12 & 4 \\ 15 & -13 & 6 \end{bmatrix}$ *as the sum of a lower triangular matrix and an upper triangular matrix with zero leading diagonal.*

Sol. Let $L = \begin{bmatrix} a & 0 & 0 \\ b & c & 0 \\ d & e & f \end{bmatrix}$ be the lower triangular matrix and $U = \begin{bmatrix} 0 & p & q \\ 0 & 0 & r \\ 0 & 0 & 0 \end{bmatrix}$ be the upper triangular matrix with zero leading diagonal such that

$$\begin{bmatrix} 2 & 5 & -7 \\ -9 & 12 & 4 \\ 15 & -13 & 6 \end{bmatrix} = \begin{bmatrix} a & 0 & 0 \\ b & c & 0 \\ d & e & f \end{bmatrix} + \begin{bmatrix} 0 & p & q \\ 0 & 0 & r \\ 0 & 0 & 0 \end{bmatrix} = \begin{bmatrix} a & p & q \\ b & c & r \\ d & e & f \end{bmatrix}$$

Equating the corresponding elements on the two sides

$$2 = a, \quad 5 = p, \quad -7 = q, \quad -9 = b, \quad 12 = c, \quad 4 = r, \quad 15 = d, \quad -13 = e, \quad 6 = f$$

$$\therefore \quad L = \begin{bmatrix} 2 & 0 & 0 \\ -9 & 12 & 0 \\ 15 & -13 & 6 \end{bmatrix} \text{ and } U = \begin{bmatrix} 0 & 5 & -7 \\ 0 & 0 & 4 \\ 0 & 0 & 0 \end{bmatrix}.$$

1.7. MATRIX MULTIPLICATION

Two matrices A and B are said to be conformable for the product AB **(in this very order of A and B)** if the number of columns in A (called the pre-factor) is equal to the number of rows in B (called the post-factor).

Thus, if the orders of A and B are $m \times n$ and $p \times q$ respectively, then

(*i*) AB is defined if number of columns in A = number of rows in B, *i.e.*, if $n = p$

(*ii*) BA is defined if number of columns in B = number of rows in A, *i.e.*, if $q = m$.

Let $A = [a_{ij}]_{m \times n}$, and $B = [b_{ij}]_{n \times p}$ be two matrices conformable for the product AB, then AB is defined as the matrix $C = [c_{ij}]_{m \times p}$, where $c_{ij} = a_{i1} b_{1j} + a_{i2} b_{2j} + \ldots + a_{in} b_{nj} = \sum_{k=1}^{n} a_{ik} b_{kj}$

i.e., $(i, j)^{\text{th}}$ element of AB = sum of the products of the elements of ith row of A with the corresponding elements of jth column of B.

The rule for multiplication of two conformable matrices is called **row-by-column method.**

Consider $A = \begin{bmatrix} a_{11} & a_{12} & a_{13} \\ a_{21} & a_{22} & a_{23} \\ a_{31} & a_{32} & a_{33} \end{bmatrix}$ and $B = \begin{bmatrix} b_{11} & b_{12} \\ b_{21} & b_{22} \\ b_{31} & b_{32} \end{bmatrix}$

Orders of A and B are 3 × 3 and 3 × 2 respectively. AB is defined and is of order 3 × 2.

$$AB = C = \begin{bmatrix} c_{11} & c_{12} \\ c_{21} & c_{22} \\ c_{31} & c_{32} \end{bmatrix}$$

where c_{11} = sum of products of elements of 1st row of A and 1st column of B

$$= [a_{11}\ a_{12}\ a_{13}] \begin{bmatrix} b_{11} \\ b_{21} \\ b_{31} \end{bmatrix} = a_{11}b_{11} + a_{12}b_{21} + a_{13}b_{31}$$

c_{12} = sum of products of elements of 1st row of A and 2nd column of B

$$= [a_{11}\ a_{12}\ a_{13}] \begin{bmatrix} b_{12} \\ b_{22} \\ b_{32} \end{bmatrix} = a_{11}b_{12} + a_{12}b_{22} + a_{13}b_{32}$$

c_{21} = sum of products of elements of 2nd row of A and 1st column of B

$$= [a_{21}\ a_{22}\ a_{23}] \begin{bmatrix} b_{11} \\ b_{21} \\ b_{31} \end{bmatrix} = a_{21}b_{11} + a_{22}b_{21} + a_{23}b_{31}$$

c_{22} = sum of products of elements of 2nd row of A and 2nd column of B

$$= [a_{21}\ a_{22}\ a_{23}] \begin{bmatrix} b_{12} \\ b_{22} \\ b_{32} \end{bmatrix} = a_{21}b_{12} + a_{22}b_{22} + a_{23}b_{32}$$

c_{31} = sum of products of elements of 3rd row of A and 1st column of B

$$= [a_{31}\ a_{32}\ a_{33}] \begin{bmatrix} b_{11} \\ b_{21} \\ b_{31} \end{bmatrix} = a_{31}b_{11} + a_{32}b_{21} + a_{33}b_{31}$$

c_{32} = sum of products of elements of 3rd row of A and 2nd column of B

$$= [a_{31}\ a_{32}\ a_{33}] \begin{bmatrix} b_{12} \\ b_{22} \\ b_{32} \end{bmatrix} = a_{31}b_{12} + a_{32}b_{22} + a_{33}b_{32}$$

Thus,

$$AB = \begin{bmatrix} a_{11}b_{11} + a_{12}b_{21} + a_{13}b_{31} & a_{11}b_{12} + a_{12}b_{22} + a_{13}b_{32} \\ a_{21}b_{11} + a_{22}b_{21} + a_{23}b_{31} & a_{21}b_{12} + a_{22}b_{22} + a_{23}b_{32} \\ a_{31}b_{11} + a_{32}b_{21} + a_{33}b_{31} & a_{31}b_{12} + a_{32}b_{22} + a_{33}b_{32} \end{bmatrix}$$

Example 3. *If* $A = \begin{bmatrix} 1 & -2 & 3 \\ 2 & 3 & -1 \\ -3 & 1 & 2 \end{bmatrix}$ *and* $B = \begin{bmatrix} 1 & 0 & 2 \\ 0 & 1 & 2 \\ 1 & 2 & 0 \end{bmatrix}$, *Obtain the products AB and BA and show that AB ≠ BA.*

Sol. $AB = \begin{bmatrix} 1 & -2 & 3 \\ 2 & 3 & -1 \\ -3 & 1 & 2 \end{bmatrix} \begin{bmatrix} 1 & 0 & 2 \\ 0 & 1 & 2 \\ 1 & 2 & 0 \end{bmatrix}$

$$= \begin{bmatrix} 1(1)-2(0)+3(1) & 1(0)-2(1)+3(2) & 1(2)-2(2)+3(0) \\ 2(1)+3(0)-1(1) & 2(0)+3(1)-1(2) & 2(2)+3(2)-1(0) \\ -3(1)+1(0)+2(1) & -3(0)+1(1)+2(2) & -3(2)+1(2)+2(0) \end{bmatrix} = \begin{bmatrix} 4 & 4 & -2 \\ 1 & 1 & 10 \\ -1 & 5 & -4 \end{bmatrix}$$

$$BA = \begin{bmatrix} 1 & 0 & 2 \\ 0 & 1 & 2 \\ 1 & 2 & 0 \end{bmatrix} \begin{bmatrix} 1 & -2 & 3 \\ 2 & 3 & -1 \\ -3 & 1 & 2 \end{bmatrix}$$

$$= \begin{bmatrix} 1(1)+0(2)+2(-3) & 1(-2)+0(3)+2(1) & 1(3)+0(-1)+2(2) \\ 0(1)+1(2)+2(-3) & 0(-2)+1(3)+2(1) & 0(3)+1(-1)+2(2) \\ 1(1)+2(2)+0(-3) & 1(-2)+2(3)+0(1) & 1(3)+2(-1)+0(2) \end{bmatrix} = \begin{bmatrix} -5 & 0 & 7 \\ -4 & 5 & 3 \\ 5 & 4 & 1 \end{bmatrix}$$

Orders of AB and BA are the same but their corresponding elements are not equal.

Hence $AB \neq BA$.

1.8. PROPERTIES OF MATRIX MULTIPLICATION

(*i*) Matrix multiplication is not commutative in general *i.e.*, $AB \neq BA$.

(*ii*) Matrix multiplication is associative *i.e.*, $(AB)C = A(BC)$.

(*iii*) Matrix multiplication is distributive with respect to matrix addition *i.e.*,

$$A(B + C) = AB + AC.$$

(*iv*) If A and I are square matrices of the same order, then $AI = IA = A$.

(*v*) If A is a square matrix of order n, then $A \times A = A^2$, $A \times A \times A = A^3$, $A \times A \times A \ldots m$ times $= A^m$ are all square matrices of order n.

Also, we define $A^\circ = I$.

(*vi*) For any positive integer n, $I^n = I$.

Example 4. *Evaluate $A^2 - 3A + 9I$, if I is the unit matrix of order 3 and* $A = \begin{bmatrix} 1 & -2 & 3 \\ 2 & 3 & -1 \\ -3 & 1 & 2 \end{bmatrix}$.

Sol. $$A^2 = A \times A = \begin{bmatrix} 1 & -2 & 3 \\ 2 & 3 & -1 \\ -3 & 1 & 2 \end{bmatrix} \begin{bmatrix} 1 & -2 & 3 \\ 2 & 3 & -1 \\ -3 & 1 & 2 \end{bmatrix} = \begin{bmatrix} -12 & -5 & 11 \\ 11 & 4 & 1 \\ -7 & 11 & -6 \end{bmatrix}$$

$$\therefore \quad A^2 - 3A + 9I_3 = \begin{bmatrix} -12 & -5 & 11 \\ 11 & 4 & 1 \\ -7 & 11 & -6 \end{bmatrix} - 3\begin{bmatrix} 1 & -2 & 3 \\ 2 & 3 & -1 \\ -3 & 1 & 2 \end{bmatrix} + 9\begin{bmatrix} 1 & 0 & 0 \\ 0 & 1 & 0 \\ 0 & 0 & 1 \end{bmatrix}$$

$$= \begin{bmatrix} -12 & -5 & 11 \\ 11 & 4 & 1 \\ -7 & 11 & -6 \end{bmatrix} - \begin{bmatrix} 3 & -6 & 9 \\ 6 & 9 & -3 \\ -9 & 3 & 6 \end{bmatrix} + \begin{bmatrix} 9 & 0 & 0 \\ 0 & 9 & 0 \\ 0 & 0 & 9 \end{bmatrix} = \begin{bmatrix} -6 & 1 & 2 \\ 5 & 4 & 4 \\ 2 & 8 & -3 \end{bmatrix}.$$

1.9. MATHEMATICAL INDUCTION

Mathematical induction is a very useful device for proving results for all positive integers. If the result to be proved involves n, where n is a positive integer, then the proof by mathematical induction consists of the following two steps:

Step 1. Verify the result for $n = 1$.

Step 2. Assume the result to be true for $n = k$ and **then prove** that it is true for $n = k + 1$.

Now the result is true for $n = 1$

Using step 2, the result is true for $n = 1 + 1 = 2$

$\Rightarrow$ The result is true for $n = 2 + 1 = 3$

$\Rightarrow$ The result is true for $n = 3 + 1 = 4$ and so on.

Hence, the result is true for all positive integers n.

Example 5. *By mathematical induction, prove that if*

$$A = \begin{bmatrix} 3 & -4 \\ 1 & -1 \end{bmatrix}, \text{ then } A^n = \begin{bmatrix} 1+2n & -4n \\ n & 1-2n \end{bmatrix} \text{ where } n \text{ is any positive integer.}$$

Sol. We prove the result by mathematical induction.

When $n = 1$, $A^n = \begin{bmatrix} 1+2n & -4n \\ n & 1-2n \end{bmatrix}$

$$\Rightarrow \quad A^1 = \begin{bmatrix} 1+2.1 & -4.1 \\ 1 & 1-2.1 \end{bmatrix} = \begin{bmatrix} 3 & -4 \\ 1 & -1 \end{bmatrix} = A$$

$\Rightarrow$ The result is true when $n = 1$.

Let us assume that the result is true for any positive integer k.

i.e., let $$A^k = \begin{bmatrix} 1+2k & -4k \\ k & 1-2k \end{bmatrix} \quad \text{...(1)}$$

Now, $$A^{k+1} = A^k . A = \begin{bmatrix} 1+2k & -4k \\ k & 1-2k \end{bmatrix}\begin{bmatrix} 3 & -4 \\ 1 & -1 \end{bmatrix} \quad \text{[Using (1)]}$$

$$= \begin{bmatrix} 3(1+2k)-4k & -4(1+2k)+4k \\ 3k+1-2k & -4k-1+2k \end{bmatrix} = \begin{bmatrix} 3+2k & -4-4k \\ 1+k & -1-2k \end{bmatrix}$$

$$= \begin{bmatrix} 1+2(k+1) & -4(k+1) \\ k+1 & 1-2(k+1) \end{bmatrix}$$

$\Rightarrow$ The result is true for $n = k + 1$

Hence, by mathematical induction, the result is true for all positive integers n.

1.10. TRANSPOSE OF A MATRIX

Given a matrix A, then the matrix obtained from A by changing its rows into columns and columns into rows is called the **transpose** of A and is denoted by A′ or A^t.

$$\text{For example, if } A = \begin{bmatrix} 1 & 0 & 2 & 5 \\ 2 & -1 & 3 & 7 \end{bmatrix}, \text{ then } A' = \begin{bmatrix} 1 & 2 \\ 0 & -1 \\ 2 & 3 \\ 5 & 7 \end{bmatrix}$$

Clearly (*i*) if the order of A is $m \times n$ then order of A′ is $n \times m$.

(*ii*) $(i, j)^{th}$ element of A′ = $(j, i)^{th}$ element of A.

In symbols, if $A = [a_{ij}]_{m \times n}$, then $A' = [b_{ij}]_{n \times m}$, where $b_{ij} = a_{ij}$.

1.11. PROPERTIES OF TRANSPOSE OF A MATRIX

If A′ and B′ denote the transposes of A and B respectively, then

(*i*) **(A′)′ = A** *i.e.,* **the transpose of the transpose of a matrix is the matrix itself.**

(ii) **(A + B)′ = A′ + B′** *i.e.*, **the transpose of the sum of two matrices is equal to the sum of their transposes.**

(iii) **(AB)′ = B′A′** *i.e.*, **the transpose of the product of two matrices is equal to the product of their transposes taken in the reverse order.**

1.12. SYMMETRIC MATRIX

A square matrix $A = \{a_{ij}\}$ is said to be symmetric if **A′ = A** *i.e.*, if the transpose of the matrix is equal to the matrix itself.

Thus, for a symmetric matrix $A = [a_{ij}]$, $a_{ij} = a_{ji}$

For example, $\begin{bmatrix} -1 & 2 & 3 \\ 2 & 5 & 6 \\ 3 & 6 & 4 \end{bmatrix}$, $\begin{bmatrix} a & h & g \\ h & b & f \\ g & f & c \end{bmatrix}$ are symmetric matrices.

1.13. SKEW-SYMMETRIC MATRIX (*OR* ANTI-SYMMETRIC MATRIX)

A square matrix $A = \{a_{ij}\}$ is said to be skew-symmetric if **A′ = – A** *i.e.*, if the transpose of the matrix is equal to the negative of the matrix.

Thus, for a skew-symmetric matrix $A = [a_{ij}]$, $a_{ij} = -a_{ji}$.

Putting $j = i$, $a_{ii} = -a_{ii} \Rightarrow 2a_{ii} = 0$ or $a_{ii} = 0$ for all i.

Thus, all diagonal elements of a skew-symmetric matrix are zero.

For example, $\begin{bmatrix} 0 & 2 & -3 \\ -2 & 0 & 1 \\ 3 & -1 & 0 \end{bmatrix}$, $\begin{bmatrix} 0 & h & -g \\ -h & 0 & f \\ g & -f & 0 \end{bmatrix}$ are skew-symmetric matrices.

1.14. EVERY SQUARE MATRIX CAN UNIQUELY BE EXPRESSED AS THE SUM OF A SYMMETRIC MATRIX *AND* A SKEW-SYMMETRIC MATRIX

Let A be any square matrix.

Consider $B = \frac{1}{2}(A + A')$ and $C = \frac{1}{2}(A - A')$

$$B + C = \frac{1}{2}(A + A') + \frac{1}{2}(A - A') = A$$

Also $B' = \frac{1}{2}(A + A')' = \frac{1}{2}[A' + (A')'] = \frac{1}{2}(A' + A) = B$

$\Rightarrow$ B is a symmetric matrix.

$$C' = \frac{1}{2}(A - A')' = \frac{1}{2}[A' - (A')'] = \frac{1}{2}[A' - A] = -\frac{1}{2}(A - A') = -C$$

$\Rightarrow$ C is a skew-symmetric matrix.

Hence **every square matrix A can be expressed as A = B + C,** where $B = \frac{1}{2}(A + A')$ is a symmetric matrix and $C = \frac{1}{2}(A - A')$ is a skew-symmetric matrix.

To prove that there is only one way in which A can be expressed as the sum of a symmetric matrix and a skew-symmetric matrix, suppose

$$A = P + Q \quad ...(1)$$

is another such representation in which P is a symmetric matrix, and Q is a skew-symmetric matrix.

i.e., $P' = P$ and $Q' = -Q$

Then $A' = (P + Q)' = P' + Q' = P - Q$ ...(2)

From (1) and (2), $A + A' = 2P$ and $A - A' = 2Q$

$\therefore$ $P = \frac{1}{2}(A + A') = B$ and $Q = \frac{1}{2}(A - A') = C$ so that $P + Q = B + C$

Hence the result.

1.15. ORTHOGONAL MATRIX

A square matrix A is called an orthogonal matrix if $AA' = A'A = I$ or $A' = A^{-1}$.

Note. If A and B are any two orthogonal matrices, AB will also be an orthogonal matrix.

1.16. NILPOTENT MATRIX

A square matrix A is said to be nilpotent if $A^2 = O$ *e.g.* $A = \begin{bmatrix} 0 & 1 \\ 0 & 0 \end{bmatrix}$

It will be of index p if p is the least positive integer such that $A^p = O$.

1.17. IDEMPOTENT MATRIX

A square matrix A is idempotent if $A^2 = A$ *e.g.* $A = \begin{bmatrix} 1 & 0 \\ 0 & 1 \end{bmatrix}$

A square matrix A is said to be idempotent of period n if n is the least positive integer such that $A^{n+1} = A$.

1.18. INVOLUTARY MATRIX

A square matrix A is said to be involutary if $A^2 = I$ where I is a unit matrix.

e.g. $$A = \begin{bmatrix} -5 & -8 & 0 \\ 3 & 5 & 0 \\ 1 & 2 & -1 \end{bmatrix}.$$

Example 6. *If A and B are symmetric matrices, prove that AB – BA is a skew-symmetric matrix.*

Sol. A and B are symmetric matrices.

$\Rightarrow$ $A' = A$ and $B' = B$ ...(1)

Now, $(AB - BA)' = (AB)' - (BA)'$ $[\because (A - B)' = A' - B']$

$= B'A' - A'B'$ $[\because (AB)' = B'A']$

$= BA - AB$ [Using (1)]

$= -(AB - BA)$

$\therefore$ (AB – BA) is a skew-symmetric matrix.

Example 7. *Express the following matrix as the sum of a symmetric and a skew-symmetric matrix,*

$$\begin{bmatrix} -1 & 7 & 1 \\ 2 & 3 & 4 \\ 5 & 0 & 5 \end{bmatrix}.$$

Sol. Given matrix is $A = \begin{bmatrix} -1 & 7 & 1 \\ 2 & 3 & 4 \\ 5 & 0 & 5 \end{bmatrix}$ $\therefore$ $A' = \begin{bmatrix} -1 & 2 & 5 \\ 7 & 3 & 0 \\ 1 & 4 & 5 \end{bmatrix}$

Now, $$B = \frac{1}{2}(A + A') = \begin{bmatrix} -1 & 9/2 & 3 \\ 9/2 & 3 & 2 \\ 3 & 2 & 5 \end{bmatrix}$$

$$C = \frac{1}{2}(A - A') = \begin{bmatrix} 0 & 5/2 & -2 \\ -5/2 & 0 & 2 \\ 2 & -2 & 0 \end{bmatrix}$$

$$\therefore \quad A = B + C = \begin{bmatrix} -1 & 9/2 & 3 \\ 9/2 & 3 & 2 \\ 3 & 2 & 5 \end{bmatrix} + \begin{bmatrix} 0 & 5/2 & -2 \\ -5/2 & 0 & 2 \\ 2 & -2 & 0 \end{bmatrix}$$

where B is symmetric and C is a skew-symmetric matrix.

1.19. ADJOINT OF A SQUARE MATRIX

The adjoint of a square matrix is the transpose of the matrix obtained by replacing each element of A by its co-factor in | A |. Adjoint of A is briefly written as adj. A.

Thus, if $A = \begin{bmatrix} a_1 & b_1 & c_1 \\ a_2 & b_2 & c_2 \\ a_3 & b_3 & c_3 \end{bmatrix}$ then adj. A = transpose of $\begin{bmatrix} A_1 & B_1 & C_1 \\ A_2 & B_2 & C_2 \\ A_3 & B_3 & C_3 \end{bmatrix} = \begin{bmatrix} A_1 & A_2 & A_3 \\ B_1 & B_2 & B_3 \\ C_1 & C_2 & C_3 \end{bmatrix}$

where the capital letters denote the co-factors of corresponding small letters in

$$|A| = \begin{vmatrix} a_1 & b_1 & c_1 \\ a_2 & b_2 & c_2 \\ a_3 & b_3 & c_3 \end{vmatrix}.$$

Note. If A is an n-rowed square matrix then

$$A\,(\text{adj. } A) = (\text{adj. } A)\,A = |A|\, I_n.$$

1.19.1. Properties of Adjoint

(1) If A and B are two non-singular square matrices of the same order then

$$\text{adj. }(AB) = (\text{adj. } B)(\text{adj. } A).$$

(2) If A is a non singular matrix of order n then,

(*i*) $|\text{adj. } A| = |A|^{n-1}$ (*ii*) $\text{adj. }(\text{adj. } A) = |A|^{n-2} A$ (*iii*) $\text{adj. } A^T = (\text{adj. } A)^T$.

1.20. INVERSE (*or* Reciprocal) OF A SQUARE MATRIX

Let A be an n-rowed square matrix. If there exists an n-rowed square matrix B such that

$$AB = BA = I$$

then the matrix **A is said to be invertible** and B is called the **inverse** (or **reciprocal**) of A. The inverse of a square matrix if it exists is unique. The necessary and sufficient condition for a square matrix A to possess inverse is that A is non-singular *i.e.*, $|A| \neq 0$.

Note 1. Inverse of A is denoted by A^{-1}, thus $B = A^{-1}$ and $AA^{-1} = A^{-1}A = I$.

Note 2. $A^{-1} = \dfrac{\text{adj. } A}{|A|}$; $|A| \neq 0$. **Note 3.** $|A^{-1}| = |A|^{-1}$.

1.21. PROPERTIES OF INVERSE

(1) If A is invertible then so is A^{-1} and $(A^{-1})^{-1} = A$.

(2) If A and B are two non-singular square matrices of the same order then

$$(AB)^{-1} = B^{-1} A^{-1}.$$ (*U.K.T.U. 2011*)

(3) If A is a non-singular square matrix then so is A′ and $(A')^{-1} = (A^{-1})'$.

Example 8. *If* $A = \begin{bmatrix} 3 & -3 & 4 \\ 2 & -3 & 4 \\ 0 & -1 & 1 \end{bmatrix}$, *(i) find* A^{-1} *(ii) show that* $A^3 = A^{-1}$.

Sol. (*i*)

$$|A| = \begin{vmatrix} 3 & -3 & 4 \\ 2 & -3 & 4 \\ 0 & -1 & 1 \end{vmatrix} = 1 \quad \therefore \quad A^{-1} \text{ exists}$$

$$\text{adj. } A = \begin{bmatrix} 1 & -1 & 0 \\ -2 & 3 & -4 \\ -2 & 3 & -3 \end{bmatrix}$$

$$\therefore \quad A^{-1} = \frac{\text{adj. } A}{|A|} = \begin{bmatrix} 1 & -1 & 0 \\ -2 & 3 & -4 \\ -2 & 3 & -3 \end{bmatrix}$$

(*ii*)

$$A^2 = A \times A = \begin{bmatrix} 3 & -3 & 4 \\ 2 & -3 & 4 \\ 0 & -1 & 1 \end{bmatrix} \begin{bmatrix} 3 & -3 & 4 \\ 2 & -3 & 4 \\ 0 & -1 & 1 \end{bmatrix} = \begin{bmatrix} 3 & -4 & 4 \\ 0 & -1 & 0 \\ -1 & 2 & -3 \end{bmatrix}$$

$$A^4 = A^2 \times A^2 = \begin{bmatrix} 3 & -4 & 4 \\ 0 & -1 & 0 \\ -1 & 2 & -3 \end{bmatrix} \begin{bmatrix} 3 & -4 & 4 \\ 0 & -1 & 0 \\ -1 & 2 & -3 \end{bmatrix} = \begin{bmatrix} 1 & 0 & 0 \\ 0 & 1 & 0 \\ 0 & 0 & 1 \end{bmatrix} = I$$

Now, $A \cdot A^3 = A^3 \cdot A = I \Rightarrow A^3 = A^{-1}$.

TEST YOUR KNOWLEDGE

1. If A is a square matrix, prove that

(*i*) A + A′ is symmetric. (*ii*) A – A′ is skew-symmetric.

2. Prove that

(*i*) if A, B are symmetric, then so is A + B (*ii*) if A, B are skew-symmetric, then so is A + B

(*iii*) if A is a square matrix, then AA′ and A′A are both symmetric.

(*iv*) if A is symmetric, then B′ AB is symmetric.

3. If $A = \begin{bmatrix} 1 & 2 & 1 \\ a & 0 & 4 \\ 1 & 1 & 1 \end{bmatrix}$ and adj. (adj. A) = A, find a. *(M.T.U. 2012)*

4. If $A = \begin{bmatrix} 1 & 2 & -1 \\ 3 & 0 & 2 \\ 4 & 5 & 0 \end{bmatrix}$ and $B = \begin{bmatrix} 1 & 0 & 0 \\ 2 & 1 & 0 \\ 0 & 1 & 3 \end{bmatrix}$, verify that $(AB)' = B'A'$.

5. Prove that the following matrices are orthogonal.

(i) $\frac{1}{3}\begin{bmatrix} -2 & 1 & 2 \\ 2 & 2 & 1 \\ 1 & -2 & 2 \end{bmatrix}$ *(ii)* $\frac{1}{9}\begin{bmatrix} -8 & 4 & 1 \\ 1 & 4 & -8 \\ 4 & 7 & 4 \end{bmatrix}$

6. Express each of the following matrices as the sum of a symmetric and a skew-symmetric matrix:

(i) $\begin{bmatrix} 4 & 2 & -3 \\ 1 & 3 & -6 \\ -5 & 0 & -7 \end{bmatrix}$ *(ii)* $\begin{bmatrix} a & a & b \\ c & b & b \\ c & a & c \end{bmatrix}$

Answers

3. 3

6. *(i)* $\begin{bmatrix} 4 & \frac{3}{2} & -4 \\ \frac{3}{2} & 3 & -3 \\ -4 & -3 & -7 \end{bmatrix} + \begin{bmatrix} 0 & \frac{1}{2} & 1 \\ -\frac{1}{2} & 0 & -3 \\ -1 & 3 & 0 \end{bmatrix}$

(ii) $\begin{bmatrix} a & \frac{1}{2}(a+c) & \frac{1}{2}(b+c) \\ \frac{1}{2}(a+c) & b & \frac{1}{2}(a+b) \\ \frac{1}{2}(b+c) & \frac{1}{2}(a+b) & c \end{bmatrix} + \begin{bmatrix} 0 & \frac{1}{2}(a-c) & \frac{1}{2}(b-c) \\ \frac{1}{2}(c-a) & 0 & \frac{1}{2}(b-a) \\ \frac{1}{2}(c-b) & \frac{1}{2}(a-b) & 0 \end{bmatrix}$.

1.22. ELEMENTARY TRANSFORMATIONS (*or* Operations)

Any one of the following operations on a matrix is called an elementary transformation (or E-operation).

(i) **Interchange of two rows or two columns.**

The interchange of ith and j^{th} rows is denoted by R_{ij}.

The interchange of ith and j^{th} columns is denoted by C_{ij}.

(ii) **Multiplication of (each element of) a row or column by a non-zero number k.**

The multiplication of ith row by k is denoted by $R_i(k)$.

The multiplication of ith column by k is denoted by $C_i(k)$.

(iii) **Addition of k times the elements of a row (or column) to the corresponding elements of another row (or column), k ≠ 0.**

The addition of k times the jth row to the ith row is denoted by $R_{ij}(k)$.

The addition of k times the jth column to the ith column is denoted by $C_{ij}(k)$.

If a matrix B is obtained from a matrix A by one or more E-operations, then B is said to be **equivalent** to A. Two equivalent matrices A and B are written as A ~ B.

1.23. ELEMENTARY MATRICES

The matrix obtained from a unit matrix I by subjecting it to one of the E-operations is called an elementary matrix.

e.g. Let $I = \begin{bmatrix} 1 & 0 & 0 \\ 0 & 1 & 0 \\ 0 & 0 & 1 \end{bmatrix}$

Operating R_{23} or C_{23} on I, we get the elementary matrix $\begin{bmatrix} 1 & 0 & 0 \\ 0 & 0 & 1 \\ 0 & 1 & 0 \end{bmatrix}$. It is denoted by E_{23}.

1.24. THEOREMS ON THE EFFECT OF E-OPERATIONS ON MATRICES

(a) *Any E-row operation on the product of two matrices is equivalent to the same E-row operation on the pre-factor.*

If the E-row operation is denoted by R, then R(AB) = R(A) . B.

(b) *Any E-column operation on the product of two matrices is equivalent to the same E-column operation on the post-factor.*

If the E-column operation is denoted by C, then C(AB) = A . C(B).

(c) *Every E-row operation on a matrix is equivalent to pre-multiplication by the corresponding E-matrix.*

Thus, the effect of E-row operation R_{ij} on $A = E_{ij} . A$

The effect of E-row operation $R_i(k)$ on $A = E_i(k) . A$

The effect of E-row operation $R_{ij}(k)$ on $A = E_{ij}(k) . A$.

(d) *Every E-column operation on a matrix is equivalent to post-multiplication by the corresponding E-matrix.*

Thus, the effect of E-column operation C_{ij} on $A = A . E_{ij}$

The effect of E-column operation $C_i(k)$ on $A = A . \{E_i(k)\}$

The effect of E-column operation $C_{ij}(k)$ on $A = A . E_{ij}'(k)$.

(e) The inverse of an elementary matrix is an elementary matrix.

ILLUSTRATIVE EXAMPLES

Example 1. *Transform* $\begin{bmatrix} 1 & 3 & 3 \\ 2 & 4 & 10 \\ 3 & 8 & 4 \end{bmatrix}$ *into a unit matrix by using elementary transformations.*

(U.K.T.U. 2011)

Sol. We have $\begin{bmatrix} 1 & 3 & 3 \\ 2 & 4 & 10 \\ 3 & 8 & 4 \end{bmatrix}$

Operating $R_{21}(-2)$, $R_{31}(-3)$

$$\sim \begin{bmatrix} 1 & 3 & 3 \\ 0 & -2 & 4 \\ 0 & -1 & -5 \end{bmatrix}$$

Operating $R_2(-\frac{1}{2})$

$$\sim \begin{bmatrix} 1 & 3 & 3 \\ 0 & 1 & -2 \\ 0 & -1 & -5 \end{bmatrix}$$

Operating $R_{12}(-3)$, $R_{32}(1)$

$$\sim \begin{bmatrix} 1 & 0 & 9 \\ 0 & 1 & -2 \\ 0 & 0 & -7 \end{bmatrix}$$

Operating $R_3(-\frac{1}{7})$

$$\sim \begin{bmatrix} 1 & 0 & 9 \\ 0 & 1 & -2 \\ 0 & 0 & 1 \end{bmatrix}$$

Operating $R_{13}(-9)$, $R_{23}(2)$

$$\sim \begin{bmatrix} 1 & 0 & 0 \\ 0 & 1 & 0 \\ 0 & 0 & 1 \end{bmatrix}.$$

Example 2. *Reduce the matrix* $\begin{bmatrix} -1 & 2 & -2 \\ 1 & 2 & 1 \\ -1 & -1 & 0 \end{bmatrix}$ *to the upper triangular form by using elementary transformations.*

Sol. We have

$$\begin{bmatrix} -1 & 2 & -2 \\ 1 & 2 & 1 \\ -1 & -1 & 0 \end{bmatrix}$$

Operating $R_{21}(1)$, $R_{31}(-1)$

$$\sim \begin{bmatrix} -1 & 2 & -2 \\ 0 & 4 & -1 \\ 0 & -3 & 2 \end{bmatrix}$$

Operating $R_{32}(1)$

$$\sim \begin{bmatrix} -1 & 2 & -2 \\ 0 & 4 & -1 \\ 0 & 1 & 1 \end{bmatrix}$$

Operating $R_{32}(-\frac{1}{4})$

$$\sim \begin{bmatrix} -1 & 2 & -2 \\ 0 & 4 & -1 \\ 0 & 0 & 5/4 \end{bmatrix}$$

1.25. INVERSE OF MATRIX BY E-OPERATIONS (Gauss-Jordan Method)

The elementary row transformations which reduce a square matrix A to the unit matrix, when applied to the unit matrix, gives the inverse matrix A^{-1}.

To compute the inverse of a matrix, we use the concept of equivalent matrices.

If we are to find out the inverse of a non-singular square matrix A, we first write A as equivalent to I, a unit matrix of the same order.

$$A \sim I$$

Then we apply elementary row operations on them. The objective is to reduce A to I. As soon as this is achieved, the other matrix gives A^{-1}.

$$I \sim A^{-1}$$

This is an elegant way of determining the inverse or reciprocal of a matrix A.

Example 3. *Employing elementary row transformations, find the inverse of the matrix*

$$A = \begin{bmatrix} 0 & 1 & 2 \\ 1 & 2 & 3 \\ 3 & 1 & 1 \end{bmatrix}.$$ (A.K.T.U. 2018)

Sol. Let $A \sim I$

$$\begin{bmatrix} 0 & 1 & 2 \\ 1 & 2 & 3 \\ 3 & 1 & 1 \end{bmatrix} \sim \begin{bmatrix} 1 & 0 & 0 \\ 0 & 1 & 0 \\ 0 & 0 & 1 \end{bmatrix}$$

Operating R_{12}

$$\begin{bmatrix} 1 & 2 & 3 \\ 0 & 1 & 2 \\ 3 & 1 & 1 \end{bmatrix} \sim \begin{bmatrix} 0 & 1 & 0 \\ 1 & 0 & 0 \\ 0 & 0 & 1 \end{bmatrix}$$

Operating $R_{31}(-3)$

$$\begin{bmatrix} 1 & 2 & 3 \\ 0 & 1 & 2 \\ 0 & -5 & -8 \end{bmatrix} \sim \begin{bmatrix} 0 & 1 & 0 \\ 1 & 0 & 0 \\ 0 & -3 & 1 \end{bmatrix}$$

Operating $R_{12}(-2)$, $R_{32}(5)$

$$\begin{bmatrix} 1 & 0 & -1 \\ 0 & 1 & 2 \\ 0 & 0 & 2 \end{bmatrix} \sim \begin{bmatrix} -2 & 1 & 0 \\ 1 & 0 & 0 \\ 5 & -3 & 1 \end{bmatrix}$$

Operating $R_{13}(\frac{1}{2})$, $R_{23}(-1)$

$$\begin{bmatrix} 1 & 0 & 0 \\ 0 & 1 & 0 \\ 0 & 0 & 2 \end{bmatrix} \sim \begin{bmatrix} \frac{1}{2} & -\frac{1}{2} & \frac{1}{2} \\ -4 & 3 & -1 \\ 5 & -3 & 1 \end{bmatrix}$$

Operating $R_3(\frac{1}{2})$

$$\begin{bmatrix} 1 & 0 & 0 \\ 0 & 1 & 0 \\ 0 & 0 & 1 \end{bmatrix} \sim \begin{bmatrix} \frac{1}{2} & -\frac{1}{2} & \frac{1}{2} \\ -4 & 3 & -1 \\ 5/2 & -3/2 & 1/2 \end{bmatrix}$$

$$I \sim A^{-1}$$

$$\therefore \qquad A^{-1} = \begin{bmatrix} \frac{1}{2} & -\frac{1}{2} & \frac{1}{2} \\ -4 & 3 & -1 \\ 5/2 & -3/2 & 1/2 \end{bmatrix}.$$

Example 4. *Find the inverse of the matrix M by applying elementary transformations*

$$M = \begin{bmatrix} 0 & 2 & 1 & 3 \\ 1 & 1 & -1 & -2 \\ 1 & 2 & 0 & 1 \\ -1 & 1 & 2 & 6 \end{bmatrix}.$$

Sol. Let $\qquad M \sim I_4$

$$\begin{bmatrix} 0 & 2 & 1 & 3 \\ 1 & 1 & -1 & -2 \\ 1 & 2 & 0 & 1 \\ -1 & 1 & 2 & 6 \end{bmatrix} \sim \begin{bmatrix} 1 & 0 & 0 & 0 \\ 0 & 1 & 0 & 0 \\ 0 & 0 & 1 & 0 \\ 0 & 0 & 0 & 1 \end{bmatrix}$$

Operating R_{12}

$$\begin{bmatrix} 1 & 1 & -1 & -2 \\ 0 & 2 & 1 & 3 \\ 1 & 2 & 0 & 1 \\ -1 & 1 & 2 & 6 \end{bmatrix} \sim \begin{bmatrix} 0 & 1 & 0 & 0 \\ 1 & 0 & 0 & 0 \\ 0 & 0 & 1 & 0 \\ 0 & 0 & 0 & 1 \end{bmatrix}$$

Operating $R_{31}(-1)$, $R_{41}(1)$

$$\begin{bmatrix} 1 & 1 & -1 & -2 \\ 0 & 2 & 1 & 3 \\ 0 & 1 & 1 & 3 \\ 0 & 2 & 1 & 4 \end{bmatrix} \sim \begin{bmatrix} 0 & 1 & 0 & 0 \\ 1 & 0 & 0 & 0 \\ 0 & -1 & 1 & 0 \\ 0 & 1 & 0 & 1 \end{bmatrix}$$

Operating $R_{42}(-1)$, $R_{13}(1)$

$$\begin{bmatrix} 1 & 2 & 0 & 1 \\ 0 & 2 & 1 & 3 \\ 0 & 1 & 1 & 3 \\ 0 & 0 & 0 & 1 \end{bmatrix} \sim \begin{bmatrix} 0 & 0 & 1 & 0 \\ 1 & 0 & 0 & 0 \\ 0 & -1 & 1 & 0 \\ -1 & 1 & 0 & 1 \end{bmatrix}$$

Operating $R_{23}(-1)$

$$\begin{bmatrix} 1 & 2 & 0 & 1 \\ 0 & 1 & 0 & 0 \\ 0 & 1 & 1 & 3 \\ 0 & 0 & 0 & 1 \end{bmatrix} \sim \begin{bmatrix} 0 & 0 & 1 & 0 \\ 1 & 1 & -1 & 0 \\ 0 & -1 & 1 & 0 \\ -1 & 1 & 0 & 1 \end{bmatrix}$$

Operating $R_{12}(-2)$, $R_{32}(-1)$

$$\begin{bmatrix} 1 & 0 & 0 & 1 \\ 0 & 1 & 0 & 0 \\ 0 & 0 & 1 & 3 \\ 0 & 0 & 0 & 1 \end{bmatrix} \sim \begin{bmatrix} -2 & -2 & 3 & 0 \\ 1 & 1 & -1 & 0 \\ -1 & -2 & 2 & 0 \\ -1 & 1 & 0 & 1 \end{bmatrix}$$

Operating $R_{14}(-1), R_{34}(-3)$

$$\begin{bmatrix} 1 & 0 & 0 & 0 \\ 0 & 1 & 0 & 0 \\ 0 & 0 & 1 & 0 \\ 0 & 0 & 0 & 1 \end{bmatrix} \sim \begin{bmatrix} -1 & -3 & 3 & -1 \\ 1 & 1 & -1 & 0 \\ 2 & -5 & 2 & -3 \\ -1 & 1 & 0 & 1 \end{bmatrix}$$

$\Rightarrow \quad I \sim M^{-1}$

Hence, $$M^{-1} = \begin{bmatrix} -1 & -3 & 3 & -1 \\ 1 & 1 & -1 & 0 \\ 2 & -5 & 2 & -3 \\ -1 & 1 & 0 & 1 \end{bmatrix}$$

TEST YOUR KNOWLEDGE

1. Reduce the matrix A to triangular form, $A = \begin{bmatrix} 1 & 2 & 3 \\ 2 & 5 & 7 \\ 3 & 1 & 2 \end{bmatrix}$.

2. If X, Y are non-singular matrices and $B = \begin{bmatrix} X & O \\ O & Y \end{bmatrix}$, show that $B^{-1} = \begin{bmatrix} X^{-1} & O \\ O & Y^{-1} \end{bmatrix}$ where O is a null matrix.

3. Find the inverse of the matrix $\begin{bmatrix} 1 & 0 & k \\ 0 & 1 & 0 \\ 0 & 0 & 1 \end{bmatrix}\begin{bmatrix} 1 & 0 & 0 \\ 0 & 1 & s \\ 0 & 0 & 1 \end{bmatrix}\begin{bmatrix} 1 & 0 & 0 \\ 0 & p & 0 \\ 0 & 0 & 1 \end{bmatrix}$ without first evaluating the product.

4. Find the inverse of the following matrices by using elementary row operations:

(i) $\begin{bmatrix} 2 & 0 & -1 \\ 5 & 1 & 0 \\ 0 & 1 & 3 \end{bmatrix}$ (ii) $\begin{bmatrix} 2 & 1 & -1 \\ 0 & 2 & 1 \\ 5 & 2 & -3 \end{bmatrix}$ (iii) $\begin{bmatrix} 8 & 4 & 3 \\ 2 & 1 & 1 \\ 1 & 2 & 1 \end{bmatrix}$

(iv) $\begin{bmatrix} 3 & -1 & 1 \\ -15 & 6 & -5 \\ 5 & -2 & 2 \end{bmatrix}$ (v) $\begin{bmatrix} 1 & 2 & 1 \\ 3 & 2 & 3 \\ 1 & 1 & 2 \end{bmatrix}$ (vi) $\begin{bmatrix} 1 & 0 & -1 \\ 3 & 4 & 5 \\ 0 & -6 & -7 \end{bmatrix}$

(vii) $\begin{bmatrix} 3 & -3 & 4 \\ 2 & -3 & 4 \\ 0 & -1 & 1 \end{bmatrix}$ *(A.K.T.U. 2014, 2019)* (viii) $\begin{bmatrix} 1 & 2 & 5 \\ 2 & 3 & 1 \\ -1 & 1 & 1 \end{bmatrix}$ (ix) $\begin{bmatrix} 1 & 2 & -2 \\ -1 & 3 & 0 \\ 0 & -2 & 1 \end{bmatrix}$

(x) $\begin{bmatrix} 1 & 3 & 3 \\ 1 & 4 & 3 \\ 1 & 3 & 4 \end{bmatrix}$ (xi) $\begin{bmatrix} 1 & 2 & 1 \\ 3 & 1 & 2 \\ 0 & 1 & 2 \end{bmatrix}$ (xii) $\begin{bmatrix} 1 & -1 & 1 \\ 4 & 1 & 0 \\ 8 & 1 & 1 \end{bmatrix}$.

(xiii) $\begin{bmatrix} 1 & 2 & 3 \\ 2 & 4 & 5 \\ 3 & 5 & 6 \end{bmatrix}$. (xiv) $\begin{bmatrix} 2 & 3 & 4 \\ 4 & 3 & 1 \\ 1 & 2 & 4 \end{bmatrix}$ *(A.K.T.U. 2021)* (xv) $\begin{bmatrix} 2 & 4 & 3 & 2 \\ 3 & 6 & 5 & 2 \\ 2 & 5 & 2 & -3 \\ 4 & 5 & 14 & 14 \end{bmatrix}$

(xvi) $\begin{bmatrix} \frac{1}{3} & \frac{1}{5} & \frac{1}{7} \\ \frac{1}{5} & \frac{1}{7} & \frac{1}{11} \\ \frac{1}{7} & \frac{1}{11} & \frac{1}{13} \end{bmatrix}$ (xvii) $\begin{bmatrix} i & -1 & 2i \\ 2 & 0 & 2 \\ -1 & 0 & 1 \end{bmatrix}$

5. Employing elementary row transformations, find the inverse of the following non-singular matrices:

$$(i)\ \begin{bmatrix} 1 & 2 & 3 & 1 \\ 1 & 3 & 3 & 2 \\ 2 & 4 & 3 & 3 \\ 1 & 1 & 1 & 1 \end{bmatrix} \qquad (ii)\ \begin{bmatrix} 1 & -1 & 0 & 2 \\ 0 & 1 & 1 & -1 \\ 2 & 1 & 2 & 1 \\ 3 & -2 & 1 & 6 \end{bmatrix} \qquad (iii)\ \begin{bmatrix} 2 & 1 & -1 & 2 \\ 1 & 3 & 2 & -3 \\ -1 & 2 & 1 & -1 \\ 2 & -3 & -1 & 4 \end{bmatrix}$$

Answers

1. $\begin{bmatrix} 1 & 2 & 3 \\ 0 & 1 & 1 \\ 0 & 0 & -2 \end{bmatrix}$ 3. $\begin{bmatrix} 1 & 0 & -k \\ 0 & 1/p & -s/p \\ 0 & 0 & 1 \end{bmatrix}$

4. (i) $\begin{bmatrix} 3 & -1 & 1 \\ -15 & 6 & -5 \\ 5 & -2 & 2 \end{bmatrix}$ (ii) $\begin{bmatrix} 8 & -1 & -3 \\ -5 & 1 & 2 \\ 10 & -1 & -4 \end{bmatrix}$ (iii) $\dfrac{1}{3}\begin{bmatrix} 1 & -2 & -1 \\ 1 & -5 & 2 \\ -3 & 12 & 0 \end{bmatrix}$

(iv) $\begin{bmatrix} 2 & 0 & -1 \\ 5 & 1 & 0 \\ 0 & 1 & 3 \end{bmatrix}$ (v) $\begin{bmatrix} -1/4 & 3/4 & -1 \\ 3/4 & -1/4 & 0 \\ -1/4 & -1/4 & 1 \end{bmatrix}$ (vi) $\begin{bmatrix} 1/10 & 3/10 & 1/5 \\ 21/20 & -7/20 & -2/5 \\ -9/10 & -3/10 & 1/5 \end{bmatrix}$

(vii) $\begin{bmatrix} 1 & -1 & 0 \\ -2 & 3 & -4 \\ -2 & 3 & -3 \end{bmatrix}$ (viii) $\dfrac{1}{21}\begin{bmatrix} 2 & 3 & -13 \\ -3 & 6 & 9 \\ 5 & -3 & -1 \end{bmatrix}$ (ix) $\begin{bmatrix} 3 & 2 & 6 \\ 1 & 1 & 2 \\ 2 & 2 & 5 \end{bmatrix}$

(x) $\begin{bmatrix} 7 & -3 & -3 \\ -1 & 1 & 0 \\ -1 & 0 & 1 \end{bmatrix}$ (xi) $\dfrac{1}{9}\begin{bmatrix} 0 & 3 & -3 \\ 6 & -2 & -1 \\ -1 & 1 & -5 \end{bmatrix}$ (xii) $\begin{bmatrix} 1 & 2 & -1 \\ -4 & -7 & 4 \\ -4 & -9 & 5 \end{bmatrix}$

(xiii) $\begin{bmatrix} 1 & -3 & 2 \\ -3 & 3 & -1 \\ 2 & -1 & 0 \end{bmatrix}$ (xiv) $\dfrac{1}{5}\begin{bmatrix} -10 & 4 & 9 \\ 15 & -4 & -14 \\ -5 & 1 & 6 \end{bmatrix}$

(xv) $\begin{bmatrix} -23 & 29 & -64/5 & -18/5 \\ 10 & -12 & 26/5 & 7/5 \\ 1 & -2 & 6/5 & 2/5 \\ 2 & -2 & 3/5 & 1/5 \end{bmatrix}$ (xvi) $\dfrac{1}{2238}\begin{bmatrix} 55125 & -48510 & -45045 \\ -48510 & 105875 & -35035 \\ -45045 & -35035 & 154154 \end{bmatrix}$

(xvii) $\dfrac{1}{4}\begin{bmatrix} 0 & 1 & -2 \\ -4 & 3i & 2i \\ 0 & 1 & 2 \end{bmatrix}$

5. (i) $\begin{bmatrix} 1 & -2 & 1 & 0 \\ 1 & -2 & 2 & -3 \\ 0 & 1 & -1 & 1 \\ -2 & 3 & -2 & 3 \end{bmatrix}$ (ii) $\begin{bmatrix} 2 & -1 & 1 & -1 \\ -5 & -3 & 1 & 1 \\ 2 & 3 & -1 & 0 \\ -3 & -1 & 0 & 1 \end{bmatrix}$ (iii) $\dfrac{1}{18}\begin{bmatrix} 2 & 5 & -7 & 1 \\ 5 & -1 & 5 & -2 \\ -7 & 5 & 11 & 10 \\ 1 & -2 & 10 & 5 \end{bmatrix}$

1.26. RANK OF A MATRIX

Let A be any $m \times n$ matrix. It has square sub-matrices of different orders. The determinants of these square sub-matrices are called minors of A.

A matrix A is said to be of rank r if

(i) it has at least one non-zero minor of order r.

(ii) all the minors of order $(r + 1)$ or higher than r are zero.

Symbolically, **rank** of A = r is written as $\rho(A) = r$.

From the definition of the rank of a matrix A, it follows that:

(*i*) If A is a null matrix, then $\rho(A) = 0$. [$\because$ every minor of A has zero value.]

(*ii*) If A is not a null matrix, then $\rho(A) \geq 1$.

(*iii*) If A is a non-singular $n \times n$ matrix, then $\rho(A) = n$.

[$\because$ $|A| \neq 0$ is the largest minor of A.]

If I_n is the $n \times n$ unit matrix, then $|I_n| = 1 \neq 0 \Rightarrow \rho(I_n) = n$.

(*iv*) If A is an $m \times n$ matrix, then $\rho(A) \leq$ minimum of m and n.

(*v*) If all minors of order r are equal to zero, then $\rho(A) < r$.

1.27. METHODS OF FINDING RANK

To determine the rank of a matrix A, we adopt the following different methods:

(1) Start with the highest order minor (or minors) of A. Let their order be r. **If any one of them** is non-zero, then $\rho(A) = r$.

If **all of them** are zero, start with minors of next lower order $(r - 1)$ and so on till you get a non-zero minor. The order of that minor is the rank of A.

This method **usually** involves a lot of computational work since we have to evaluate several determinants.

(2) **Normal form method.** If A is an $m \times n$ matrix and by a series of elementary (*row* or *column* or *both*) operations, it can be put into one of the following forms (called normal or canonical forms):

$$\begin{bmatrix} I_r & \vdots & O \\ \cdots & \cdots & \cdots \\ O & \vdots & O \end{bmatrix}, \begin{bmatrix} I_r \\ \cdots \\ O \end{bmatrix}, [I_r \;\vdots\; O], [I_r], \text{ where } I_r \text{ is the unit matrix of order } r.$$

Since the rank of a matrix is not changed as a result of elementary transformations, it follows that

$$\rho(A) = r. \qquad [\because \ r^{th} \text{ order minor } |I_r| = 1 \neq 0]$$

This method is also called **sweep out method** or **pivotal method.**

Procedure to obtain normal form

(*i*) Interchange rows (or columns) to obtain a non-zero element in I row and I column of given matrix.

(*ii*) Make this non-zero element as 1.

(*iii*) Obtain zeros in the remainder of I row and I column.

(*iv*) Repeat the above three steps starting with element in II row and II column.

(*v*) Continue the process down the main diagonal either until the end of diagonal is reached or all the remaining elements of matrix are zero.

(3) **Echelon form method.** In this form of the matrix, each of the first r elements of the leading diagonal is non-zero and every element below this diagonal/r^{th} row is zero. A matrix is reduced in Echelon form as

(*i*) The first non-zero element in a row should be unity (if possible).

(*ii*) All the non-zero rows, if any, precede the zero rows.

The rank of the matrix is equal to the number of non-zero diagonal elements or the number of non zero rows (as the case may be) when it has been reduced to Echelon form.

In other words, a matrix $A = [a_{ij}]$ is an Echelon matrix or is said to be in Echelon form if the number of zeros preceding the first non-zero entry of a row increases row by row until only zero rows remain.

In *row reduced Echelon form* (row canonical form) of the matrix, the first non-zero entry of a row is unity and are the only non-zero entry in their respective columns.

Further, any matrix can be transformed to a column Echelon form (CEF) by sequence of elementary column transformations. We say that a matrix is in CEF if

(*i*) all zero columns are on the right

(*ii*) the first (if we go down from the top) non-zero entry in each column is 1.

(*iii*) if $j > i$, then the leading one of the column c_j appears below that of c_i.

e.g., $$\begin{bmatrix} 1 & 0 & 0 \\ 0 & 1 & 0 \\ 3 & 2 & 0 \end{bmatrix}, \begin{bmatrix} 0 & 0 \\ 1 & 0 \\ 2 & 1 \end{bmatrix}, \begin{bmatrix} 0 & 0 & 0 \\ 1 & 0 & 0 \end{bmatrix}$$

A matrix is said to be in a column reduced Echelon form if it is in CEF and additionally, any row containing the leading one of a column consists of all zeros except this leading one. In above examples, first and third matrices are in column reduced echelon form.

1.28. FOR AN *m* × *n* MATRIX A, TO FIND SQUARE MATRICES P *AND* Q OF ORDERS *m AND n* RESPECTIVELY, SUCH THAT PAQ IS IN THE NORMAL FORM

Method. Write A = IAI.

Reduce the matrix on LHS to normal form by affecting elementary row and/or column transformations.

Every elementary row transformation on A must be accompanied by the same transformation on the pre-factor on RHS.

Every elementary column transformation on A must be accompanied by the same transformation on the post-factor on RHS.

1.29. RANK-NULLITY THEOREM *(A.K.T.U. 2021)*

If A is an $m \times n$ order matrix over some field then

$$\text{rank (A)} + \text{nul (A)} = n$$

This applies to linear maps also. Let V be a finite dimensional vector space and W be a vector space over some field. Let $T: V \to W$ be a linear map, then

$$\text{rank (T)} + \text{nul (T)} = \text{dim (V)}$$

where the rank of T is the dimension of the image of T and nullity of T is the dimension of the kernel of T.

ILLUSTRATIVE EXAMPLES

Example 1. *Find the rank of the matrix*

(*i*) $\begin{bmatrix} 1 & 2 & 3 & 4 \\ -2 & 0 & 5 & 7 \end{bmatrix}$ (*ii*) $\begin{bmatrix} 1 & 2 & 3 \\ 1 & 4 & 2 \\ 2 & 6 & 5 \end{bmatrix}$

(*iii*) $\begin{bmatrix} 2 & 3 & 4 & -1 \\ 5 & 2 & 0 & -1 \\ -4 & 5 & 12 & -1 \end{bmatrix}$ (*iv*) $\begin{bmatrix} 2 & 3 & -1 & -1 \\ 1 & -1 & -2 & -4 \\ 3 & 1 & 3 & -2 \\ 6 & 3 & 0 & -7 \end{bmatrix}$.

Sol. (*i*) Here, $A = \begin{bmatrix} 1 & 2 & 3 & 4 \\ -2 & 0 & 5 & 7 \end{bmatrix}$.

Operating R_{21} (2)

$$\sim \begin{bmatrix} 1 & 2 & 3 & 4 \\ 0 & 4 & 11 & 15 \end{bmatrix}$$

which is Echelon form.

$\therefore$ $\rho(A)$ = number of non-zero rows = 2

(*ii*) Here, $A = \begin{bmatrix} 1 & 2 & 3 \\ 1 & 4 & 2 \\ 2 & 6 & 5 \end{bmatrix}$

$\because$ $| A | = 0 \quad \therefore \quad \rho(A) \neq 3$

A 2×2 order minor is $\begin{vmatrix} 1 & 2 \\ 1 & 4 \end{vmatrix} = 2 \neq 0$

$\therefore$ $\rho(A) = 2$ | By definition

(*iii*) Here, $A = \begin{bmatrix} 2 & 3 & 4 & -1 \\ 5 & 2 & 0 & -1 \\ -4 & 5 & 12 & -1 \end{bmatrix}$

Operating R_{21} (– 2)

$$\sim \begin{bmatrix} 2 & 3 & 4 & -1 \\ 1 & -4 & -8 & 1 \\ -4 & 5 & 12 & -1 \end{bmatrix}$$

Operating R_{12}

$$\sim \begin{bmatrix} 1 & -4 & -8 & 1 \\ 2 & 3 & 4 & -1 \\ -4 & 5 & 12 & -1 \end{bmatrix}$$

Operating R_{21} (– 2), R_{31} (4)

$$\sim \begin{bmatrix} 1 & -4 & -8 & 1 \\ 0 & 11 & 20 & -3 \\ 0 & -11 & -20 & 3 \end{bmatrix}$$

Operating R_{32} (1)

$$\sim \begin{bmatrix} 1 & -4 & -8 & 1 \\ 0 & 11 & 20 & -3 \\ 0 & 0 & 0 & 0 \end{bmatrix}$$

which is Echelon form.

$\therefore$ $\rho(A) = 2$

(*iv*) Here $A = \begin{bmatrix} 2 & 3 & -1 & -1 \\ 1 & -1 & -2 & -4 \\ 3 & 1 & 3 & -2 \\ 6 & 3 & 0 & -7 \end{bmatrix}$ is a 4×4 matrix. $\therefore$ $\rho(A) \leq 4$.

Operating R_{12}

$$A \sim \begin{bmatrix} 1 & -1 & -2 & -4 \\ 2 & 3 & -1 & -1 \\ 3 & 1 & 3 & -2 \\ 6 & 3 & 0 & -7 \end{bmatrix}$$

Operating $C_{21}(1)$, $C_{31}(2)$, $C_{41}(4)$

$$\sim \begin{bmatrix} 1 & 0 & 0 & 0 \\ 2 & 5 & 3 & 7 \\ 3 & 4 & 9 & 10 \\ 6 & 9 & 12 & 17 \end{bmatrix}$$

Operating $R_{21}(-2)$, $R_{31}(-3)$, $R_{41}(-6)$

$$\sim \begin{bmatrix} 1 & 0 & 0 & 0 \\ 0 & 5 & 3 & 7 \\ 0 & 4 & 9 & 10 \\ 0 & 9 & 12 & 17 \end{bmatrix}$$

Operating $R_{23}(-1)$, $R_{43}(-2)$

$$\sim \begin{bmatrix} 1 & 0 & 0 & 0 \\ 0 & 1 & -6 & -3 \\ 0 & 4 & 9 & 10 \\ 0 & 1 & -6 & -3 \end{bmatrix}$$

Operating $C_{32}(6)$, $C_{42}(3)$

$$\sim \begin{bmatrix} 1 & 0 & 0 & 0 \\ 0 & 1 & 0 & 0 \\ 0 & 4 & 33 & 22 \\ 0 & 1 & 0 & 0 \end{bmatrix}$$

Operating $R_{32}(-4)$, $R_{42}(-1)$

$$\sim \begin{bmatrix} 1 & 0 & 0 & 0 \\ 0 & 1 & 0 & 0 \\ 0 & 0 & 33 & 22 \\ 0 & 0 & 0 & 0 \end{bmatrix}$$

Operating $C_3\left(\frac{1}{33}\right)$

$$\sim \begin{bmatrix} 1 & 0 & 0 & 0 \\ 0 & 1 & 0 & 0 \\ 0 & 0 & 1 & 22 \\ 0 & 0 & 0 & 0 \end{bmatrix}$$

Operating $C_{43}(-22)$

$$\sim \left[\begin{array}{ccc:c} 1 & 0 & 0 & 0 \\ 0 & 1 & 0 & 0 \\ 0 & 0 & 1 & 0 \\ \hdashline 0 & 0 & 0 & 0 \end{array}\right] = \left[\begin{array}{c:c} I_3 & O_{3\times 1} \\ \hdashline O_{1\times 3} & O_{1\times 1} \end{array}\right]$$

which is normal form.

$\therefore \quad \rho(A) = 3.$

Example 2. *Use elementary transformations to reduce the following matrix A to triangular form and hence find the rank of A.*

$$A = \begin{bmatrix} 2 & 3 & -1 & -1 \\ 1 & -1 & -2 & -4 \\ 3 & 1 & 3 & -2 \\ 6 & 3 & 0 & -7 \end{bmatrix}$$

Sol. We have,

$$A = \begin{bmatrix} 2 & 3 & -1 & -1 \\ 1 & -1 & -2 & -4 \\ 3 & 1 & 3 & -2 \\ 6 & 3 & 0 & -7 \end{bmatrix}$$

Operating R_{12}

$$\sim \begin{bmatrix} 1 & -1 & -2 & -4 \\ 2 & 3 & -1 & -1 \\ 3 & 1 & 3 & -2 \\ 6 & 3 & 0 & -7 \end{bmatrix}$$

Operating $R_{21}(-2)$, $R_{31}(-3)$, $R_{41}(-6)$

$$\sim \begin{bmatrix} 1 & -1 & -2 & -4 \\ 0 & 5 & 3 & 7 \\ 0 & 4 & 9 & 10 \\ 0 & 9 & 12 & 17 \end{bmatrix}$$

Operating $R_{23}(-1)$

$$\sim \begin{bmatrix} 1 & -1 & -2 & -4 \\ 0 & 1 & -6 & -3 \\ 0 & 4 & 9 & 10 \\ 0 & 9 & 12 & 17 \end{bmatrix}$$

Operating $R_{32}(-4)$, $R_{42}(-9)$.

$$\sim \begin{bmatrix} 1 & -1 & -2 & -4 \\ 0 & 1 & -6 & -3 \\ 0 & 0 & 33 & 22 \\ 0 & 0 & 66 & 44 \end{bmatrix}$$

Operating $R_{43}(-2)$

$$\sim \begin{bmatrix} 1 & -1 & -2 & -4 \\ 0 & 1 & -6 & -3 \\ 0 & 0 & 33 & 22 \\ 0 & 0 & 0 & 0 \end{bmatrix}$$

which is upper triangular form.

$\therefore \qquad \rho(A) = 3.$

Example 3. *Find the rank of matrix*

$$A = \begin{bmatrix} 2 & 3 & -2 & 4 \\ 3 & -2 & 1 & 2 \\ 3 & 2 & 3 & 4 \\ -2 & 4 & 0 & 5 \end{bmatrix}$$

Sol. We have

$$A = \begin{bmatrix} 2 & 3 & -2 & 4 \\ 3 & -2 & 1 & 2 \\ 3 & 2 & 3 & 4 \\ -2 & 4 & 0 & 5 \end{bmatrix}$$

Operating $R_{21}(-1)$

$$\sim \begin{bmatrix} 2 & 3 & -2 & 4 \\ 1 & -5 & 3 & -2 \\ 3 & 2 & 3 & 4 \\ -2 & 4 & 0 & 5 \end{bmatrix}$$

Operating R_{12}

$$\sim \begin{bmatrix} 1 & -5 & 3 & -2 \\ 2 & 3 & -2 & 4 \\ 3 & 2 & 3 & 4 \\ -2 & 4 & 0 & 5 \end{bmatrix}$$

Operating $R_{21}(-2)$, $R_{31}(-3)$, $R_{41}(2)$

$$\sim \begin{bmatrix} 1 & -5 & 3 & -2 \\ 0 & 13 & -8 & 8 \\ 0 & 17 & -6 & 10 \\ 0 & -6 & 6 & 1 \end{bmatrix}$$

Operating $R_{24}(2)$

$$\sim \begin{bmatrix} 1 & -5 & 3 & -2 \\ 0 & 1 & 4 & 10 \\ 0 & 17 & -6 & 10 \\ 0 & -6 & 6 & 1 \end{bmatrix}$$

Operating $R_{32}(-17)$, $R_{42}(6)$

$$\sim \begin{bmatrix} 1 & -5 & 3 & -2 \\ 0 & 1 & 4 & 10 \\ 0 & 0 & -74 & -160 \\ 0 & 0 & 30 & 61 \end{bmatrix}$$

Operating $R_{34}\left(\frac{5}{2}\right)$

$$\sim \begin{bmatrix} 1 & -5 & 3 & -2 \\ 0 & 1 & 4 & 10 \\ 0 & 0 & 1 & -15/2 \\ 0 & 0 & 30 & 61 \end{bmatrix}$$

Operating $R_{43}(-30)$

$$\sim \begin{bmatrix} 1 & -5 & 3 & -2 \\ 0 & 1 & 4 & 10 \\ 0 & 0 & 1 & -15/2 \\ 0 & 0 & 0 & 286 \end{bmatrix}$$

which is Echelon form.

$\therefore \qquad \rho(A) = 4.$

Example 4. (*i*) *Reduce the matrix A to its normal form when*

$$A = \begin{bmatrix} 1 & 2 & -1 & 4 \\ 2 & 4 & 3 & 4 \\ 1 & 2 & 3 & 4 \\ -1 & -2 & 6 & -7 \end{bmatrix}.$$

Hence find the rank of A. (A.K.T.U. 2019)

(*ii*) *Reduce the matrix A to its normal form and hence find its rank, where*

$$A = \begin{bmatrix} 2 & 1 & -3 & -6 \\ 3 & -3 & 1 & 2 \\ 1 & 1 & 1 & 2 \end{bmatrix}.$$

Sol. (*i*)

Pivot element ← $\boxed{1}$

$$A = \begin{bmatrix} \boxed{1} & 2 & -1 & 4 \\ 2 & 4 & 3 & 4 \\ 1 & 2 & 3 & 4 \\ -1 & -2 & 6 & -7 \end{bmatrix}$$

Operating $R_{21}(-2)$, $R_{31}(-1)$, $R_{41}(1)$

$$\sim \begin{bmatrix} 1 & 2 & -1 & 4 \\ 0 & 0 & 5 & -4 \\ 0 & 0 & 4 & 0 \\ 0 & 0 & 5 & -3 \end{bmatrix}$$

Operating $C_{21}(-2)$, $C_{31}(1)$, $C_{41}(-4)$

$$\sim \begin{bmatrix} 1 & 0 & 0 & 0 \\ 0 & 0 & 5 & -4 \\ 0 & 0 & 4 & 0 \\ 0 & 0 & 5 & -3 \end{bmatrix}$$

Operating C_{32}

$$\sim \begin{bmatrix} 1 & 0 & 0 & 0 \\ 0 & 5 & 0 & -4 \\ 0 & 4 & 0 & 0 \\ 0 & 5 & 0 & -3 \end{bmatrix}$$

Operating $R_{23}(-1)$

Pivot element ←

$$\sim \begin{bmatrix} 1 & 0 & 0 & 0 \\ 0 & \boxed{1} & 0 & -4 \\ 0 & 4 & 0 & 0 \\ 0 & 5 & 0 & -3 \end{bmatrix}$$

Operating $R_{32}(-4)$, $R_{42}(-5)$

$$\sim \begin{bmatrix} 1 & 0 & 0 & 0 \\ 0 & 1 & 0 & -4 \\ 0 & 0 & 0 & 16 \\ 0 & 0 & 0 & 17 \end{bmatrix}$$

Operating $C_{42}(4)$

$$\sim \begin{bmatrix} 1 & 0 & 0 & 0 \\ 0 & 1 & 0 & 0 \\ 0 & 0 & 0 & 16 \\ 0 & 0 & 0 & 17 \end{bmatrix}$$

Operating C_{43}

$$\sim \begin{bmatrix} 1 & 0 & 0 & 0 \\ 0 & 1 & 0 & 0 \\ 0 & 0 & 16 & 0 \\ 0 & 0 & 17 & 0 \end{bmatrix}$$

Operating $R_{34}\left(-\frac{15}{17}\right)$

$$\sim \begin{bmatrix} 1 & 0 & 0 & 0 \\ 0 & 1 & 0 & 0 \\ 0 & 0 & \boxed{1} & 0 \\ 0 & 0 & 17 & 0 \end{bmatrix}$$

Pivot element ←

Operating $R_{43}(-17)$

$$\sim \left[\begin{array}{ccc:c} 1 & 0 & 0 & 0 \\ 0 & 1 & 0 & 0 \\ 0 & 0 & 1 & 0 \\ \hdashline 0 & 0 & 0 & 0 \end{array}\right]$$

$$\sim \left[\begin{array}{c:c} I_3 & O \\ \hdashline O & O \end{array}\right] \text{ which is normal form.}$$

Hence $\rho(A) = 3$.

(*ii*) $$A = \begin{bmatrix} 2 & 1 & -3 & -6 \\ 3 & -3 & 1 & 2 \\ 1 & 1 & 1 & 2 \end{bmatrix}$$

Operating R_{13}

$$\sim \begin{bmatrix} 1 & 1 & 1 & 2 \\ 3 & -3 & 1 & 2 \\ 2 & 1 & -3 & -6 \end{bmatrix}$$

Operating $R_{21}(-3)$, $R_{31}(-2)$

$$\sim \begin{bmatrix} 1 & 1 & 1 & 2 \\ 0 & -6 & -2 & -4 \\ 0 & -1 & -5 & -10 \end{bmatrix}$$

Operating $C_{21}(-1)$, $C_{31}(-1)$, $C_{41}(-2)$

$$\sim \begin{bmatrix} 1 & 0 & 0 & 0 \\ 0 & -6 & -2 & -4 \\ 0 & -1 & -5 & -10 \end{bmatrix}$$

Operating $R_2(-1)$, $R_3(-1)$

$$\sim \begin{bmatrix} 1 & 0 & 0 & 0 \\ 0 & 6 & 2 & 4 \\ 0 & 1 & 5 & 10 \end{bmatrix}$$

Operating R_{23}

$$\sim \begin{bmatrix} 1 & 0 & 0 & 0 \\ 0 & 1 & 5 & 10 \\ 0 & 6 & 2 & 4 \end{bmatrix}$$

Operating $R_{32}(-6)$

$$\sim \begin{bmatrix} 1 & 0 & 0 & 0 \\ 0 & 1 & 5 & 10 \\ 0 & 0 & -28 & -56 \end{bmatrix}$$

Operating $C_{32}(-5)$, $C_{42}(-10)$

$$\sim \begin{bmatrix} 1 & 0 & 0 & 0 \\ 0 & 1 & 0 & 0 \\ 0 & 0 & -28 & -56 \end{bmatrix}$$

Operating $R_3\left(-\dfrac{1}{28}\right)$

$$\sim \begin{bmatrix} 1 & 0 & 0 & 0 \\ 0 & 1 & 0 & 0 \\ 0 & 0 & 1 & 2 \end{bmatrix}$$

Operating $C_{43}(-2)$

$$\sim \left[\begin{array}{ccc:c} 1 & 0 & 0 & 0 \\ 0 & 1 & 0 & 0 \\ 0 & 0 & 1 & 0 \end{array}\right]$$

$$\sim [I_3 \;\vdots\; O] \quad \text{which is normal form.}$$

Hence $\rho(A) = 3$.

Example 5. *For the matrix* $A = \begin{bmatrix} 1 & 1 & 2 \\ 1 & 2 & 3 \\ 0 & -1 & -1 \end{bmatrix}$, *find non-singular matrices P and Q such that PAQ is in the normal form.*

Sol. We write $A = I_3 A I_3$

$$\Rightarrow \quad \begin{bmatrix} 1 & 1 & 2 \\ 1 & 2 & 3 \\ 0 & -1 & -1 \end{bmatrix} = \begin{bmatrix} 1 & 0 & 0 \\ 0 & 1 & 0 \\ 0 & 0 & 1 \end{bmatrix} A \begin{bmatrix} 1 & 0 & 0 \\ 0 & 1 & 0 \\ 0 & 0 & 1 \end{bmatrix}$$

Operating $C_{21}(-1)$, $C_{31}(-2)$ (subjecting the *post-factor* on RHS to same operations)

$$\begin{bmatrix}1 & 0 & 0\\ 1 & 1 & 1\\ 0 & -1 & -1\end{bmatrix} = \begin{bmatrix}1 & 0 & 0\\ 0 & 1 & 0\\ 0 & 0 & 1\end{bmatrix} A \begin{bmatrix}1 & -1 & -2\\ 0 & 1 & 0\\ 0 & 0 & 1\end{bmatrix}$$

Operating $R_{21}(-1)$ (subjecting the *pre-factor* on RHS to same operation)

$$\begin{bmatrix}1 & 0 & 0\\ 0 & 1 & 1\\ 0 & -1 & -1\end{bmatrix} = \begin{bmatrix}1 & 0 & 0\\ -1 & 1 & 0\\ 0 & 0 & 1\end{bmatrix} A \begin{bmatrix}1 & -1 & -2\\ 0 & 1 & 0\\ 0 & 0 & 1\end{bmatrix}$$

Operating $C_{32}(-1)$ (subjecting the *post-factor* on RHS to same operation)

$$\begin{bmatrix}1 & 0 & 0\\ 0 & 1 & 0\\ 0 & -1 & 0\end{bmatrix} = \begin{bmatrix}1 & 0 & 0\\ -1 & 1 & 0\\ 0 & 0 & 1\end{bmatrix} A \begin{bmatrix}1 & -1 & -1\\ 0 & 1 & -1\\ 0 & 0 & 1\end{bmatrix}$$

Operating $R_{32}(1)$ (subjecting the *pre-factor* on RHS to same operation)

$$\begin{bmatrix}1 & 0 & 0\\ 0 & 1 & 0\\ 0 & 0 & 0\end{bmatrix} = \begin{bmatrix}1 & 0 & 0\\ -1 & 1 & 0\\ -1 & 1 & 0\end{bmatrix} A \begin{bmatrix}1 & -1 & -1\\ 0 & 1 & -1\\ 0 & 0 & 1\end{bmatrix}$$

$$\Rightarrow \quad \begin{bmatrix}I_2 & O_{2\times 1}\\ O_{1\times 2} & O_{1\times 1}\end{bmatrix} = PAQ \quad \text{where} \quad P = \begin{bmatrix}1 & 0 & 0\\ -1 & 1 & 0\\ -1 & 1 & 1\end{bmatrix} \text{ and } Q = \begin{bmatrix}1 & -1 & -1\\ 0 & 1 & -1\\ 0 & 0 & 1\end{bmatrix}.$$

Example 6. *If* $A = \begin{bmatrix}3 & -3 & 4\\ 2 & -3 & 4\\ 0 & -1 & 1\end{bmatrix}$, *determine two non-singular matrices P and Q such that PAQ = I. Hence find* A^{-1}.

Sol. Let $\quad A = I_3 A I_3$

$$\begin{bmatrix}3 & -3 & 4\\ 2 & -3 & 4\\ 0 & -1 & 1\end{bmatrix} = \begin{bmatrix}1 & 0 & 0\\ 0 & 1 & 0\\ 0 & 0 & 1\end{bmatrix} A \begin{bmatrix}1 & 0 & 0\\ 0 & 1 & 0\\ 0 & 0 & 1\end{bmatrix}$$

Operating $R_{12}(-1)$

$$\begin{bmatrix}1 & 0 & 0\\ 2 & -3 & 4\\ 0 & -1 & 1\end{bmatrix} = \begin{bmatrix}1 & -1 & 0\\ 0 & 1 & 0\\ 0 & 0 & 1\end{bmatrix} A \begin{bmatrix}1 & 0 & 0\\ 0 & 1 & 0\\ 0 & 0 & 1\end{bmatrix}$$

Operating $R_{21}(-2)$

$$\begin{bmatrix}1 & 0 & 0\\ 0 & -3 & 4\\ 0 & -1 & 1\end{bmatrix} = \begin{bmatrix}1 & -1 & 0\\ -2 & 3 & 0\\ 0 & 0 & 1\end{bmatrix} A \begin{bmatrix}1 & 0 & 0\\ 0 & 1 & 0\\ 0 & 0 & 1\end{bmatrix}$$

Operating $R_{23}(-4)$

$$\begin{bmatrix}1 & 0 & 0\\ 0 & 1 & 0\\ 0 & -1 & 1\end{bmatrix} = \begin{bmatrix}1 & -1 & 0\\ -2 & 3 & -4\\ 0 & 0 & 1\end{bmatrix} A \begin{bmatrix}1 & 0 & 0\\ 0 & 1 & 0\\ 0 & 0 & 1\end{bmatrix}$$

Operating $C_{23}(1)$

$$\begin{bmatrix} 1 & 0 & 0 \\ 0 & 1 & 0 \\ 0 & 0 & 1 \end{bmatrix} = \begin{bmatrix} 1 & -1 & 0 \\ -2 & 3 & -4 \\ 0 & 0 & 1 \end{bmatrix} A \begin{bmatrix} 1 & 0 & 0 \\ 0 & 1 & 0 \\ 0 & 1 & 1 \end{bmatrix}$$

$\Rightarrow \qquad I = PAQ$

where $\qquad P = \begin{bmatrix} 1 & -1 & 0 \\ -2 & 3 & -4 \\ 0 & 0 & 1 \end{bmatrix}$ and $Q = \begin{bmatrix} 1 & 0 & 0 \\ 0 & 1 & 0 \\ 0 & 1 & 1 \end{bmatrix}$

Now, $\quad PAQ = I$

Pre-multiplying by P^{-1},

$$P^{-1} PAQ = P^{-1}$$

$$AQ = P^{-1} \qquad | \because \; P^{-1}P = I$$

Post-multiplying by P,

$$AQP = I$$

Pre-multiplying by A^{-1},

$$QP = A^{-1}$$

$$\therefore \qquad A^{-1} = \begin{bmatrix} 1 & 0 & 0 \\ 0 & 1 & 0 \\ 0 & 1 & 1 \end{bmatrix} \begin{bmatrix} 1 & -1 & 0 \\ -2 & 3 & -4 \\ 0 & 0 & 1 \end{bmatrix} = \begin{bmatrix} 1 & -1 & 0 \\ -2 & 3 & -4 \\ -2 & 3 & -3 \end{bmatrix}.$$

Example 7. *Prove that the points (x_1, y_1), (x_2, y_2) and (x_3, y_3) are collinear if and only if the rank of the matrix $\begin{bmatrix} x_1 & y_1 & 1 \\ x_2 & y_2 & 1 \\ x_3 & y_3 & 1 \end{bmatrix}$ is less than three.*

Sol. The condition is necessary

We are given that the points (x_i, y_i); $i = 1, 2, 3$ are collinear. We are to prove that the rank of matrix $\begin{bmatrix} x_1 & y_1 & 1 \\ x_2 & y_2 & 1 \\ x_3 & y_3 & 1 \end{bmatrix}$ is less than three.

Since the points are collinear, therefore, the area of the triangle formed by these points is zero.

Hence $\quad \dfrac{1}{2}\begin{vmatrix} x_1 & y_1 & 1 \\ x_2 & y_2 & 1 \\ x_3 & y_3 & 1 \end{vmatrix} = 0 \qquad \Rightarrow \qquad \begin{vmatrix} x_1 & y_1 & 1 \\ x_2 & y_2 & 1 \\ x_3 & y_3 & 1 \end{vmatrix} = 0$

$\Rightarrow$ The rank of matrix $\begin{bmatrix} x_1 & y_1 & 1 \\ x_2 & y_2 & 1 \\ x_3 & y_3 & 1 \end{bmatrix}$ is less than 3.

Hence the condition is necessary.

The condition is sufficient.

We are given that the rank of the matrix $\begin{bmatrix} x_1 & y_1 & 1 \\ x_2 & y_2 & 1 \\ x_3 & y_3 & 1 \end{bmatrix}$ is less than 3 and we are to prove that the points (x_i, y_i) ; $i = 1, 2, 3$ are collinear. Since rank of matrix $\begin{bmatrix} x_1 & y_1 & 1 \\ x_2 & y_2 & 1 \\ x_3 & y_3 & 1 \end{bmatrix}$ is less than 3,

hence $$\begin{vmatrix} x_1 & y_1 & 1 \\ x_2 & y_2 & 1 \\ x_3 & y_3 & 1 \end{vmatrix} = 0.$$

$$\Rightarrow \quad \frac{1}{2}\begin{vmatrix} x_1 & y_1 & 1 \\ x_2 & y_2 & 1 \\ x_3 & y_3 & 1 \end{vmatrix} = 0$$

Hence the given points are collinear.

∴ The condition is sufficient.

Example 8. *Show that:*

(i) rank (AA′) = rank (A) *(ii) rank (A) = rank (AA*).*

Sol. (*i*) Let B = AA′, then

$\rho(B) = \rho(AA')$

$\leq \rho(A)$ | ∵ Rank of a product of two matrices cannot exceed the rank of either matrix.

∴ $\rho(B) \leq \rho(A)$ ...(1)

Now, $B = AA'$

⇒ $A^{-1}B = A'$

∴ $\rho(A) = \rho(A') = \rho(A^{-1}B)$ | ∵ A matrix and its transpose have the same rank.

$\leq \rho(B)$

∴ $\rho(A) \leq \rho(B)$ ...(2)

From (1) and (2), $\rho(A) = \rho(B) = \rho(AA')$.

(*ii*) Let $C = AA^*$

∴ $\rho(C) = \rho(AA^*) \leq \rho(A)$

∴ $\rho(C) \leq \rho(A)$ ...(1)

Again, $C = AA^*$

⇒ $A^{-1}C = A^*$

Now, $\rho(A) = \rho(A^*) = \rho(A^{-1}C)$ | ∵ $\rho(A) = \rho(A^*)$

$\leq \rho(C)$

∴ $\rho(A) \leq \rho(C)$ ...(2)

Combining (1) and (2), we get

$\rho(A) = \rho(C)$

⇒ $\rho(A) = \rho(AA^*)$.

TEST YOUR KNOWLEDGE

1. Find the rank of the matrices

(i) $\begin{bmatrix} 2 & -1 & 0 & 5 \\ 0 & 3 & 1 & 4 \end{bmatrix}$ (ii) $\begin{bmatrix} 1 & 2 & 3 \\ 2 & 4 & 7 \\ 3 & 6 & 10 \end{bmatrix}$ (iii) $\begin{bmatrix} 1 & 3 & 4 & 5 \\ 1 & 2 & 6 & 7 \\ 1 & 5 & 0 & 10 \end{bmatrix}$ (iv) $\begin{bmatrix} 1 & 2 & 3 & 0 \\ 2 & 4 & 3 & 2 \\ 3 & 2 & 1 & 3 \\ 6 & 8 & 7 & 5 \end{bmatrix}$.

2. Find the rank of

(i) $\begin{bmatrix} 3 & -1 & 2 \\ -6 & 2 & 4 \\ -3 & 1 & 2 \end{bmatrix}$ (ii) $\begin{bmatrix} 1 & -1 & 2 & -3 \\ 4 & 1 & 0 & 2 \\ 0 & 3 & 0 & 4 \\ 0 & 1 & 0 & 2 \end{bmatrix}$ (iii) $\begin{bmatrix} 0 & 1 & -3 & -1 \\ 1 & 0 & 1 & 1 \\ 3 & 1 & 0 & 2 \\ 1 & 1 & -2 & 0 \end{bmatrix}$.

(U.K.T.U. 2012)

3. Find the rank of the following matrices:

(i) $\begin{bmatrix} 3 & 11 & 1 & 5 \\ 5 & 13 & -1 & -11 \\ -2 & 2 & 4 & -8 \end{bmatrix}$ (ii) $\begin{bmatrix} 1^2 & 2^2 & 3^2 & 4^2 \\ 2^2 & 3^2 & 4^2 & 5^2 \\ 3^2 & 4^2 & 5^2 & 6^2 \\ 4^2 & 5^2 & 6^2 & 7^2 \end{bmatrix}$ (iii) $\begin{bmatrix} 5 & 3 & 14 & 4 \\ 0 & 1 & 3 & 1 \\ 1 & -1 & 2 & 0 \end{bmatrix}$.

4. (i) Reduce the matrix $A = \begin{bmatrix} 1 & 1 & -1 & 1 \\ -1 & 1 & -3 & -3 \\ 1 & 0 & 1 & 2 \\ 1 & -1 & 3 & 3 \end{bmatrix}$ to column echelon form and find its rank.

(ii) Reduce A to Echelon form and then to its row canonical form

where $A = \begin{bmatrix} 1 & 3 & -1 & 2 \\ 0 & 11 & -5 & 3 \\ 2 & -5 & 3 & 1 \\ 4 & 1 & 1 & 5 \end{bmatrix}$

Hence find the rank of A. (U.P.T.U. 2015)

5. Using elementary transformation, reduce the following matrices to the canonical form:

(i) $\begin{bmatrix} 0 & 0 & 0 & 0 & 0 \\ 0 & 1 & 2 & 3 & 4 \\ 0 & 2 & 3 & 4 & 1 \\ 0 & 3 & 4 & 1 & 2 \end{bmatrix}$ (ii) $\begin{bmatrix} 0 & 4 & -12 & 8 & 9 \\ 0 & 2 & -6 & 2 & 5 \\ 0 & 1 & -3 & 6 & 4 \\ 0 & -8 & 24 & 3 & 1 \end{bmatrix}$ (iii) $\begin{bmatrix} 1 & 2 & 0 & -1 \\ 3 & 4 & 1 & 2 \\ -2 & 3 & 2 & 5 \end{bmatrix}$

(iv) $\begin{bmatrix} 1 & 2 & 3 & 4 \\ 3 & 4 & 1 & 2 \\ 4 & 3 & 1 & 2 \end{bmatrix}$ (v) $\begin{bmatrix} 1 & 2 & -1 & 4 \\ 2 & 3 & 3 & 4 \\ 1 & 2 & 3 & 4 \\ -1 & -2 & 6 & -7 \end{bmatrix}$. (U.K.T.U. 2010)

6. (i) Find non-singular matrices P and Q such that PAQ is in the normal form for the matrix

$$A = \begin{bmatrix} 1 & -1 & -1 \\ 1 & 1 & 1 \\ 3 & 1 & 1 \end{bmatrix}$$

(ii) Find non-singular matrices P and Q such that PAQ is in normal form of the matrix and hence find the rank of matrix $A = \begin{bmatrix} 1 & 2 & 3 & -2 \\ 2 & -2 & 1 & 3 \\ 3 & 0 & 4 & 1 \end{bmatrix}$ (U.P.T.U. 2015) (*iii*) **Obtain a matrix N in the normal form equivalent to** $A = \begin{bmatrix} 0 & 0 & 0 & 0 & 0 \\ 0 & 4 & 5 & 0 & 0 \\ 0 & 9 & 1 & -1 & 2 \\ 0 & 10 & 0 & 1 & 11 \end{bmatrix}$

Hence find non-singular matrices P and Q such that PAQ = N.

(*iv*) **Find non-singular matrices P and Q such that PAQ is in normal form for the matrix A where**

$$A = \begin{bmatrix} 1 & 1 & 2 \\ 1 & 2 & 3 \\ 0 & 1 & 1 \end{bmatrix}$$ (*A.K.T.U. 2021*)

7. Find the rank of the following matrices using elementary transformations:

(*i*) $\begin{bmatrix} 1 & -3 & 1 & 2 \\ 0 & 1 & 2 & 3 \\ 3 & 4 & 1 & -2 \end{bmatrix}$ (*ii*) $\begin{bmatrix} 1 & 2 & 3 & 4 & 5 \\ 2 & 3 & 4 & 5 & 1 \\ 5 & 8 & 11 & 14 & 7 \end{bmatrix}$ (*iii*) $\begin{bmatrix} 3 & 4 & 1 & 1 \\ 2 & 4 & 3 & 6 \\ -1 & -2 & 6 & 4 \\ 1 & -1 & 2 & -3 \end{bmatrix}$

(*G.B.T.U. 2010*)

(*iv*) $\begin{bmatrix} 1 & 2 & 3 & 1 \\ 2 & 4 & 6 & 2 \\ 1 & 2 & 3 & 2 \end{bmatrix}$ (*v*) $\begin{bmatrix} -2 & -1 & 3 & -1 \\ 1 & 2 & -3 & -1 \\ 1 & 0 & 1 & 1 \\ 0 & 1 & 1 & -1 \end{bmatrix}$. (*A.K.T.U. 2018*)

8. Find the rank of the following matrices by reducing it to normal form (*or* canonical form):

(*i*) $\begin{bmatrix} 1 & 2 & -1 & 3 \\ 4 & 1 & 2 & 1 \\ 3 & -1 & 1 & 2 \\ 1 & 2 & 0 & 1 \end{bmatrix}$ (*ii*) $\begin{bmatrix} 1 & 2 & 3 & 2 \\ 2 & 3 & 5 & 1 \\ 1 & 3 & 4 & 5 \end{bmatrix}$ (*iii*) $\begin{bmatrix} 1 & 3 & 2 & 5 & 1 \\ 2 & 2 & -1 & 6 & 3 \\ 1 & 1 & 2 & 3 & -1 \\ 0 & 2 & 5 & 2 & -3 \end{bmatrix}$

(*iv*) $\begin{bmatrix} 3 & 2 & -1 \\ 4 & 2 & 6 \\ 7 & 4 & 5 \end{bmatrix}$ (*v*) $\begin{bmatrix} 0 & 1 & 2 & -2 \\ 4 & 0 & 2 & 6 \\ 2 & 1 & 3 & 1 \end{bmatrix}$ (*vi*) $\begin{bmatrix} 1 & 2 & -1 & 3 \\ 2 & 4 & -4 & 7 \\ -1 & -2 & -1 & -2 \end{bmatrix}$

(*A.K.T.U. 2016*)

(*vii*) $\begin{bmatrix} 5 & 3 & 14 & 4 \\ 0 & 1 & 2 & 1 \\ 1 & -1 & 2 & 0 \end{bmatrix}$ (*viii*) $\begin{bmatrix} 1 & 2 & 1 & 0 \\ -2 & 4 & 3 & 0 \\ 1 & 0 & 2 & -8 \end{bmatrix}$ (*ix*) $\begin{bmatrix} 0 & 1 & 2 & -1 \\ 1 & 2 & 1 & 1 \\ 3 & 1 & 0 & 2 \\ 1 & 1 & -2 & 0 \end{bmatrix}$.

(*G.B.T.U. 2011*) (*U.P.T.U. 2015*) (*A.K.T.U. 2018*)

(*x*) $\begin{bmatrix} 6 & 1 & 3 & 8 \\ 4 & 2 & 6 & -1 \\ 10 & 3 & 9 & 7 \\ 16 & 4 & 12 & 15 \end{bmatrix}$ (*xi*) $\begin{bmatrix} 1 & 1 & 2 & 3 \\ 1 & 3 & 0 & 3 \\ 1 & -2 & -3 & -3 \\ 1 & 1 & 2 & 3 \end{bmatrix}$ (*xii*) $\begin{bmatrix} -1 & -3 & 3 & -1 \\ 1 & 1 & -1 & 0 \\ 2 & -5 & 2 & -3 \end{bmatrix}$

(*U.P.T.U. 2014*)

9. Find the rank of the matrix $A = \begin{bmatrix} 1 & 1 & 1 \\ a & b & c \\ a^3 & b^3 & c^3 \end{bmatrix}$ **;** a, b, c **being all real.**

10. Show that the rank of a skew-symmetric matrix cannot be unity.

11. (*i*) **Find all values of μ for which rank of the matrix**

$$A = \begin{bmatrix} \mu & -1 & 0 & 0 \\ 0 & \mu & -1 & 0 \\ 0 & 0 & \mu & -1 \\ -6 & 11 & -6 & 1 \end{bmatrix}$$ is equal to 3.

(*ii*) Find the value of P for which the matrix

$$A = \begin{bmatrix} 3 & P & P \\ P & 3 & P \\ P & P & 3 \end{bmatrix} \text{ is of rank 1.}$$

(*M.T.U. 2012*)

12. Under what condition, the rank of the following matrix A is 3? Is it possible for the rank to be 1? Why?

$$A = \begin{bmatrix} 2 & 4 & 2 \\ 3 & 1 & 2 \\ 1 & 0 & x \end{bmatrix}.$$

13. If $A = \begin{bmatrix} 1 & 5 & 4 \\ 0 & 3 & 2 \\ 2 & 3 & 10 \end{bmatrix}$, $B = \begin{bmatrix} 1 & 1 & 1 \\ 2 & 2 & 2 \\ 3 & 3 & 3 \end{bmatrix}$, find rank of A, rank of B, rank of A + B and rank of AB.

14. Show that if A is a non-zero column matrix and B is a non-zero row matrix then $\rho(AB) = 1$.

15. (*i*) Show that the rank of the transpose of a matrix is the same as that of the original matrix.

(*ii*) Prove that for a $m \times n$ matrix, whose every element is 1, the rank is one.

16. Find the rank of the matrix

$$A = \begin{bmatrix} 1 & 3 & 4 & 2 \\ 2 & -1 & 3 & 2 \\ 3 & -5 & 2 & 2 \\ 6 & -3 & 8 & 6 \end{bmatrix}$$

by reducing it to normal form. (*A.K.T.U. 2020*)

Answers

1. (*i*) 2 (*ii*) 2 (*iii*) 3 (*iv*) 3

2. (*i*) 2 (*ii*) 4 (*iii*) 2

3. (*i*) 3 (*ii*) 3 (*iii*) 3

4. (*i*) 2 (*ii*) 2

5. (*i*) $\begin{bmatrix} I_3 & \vdots & O_{3\times 2} \\ \cdots & & \cdots \\ O_{1\times 3} & \vdots & O_{1\times 2} \end{bmatrix}$ (*ii*) $\begin{bmatrix} I_3 & \vdots & O_{3\times 2} \\ \cdots & & \cdots \\ O_{1\times 3} & \vdots & O_{1\times 2} \end{bmatrix}$ (*iii*) $[I_3 : O_{3\times 1}]$

(*iv*) $[I_3 : O_{3\times 1}]$ (*v*) $\begin{bmatrix} I_3 & \vdots & O_{3\times 1} \\ \cdots & & \cdots \\ O_{1\times 3} & \vdots & O_{1\times 1} \end{bmatrix}$

6. (*i*) $P = \begin{bmatrix} 1 & 0 & 0 \\ -\frac{1}{2} & \frac{1}{2} & 0 \\ \frac{1}{2} & 1 & -\frac{1}{2} \end{bmatrix}$, $Q = \begin{bmatrix} 1 & 1 & 0 \\ 0 & 1 & -1 \\ 0 & 0 & 1 \end{bmatrix}$. (*ii*) $P = \begin{bmatrix} 1 & 0 & 0 \\ -2 & 1 & 0 \\ -1 & -1 & 1 \end{bmatrix}$, $Q = \begin{bmatrix} 1 & \frac{1}{3} & \frac{4}{15} & -\frac{1}{21} \\ 0 & -\frac{1}{6} & \frac{1}{6} & \frac{1}{6} \\ 0 & 0 & -\frac{1}{5} & 0 \\ 0 & 0 & 0 & \frac{1}{7} \end{bmatrix}$; $\rho(A) = 2$

(*iii*) $P = \begin{bmatrix} 0 & 0 & 1 & 0 \\ 0 & -1 & 5 & -4 \\ 0 & 10 & -50 & 41 \\ 1 & 0 & 0 & 0 \end{bmatrix}$, $Q = \begin{bmatrix} 0 & 0 & 0 & 1 & 0 \\ 0 & 1 & 0 & 0 & 34 \\ 1 & -9 & 1 & 0 & -659 \\ 0 & 0 & 1 & 0 & -351 \\ 0 & 0 & 0 & 0 & 1 \end{bmatrix}$

(*iv*) $P = \begin{bmatrix} 1 & 0 & 0 \\ -1 & 1 & 0 \\ 1 & -1 & 1 \end{bmatrix}$, $Q = \begin{bmatrix} 1 & -1 & -1 \\ 0 & 1 & -1 \\ 0 & 0 & 1 \end{bmatrix}$

7. (*i*) 3 (*ii*) 2 (*iii*) 4 (*iv*) 2 (*v*) 3

8. (*i*) 3 (*ii*) 2 (*iii*) 3 (*iv*) 2 (*v*) 2
(*vi*) 2 (*vii*) 3 (*viii*) 3 (*ix*) 4 (*x*) 2
(*xi*) 3 (*xii*) 3

9. $\rho(A) = 3$ if $a \neq b \neq c$ and $a + b + c \neq 0$; $\rho(A) = 2$ if $a \neq b \neq c$ and $a + b + c = 0$
Also $\rho(A) = 2$ if $a = b \neq c$ while $\rho(A) = 1$ if $a = b = c$

11. (*i*) $\mu = 1, 2, 3$ (*ii*) $P = 3$ **12.** $x = 3/5$; No

13. $\rho(A) = 3$, $\rho(B) = 1$, $\rho(A + B) = 3$, $\rho(AB) = 1$.

16. 3

1.30. SOLUTION OF A SYSTEM OF LINEAR EQUATIONS

Consider the system of equations $\left.\begin{aligned} a_1x + b_1y + c_1z = d_1 \\ a_2x + b_2y + c_2z = d_2 \\ a_3x + b_3y + c_3z = d_3 \end{aligned}\right\}$ (3 equations in 3 unknowns)

In matrix notation, these equations can be written as

$$\begin{bmatrix} a_1 & b_1 & c_1 \\ a_2 & b_2 & c_2 \\ a_3 & b_3 & c_3 \end{bmatrix}\begin{bmatrix} x \\ y \\ z \end{bmatrix} = \begin{bmatrix} d_1 \\ d_2 \\ d_3 \end{bmatrix}$$

or $$AX = B$$

where $A = \begin{bmatrix} a_1 & b_1 & c_1 \\ a_2 & b_2 & c_2 \\ a_3 & b_3 & c_3 \end{bmatrix}$ is called the co-efficient matrix, $X = \begin{bmatrix} x \\ y \\ z \end{bmatrix}$ is the column matrix of unknowns, $B = \begin{bmatrix} d_1 \\ d_2 \\ d_3 \end{bmatrix}$ is the column matrix of constants.

If $d_1 = d_2 = d_3 = 0$, then $B = O$ and the matrix equation $AX = B$ reduces to $AX = O$. Such a system of equations is called a system of **homogeneous linear equations.**
If at least one of d_1, d_2, d_3 is non-zero, then $B \neq O$.
Such a system of equations is called a system of **non-homogeneous linear equations.**

Solving the matrix equation $AX = B$ means finding X, *i.e.*, finding a column matrix $\begin{bmatrix} \alpha \\ \beta \\ \gamma \end{bmatrix}$ such that $X = \begin{bmatrix} x \\ y \\ z \end{bmatrix} = \begin{bmatrix} \alpha \\ \beta \\ \gamma \end{bmatrix}$. Then $x = \alpha$, $y = \beta$, $z = \gamma$.

The matrix equation $AX = B$ need not always have a solution. It may have no solution or a unique solution or an infinite number of solutions.

A system of equations having **no solution** is called an **inconsistent** system of equations.

A system of equations having **one or more solution** is called **a consistent** system of equations.

For a system of non-homogeneous linear equations AX = B.

(*i*) if $\rho[A : B] \neq \rho(A)$, the system is inconsistent.

(*ii*) if $\rho[A : B] = \rho(A)$ = number of unknowns, the system has a unique solution.

(*iii*) $\rho[A : B] = \rho(A) <$ number of unknowns, the system has an infinite number of solutions.

The matrix [A : B] in which the elements of A and B are written side by side is called the **augmented matrix.**

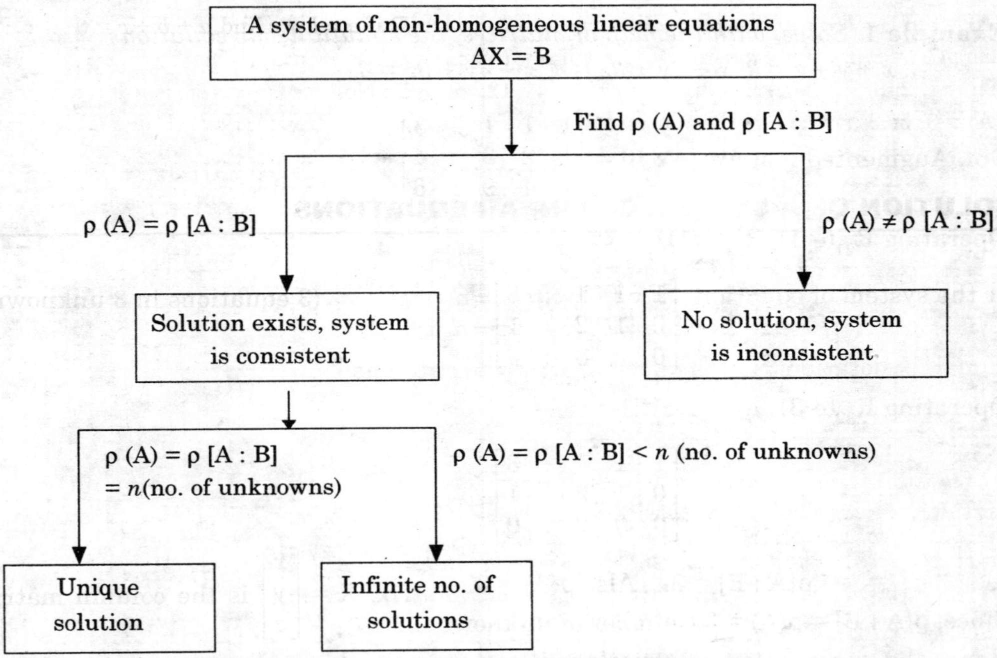

For a system of homogeneous linear equations AX = O

(*i*) X = O is always a solution. This solution in which each unknown has the value zero is called the **Null Solution** or the **Trivial Solution.** Thus, a homogeneous system is always consistent.

A system of homogeneous linear equations has either the trivial solution or an infinite number of solutions.

(*ii*) if ρ (A) = number of unknowns, the system has only the trivial solution.

(*iii*) if ρ (A) < number of unknowns, the system has an infinite number of non-trivial solutions.

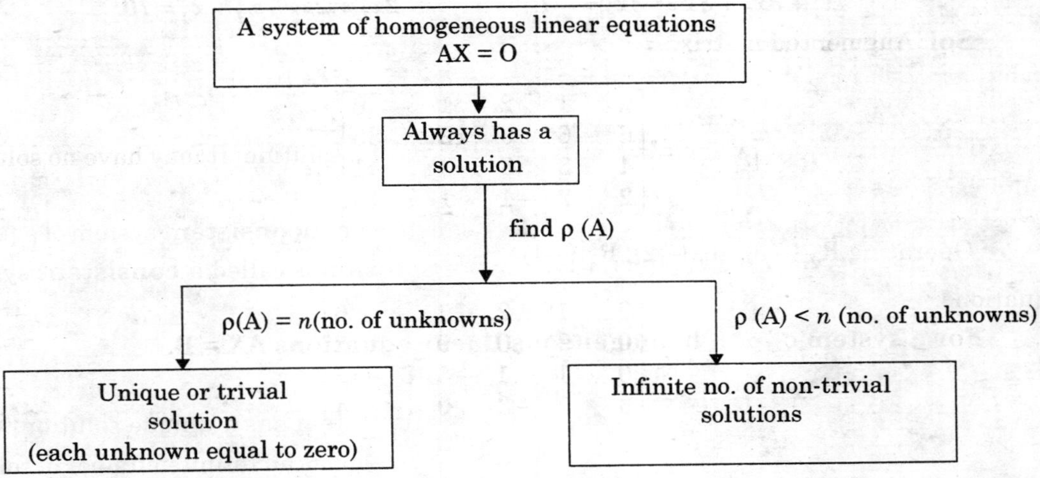

ILLUSTRATIVE EXAMPLES

Example 1. *Solve, with the help of matrices, the simultaneous equations:*

$$x + y + z = 3,\ x + 2y + 3z = 4,\ x + 4y + 9z = 6.$$

Sol. Augmented matrix $[A : B] = \begin{bmatrix} 1 & 1 & 1 & \vdots & 3 \\ 1 & 2 & 3 & \vdots & 4 \\ 1 & 4 & 9 & \vdots & 6 \end{bmatrix}$

Operating $R_{21}(-1)$, $R_{31}(-1)$

$$\sim \begin{bmatrix} 1 & 1 & 1 & \vdots & 3 \\ 0 & 1 & 2 & \vdots & 1 \\ 0 & 3 & 8 & \vdots & 3 \end{bmatrix}$$

Operating $R_{32}(-3)$

$$\sim \begin{bmatrix} 1 & 1 & 1 & \vdots & 3 \\ 0 & 1 & 2 & \vdots & 1 \\ 0 & 0 & 2 & \vdots & 0 \end{bmatrix}$$

$\therefore$ $\rho[A : B] = 3$. Also $\rho(A) = 3$.

Since, $\rho[A : B] = \rho(A) = 3$ (number of unknowns).

Hence the given system of equations is consistent and has a unique solution.

Equivalent system of equations is

$$x + y + z = 3$$
$$y + 2z = 1$$
$$2z = 0$$

$\Rightarrow$ $x = 2,\ y = 1,\ z = 0.$

Example 2. *Solve the system of equations using matrix method:*

$$2x_1 + x_2 + 2x_3 + x_4 = 6, \qquad 6x_1 - 6x_2 + 6x_3 + 12x_4 = 36,$$
$$4x_1 + 3x_2 + 3x_3 - 3x_4 = -1, \qquad 2x_1 + 2x_2 - x_3 + x_4 = 10.$$

Sol. Augmented matrix,

$$[A : B] = \begin{bmatrix} 2 & 1 & 2 & 1 & \vdots & 6 \\ 6 & -6 & 6 & 12 & \vdots & 36 \\ 4 & 3 & 3 & -3 & \vdots & -1 \\ 2 & 2 & -1 & 1 & \vdots & 10 \end{bmatrix}$$

Operating $R_{21}(-3)$, $R_{31}(-2)$, $R_{41}(-1)$

$$\sim \begin{bmatrix} 2 & 1 & 2 & 1 & \vdots & 6 \\ 0 & -9 & 0 & 9 & \vdots & 18 \\ 0 & 1 & -1 & -5 & \vdots & -13 \\ 0 & 1 & -3 & 0 & \vdots & 4 \end{bmatrix}$$

Operating $R_2\left(-\frac{1}{9}\right)$

$$\sim \left[\begin{array}{cccc:c} 2 & 1 & 2 & 1 & 6 \\ 0 & 1 & 0 & -1 & -2 \\ 0 & 1 & -1 & -5 & -13 \\ 0 & 1 & -3 & 0 & 4 \end{array}\right]$$

Operating $R_{32}(-1)$, $R_{42}(-1)$

$$\sim \left[\begin{array}{cccc:c} 2 & 1 & 2 & 1 & 6 \\ 0 & 1 & 0 & -1 & -2 \\ 0 & 0 & -1 & -4 & -11 \\ 0 & 0 & -3 & 1 & 6 \end{array}\right]$$

Operating $R_{43}(-3)$

$$\sim \left[\begin{array}{cccc:c} 2 & 1 & 2 & 1 & 6 \\ 0 & 1 & 0 & -1 & -2 \\ 0 & 0 & -1 & -4 & -11 \\ 0 & 0 & 0 & 13 & 39 \end{array}\right]$$

which is echelon form

$\therefore$ $\rho[A : B] = 4$. Also $\rho(A) = 4$.

Since $\rho[A : B] = \rho(A) = 4$ (number of variables).

Hence the system of equations is consistent and has a unique solution.

Equivalent system of equations is

$$2x_1 + x_2 + 2x_3 + x_4 = 6$$
$$x_2 - x_4 = -2$$
$$-x_3 - 4x_4 = -11$$
$$13x_4 = 39$$

On solving, we get $x_1 = 2, x_2 = 1, x_3 = -1, x_4 = 3$.

Example 3. *Using matrix method, show that the equations*

$$3x + 3y + 2z = 1,\ x + 2y = 4,\ 10y + 3z = -2,\ 2x - 3y - z = 5$$

are consistent and hence obtain the solutions for x, y and z. [U.K.T.U. 2010]

Sol. Augmented matrix

$$[A : B] = \left[\begin{array}{ccc:c} 3 & 3 & 2 & 1 \\ 1 & 2 & 0 & 4 \\ 0 & 10 & 3 & -2 \\ 2 & -3 & -1 & 5 \end{array}\right]$$

Operating R_{12}

$$\sim \left[\begin{array}{ccc:c} 1 & 2 & 0 & 4 \\ 3 & 3 & 2 & 1 \\ 0 & 10 & 3 & -2 \\ 2 & -3 & -1 & 5 \end{array}\right]$$

Operating $R_{21}(-3)$, $R_{41}(-2)$

$$\sim \begin{bmatrix} 1 & 2 & 0 & : & 4 \\ 0 & -3 & 2 & : & -11 \\ 0 & 10 & 3 & : & -2 \\ 0 & -7 & -1 & : & -3 \end{bmatrix}$$

Operating $R_{32}(3)$, $R_{42}(-2)$

$$\sim \begin{bmatrix} 1 & 2 & 0 & : & 4 \\ 0 & -3 & 2 & : & -11 \\ 0 & 1 & 9 & : & -35 \\ 0 & -1 & -5 & : & 19 \end{bmatrix}$$

Operating $R_{13}(-2)$, $R_{23}(3)$, $R_{43}(1)$

$$\sim \begin{bmatrix} 1 & 0 & -18 & : & 74 \\ 0 & 0 & 29 & : & -116 \\ 0 & 1 & 9 & : & -35 \\ 0 & 0 & 4 & : & -16 \end{bmatrix}$$

Operating R_{23}, $R_4\left(\frac{1}{4}\right)$

$$\sim \begin{bmatrix} 1 & 0 & -18 & : & 74 \\ 0 & 1 & 9 & : & -35 \\ 0 & 0 & 29 & : & -116 \\ 0 & 0 & 1 & : & -4 \end{bmatrix}$$

Operating $R_{43}\left(-\frac{1}{29}\right)$

$$\sim \begin{bmatrix} 1 & 0 & -18 & 74 \\ 0 & 1 & 9 & -35 \\ 0 & 0 & 29 & -116 \\ 0 & 0 & 0 & 0 \end{bmatrix}$$

which is echelon form.

$\therefore$ $\rho[A : B] = 3$ and $\rho(A) = 3$

Since $\rho[A : B] = \rho(A) = 3$ (number of variables).

$\therefore$ The given system of equations is consistent and has a unique solution.

Equivalent system of equations is

$$x - 18z = 74$$
$$y + 9z = -35$$
$$29z = -116$$

$\Rightarrow$ $x = 2, y = 1, z = -4.$

Example 4. *Test the consistency of following system of linear equations and hence find the solution, if exists:*

(*i*)
$$4x_1 - x_2 = 12$$
$$-x_1 + 5x_2 - 2x_3 = 0$$
$$-2x_2 + 4x_3 = -8$$

(*ii*)
$$7x_1 + 2x_2 + 3x_3 = 16$$
$$2x_1 + 11x_2 + 5x_3 = 25$$
$$x_1 + 3x_2 + 4x_3 = 13$$

Sol. (*i*) Augmented matrix

$$[A : B] = \left[\begin{array}{ccc:c} 4 & -1 & 0 & 12 \\ -1 & 5 & -2 & 0 \\ 0 & -2 & 4 & -8 \end{array}\right]$$

Operating $R_{12}(3)$

$$\sim \left[\begin{array}{ccc:c} 1 & 14 & -6 & 12 \\ -1 & 5 & -2 & 0 \\ 0 & -2 & 4 & -8 \end{array}\right]$$

Operating R_{21}

$$\sim \left[\begin{array}{ccc:c} 1 & 14 & -6 & 12 \\ 0 & 19 & -8 & 12 \\ 0 & -2 & 4 & -8 \end{array}\right]$$

Operating R_{23} (9)

$$\sim \left[\begin{array}{ccc:c} 1 & 14 & -6 & 12 \\ 0 & 1 & 28 & -60 \\ 0 & -2 & 4 & -8 \end{array}\right]$$

Operating R_{32} (2)

$$\sim \left[\begin{array}{ccc:c} 1 & 14 & -6 & 12 \\ 0 & 1 & 28 & -60 \\ 0 & 0 & 60 & -128 \end{array}\right]$$

which is echelon form.

$\therefore \quad \rho(A : B) = 3$ and $\rho(A) = 3$

Since $\rho(A) = \rho(A : B) = 3$ = number of variables,

$\therefore$ The given system of equations is consistent and has a unique solution.

Equivalent system of equations is

$$x_1 + 14\, x_2 - 6x_3 = 12$$
$$x_2 + 28x_3 = -60$$
$$60x_3 = -128$$

$$\Rightarrow \qquad x_3 = -\frac{32}{15},\ x_2 = -\frac{4}{15},\ x_1 = \frac{44}{15}$$

(*ii*) Augmented matrix

$$[A : B] = \left[\begin{array}{ccc:c} 7 & 2 & 3 & 16 \\ 2 & 11 & 5 & 25 \\ 1 & 3 & 4 & 13 \end{array}\right]$$

Operating R_{13}

$$\sim \left[\begin{array}{ccc:c} 1 & 3 & 4 & 13 \\ 2 & 11 & 5 & 25 \\ 7 & 2 & 3 & 16 \end{array}\right]$$

Operating $R_{21}(-2)$, $R_{31}(-7)$

$$\sim \left[\begin{array}{ccc:c} 1 & 3 & 4 & 13 \\ 0 & 5 & -3 & -1 \\ 0 & -19 & -25 & -75 \end{array}\right]$$

Operating $R_2(4)$

$$\sim \begin{bmatrix} 1 & 3 & 4 & : & 13 \\ 0 & 20 & -12 & : & -4 \\ 0 & -19 & -25 & : & -75 \end{bmatrix}$$

Operating $R_{23}(1)$

$$\sim \begin{bmatrix} 1 & 3 & 4 & : & 13 \\ 0 & 1 & -37 & : & -79 \\ 0 & -19 & -25 & : & -75 \end{bmatrix}$$

Operating R_{32} (19)

$$\sim \begin{bmatrix} 1 & 3 & 4 & : & 13 \\ 0 & 1 & -37 & : & -79 \\ 0 & 0 & -728 & : & 1576 \end{bmatrix}$$

which is echelon form

$\therefore \quad \rho(A : B) = 3 \quad \text{and} \quad \rho(A) = 3$

Since $\rho(A) = \rho(A : B) = 3$ = number of variables, hence the given system of linear equations is consistent.

Equivalent system of equations is

$$x_1 + 3x_2 + 4x_3 = 13$$
$$x_2 - 37x_3 = -79$$
$$-728\, x_3 = 1576$$

$$\Rightarrow \qquad x_3 = \frac{197}{91},\ x_2 = \frac{100}{91},\ x_1 = \frac{95}{91}$$

Example 5. *Show that the system of equations*

$x + y + z = -3$, $3x + y - 2z = -2$, $2x + 4y + 7z = 7$ is not consistent.

Sol. Augmented matrix,

$$[A : B] = \begin{bmatrix} 1 & 1 & 1 & : & -3 \\ 3 & 1 & -2 & : & -2 \\ 2 & 4 & 7 & : & 7 \end{bmatrix}$$

Operating $R_{21}(-3)$, $R_{31}(-2)$

$$\sim \begin{bmatrix} 1 & 1 & 1 & : & -3 \\ 0 & -2 & -5 & : & 7 \\ 0 & 2 & 5 & : & 13 \end{bmatrix}$$

Operating R_{32} (1)

$$\sim \begin{bmatrix} 1 & 1 & 1 & : & -3 \\ 0 & -2 & -5 & : & 7 \\ 0 & 0 & 0 & : & 20 \end{bmatrix}$$

which is echelon form.

$\therefore \quad \rho[A : B] = 3 \quad \text{and} \quad \rho(A) = 2$

We observe that

$$\rho(A) \neq \rho(B)$$

Hence the system of given equations is inconsistent and has no solution.

Example 6. *Investigate for consistency of the following equations and if possible, find the solutions:*

$$4x - 2y + 6z = 8$$
$$x + y - 3z = -1$$
$$15x - 3y + 9z = 21.$$

Sol. Augmented matrix,

$$[A : B] = \begin{bmatrix} 4 & -2 & 6 & : & 8 \\ 1 & 1 & -3 & : & -1 \\ 15 & -3 & 9 & : & 21 \end{bmatrix}$$

Operating R_{12}

$$\sim \begin{bmatrix} 1 & 1 & -3 & : & -1 \\ 4 & -2 & 6 & : & 8 \\ 15 & -3 & 9 & : & 21 \end{bmatrix}$$

Operating $R_{21}(-4)$, $R_{31}(-15)$

$$\sim \begin{bmatrix} 1 & 1 & -3 & : & -1 \\ 0 & -6 & 18 & : & 12 \\ 0 & -18 & 54 & : & 36 \end{bmatrix}$$

Operating $R_{32}(-3)$

$$\sim \begin{bmatrix} 1 & 1 & -3 & : & -1 \\ 0 & -6 & 18 & : & 12 \\ 0 & 0 & 0 & : & 0 \end{bmatrix}$$

$\therefore \qquad \rho[A : B] = 2 \quad \text{and} \quad \rho(A) = 2$

Since $\rho[A : B] = \rho(A) = 2$ (< number of variables).

Hence the given system of equations is consistent but has infinite number of solutions. Equivalent system of equations is

$$x + y - 3z = -1 \qquad ...(1)$$
$$-6y + 18z = 12 \qquad ...(2)$$

Let $z = k$ then from (2), $y = 3k - 2$

From (1), $x + 3k - 2 - 3k = -1 \Rightarrow x = 1$

Hence the solutions are

$$x = 1, y = 3k - 2 \quad \text{and} \quad z = k \text{ where } k \text{ is arbitrary.}$$

Note. If the system of equations reduces to only one equation, we let any two unknowns as arbitrary constants.

Example 7. (*i*) *Apply the matrix method to solve the system of equations*

$x + 2y - z = 3,\ 3x - y + 2z = 1,\ 2x - 2y + 3z = 2,\ x - y + z = -1.$ (U.P.T.U. 2014)

(*ii*) *Test for the consistency of the following system of equations:*

$$x_1 + 2x_2 + 3x_3 + 4x_4 = 5$$
$$6x_1 + 7x_2 + 8x_3 + 9x_4 = 10$$

$$11x_1 + 12x_2 + 13x_3 + 14x_4 = 15$$
$$16x_1 + 17x_2 + 18x_3 + 19x_4 = 20$$
$$21x_1 + 22x_2 + 23x_3 + 24x_4 = 25.$$

Sol. (*i*) Augmented matrix,

$$[A : B] = \begin{bmatrix} 1 & 2 & -1 & \vdots & 3 \\ 3 & -1 & 2 & \vdots & 1 \\ 2 & -2 & 3 & \vdots & 2 \\ 1 & -1 & 1 & \vdots & -1 \end{bmatrix}$$

Operating $R_{21}(-3)$, $R_{31}(-2)$, $R_{41}(-1)$

$$\sim \begin{bmatrix} 1 & 2 & -1 & \vdots & 3 \\ 0 & -7 & 5 & \vdots & -8 \\ 0 & -6 & 5 & \vdots & -4 \\ 0 & -3 & 2 & \vdots & -4 \end{bmatrix}$$

Operating $R_{32}(-1)$

$$\sim \begin{bmatrix} 1 & 2 & -1 & \vdots & 3 \\ 0 & -7 & 5 & \vdots & -8 \\ 0 & 1 & 0 & \vdots & 4 \\ 0 & -3 & 2 & \vdots & -4 \end{bmatrix}$$

Operating R_{23}

$$\sim \begin{bmatrix} 1 & 2 & -1 & \vdots & 3 \\ 0 & 1 & 0 & \vdots & 4 \\ 0 & -7 & 5 & \vdots & -8 \\ 0 & -3 & 2 & \vdots & -4 \end{bmatrix}$$

Operating $R_{32}(7)$, $R_{42}(3)$

$$\sim \begin{bmatrix} 1 & 2 & -1 & \vdots & 3 \\ 0 & 1 & 0 & \vdots & 4 \\ 0 & 0 & 5 & \vdots & 20 \\ 0 & 0 & 2 & \vdots & 8 \end{bmatrix}$$

Operating $R_{43}(-2/5)$

$$\sim \begin{bmatrix} 1 & 2 & -1 & \vdots & 3 \\ 0 & 1 & 0 & \vdots & 4 \\ 0 & 0 & 5 & \vdots & 20 \\ 0 & 0 & 0 & \vdots & 0 \end{bmatrix}$$

which is echelon form.

$\therefore$ $\rho(A : B) = 3$ and $\rho(A) = 3$

Since $\rho(A : B) = \rho(A) = 3$ (= no. of variables).

Hence the given system of equations is consistent and has a unique solution.

Equivalent system of equations is

$$x + 2y - z = 3 \quad \text{...(1)}$$
$$y = 4 \quad \text{...(2)}$$
$$5z = 20. \quad \text{...(3)}$$

On solving, we get $x = -1, y = 4, z = 4$.

(*ii*) Augmented matrix,

$$[A : B] = \begin{bmatrix} 1 & 2 & 3 & 4 & : & 5 \\ 6 & 7 & 8 & 9 & : & 10 \\ 11 & 12 & 13 & 14 & : & 15 \\ 16 & 17 & 18 & 19 & : & 20 \\ 21 & 22 & 23 & 24 & : & 25 \end{bmatrix}$$

Operating $R_{21}(-1), R_{31}(-1), R_{41}(-1), R_{51}(-1)$

$$\sim \begin{bmatrix} 1 & 2 & 3 & 4 & : & 5 \\ 5 & 5 & 5 & 5 & : & 5 \\ 10 & 10 & 10 & 10 & : & 10 \\ 15 & 15 & 15 & 15 & : & 15 \\ 20 & 20 & 20 & 20 & : & 20 \end{bmatrix}$$

Operating $R_{32}(-2), R_{42}(-3), R_{52}(-4)$

$$\sim \begin{bmatrix} 1 & 2 & 3 & 4 & : & 5 \\ 5 & 5 & 5 & 5 & : & 5 \\ 0 & 0 & 0 & 0 & : & 0 \\ 0 & 0 & 0 & 0 & : & 0 \\ 0 & 0 & 0 & 0 & : & 0 \end{bmatrix}$$

Operating $R_{21}(-5)$

$$\sim \begin{bmatrix} 1 & 2 & 3 & 4 & : & 5 \\ 0 & -5 & -10 & -15 & : & -20 \\ 0 & 0 & 0 & 0 & : & 0 \\ 0 & 0 & 0 & 0 & : & 0 \\ 0 & 0 & 0 & 0 & : & 0 \end{bmatrix}$$

which is echelon form.

$\therefore$ $\rho(A : B) = 2$ and $\rho(A) = 2$

Since $\rho(A : B) = \rho(A) = 2$ (< no. of variables)

$\therefore$ The given system of equations is consistent and has infinite number of solutions.

Example 8. *Investigate, for what values of λ and μ do the system of equations*

$$x + y + z = 6, \quad x + 2y + 3z = 10, \quad x + 2y + \lambda z = \mu$$

have (i) no solution (ii) unique solution (iii) infinite solutions? (A.K.T.U. 2016, 2018)

Sol. Augmented matrix $[A : B] = \begin{bmatrix} 1 & 1 & 1 & : & 6 \\ 1 & 2 & 3 & : & 10 \\ 1 & 2 & \lambda & : & \mu \end{bmatrix}$

Operating $R_{21}(-1), R_{31}(-1)$

$$\sim \begin{bmatrix} 1 & 1 & 1 & : & 6 \\ 0 & 1 & 2 & : & 4 \\ 0 & 1 & \lambda - 1 & : & \mu - 6 \end{bmatrix}$$

Operating $R_{32}(-1)$

$$\sim \left[\begin{array}{ccc:c} 1 & 1 & 1 & 6 \\ 0 & 1 & 2 & 4 \\ 0 & 0 & \lambda-3 & \mu-10 \end{array}\right]$$

Case I. If $\lambda = 3, \mu \neq 10$

$$\rho(A) = 2, \rho[A:B] = 3$$

$\because$ $\rho(A) \neq \rho[A:B]$

$\therefore$ The system has **no solution.**

Case II. If $\lambda \neq 3$, μ may have any value

$$\rho(A) = \rho[A:B] = 3 = \text{number of unknowns.}$$

$\therefore$ The system has **unique solution.**

Case III. If $\lambda = 3, \mu = 10$

$$\rho(A) = \rho[A:B] = 2 < \text{number of unknowns.}$$

$\therefore$ The system has an **infinite number of solutions.**

Example 9. *Determine the values of λ and μ such that the system $2x - 5y + 2z = 8$, $2x + 4y + 6z = 5$, $x + 2y + \lambda z = \mu$ has (i) no solution (ii) a unique solution (iii) infinite number of solutions.*

Sol. Augmented matrix,

$$[A:B] = \left[\begin{array}{ccc:c} 2 & -5 & 2 & 8 \\ 2 & 4 & 6 & 5 \\ 1 & 2 & \lambda & \mu \end{array}\right]$$

Operating $R_{21}(-1)$, $R_{31}\left(-\frac{1}{2}\right)$

$$\sim \left[\begin{array}{ccc:c} 2 & -5 & 2 & 8 \\ 0 & 9 & 4 & -3 \\ 0 & 9/2 & \lambda-1 & \mu-4 \end{array}\right]$$

Operating $R_{32}\left(-\frac{1}{2}\right)$

$$\sim \left[\begin{array}{ccc:c} 2 & -5 & 2 & 8 \\ 0 & 9 & 4 & -3 \\ 0 & 0 & \lambda-3 & \mu-5/2 \end{array}\right]$$

Case I. If $\lambda = 3, \mu \neq 5/2$

$$\rho(A) = 2, \rho(A:B) = 3$$

$\because$ $\rho(A) \neq \rho(A:B)$

$\therefore$ The system has **no solution.**

Case II. If $\lambda \neq 3$, μ may have any value

$$\rho(A) = \rho(A:B) = 3 (= \text{no. of variables}).$$

$\therefore$ The system has a **unique solution.**

Case III. If $\lambda = 3, \mu = 5/2$,

$$\rho(A) = \rho(A:B) = 2 (< \text{no. of variables}).$$

$\therefore$ The system has an **infinite no. of solutions.**

Example 10. *Determine the values of a and b for which the system* $\begin{bmatrix} 3 & -2 & 1 \\ 5 & -8 & 9 \\ 2 & 1 & a \end{bmatrix}\begin{bmatrix} x \\ y \\ z \end{bmatrix} = \begin{bmatrix} b \\ 3 \\ -1 \end{bmatrix}$ *has*

(i) a unique solution (ii) no solution (iii) infinitely many solutions.

Sol. Augmented matrix $[A : B] = \left[\begin{array}{ccc:c} 3 & -2 & 1 & b \\ 5 & -8 & 9 & 3 \\ 2 & 1 & a & -1 \end{array}\right]$

Operating $R_{13}(-1)$

$$\sim \left[\begin{array}{ccc:c} 1 & -3 & 1-a & b+1 \\ 5 & -8 & 9 & 3 \\ 2 & 1 & a & -1 \end{array}\right]$$

Operating $R_{21}(-5)$, $R_{31}(-2)$

$$\sim \left[\begin{array}{ccc:c} 1 & -3 & 1-a & b+1 \\ 0 & 7 & 4+5a & -5b-2 \\ 0 & 7 & -2+3a & -2b-3 \end{array}\right]$$

Operating $R_{32}(-1)$

$$\sim \left[\begin{array}{ccc:c} 1 & -3 & 1-a & b+1 \\ 0 & 7 & 4+5a & -5b-2 \\ 0 & 0 & -6-2a & 3b-1 \end{array}\right]$$

Case I. If $a \neq -3$, b may have any value

$\rho(A) = \rho[A : B] = 3$(= no. of variables).

$\therefore$ The system has a unique solution.

Case II. If $a = -3$ and $b \neq \frac{1}{3}$ then $\rho(A) = 2$ and $\rho(A : B) = 3$

$\therefore$ $\rho(A) \neq \rho[A : B]$

Hence the system has no solution.

Case III. If $a = -3$ and $b = \frac{1}{3}$, then

$\rho(A) = \rho[A : B] = 2$(< no. of variables)

$\therefore$ The given system has infinitely many solutions.

Example 11. *Test whether the following system of equations possess a non-trivial solution:*

$$x_1 + x_2 + 2x_3 + 3x_4 = 0$$
$$3x_1 + 4x_2 + 7x_3 + 10x_4 = 0$$
$$5x_1 + 7x_2 + 11x_3 + 17x_4 = 0$$
$$6x_1 + 8x_2 + 13x_3 + 16x_4 = 0.$$

Sol. The given system is a homogeneous linear system of the form AX = O

Coefficient matrix,

$$A = \begin{bmatrix} 1 & 1 & 2 & 3 \\ 3 & 4 & 7 & 10 \\ 5 & 7 & 11 & 17 \\ 6 & 8 & 13 & 16 \end{bmatrix}$$

Operating $R_{21}(-3)$, $R_{31}(-5)$, $R_{41}(-6)$

$$\sim \begin{bmatrix} 1 & 1 & 2 & 3 \\ 0 & 1 & 1 & 1 \\ 0 & 2 & 1 & 2 \\ 0 & 2 & 1 & -2 \end{bmatrix}$$

Operating $R_{32}(-2)$, $R_{42}(-2)$

$$\sim \begin{bmatrix} 1 & 1 & 2 & 3 \\ 0 & 1 & 1 & 1 \\ 0 & 0 & -1 & 0 \\ 0 & 0 & -1 & -4 \end{bmatrix}$$

Operating $R_{43}(-1)$

$$\sim \begin{bmatrix} 1 & 1 & 2 & 3 \\ 0 & 1 & 1 & 1 \\ 0 & 0 & -1 & 0 \\ 0 & 0 & 0 & -4 \end{bmatrix}$$

$\therefore$ $\rho(A) = 4$(= no. of variables),

Hence the given homogeneous system of equations has trivial solution.

$\therefore$ $x_1 = 0, x_2 = 0, x_3 = 0$ and $x_4 = 0$.

Example 12. *Solve the equations using matrix method:*

$x_1 + 3x_2 + 2x_3 = 0,\ 2x_1 - x_2 + 3x_3 = 0,\ 3x_1 - 5x_2 + 4x_3 = 0,\ x_1 + 17x_2 + 4x_3 = 0.$

Sol. Coefficient matrix,

$$A = \begin{bmatrix} 1 & 3 & 2 \\ 2 & -1 & 3 \\ 3 & -5 & 4 \\ 1 & 17 & 4 \end{bmatrix}$$

Operating $R_{21}(-2)$, $R_{31}(-3)$, $R_{41}(-1)$

$$\sim \begin{bmatrix} 1 & 3 & 2 \\ 0 & -7 & -1 \\ 0 & -14 & -2 \\ 0 & 14 & 2 \end{bmatrix}$$

Operating $R_{32}(-2)$, $R_{42}(2)$

$$\sim \begin{bmatrix} 1 & 3 & 2 \\ 0 & -7 & -1 \\ 0 & 0 & 0 \\ 0 & 0 & 0 \end{bmatrix}$$

$\rho(A) = 2$ (< no. of unknowns)

$\therefore$ The system has an infinite number of non-trivial solutions given by

$$x_1 + 3x_2 + 2x_3 = 0 \quad \text{...(1)}$$

$$-7x_2 - x_3 = 0 \quad \text{...(2)}$$

Let $x_2 = k$ then from (2), $x_3 = -7k$

$\therefore$ From (1), $x_1 + 3k - 14k = 0 \Rightarrow x_1 = 11k$

Hence, $x_1 = 11k, x_2 = k, x_3 = -7k$ where k is arbitrary. Different values of k give different solutions.

Example 13. *Find the values of k for which the system of equations*

$$(3k - 8)\,x + 3y + 3z = 0$$
$$3x + (3k - 8)y + 3z = 0$$
$$3x + 3y + (3k - 8)z = 0$$

has a non-trivial solution. (A.K.T.U. 2022)

Sol. For the given system of equations to have a non-trivial solution,

$$\rho(A) < 3$$

where $A = \begin{bmatrix} 3k-8 & 3 & 3 \\ 3 & 3k-8 & 3 \\ 3 & 3 & 3k-8 \end{bmatrix}$ is the coefficient matrix.

For this, $|A| = 0$

$$\Rightarrow \quad \begin{vmatrix} 3k-8 & 3 & 3 \\ 3 & 3k-8 & 3 \\ 3 & 3 & 3k-8 \end{vmatrix} = 0$$

Operating $C_1 + (C_2 + C_3)$,

$$\begin{vmatrix} 3k-2 & 3 & 3 \\ 3k-2 & 3k-8 & 3 \\ 3k-2 & 3 & 3k-8 \end{vmatrix} = 0$$

$$\Rightarrow \quad (3k-2)\begin{vmatrix} 1 & 3 & 3 \\ 1 & 3k-8 & 3 \\ 1 & 3 & 3k-8 \end{vmatrix} = 0$$

Operating $R_2 \to R_2 - R_1$, $R_3 \to R_3 - R_1$, we get

$$(3k-2)\begin{vmatrix} 1 & 3 & 3 \\ 0 & 3k-11 & 0 \\ 0 & 0 & 3k-11 \end{vmatrix} = 0$$

$$\Rightarrow \quad (3k-2)\,(3k-11)^2 = 0$$

$$\Rightarrow \quad k = 2/3,\ 11/3,\ 11/3.$$

Example 14. *Find values of λ for which the following system of equations is consistent and has non-trivial solutions. Solve equations for all such values of λ:*

$$(\lambda - 1)\,x + (3\lambda + 1)\,y + 2\lambda z = 0$$
$$(\lambda - 1)\,x + (4\lambda - 2)\,y + (\lambda + 3)\,z = 0$$
$$2x + (3\lambda + 1)y + 3(\lambda - 1)\,z = 0.$$

Sol. For the given system of equations to have a non-trivial solution,

$\rho(A) < 3$ where $A = \begin{bmatrix} \lambda-1 & 3\lambda+1 & 2\lambda \\ \lambda-1 & 4\lambda-2 & \lambda+3 \\ 2 & 3\lambda+1 & 3(\lambda-1) \end{bmatrix}$

For this, $|A| = 0$

$$\begin{vmatrix} \lambda-1 & 3\lambda+1 & 2\lambda \\ \lambda-1 & 4\lambda-2 & \lambda+3 \\ 2 & 3\lambda+1 & 3(\lambda-1) \end{vmatrix} = 0$$

Operating $R_1 \to R_1 - R_2$

$$\begin{vmatrix} 0 & -\lambda+3 & \lambda-3 \\ \lambda-1 & 4\lambda-2 & \lambda+3 \\ 2 & 3\lambda+1 & 3\lambda-3 \end{vmatrix} = 0$$

Operating $C_2 \to C_2 + C_3$

$$\begin{vmatrix} 0 & 0 & \lambda-3 \\ \lambda-1 & 5\lambda+1 & \lambda+3 \\ 2 & 6\lambda-2 & 3\lambda-3 \end{vmatrix} = 0$$

$$\Rightarrow \quad (\lambda-3)\,[(\lambda-1)(6\lambda-2) - 2(5\lambda+1)] = 0$$

$$\Rightarrow \quad \lambda = 0, 3$$

Putting $\lambda = 0$, we get

$$A = \begin{bmatrix} -1 & 1 & 0 \\ -1 & -2 & 3 \\ 2 & 1 & -3 \end{bmatrix}$$

Operating $R_{21}(-1)$, $R_{31}(2)$

$$\sim \begin{bmatrix} -1 & 1 & 0 \\ 0 & -3 & 3 \\ 0 & 3 & -3 \end{bmatrix}$$

Operating $R_{32}(1)$

$$\sim \begin{bmatrix} -1 & 1 & 0 \\ 0 & -3 & 3 \\ 0 & 0 & 0 \end{bmatrix}$$

Equivalent system of equations is

$$-x + y = 0 \qquad \ldots(1)$$

$$-3y + 3z = 0 \qquad \ldots(2)$$

Let $z = k_1$ then from (2), $y = k_1$

From (1), $x = k_1$

$\therefore$ Infinite solutions are given by $x = k_1, y = k_1, z = k_1$ where k_1 is arbitrary.

Putting $\lambda = 3$, we get

$$A = \begin{bmatrix} 2 & 10 & 6 \\ 2 & 10 & 6 \\ 2 & 10 & 6 \end{bmatrix}$$

Operating $R_{21}(-1)$, $R_{31}(-1)$

$$\sim \begin{bmatrix} 2 & 10 & 6 \\ 0 & 0 & 0 \\ 0 & 0 & 0 \end{bmatrix}$$

Equivalent system of equations is

$$2x + 10y + 6z = 0 \qquad \ldots(3)$$

Let $z = k_2$, $y = k_3$ then from (3),

$$x = -5k_3 - 3k_2$$

Hence infinite solutions are given by

$$x = -5k_3 - 3k_2, \; y = k_3, \; z = k_2$$

where k_2 and k_3 are arbitrary.

Example 15. *Show that the homogeneous system of equations*

$$x + y \cos\gamma + z \cos\beta = 0$$
$$x \cos\gamma + y + z \cos\alpha = 0$$
$$x \cos\beta + y \cos\alpha + z = 0$$

has non-trivial solution if $\alpha + \beta + \gamma = 0$.

Sol. If the system has only non-trivial solutions, then

$$\begin{vmatrix} 1 & \cos\gamma & \cos\beta \\ \cos\gamma & 1 & \cos\alpha \\ \cos\beta & \cos\alpha & 1 \end{vmatrix} = 0$$

$$\Rightarrow \quad 1 - \cos^2\alpha + \cos\gamma(\cos\alpha\cos\beta - \cos\gamma) + \cos\beta(\cos\gamma\cos\alpha - \cos\beta) = 0$$

$$\Rightarrow \quad \sin^2\alpha - \cos^2\beta - \cos^2\gamma + 2\cos\alpha\cos\beta\cos\gamma = 0$$

$$\Rightarrow \quad -(\cos^2\beta - \sin^2\alpha) - \cos^2\gamma + 2\cos\alpha\cos\beta\cos\gamma = 0$$

$$\Rightarrow \quad -\cos(\alpha+\beta)\cos(\beta-\alpha) - \cos^2\gamma + 2\cos\alpha\cos\beta\cos\gamma = 0$$

$$|\text{ if } \alpha + \beta + \gamma = 0$$

$$\Rightarrow \quad -\cos(-\gamma)\cos(\beta-\alpha) - \cos^2\gamma + 2\cos\alpha\cos\beta\cos\gamma = 0$$

$$\Rightarrow \quad -\cos\gamma[\cos(\beta-\alpha) + \cos(\beta+\alpha)] + 2\cos\alpha\cos\beta\cos\gamma = 0$$

$$\Rightarrow \quad -2\cos\beta\cos\alpha\cos\gamma + 2\cos\alpha\cos\beta\cos\gamma = 0$$

which is true.

Hence the given homogeneous system of equations has non-trivial solution if $\alpha + \beta + \gamma = 0$.

Example 16. *Find the value of* λ *such that the following equations have unique solution:*

$$\lambda x + 2y - 2z - 1 = 0,\ 4x + 2\lambda y - z - 2 = 0,\ 6x + 6y + \lambda z - 3 = 0$$

and use matrix method to solve these equations when $\lambda = 2$. **(M.T.U. 2013)**

Sol. The system will have a unique solution if coeff. matrix A is non-singular.

i.e.,
$$\begin{vmatrix} \lambda & 2 & -2 \\ 4 & 2\lambda & -1 \\ 6 & 6 & \lambda \end{vmatrix} \neq 0$$

$$\Rightarrow \quad (\lambda - 2)(\lambda^2 + 2\lambda + 15) \neq 0 \quad \Rightarrow \quad \lambda \neq 2$$

The solution can be found by Cramer's rule.

In case when $\lambda = 2$.

$$\text{Augmented matrix, } [A : B] = \left[\begin{array}{ccc:c} 2 & 2 & -2 & 1 \\ 4 & 4 & -1 & 2 \\ 6 & 6 & 2 & 3 \end{array}\right]$$

Operating $R_1(\frac{1}{2}), R_2(\frac{1}{4}), R_3(\frac{1}{6})$

$$\sim \left[\begin{array}{ccc:c} 1 & 1 & -1 & 1/2 \\ 1 & 1 & -1/4 & 1/2 \\ 1 & 1 & 1/3 & 1/2 \end{array}\right]$$

Operating $R_{21}(-1), R_{31}(-1)$

$$\sim \left[\begin{array}{ccc:c} 1 & 1 & -1 & 1/2 \\ 0 & 0 & 3/4 & 0 \\ 0 & 0 & 4/3 & 0 \end{array}\right]$$

Operating $R_2(\frac{4}{3})$, $R_3(\frac{3}{4})$

$$\sim \begin{bmatrix} 1 & 1 & -1 & : & 1/2 \\ 0 & 0 & 1 & : & 0 \\ 0 & 0 & 1 & : & 0 \end{bmatrix}$$

$\therefore \qquad \rho[A : B] = 2 = \rho(A) \; (< 3)$

Equivalent system of equations is

$$x + y - z = \frac{1}{2} \quad \text{and} \quad z = 0$$

$$\therefore \qquad x + y = \frac{1}{2}$$

Let $y = k_1$ then $x = \frac{1}{2} - k_1$

Hence solutions are $x = \frac{1}{2} - k_1, y = k_1$ and $z = 0$ where k_1 is arbitrary.

Example 17. *Show that the equations*

$$-2x + y + z = a$$
$$x - 2y + z = b$$
$$x + y - 2z = c$$

have no solution unless a + b + c = 0. In which case they have infinitely many solutions? Find these solutions when a = 1, b = 1, c = – 2.

Sol. Augmented matrix,

$$[A : B] = \begin{bmatrix} -2 & 1 & 1 & : & a \\ 1 & -2 & 1 & : & b \\ 1 & 1 & -2 & : & c \end{bmatrix} \qquad | \rho(A) = 2$$

Operating R_{13}

$$\sim \begin{bmatrix} 1 & 1 & -2 & : & c \\ 1 & -2 & 1 & : & b \\ -2 & 1 & 1 & : & a \end{bmatrix}$$

Operating $R_{21}(-1)$, $R_{31}(2)$

$$\sim \begin{bmatrix} 1 & 1 & -2 & : & c \\ 0 & -3 & 3 & : & b-c \\ 0 & 3 & -3 & : & a+2c \end{bmatrix}$$

Operating $R_{32}(1)$

$$\sim \begin{bmatrix} 1 & 1 & -2 & : & c \\ 0 & -3 & 3 & : & b-c \\ 0 & 0 & 0 & : & a+b+c \end{bmatrix}.$$

Case I. If $a + b + c \neq 0$

$$\rho[A : B] = 3 \neq \rho(A)$$

where A is the coefficient matrix.

Hence the system, being inconsistent, have no solution.

Case II. If $a + b + c = 0$

$$\rho[A : B] = 2 = \rho(A) \qquad (< 3)$$

Hence the system has infinite number of solutions.

Equivalent system of equations is

$$x + y - 2z = -2 \qquad | \text{ Putting } b = 1, c = -2$$
$$-3y + 3z = 3$$

Let $z = k$, k being an arbitrary constant.

$$y = k - 1$$
$$x = k - 1$$

Hence the solutions are $x = k - 1, y = k - 1, z = k$.

Example 18. *For what values of k, the equations $x + y + z = 1$, $2x + y + 4z = k$, $4x + y + 10z = k^2$ have a solution and solve them completely in each case.*

Sol. Augmented matrix,

$$[A : B] = \begin{bmatrix} 1 & 1 & 1 & : & 1 \\ 2 & 1 & 4 & : & k \\ 4 & 1 & 10 & : & k^2 \end{bmatrix}$$

Operating $R_{21}(-2)$, $R_{31}(-4)$

$$\sim \begin{bmatrix} 1 & 1 & 1 & : & 1 \\ 0 & -1 & 2 & : & k-2 \\ 0 & -3 & 6 & : & k^2-4 \end{bmatrix}$$

Operating $R_{32}(-3)$

$$\sim \begin{bmatrix} 1 & 1 & 1 & : & 1 \\ 0 & -1 & 2 & : & k-2 \\ 0 & 0 & 0 & : & k^2-3k+2 \end{bmatrix}$$

If the given system has solutions, then

$$k^2 - 3k + 2 = 0 \Rightarrow k = 1, 2$$

Case I. When $k = 1$, we have

$$x + y + z = 1 \qquad ...(1)$$
$$-y + 2z = -1 \qquad ...(2)$$

Let $y = k$ then $z = \dfrac{k-1}{2}$

From (1), $x + k + \dfrac{k-1}{2} = 1 \Rightarrow x = \dfrac{3}{2}(1 - k)$

Hence solutions are $\quad x = \dfrac{3}{2}(1-k), y = k, z = \dfrac{1}{2}(k-1)$ where k is arbitrary.

Case II. When $k = 2$, we have

$$x + y + z = 1 \qquad ...(3)$$
$$-y + 2z = 0 \qquad ...(4)$$

Let $z = \lambda$ then from (4), $y = 2\lambda$

From (3), $\quad x = 1 - 3\lambda$

Hence solutions are

$$x = 1 - 3\lambda, y = 2\lambda, z = \lambda \quad \text{where } \lambda \text{ is arbitrary.}$$

TEST YOUR KNOWLEDGE

1. (*i*) Test the consistency of the following system of equations:

$$5x + 3y + 7z = 4,\ 3x + 26y + 2z = 9,\ 7x + 2y + 11z = 5.$$

(*ii*) Test for the consistency of the following system of equations:

$$\begin{bmatrix} 3 & 4 & 5 & 6 \\ 4 & 5 & 6 & 7 \\ 5 & 6 & 7 & 8 \\ 10 & 11 & 12 & 13 \\ 15 & 16 & 17 & 18 \end{bmatrix} \begin{bmatrix} x_1 \\ x_2 \\ x_3 \\ x_4 \end{bmatrix} = \begin{bmatrix} 7 \\ 8 \\ 9 \\ 14 \\ 19 \end{bmatrix}$$

(*iii*) Show that the equations $2x + 6y + 11 = 0$, $6x + 20y - 6z + 3 = 0$ and $6y - 18z + 1 = 0$ are not consistent. (*U.K.T.U. 2011*)

2. Solve the following system of equations by matrix method:

(*i*) $x + y + z = 8,\ x - y + 2z = 6,\ 3x + 5y - 7z = 14$

(*ii*) $x + y + z = 6,\ x - y + 2z = 5,\ 3x + y + z = 8$

(*iii*) $x + 2y + 3z = 1,\ 2x + 3y + 2z = 2,\ 3x + 3y + 4z = 1.$

3. (*i*) Test the consistency and hence solve the following set of equations:

$$x_1 + 2x_2 + x_3 = 2,\ 3x_1 + x_2 - 2x_3 = 1,\ 4x_1 - 3x_2 - x_3 = 3,\ 2x_1 + 4x_2 + 2x_3 = 4.$$

(*ii*) Solve the system of linear equations using matrix method:

$$x + 2y + 3z = 5$$
$$7x + 11y + 13z = 17$$
$$19x + 23y + 29z = 31$$

(*iii*) Test for consistency and solve the following system of equations:

$$2x - y + 3z = 8$$
$$-x + 2y + z = 4$$
$$3x + y - 4z = 0.$$

(*U.P.T.U. 2014*)

4. (*i*) Test for consistency, the equations $2x - 3y + 7z = 5,\ 3x + y - 3z = 13,\ 2x + 19y - 47z = 32.$

(*ii*) Verify that the following system of equations is inconsistent:

$$x + 2y + 2z = 1,\ 2x + y + z = -2,\ 3x + 2y + 2z = 3,\ y + z = 0.$$

(*iii*) Test for consistency of the equations:

$$2x - 3y + 7z = 5$$
$$3x + y - 3z = 13$$
$$2x + 19y - 47z = 32$$

(*iv*) Test the consistency and hence, solve the following set of equations:

$$10y + 3z = 0$$
$$3x + 3y + z = 1$$
$$2x - 3y - z = 5$$
$$x + 2y = 4$$

(*A.K.T.U. 2018*)

5. (*i*) Apply rank test to examine if the following system of equations is consistent, solve them.

$$2x + 4y - z = 9,\ 3x - y + 5z = 5,\ 8x + 2y + 9z = 19.$$

(*ii*) Test the consistency for the following system of equations and if system is consistent, solve them:

$$x + y + z = 6,\ x + 2y + 3z = 14,\ x + 4y + 7z = 30$$

(*U.K.T.U. 2012*)

6. Show that if $\lambda \neq -5$, the system of equations $3x - y + 4z = 3,\ x + 2y - 3z = -2,\ 6x + 5y + \lambda z = -3$ have a unique solution. If $\lambda = -5$, show that the equations are consistent. Determine the solutions in each case.

7. For what values of λ, the equations

$$x + y + z = 1,\ x + 2y + 4z = \lambda,\ x + 4y + 10z = \lambda^2$$

have a solution and solve them completely in each case.

8. (*i*) Verify that the following set of equations has a non-trivial solution:

$$x + 3y - 2z = 0,\ 2x - y + 4z = 0,\ x - 11y + 14z = 0.$$

(*ii*) Show that the following system of equations:

$x + 2y - 2u = 0$, $2x - y - u = 0$, $x + 2z - u = 0$, $4x - y + 3z - u = 0$ do not have a non-trivial solution.

9. Find the values of λ for which the equations

$$x + (\lambda + 4)y + (4\lambda + 2)z = 0$$
$$x + 2(\lambda + 1)y + (3\lambda + 4)z = 0$$
$$2x + 3\lambda y + (3\lambda + 4)z = 0$$

have a non-trivial solution. Also find the solution in each case. (*U.P.T.U. 2015*)

10. (*i*) Find the values of λ for which the equations

$$(11 - \lambda)x - 4y - 7z = 0$$
$$7x - (\lambda + 2)y - 5z = 0$$
$$10x - 4y - (6 + \lambda)z = 0$$

possess a non-trivial solution. For these values of λ, find the solution also.

(*ii*) For what values of λ, the system of equations

$$2x - 2y + z = \lambda x,\ 2x - 3y + 2z = \lambda y,\ -x + 2y + 0z = \lambda z$$

possess a non-trivial solution? Obtain its general solution.

11. Determine the values of λ for which the following set of equations may possess non-trivial solution $3x_1 + x_2 - \lambda x_3 = 0$, $4x_1 - 2x_2 - 3x_3 = 0$, $2\lambda x_1 + 4x_2 + \lambda x_3 = 0$. For each permissible value of λ, determine the general solution.

12. Find the value of k so that the equations $x + y + 3z = 0$, $4x + 3y + kz = 0$, $2x + y + 2z = 0$ have a non-trivial solution.

13. Determine 'b' such that the system of homogeneous equations

$$2x + y + 2z = 0$$
$$x + y + 3z = 0$$
$$4x + 3y + bz = 0$$

has (*i*) trivial solution (*ii*) non-trivial solutions. Find the non-trivial solutions using matrix method.

14. (*i*) Show that the system of equations $3x + 4y + 5z = a$, $4x + 5y + 6z = b$, $5x + 6y + 7z = c$ does not have a solution unless $a + c = 2b$.

(*ii*) Find the values of a and b for which the equations $x + ay + z = 3$, $x + 2y + 2z = b$, $x + 5y \quad z = 9$ are consistent. When will these equations have a unique solution?

(*iii*) Find for what value of λ and μ, the system of linear equations

$$x + y + z = 6,\ x + 2y + 5z = 10,\ 2x + 3y + \lambda z = \mu$$

has no solution and also find the solution when $\lambda = 2$ and $\mu = 10$. (*U.K.T.U. 2011*)

15. For what values of a and b do the equations $x + 2y + 3z = 6$, $x + 3y + 5z = 9$, $2x + 5y + az = b$ have (*i*) no solution (*ii*) a unique solution (*iii*) more than one solution?

16. Investigate the values of λ and μ so that the equations $2x + 3y + 5z = 9$, $7x + 3y - 2z = 8$, $2x + 3y + \lambda z = \mu$ have (*a*) no solution, (*b*) a unique solution and (*c*) an infinite number of solutions.

17. Solve by calculating the inverse by elementary row operations:

$$x_1 + x_2 + x_3 + x_4 = 0$$
$$x_1 + x_2 + x_3 - x_4 = 4$$
$$x_1 + x_2 - x_3 + x_4 = -4$$
$$x_1 - x_2 + x_3 + x_4 = 2$$

(*U.P.T.U. 2015*)

18. For what values of λ and μ, the system of linear equations:

$$x + y + z = 6$$
$$x + 2y + 5z = 10$$
$$2x + 3y + \lambda z = \mu$$

has

(*i*) a unique solution (*ii*) no solution (*iii*) infinite solutions

Also find the solution for $\lambda = 2$ and $\mu = 8$. (*A.K.T.U. 2020*)

Answers

1. (*i*) Consistent (*ii*) Consistent with many solutions.

2. (*i*) $x = 5, y = \frac{5}{3}, z = \frac{4}{3}$ (*ii*) $x = 1, y = 2, z = 3$ (*iii*) $x = -\frac{3}{7}, y = \frac{8}{7}, z = -\frac{2}{7}$

3. (*i*) $x_1 = 1, x_2 = 0, x_3 = 1$ (*ii*) $x = -\frac{35}{18}, y = \frac{2}{9}, z = \frac{13}{6}$ (*iii*) $x = 2, y = 2, z = 2.$

4. (*i*) Inconsistent (*iii*) inconsistent (*iv*) inconsistent, no solution exists

5. (*i*) $x = -\frac{19}{14}k + \frac{29}{14}, y = \frac{13}{14}k + \frac{17}{14}, z = k$ where k is arbitrary.

(*ii*) $x = k - 2, y = 8 - 2k, z = k$, where k is arbitrary.

6. $\lambda \neq -5, x = \frac{4}{7}, y = -\frac{9}{7}, z = 0$; $\lambda = -5, x = \frac{4 - 5k}{7}, y = -\frac{13k - 9}{7}, z = k$ where k is arbitrary.

7. $\lambda = 1, 2$; when $\lambda = 1, x = 1 + 2k, y = -3k$ and $z = k$ when $\lambda = 2, x = 2k, y = 1 - 3k, z = k$; k is arbitrary.

9. $\lambda = 2, x = 0, y = -5k, z = 3k$; $\lambda = -2, x = 4k, y = k, z = k.$

10. (*i*) $\lambda = 0, 1, 2$; when $\lambda = 0$, solution is (k, k, k)

when $\lambda = 1$, solution is $(k, -k, 2k)$; when $\lambda = 2$, solution is $(2k, k, 2k)$.

(*ii*) $\lambda = 1, -3$

when $\lambda = 1, x = 2k_1 - k_2, y = k_1, z = k_2$; when $\lambda = -3, x = -k, y = -2k, z = k.$

11. $\lambda = 1, -9$. For $\lambda = 1$, solution is $x = k, y = -k, z = 2k$. For $\lambda = -9$, solution is $x = 3k, y = 9k, z = -2k$; k is arbitrary.

12. $k = 8$

13. (*i*) $b \neq 8$ (*ii*) $b = 8$; $x = k, y = -4k, z = k$, k is arbitrary.

14. (*ii*) If $a = -1, b = 6$, equations will be consistent and have infinitely many solutions.

If $a = -1, b \neq 6$, equations will be inconsistent.

If $a \neq -1$, b is arbitrary, equations will be consistent and have a unique solution.

(*iii*) $\lambda = 6, \mu \neq 16; x = \frac{13}{2}, y = -2, z = \frac{3}{2}.$

15. (*i*) $a = 8, b \neq 15$ (*ii*) $a \neq 8$, b may have any value (*iii*) $a = 8, b = 15$

16. (*a*) $\lambda = 5, \mu \neq 9$ (*b*) $\lambda \neq 5$, μ arbitrary (*c*) $\lambda = 5, \mu = 9$

17. $x_1 = 1, x_2 = -1, x_3 = 2, x_4 = -2.$

18. (*i*) $\lambda \neq 6$, μ is arbitrary (*ii*) $\lambda = 6, \mu \neq 16$ (*iii*) $\lambda = 6, \mu = 16$

when $\lambda = 2, \mu = 8$, solution is $x = 8, y = -4$ and $z = 2$.

1.31. VECTORS

Any ordered n-tuple of numbers is called an *n-vector*. By an ordered n-tuple, we mean a set consisting of n numbers in which the place of each number is fixed. If $x_1, x_2, ..., x_n$ be any n numbers then the ordered n-tuple $X = (x_1, x_2, ..., x_n)$ is called an n-vector. Thus the coordinates of a point in space can be represented by a 3-vector (x, y, z). Similarly $(1, 0, 2, -1)$ and $(2, 7, 5, -3)$ are 4-vectors. The n numbers $x_1, x_2, ..., x_n$ are called the components of the n-vector $X = (x_1, x_2, ..., x_n)$. A vector may be written either as a *row vector* or as a *column vector*. If A is a matrix of order $m \times n$, then each row of A will be an n-vector and each column of A will be an m-vector. A vector whose components are all zero is called a zero vector and is denoted by O. Thus O = (0, 0, 0, ..., 0).

Let $X = (x_1, x_2, ..., x_n)$ and $Y = (y_1, y_2, ..., y_n)$ be two vectors.

Then X = Y if and only if their corresponding components are equal.

i.e., if $x_i = y_i$, for $i = 1, 2, ..., n$

$$X + Y = (x_1 + y_1, x_2 + y_2, ..., x_n + y_n)$$

If k be a scalar, then $kX = (kx_1, kx_2, ..., kx_n)$.

n, n-vectors $e_1 = [1, 0, 0, ..., 0]$, $e_2 = [0, 1, 0, 0, ..., 0]$, ..., $e_n = [0, 0, ..., 1]$ are called fundamental unit vectors or elementary vectors.

1.32. LINEAR DEPENDENCE *AND* LINEAR INDEPENDENCE OF VECTORS

A set of r n-vectors $X_1, X_2, ..., X_r$ is said to be *linearly dependent* if there exist r scalars (numbers) $k_1, k_2, ..., k_r$ not all zero, such that

$$k_1X_1 + k_2X_2 + ... + k_rX_r = O$$

A set of r n-vectors $X_1, X_2, ..., X_r$ is said to be *linearly independent* if every relation of the type

$$k_1X_1 + k_2X_2 + ... + k_rX_r = O \text{ implies } k_1 = k_2 = ... = k_r = 0$$

To test the linear dependence of r given vectors, write them as row vectors. Add suitable multiples of one vector to the others so that the resulting $(r - 1)$ vectors have their first component zero. Choose anyone of these $(r - 1)$ vectors and add its multiples to the others so that the resulting $(r - 2)$ vectors have their second component zero. In this way continue, reducing the successive components to zero. If the final reduction gives a vector all of whose components are zero, then the original vectors are linearly dependent. However, if the final reduction gives a vector all of whose components are not zero, then the original vectors are linearly independent.

Note 1. If a set of vectors is linearly dependent, then at least one member of the set can be expressed as a linear combination of the remaining vectors.

Note 2. The m, n-dimensional vectors $X_1, X_2, X_3, ..., X_m$ are linearly dependent if the rank of the matrix $[X_1, X_2, X_3, ..., X_m]$ with the given vectors as columns is less than m.

1.32.1. Procedure to test linear dependence of vectors

Step 1. Construct coefficient matrix A with elements of given vectors as columns.

Step 2. Find $\rho(A)$.

Step 3. If $\rho(A)$ = no. of vectors then given set of vectors is linearly independent and if $\rho(A)$ < no. of vectors then given set of vectors is linearly dependent.

1.33. ANY NON-EMPTY SUBSET OF A LINEARLY INDEPENDENT SET OF VECTORS IS LINEARLY INDEPENDENT

Let $S_1 = \{X_1, X_2, X_3, ..., X_n\}$

$S_2 = \{X_1, X_2, X_3, ..., X_r\}$ where $r < n$

Clearly, S_2 is a subset of S_1.

Let $a_1X_1 + a_2X_2 + a_3X_3 + ... + a_rX_r = O$ ...(1)

$\Rightarrow$ $a_1X_1 + a_2X_2 + a_3X_3 + ... + a_rX_r + 0.X_{r+1} + ... + 0.X_n = O$ ...(2)

But S_1 is linearly independent

$\therefore$ From (2), $a_1 = a_2 = a_3 = ... = a_r = 0$

Hence by eqn. (1), it is true only if all a's are zero.

$\therefore$ Set S_2 is also linearly independent.

1.34. ANY SUPER SET OF A LINEARLY DEPENDENT SET OF VECTORS IS LINEARLY DEPENDENT

Let the set $\{X_1, X_2, X_3, ..., X_r\}$ be linearly dependent set, therefore $\exists$ r scalars $a_1, a_2, a_3, ..., a_r$ not all zero such that

$$a_1X_1 + a_2X_2 + a_3X_3 + ... + a_rX_r = O \quad ...(1)$$

$$\Rightarrow \quad a_1X_1 + a_2X_2 + a_3X_3 + ... + a_rX_r + 0.X_{r+1} + ... + 0.X_n = O \quad ...(2)$$

where all a's are not zero.

Hence by definition, the set $\{X_1, X_2, ..., X_n\}$ is linearly dependent.

ILLUSTRATIVE EXAMPLES

Example 1. *If $X_1 = [1, 0, 0]$, $X_2 = [0, 1, 0]$, $X_3 = [0, 0, 1]$, find*

(i) $X_1 + X_2 + X_3$ *(ii)* $2X_1 + 3X_2 - X_3$.

Sol. *(i)* $X_1 + X_2 + X_3 = [1, 0, 0] + [0, 1, 0] + [0, 0, 1] = [1, 1, 1]$

(ii) $2X_1 + 3X_2 - X_3 = 2\,[1, 0, 0] + 3\,[0, 1, 0] - [0, 0, 1]$

$= [2, 0, 0] + [0, 3, 0] - [0, 0, 1] = [2, 3, -1]$.

Example 2. *Show that the vectors $X_1 = [1, 2, 3]$, $X_2 = [2, -2, 0]$ form a linearly independent set.*

Sol. Coefficient matrix

$$A = \begin{bmatrix} 1 & 2 \\ 2 & -2 \\ 3 & 0 \end{bmatrix}$$

Operating $R_{21}(-2)$, $R_{31}(-3)$.

$$\sim \begin{bmatrix} 1 & 2 \\ 0 & -6 \\ 0 & -6 \end{bmatrix}$$

Operating $R_{32}\ (-1)$

$$\sim \begin{bmatrix} 1 & 2 \\ 0 & -6 \\ 0 & 0 \end{bmatrix}$$

$\therefore \qquad \rho(A) = 2(= \text{no. of vectors})$

Hence the given set of vectors is linearly independent.

Example 3. *Prove that the vectors $\eta_1 = [1, 2, 3]$, $\eta_2 = [3, -2, 1]$ and $\eta_3 = [1, -6, -5]$ form a linearly dependent system. Also find the relation between them.*

Sol. Coefficient matrix,

$$A = \begin{bmatrix} 1 & 3 & 1 \\ 2 & -2 & -6 \\ 3 & 1 & -5 \end{bmatrix}$$

Operating $R_{21}(-2)$, $R_{31}(-3)$

$$\sim \begin{bmatrix} 1 & 3 & 1 \\ 0 & -8 & -8 \\ 0 & -8 & -8 \end{bmatrix}$$

Operating $R_{32}(-1)$

$$\sim \begin{bmatrix} 1 & 3 & 1 \\ 0 & -8 & -8 \\ 0 & 0 & 0 \end{bmatrix}$$

$\therefore$ $\rho(A) = 2$ (< 3, no. of vectors)

Hence the given vectors are linearly dependent.

Let the relation between the vectors be

$$a_1\eta_1 + a_2\eta_2 + a_3\eta_3 = 0 \quad ...(1)$$

where a_1, a_2, a_3 are scalars given by

$$a_1 + 3a_2 + a_3 = 0 \quad ...(2)$$

$$-8a_2 - 8a_3 = 0 \quad ...(3)$$

From (3), $a_2 + a_3 = 0$

$\Rightarrow$ $\frac{a_2}{1} = \frac{a_3}{-1} = k$ (say)

$\therefore$ $a_2 = k$ and $a_3 = -k$

From (2), $a_1 + 3k - k = 0$

$a_1 = -2k$

Substituting these values in eqn. (1), we get

$$-2k\eta_1 + k\eta_2 - k\eta_3 = 0$$

$\Rightarrow$ $\eta_2 = 2\eta_1 + \eta_3$ $(\because k \neq 0)$

Example 4. *Show that the vectors $X_1 = (1, 2, 4)$, $X_2 = (2, -1, 3)$, $X_3 = (0, 1, 2)$ and $X_4 = (-3, 7, 2)$ are linearly dependent and find the relation between them.* (U.K.T.U. 2012)

Sol. Coefficient matrix

$$A = \begin{bmatrix} 1 & 2 & 0 & -3 \\ 2 & -1 & 1 & 7 \\ 4 & 3 & 2 & 2 \end{bmatrix}$$

Operating $R_{21}(-2)$, $R_{31}(-4)$

$$\sim \begin{bmatrix} 1 & 2 & 0 & -3 \\ 0 & -5 & 1 & 13 \\ 0 & -5 & 2 & 14 \end{bmatrix}$$

Operating $R_{32}(-1)$

$$\sim \begin{bmatrix} 1 & 2 & 0 & -3 \\ 0 & -5 & 1 & 13 \\ 0 & 0 & 1 & 1 \end{bmatrix}$$

$\therefore$ $\rho(A) = 3$ (< no. of vectors)

Hence the given set of vectors is linearly dependent.

Let the relation between the vectors be

$$a_1X_1 + a_2X_2 + a_3X_3 + a_4X_4 = O \quad ...(1)$$

where a_1, a_2, a_3 and a_4 are scalars given by

$$a_1 + 2a_2 - 3a_4 = 0 \quad ...(2)$$

$$-5a_2 + a_3 + 13a_4 = 0 \quad ...(3)$$

$$a_3 + a_4 = 0 \quad ...(4)$$

From (4), $\frac{a_3}{1} = \frac{a_4}{-1} = k$ (say)

so that, $a_3 = k$ and $a_4 = -k$

From (3), $-5a_2 + k - 13k = 0 \Rightarrow a_2 = -\frac{12}{5}k$

From (2), $a_1 - \frac{24}{5}k + 3k = 0 \Rightarrow a_1 = \frac{9}{5}k$

Substituting these values in (1), we get

$$\frac{9}{5}kX_1 - \frac{12}{5}kX_2 + kX_3 - kX_4 = O$$

$$\Rightarrow \quad 9X_1 - 12X_2 + 5X_3 - 5X_4 = O \quad (\because \; k \neq 0)$$

Example 5. *Find whether or not the following set of vectors are linearly dependent or independent.*

(i) [1, 1, 1, 1], [0, 1, 1, 1], [0, 0, 1, 1], [0, 0, 0, 1]

(ii) [1, 1, 0], [1, 0, 1], [0, 1, 1] *(iii)* [1, – 2], [2, 1], [3, 2].

Sol. *(i)* Coefficient matrix,

$$A = \begin{bmatrix} 1 & 0 & 0 & 0 \\ 1 & 1 & 0 & 0 \\ 1 & 1 & 1 & 0 \\ 1 & 1 & 1 & 1 \end{bmatrix}$$

Operating $R_{21}(-1)$, $R_{31}(-1)$, $R_{41}(-1)$

$$\sim \begin{bmatrix} 1 & 0 & 0 & 0 \\ 0 & 1 & 0 & 0 \\ 0 & 1 & 1 & 0 \\ 0 & 1 & 1 & 1 \end{bmatrix}$$

Operating $R_{32}(-1)$, $R_{42}(-1)$

$$\sim \begin{bmatrix} 1 & 0 & 0 & 0 \\ 0 & 1 & 0 & 0 \\ 0 & 0 & 1 & 0 \\ 0 & 0 & 1 & 1 \end{bmatrix}$$

Operating $R_{43}(-1)$

$$\sim \begin{bmatrix} 1 & 0 & 0 & 0 \\ 0 & 1 & 0 & 0 \\ 0 & 0 & 1 & 0 \\ 0 & 0 & 0 & 1 \end{bmatrix}$$

$\therefore$ $\rho(A) = 4$ (= no. of vectors)

Hence the given set of vectors is linearly independent.

(ii) Coefficient matrix,

$$A = \begin{bmatrix} 1 & 1 & 0 \\ 1 & 0 & 1 \\ 0 & 1 & 1 \end{bmatrix}$$

Operating $R_{21}(-1)$

$$\sim \begin{bmatrix} 1 & 1 & 0 \\ 0 & -1 & 1 \\ 0 & 1 & 1 \end{bmatrix}$$

Operating $R_{32}(1)$

$$\sim \begin{bmatrix} 1 & 1 & 0 \\ 0 & -1 & 1 \\ 0 & 0 & 2 \end{bmatrix}$$

$\therefore$ $\rho(A) = 3$ (= no. of vectors)

Hence the given set of vectors is linearly independent.

(*iii*) Coefficient matrix,

$$A = \begin{bmatrix} 1 & 2 & 3 \\ -2 & 1 & 2 \end{bmatrix}$$

Operating $R_{21}(2)$

$$\sim \begin{bmatrix} 1 & 2 & 3 \\ 0 & 5 & 8 \end{bmatrix}$$

$\therefore$ $\rho(A) = 2$ (< 3, no. of vectors)

Hence the given set of vectors is linearly dependent.

Example 6. *Find the value of λ for which the vectors $(1, -2, \lambda)$, $(2, -1, 5)$ and $(3, -5, 7\lambda)$ are linearly dependent.*

Sol. Coefficient matrix

$$A = \begin{bmatrix} 1 & 2 & 3 \\ -2 & -1 & -5 \\ \lambda & 5 & 7\lambda \end{bmatrix}$$

The given vectors will be linearly dependent when $\rho(A) <$ no. of vectors (*i.e.* 3)

Now, $\rho(A) < 3$ is possible only if $|A| = 0$

Hence, $$|A| = \begin{vmatrix} 1 & 2 & 3 \\ -2 & -1 & -5 \\ \lambda & 5 & 7\lambda \end{vmatrix} = 0$$

$$\Rightarrow \quad 1(-7\lambda + 25) - 2(-14\lambda + 5\lambda) + 3(-10 + \lambda) = 0$$

$$\Rightarrow \quad 14\lambda = 5$$

$$\Rightarrow \quad \lambda = \frac{5}{14}.$$

Example 7. *If $X_1 = [3, 1, -4]$, $X_2 = [2, 2, -3]$, $X_3 = [0, -4, 1]$, then show that*

(i) the vectors X_1 and X_2 are linearly independent

and (ii) the vectors X_1, X_2 and X_3 are linearly dependent.

Sol. (*i*) Coefficient matrix

$$A = \begin{bmatrix} 3 & 2 \\ 1 & 2 \\ -4 & -3 \end{bmatrix}$$

Operating R_{12}

$$\sim \begin{bmatrix} 1 & 2 \\ 3 & 2 \\ -4 & -3 \end{bmatrix}$$

Operating $R_{21}(-3)$, $R_{31}(4)$

$$\sim \begin{bmatrix} 1 & 2 \\ 0 & -4 \\ 0 & 5 \end{bmatrix}$$

Operating $R_{32}\left(\frac{5}{4}\right)$

$$\sim \begin{bmatrix} 1 & 2 \\ 0 & -4 \\ 0 & 0 \end{bmatrix}$$

$\therefore$ $\rho(A) = 2$ (= no. of vectors)

Hence the vectors X_1 and X_2 are linearly independent.

(*ii*) Coefficient matrix

$$A = \begin{bmatrix} 3 & 2 & 0 \\ 1 & 2 & -4 \\ -4 & -3 & 1 \end{bmatrix}$$

Operating R_{12}

$$\sim \begin{bmatrix} 1 & 2 & -4 \\ 3 & 2 & 0 \\ -4 & -3 & 1 \end{bmatrix}$$

Operating $R_{21}(-3)$, $R_{31}(4)$

$$\sim \begin{bmatrix} 1 & 2 & -4 \\ 0 & -4 & 12 \\ 0 & 5 & -15 \end{bmatrix}$$

Operating R_{32} (5/4)

$$\sim \begin{bmatrix} 1 & 2 & -4 \\ 0 & -4 & 12 \\ 0 & 0 & 0 \end{bmatrix}$$

$\therefore$ $\rho(A) = 2$ (< 3, no. of vectors)

Hence the vectors X_1, X_2 and X_3 are linearly dependent.

TEST YOUR KNOWLEDGE

1. For the vectors X = [3, – 9, 12], and Y = [– 4, 12, – 16] show that 4X + 3Y = O.
2. Show that the elementary unit vectors $E_1 = [1, 0, 0, ..., 0]$, $E_2 = [0, 1, 0, ..., 0]$, ..., $E_n = [0, 0, 0, ..., 1]$ are linearly independent.

3. (*i*) Show that the column vectors of $A = \begin{bmatrix} 1 & 0 & 0 \\ 6 & 2 & 1 \\ 4 & 3 & 2 \end{bmatrix}$ are linearly independent. (*A.K.T.U. 2020*)

(*ii*) Show that row vectors of the matrix $\begin{bmatrix} 1 & 2 & -2 \\ -1 & 3 & 0 \\ 0 & -2 & 1 \end{bmatrix}$ are linearly independent.

(*G.B.T.U. 2010*)

4. Show that the vectors $X_1 = [a_1, b_1]$, $X_2 = [a_2, b_2]$ are linearly dependent iff $a_1b_2 - a_2b_1 = 0$.
5. Prove that the four vectors $X_1 = [1, 0, 0]$, $X_2 = [0, 1, 0]$, $X_3 = [0, 0, 1]$, $X_4 = [1, 1, 1]$ in V_3 (*c*) form a linearly dependent set but any three of them are linearly independent.
6. Show that the vectors $X_1 = [2, 3, 1, -1]$, $X_2 = [2, 3, 1, -2]$, $X_3 = [4, 6, 2, 1]$ are linearly dependent. Express one of the vectors as a linear combination of the others.
7. Examine the linear dependence or independence of the following set of vectors:

   (*i*) [3, 2, 4], [1, 0, 2], [1, – 1, – 1] (*ii*) [2, 3, 1, – 3], [2, 3, 1, – 2], [4, 6, 2, – 3]

   (*iii*) [1, 1, 0], [3, 1, 3], [5, 3, 3] (*iv*) [1, 2, 1, 2], [0, 1, 2, 0], [1, 4, 3, 2].
8. Show using a matrix that the set of vectors $X = [1, 2, -3, 4]$, $Y = [3, -1, 2, 1]$, $Z = [1, -5, 8, -7]$ is linearly dependent. Determine a maximum subset of linearly independent vectors and express the others as a linear combination of these.
9. If the vectors $(0, 1, a)$, $(1, a, 1)$ and $(a, 1, 0)$ are linearly dependent, then find the value of a.
10. Show that the vectors $x_1 = (1, -1, 1)$; $x_2 = (2, 1, 1)$, $x_3 = (3, 0, 2)$ are linearly dependent. Find the relation between them. (*M.T.U. 2013 ; U.K.T.U. 2011*)
11. Examine the following vectors for linear dependence and find the relation between them, if possible: $X_1 = (1, 1, -1, 1)$, $X_2 = (1, -1, -2, -1)$, $X_3 = (3, 1, 0, 1)$. (*G.B.T.U. 2013*)

Answers

6. $5X_1 - 3X_2 - X_3 = O$
7. (*i*) linearly independent (*ii*) linearly dependent (*iii*) linearly dependent (*iv*) linearly independent.
8. 2; $X_3 = -2X_1 + X_2$ **9.** $a = 0, \pm\sqrt{2}$ **10.** $x_1 + x_2 = x_3$
11. linearly dependent; $2X_1 + X_2 = X_3$.

1.35. COMPLEX MATRICES

If *at least one* element of a matrix is a complex number $a + ib$, where a, b are real and $i = \sqrt{-1}$, then the matrix is called a *complex matrix.*

The matrix obtained by replacing the elements of a complex matrix A by the corresponding conjugate complex numbers is called the *conjugate of the matrix* A and is denoted by $\overline{A}$.

Thus, if $$A = \begin{bmatrix} 2+3i & -7i \\ 5 & 1-i \end{bmatrix}, \text{ then } \overline{A} = \begin{bmatrix} 2-3i & 7i \\ 5 & 1+i \end{bmatrix}$$

It is easy to see that the *conjugate of the transpose of* A *i.e.,* $(\overline{A'})$ and the *transposed conjugate of* A *i.e.,* $(\overline{A})'$ are equal. Each of them is denoted by A^*.

Thus, $$(\overline{A'}) = (\overline{A})' = A^*.$$

A square matrix A = $[a_{ij}]$ is said to be *Hermitian* if A* = A or $a_{ij} = \overline{a}_{ji}$.

A square matrix A = $[a_{ij}]$ is said to be *skew-Hermitian* if A* = – A or $a_{ij} = -\overline{a}_{ji}$.

In a Hermitian matrix, the diagonal elements are all real, while every other element is the conjugate complex of the element in the transposed position. For example,

$$A = \begin{bmatrix} 5 & 2+i & -3i \\ 2-i & -3 & 1-i \\ 3i & 1+i & 0 \end{bmatrix} \text{ is a Hermitian matrix.}$$

In a skew-Hermitian matrix, the diagonal elements are zero or purely imaginary numbers of the form $i\beta$, where β is real. Every other element is the negative of the conjugate complex of the element in the transposed position.

$$\text{For example, } B = \begin{bmatrix} 3i & 1+i & 7 \\ -1+i & 0 & -2-i \\ -7 & 2-i & -i \end{bmatrix} \text{ is a skew-Hermitian matrix.}$$

A square matrix A is said to be *unitary* if AA* = I = A*A.

The determinant of a unitary matrix is of unit modulus. For a matrix to be unitary, it must be non-singular.

A square matrix A is said to be *orthogonal* if AA′ = I = A′A or $A' = A^{-1}$.

Note. The following results hold:

(*i*) $\overline{(\overline{A})} = A$ (*ii*) $\overline{A+B} = \overline{A} + \overline{B}$ (*iii*) $\overline{\lambda A} = \overline{\lambda}\,\overline{A}$

(*iv*) $\overline{AB} = \overline{A}\,\overline{B}$ (*v*) $(A^*)^* = A$ (*vi*) $(A + B)^* = A^* + B^*$

(*vii*) $(\lambda A)^* = \overline{\lambda} A^*$ (*viii*) $(AB)^* = B^*A^*$.

ILLUSTRATIVE EXAMPLES

Example 1. *If* $A = \begin{bmatrix} 2+i & 3 & -1+3i \\ -5 & i & 4-2i \end{bmatrix}$, *verify that A*A is a Hermitian matrix where A* is the conjugate transpose of A.*

Sol.
$$A' = \begin{bmatrix} 2+i & -5 \\ 3 & i \\ -1+3i & 4-2i \end{bmatrix}$$

$$\therefore \quad A^* = (\overline{A'}) = \begin{bmatrix} 2-i & -5 \\ 3 & -i \\ -1-3i & 4+2i \end{bmatrix}$$

Now,
$$A^*A = \begin{bmatrix} 2-i & -5 \\ 3 & -i \\ -1-3i & 4+2i \end{bmatrix} \begin{bmatrix} 2+i & 3 & -1+3i \\ -5 & i & 4-2i \end{bmatrix}$$

$$= \begin{bmatrix} 30 & 6-8i & -19+17i \\ 6+8i & 10 & -5+5i \\ -19-17i & -5-5i & 30 \end{bmatrix} = B \text{ (say)}$$

$$\therefore \qquad B' = \begin{bmatrix} 30 & 6+8i & -19-17i \\ 6-8i & 10 & -5-5i \\ -19+17i & -5+5i & 30 \end{bmatrix}$$

Now, $$B^* = (\overline{B'}) = \begin{bmatrix} 30 & 6-8i & -19+17i \\ 6+8i & 10 & -5+5i \\ -19-17i & -5-5i & 30 \end{bmatrix} = B$$

$\Rightarrow$ **B = A*A is a Hermitian matrix.**

Example 2. *Show that the matrix*

$$\begin{bmatrix} \alpha+i\gamma & -\beta+i\delta \\ \beta+i\delta & \alpha-i\gamma \end{bmatrix} \textit{ is unitary if } \alpha^2+\beta^2+\gamma^2+\delta^2 = 1.$$

(A.K.T.U. 2018)

Sol. Let $$A = \begin{bmatrix} \alpha+i\gamma & -\beta+i\delta \\ \beta+i\delta & \alpha-i\gamma \end{bmatrix}$$

$$\therefore \qquad \overline{A} = \begin{bmatrix} \alpha-i\gamma & -\beta-i\delta \\ \beta-i\delta & \alpha+i\gamma \end{bmatrix}$$

$$\Rightarrow \qquad A^* = (\overline{A})' = \begin{bmatrix} \alpha-i\gamma & \beta-i\delta \\ -\beta-i\delta & \alpha+i\gamma \end{bmatrix}$$

For a square matrix A to be unitary,

$$AA^* = I = A^*A \qquad ...(1)$$

Now, $$AA^* = \begin{bmatrix} \alpha^2+\beta^2+\gamma^2+\delta^2 & 0 \\ 0 & \alpha^2+\beta^2+\gamma^2+\delta^2 \end{bmatrix}$$

Also, $$A^*A = \begin{bmatrix} \alpha^2+\beta^2+\gamma^2+\delta^2 & 0 \\ 0 & \alpha^2+\beta^2+\gamma^2+\delta^2 \end{bmatrix}$$

Eqn. (1) is satisfied only when

$$\alpha^2+\beta^2+\gamma^2+\delta^2 = 1.$$

Example 3. *If A and B are Hermitian, show that AB – BA is skew-Hermitian.*

Sol. A and B are Hermitian $\Rightarrow$ $A^* = A$ and $B^* = B$

Now, $$(AB - BA)^* = (AB)^* - (BA)^*$$
$$= B^*A^* - A^*B^* = BA - AB = -(AB - BA)$$

$\Rightarrow$ AB – BA is skew-Hermitian.

Example 4. *If A is a skew-Hermitian matrix, then show that iA is Hermitian.*

(M.T.U. 2013)

Sol. A is a skew-Hermitian matrix $\Rightarrow$ $A^* = -A$

Now, $$(iA)^* = \bar{i}A^* = (-i)(-A) = iA$$

$\Rightarrow$ iA is a Hermitian matrix.

Example 5. *Show that every square matrix is expressible as the sum of a Hermitian matrix and a skew-Hermitian matrix.*

Sol. Let A be any square matrix.

Since $(A + A^*)^* = A^* + (A^*)^* = A^* + A = A + A^*$

and $(A - A^*)^* = A^* - (A^*)^* = A^* - A = -(A - A^*)$

$\therefore$ $A + A^*$ is Hermitian and $A - A^*$ is skew-Hermitian.

Now, $$A = \frac{1}{2}(A + A^*) + \frac{1}{2}(A - A^*) = P + Q \text{ (say)}$$

where P is Hermitian and Q is skew-Hermitian. Thus, every square matrix can be expressed as the sum of a Hermitian matrix and a skew-Hermitian matrix.

Example 6. *If* $N = \begin{bmatrix} 0 & 1+2i \\ -1+2i & 0 \end{bmatrix}$, *obtain the matrix* $(I - N)(I + N)^{-1}$, *and show that it is unitary.*

Sol. $$I - N = \begin{bmatrix} 1 & 0 \\ 0 & 1 \end{bmatrix} - \begin{bmatrix} 0 & 1+2i \\ -1+2i & 0 \end{bmatrix} = \begin{bmatrix} 1 & -1-2i \\ 1-2i & 1 \end{bmatrix}$$

$$I + N = \begin{bmatrix} 1 & 0 \\ 0 & 1 \end{bmatrix} + \begin{bmatrix} 0 & 1+2i \\ -1+2i & 0 \end{bmatrix} = \begin{bmatrix} 1 & 1+2i \\ -1+2i & 1 \end{bmatrix}$$

$$|I + N| = \begin{vmatrix} 1 & 1+2i \\ -1+2i & 1 \end{vmatrix} = 1 - (4i^2 - 1) = 6$$

$$(I + N)^{-1} = \frac{1}{|I+N|} \text{ adj } (I + N) = \frac{1}{6}\begin{bmatrix} 1 & -1-2i \\ 1-2i & 1 \end{bmatrix}$$

$$\therefore \quad (I - N)(I + N)^{-1} = \begin{bmatrix} 1 & -1-2i \\ 1-2i & 1 \end{bmatrix} \frac{1}{6}\begin{bmatrix} 1 & -1-2i \\ 1-2i & 1 \end{bmatrix}$$

$$= \frac{1}{6}\begin{bmatrix} -4 & -2-4i \\ 2-4i & -4 \end{bmatrix} = A \text{ (say)}$$

$$A' = \frac{1}{6}\begin{bmatrix} -4 & 2-4i \\ -2-4i & -4 \end{bmatrix}$$

$$\overline{(A')} = A^* = \frac{1}{6}\begin{bmatrix} -4 & 2+4i \\ -2+4i & -4 \end{bmatrix}$$

$$A^*A = \frac{1}{6}\begin{bmatrix} -4 & 2+4i \\ -2+4i & -4 \end{bmatrix} \frac{1}{6}\begin{bmatrix} -4 & -2-4i \\ 2-4i & -4 \end{bmatrix}$$

$$= \frac{1}{36}\begin{bmatrix} 36 & 0 \\ 0 & 36 \end{bmatrix} = \begin{bmatrix} 1 & 0 \\ 0 & 1 \end{bmatrix} = I$$

$\Rightarrow$ $A = (I - N)(I + N)^{-1}$ is unitary.

TEST YOUR KNOWLEDGE

1. Show that the matrix A is Hermitian and iA is Skew-Hermitian where A is

(i) $\begin{bmatrix} 2 & 3-4i \\ 3+4i & 2 \end{bmatrix}$ (*U.K.T.U. 2012*) (ii) $\begin{bmatrix} 3 & 5+2i & -3 \\ 5-2i & 7 & 4i \\ -3 & -4i & 5 \end{bmatrix}$

(iii) $\begin{bmatrix} -1 & 2+i & 5-3i \\ 2-i & 7 & 5i \\ 5+3i & -5i & 2 \end{bmatrix}$ (iv) $\begin{bmatrix} 2 & 3+2i & -4 \\ 3-2i & 5 & 6i \\ -4 & -6i & 3 \end{bmatrix}$ (*U.P.T.U. 2014*)

2. (i) Express the matrix $A = \begin{bmatrix} i & 2-3i & 4+5i \\ 6+i & 0 & 4-5i \\ -i & 2-i & 2+i \end{bmatrix}$, as a sum of Hermitian and Skew-Hermitian matrix. (*G.B.T.U. 2010*)

(ii) Express the Hermitian matrix $A = \begin{bmatrix} 1 & -i & 1+i \\ i & 0 & 2-3i \\ 1-i & 2+3i & 2 \end{bmatrix}$ as $P + iQ$ where P is a real symmetric and Q is a real skew-symmetric matrix. (*G.B.T.U. 2012*)

3. Show that $A = \begin{bmatrix} i & 0 & 0 \\ 0 & 0 & i \\ 0 & i & 0 \end{bmatrix}$ is skew-Hermitian and also unitary. (*M.T.U. 2012*)

4. (i) If A is any square matrix, prove that $A + A^*$, AA^*, A^*A are all Hermitian and $A - A^*$ is Skew-Hermitian.

(ii) If A, B are Hermitian or Skew-Hermitian, then so is A + B.

(iii) Show that the matrix B^*AB is Hermitian or Skew-Hermitian according as A is Hermitian or Skew-Hermitian.

(iv) If A is a Hermitian matrix, then show that iA is a Skew-Hermitian matrix.

5. (i) Define unitary matrix. Show that the following matrix is unitary:

$$\frac{1}{\sqrt{3}}\begin{bmatrix} 1 & 1+i \\ 1-i & -1 \end{bmatrix}.$$ (*A.K.T.U. 2021*)

(ii) Prove that $\frac{1}{2}\begin{bmatrix} 1+i & -1+i \\ 1+i & 1-i \end{bmatrix}$ is a unitary matrix.

(iii) Show that $A = \frac{1}{\sqrt{3}}\begin{bmatrix} 1 & 1 & 1 \\ 1 & \omega & \omega^2 \\ 1 & \omega^2 & \omega \end{bmatrix}$ is a unitary matrix, where ω is complex cube root of unity. (*A.K.T.U. 2017*)

6. Verify that the matrix $A = \begin{bmatrix} \frac{1+i}{2} & \frac{1-i}{2} \\ \frac{1-i}{2} & \frac{1+i}{2} \end{bmatrix}$ have eigen values with unit modulus.

Answers

2. (*i*) $\begin{bmatrix} 0 & 4-2i & 2+3i \\ 4+2i & 0 & 3-2i \\ 2-3i & 3+2i & 2 \end{bmatrix} + \begin{bmatrix} i & -2-i & 2+2i \\ 2-i & 0 & 1-3i \\ -2+2i & -1-3i & i \end{bmatrix}$ (*ii*) $\begin{bmatrix} 1 & 0 & 1 \\ 0 & 0 & 2 \\ 1 & 2 & 2 \end{bmatrix} + \begin{bmatrix} 0 & -i & i \\ i & 0 & -3i \\ -i & 3i & 0 \end{bmatrix}$

1.36. CHARACTERISTIC EQUATION

If A is a square matrix of order n, we can form the matrix $A - \lambda I$, where λ is a scalar and I is the unit matrix of order n. The determinant of this matrix equated to zero, *i.e.*,

$$|A - \lambda I| = \begin{vmatrix} a_{11} - \lambda & a_{12} & \dots & a_{1n} \\ a_{21} & a_{22} - \lambda & \dots & a_{2n} \\ \dots & \dots & \dots & \dots \\ a_{n1} & a_{n2} & \dots & a_{nn} - \lambda \end{vmatrix} = 0 \text{ is called the } \textit{characteristic equation} \text{ of A.}$$

On expanding the determinant, the characteristic equation can be written as a polynomial equation of degree n in λ of the form $(-1)^n \lambda^n + k_1 \lambda^{n-1} + k_2 \lambda^{n-2} + \dots + k_n = 0$.

The roots of this equation are called the *characteristic roots* or *latent roots* or *eigen values* of A.

The set of eigen values of a square matrix A is called the *spectrum* of A.

Note. The sum of the eigen values of a matrix A is equal to trace of A.

[The trace of a square matrix is the sum of its diagonal elements.]

1.37. EIGEN VECTORS

Consider the linear transformation $Y = AX$ ...(1)

which transforms the column vector X into the column vector Y. In practice, we are often required to find those vectors X which transform into scalar multiples of themselves.

Let X be such a vector which transforms into λX (λ being a non-zero scalar) by the transformation (1).

Then $\qquad Y = \lambda X$ ...(2)

From (1) and (2), $\qquad AX = \lambda X \Rightarrow AX - \lambda IX = O \Rightarrow (A - \lambda I)X = O$ ...(3)

This matrix equation gives n homogeneous linear equations

$$\left.\begin{array}{llll} (a_{11} - \lambda)x_1 & + a_{12} x_2 & + \dots + a_{1n} x_n = 0 \\ a_{21} x_1 & + (a_{22} - \lambda)x_2 & + \dots + a_{2n} x_n = 0 \\ \dots & \dots & \dots \\ a_{n1} x_1 & + a_{n2} x_2 & + \dots + (a_{nn} - \lambda)x_n = 0 \end{array}\right\} \quad \text{...(4)}$$

These equations will have a non-trivial solution only if the co-efficient matrix $A - \lambda I$ is singular

i.e., if $\qquad |A - \lambda I| = 0$ ...(5)

This is the characteristic equation of the matrix A and has n roots which are the eigen values of A. Corresponding to each root of (5), the homogeneous system (3) has a non-zero solution

$$X = \begin{bmatrix} x_1 \\ x_2 \\ \vdots \\ x_n \end{bmatrix}$$ which is called an eigen vector or latent vector.

Note. If X is a solution of (3), then so is kX, where k is an arbitrary constant. Thus, the eigen vector corresponding to an eigen value is not unique.

1.38. THE CHARACTERISTIC ROOTS OF A UNITARY MATRIX ARE OF UNIT MODULUS

(U.K.T.U. 2010, 2012)

Let A be a unitary matrix so that

$$A^*A = I = AA^* \quad \ldots(1)$$

If λ is a characteristic root of the matrix A and X is its latent vector, then we have

$$AX = \lambda X \quad \ldots(2)$$

Taking transpose conjugate of (2), we obtain

$$(AX)^* = (\lambda X)^*$$

$$X^*A^* = \bar{\lambda} X^* \quad \ldots(3) \quad | \because \lambda^* = \bar{\lambda}$$

On multiplying (2) and (3), we get

$$(X^*A^*)(AX) = (\bar{\lambda} X^*)(\lambda X)$$

$$\Rightarrow \quad X^*(A^*A)X = \lambda\bar{\lambda}(X^*X)$$

$$\Rightarrow \quad X^*X = \lambda\bar{\lambda}(X^*X) \quad | \text{ Using (1)}$$

$$\Rightarrow \quad (1 - \lambda\bar{\lambda}) X^*X = O \quad \ldots(4)$$

Since X is a characteristic vector, $X \neq O$

Consequently, $X^*X \neq O$

Hence equation (4) gives

$$1 - \lambda\bar{\lambda} = 0$$

$$\Rightarrow \quad \lambda\bar{\lambda} = 1$$

$$\Rightarrow \quad |\lambda|^2 = 1 \quad \Rightarrow \quad |\lambda| = 1$$

Hence the characteristic roots of a unitary matrix are of unit modulus.

1.39. THE LATENT ROOTS OF A HERMITIAN MATRIX ARE ALL REAL

(U.K.T.U. 2011)

Let λ be the characteristic or latent root of a Hermitian matrix A. Then $\exists$ a non-zero latent vector X such that

$$AX = \lambda X \quad \ldots(1)$$

Pre-multiplying both sides of (1) by X*, we get

$$X^*AX = X^*\lambda X \quad \ldots(2)$$

Transpose conjugate of (2) gives

$$(X^*AX)^* = (\lambda X^*X)^*$$

$$\Rightarrow \quad X^*A^*(X^*)^* = X^*(X^*)^* \lambda^* \quad | \text{ By reversal law}$$

$$\Rightarrow \quad X^*A^*X = X^*X\bar{\lambda} \quad | \because \lambda^* = \bar{\lambda}$$

But A is a Hermitian matrix so that $A^* = A$

Hence above equation becomes

$$X^*AX = \bar{\lambda}\, X^*X \qquad ...(3)$$

From (2) and (3), we have

$$X^*(\lambda X) = \bar{\lambda}\, X^*X$$

$$\Rightarrow \quad (\lambda - \bar{\lambda})X^*X = O \qquad ...(4)$$

Since X is a non-zero latent vector

$$\therefore \quad X^*X \neq O$$

Hence from (4), we have

$$\lambda - \bar{\lambda} = 0 \quad \Rightarrow \quad \lambda = \bar{\lambda}$$

which is possible only when λ is real.

Hence the latent roots of a Hermitian matrix are all real.

1.40. THE CHARACTERISTIC ROOTS OF A SKEW-HERMITIAN MATRIX IS EITHER ZERO *OR* PURELY AN IMAGINARY NUMBER

Since A is a Skew-Hermitian matrix

$\therefore$ iA is Hermitian matrix.

Let λ be a characteristic root of A.

Then, $AX = \lambda X \quad \Rightarrow \quad (iA)X = (i\lambda)X$

$\Rightarrow$ $i\lambda$ is a characteristic root of matrix iA.

But iA is a Hermitian matrix.

Therefore, $i\lambda$ should be real.

Hence, λ is either zero or purely imaginary.

1.41. THE CHARACTERISTIC ROOTS OF AN IDEMPOTENT MATRIX ARE EITHER ZERO *OR* UNITY

Since A is an idempotent matrix. $\therefore$ $A^2 = A$.

Let X be a latent vector of the matrix A corresponding to the latent root λ so that

$$AX = \lambda X \qquad ...(1)$$

$$\Rightarrow \quad (A - \lambda I)X = O \quad \text{such that} \quad X \neq O$$

Pre-multiplying (1) by A

$$A(AX) = A(\lambda X) = \lambda(AX)$$

$$\Rightarrow \quad (AA)X = \lambda(\lambda X) \qquad | \text{ by } (1)$$

$$\Rightarrow \quad A^2X = \lambda^2 X \quad \Rightarrow \quad AX = \lambda^2 X \qquad | \because A^2 = A$$

$$\Rightarrow \quad \lambda X = \lambda^2 X \qquad | \text{ by } (1)$$

$$\Rightarrow \quad (\lambda^2 - \lambda)X = O \quad \Rightarrow \quad \lambda^2 - \lambda = 0 \qquad (\text{Since } X \neq O)$$

$$\Rightarrow \quad \lambda(\lambda - 1) = 0$$

$$\Rightarrow \quad \lambda = 0, 1.$$

ILLUSTRATIVE EXAMPLES

Example 1. *Show that if $\lambda_1, \lambda_2, ..., \lambda_n$ are the latent roots of the matrix A, then A^3 has the latent roots $\lambda_1^3, \lambda_2^3, ..., \lambda_n^3$.*

Sol. Let λ be a latent root of the matrix A. Then there exists a non-zero vector X such that

$$AX = \lambda X \qquad ...(1)$$

$\Rightarrow$ $A^2(AX) = A^2(\lambda X)$ $\Rightarrow$ $A^3X = \lambda(A^2X)$

But $A^2X = A(AX) = A(\lambda X)$ [Using (1)]

$= \lambda(AX) = \lambda(\lambda X) = \lambda^2 X$

$\therefore$ $A^3X = \lambda(\lambda^2 X) = \lambda^3 X$

$\Rightarrow$ λ^3 is a latent root of A^3.

$\therefore$ If $\lambda_1, \lambda_2, ..., \lambda_n$ are the latent roots of A, then $\lambda_1^3, \lambda_2^3, ..., \lambda_n^3$ are the latent roots of A^3.

Example 2. *If $\lambda_1, \lambda_2, ..., \lambda_n$ are eigen values of A, then find eigen values of the matrix $(A - \lambda I)^2$.*

Sol. $(A - \lambda I)^2 = A^2 - 2\lambda AI + \lambda^2 I^2$

$= A^2 - 2\lambda A + \lambda^2 I$

Eigen values of A^2 are $\lambda_1^2, \lambda_2^2, ..., \lambda_n^2$.

Eigen values of $2\lambda A$ are $2\lambda\lambda_1, 2\lambda\lambda_2, ..., 2\lambda\lambda_n$.

Eigen values of $\lambda^2 I$ are λ^2

$\therefore$ Eigen values of $(A - \lambda I)^2$ are

$$\lambda_1^2 - 2\lambda\lambda_1 + \lambda^2, \lambda_2^2 - 2\lambda\lambda_2 + \lambda^2, ..., \lambda_n^2 - 2\lambda\lambda_n + \lambda^2.$$

or $$(\lambda_1 - \lambda)^2, (\lambda_2 - \lambda)^2, ..., (\lambda_n - \lambda)^2.$$

Example 3. *If λ is an eigen value of a non-singular matrix A, show that*

(i) λ^{-1} is an eigen value of A^{-1}. *(ii) $\frac{|A|}{\lambda}$ is an eigen value of adj. A.*

Sol. *(i)* λ is an eigen value of A.

$\Rightarrow$ There exists a non-zero matrix X such that $AX = \lambda X$

$\Rightarrow$ $X = A^{-1}(\lambda X)$

$\Rightarrow$ $X = \lambda(A^{-1}X)$

$\Rightarrow$ $\frac{1}{\lambda}X = A^{-1}X$

$\Rightarrow$ $A^{-1}X = \lambda^{-1}X$

$\Rightarrow$ λ^{-1} is an eigen value of A^{-1}.

(ii) λ is an eigen value of A.

$\Rightarrow$ There exists a non-zero matrix X such that $AX = \lambda X$

$\Rightarrow$ $(\text{adj. } A)(AX) = (\text{adj. } A)(\lambda X)$

$\Rightarrow$ $\{(\text{adj. } A)A\} X = \lambda (\text{adj. } A) X$

$\Rightarrow$ $|A| IX = \lambda (\text{adj. } A) X$ [$\because$ $(\text{adj. } A) A = |A| I$]

$\Rightarrow$ $|A| X = \lambda (\text{adj. } A) X$

$\Rightarrow$ $\frac{|A|}{\lambda} X = (\text{adj. } A) X$

$\Rightarrow$ $(\text{adj. } A) X = \frac{|A|}{\lambda} X$

$\Rightarrow$ $\frac{|A|}{\lambda}$ is an eigen value of adj. A.

Example 4. *Find the eigen values of the following matrices:*

(*i*) $A = \begin{bmatrix} 1 & 0 & 4 \\ 0 & 2 & 0 \\ 3 & 1 & -3 \end{bmatrix}$ (*ii*) $A = \begin{bmatrix} 2 & 5 & 7 \\ 5 & 3 & 1 \\ 7 & 0 & 2 \end{bmatrix}$

Sol. (*i*) Characteristic equation is

$$| A - \lambda I | = 0$$

$$\Rightarrow \begin{vmatrix} 1-\lambda & 0 & 4 \\ 0 & 2-\lambda & 0 \\ 3 & 1 & -3-\lambda \end{vmatrix} = 0$$

$$\Rightarrow \lambda^3 - 19\lambda + 30 = 0$$

$$\Rightarrow \lambda = 3, -5, 2$$

(*ii*) Characteristic equation is

$$| A - \lambda I | = 0$$

$$\Rightarrow \begin{vmatrix} 2-\lambda & 5 & 7 \\ 5 & 3-\lambda & 1 \\ 7 & 0 & 2-\lambda \end{vmatrix} = 0$$

$$\Rightarrow \lambda^3 - 7\lambda^2 - 58\lambda + 150 = 0$$

$$\Rightarrow \lambda = 11.026, -6.215, 2.1888$$

Example 5. (*i*) *Obtain the eigen value of* A^3 *where* $A = \begin{bmatrix} 3 & 2 \\ 1 & 2 \end{bmatrix}$.

(*ii*) *Two eigen values of the matrix* $A = \begin{bmatrix} 2 & 2 & 1 \\ 1 & 3 & 1 \\ 1 & 2 & 2 \end{bmatrix}$ *are equal to 1 each. Find the eigen values of* A^{-1}.

Sol. (*i*) Characteristic equation of A is

$$| A - \lambda I | = 0$$

$$\begin{vmatrix} 3-\lambda & 2 \\ 1 & 2-\lambda \end{vmatrix} = 0$$

$$\Rightarrow \lambda = 1, 4$$

So, the eigen values of A^3 are 1^3, 4^3 *i.e.,* 1, 64.

(*ii*) Let λ be the third eigen value then,

$$\lambda + 1 + 1 = 2 + 3 + 2$$

$$\Rightarrow \lambda = 5$$

The three eigen values of A are 1, 1, 5.

$\therefore$ The eigen values of A^{-1} are $1, 1, \frac{1}{5}$.

Example 6. *Show that for any square matrix A,*

(*i*) *A and A′ have same set of eigen values.*

(*ii*) *The product of all eigen values of A is equal to determinant(A).*

Sol. Let A be a square matrix.

(*i*) The characteristic equation of A is $| A - \lambda I | = 0$ ...(1)

Let A′ be the transpose of A.

Then the characteristic equation of A′ will be

$$| A' - \lambda I | = 0 \qquad ...(2)$$

Since the interchange of rows and columns does not alter the value of the determinant we have,

$$| A' - \lambda I | = | A - \lambda I |$$

$$| \because \; | A - \lambda I | = | A - \lambda I |' = | A' - \lambda I' | = | A' - \lambda I | \text{ as } I' = I$$

Hence the eigen values of matrix A and its transpose A′ are identical.

(*ii*) Let $A = [a_{ij}]_{n \times n}$ be a given square matrix and $\lambda_1, \lambda_2, \lambda_3, ..., \lambda_n$ be its eigen values. If $\phi(\lambda)$ be the characteristic polynomial then,

$$\phi(\lambda) = | A - \lambda I |$$

$$= \begin{vmatrix} a_{11} - \lambda & a_{12} & \dots\dots & a_{1n} \\ a_{21} & a_{22} - \lambda & \dots\dots & a_{2n} \\ \vdots & \vdots & \vdots & \vdots \\ a_{n1} & a_{n2} & \dots\dots & a_{nn} - \lambda \end{vmatrix}$$

$$= (-1)^n \{\lambda^n + p_1 \lambda^{n-1} + p_2 \lambda^{n-2} + ... + p_n\}$$

$$= (-1)^n \{(\lambda - \lambda_1)(\lambda - \lambda_2)(\lambda - \lambda_3) ... (\lambda - \lambda_n)\}$$

Putting $\lambda = 0$, we get

$$\phi(0) = (-1)^n (-1)^n \lambda_1 \lambda_2 \lambda_3 ... \lambda_n$$

$$| A | = \lambda_1 \lambda_2 \lambda_3 ... \lambda_n$$

Hence the product of all eigen values of A is equal to determinant(A).

Example 7. *Show that for a square matrix,*

(*i*) *There are infinitely many eigen vectors corresponding to a single eigen value.*

(*ii*) *Every eigen vector corresponds to a unique eigen value.*

Sol. (*i*) Let X be a characteristic vector of a square matrix A corresponding to a single eigen value λ. Then we have,

$$AX = \lambda X$$

Let *k* be any non-zero scalar. Then,

$$k(AX) = k(\lambda X) \quad \Rightarrow \quad A(kX) = \lambda(kX)$$

Therefore, *k*X is also a characteristic vector of A corresponding to the same characteristic root λ.

Since *k* is any non-zero scalar, ∃ infinitely many eigen vectors corresponding to a single eigen value.

(*ii*) Let there exist two distinct eigen values λ_1 and λ_2 corresponding to an eigen vector X of a square matrix A. Then, we have

$$AX = \lambda_1 X \qquad | \lambda_1 \neq \lambda_2$$

$$AX = \lambda_2 X$$

$$\therefore \quad AX = \lambda_1 X = \lambda_2 X$$

$$\Rightarrow \quad (\lambda_1 - \lambda_2)X = O$$

$$\Rightarrow \quad X = O \qquad | \because \; \lambda_1 - \lambda_2 \neq 0$$

which is impossible since X is a non-zero vector. Hence every eigen vector corresponds to a unique eigen value.

Example 8. *Find the eigen values and eigen vectors of the matrix* $A = \begin{bmatrix} 1 & -2 \\ -5 & 4 \end{bmatrix}$.

Sol. The characteristic equation of the given matrix is

$$|A - \lambda I| = 0$$

or $$\begin{vmatrix} 1-\lambda & -2 \\ -5 & 4-\lambda \end{vmatrix} = 0$$

$\Rightarrow$ $$\lambda^2 - 5\lambda - 6 = 0$$

$\Rightarrow$ $$\lambda = 6, -1.$$

Thus, the eigen values of A are 6, – 1.

Corresponding to $\lambda = 6$, the eigen vectors are given by

$$(A - 6I)X_1 = O$$

or $$\begin{bmatrix} 1-6 & -2 \\ -5 & 4-6 \end{bmatrix}\begin{bmatrix} x_1 \\ x_2 \end{bmatrix} = O$$

or $$\begin{bmatrix} -5 & -2 \\ -5 & -2 \end{bmatrix}\begin{bmatrix} x_1 \\ x_2 \end{bmatrix} = O$$

We get only one independent equation $-5x_1 - 2x_2 = 0$

$\Rightarrow$ $$\frac{x_1}{2} = \frac{x_2}{-5} = k_1 \text{ (say)}$$

$$x_1 = 2k_1, \quad x_2 = -5k_1$$

$\therefore$ The eigen vectors are $X_1 = k_1 \begin{bmatrix} 2 \\ -5 \end{bmatrix}$

Corresponding to $\lambda = -1$, the eigen vectors are given by

$$(A + I)X_2 = O$$

$\Rightarrow$ $$\begin{bmatrix} 2 & -2 \\ -5 & 5 \end{bmatrix}\begin{bmatrix} x_1 \\ x_2 \end{bmatrix} = \begin{bmatrix} 0 \\ 0 \end{bmatrix}$$

$\Rightarrow$ $$x_1 - x_2 = 0$$

$\Rightarrow$ $$\frac{x_1}{1} = \frac{x_2}{1} = k_2 \text{ (say)}$$

$$x_1 = k_2, x_2 = k_2$$

$\therefore$ The eigen vectors are $X_2 = k_2 \begin{bmatrix} 1 \\ 1 \end{bmatrix}$.

Example 9. *Find the eigen values and eigen vectors of the matrix* $A = \begin{bmatrix} -2 & 2 & -3 \\ 2 & 1 & -6 \\ -1 & -2 & 0 \end{bmatrix}$.

[U.K.T.U. 2011; G.B.T.U. (SUM) 2010]

Sol. The characteristic equation of the given matrix is

$$|A - \lambda I| = 0$$

or $$\begin{vmatrix} -2-\lambda & 2 & -3 \\ 2 & 1-\lambda & -6 \\ -1 & -2 & -\lambda \end{vmatrix} = 0$$

or $$(-2-\lambda)[-\lambda(1-\lambda) - 12] - 2[-2\lambda - 6] - 3[-4 + 1(1-\lambda)] = 0$$

or $$\lambda^3 + \lambda^2 - 21\lambda - 45 = 0$$

By trial, $\lambda = -3$ satisfies it.

$\therefore \quad (\lambda + 3)(\lambda^2 - 2\lambda - 15) = 0 \Rightarrow (\lambda + 3)(\lambda + 3)(\lambda - 5) = 0 \Rightarrow \lambda = -3, -3, 5$

Thus, the eigen values of A are $-3, -3, 5$.

Corresponding to $\lambda = -3$, the eigen vectors are given by

$$(A + 3I)\, X_1 = O$$

or
$$\begin{bmatrix} -1 & 2 & -3 \\ 2 & 4 & -6 \\ -1 & -2 & 3 \end{bmatrix} \begin{bmatrix} x_1 \\ x_2 \\ x_3 \end{bmatrix} = O$$

We get only one independent equation $x_1 + 2x_2 - 3x_3 = 0$

Let $x_3 = k_1$, $x_2 = k_2$ then $x_1 = 3k_1 - 2k_2$

$\therefore$ The eigen vectors are given by

$$X_1 = \begin{bmatrix} 3k_1 - 2k_2 \\ k_2 \\ k_1 \end{bmatrix} = k_1 \begin{bmatrix} 3 \\ 0 \\ 1 \end{bmatrix} + k_2 \begin{bmatrix} -2 \\ 1 \\ 0 \end{bmatrix}$$

Corresponding to $\lambda = 5$, the eigen vectors are given by $(A - 5I)\, X_2 = O$

$$\Rightarrow \quad \begin{bmatrix} -7 & 2 & -3 \\ 2 & -4 & -6 \\ -1 & -2 & -5 \end{bmatrix} \begin{bmatrix} x_1 \\ x_2 \\ x_3 \end{bmatrix} = \begin{bmatrix} 0 \\ 0 \\ 0 \end{bmatrix}$$

$$\Rightarrow \quad -7x_1 + 2x_2 - 3x_3 = 0$$

$$x_1 - 2x_2 - 3x_3 = 0$$

$$-x_1 - 2x_2 - 5x_3 = 0$$

From first two equations,

$$\frac{x_1}{10 - 6} = \frac{x_2}{3 + 5} = \frac{x_3}{-2 - 2}$$

$$\Rightarrow \quad \frac{x_1}{1} = \frac{x_2}{2} = \frac{x_3}{-1} = k_3 \text{ (say)}$$

$$\therefore \quad x_1 = k_3, \; x_2 = 2k_3, \; x_3 = -k_3$$

Hence the eigen vectors are given by

$$X_2 = k_3 \begin{bmatrix} 1 \\ 2 \\ -1 \end{bmatrix}.$$

Example 10. *Find the eigen values and eigen vectors of the following matrices:*

(*i*) $A = \begin{bmatrix} 8 & -6 & 2 \\ -6 & 7 & -4 \\ 2 & -4 & 3 \end{bmatrix}$ (G.B.T.U. 2011) (*ii*) $A = \begin{bmatrix} 6 & -2 & 2 \\ -2 & 3 & -1 \\ 2 & -1 & 3 \end{bmatrix}$

(*iii*) $A = \begin{bmatrix} 3 & 1 & 4 \\ 0 & 2 & 6 \\ 0 & 0 & 5 \end{bmatrix}$. (U.K.T.U. 2010)

Sol. (*i*) The characteristic equation is

$$| A - \lambda I | = 0$$

$$\Rightarrow \quad \begin{vmatrix} 8-\lambda & -6 & 2 \\ -6 & 7-\lambda & -4 \\ 2 & -4 & 3-\lambda \end{vmatrix} = 0$$

$$\Rightarrow \quad (8-\lambda)\{(7-\lambda)(3-\lambda)-16\}+6\{(-6)(3-\lambda)+8\}+2\{24-2(7-\lambda)\}=0$$

$$\Rightarrow \quad -\lambda^3+18\lambda^2-45\lambda=0$$

$$\Rightarrow \quad \lambda(\lambda-3)(\lambda-15)=0 \quad \Rightarrow \quad \lambda = 0, 3, 15$$

Hence the eigen values are 0, 3, 15.

The eigen vector corresponding to $\lambda_1 = 0$ is given by

$$(A-\lambda_1 I)X_1 = O \quad \text{where} \quad X_1 = \begin{bmatrix} x_1 \\ x_2 \\ x_3 \end{bmatrix}$$

$$\Rightarrow \quad \begin{bmatrix} 8 & -6 & 2 \\ -6 & 7 & -4 \\ 2 & -4 & 3 \end{bmatrix}\begin{bmatrix} x_1 \\ x_2 \\ x_3 \end{bmatrix} = \begin{bmatrix} 0 \\ 0 \\ 0 \end{bmatrix}$$

$$\Rightarrow \quad 8x_1 - 6x_2 + 2x_3 = 0 \qquad \text{...(1)}$$

$$-6x_1 + 7x_2 - 4x_3 = 0 \qquad \text{...(2)}$$

$$2x_1 - 4x_2 + 3x_3 = 0 \qquad \text{...(3)}$$

Solving eqns. (1) and (2) by cross-multiplication,

$$\frac{x_1}{24-14} = \frac{x_2}{-12+32} = \frac{x_3}{56-36}$$

$$\Rightarrow \quad \frac{x_1}{10} = \frac{x_2}{20} = \frac{x_3}{20}$$

$$\Rightarrow \quad \frac{x_1}{1} = \frac{x_2}{2} = \frac{x_3}{2} = k_1 \quad \text{(say)} \quad \text{where} \quad k_1 \neq 0$$

$$x_1 = k_1$$

$$x_2 = 2k_1$$

$$x_3 = 2k_1 \quad \text{which satisfy equation (3) also.}$$

$\therefore$ The required eigen vector X_1 corresponding to eigen value $\lambda_1 = 0$ is given by

$$X_1 = \begin{bmatrix} k_1 \\ 2k_1 \\ 2k_1 \end{bmatrix} = k_1 \begin{bmatrix} 1 \\ 2 \\ 2 \end{bmatrix}$$

The eigen vector X_2 corresponding to $\lambda_2 = 3$ is given by

$$(A-\lambda_2 I)X_2 = O \quad \text{where} \quad X_2 = \begin{bmatrix} x_1 \\ x_2 \\ x_3 \end{bmatrix}$$

$$\Rightarrow \quad \begin{bmatrix} 5 & -6 & 2 \\ -6 & 4 & -4 \\ 2 & -4 & 0 \end{bmatrix}\begin{bmatrix} x_1 \\ x_2 \\ x_3 \end{bmatrix} = \begin{bmatrix} 0 \\ 0 \\ 0 \end{bmatrix}$$

$$\Rightarrow \quad 5x_1 - 6x_2 + 2x_3 = 0 \qquad \text{...(4)}$$

$$-6x_1 + 4x_2 - 4x_3 = 0 \qquad \text{...(5)}$$

$$2x_1 - 4x_2 = 0 \qquad \text{...(6)}$$

Solving eqns. (4) and (5) by cross-multiplication, we get

$$\frac{x_1}{24-8} = \frac{x_2}{-12+20} = \frac{x_3}{20-36}$$

$$\Rightarrow \quad \frac{x_1}{16} = \frac{x_2}{8} = \frac{x_3}{-16}$$

$$\Rightarrow \quad \frac{x_1}{2} = \frac{x_2}{1} = \frac{x_3}{-2} = k_2 \text{ (say)} \quad \text{where} \quad k_2 \neq 0$$

$$\therefore \quad x_1 = 2k_2, \quad x_2 = k_2, \quad x_3 = -2k_2$$

which satisfy equation (6) also.

$\therefore$ The required eigen vector X_2 is

$$X_2 = \begin{bmatrix} 2k_2 \\ k_2 \\ -2k_2 \end{bmatrix} = k_2 \begin{bmatrix} 2 \\ 1 \\ -2 \end{bmatrix}$$

The eigen vector corresponding to $\lambda_3 = 15$ is given by

$$(A - \lambda_3 I)X_3 = O \quad \text{where} \quad X_3 = \begin{bmatrix} x_1 \\ x_2 \\ x_3 \end{bmatrix}$$

$$\Rightarrow \quad \begin{bmatrix} -7 & -6 & 2 \\ -6 & -8 & -4 \\ 2 & -4 & -12 \end{bmatrix} \begin{bmatrix} x_1 \\ x_2 \\ x_3 \end{bmatrix} = \begin{bmatrix} 0 \\ 0 \\ 0 \end{bmatrix}$$

$$\Rightarrow \quad -7x_1 - 6x_2 + 2x_3 = 0 \qquad \text{...(7)}$$

$$-6x_1 - 8x_2 - 4x_3 = 0 \qquad \text{...(8)}$$

$$2x_1 - 4x_2 - 12x_3 = 0 \qquad \text{...(9)}$$

Solving equations (7) and (8) by cross-multiplication, we get

$$\frac{x_1}{40} = \frac{x_2}{-40} = \frac{x_3}{20}$$

$$\Rightarrow \quad \frac{x_1}{2} = \frac{x_2}{-2} = \frac{x_3}{1} = k_3 \text{ (say)} \quad \text{where} \quad k_3 \neq 0$$

$$\Rightarrow \quad x_1 = 2k_3, \quad x_2 = -2k_3, \quad x_3 = k_3$$

which satisfy equation (9) also.

$\therefore$ The required eigen vector X_3 is $X_3 = \begin{bmatrix} 2k_3 \\ -2k_3 \\ k_3 \end{bmatrix} = k_3 \begin{bmatrix} 2 \\ -2 \\ 1 \end{bmatrix}$

(*ii*) The characteristic equation is

$$|A - \lambda I| = 0$$

$$\Rightarrow \quad \begin{vmatrix} 6-\lambda & -2 & 2 \\ -2 & 3-\lambda & -1 \\ 2 & -1 & 3-\lambda \end{vmatrix} = 0$$

$$\Rightarrow \quad -\lambda^3 + 12\lambda^2 - 36\lambda + 32 = 0$$

$$\Rightarrow \quad \lambda = 2, 2, 8$$

Hence eigen values are 2, 2, 8.

Eigen vector corresponding to $\lambda_1 = \lambda_2 = 2$ is given by

$$(A - \lambda_1 I)X_1 = O \quad \text{where} \quad X_1 = \begin{bmatrix} x_1 \\ x_2 \\ x_3 \end{bmatrix}$$

$$\Rightarrow \quad \begin{bmatrix} 4 & -2 & 2 \\ -2 & 1 & -1 \\ 2 & -1 & 1 \end{bmatrix} \begin{bmatrix} x_1 \\ x_2 \\ x_3 \end{bmatrix} = \begin{bmatrix} 0 \\ 0 \\ 0 \end{bmatrix}$$

$$\Rightarrow \quad 4x_1 - 2x_2 + 2x_3 = 0 \quad \text{...(1)}$$

$$-2x_1 + x_2 - x_3 = 0 \quad \text{...(2)}$$

$$2x_1 - x_2 + x_3 = 0 \quad \text{...(3)}$$

These equations are equivalent to a single equation

$$2x_1 - x_2 + x_3 = 0 \quad \text{...(4)}$$

Let $x_3 = 2k_3$ and $x_2 = 2k_2$ then from (4)

$$2x_1 - 2k_2 + 2k_3 = 0 \quad \Rightarrow \quad x_1 = k_2 - k_3$$

$\therefore$ Required eigen vector corresponding to $\lambda_1 = 2 = \lambda_2$ is

$$X_1 = \begin{bmatrix} k_2 - k_3 \\ 2k_2 \\ 2k_3 \end{bmatrix} = k_2 \begin{bmatrix} 1 \\ 2 \\ 0 \end{bmatrix} + k_3 \begin{bmatrix} -1 \\ 0 \\ 2 \end{bmatrix}$$

Similarly, the eigen vector corresponding to $\lambda_3 = 8$ is given by,

$$(A - \lambda_3 I)X_3 = O \quad \text{where} \quad X_3 = \begin{bmatrix} x_1 \\ x_2 \\ x_3 \end{bmatrix}$$

$$\Rightarrow \quad (A - 8I)X_3 = O$$

$$\Rightarrow \quad \begin{bmatrix} -2 & -2 & 2 \\ -2 & -5 & -1 \\ 2 & -1 & -5 \end{bmatrix} \begin{bmatrix} x_1 \\ x_2 \\ x_3 \end{bmatrix} = \begin{bmatrix} 0 \\ 0 \\ 0 \end{bmatrix}$$

$$\Rightarrow \quad -2x_1 - 2x_2 + 2x_3 = 0 \quad \text{...(5)}$$

$$-2x_1 - 5x_2 - x_3 = 0 \quad \text{...(6)}$$

$$2x_1 - x_2 - 5x_3 = 0 \quad \text{...(7)}$$

Solving equations (5) and (6) by cross-multiplication, we get

$$\frac{x_1}{12} = \frac{x_2}{-6} = \frac{x_3}{6}$$

$$\Rightarrow \quad \frac{x_1}{2} = \frac{x_2}{-1} = \frac{x_3}{1} = k_1 \text{ (say)}$$

$$x_1 = 2k_1, \quad x_2 = -k_1, \quad x_3 = k_1$$

which satisfy equation (7) also.

$\therefore$ Required eigen vector corresponding to $\lambda_3 = 8$ is $X_3 = \begin{bmatrix} 2k_1 \\ -k_1 \\ k_1 \end{bmatrix} = k_1 \begin{bmatrix} 2 \\ -1 \\ 1 \end{bmatrix}$

(*iii*) The characteristic equation is

$$| A - \lambda I | = 0$$

$$\Rightarrow \quad \begin{vmatrix} 3-\lambda & 1 & 4 \\ 0 & 2-\lambda & 6 \\ 0 & 0 & 5-\lambda \end{vmatrix} = 0$$

$$\Rightarrow \quad (3-\lambda)(2-\lambda)(5-\lambda) = 0 \quad \Rightarrow \quad \lambda = 3, 2, 5$$

Hence eigen values are 3, 2, 5.

Eigen vector X_1 corresponding to eigen value $\lambda_1 = 3$ is

$$(A - 3I)X_1 = O \quad \text{where} \quad X_1 = [x_1 \quad x_2 \quad x_3]'$$

$$\Rightarrow \quad \begin{bmatrix} 0 & 1 & 4 \\ 0 & -1 & 6 \\ 0 & 0 & 2 \end{bmatrix} \begin{bmatrix} x_1 \\ x_2 \\ x_3 \end{bmatrix} = \begin{bmatrix} 0 \\ 0 \\ 0 \end{bmatrix}$$

$$\Rightarrow \quad x_2 + 4x_3 = 0$$

$$-x_2 + 6x_3 = 0$$

$$2x_3 = 0$$

$$\Rightarrow \quad x_3 = 0, \quad x_2 = 0$$

$\therefore$ **The characteristic vector corresponding to eigen value $\lambda_1 = 3$ is given by**

$$X_1 = \begin{bmatrix} x_1 \\ x_2 \\ x_3 \end{bmatrix} = \begin{bmatrix} k_1 \\ 0 \\ 0 \end{bmatrix} \quad \text{where} \quad x_1 = k_1, \quad k_1 \neq 0$$

When $\lambda = 2$, let X_2 be the eigen vector then,

$$(A - 2I)X_2 = O \quad \text{where} \quad X_2 = [x_1 \quad x_2 \quad x_3]'$$

$$\Rightarrow \quad \begin{bmatrix} 1 & 1 & 4 \\ 0 & 0 & 6 \\ 0 & 0 & 3 \end{bmatrix} \begin{bmatrix} x_1 \\ x_2 \\ x_3 \end{bmatrix} = \begin{bmatrix} 0 \\ 0 \\ 0 \end{bmatrix}$$

$$\Rightarrow \quad x_1 + x_2 + 4x_3 = 0$$

$$6x_3 = 0$$

$$3x_3 = 0 \quad \Rightarrow \quad x_3 = 0$$

$$\therefore \quad x_1 + x_2 = 0$$

or

$$x_1 = -x_2$$

$$\frac{x_1}{1} = \frac{x_2}{-1} = \frac{x_3}{0} = k_2 \text{ (say)}, \quad \text{where} \quad k_2 \neq 0$$

$$\therefore \quad x_1 = k_2, \quad x_2 = -k_2, \quad x_3 = 0$$

$\therefore$ **Required eigen vector is** $X_2 = \begin{bmatrix} k_2 \\ -k_2 \\ 0 \end{bmatrix} = k_2 \begin{bmatrix} 1 \\ -1 \\ 0 \end{bmatrix}$

When $\lambda = 5$, let X_3 be the eigen vector then

$$(A - 5I)X_3 = O \quad \text{where} \quad X_3 = [x_1 \quad x_2 \quad x_3]'$$

$$\Rightarrow \quad \begin{bmatrix} -2 & 1 & 4 \\ 0 & -3 & 6 \\ 0 & 0 & 0 \end{bmatrix} \begin{bmatrix} x_1 \\ x_2 \\ x_3 \end{bmatrix} = \begin{bmatrix} 0 \\ 0 \\ 0 \end{bmatrix}$$

$$\Rightarrow \quad -2x_1 + x_2 + 4x_3 = 0$$

$$-3x_2 + 6x_3 = 0$$

Solving above equations by cross-multiplication, we get

$$\frac{x_1}{6+12} = \frac{x_2}{12} = \frac{x_3}{6}$$

$\Rightarrow \quad \frac{x_1}{3} = \frac{x_2}{2} = \frac{x_3}{1} = k_3$ (say) where $k_3 \neq 0$

$\Rightarrow \quad x_1 = 3k_3, \quad x_2 = 2k_3, \quad x_3 = k_3$

$\therefore$ Required eigen vector is $X_3 = \begin{bmatrix} 3k_3 \\ 2k_3 \\ k_3 \end{bmatrix} = k_3 \begin{bmatrix} 3 \\ 2 \\ 1 \end{bmatrix}$.

Example 11. *Prove that*

(i) 0 is a characteristic root of a matrix if and only if the matrix is singular.

(ii) The characteristic root of a real symmetric matrix are all real.

(iii) If A and B are square matrices of same type and if P be invertible then A and $P^{-1}AP$ have same eigen values. [M.T.U. (SUM) 2011]

(iv) The sum of the eigen values of a square matrix is equal to the sum of the elements of its principal diagonal.

Sol. (*i*) The characteristic root of a matrix A is given by $|A - \lambda I| = 0$

If $\lambda = 0$, then it gives $|A| = 0$

$\Rightarrow$ A is singular.

Again if matrix A is singular, then

$$|A - \lambda I| = 0$$

$$\Rightarrow \quad |A| - \lambda |I| = 0$$

$$\Rightarrow \quad 0 - \lambda \cdot 1 = 0$$

$$\Rightarrow \quad \lambda = 0.$$

(*ii*) Let A be a real symmetric matrix

$\because$ A is real

$$\therefore \quad \overline{A} = A$$

$$\Rightarrow \quad (\overline{A})' = A'$$

$$\Rightarrow \quad A^* = A \qquad | \because \text{ A is symmetric}$$

$\Rightarrow$ A is Hermitian.

Hence the characteristic roots of A are all real.

(*iii*) Let $B = P^{-1}AP$

then

$$B - \lambda I = P^{-1}AP - \lambda I$$

$$= P^{-1}AP - P^{-1}\lambda IP = P^{-1}(A - \lambda I)P$$

$$\therefore \quad |B - \lambda I| = |P^{-1}(A - \lambda I)P| = |P^{-1}|\,|A - \lambda I|\,|P|$$

$$= |A - \lambda I|\,|P^{-1}|\,|P| = |A - \lambda I|\,|P^{-1}P|$$

$$= |A - \lambda I|\,|I| = |A - \lambda I| \qquad | \because \; |I| = 1$$

Hence matrices A and $P^{-1}AP$ have the same characteristic roots.

(*iv*) Let $A = [a_{ij}]_{3 \times 3}$ be a square matrix of order 3. The characteristic equation is

$$|A - \lambda I| = 0$$

$$\Rightarrow \begin{vmatrix} a_{11}-\lambda & a_{12} & a_{13} \\ a_{21} & a_{22}-\lambda & a_{23} \\ a_{31} & a_{32} & a_{33}-\lambda \end{vmatrix} = 0$$

$$\Rightarrow \quad -\lambda^3 + \lambda^2 (a_{11} + a_{22} + a_{33}) - \ldots = 0 \qquad \ldots(1)$$

But we know that,

$$| A - \lambda I | = (-1)^3 (\lambda - \lambda_1)(\lambda - \lambda_2)(\lambda - \lambda_3)$$
$$= -\lambda^3 + \lambda^2 (\lambda_1 + \lambda_2 + \lambda_3) - \ldots \qquad \ldots(2)$$

Comparing equations (1) and (2), we get

$$\lambda_1 + \lambda_2 + \lambda_3 = a_{11} + a_{22} + a_{33}.$$

TEST YOUR KNOWLEDGE

1. Find the eigen values and corresponding eigen vectors of the following matrices:

(*i*) $\begin{bmatrix} -5 & 2 \\ 2 & -2 \end{bmatrix}$ (*ii*) $\begin{bmatrix} 5 & 4 \\ 1 & 2 \end{bmatrix}$

(*iii*) $\begin{bmatrix} 1 & 2 & 2 \\ 0 & 2 & 1 \\ -1 & 2 & 2 \end{bmatrix}$ (*iv*) $\begin{bmatrix} a & h & g \\ 0 & b & 0 \\ 0 & 0 & c \end{bmatrix}$

(*v*) $\begin{bmatrix} 1 & 0 & 0 \\ 0 & 2 & 1 \\ 2 & 0 & 3 \end{bmatrix}$ (*vi*) $\begin{bmatrix} 1 & 0 & -1 \\ 1 & 2 & 1 \\ 2 & 2 & 3 \end{bmatrix}$

(*vii*) $\begin{bmatrix} 1 & 2 & 3 \\ 3 & 1 & 2 \\ 2 & 3 & 1 \end{bmatrix}$ (*viii*) $\begin{bmatrix} 1 & 1 & 0 \\ 0 & 1 & 1 \\ 1 & 0 & 1 \end{bmatrix}$

(*ix*) $\begin{bmatrix} 1 & 1 & 3 \\ 1 & 5 & 1 \\ 3 & 1 & 1 \end{bmatrix}$ (*U.P.T.U. 2015*) (*x*) $\begin{bmatrix} 2 & -1 & 1 \\ -1 & 2 & -1 \\ 1 & -1 & 2 \end{bmatrix}$ (*A.K.T.U. 2018*)

(*xi*) $A = \begin{bmatrix} 2 & 1 & 1 \\ 2 & 3 & 2 \\ 3 & 3 & 4 \end{bmatrix}$ (*A.K.T.U. 2022*) (*xii*) $A = \begin{bmatrix} 2 & 0 & 1 \\ 0 & 3 & 0 \\ 1 & 0 & 2 \end{bmatrix}$ (*A.K.T.U. 2021*)

2. (*i*) Find the characteristic equation of the matrix $A = \begin{bmatrix} 1 & 2 & 2 \\ 0 & 2 & 1 \\ -1 & 2 & 2 \end{bmatrix}$.

(*ii*) Find the eigen values of the matrix $A = \begin{bmatrix} 2 & -3 & 1 \\ 3 & 1 & 3 \\ -5 & 2 & -4 \end{bmatrix}$. (*U.K.T.U. 2011*)

(*iii*) Find the eigen vectors for the matrix $A = \begin{bmatrix} 1 & -1 \\ 2 & -1 \end{bmatrix}$. (*U.P.T.U. 2014*)

3. Find the eigen values of $3A^3 + 5A^2 - 6A + 2I$ where $A = \begin{bmatrix} 1 & 2 & -3 \\ 0 & 3 & 2 \\ 0 & 0 & -2 \end{bmatrix}$.

4. Find the eigen value of the matrix $\begin{bmatrix} 4 & 2 \\ 2 & 4 \end{bmatrix}$ corresponding to the eigen vector $\begin{bmatrix} 101 \\ 101 \end{bmatrix}$.

(*A.K.T.U. 2017*)

5. (*i*) Show that if λ is a characteristic root of the matrix A then $\lambda + k$ is a characteristic root of the matrix $A + kI$.

(*ii*) Show that if λ_i $(1 \le i \le n)$ are the eigen values of a square matrix A then A^m has the eigen values λ_i^m $(1 \le i \le n)$, m being a positive integer.

(*iii*) Prove that the characteristic roots of a diagonal matrix are the diagonal elements of the matrix.

6. Verify that the matrices $X = \begin{bmatrix} 0 & h & g \\ h & 0 & f \\ g & f & 0 \end{bmatrix}$, $Y = \begin{bmatrix} 0 & f & h \\ f & 0 & g \\ h & g & 0 \end{bmatrix}$, $Z = \begin{bmatrix} 0 & g & f \\ g & 0 & h \\ f & h & 0 \end{bmatrix}$ have the same characteristic equation.

7. If $a + b + c = 0$, find the characteristic roots of the matrix $A = \begin{bmatrix} a & c & b \\ c & b & a \\ b & a & c \end{bmatrix}$.

8. Prove that for matrix $A = \begin{bmatrix} 2 & 1 & 1 \\ 2 & 3 & 4 \\ -1 & -1 & -2 \end{bmatrix}$, all its eigen values are distinct and real. Hence find the corresponding eigen vectors.

9. Show that the matrix $\begin{bmatrix} 3 & 10 & 5 \\ -2 & -3 & -4 \\ 3 & 5 & 7 \end{bmatrix}$ has less than three linearly independent eigen vectors. Also find them. (*U.P.T.U. 2014*)

10. If $M = \begin{bmatrix} 0 & 1 & 1 \\ 1 & 0 & 1 \\ 1 & 1 & 0 \end{bmatrix}$ and $P = \dfrac{1}{2}\begin{bmatrix} b+c & c-a & b-a \\ c-b & c+a & a-b \\ b-c & a-c & a+b \end{bmatrix}$, show that the transform of P by M *i.e.*, MPM^{-1} is a diagonal matrix and hence find the eigen values of P.

11. Find characteristic equation and eigen values of the matrix $A = \begin{bmatrix} 3 & 2 & 2 & -4 \\ 2 & 3 & 2 & -1 \\ 1 & 1 & 2 & -1 \\ 2 & 2 & 2 & -1 \end{bmatrix}$.

12. If $A = \begin{bmatrix} 4 & -1 & -1 \\ -1 & 4 & -1 \\ -1 & -1 & 4 \end{bmatrix}$ and $B = I - \dfrac{1}{4}A$, then show that $\mu_i = 1 - \dfrac{1}{4}\lambda_i$, where λ_i and μ_i are the eigen values of A and B respectively.

Answers

1. (*i*) $-1, -6$; $k_1 \begin{bmatrix} 1 \\ 2 \end{bmatrix}, k_2 \begin{bmatrix} 2 \\ -1 \end{bmatrix}$ (*ii*) $1, 6$; $k_1 \begin{bmatrix} 1 \\ -1 \end{bmatrix}, k_2 \begin{bmatrix} 4 \\ 1 \end{bmatrix}$

(*iii*) $1, 2, 2$; $k_1 \begin{bmatrix} 1 \\ \; \\ \; \end{bmatrix}$ $k_2 \begin{bmatrix} 2 \\ 1 \\ \; \end{bmatrix}$ (*iv*) a, b, c ; $k_1 \begin{bmatrix} 1 \\ 0 \\ 0 \end{bmatrix}, k_2 \begin{bmatrix} h \\ b-a \\ 0 \end{bmatrix}, k_3 \begin{bmatrix} g \\ 0 \\ c-a \end{bmatrix}$

(*v*) $1, 2, 3$; $k_1 \begin{bmatrix} -1 \\ -1 \\ 1 \end{bmatrix}, k_2 \begin{bmatrix} 0 \\ 1 \\ 0 \end{bmatrix}, k_3 \begin{bmatrix} 0 \\ 1 \\ 1 \end{bmatrix}$ (*vi*) $1, 2, 3$; $k_1 \begin{bmatrix} 1 \\ -1 \\ 0 \end{bmatrix}, k_2 \begin{bmatrix} 2 \\ -1 \\ -2 \end{bmatrix}, k_3 \begin{bmatrix} 1 \\ -1 \\ -2 \end{bmatrix}$

(vii) $6, -\frac{3}{2} \pm \frac{\sqrt{3}}{2}i$; $k_1 \begin{bmatrix} 1 \\ 1 \\ 1 \end{bmatrix}$, $k_2 \begin{bmatrix} -\frac{7}{2} + \frac{3\sqrt{3}}{2}i \\ 4 + \sqrt{3}\,i \\ -\frac{11}{2} + \sqrt{3}\,i \end{bmatrix}$, $k_3 \begin{bmatrix} -\frac{7}{2} - \frac{3\sqrt{3}}{2}i \\ 4 - \sqrt{3}\,i \\ \frac{9}{2} + \sqrt{3}\,i \end{bmatrix}$

(viii) $2, \frac{1}{2} \pm \frac{\sqrt{3}}{2}i$; $k_1 \begin{bmatrix} 1 \\ 1 \\ 1 \end{bmatrix}$, $k_2 \begin{bmatrix} 1 \\ -\frac{1}{2} + \frac{\sqrt{3}}{2}i \\ -\frac{3}{2} + \sqrt{3}\,i \end{bmatrix}$, $k_3 \begin{bmatrix} 1 \\ -\frac{1}{2} - \frac{\sqrt{3}}{2}i \\ \frac{1}{2} + \sqrt{3}\,i \end{bmatrix}$

(ix) $-2, 6, 3$; $k_1 \begin{bmatrix} -1 \\ 0 \\ 1 \end{bmatrix}$, $k_2 \begin{bmatrix} 1 \\ 2 \\ 1 \end{bmatrix}$, $k_3 \begin{bmatrix} 1 \\ -1 \\ 1 \end{bmatrix}$ (x) $1, 1, 4$; $k_1 \begin{bmatrix} 1 \\ 1 \\ 0 \end{bmatrix}$, $k_2 \begin{bmatrix} -1 \\ 0 \\ 1 \end{bmatrix}$, $k_3 \begin{bmatrix} 1 \\ -1 \\ 1 \end{bmatrix}$

(xi) $1, 1, 7$; $k_2 \begin{bmatrix} -1 \\ 1 \\ 0 \end{bmatrix} + k_3 \begin{bmatrix} -1 \\ 0 \\ 1 \end{bmatrix}$, $k_1 \begin{bmatrix} 1 \\ 2 \\ 3 \end{bmatrix}$ (xii) $1, 3, 3$; $k_1 \begin{bmatrix} 1 \\ 0 \\ -1 \end{bmatrix}$, $k_2 \begin{bmatrix} 0 \\ 1 \\ 0 \end{bmatrix} + k_3 \begin{bmatrix} 1 \\ 0 \\ 1 \end{bmatrix}$

2. (i) $\lambda^3 - 5\lambda^2 + 8\lambda - 4 = 0$ (ii) $0, 1, -2$ (iii) $k_1 \begin{bmatrix} 1 \\ 1-i \end{bmatrix}$, $k_2 \begin{bmatrix} 1 \\ 1+i \end{bmatrix}$

3. 4, 110, 10 **4.** 6

7. $\lambda = 0, \pm \left[\frac{3}{2}(a^2 + b^2 + c^2)\right]^{1/2}$

8. $3, -1, 1$; $k_1 \begin{bmatrix} 2 \\ 3 \\ -1 \end{bmatrix}$, $k_2 \begin{bmatrix} 0 \\ -1 \\ 1 \end{bmatrix}$, $k_3 \begin{bmatrix} 1 \\ -1 \\ 0 \end{bmatrix}$ **9.** $2, 2, 3$; $k_1 \begin{bmatrix} 5 \\ 2 \\ -5 \end{bmatrix}$, $k_2 \begin{bmatrix} 1 \\ 1 \\ -2 \end{bmatrix}$

10. a, b, c **11.** $\lambda^4 - 7\lambda^3 + 17\lambda^2 - 17\lambda + 6 = 0$; $\lambda = 1, 1, 2, 3$

1.42. CAYLEY-HAMILTON THEOREM

Every square matrix satisfies its own characteristic equation.

Let $A = [a_{ij}]_{n \times n}$ be a square matrix

then,
$$A = \begin{bmatrix} a_{11} & a_{12} & \cdots & a_{1n} \\ a_{21} & a_{22} & \cdots & a_{2n} \\ \vdots & \vdots & \vdots & \vdots \\ a_{n1} & a_{n2} & \cdots & a_{nn} \end{bmatrix}_{n \times n}$$

Let the characteristic polynomial of A be $\phi(\lambda)$

then,
$$\phi(\lambda) = A - \lambda I$$
$$= \begin{bmatrix} a_{11} - \lambda & a_{12} & \cdots & a_{1n} \\ a_{21} & a_{22} - \lambda & \cdots & a_{2n} \\ \vdots & \vdots & & \vdots \\ a_{n1} & a_{n2} & \cdots & a_{nn} - \lambda \end{bmatrix}$$

Therefore, the characteristic equation is
$$| A - \lambda I | = 0$$

$$\Rightarrow \quad p_0\lambda^n + p_1\lambda^{n-1} + p_2\lambda^{n-2} + \cdots + p_n = 0$$

We have to prove that

$$p_0A^n + p_1A^{n-1} + p_2A^{n-2} + \cdots + p_nI = O$$

Since the elements of $(A - \lambda I)$ are atmost of the first degree in λ, the elements of adj. $(A - \lambda I)$ are atmost of degree $(n - 1)$ in λ.

Thus, adj. $(A - \lambda I)$ can be written as a matrix polynomial in λ given by

$$\text{adj. } (A - \lambda I) = B_0\lambda^{n-1} + B_1\lambda^{n-2} + \cdots + B_{n-1}$$

where $B_0, B_1, ..., B_{n-1}$ are square matrices of order n, their elements being polynomial in λ.

We know that

$$(A - \lambda I) \text{ adj. } (A - \lambda I) = |A - \lambda I| I$$

$$\Rightarrow \quad (A - \lambda I)(B_0\lambda^{n-1} + B_1\lambda^{n-2} + \cdots + B_{n-1}) = (p_0\lambda^n + p_1\lambda^{n-1} + p_2\lambda^{n-2} + \cdots + p_n)I \qquad ...(1)$$

This is an identity in scalar λ.

Equating the coefficients of like powers of λ on both sides, we get

$$-B_0 = p_0I$$
$$AB_0 - B_1 = p_1I$$
$$AB_1 - B_2 = p_2I$$
$$\vdots \qquad \vdots$$
$$AB_{r-1} - B_r = p_rI$$
$$\vdots$$
$$AB_{n-1} = p_nI$$

Pre-multiplying these equations by $A^n, A^{n-1}, A^{n-2}, ..., A, I$ respectively and adding, we get

$$O = p_0A^n + p_1A^{n-1} + p_2A^{n-2} + \cdots + p_{n-1}A + p_nI \qquad ...(2)$$

which proves the theorem.

Note 1. Pre-multiplying equation (2) by A^{-1}, we have

$$O = p_0A^{n-1} + p_1A^{n-2} + p_2A^{n-3} + ... + p_{n-1}I + p_nA^{-1} \qquad ...(3)$$

$$\Rightarrow \quad A^{-1} = -\frac{1}{p_n}[p_0A^{n-1} + p_1A^{n-2} + p_2A^{n-3} + ... + p_{n-1}I]$$

Thus, Cayley-Hamilton theorem gives another method for computing the inverse of a square matrix. Since this method expresses the inverse of a matrix of order n in terms of $(n - 1)$ powers of A, it is most suitable for computing inverses of large matrices.

Note 2. Again pre-multiplying eqn. (3) by A^{-1}, we have

$$O = p_0A^{n-2} + p_1A^{n-3} + p_2A^{n-4} + ... + p_{n-1}A^{-1} + p_nA^{-2}$$

$$\Rightarrow \quad A^{-2} = -\frac{1}{p_n}[p_0A^{n-2} + p_1A^{n-3} + p_2A^{n-4} + ... + p_{n-1}A^{-1}]$$

Hence, by using this theorem, we can also find A^{-2} provided A^{-1} is known and hence $A^{-3}, A^{-4}, ...$ etc.

Note 3. If m is a positive integer such that $m > n$ then any positive integral power A^m of A is linearly expressible in terms of those of lower degree.

ILLUSTRATIVE EXAMPLES

Example 1. *Verify Cayley-Hamilton theorem for the matrix* $A = \begin{bmatrix} 2 & -1 & 1 \\ -1 & 2 & -1 \\ 1 & -1 & 2 \end{bmatrix}$.

Hence compute A^{-1}. *Also evaluate* $A^6 - 6A^5 + 9A^4 - 2A^3 - 12A^2 + 23A - 9I$.

(A.K.T.U. 2016, 2018)

Sol. The characteristic equation of A is

$$| A - \lambda I | = 0$$

i.e.,
$$\begin{vmatrix} 2-\lambda & -1 & 1 \\ -1 & 2-\lambda & -1 \\ 1 & -1 & 2-\lambda \end{vmatrix} = 0$$

or
$$\lambda^3 - 6\lambda^2 + 9\lambda - 4 = 0 \quad \text{(on simplification)}$$

To verify Cayley-Hamilton theorem, we have to show that

$$A^3 - 6A^2 + 9A - 4I = O \qquad ...(1)$$

Now,
$$A^2 = \begin{bmatrix} 2 & -1 & 1 \\ -1 & 2 & -1 \\ 1 & -1 & 2 \end{bmatrix}\begin{bmatrix} 2 & -1 & 1 \\ -1 & 2 & -1 \\ 1 & -1 & 2 \end{bmatrix} = \begin{bmatrix} 6 & -5 & 5 \\ -5 & 6 & -5 \\ 5 & -5 & 6 \end{bmatrix}$$

$$A^3 = A^2 \times A = \begin{bmatrix} 6 & -5 & 5 \\ -5 & 6 & -5 \\ 5 & -5 & 6 \end{bmatrix}\begin{bmatrix} 2 & -1 & 1 \\ -1 & 2 & -1 \\ 1 & -1 & 2 \end{bmatrix} = \begin{bmatrix} 22 & -22 & -21 \\ -21 & 22 & -21 \\ 21 & -21 & 22 \end{bmatrix}$$

$\therefore$ $A^3 - 6A^2 + 9A - 4I$

$$= \begin{bmatrix} 22 & -22 & -21 \\ -21 & 22 & -21 \\ 21 & -21 & 22 \end{bmatrix} - 6\begin{bmatrix} 6 & -5 & 5 \\ -5 & 6 & -5 \\ 5 & -5 & 6 \end{bmatrix} + 9\begin{bmatrix} 2 & -1 & 1 \\ -1 & 2 & -1 \\ 1 & -1 & 2 \end{bmatrix} - 4\begin{bmatrix} 1 & 0 & 0 \\ 0 & 1 & 0 \\ 0 & 0 & 1 \end{bmatrix}$$

$$= \begin{bmatrix} 0 & 0 & 0 \\ 0 & 0 & 0 \\ 0 & 0 & 0 \end{bmatrix} = O$$

This verifies Cayley-Hamilton theorem.

Now, pre-multiplying both sides of (1) by A^{-1}, we have

$$A^2 - 6A + 9I - 4A^{-1} = O$$

$$\Rightarrow \quad 4A^{-1} = A^2 - 6A + 9I$$

$$\Rightarrow \quad 4A^{-1} = \begin{bmatrix} 6 & -5 & 5 \\ -5 & 6 & -5 \\ 5 & -5 & 6 \end{bmatrix} - 6\begin{bmatrix} 2 & -1 & 1 \\ -1 & 2 & -1 \\ 1 & -1 & 2 \end{bmatrix} + 9\begin{bmatrix} 1 & 0 & 0 \\ 0 & 1 & 0 \\ 0 & 0 & 1 \end{bmatrix} = \begin{bmatrix} 3 & 1 & -1 \\ 1 & 3 & 1 \\ -1 & 1 & 3 \end{bmatrix}$$

$$\therefore \quad A^{-1} = \frac{1}{4}\begin{bmatrix} 3 & 1 & -1 \\ 1 & 3 & 1 \\ -1 & 1 & 3 \end{bmatrix}.$$

Also, $A^6 - 6A^5 + 9A^4 - 2A^3 - 12A^2 + 23A - 9I$

$= A^3 (A^3 - 6A^2 + 9A - 4I) + 2(A^3 - 6A^2 + 9A - 4I) + 5A - I = 5A - I$ | Using (1)

$$= 5\begin{bmatrix} 2 & -1 & 1 \\ -1 & 2 & -1 \\ 1 & -1 & 2 \end{bmatrix} - \begin{bmatrix} 1 & 0 & 0 \\ 0 & 1 & 0 \\ 0 & 0 & 1 \end{bmatrix} = \begin{bmatrix} 9 & -5 & 5 \\ -5 & 9 & -5 \\ 5 & -5 & 9 \end{bmatrix}$$

Example 2. *Find the characteristic equation of the matrix* $A = \begin{bmatrix} 2 & 1 & 1 \\ 0 & 1 & 0 \\ 1 & 1 & 2 \end{bmatrix}$ *and hence, compute* A^{-1}. *Also find the matrix represented by* $A^8 - 5A^7 + 7A^6 - 3A^5 + A^4 - 5A^3 + 8A^2 - 2A + I$.
[A.K.T.U. 2022]

Sol. The characteristic equation of A is

$$|A - \lambda I| = \begin{vmatrix} 2-\lambda & 1 & 1 \\ 0 & 1-\lambda & 0 \\ 1 & 1 & 2-\lambda \end{vmatrix} = 0$$

or
$$\lambda^3 - 5\lambda^2 + 7\lambda - 3 = 0$$

By Cayley-Hamilton theorem, $A^3 - 5A^2 + 7A - 3I = O$ ...(1)

Pre-multiplying (1) by A^{-1}, we get

$$A^{-1}A^3 - 5A^{-1}A^2 + 7A^{-1}A - 3A^{-1}I = A^{-1}O$$

$$\Rightarrow \quad A^2 - 5A + 7I - 3A^{-1} = O$$

$$\Rightarrow \quad 3A^{-1} = A^2 - 5A + 7I$$

$$\Rightarrow \quad A^{-1} = \frac{1}{3}(A^2 - 5A + 7I) \quad ...(2)$$

Now, $$A^2 = A\,.\,A = \begin{bmatrix} 2 & 1 & 1 \\ 0 & 1 & 0 \\ 1 & 1 & 2 \end{bmatrix}\begin{bmatrix} 2 & 1 & 1 \\ 0 & 1 & 0 \\ 1 & 1 & 2 \end{bmatrix} = \begin{bmatrix} 5 & 4 & 4 \\ 0 & 1 & 0 \\ 4 & 4 & 5 \end{bmatrix}$$

From (2), $$3A^{-1} = \begin{bmatrix} 5 & 4 & 4 \\ 0 & 1 & 0 \\ 4 & 4 & 5 \end{bmatrix} - \begin{bmatrix} 10 & 5 & 5 \\ 0 & 5 & 0 \\ 5 & 5 & 10 \end{bmatrix} + \begin{bmatrix} 7 & 0 & 0 \\ 0 & 7 & 0 \\ 0 & 0 & 7 \end{bmatrix} = \begin{bmatrix} 2 & -1 & -1 \\ 0 & 3 & 0 \\ -1 & -1 & 2 \end{bmatrix}$$

$$\Rightarrow \quad A^{-1} = \frac{1}{3}\begin{bmatrix} 2 & -1 & -1 \\ 0 & 3 & 0 \\ -1 & -1 & 2 \end{bmatrix}$$

Now, $A^8 - 5A^7 + 7A^6 - 3A^5 + A^4 - 5A^3 + 8A^2 - 2A + I$

$$= A^5(A^3 - 5A^2 + 7A - 3I) + A(A^3 - 5A^2 + 7A - 3I) + (A^2 + A + I)$$

$$= A^2 + A + I \quad \text{[Using (1)]}$$

$$= \begin{bmatrix} 2 & 1 & 1 \\ 0 & 1 & 0 \\ 1 & 1 & 2 \end{bmatrix}\begin{bmatrix} 2 & 1 & 1 \\ 0 & 1 & 0 \\ 1 & 1 & 2 \end{bmatrix} + \begin{bmatrix} 2 & 1 & 1 \\ 0 & 1 & 0 \\ 1 & 1 & 2 \end{bmatrix} + \begin{bmatrix} 1 & 0 & 0 \\ 0 & 1 & 0 \\ 0 & 0 & 1 \end{bmatrix}$$

$$= \begin{bmatrix} 5 & 4 & 4 \\ 0 & 1 & 0 \\ 4 & 4 & 5 \end{bmatrix} + \begin{bmatrix} 2 & 1 & 1 \\ 0 & 1 & 0 \\ 1 & 1 & 2 \end{bmatrix} + \begin{bmatrix} 1 & 0 & 0 \\ 0 & 1 & 0 \\ 0 & 0 & 1 \end{bmatrix} = \begin{bmatrix} 8 & 5 & 5 \\ 0 & 3 & 0 \\ 5 & 5 & 8 \end{bmatrix}.$$

Example 3. *Given* $A = \begin{bmatrix} 1 & 2 & -1 \\ 0 & 1 & -1 \\ 3 & -1 & 1 \end{bmatrix}$ *find Adj. A by using Cayley-Hamilton theorem.*

Sol. The characteristic equation of the given matrix A is

$$|A - \lambda I| = 0$$

$$\Rightarrow \quad \begin{vmatrix} 1-\lambda & 2 & -1 \\ 0 & 1-\lambda & -1 \\ 3 & -1 & 1-\lambda \end{vmatrix} = 0$$

$\Rightarrow$ $(1-\lambda)[(1-\lambda)^2 - 1] - 2[3] - 1[-3(1-\lambda)] = 0$

$\Rightarrow$ $\lambda^3 - 3\lambda^2 + 5\lambda + 3 = 0$ (On simplification)

By Cayley-Hamilton theorem, A should satisfy

$$A^3 - 3A^2 + 5A + 3I = O$$

Pre-multiplying by A^{-1}, we get

$$A^2 - 3A + 5I + 3A^{-1} = O$$

$$\Rightarrow \quad A^{-1} = -\frac{1}{3}(A^2 - 3A + 5I) \quad ...(1)$$

Now, $A^2 = A \cdot A = \begin{bmatrix} 1 & 2 & -1 \\ 0 & 1 & -1 \\ 3 & -1 & 1 \end{bmatrix}\begin{bmatrix} 1 & 2 & -1 \\ 0 & 1 & -1 \\ 3 & -1 & 1 \end{bmatrix} = \begin{bmatrix} -2 & 5 & -4 \\ -3 & 2 & -2 \\ 6 & 4 & -1 \end{bmatrix}$

$$3A = \begin{bmatrix} 3 & 6 & -3 \\ 0 & 3 & -3 \\ 9 & -3 & 3 \end{bmatrix}$$

$\therefore$ From (1), $A^{-1} = -\frac{1}{3}\left\{\begin{bmatrix} -2 & 5 & -4 \\ -3 & 2 & -2 \\ 6 & 4 & -1 \end{bmatrix} - \begin{bmatrix} 3 & 6 & -3 \\ 0 & 3 & -3 \\ 9 & -3 & 3 \end{bmatrix} + \begin{bmatrix} 5 & 0 & 0 \\ 0 & 5 & 0 \\ 0 & 0 & 5 \end{bmatrix}\right\}$

$$= -\frac{1}{3}\begin{bmatrix} 0 & -1 & -1 \\ -3 & 4 & 1 \\ -3 & 7 & 1 \end{bmatrix}$$

We know that, $A^{-1} = \dfrac{\text{Adj. } A}{|A|}$

$\therefore$ Adj. $A = A^{-1}\,|A|$

Now, $|A| = \begin{vmatrix} 1 & 2 & -1 \\ 0 & 1 & -1 \\ 3 & -1 & 1 \end{vmatrix} = -3$

$\therefore$ Adj. $A = (-3)\left(-\frac{1}{3}\right)\begin{bmatrix} 0 & -1 & -1 \\ -3 & 4 & 1 \\ -3 & 7 & 1 \end{bmatrix} = \begin{bmatrix} 0 & -1 & -1 \\ -3 & 4 & 1 \\ -3 & 7 & 1 \end{bmatrix}$.

Example 4. *If* $A = \begin{bmatrix} 1 & 0 & 0 \\ 1 & 0 & 1 \\ 0 & 1 & 0 \end{bmatrix}$, *find* A^{-1} *and* A^4 *using Cayley-Hamilton's theorem. Also show that for every integer* $n \geq 3$, $A^n = A^{n-2} + A^2 - I$. [U.P.T.U. 2015]

Sol. Characteristic equation is

$$|A - \lambda I| = 0$$

$$\Rightarrow \quad \begin{vmatrix} 1-\lambda & 0 & 0 \\ 1 & -\lambda & 1 \\ 0 & 1 & -\lambda \end{vmatrix} = 0$$

$$\Rightarrow \quad (1-\lambda)(\lambda^2 - 1) = 0$$

$$\Rightarrow \quad \lambda^3 - \lambda^2 - \lambda + 1 = 0$$

Therefore, by Cayley-Hamilton theorem, we have

$$A^3 - A^2 - A + I = O \quad ...(1)$$

Premultiplying eqn. (1) by A^{-1}, we get

$$A^2 - A - I + A^{-1} = O$$

$$\Rightarrow \quad A^{-1} = I + A - A^2$$

$$\Rightarrow \quad A^{-1} = \begin{bmatrix} 1 & 0 & 0 \\ 0 & 1 & 0 \\ 0 & 0 & 1 \end{bmatrix} + \begin{bmatrix} 1 & 0 & 0 \\ 1 & 0 & 1 \\ 0 & 1 & 0 \end{bmatrix} - \begin{bmatrix} 1 & 0 & 0 \\ 1 & 1 & 0 \\ 1 & 0 & 1 \end{bmatrix} = \begin{bmatrix} 1 & 0 & 0 \\ 0 & 0 & 1 \\ -1 & 1 & 0 \end{bmatrix}$$

Again Premultiplying eqn. (1) by A, we get

$$A^4 - A^3 - A^2 + A = O$$

$$\Rightarrow \quad A^4 = A^3 + A^2 - A$$

$$\Rightarrow \quad A^4 = \begin{bmatrix} 1 & 0 & 0 \\ 2 & 0 & 1 \\ 1 & 1 & 0 \end{bmatrix} + \begin{bmatrix} 1 & 0 & 0 \\ 1 & 1 & 0 \\ 1 & 0 & 1 \end{bmatrix} - \begin{bmatrix} 1 & 0 & 0 \\ 1 & 0 & 1 \\ 0 & 1 & 0 \end{bmatrix} = \begin{bmatrix} 1 & 0 & 0 \\ 2 & 1 & 0 \\ 2 & 0 & 1 \end{bmatrix}$$

Now, again from equation (1), we write

$$A^3 - A^2 = A - I$$

Premultiplying both sides successively by A, we obtain

$$A^4 - A^3 = A^2 - A$$

$$A^5 - A^4 = A^3 - A^2$$

....

....

$$A^{n-1} - A^{n-2} = A^{n-3} - A^{n-4}$$

$$A^n - A^{n-1} = A^{n-2} - A^{n-3}$$

Adding these equations, we get

$$A^n - A^2 = A^{n-2} - I$$

$$\Rightarrow \quad A^n = A^{n-2} + A^2 - I;\ n \geq 3$$

Example 5. *Find the eigen values of A and using Cayley-Hamilton theorem, find* A^n *(n is a positive integer) ; given that* $A = \begin{bmatrix} 1 & 2 \\ 4 & 3 \end{bmatrix}$.

Sol. Characteristic equation is

$$|A - \lambda I| = 0$$

$$\Rightarrow \quad \begin{vmatrix} 1-\lambda & 2 \\ 4 & 3-\lambda \end{vmatrix} = 0$$

$$\Rightarrow \quad \lambda^2 - 4\lambda - 5 = 0$$

$$\Rightarrow \quad \lambda = -1, 5$$

$\therefore$ Eigen values of A are – 1, 5.

Let $\lambda^n \equiv (\lambda^2 - 4\lambda - 5)\, Q(\lambda) + (a\lambda + b)$ ...(1) | where $Q(\lambda)$ is quotient

Put $\lambda = -1$, $(-1)^n = -a + b$ ...(2)

Put $\lambda = 5$, $5^n = 5a + b$ ...(3)

Solving (2) and (3), we get

$$a = \frac{5^n - (-1)^n}{6}, \quad b = \frac{5^n + 5(-1)^n}{6}$$

Replacing λ by matrix A in (1), we get

$$A^n = (A^2 - 4A - 5I)\,Q(A) + (aA + bI)$$
$$= O + aA + bI \qquad | \text{by Cayley-Hamilton theorem}$$
$$= aA + bI$$
$$= \left\{\frac{5^n - (-1)^n}{6}\right\}\begin{bmatrix} 1 & 2 \\ 4 & 3 \end{bmatrix} + \left\{\frac{5^n + 5(-1)^n}{6}\right\}\begin{bmatrix} 1 & 0 \\ 0 & 1 \end{bmatrix}.$$

TEST YOUR KNOWLEDGE

1. Verify Cayley-Hamilton theorem for the matrices:

(*i*) $A = \begin{bmatrix} 1 & 2 \\ 2 & -1 \end{bmatrix}$

(*ii*) $A = \begin{bmatrix} 2 & 2 & 1 \\ 0 & 1 & -1 \\ 3 & -1 & 1 \end{bmatrix}$

(*iii*) $A = \begin{bmatrix} 3 & 0 & 1 \\ 0 & 2 & 0 \\ 0 & 0 & 1 \end{bmatrix}$

(*iv*) $A = \begin{bmatrix} 1 & 0 & -4 \\ 0 & 5 & 4 \\ -4 & 4 & 3 \end{bmatrix}$

(*v*) $A = \begin{bmatrix} 2 & 3 \\ 3 & 5 \end{bmatrix}$

(*vi*) $A = \begin{bmatrix} 4 & 0 & 1 \\ 0 & 1 & 2 \\ 1 & 0 & 1 \end{bmatrix}$ (*A.K.T.U. 2020*)

(*vii*) $A = \begin{bmatrix} 1 & 2 & 3 \\ 2 & 4 & 5 \\ 3 & 5 & 6 \end{bmatrix}$ (*A.K.T.U. 2015, 2019*)

(*viii*) $A = \begin{bmatrix} 1 & 3 & 7 \\ 4 & 2 & 3 \\ 1 & 2 & 1 \end{bmatrix}$

(*ix*) $A = \begin{bmatrix} 2 & -1 & 1 \\ -1 & 2 & -1 \\ 1 & -2 & 2 \end{bmatrix}$

(*x*) $A = \begin{bmatrix} 7 & 2 & -2 \\ -6 & -1 & 2 \\ 6 & 2 & -1 \end{bmatrix}$ (*U.K.T.U. 2012*)

(*xi*) $A = \begin{bmatrix} 1 & 1 & 2 \\ 3 & 1 & 1 \\ 2 & 3 & 1 \end{bmatrix}$ (*U.P.T.U. 2014*)

Hence find A^{-1} in each case.

2. Verify Cayley-Hamilton theorem for:

(*i*) $\begin{bmatrix} 1 & 2 \\ 2 & 1 \end{bmatrix}$

(*ii*) $\begin{bmatrix} 1 & -2 & 3 \\ 2 & 4 & -2 \\ -1 & 1 & 2 \end{bmatrix}$

(*iii*) $\begin{bmatrix} 3 & 4 & 1 \\ 2 & 1 & 6 \\ -1 & 4 & 7 \end{bmatrix}$

(*iv*) $\begin{bmatrix} 2 & 1 & 1 \\ 0 & 1 & 0 \\ 1 & 1 & 2 \end{bmatrix}$ (*U.P.T.U. 2014*)

3. Using Cayley-Hamilton theorem, find the inverse of:

(*i*) $\begin{bmatrix} 7 & -1 & 3 \\ 6 & 1 & 4 \\ 2 & 4 & 8 \end{bmatrix}$

(*ii*) $\begin{bmatrix} 4 & 3 & 1 \\ 2 & 1 & -2 \\ 1 & 2 & 1 \end{bmatrix}$

4. Show that the matrix $A = \begin{bmatrix} 1 & 2 & 0 \\ 2 & -1 & 0 \\ 0 & 0 & -1 \end{bmatrix}$ satisfies its own characteristic equation and hence or otherwise obtain A^{-2}.

5. If $A = \begin{bmatrix} 1 & 2 \\ -1 & 3 \end{bmatrix}$, use Cayley-Hamilton Theorem to express $A^6 - 4A^5 + 8A^4 - 12A^3 + 14A^2$ as a linear polynomial in A.

6. Express $2A^5 - 3A^4 + A^2 - 4I$ as a linear polynomial in A where $A = \begin{bmatrix} 3 & 1 \\ -1 & 2 \end{bmatrix}$. *(A.K.T.U. 2017)*

7. If $A = \begin{bmatrix} -3 & 2 \\ -1 & 0 \end{bmatrix}$ then evaluate the expression $A + 5I + 2A^{-1}$. *(A.K.T.U. 2017)*

8. Using Cayley-Hamilton theorem, find the inverse of the matrix

$$A = \begin{bmatrix} 1 & 2 & 3 \\ 2 & 4 & 5 \\ 3 & 5 & 6 \end{bmatrix}$$

Also express the polynomial $B = A^8 - 11A^7 - 4A^6 + A^5 + A^4 - 11A^3 - 3A^2 + 2A + I$ as a quadratic polynomial in A and hence find B. *(A.K.T.U. 2019)*

Answers

1. (*i*) $A^{-1} = \frac{1}{5}\begin{bmatrix} 1 & 2 \\ 2 & -1 \end{bmatrix}$ (*ii*) $A^{-1} = \frac{1}{9}\begin{bmatrix} 0 & 3 & 3 \\ 3 & 1 & -2 \\ 3 & -8 & -2 \end{bmatrix}$

(*iii*) $A^{-1} = \frac{1}{6}\begin{bmatrix} 11 & 0 & 1 \\ 0 & 9 & 0 \\ 0 & 0 & 9 \end{bmatrix}$ (*iv*) $A^{-1} = \frac{1}{81}\begin{bmatrix} 1 & 16 & -20 \\ 16 & 13 & 4 \\ -20 & 4 & -5 \end{bmatrix}$

(*v*) $A^{-1} = \begin{bmatrix} 5 & -3 \\ -3 & 2 \end{bmatrix}$ (*vi*) $A^{-1} = \frac{1}{3}\begin{bmatrix} 1 & 0 & -1 \\ 2 & 3 & -8 \\ -1 & 0 & 4 \end{bmatrix}$

(*vii*) $A^{-1} = \begin{bmatrix} 1 & -3 & 2 \\ -3 & 3 & -1 \\ 2 & -1 & 0 \end{bmatrix}$ (*viii*) $A^{-1} = \frac{1}{35}\begin{bmatrix} -4 & 11 & -5 \\ -1 & -6 & 25 \\ -6 & 1 & -10 \end{bmatrix}$

(*ix*) $A^{-1} = \frac{1}{3}\begin{bmatrix} 2 & 0 & -1 \\ 1 & 3 & 1 \\ 0 & 3 & 3 \end{bmatrix}$ (*x*) $A^{-1} = \frac{1}{3}\begin{bmatrix} -3 & -2 & 2 \\ 6 & 5 & -2 \\ -6 & -2 & 5 \end{bmatrix}$

(*xi*) $A^{-1} = \frac{1}{11}\begin{bmatrix} -2 & 5 & -1 \\ -1 & -3 & 5 \\ 7 & -1 & -2 \end{bmatrix}$

3. (*i*) $\frac{1}{50}\begin{bmatrix} -8 & 20 & -7 \\ -40 & 50 & -10 \\ 22 & -30 & 13 \end{bmatrix}$ (*ii*) $\frac{1}{11}\begin{bmatrix} 5 & -1 & -7 \\ -4 & 3 & 10 \\ 3 & -5 & -2 \end{bmatrix}$

4. $A^{-2} = \begin{bmatrix} 1 & 0 & 0 \\ 0 & 1 & 0 \\ 0 & 0 & 5 \end{bmatrix}$ **5.** $-4A + 5I$

6. 138A – 403I **7.** 2I

8. $\begin{bmatrix} 1 & -3 & 2 \\ -3 & 3 & -1 \\ 2 & -1 & 0 \end{bmatrix}$; $B = A^2 + A + I = \begin{bmatrix} 16 & 27 & 34 \\ 27 & 50 & 61 \\ 34 & 61 & 77 \end{bmatrix}$

1.43. SIMILARITY TRANSFORMATION

Let A and B be two square matrices of order n. Then B is said to be similar to A if there exists a non-singular matrix M such that

$$B = M^{-1} AM \qquad ...(1)$$

Equation (1) is called a similarity transformation.

1.44. DIAGONALISATION OF A MATRIX

Diagonalisation of a matrix A is the process of reduction of A to a diagonal form.

If A is related to D by a similarity transformation such that $D = M^{-1} AM$ then A is reduced to the diagonal matrix D through modal matrix M. D is also called **spectral matrix** of A.

1.45. REDUCTION OF A MATRIX TO DIAGONAL FORM

If a square matrix A of order n has n linearly independent eigen vectors, then a matrix B can be found such that $B^{-1}AB$ is a diagonal matrix.

Note. The matrix B which diagonalises A is called the **modal matrix** of A and is obtained by grouping the eigen vectors of A into a square matrix.

ILLUSTRATIVE EXAMPLES

Example 1. *Reduce the matrix* $A = \begin{bmatrix} 1 & -1 & 2 \\ 0 & 2 & -1 \\ 0 & 0 & 3 \end{bmatrix}$ *to diagonal form by similarity transformation. Hence find* A^3.

Sol. Characteristic equation is

$$\begin{vmatrix} 1-\lambda & -1 & 2 \\ 0 & 2-\lambda & -1 \\ 0 & 0 & 3-\lambda \end{vmatrix} = 0$$

$$\Rightarrow \qquad \lambda = 1, 2, 3$$

Hence eigen values of A are 1, 2, 3.

Corresponding to $\lambda = 1$, let $X_1 = \begin{bmatrix} x_1 \\ x_2 \\ x_3 \end{bmatrix}$ be the eigen vector then,

$$(A - I) X_1 = O$$

$$\begin{bmatrix} 0 & -1 & 2 \\ 0 & 1 & -1 \\ 0 & 0 & 2 \end{bmatrix} \begin{bmatrix} x_1 \\ x_2 \\ x_3 \end{bmatrix} = \begin{bmatrix} 0 \\ 0 \\ 0 \end{bmatrix}$$

$$\Rightarrow \quad -x_2 + 2x_3 = 0$$
$$x_2 - x_3 = 0$$
$$2x_3 = 0$$
$$\therefore \quad x_1 = k_1, x_2 = 0 = x_3$$
$$\therefore \quad X_1 = k_1 \begin{bmatrix} 1 \\ 0 \\ 0 \end{bmatrix}$$

Corresponding to $\lambda = 2$, let $X_2 = \begin{bmatrix} x_1 \\ x_2 \\ x_3 \end{bmatrix}$ be the eigen vector then,

$$(A - 2I) X_2 = O$$
$$\begin{bmatrix} -1 & -1 & 2 \\ 0 & 0 & -1 \\ 0 & 0 & 1 \end{bmatrix} \begin{bmatrix} x_1 \\ x_2 \\ x_3 \end{bmatrix} = \begin{bmatrix} 0 \\ 0 \\ 0 \end{bmatrix}$$
$$\Rightarrow \quad -x_1 - x_2 + 2x_3 = 0$$
$$-x_3 = 0$$
$$x_3 = 0$$
$$\therefore \quad x_1 = k_2, x_2 = -k_2, x_3 = 0$$
$$\therefore \quad X_2 = k_2 \begin{bmatrix} 1 \\ -1 \\ 0 \end{bmatrix}$$

Corresponding to $\lambda = 3$, let $X_3 = \begin{bmatrix} x_1 \\ x_2 \\ x_3 \end{bmatrix}$ be the eigen vector then,

$$(A - 3I) X_3 = O$$
$$\Rightarrow \quad \begin{bmatrix} -2 & -1 & 2 \\ 0 & -1 & -1 \\ 0 & 0 & 0 \end{bmatrix} \begin{bmatrix} x_1 \\ x_2 \\ x_3 \end{bmatrix} = \begin{bmatrix} 0 \\ 0 \\ 0 \end{bmatrix}$$
$$\Rightarrow \quad -2x_1 - x_2 + 2x_3 = 0$$
$$-x_2 - x_3 = 0$$
$$\therefore \quad x_2 = k_3, \; x_3 = -k_3, x_1 = \frac{-3}{2} k_3$$
$$\therefore \quad X_3 = k_3 \begin{bmatrix} 3 \\ -2 \\ 2 \end{bmatrix}$$

Hence modal matrix is $M = \begin{bmatrix} 1 & 1 & 3 \\ 0 & -1 & -2 \\ 0 & 0 & 2 \end{bmatrix}$; $|M| = -2$

$$\text{Adj. } M = \begin{bmatrix} -2 & -2 & 1 \\ 0 & 2 & 2 \\ 0 & 0 & -1 \end{bmatrix}$$

$$\therefore \qquad M^{-1} = \frac{\text{Adj. M}}{|M|} = \begin{bmatrix} 1 & 1 & -1/2 \\ 0 & -1 & -1 \\ 0 & 0 & 1/2 \end{bmatrix}$$

$$M^{-1}AM = \begin{bmatrix} 1 & 1 & -1/2 \\ 0 & -1 & -1 \\ 0 & 0 & 1/2 \end{bmatrix} \begin{bmatrix} 1 & -1 & 2 \\ 0 & 2 & -1 \\ 0 & 0 & 3 \end{bmatrix} \begin{bmatrix} 1 & 1 & 3 \\ 0 & -1 & -2 \\ 0 & 0 & 2 \end{bmatrix} = \begin{bmatrix} 1 & 0 & 0 \\ 0 & 2 & 0 \\ 0 & 0 & 3 \end{bmatrix}$$

Now, since, $\qquad D = M^{-1}AM$

$\Rightarrow \qquad A = MDM^{-1}$

$\qquad A^2 = (MDM^{-1})(MDM^{-1}) = MD^2M^{-1} \qquad | \because \quad M^{-1}M = I$

Similarly, $\qquad A^3 = MD^3M^{-1}$

$$= \begin{bmatrix} 1 & 1 & 3 \\ 0 & -1 & -2 \\ 0 & 0 & 2 \end{bmatrix} \begin{bmatrix} 1 & 0 & 0 \\ 0 & 8 & 0 \\ 0 & 0 & 27 \end{bmatrix} \begin{bmatrix} 1 & 1 & -1/2 \\ 0 & -1 & -1 \\ 0 & 0 & 1/2 \end{bmatrix} = \begin{bmatrix} 1 & -7 & 32 \\ 0 & 8 & -19 \\ 0 & 0 & 27 \end{bmatrix}.$$

Example 2. *Reduce the matrix* $A = \begin{bmatrix} -1 & 2 & -2 \\ 1 & 2 & 1 \\ -1 & -1 & 0 \end{bmatrix}$ *to the diagonal form.*

(U.K.T.U. 2011)

Sol. The characteristic equation of A is

$$| A - \lambda I | = \begin{vmatrix} -1-\lambda & 2 & -2 \\ 1 & 2-\lambda & 1 \\ -1 & -1 & -\lambda \end{vmatrix} = 0$$

or $\qquad \lambda^3 - \lambda^2 - 5\lambda + 5 = 0$ (on simplification)

Solving, we have $\lambda = 1, \pm \sqrt{5}$.

When $\lambda = 1$, the corresponding eigen vector is given by

$$-2x_1 + 2y_1 - 2z_1 = 0$$
$$x_1 + y_1 + z_1 = 0$$
$$x_1 - y_1 + z_1 = 0$$

Solving the last two $\dfrac{x_1}{2} = \dfrac{y_1}{0} = \dfrac{z_1}{-2}$ giving the eigen vector $k_1 \begin{bmatrix} 1 \\ 0 \\ -1 \end{bmatrix}$.

When $\lambda = \sqrt{5}$, the corresponding eigen vector is given by

$$(-1 - \sqrt{5})x_1 + 2y_1 - 2z_1 = 0$$
$$x_1 + (2 - \sqrt{5})y_1 + z_1 = 0$$
$$-x_1 - y_1 - \sqrt{5}z_1 = 0$$

Solving the last two $\dfrac{x_1}{6 - 2\sqrt{5}} = \dfrac{y_1}{-1 + \sqrt{5}} = \dfrac{z_1}{1 - \sqrt{5}}$

or $\qquad \dfrac{x_1}{(\sqrt{5} - 1)^2} = \dfrac{y_1}{\sqrt{5} - 1} = \dfrac{z_1}{1 - \sqrt{5}} \quad$ or $\quad \dfrac{x_1}{\sqrt{5} - 1} = \dfrac{y_1}{1} = \dfrac{z_1}{-1}$

giving the eigen vector $k_2 \begin{bmatrix} \sqrt{5} - 1 \\ 1 \\ -1 \end{bmatrix}$

Similarly, the eigen vector corresponding to $\lambda = -\sqrt{5}$ is $k_3 \begin{bmatrix} \sqrt{5}+1 \\ -1 \\ 1 \end{bmatrix}$

$\therefore$ The modal matrix $M = \begin{bmatrix} 1 & \sqrt{5}-1 & \sqrt{5}+1 \\ 0 & 1 & -1 \\ -1 & -1 & 1 \end{bmatrix}$

and $D = \begin{bmatrix} \lambda_1 & 0 & 0 \\ 0 & \lambda_2 & 0 \\ 0 & 0 & \lambda_3 \end{bmatrix} = \begin{bmatrix} 1 & 0 & 0 \\ 0 & \sqrt{5} & 0 \\ 0 & 0 & -\sqrt{5} \end{bmatrix}$

is the diagonal matrix, obtained by diagonalising A.

Example 3. *The eigen vectors of a 3 × 3 matrix A corresponding to the eigen values 1, 1, 3 are $[1, 0, -1]^T$, $[0, 1, -1]^T$ and $[1, 1, 0]^T$ respectively. Find the matrix A.*

Sol. Modal matrix, $P = \begin{bmatrix} 1 & 0 & 1 \\ 0 & 1 & 1 \\ -1 & -1 & 0 \end{bmatrix}$

Spectral matrix, $D = \begin{bmatrix} 1 & 0 & 0 \\ 0 & 1 & 0 \\ 0 & 0 & 3 \end{bmatrix}$

We find by elementary operations,

$$P^{-1} = \frac{1}{2}\begin{bmatrix} 1 & -1 & -1 \\ -1 & 1 & -1 \\ 1 & 1 & 1 \end{bmatrix}$$

Therefore, $A = PDP^{-1}$

$$= \frac{1}{2}\begin{bmatrix} 1 & 0 & 1 \\ 0 & 1 & 1 \\ -1 & -1 & 0 \end{bmatrix}\begin{bmatrix} 1 & 0 & 0 \\ 0 & 1 & 0 \\ 0 & 0 & 3 \end{bmatrix}\begin{bmatrix} 1 & -1 & -1 \\ -1 & 1 & -1 \\ 1 & 1 & 1 \end{bmatrix} = \begin{bmatrix} 2 & 1 & 1 \\ 1 & 2 & 1 \\ 0 & 0 & 1 \end{bmatrix}$$

Example 4. *The matrix $A = \begin{bmatrix} a & h \\ h & b \end{bmatrix}$ is transformed to the diagonal form $D = T^{-1}AT$ where $T = \begin{bmatrix} \cos\theta & \sin\theta \\ -\sin\theta & \cos\theta \end{bmatrix}$. Find the value of θ which gives the diagonal transformation.*

Sol. $T = \begin{bmatrix} \cos\theta & \sin\theta \\ -\sin\theta & \cos\theta \end{bmatrix}$ $\therefore$ $T^{-1} = \begin{bmatrix} \cos\theta & -\sin\theta \\ \sin\theta & \cos\theta \end{bmatrix}$

Now, $T^{-1}AT = \begin{bmatrix} \cos\theta & -\sin\theta \\ \sin\theta & \cos\theta \end{bmatrix}\begin{bmatrix} a & h \\ h & b \end{bmatrix}\begin{bmatrix} \cos\theta & \sin\theta \\ -\sin\theta & \cos\theta \end{bmatrix}$

$$= \begin{bmatrix} a\cos\theta - h\sin\theta & h\cos\theta - b\sin\theta \\ a\sin\theta + h\cos\theta & h\sin\theta + b\cos\theta \end{bmatrix}\begin{bmatrix} \cos\theta & \sin\theta \\ -\sin\theta & \cos\theta \end{bmatrix}$$

$$= \begin{bmatrix} a\cos^2\theta - 2h\sin\theta\cos\theta + b\sin^2\theta & (a-b)\sin\theta\cos\theta - h\sin^2\theta + h\cos^2\theta \\ (a-b)\sin\theta\cos\theta + h\cos^2\theta - h\sin^2\theta & a\sin^2\theta + 2h\sin\theta\cos\theta + b\cos^2\theta \end{bmatrix}$$

$$= \begin{bmatrix} a\cos^2\theta - h\sin 2\theta + b\sin^2\theta & (a-b)\sin\theta\cos\theta + h\cos 2\theta \\ (a-b)\sin\theta\cos\theta + h\cos 2\theta & a\sin^2\theta + h\sin 2\theta + b\cos^2\theta \end{bmatrix}$$

But $T^{-1}AT$ is a diagonal matrix

$$\therefore \quad (a-b)\sin\theta\cos\theta + h\cos 2\theta = 0$$

$$\Rightarrow \quad \left(\frac{a-b}{2}\right)\sin 2\theta + h\cos 2\theta = 0$$

$$\Rightarrow \quad \tan 2\theta = \frac{-2h}{a-b} = \frac{2h}{b-a}$$

$$\Rightarrow \quad \theta = \frac{1}{2}\tan^{-1}\left(\frac{2h}{b-a}\right).$$

TEST YOUR KNOWLEDGE

1. Diagonalise the following matrices:

(i) $\begin{bmatrix} -1 & 1 & 2 \\ 0 & -2 & 1 \\ 0 & 0 & -3 \end{bmatrix}$ (ii) $\begin{bmatrix} 9 & -1 & -9 \\ 3 & -1 & 3 \\ -7 & 1 & -7 \end{bmatrix}$ (iii) $\begin{bmatrix} 1 & -6 & -4 \\ 0 & 4 & 2 \\ 0 & -6 & -3 \end{bmatrix}$

(iv) $\begin{bmatrix} 3 & 5 \\ 7 & 31 \end{bmatrix}$ (v) $\begin{bmatrix} 1 & 6 & 1 \\ 1 & 2 & 0 \\ 0 & 0 & 3 \end{bmatrix}$ (vi) $\begin{bmatrix} 3 & -1 & 1 \\ -1 & 5 & -1 \\ 1 & -1 & 3 \end{bmatrix}$.

2. (i) Find a matrix P which diagonalises the matrix $A = \begin{bmatrix} 4 & 1 \\ 2 & 3 \end{bmatrix}$. Verify $P^{-1}AP = D$ where D is the diagonal matrix.

(ii) Let $A = \begin{bmatrix} 6 & -2 & 2 \\ -2 & 3 & -1 \\ 2 & -1 & 3 \end{bmatrix}$. Find matrix P such that $P^{-1}AP$ is a diagonal matrix.

(iii) For the matrix $A = \begin{bmatrix} 4 & 1 & 0 \\ 1 & 4 & 1 \\ 0 & 1 & 4 \end{bmatrix}$, determine a matrix P such that $P^{-1}AP$ is a diagonal matrix.

[*M.T.U. (SUM) 2011*]

3. Reduce the matrix $P = \begin{bmatrix} 1 & 2 & -2 \\ 1 & 2 & 1 \\ -1 & -1 & 0 \end{bmatrix}$ to diagonal form. (*A.K.T.U. 2017*)

4. Diagonalise the matrix $A = \begin{bmatrix} 1 & 1 & 1 \\ 0 & 2 & 1 \\ -4 & 4 & 3 \end{bmatrix}$ and hence find A^8.

5. Show that the matrix $A = \begin{bmatrix} 3 & 1 & -1 \\ -2 & 1 & 2 \\ 0 & 1 & 2 \end{bmatrix}$ is diagonalisable. Hence, find P such that $P^{-1}AP$ is a diagonal matrix. [*M.T.U. 2012*]

[Hint. A is diagonalisable iff it has 3 linearly independent eigen vectors]

6. Find the eigen values and the corresponding eigen vectors of the following matrix and hence diagonalise it. [*G.B.T.U. (C.O.) 2011*]

$$\begin{bmatrix} 2 & 0 & 1 \\ 0 & 3 & 0 \\ 1 & 0 & 2 \end{bmatrix}$$

7. Show that the matrix $A = \begin{bmatrix} 3 & 10 & 5 \\ -2 & -3 & -4 \\ 3 & 5 & 7 \end{bmatrix}$ has less than three linearly independent eigen vectors.

Is it possible to obtain a similarity transformation that will diagonalise this matrix? (*U.P.T.U. 2014*)

8. Find the square matrix A whose eigen values are 1, 2 and 3 and their corresponding eigen vectors are $[1, 0, -1]^t$, $[0, 1, 0]^t$ and $[1, 0, 1]^t$ respectively. (*M.T.U. 2013*)

9. Diagonalise the unitary matrix $A = \dfrac{1}{\sqrt{2}} \begin{bmatrix} 1 & i \\ -i & -1 \end{bmatrix}$. (*U.P.T.U. 2014*)

Answers

1. (*i*) $\begin{bmatrix} 1 & 1 & 1 \\ 0 & -1 & 2 \\ 0 & 0 & -2 \end{bmatrix}$ (*ii*) $\begin{bmatrix} 0 & 1 & 4 \\ 1 & 0 & 1 \\ -1 & -1 & -3 \end{bmatrix}$ (*iii*) $\begin{bmatrix} 0 & 0 & 0 \\ 0 & 1 & 0 \\ 0 & 0 & 1 \end{bmatrix}$

(*iv*) $\begin{bmatrix} 32.199 & 0 \\ 0 & 1.801 \end{bmatrix}$ (*v*) $\begin{bmatrix} -1 & 0 & 0 \\ 0 & 3 & 0 \\ 0 & 0 & 4 \end{bmatrix}$ (*vi*) $\begin{bmatrix} 2 & 0 & 0 \\ 0 & 3 & 0 \\ 0 & 0 & 6 \end{bmatrix}$

2. (*i*) $P = \begin{bmatrix} 1 & 1 \\ -2 & 1 \end{bmatrix}$ (*ii*) $\begin{bmatrix} 2 & 1 & -1 \\ -1 & 2 & 0 \\ 1 & 0 & 2 \end{bmatrix}$ (*iii*) $\begin{bmatrix} -1 & 1 & 1 \\ 0 & \sqrt{2} & -\sqrt{2} \\ 1 & 1 & 1 \end{bmatrix}$

3. $\begin{bmatrix} 3 & 0 & 0 \\ 0 & -1 & 0 \\ 0 & 0 & 1 \end{bmatrix}$

4. $\begin{bmatrix} 1 & 0 & 0 \\ 0 & 2 & 0 \\ 0 & 0 & 3 \end{bmatrix}$; $\begin{bmatrix} -12099 & 12355 & 6305 \\ -12100 & 12355 & 6305 \\ -13120 & 13120 & 6561 \end{bmatrix}$

5. $P = \begin{bmatrix} 1 & 1 & 0 \\ -1 & 0 & 1 \\ 1 & 1 & 1 \end{bmatrix}$

6. 1, 3, 3 ; $k_1 \begin{bmatrix} 1 \\ 0 \\ -1 \end{bmatrix}$, $k_2 \begin{bmatrix} 1 \\ 0 \\ 1 \end{bmatrix}$; $k_3 \begin{bmatrix} 0 \\ 1 \\ 0 \end{bmatrix}$; $\begin{bmatrix} 1 & 0 & 0 \\ 0 & 3 & 0 \\ 0 & 0 & 3 \end{bmatrix}$

7. No

8. $A = \begin{bmatrix} 2 & 0 & 1 \\ 0 & 2 & 0 \\ 1 & 0 & 2 \end{bmatrix}$

9. $\begin{bmatrix} 1 & 0 \\ 0 & -1 \end{bmatrix}$.

1.46. APPLICATION OF MATRICES TO ENGINEERING PROBLEMS

1. Electric circuits with resistors and applied voltages. Consider a simple electric circuit involving 5 resistors and 3 applied voltages as shown below:

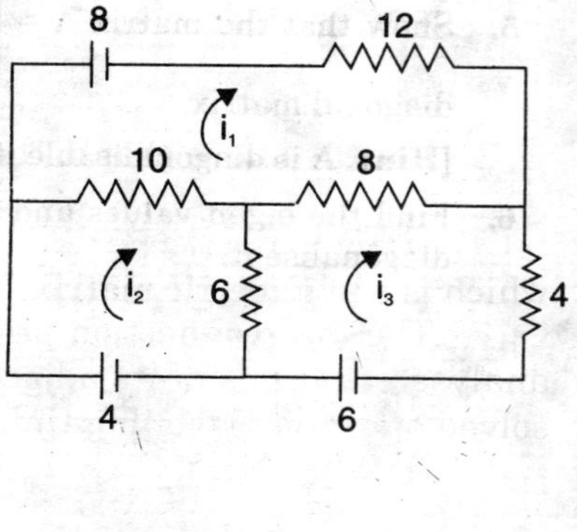

The directions of the currents i_1, i_2 and i_3 flowing in each branch of the circuit are shown by arrows. The currents can be determined by an application of Ohm's law and the Kirchhoff laws.

An application of these laws to the adjoining circuit where the potentials are in volts, the resistances are in ohms, and the currents are in amperes leads to the following set of simultaneous equations:

$$12i_1 + 10(i_1 - i_2) + 8(i_1 - i_3) = 8$$
$$10(i_2 - i_1) + 6(i_2 - i_3) = 4$$
$$8\ (i_3 - i_1) + 6(i_3 - i_2) + 4i_3 = 6$$

After collecting terms, this system can be written as the matrix equation AX = B with

$$A = \begin{bmatrix} 30 & -10 & -8 \\ -10 & 16 & -6 \\ -8 & -6 & 18 \end{bmatrix}, X = \begin{bmatrix} i_1 \\ i_2 \\ i_3 \end{bmatrix}, B = \begin{bmatrix} 8 \\ 4 \\ 6 \end{bmatrix}$$

The directions assumed for the currents i_1, i_2, i_3 are shown by the arrows in above figure but if after the system of equations is solved, the value of the current is found to be negative, the direction of its arrow must be reversed.

When many coupled circuits are involved, a matrix approach is useful and it then becomes necessary to develop a method for solving for X, the matrix equation AX = B, the elements of which are the required currents.

2. Graph Theory. Matrices play an important role in combinatorial problems. These problems involve dealing with the possible arrangements of situations of various different kinds and computing the number and properties of such arrangements.

Consider the delivery from a storage depot of a consumable product to a group of supermarkets in a large city where it is important that daily deliveries be made as rapidly as possible. One possibility involves a delivery truck making a delivery to each supermarket in turn and returning to the storage depot between each delivery before setting out on the next delivery.

An alternative is to travel between supermarkets after each delivery without returning to the storage depot. The question is which approach to routing is the best and how it is to be determined.

A typical situation is illustrated in the adjoining figure in which supermarkets numbered 1 to 5 are involved with circles representing supermarkets and lines, arcs representing the routes.

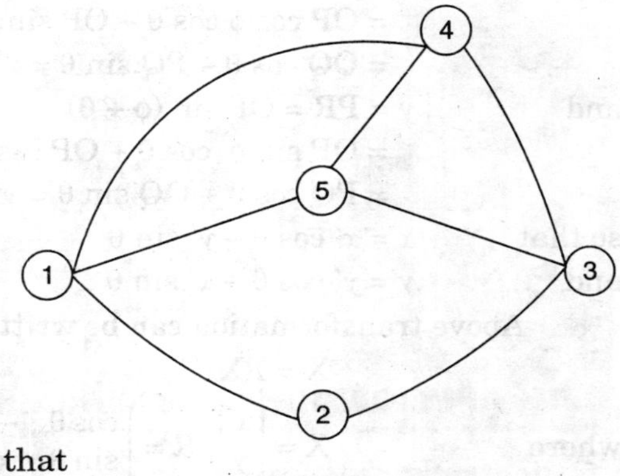

Adjoining representation is called a graph which is regarded as a set of points represented by the circles called vertices of the graph and edges of the graph represented by the lines and arcs.

A special type of matrix associated with such a graph is an adjacency matrix *i.e.*, a matrix whose only entries are 0 or 1.

The adjacency matrix $A = [a_{ij}]$ is formed such that

$$a_{ij} = \begin{cases} 1, \text{ if vertices } i \text{ and } j \text{ are joined by an edge} \\ 0, \text{ otherwise} \end{cases}$$

For the above graph, the adjacency matrix is $A = \begin{bmatrix} 0 & 1 & 0 & 1 & 1 \\ 1 & 0 & 1 & 0 & 0 \\ 0 & 1 & 0 & 1 & 1 \\ 1 & 0 & 1 & 0 & 1 \\ 1 & 0 & 1 & 1 & 0 \end{bmatrix}$

which is a symmetric matrix.

The interconnection properties of the problem represented by the above graph can be analysed in terms of its adjacency matrix A. The optimum routing problem can then be resolved once the travelling times along roads (or lines) are known.

3. Computer Graphics. The simplest operations in computer graphics involve copying a picture to a different location, rotating a picture about a fixed point and scaling a picture where the scaling can be different in the horizontal and vertical directions. These operations are called, respectively, a translation, a rotation and a scaling of the picture.

Operations of this nature can all be represented in terms of matrices and they involve linear transformations of the original picture.

Translation. A translation of a two dimensional figure involves copying it to a different location without either rotating it or changing its horizontal and vertical scales.

The adjoining figure shows the original cartesian axes $O(x, y)$ and the shifted axes $O'(x', y')$ where the respective axes remain parallel to their original directions and the origin O' is located at the point (h, k) relative to the $O(x, y)$ axes.

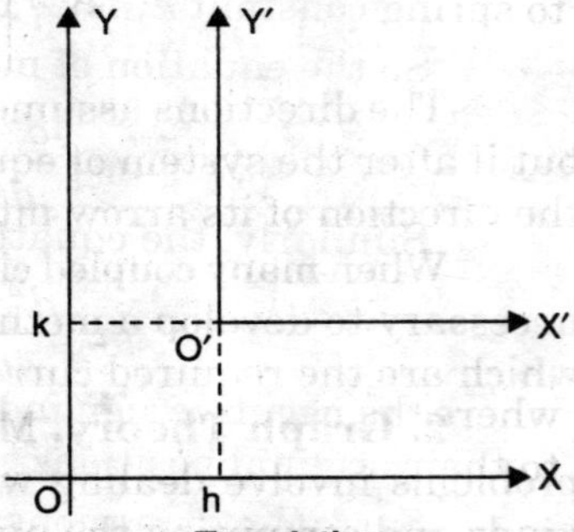

Translation

The relationship between the two sets of coordinates is given by

$$x = x' + h \quad \text{and} \quad y = y' + k$$

If $X = \begin{bmatrix} x \\ y \end{bmatrix}$, $X' = \begin{bmatrix} x' \\ y' \end{bmatrix}$ and $B = \begin{bmatrix} h \\ k \end{bmatrix}$, the coordinate transformation can be written in matrix form as $X = X' + B$ where matrix B represents the translation.

Rotation. Consider a rotation of the coordinate axes through an angle θ where $P(x, y)$ is an arbitrary point. The coordinates of P in the (x, y) reference frame and the (x', y') reference frame are related as

$$x = OR = OP \cos(\phi + \theta)$$
$$= OP \cos\phi \cos\theta - OP \sin\phi \sin\theta$$
$$= OQ \cos\theta - PQ \sin\theta = x' \cos\theta - y' \sin\theta$$

and
$$y = PR = OP \sin(\phi + \theta)$$
$$= OP \sin\phi \cos\theta + OP \cos\phi \sin\theta$$
$$= PQ \cos\theta + OQ \sin\theta = y' \cos\theta + x' \sin\theta$$

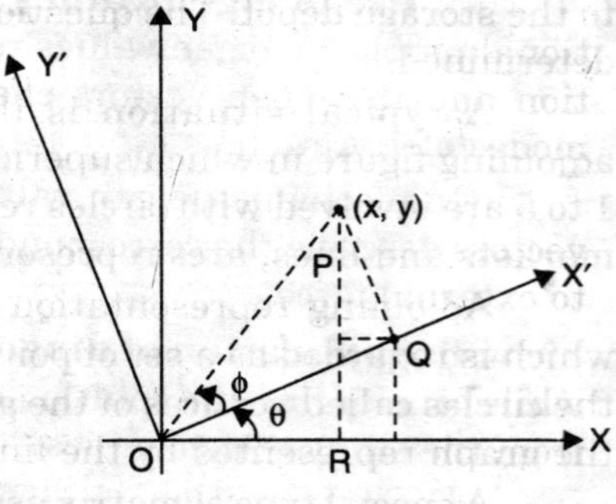

so that $x = x' \cos\theta - y' \sin\theta$

and $y = y' \cos\theta + x' \sin\theta$

Above transformation can be written as

$$X = RX'$$

where $X = \begin{bmatrix} x \\ y \end{bmatrix}$, $R = \begin{bmatrix} \cos\theta & -\sin\theta \\ \sin\theta & \cos\theta \end{bmatrix}$, $X' = \begin{bmatrix} x' \\ y' \end{bmatrix}$

Scaling. If S is a matrix of the form

$$S = \begin{bmatrix} k_x & 0 \\ 0 & k_y \end{bmatrix}$$

where k_x and k_y are positive constants and $X' = SX$, it follows that

$$x = k_x x' \quad \text{and} \quad y = k_y y'$$

showing that x is obtained by scaling x' by k_x, while y is obtained by scaling y' by k_y.

4. A compound mass-spring system. The adjoining figure represents a mass m_1 suspended from a rigid support by a spring of negligible mass with spring constant k_1 and a mass m_2 suspended from mass m_1 by a spring of negligible mass with spring constant k_2. The vertical displacement of m_1 from its equilibrium position is x_1 and the vertical displacement of

m_2 from its equilibrium position is x_2. Each spring is assumed to be linearly elastic so the restoring force exerted by a spring is equal to the product of the displacement from its equilibrium position and the spring constant.

The product of the mass m_1 and its acceleration is $m_1 \dfrac{d^2x_1}{dt^2}$ and the restoring force due to spring constant k_1 is $k_1 x_1$ while the restoring force due to spring constant k_2 is $k_2(x_1 - x_2)$.

So the equation of motion of m_1 is

$$m_1 \frac{d^2x_1}{dt^2} = -k_1x_1 - k_2(x_1 - x_2) \quad \text{...(1)}$$

Similarly, the equation of motion of m_2 is

$$m_2 \frac{d^2x_2}{dt^2} = -k_2(x_2 - x_1) \quad \text{...(2)}$$

where the negative sign indicates that the springs act to restore the masses to their original positions.

System of equations (1) and (2) can be written as the matrix differential equation $\ddot{X} + AX = 0$ where

$$\ddot{X} = \begin{bmatrix} \dfrac{d^2x_1}{dt^2} \\ \dfrac{d^2x_2}{dt^2} \end{bmatrix}, \; A = \begin{bmatrix} \dfrac{k_1 + k_2}{m_1} & \dfrac{-k_2}{m_1} \\ \dfrac{-k_2}{m_2} & \dfrac{k_2}{m_2} \end{bmatrix}, \; X = \begin{bmatrix} x_1 \\ x_2 \end{bmatrix}$$

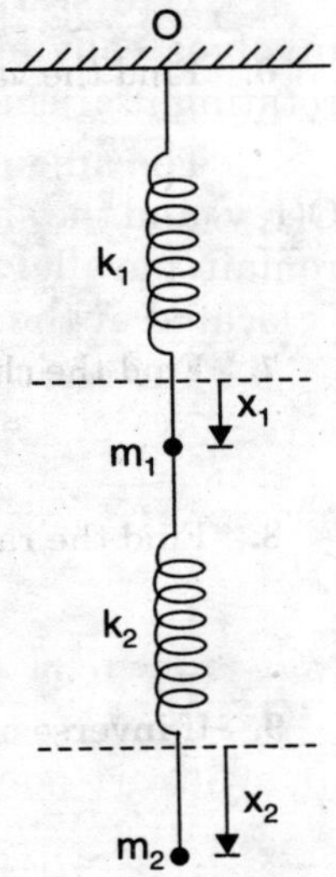

A compound mass spring system

5. Use of eigen values and eigen vectors. While studying vibrational problems, eigen values represent fundamental frequencies of vibration and the eigen vectors characterise the corresponding fundamental modes of vibration.

In mechanics, eigen values represent the principal stresses in a solid body while eigen vectors describe the corresponding principal axes of stress caused by the body being subjected to external forces.

6. Application of diagonalization. An important application of diagonalization arises while considering the solution of linear systems of ordinary differential equations which arise from the study of mechanical, electrical and chemical reaction problems.

Diagonalization is also an important tool when working with partial differential equations, different types of which describe the temperature distribution in a metal, electromagnetic wave propagation and diffusion processes.

ASSIGNMENT I

1. The matrix $A = \begin{bmatrix} 2 & 0 & -1 \\ 5 & 1 & 0 \\ 0 & 1 & 3 \end{bmatrix}$ satisfies the matrix equation $A^3 - 6A^2 + 11A - I = O$ where I is an identity matrix of order 3. Find A^{-1}. *(M.T.U. 2012)*

2. Show that the matrix $A = \begin{bmatrix} 3 & 1 & -1 \\ -2 & 1 & 2 \\ 0 & 1 & 2 \end{bmatrix}$ is diagonalizable. *(M.T.U. 2012)*

3. If $A = \begin{bmatrix} \alpha & \beta \\ \gamma & \delta \end{bmatrix}$ and $\alpha\delta - \beta\gamma = 1$ then find A^{-1}. [M.T.U. (SUM) 2011]

4. Determine the type of complex matrix $A = \frac{1}{\sqrt{3}} \begin{bmatrix} 1 & 1+i \\ 1-i & -1 \end{bmatrix}$. [A.K.T.U. 2021]

5. Calculate the inverse of the matrix $A = \begin{bmatrix} 1 & 2 \\ 5 & 7 \end{bmatrix}$ (G.B.T.U. 2012)

6. Find the value of 'b' for which the rank of the matrix

$$A = \begin{bmatrix} 1 & 5 & 4 \\ 0 & 3 & 2 \\ b & 13 & 10 \end{bmatrix} \text{ is 2.}$$

7. Find the characteristic values of the matrix $\begin{bmatrix} 4 & 1 \\ 2 & 3 \end{bmatrix}$.

8. Find the rank of matrix $A = \begin{bmatrix} 2 & -1 & 3 & 1 \\ 1 & 4 & -2 & 1 \\ 5 & 2 & 4 & 3 \end{bmatrix}$.

9. If inverse of $\begin{bmatrix} 1 & 2 & -2 \\ -1 & 3 & 0 \\ 0 & -2 & 1 \end{bmatrix}$ is $\begin{bmatrix} 3 & 2 & 6 \\ 1 & 1 & k \\ 2 & 2 & 5 \end{bmatrix}$ then find k.

10. If the two eigen values of $\begin{bmatrix} 8 & -6 & 2 \\ -6 & 7 & -4 \\ 2 & -4 & 3 \end{bmatrix}$ are 3 and 15 then find the third eigen value.

11. (*i*) Find the product of the eigen values of the matrix $\begin{bmatrix} 1 & 0 & 0 \\ 0 & 3 & -1 \\ 0 & -1 & 3 \end{bmatrix}$.

(*ii*) Find the sum and product of eigen values of the matrix: $\begin{bmatrix} 1 & 6 & 1 \\ 1 & 2 & 0 \\ 0 & 0 & 3 \end{bmatrix}$ (M.T.U. 2013)

12. Find the rank of matrix $\begin{bmatrix} 3 & 4 & 5 & 6 & 7 \\ 4 & 5 & 6 & 7 & 8 \\ 5 & 6 & 7 & 8 & 9 \\ 10 & 11 & 12 & 13 & 14 \end{bmatrix}$.

13. Find the rank of matrices:

(*i*) $\begin{bmatrix} 1 & 1 & -1 \\ 2 & -3 & 4 \\ 3 & -2 & 3 \end{bmatrix}$ (*ii*) $\begin{bmatrix} 1 & 3 & 5 \\ 7 & 11 & 13 \\ 17 & 19 & 23 \end{bmatrix}$.

14. Find the eigen values of matrices:

(*i*) $\begin{bmatrix} 2 & \sqrt{2} \\ \sqrt{2} & 2 \end{bmatrix}$ (*ii*) $\begin{bmatrix} 1 & 1 \\ 1 & 1 \end{bmatrix}$. (U.P.T.U. 2014)

15. (*i*) If $A = \begin{bmatrix} 1 & 2 & 3 \\ 0 & 2 & 5 \\ 0 & 0 & 3 \end{bmatrix}$, find the eigen values of A^{-1}.

(*ii*) Find the eigen values of A^2 where $A = \begin{bmatrix} 1 & 0 & 0 \\ 2 & 3 & 0 \\ 3 & 4 & 5 \end{bmatrix}$. (*U.P.T.U. 2014*)

(*iii*) If $A = \begin{bmatrix} -1 & 0 & 0 \\ 2 & -3 & 0 \\ 1 & 4 & -2 \end{bmatrix}$, find the eigen values of A^2. (*A.K.T.U. 2016*)

16. Find the rank of the following matrix by reducing into echelon form: $\begin{bmatrix} 3 & 1 & 2 \\ 6 & 2 & 4 \\ 3 & 1 & 2 \end{bmatrix}$.

(*U.P.T.U. 2014*)

17. Find the rank of matrices:

(*i*) $\begin{bmatrix} 4 & 2 & 3 \\ 8 & 4 & 6 \\ -2 & -1 & -1.5 \end{bmatrix}$ (*ii*) $\begin{bmatrix} 1 & 1 & 0 \\ 0 & 1 & 1 \\ 1 & 0 & 1 \end{bmatrix}$. (*U.P.T.U. 2014*)

18. If $A \begin{bmatrix} 0 & 1 \\ 2 & -1 \end{bmatrix} = \begin{bmatrix} 2 & 1 \\ -1 & 0 \end{bmatrix}$ where $A = \begin{bmatrix} a & b \\ c & d \end{bmatrix}$, then find A.

19. Find the characteristic equation of the matrix $A = \begin{bmatrix} 0 & 1 & 0 & 0 \\ 0 & 0 & 0 & 1 \\ 0 & 0 & 1 & 0 \\ 1 & 0 & 0 & 0 \end{bmatrix}$.

20. Find the eigen values of $\begin{bmatrix} 1 & 1 & 1 \\ 1 & 1 & 1 \\ 1 & 1 & 1 \end{bmatrix}$. (*G.B.T.U. 2011, 2012*)

21. Find the rank of matrix $\begin{bmatrix} 1 & 1 & 1 \\ 1 & -1 & 0 \\ 1 & 1 & 1 \end{bmatrix}$ (*G.B.T.U. 2011*)

22. Let A be a 3 × 3 matrix with real entries such that det (A) = 6 and the trace of A is 0. If | A + I | = 0 where I denotes the 3 × 3 identity matrix then find the eigen values of A.

23. Suppose the matrix $A = \begin{bmatrix} 40 & -29 & -11 \\ -18 & 30 & -12 \\ 26 & 24 & -50 \end{bmatrix}$ has a certain complex number $\lambda \neq 0$ as an eigen value.

What are the other eigen values of A?

24. (*i*) Prove that $A = \begin{bmatrix} 4 & 1-3i \\ 1+3i & 7 \end{bmatrix}$ is Hermitian matrix. Find its eigen values.

(*ii*) Verify whether the matrix $A = \begin{bmatrix} 0 & i & 3 \\ -7 & 0 & 5i \\ 3i & 1 & 0 \end{bmatrix}$ is a Hermitian matrix? (*U.P.T.U. 2014*)

25. Find the eigen values and eigen vectors of $A = \begin{bmatrix} i & 0 & 0 \\ 0 & 0 & i \\ 0 & i & 0 \end{bmatrix}$.

26. Determine a, b, c so that A is orthogonal where $A = \begin{bmatrix} 0 & 2b & c \\ a & b & -c \\ a & -b & c \end{bmatrix}$.

27. If $\alpha_1, \alpha_2, \alpha_3, ..., \alpha_n$ are the characteristic roots of the n-square matrix A and k is a scalar, prove that the characteristic roots of $[A - kI]$ are $\alpha_1 - k, \alpha_2 - k, \alpha_3 - k, ..., \alpha_n - k$. [*M.T.U. (C.O.) 2013*]

28. Explain the working rule to find the inverse of a matrix A by elementary row or column transformations. [*M.T.U. (C.O.) 2013*]

29. The Eigen values of a matrix A are 2,3,1 then find the Eigen values of $A^{-1} + A^2$. (*A.K.T.U. 2018*)

30. Examine whether the vectors $x_1 = [3, 1, 1]$, $x_2 = [2, 0, 1]$, $x_3 = [4, 2, 1]$ are linearly independent? (*A.K.T.U. 2016*)

31. Reduce the matrix $\begin{bmatrix} 1 & 1 & 1 \\ 3 & 1 & 1 \end{bmatrix}$ into normal form and find its rank. (*A.K.T.U. 2018, 2022*)

32. For what value of 'x', the eigen values of the given matrix A are real

$$A = \begin{bmatrix} 10 & 5+i & 4 \\ x & 20 & 2 \\ 4 & 2 & -10 \end{bmatrix}$$ (*A.K.T.U. 2017*)

33. For the given matrix $A = \begin{bmatrix} -5 & -3 \\ 2 & 0 \end{bmatrix}$ and $I = \begin{bmatrix} 1 & 0 \\ 0 & 1 \end{bmatrix}$ prove that $A^3 = 19A + 30I$. (*A.K.T.U. 2017*)

34. If the matrix $A = \begin{bmatrix} -1 & 2 & 3 \\ 0 & 3 & 5 \\ 0 & 0 & -2 \end{bmatrix}$ then find the Eigen value of $A^3 + 5A + 8I$. (*A.K.T.U. 2022*)

35. State Rank-Nullity theorem. (*A.K.T.U. 2021*)

36. Show that vectors (1, 6, 4), (0, 2, 3) and (0, 1, 2) are linearly independent. (*A.K.T.U. 2020*)

37. Find the value of 'b' so that rank of

$$A = \begin{bmatrix} 2 & 4 & 2 \\ 3 & 1 & 2 \\ 1 & 0 & b \end{bmatrix}$$

is 2. (*A.K.T.U. 2020*)

38. Find the rank of the matrix $\begin{bmatrix} 2 & 2 & 2 \\ 2 & 2 & 2 \\ 2 & 2 & 2 \end{bmatrix}$. (*A.K.T.U. 2019*)

39. If the Eigen values of matrix A are 1, 1, 1 then find the Eigen values of $A^2 + 2A + 3I$. (*A.K.T.U. 2019*)

Answers

1. $A^{-1} = \begin{bmatrix} 3 & -1 & 1 \\ -15 & 6 & -5 \\ 5 & -2 & 2 \end{bmatrix}$ **3.** $A^{-1} = \begin{bmatrix} \delta & -\beta \\ -\gamma & \alpha \end{bmatrix}$ **4.** unitary **5.** $-\dfrac{1}{3}\begin{bmatrix} 7 & -2 \\ -5 & 1 \end{bmatrix}$

6. 2 **7.** 2, 5 **8.** 2 **9.** 2

10. 0 **11.** (*i*) 8 (*ii*) Sum = 6, Product = – 12

12. 2

13. (*i*) 2 (*ii*) 3 **14.** (*i*) $1 \pm \sqrt{6}$ (*ii*) 0, 2

15. (*i*) $1, \dfrac{1}{2}, \dfrac{1}{3}$ (*ii*) 1, 9, 25 (*iii*) 1, 9, 4 **16.** 1

17. (*i*) 1 (*ii*) 3

18. $A = \begin{bmatrix} 2 & 1 \\ -\dfrac{1}{2} & -\dfrac{1}{2} \end{bmatrix}$

19. $\lambda^4 - \lambda^3 - \lambda + 1 = 0$

20. 0, 0, 3

21. 2

22. – 1, – 2, 3

23. 0, 20 – λ

24. (*i*) 9, 2 (*ii*) No

25. $-i, i, i\ ; k_1 \begin{bmatrix} 0 \\ 1 \\ -1 \end{bmatrix}, k_2 \begin{bmatrix} 0 \\ 1 \\ 1 \end{bmatrix}, k_3 \begin{bmatrix} 1 \\ 0 \\ 0 \end{bmatrix}$

26. $b = \pm \dfrac{1}{\sqrt{6}}, c = \pm \dfrac{1}{\sqrt{3}}, a = \pm \dfrac{1}{\sqrt{2}}$

29. $\dfrac{9}{2}, \dfrac{28}{3}, 2$

30. No

31. $\left[\begin{array}{cc|c} 1 & 0 & 0 \\ 0 & 1 & 0 \end{array}\right]$ or $[I_2 \ \vdots\ O]$; 2

32. $x = 5 - i$

34. 2, 50, –10

37. $\dfrac{3}{5}$

38. 1

39. 6, 6, 6

MODULE 2

Differential Calculus – I

2.1. SUCCESSIVE DIFFERENTIATION

If y is a function of x, its derivative $\frac{dy}{dx}$ is itself a function of x. In general, we assume that it also possesses a derivative if it is differentiated further. The derivative $\frac{dy}{dx}$ is called the first differential co-efficient or first derivative of y w.r.t. x. The differential co-efficient of $\frac{dy}{dx}$, *i.e.*, $\frac{d}{dx}\left(\frac{dy}{dx}\right)$ is called second differential co-efficient or **second derivative** of y w.r.t. x, which we denote as $\frac{d^2y}{dx^2}$ [read as *dee two y* over *dee x square*]. In general, the nth differential co-efficient of y w.r.t. x is denoted by $\frac{d^n y}{dx^n}$. The process of finding the differential co-efficient of a function again and again is called **Successive Differentiation.**

Thus, if $y = f(x)$, the successive differential co-efficients of $f(x)$ are

$$\frac{dy}{dx}, \frac{d^2y}{dx^2}, \frac{d^3y}{dx^3}, \ldots, \frac{d^n y}{dx^n}.$$

These are also denoted by :

(*i*) $y_1, y_2, y_3, \ldots, y_n$. (*ii*) $y', y'', y''', \ldots, y^{(n)}$.

(*iii*) $Dy, D^2y, D^3y, \ldots, D^n y$. (*iv*) $f'(x), f''(x), f'''(x), \ldots, f^n(x)$.

ILLUSTRATIVE EXAMPLES

Example 1. *Find* $\frac{d^3y}{dx^3}$, *when* $y = 4x^3 + 4x + 2$.

Sol. Here, $y = 4x^3 + 4x + 2$

Differentiating w.r.t. x, we get

$$\frac{dy}{dx} = 12x^2 + 4$$

Again, $\frac{d^2y}{dx^2} = 24x$ and $\frac{d^3y}{dx^3} = 24.$

Example 2. *If $y = a \cos (\log x) + b \sin (\log x)$, prove that $x^2y_2 + xy_1 + y = 0$.*

Sol. $y = a \cos (\log x) + b \sin (\log x)$ ...(1)

Differentiating w.r.t. x,

$$y_1 = - a \sin (\log x) \cdot \frac{1}{x} + b \cos (\log x) \cdot \frac{1}{x}$$

Multiplying every term by x,

$$xy_1 = - a \sin (\log x) + b \cos (\log x)$$

Again differentiating both sides w.r.t. x

$$\frac{d}{dx}(xy_1) = - a \cos (\log x) \cdot \frac{1}{x} - b \sin (\log x) \cdot \frac{1}{x}$$

or $$x \cdot y_2 + y_1 \cdot 1 = - \frac{[a \cos (\log x) + b \sin (\log x)]}{x}$$

Cross-multiplying $x^2y_2 + xy_1 = - y$ [By (1)]

or $$x^2y_2 + xy_1 + y = 0.$$

Example 3. *If $y = \sin (a \sin^{-1} x)$, prove that $(1 - x^2) \frac{d^2y}{dx^2} - x \frac{dy}{dx} = - a^2y$.*

Sol. $y = \sin (a \sin^{-1} x)$ ...(1)

$\therefore$ $$\frac{dy}{dx} = \cos (a \sin^{-1} x) \frac{a}{\sqrt{1-x^2}}$$

$\Rightarrow$ $$\sqrt{1-x^2} \frac{dy}{dx} = a \cos (a \sin^{-1} x)$$

Squaring both sides,

$$(1 - x^2) \left(\frac{dy}{dx}\right)^2 = a^2 \cos^2 (a \sin^{-1} x)$$

or $$(1 - x^2) \left(\frac{dy}{dx}\right)^2 = a^2 [1 - \sin^2 (a \sin^{-1} x)]$$ $| \because \cos^2 \theta = 1 - \sin^2 \theta$

or $$(1 - x^2) \left(\frac{dy}{dx}\right)^2 = a^2 (1 - y^2)$$ [By (1)]

Again differentiating w.r.t. x, we have

$$(1 - x^2)\, 2 \frac{dy}{dx} \frac{d^2y}{dx^2} - 2x \left(\frac{dy}{dx}\right)^2 = - 2a^2\, y \frac{dy}{dx}$$

Dividing every term by $2 \frac{dy}{dx}$, we get

$$(1 - x^2) \frac{d^2y}{dx^2} - x \frac{dy}{dx} = - a^2y.$$

TEST YOUR KNOWLEDGE

1. (i) If $y = x^3 \log x$, prove that $y_4 = \dfrac{6}{x}$.

   (ii) If $y = \dfrac{\log x}{x}$, prove that $\dfrac{d^2y}{dx^2} = \dfrac{2 \log x - 3}{x^3}$.

2. If $y = A \sin nx + B \cos nx$, prove that $\dfrac{d^2y}{dx^2} + n^2y = 0$.

3. (i) If $y = (\sin^{-1} x)^2$, prove that $(1 - x^2) \dfrac{d^2y}{dx^2} - x\dfrac{dy}{dx} = 2$.

   [Hint. Differentiate w.r.t. x ; cross-multiply and square again, differentiate.]

   (ii) If $y = \tan^{-1} x$, prove that $(1 + x^2) \dfrac{d^2y}{dx^2} + 2x\dfrac{dy}{dx} = 0$.

   (iii) If $y = (\tan^{-1} x)^2$, prove that $(x^2 + 1)^2 y_2 + 2x(x^2 + 1)y_1 = 2$.

4. (i) If $y = ae^{mx} + be^{-mx}$, prove that $\dfrac{d^2y}{dx^2} - m^2y = 0$.

   (ii) If $y = e^{ax} \sin bx$, prove that $\dfrac{d^2y}{dx^2} - 2a\dfrac{dy}{dx} + (a^2 + b^2)\, y = 0$.

5. (i) If $y = \cos^{-1} x$, prove that $y_2 (1 - x^2) - xy_1 = 0$.

   (ii) If $x = (a + bt)\, e^{-nt}$, prove that $\dfrac{d^2x}{dt^2} + 2n\dfrac{dx}{dt} + n^2x = 0$.

   (iii) If $y = e^{m \sin^{-1} x}$, prove that $(1 - x^2)\, y_2 - xy_1 = m^2y$.

6. If $y = \dfrac{\sin^{-1} x}{\sqrt{1 - x^2}}$, prove that

   (i) $(1 - x^2)\, y_1 - xy = 1$ (ii) $(1 - x^2)\, y_2 - 3xy_1 - y = 0$.

7. (a) If $y = A(x + \sqrt{1 + x^2})^n$, prove that $(x^2 + 1)\, y_2 + xy_1 - n^2y = 0$.

   (b) If $y = A\,(x + \sqrt{x^2 - 1})^n$, prove that $(x^2 - 1)\, y_2 + xy_1 - n^2y = 0$.

8. (i) If $y = \log\left(x + \sqrt{1 + x^2}\right)$, prove that $(1 + x^2) \dfrac{d^2y}{dx^2} + x\dfrac{dy}{dx} = 0$.

   (ii) If $y = [\log (x + \sqrt{1 + x^2})]^2$, prove that $(1 + x^2) \dfrac{d^2y}{dx^2} + x\dfrac{dy}{dx} = 2$.

2.2. CALCULATION OF n^{TH} ORDER DERIVATIVES

I. To find the nth differential co-efficient of x^m.

Let $y = x^m$

then $y_1 = mx^{m-1}$

$y_2 = m(m-1)\, x^{m-2}$

$y_3 = m(m-1)\,(m-2)\, x^{m-3}$

....

....

$\therefore$ $y_n = [m(m-1)\,(m-2)\, \ldots \text{ upto } n \text{ factors}] \times x^{m-n}$

$= m(m-1)\,(m-2)\, \ldots\, (m-n+1)\,.\, x^{m-n}$, where $n < m$.

[**Cor.** If m be a positive integer, and if $n = m$

then $y_m = m(m-1)\,(m-2)\, \ldots\, (m-m+1)\, x^{m-m}$

$= m(m-1)(m-2)\, \ldots\, 3.2.1. = m\,!$

i.e., $$\frac{d^m}{dx^m}(x^m) = m\,!$$

and $(m+1)^{th}$, $(m+2)^{th}$ derivative ... etc., each will be = 0.

$\therefore$ $y_m = m\,!$ which is constant and $y_{(m+1)}$ which is the derivative of y_m will be zero and so on.]

II. The nth differential co-efficient of $(ax + b)^m$, where $n < m$.

Let $y = (ax+b)^m$

then $y_1 = m(ax+b)^{m-1}\,.\,a$

$y_2 = m(m-1)\,(ax+b)^{m-2}\,.\,a^2$

$y_3 = m(m-1)\,(m-2)\,(ax+b)^{m-3}\,.\,a^3$

....

....

$\therefore$ $$y_n = m(m-1)\,(m-2)\,(m-3)\, \ldots\, (m-n+1)\,(ax+b)^{m-n}\,.\,a^n$$

[**Cor.** If $n = m$, then

$y_m = m(m-1)\,(m-2)\, \ldots\, 3.2.1.\,(ax+b)^0\, a^m = m\,!\, a^m$]

III. The nth differential co-efficient of $\dfrac{1}{ax+b}\left[x \neq -\dfrac{b}{a}\right]$

Let $y = \dfrac{1}{ax+b} = (ax+b)^{-1}$

then $y_1 = (-1)(ax+b)^{-2}\,.\,a$

$y_2 = (-1)(-2)(ax+b)^{-3}\,.\,a^2 = (-1)^2\,.\,2\,!\,(ax+b)^{-3}\,.\,a^2$

$y_3 = (-1)(-2)(-3)(ax+b)^{-4}\,.\,a^3 = (-1)^3\,.\,3\,!\,(ax+b)^{-4}\,.\,a^3$

....

....

$\therefore$ $$y_n = (-1)^n \, . \, n \, ! \, (ax+b)^{-(n+1)} \, . \, a^n$$

or $$\boxed{y_n = \frac{(-1)^n \, n \, ! \, a^n}{(ax+b)^{n+1}}}.$$

IV. The nth differential co-efficient of log $(ax + b)$.

Let $$y = \log(ax+b)$$

then $$y_1 = \frac{1}{ax+b} \, a = (ax+b)^{-1} \, . \, a$$

$$y_2 = (-1)(ax+b)^{-2} \, . \, a^2$$

$$y_3 = (-1)(-2)(ax+b)^{-3} \, . \, a^3 = (-1)^2 \, 2 \, ! \, (ax+b)^{-3} \, . \, a^3$$

....

....

$\therefore$ $$y_n = (-1)^{n-1} \, (n-1) \, ! \, (ax+b)^{-n} \, . \, a^n$$

or $$\boxed{y_n = \frac{(-1)^{n-1} \, (n-1) \, ! \, a^n}{(ax+b)^n}}$$

[**Cor.** If $a = 1; b = 0, y = \log x$, then $y_n = \dfrac{(-1)^{n-1} \, (n-1) \, !}{x^n}$]

V. The nth differential co-efficient of a^{mx}

Let $$y = a^{mx}$$

$\therefore$ $$y_1 = m \, . \, a^{mx} \, (\log a)$$

$$y_2 = m^2 \, . \, a^{mx} \, (\log a)^2$$

$$y_3 = m^3 \, . \, a^{mx} \, (\log a)^3$$

....

....

$\therefore$ $$\boxed{y_n = m^n \, . \, a^{mx} \, (\log a)^n}$$

[**Cor.** Put $a = e$, then $y = e^{mx}$ and $y_n = m^n \, e^{mx}$.]

VI. The nth differential co-efficient of sin $(ax + b)$

Let $$y = \sin(ax+b)$$

then $$y_1 = a \cos(ax+b) = a \sin\left(ax+b+\frac{\pi}{2}\right)$$ [Note this step]

$$y_2 = a^2 \cos\left(ax+b+\frac{\pi}{2}\right) = a^2 \sin\left(ax+b+\frac{\pi}{2}+\frac{\pi}{2}\right)$$

$$= a^2 \sin\left(ax+b+2 \, . \, \frac{\pi}{2}\right)$$

$$y_3 = a^3 \cos\left(ax+b+2 \, . \, \frac{\pi}{2}\right) = a^3 \sin\left(ax+b+2 \, . \, \frac{\pi}{2}+\frac{\pi}{2}\right)$$

$$= a^3 \sin\left(ax + b + 3 \cdot \frac{\pi}{2}\right)$$

....

....

$$\boldsymbol{y_n = a^n \sin\left(ax + b + n\frac{\pi}{2}\right)}$$

$$\left[\textbf{Cor. } \text{If } b = 0, y = \sin ax, y_n = a^n \sin\left(ax + n\frac{\pi}{2}\right)\right]$$

VII. The nth differential co-efficient of cos $(ax + b)$

Let $y = \cos(ax + b)$

then $y_1 = -\sin(ax + b) \cdot a = a \cos\left(ax + b + \frac{\pi}{2}\right)$ **[Note this step]**

$$y_2 = -a^2 \sin\left(ax + b + \frac{\pi}{2}\right) = a^2 \cos\left(ax + b + \frac{\pi}{2} + \frac{\pi}{2}\right)$$

$$= a^2 \cos\left(ax + b + 2 \cdot \frac{\pi}{2}\right)$$

$$y_3 = -a^3 \sin\left(ax + b + 2 \cdot \frac{\pi}{2}\right) = a^3 \cos\left(ax + b + 2 \cdot \frac{\pi}{2} + \frac{\pi}{2}\right)$$

$$= a^3 \cos\left(ax + b + 3 \cdot \frac{\pi}{2}\right)$$

....

....

$$\boxed{\boldsymbol{y_n = a^n \cos\left(ax + b + n\frac{\pi}{2}\right)}}$$

$$\left[\textbf{Cor. } \text{If } y = \cos ax, y_n = a^n \cos\left(ax + n \cdot \frac{\pi}{2}\right).\right]$$

VIII. The nth differential co-efficient of e^{ax} sin $(bx + c)$

Let $y = e^{ax} \sin(bx + c)$

then $y_1 = e^{ax} \cdot \cos(bx + c) \cdot b + \sin(bx + c) \cdot a \cdot e^{ax}$

$$= e^{ax} [a \sin(bx + c) + b \cos(bx + c)].$$

We determine two constants r and θ, to change the expression into a single sine which will enable us to make the required generalisation by putting $a = r \cos \theta$, $b = r \sin \theta$.

$\therefore$ $r = \sqrt{a^2 + b^2}$, $\theta = \tan^{-1} \dfrac{b}{a}$.

Hence, $y_1 = e^{ax} [r \cos \theta \cdot \sin(bx + c) + r \sin \theta \cdot \cos(bx + c)]$

$$= r e^{ax} \sin(bx + c + \theta).$$

Thus y_1 is obtained from y on multiplying it by the constant r and increasing the angle by the constant θ. Repeating the same rule to y_1, we have

$$y_2 = r^2 e^{ax} \sin (bx + c + 2\theta).$$

Similarly, $$y_3 = r^3 e^{ax} \sin (bx + c + 3\theta).$$

Hence, in general, $$y_n = \frac{d^n}{dx^n} [e^{ax} \sin (bx + c)] = r^n . e^{ax} \sin (bx + c + n\theta)$$

Putting values of r and θ,

$$\boldsymbol{y_n = (a^2 + b^2)^{n/2} e^{ax} . \sin \left(bx + c + n . \tan^{-1} \frac{b}{a}\right)}$$

IX. The nth differential co-efficient of $e^{ax} \cos (bx + c)$

Let $$y = e^{ax} \cos (bx + c)$$

then $$y_1 = a\, e^{ax} \cos (bx + c) - b\, e^{ax} \sin (bx + c)$$
$$= e^{ax} [a \cos (bx + c) - b \sin (bx + c)]$$

As in Art. VIII, put $a = r \cos \theta$, and $b = r \sin \theta$.

∴ $$r = \sqrt{a^2 + b^2}, \tan \theta = \frac{b}{a}, \textit{ i.e.}, \theta = \tan^{-1} \frac{b}{a}$$

∴ $$y_1 = e^{ax} r [\cos \theta . \cos (bx + c) - \sin \theta \sin (bx + c)]$$
$$= r\, e^{ax} [\cos (bx + c + \theta)].$$

Thus y_1 is obtained from y on multiplying it by the constant r and increasing the angle by the constant θ.

Repeating the same rule to y_1, we have

$$y_2 = r^2 e^{ax} \cos (bx + c + 2\theta).$$

Similarly, $$y_3 = r^3 e^{ax} \cos (bx + c + 3\theta).$$

Hence, in general, $$y_n = \frac{d^n}{dx^n} [e^{ax} \cos (bx + c)] = r^n e^{ax} . \cos (bx + c + n\theta)$$

Putting values of r and θ,

$$\boldsymbol{y_n = (a^2 + b^2)^{n/2} e^{ax} . \cos \left(bx + c + n . \tan^{-1} \frac{b}{a}\right)}$$

ILLUSTRATIVE EXAMPLES

Example 1. *Find the nth differential co-efficient of $\log (ax + x^2)$.*

Sol. Let $$y = \log (ax + x^2) = \log x(a + x) = \log x + \log (x + a)$$

Differentiating n times,

$$y_n = \frac{d^n}{dx^n} \log x + \frac{d^n}{dx^n} \log (x + a)$$

$$= \frac{(-1)^{n-1}(n-1)!\,.\,1^n}{x^n} + \frac{(-1)^{n-1}(n-1)!\,.\,1^n}{(x+a)^n} \qquad \text{[By Art. IV]}$$

$$= (-1)^{n-1}(n-1)!\left[\frac{1}{x^n} + \frac{1}{(x+a)^n}\right].$$

Example 2. *If $y = \cos^3 x$, find y_n.*

Sol. We know, $\cos 3x = 4\cos^3 x - 3\cos x$

$\therefore$ $\cos^3 x = \frac{1}{4}[\cos 3x + 3\cos x]$

i.e., $y = \frac{1}{4}[\cos 3x + 3\cos x]$

Differentiating n times,

$$y_n = \frac{1}{4}\left[\frac{d^n}{dx^n}\cos 3x + 3\frac{d^n}{dx^n}\cos x\right]$$

$$= \frac{1}{4}\left[3^n\cos\left(3x + \frac{n\pi}{2}\right) + 3.1^n\cos\left(x + \frac{n\pi}{2}\right)\right] \qquad \text{[By Art. VII]}$$

$$= \frac{1}{4}\left[3^n\cos\left(3x + \frac{n\pi}{2}\right) + 3\cos\left(x + \frac{n\pi}{2}\right)\right].$$

Example 3. *If $y = \sin 2x \sin 3x$, find y_n.*

Sol. $y = \sin 2x \sin 3x$

$= \frac{1}{2}[2\sin 3x \sin 2x] = \frac{1}{2}[\cos x - \cos 5x]$

$[\because\ 2\sin A \sin B = \cos(A - B) - \cos(A + B)]$

$$\therefore \quad y_n = \frac{1}{2}\left[\frac{d^n}{dx^n}\cos x - \frac{d^n}{dx^n}\cos 5x\right]$$

$$= \frac{1}{2}\left[\cos\left(x + \frac{n\pi}{2}\right) - 5^n\cos\left(5x + \frac{n\pi}{2}\right)\right].$$

Example 4. *Find n^{th} derivative of $\sin^2 x \cos^3 x$.*

Sol. Let $y = \sin^2 x \cos^3 x$

$= (\sin x \cos x)^2 \cos x = \frac{1}{4}\sin^2 2x \cos x$

$$= \frac{1}{4}\left(\frac{1 - \cos 4x}{2}\right)\cos x \qquad \left|\ \because \sin^2\theta = \frac{1 - \cos 2\theta}{2}\right.$$

$$= \frac{1}{8}(\cos x - \cos 4x \cos x)$$

or $$y = \frac{1}{16}[2\cos x - (\cos 5x + \cos 3x)]$$

$[\because\ 2\cos A \cos B = \cos(A + B) + \cos(A - B)]$

$$\therefore \quad y_n = \frac{1}{16}\left[2.1^n\cos\left(x + \frac{n\pi}{2}\right) - 5^n\cos\left(5x + \frac{n\pi}{2}\right) - 3^n\cos\left(3x + \frac{n\pi}{2}\right)\right]$$

$$= \frac{1}{16}\left[2\cos\left(x + \frac{n\pi}{2}\right) - 5^n\cos\left(5x + \frac{n\pi}{2}\right) - 3^n\cos\left(3x + \frac{n\pi}{2}\right)\right].$$

Example 5. *Find the nth differential co-efficient of $e^x \sin^3 x$.*

Sol. We know that, $\sin 3x = 3 \sin x - 4 \sin^3 x$

$$\therefore \quad \sin^3 x = \frac{1}{4} [3 \sin x - \sin 3x] \quad ...(1)$$

Let $$y = e^x \sin^3 x = e^x . \frac{1}{4} [3 \sin x - \sin 3x] \quad | \text{ From (1)}$$

$$= \frac{3}{4} e^x \sin x - \frac{1}{4} e^x \sin 3x$$

$$\therefore \quad y_n = \frac{3}{4} (1^2 + 1^2)^{n/2} e^x \sin (x + n \tan^{-1} \tfrac{1}{1})$$

$$- \frac{1}{4} . (1^2 + 3^2)^{n/2} e^x \sin [3x + n \tan^{-1} \tfrac{3}{1}]$$

| **Comparing to result in Art. (VIII)**

$$= \frac{3}{4} . 2^{n/2} e^x \sin \left(x + n \frac{\pi}{4}\right) - \frac{1}{4} . 10^{n/2} . e^x \sin (3x + n \tan^{-1} 3).$$

Example 6. *If* $y = \dfrac{1}{1 - 5x + 6x^2}$, *find* y_n.

Sol. $$y = \frac{1}{1 - 5x + 6x^2} = \frac{1}{(2x-1)(3x-1)} = \frac{2}{2x-1} - \frac{3}{3x-1}$$

$$\therefore \quad y_n = 2D^n \left(\frac{1}{2x-1}\right) - 3D^n \left(\frac{1}{3x-1}\right)$$

$$= 2 \left[\frac{(-1)^n \, n! (2)^n}{(2x-1)^{n+1}}\right] - 3 \left[\frac{(-1)^n \, n! (3)^n}{(3x-1)^{n+1}}\right] \quad | \text{ By Art. III}$$

$$= (-1)^n \, n! \left[\frac{(2)^{n+1}}{(2x-1)^{n+1}} - \frac{(3)^{n+1}}{(3x-1)^{n+1}}\right].$$

Example 7. *Find the nth derivative of* $\dfrac{ax+b}{cx+d}$.

Sol. Let, $$y = \frac{ax+b}{cx+d} = \frac{a}{c} + \frac{(bc - ad)}{c(cx+d)}$$

$$\therefore \quad y_n = 0 + \left(\frac{bc - ad}{c}\right) D^n \left(\frac{1}{cx+d}\right)$$

$$= \left(\frac{bc - ad}{c}\right) \frac{(-1)^n \, n! \, c^n}{(cx+d)^{n+1}} \quad | \text{ By Art. III}$$

$$= \frac{(-1)^n \, n! (bc - ad) (c)^{n-1}}{(cx+d)^{n+1}}.$$

Example 8. *If* $y = \sqrt{x+a}$, *find* y_n.

Sol. Here, $$y = (x+a)^{1/2}$$

$$\therefore \quad y_1 = \tfrac{1}{2}(x+a)^{-1/2}$$

$$y_2 = (\tfrac{1}{2})(-\tfrac{1}{2})(x+a)^{-3/2}$$

$$y_3 = \tfrac{1}{2}(-\tfrac{1}{2})(-\tfrac{3}{2})(x+a)^{-5/2} = (-1)^2 \frac{1.3}{2^3}(x+a)^{-5/2}$$

....

....

$$\therefore \quad y_n = (-1)^{n-1} \frac{1.3.5 \ldots \text{upto } (n-1) \text{ terms}}{2^n} (x+a)^{-\left(\frac{2n-1}{2}\right)}$$

$$= (-1)^{n-1} \frac{1.3 \ldots (2n-3)}{2^n} (x+a)^{-\left(\frac{2n-1}{2}\right)}.$$

Remark. If none of the formulae of the nine articles is applicable to find y_n in a problem, then we proceed as in the articles, *i.e.*, find y_1, y_2, y_3 and then generalise.

TEST YOUR KNOWLEDGE

Find the nth derivative of the following:

1. (*i*) 4^x (*ii*) $e^{2x} + e^{-2x}$ (*iii*) $\log 3x$
2. (*i*) $\dfrac{1}{(3-2x)^3}$ (*ii*) $\dfrac{1}{(bx+a)^2}$
3. (*i*) $\log \sqrt{\dfrac{2x+1}{x-2}}$ (*ii*) $\log \sqrt{\dfrac{3-2x}{5+4x}}$ (*iii*) $\log (x^2 - a^2)$ (*iv*) $\dfrac{3x+7}{x+2}$
4. (*i*) $\sin^2 x$ (*ii*) $\cos^2 x$ (*iii*) $\sin^3 x$ (*iv*) $\cos^4 x$.
5. (*i*) $\sin x \, . \cos 2x$ (*ii*) $\cos x \cos 2x$
6. (*i*) $\cos ax \sin bx$ (*ii*) $\sin x \sin 2x$ (*iii*) $\sin^3 x \cos^2 x$.
7. (*i*) $e^x \cos x$ (*ii*) $e^x \sin x$ (*iii*) $e^x \sin x \cos x$ (*iv*) $e^{3x} \sin^2 2x$
8. (*i*) $e^{ax} \cos^2 x \sin x$ (*ii*) $e^x \cos^3 x$ (*iii*) $e^x \sin^2 x \sin 2x$ (*iv*) $a^x \cos x$
9. If $\mu = \sin nx + \cos nx$, prove that $\mu_r = n^r [1 + (-1)^r \sin 2nx]^{1/2}$, where μ_r denotes the rth differential co-efficient of μ w.r.t. x. Hence show that $\mu_8(\pi) = (1/2)^{31/2}$ when $n = 1/4$. (*A.K.T.U. 2018*)

Answers

1. (*i*) $4^x [\log 4]^n$ (*ii*) $2^n [e^{2x} + (-1)^n e^{-2x}]$ (*iii*) $\dfrac{[-1]^{n-1}(n-1)!}{x^n}$.
2. (*i*) $\dfrac{(n+2)!\,.\,2^{n-1}}{[3-2x]^{n+3}}$ (*ii*) $\dfrac{[-1]^n (n+1)!\,.\,b^n}{[bx+a]^{n+2}}$.
3. (*i*) $\dfrac{[-1]^{n-1}(n-1)!}{2}\left[\dfrac{2^n}{(2x+1)^n} - \dfrac{1}{(x-2)^n}\right]$ (*ii*) $\dfrac{1}{2}(-1)^{n-1}(n-1)!\left[\dfrac{(-2)^n}{(3-2x)^n} - \dfrac{4^n}{(5+4x)^n}\right]$

   (*iii*) $(-1)^{n-1}(n-1)!\left[\dfrac{1}{(x+a)^n} + \dfrac{1}{(x-a)^n}\right]$ (*iv*) $\dfrac{(-1)^n \, n!}{(x+2)^{n+1}}$.
4. (*i*) $-2^{n-1} \cos [2x + n \, . \, \pi/2]$ (*ii*) $2^{n-1} \cos [2x + n \, . \, \pi/2]$

(*iii*) $\frac{3}{4}\sin[x + n\,.\,\pi/2] - \frac{1}{4}\,.\,3^n \sin[3x + n\,.\,\pi/2]$

(*iv*) $2^{n-1}\cos[2x + n\,.\,\pi/2] + 2^{2n-3}\cos[4x + n\,.\,\pi/2]$.

5. (*i*) $\frac{1}{2}[3^n \sin(3x + n\,.\,\pi/2) - \sin(x + n\,.\,\pi/2)]$ (*ii*) $\frac{1}{2}\left[3^n \cos\left(3x + n\frac{\pi}{2}\right) + \cos\left(x + n\frac{\pi}{2}\right)\right]$.

6. (*i*) $\frac{1}{2}\left[(a+b)^n \sin\left\{(a+b)x + \frac{n\pi}{2}\right\} + (b-a)^n \sin\left\{(b-a)x + \frac{n\pi}{2}\right\}\right]$

(*ii*) $\frac{1}{2}[\cos(x + n\,.\,\pi/2) - 3^n \cos(3x + n\,.\,\pi/2)]$

(*iii*) $\frac{1}{16}\left[2\sin\left(x + \frac{n\pi}{2}\right) - 5^n \sin\left(5x + \frac{n\pi}{2}\right) + 3^n \sin\left(3x + \frac{n\pi}{2}\right)\right]$.

7. (*i*) $2^{n/2}\,.\,e^x \cos[x + n\,.\,\pi/4]$ (*ii*) $2^{n/2}\,.\,e^x \sin[x + n\,.\,\pi/4]$

(*iii*) $\frac{1}{2}\,.\,5^{n/2} e^x \sin(2x + n\tan^{-1} 2)$ (*iv*) $\frac{1}{2}\left[3^n e^{3x} - 5^n e^{3x}\cos\left(4x + n\tan^{-1}\frac{4}{3}\right)\right]$.

8. (*i*) $\frac{1}{4}[(a^2+1)^{n/2} e^{ax}\sin(x + n\tan^{-1} 1/a) + (a^2+9)^{n/2} e^{ax}\sin(3x + n\tan^{-1} 3/a)]$

(*ii*) $\frac{3}{4}\,.\,2^{n/2}\,.\,e^x\cos[x + n\,.\,\pi/4] + \frac{1}{4}\,.\,10^{n/2} e^x \cos[3x + n\tan^{-1} 3]$

(*iii*) $\frac{1}{2}\,.\,5^{n/2}\,.\,e^x \sin[2x + n\tan^{-1} 2] - \frac{1}{4}[17]^{n/2}\,.\,e^x \sin[4x + n\tan^{-1} 4]$

(*iv*) $a^x[1 + (\log a)^2]^{n/2}\cos\left[x + n\tan^{-1}\frac{1}{\log a}\right]$.

2.3. USE OF PARTIAL FRACTIONS

In order to determine the nth derivative, of any rational function, we have to split it into partial fractions. Forming partial fractions is converse process of taking L.C.M.

To resolve a fraction into partial fractions, the degree of numerator must be less than the degree of denominator.

ILLUSTRATIVE EXAMPLES

Example 1. *Find n^{th} derivative of* (*i*) $\frac{1}{x^2 - a^2}$ (*ii*) $\frac{2x+1}{(2x-1)(2x+3)}$. (U.P.T.U. 2015)

Sol. (*i*) Let,
$$y = \frac{1}{x^2 - a^2} = \frac{1}{(x-a)(x+a)} = \frac{1}{2a}\left[\frac{1}{x-a} - \frac{1}{x+a}\right]$$

$\therefore$
$$y_n = \frac{1}{2a}\left[\frac{d^n}{dx^n}\left(\frac{1}{x-a}\right) - \frac{d^n}{dx^n}\left(\frac{1}{x+a}\right)\right]$$
$$= \frac{1}{2a}\left[\frac{(-1)^n n!\,1^n}{(x-a)^{n+1}} - \frac{(-1)^n n!\,.\,1^n}{(x+a)^{n+1}}\right]$$
$$= \frac{1}{2a}(-1)^n n!\left[\frac{1}{(x-a)^{n+1}} - \frac{1}{(x+a)^{n+1}}\right].$$

(*ii*) Let, $$y = \frac{2x+1}{(2x-1)(2x+3)} = \frac{1}{2}\left(\frac{1}{2x-1} + \frac{1}{2x+3}\right)$$

$$\therefore \quad y_n = \frac{1}{2}\left[\frac{d^n}{dx^n}\left(\frac{1}{2x-1}\right) + \frac{d^n}{dx^n}\left(\frac{1}{2x+3}\right)\right]$$

$$= \frac{1}{2}\left[\frac{(-1)^n\, n!(2)^n}{(2x-1)^{n+1}} + \frac{(-1)^n\, n!(2)^n}{(2x+3)^{n+1}}\right]$$

$$= (-1)^n\, n!\,(2)^{n-1}\left[\frac{1}{(2x-1)^{n+1}} + \frac{1}{(2x+3)^{n+1}}\right].$$

Example 2. *If* $y = \dfrac{x^2}{(x-1)^2(x+2)}$, *find* n^{th} *derivative of* y.

Sol. To split y into partial fractions.

Let $x - 1 = z$, then $$y = \frac{1}{z^2} \cdot \frac{1+2z+z^2}{3+z}$$

$$= \frac{1}{z^2}\left(\frac{1}{3} + \frac{5z}{9} + \frac{4}{9} \cdot \frac{z^2}{3+z}\right) \qquad \text{| on division}$$

$$= \frac{1}{3z^2} + \frac{5}{9z} + \frac{4}{9(3+z)} = \frac{1}{3(x-1)^2} + \frac{5}{9(x-1)} + \frac{4}{9(x+2)}$$

Hence, $$y_n = \frac{(-1)^n\,(n+1)!}{3(x-1)^{n+2}} + \frac{5(-1)^n\, n!}{9(x-1)^{n+1}} + \frac{4(-1)^n\, n!}{9(x+2)^{n+1}}.$$

Example 3. *Find the* n^{th} *derivative of* $\tan^{-1}\left(\dfrac{2x}{1-x^2}\right)$.

Sol. We have, $\tan^{-1}\left(\dfrac{2x}{1-x^2}\right) = 2\tan^{-1} x$

Hence we have to find y_n if $y = 2\tan^{-1} x$.

Now, $$y = 2\tan^{-1} x$$

$$\therefore \quad y_1 = \frac{2}{1+x^2} = \frac{2}{2i}\left[\frac{1}{x-i} - \frac{1}{x+i}\right] = \frac{1}{i}\left[\frac{1}{x-i} - \frac{1}{x+i}\right]$$

Hence, $$y_n = \frac{1}{i}(-1)^{n-1}.(n-1)!\left[\frac{1}{(x-i)^n} - \frac{1}{(x+i)^n}\right]$$

| Differentiating $(n-1)$ times

Suppose $x = r\cos\theta$, and $1 = r\sin\theta$ so that

$$r^2 = x^2 + 1 \text{ and } \theta = \tan^{-1}\left(\frac{1}{x}\right)$$

Thus, $$y_n = \frac{(-1)^{n-1}(n-1)!}{ir^n} \{(\cos\theta - i\sin\theta)^{-n} - (\cos\theta + i\sin\theta)^{-n}\}$$

$$\Rightarrow \quad y_n = \frac{(-1)^{n-1}(n-1)!}{i \cdot r^n} \cdot 2i \sin n\theta \qquad | \text{ Using De Moivre's theorem}$$

$$\Rightarrow \quad y_n = 2(-1)^{n-1}(n-1)! \cdot \sin n\theta \cdot \sin^n\theta, \qquad \text{where } \theta = \tan^{-1}\left(\frac{1}{x}\right)$$

Example 4. *Prove that the value of the nth differential co-efficient of $\frac{x^3}{x^2-1}$ for $x = 0$ is zero, if n is even; and is – (n !) if n is odd and greater than 1.*

Sol. Let $$y = \frac{x^3}{x^2-1} = x + \frac{x}{x^2-1} = x + \frac{1}{2}\left[\frac{1}{x+1} + \frac{1}{x-1}\right] \qquad \text{...(1)}$$

If $n > 1$, the second and higher derivatives of x are 0.

$$\therefore \quad y_n = \frac{1}{2}\left[\frac{(-1)^n n!}{(x-1)^{n+1}} + \frac{(-1)^n n!}{(x+1)^{n+1}}\right]$$

$$= \frac{1}{2}(-1)^n n! \left[\frac{1}{(x+1)^{n+1}} + \frac{1}{(x-1)^{n+1}}\right] \qquad \text{...(2)}$$

Case I. When n is even, putting $x = 0$ in eq. (2),

$$(y_n)_0 = \frac{(1)\, n!}{2}\left[\frac{1}{1} + \frac{1}{(-1)}\right] = \frac{n!}{2}[1-1] = 0.$$

Case II. When n is odd, putting $x = 0$ in eq. (2),

$$(y_n)_0 = \frac{(-1)\, n!}{2}\left[\frac{1}{1} + \frac{1}{(1)}\right] = -\frac{(n!)}{2}[1+1] = -n!.$$

Example 5. *If $y = x \log \frac{x-1}{x+1}$, show that $y_n = (-1)^{n-2}(n-2)!\left[\frac{x-n}{(x-1)^n} - \frac{x+n}{(x+1)^n}\right]$.*

Sol. $$y = x[\log(x-1) - \log(x+1)] \qquad \text{...(1)}$$

Differentiating y w.r.t. x, we get

$$y_1 = x\left(\frac{1}{x-1} - \frac{1}{x+1}\right) + \log(x-1) - \log(x+1)$$

$$= \frac{1}{x-1} + \frac{1}{x+1} + \log(x-1) - \log(x+1) \qquad \text{...(2)}$$

Again, differentiating (2), $(n-1)$ times w.r.t. x, we get

$$y_n = \frac{(-1)^{n-1}(n-1)!}{(x-1)^n} + \frac{(-1)^{n-1}(n-1)!}{(x+1)^n} + \frac{(-1)^{n-2}(n-2)!}{(x-1)^{n-1}} - \frac{(-1)^{n-2}(n-2)!}{(x+1)^{n-1}}$$

$$= (-1)^{n-2}(n-2)!\left[\frac{(-1)(n-1)}{(x-1)^n} + \frac{(-1)(n-1)}{(x+1)^n} + \frac{x-1}{(x-1)^n} - \frac{x+1}{(x+1)^n}\right]$$

$$= (-1)^{n-2}(n-2)!\left[\frac{x-n}{(x-1)^n} - \frac{x+n}{(x+1)^n}\right].$$

TEST YOUR KNOWLEDGE

Find the nth derivative of the following:

1. (i) $\dfrac{1}{(x+2)(x-1)}$ (ii) $\dfrac{x}{1+3x+2x^2}$ (iii) $\dfrac{1}{x^2+a^2}$. *(U.K.T.U. 2011)*

2. (i) $\dfrac{x^2}{(x+2)(2x+3)}$ *(U.K.T.U. 2011)* (ii) $\dfrac{5x+12}{x^2+5x+6}$.

3. If $y = x \log(1+x)$, prove that $y_n = \dfrac{(-1)^{n-2}(n-2)!\,(x+n)}{(1+x)^n}$.

4. Find the nth derivative of $\tan^{-1}\left(\dfrac{1+x}{1-x}\right)$.

5. If $y = \dfrac{x}{x^2+a^2}$, prove that $y_n = \dfrac{(-1)^n n!}{a^{n+1}} \sin^{n+1}\phi \cos(n+1)\phi$, where $\phi = \tan^{-1}\left(\dfrac{a}{x}\right)$.

6. If $y = (x^2-1)^m$, prove that $y_{2m} = 2m!$

Answers

1. (i) $\dfrac{(-1)^n . n!}{3}\left[\dfrac{1}{(x-1)^{n+1}} - \dfrac{1}{(x+2)^{n+1}}\right]$ (ii) $(-1)^n n!\left[\dfrac{1}{(x+1)^{n+1}} - \dfrac{2^n}{(2x+1)^{n+1}}\right]$

 (iii) $\dfrac{(-1)^n n!}{2ai}\left[\dfrac{1}{(x-ai)^{n+1}} - \dfrac{1}{(x+ai)^{n+1}}\right]$

2. (i) $\dfrac{(-1)^n n!}{2}\left[\dfrac{9.2^n}{(2x+3)^{n+1}} - \dfrac{8}{(x+2)^{n+1}}\right]$ (ii) $(-1)^n n!\left[\dfrac{2}{(x+2)^{n+1}} + \dfrac{3}{(x+3)^{n+1}}\right]$

4. $(-1)^{n-1}(n-1)!\sin^n\theta \sin n\theta$, where $\theta = \tan^{-1}\left(\dfrac{1}{x}\right)$.

2.4. LEIBNITZ THEOREM

This theorem helps us to find the nth differential co-efficient of the product of two functions in terms of the successive derivatives of the functions.

Statement. *If u and v are two functions of x, having derivatives of the nth order, then*

$$\frac{d^n}{dx^n}(uv) = u_n v + {}^nC_1 u_{n-1} v_1 + {}^nC_2 u_{n-2} v_2 + \ldots + {}^nC_r u_{n-r} v_r + \ldots + {}^nC_n uv_n,$$

where suffixes of u and v denote differentiations w.r.t. x.

Proof. We shall prove the theorem by Mathematical Induction.

Step I. Let $y = uv$

By actual differentiation, we have

$$y_1 = u_1 v + u v_1$$

and
$$y_2 = u_2 v + u_1 v_1 + u_1 v_1 + u v_2$$
$$= u_2 v + 2u_1 v_1 + u v_2 = u_2 v + {}^2C_1\, u_1 v_1 + {}^2C_2\, u v_2.$$

Thus, the theorem is true for $n = 1, 2$.

Step II. Let us assume that the theorem is true for a particular value of n, say m, so that, we have

$$y_m = u_m\, v + {}^mC_1\, u_{m-1}\, v_1 + {}^mC_2\, u_{m-2}\, v_2 + \ldots$$
$$+ {}^mC_{r-1}\, u_{m-r+1}\, v_{r-1} + {}^mC_r u_{m-r}\, v_r + \ldots + {}^mC_m\, u v_m.$$

Step III. Differentiating both sides, we have

$$y_{m+1} = u_{m+1}\, v + u_m\, v_1 + {}^mC_1\, u_m\, v_1 + {}^mC_1\, u_{m-1}\, v_2 + {}^mC_2\, u_{m-1}\, v_2$$
$$+ {}^mC_2\, u_{m-2}\, v_3 + \ldots + {}^mC_{r-1}\, u_{m-r+2}\, v_{r-1} + {}^mC_{r-1}\, u_{m-r+1}\, v_r$$
$$+ {}^mC_r\, u_{m-r+1}\, v_r + {}^mC_r\, u_{m-r}\, v_{r+1} + \ldots$$
$$+ {}^mC_m\, u_1 v_m + {}^mC_m\, u v_{m+1}.$$
$$= u_{m+1}\, v + ({}^mC_1 + 1) v_m\, v_1 + ({}^mC_2 + {}^mC_1)\, u_{m-1}\, v_2 + \ldots$$
$$+ ({}^mC_r + {}^mC_{r-1})\, u_{m-r+1}\, v_r + \ldots + {}^mC_m\, u v_{m+1}.$$

But, we know that ${}^mC_{r-1} + {}^mC_r = {}^{m+1}C_r$

Putting $r = 1, 2, 3, \ldots$

$${}^mC_0 + {}^mC_1 = {}^{m+1}C_1 \text{ or } 1 + {}^mC_1 = {}^{m+1}C_1$$
$${}^mC_1 + {}^mC_2 = {}^{m+1}C_2$$
$$\ldots\ldots\ldots\ldots\ldots\ldots$$

and
$${}^mC_m = 1 = {}^{m+1}C_{m+1}.$$

$\therefore$
$$y_{m+1} = u_{m+1}\, v + {}^{m+1}C_1\, u_m\, v_1 + {}^{m+1}C_2\, u_{m-1}\, v_2 + \ldots$$
$$+ {}^{m+1}C_r\, u_{m-r+1}\, v_r + \ldots + {}^{m+1}C_{m+1}\, u v_{m+1}$$

which is of exactly the same form as the given formula with $n = m + 1$.

$\therefore$ it is clear that if the theorem is true for $n = m$, then it is also true for the next higher value $n = m + 1$.

But in Step I, we have proved that the theorem is true for $n = 2$.

$\therefore$ it must be true for next higher values $n = 2 + 1 = 3$.

Again $\because$ the theorem is true for $n = 3$.

$\therefore$ it must be true for next higher value $n = 3 + 1$, *i.e.*, 4 and so on.

Hence the theorem is true for any positive integer n.

ILLUSTRATIVE EXAMPLES

Example 1. *Find the nth derivative of $x^2 \sin x$ at $x = 0$.*

Sol. Let $u = \sin x$ and $v = x^2$

$\therefore$ $u_n = \sin\left(x + n\dfrac{\pi}{2}\right)$ $\qquad v_1 = 2x$

$u_{n-1} = \sin\left[x + (n-1)\dfrac{\pi}{2}\right]$ $\qquad v_2 = 2$

$u_{n-2} = \sin\left[x + (n-2)\dfrac{\pi}{2}\right]$ $\qquad v_3 = 0$

Now by Leibnitz Theorem, we have

$$\frac{d^n}{dx^n}(uv) = u_n v + {}^nC_1 u_{n-1} v_1 + {}^nC_2 u_{n-2} v_2$$

or
$$\frac{d^n}{dx^n}(x^2 \sin x) = \sin\left(x + n\frac{\pi}{2}\right) . x^2 + {}^nC_1 \sin\left[x + (n-1)\frac{\pi}{2}\right] 2x + {}^nC_2 \sin\left[x + (n-2)\frac{\pi}{2}\right] 2$$

$$= x^2 \sin\left(x + \frac{n\pi}{2}\right) + 2nx \sin\left[x + (n-1)\frac{\pi}{2}\right] + n(n-1) \sin\left[x + (n-2)\frac{\pi}{2}\right].$$

At $x = 0$, $\quad y_n = n(n-1) \sin (n-2)\dfrac{\pi}{2} = (n - n^2) \sin \dfrac{n\pi}{2}$

Note. Generally, we take x^n as v.

Example 2. *Find the nth derivative of $e^x \log x$.*

Sol. Let $u = e^x$ and $v = \log x$

$\therefore$ $u_n = e^x$ $\qquad v_1 = \dfrac{1}{x}$

$u_{n-1} = u_{n-2} = u_2 = u_1 = e^x$ $\qquad v_2 = \dfrac{-1}{x^2}$

$v_3 = \dfrac{2}{x^3}$

....

$v_n = \dfrac{(-1)^{n-1}(n-1)!}{x^n}$

Now by Leibnitz Theorem, we have

$$\frac{d^n}{dx^n}(e^x \log x) = e^x \log x + {}^nC_1 e^x \left(\frac{1}{x}\right) + {}^nC_2 e^x \left(\frac{-1}{x^2}\right) + \ldots + {}^nC_n . e^x . \frac{(-1)^{n-1}(n-1)!}{x^n}$$

$$= e^x \left[\log x + \frac{n}{x} - \frac{n(n-1)}{2} . \frac{1}{x^2} + \ldots + \frac{(-1)^{n-1}(n-1)!}{x^n}\right].$$

Example 3. *If* $y = x^2 e^x$, *show that*

$$\frac{d^n y}{dx^n} = \frac{1}{2} n(n-1) \frac{d^2 y}{dx^2} - n(n-2) \frac{dy}{dx} + \frac{1}{2} (n-1)(n-2) y.$$

Sol. $y = e^x \cdot x^2$

Differentiating n times by Leibnitz Theorem, we get

$$y_n = u_n v + {}^nC_1 u_{n-1} v_1 + {}^nC_2 u_{n-2} v_2$$

or $$y_n = e^x \cdot x^2 + {}^nC_1\, e^x \cdot 2x + {}^nC_2\, e^x \cdot 2$$

$$= e^x[x^2 + 2nx + n(n-1)] \qquad ...(1)$$

Put $n = 1, 2$ in equation (1), successively

$$y_1 = e^x[x^2 + 2x] \qquad ...(2)$$

$$y_2 = e^x[x^2 + 4x + 2] \qquad ...(3)$$

Now, $$\text{RHS} = \frac{1}{2} n(n-1) \frac{d^2 y}{dx^2} - n(n-2) \frac{dy}{dx} + \frac{1}{2} (n-1)(n-2)y$$

$$= \frac{1}{2} n(n-1) \cdot e^x [x^2 + 4x + 2] - n(n-2) \cdot e^x (x^2 + 2x) + \frac{1}{2} (n-1)(n-2)x^2 e^x$$

$$= e^x \left[x^2 \left\{ \frac{n(n-1)}{2} - n(n+2) + \frac{(n-1)(n-2)}{2} \right\} + x \{2n(n-1) - 2n(n-2)\} + n(n-1) \right]$$

$$= e^x [x^2 + 2nx + n(n-1)] = y_n = \text{LHS} \qquad | \text{ From (1)}$$

Example 4. *Differentiate n times the differential equations:*

(*i*) $(1 - x^2) \frac{d^2 y}{dx^2} - x \frac{dy}{dx} + a^2 y = 0$ (*ii*) $x^2 y_2 + xy_1 + y = 0.$

Sol. (*i*) Differentiating each term of the given equation n times by Leibnitz theorem, we get

$$[y_{n+2} (1 - x^2) + {}^nC_1\, y_{n+1} (-2x) + {}^nC_2\, y_n (-2)] - [y_{n+1} (x) + {}^nC_1\, y_n (1)] + a^2 y_n = 0$$

$$\Rightarrow \quad (1 - x^2)\, y_{n+2} + (-2nx - x) y_{n+1} + \{-n(n-1) - n + a^2\}\, y_n = 0$$

$$\Rightarrow \quad (1 - x^2)\, y_{n+2} - (2n+1)\, xy_{n+1} - (n^2 - a^2)\, y_n = 0$$

(*ii*) Differentiating each term of the given equation n times by Leibnitz theorem, we get

$$[y_{n+2} (x^2) + {}^nC_1\, y_{n+1} (2x) + {}^nC_2\, y_n (2)] + [y_{n+1}(x) + {}^nC_1\, y_n(1)] + y_n = 0$$

$$\Rightarrow \quad x^2 y_{n+2} + (2nx + x)\, y_{n+1} + \{n(n-1) + n + 1\}\, y_n = 0$$

$$\Rightarrow \quad x^2 y_{n+2} + (2n+1)\, xy_{n+1} + (n^2 + 1)\, y_n = 0.$$

Example 5. *If* $y = \sin (m \sin^{-1} x)$, *then prove that*

$$(1 - x^2) y_2 - xy_1 + m^2 y = 0$$

and $$(1 - x^2) y_{n+2} - (2n+1)\, xy_{n+1} - (n^2 - m^2)\, y_n = 0.$$

Sol. $$y = \sin (m \sin^{-1} x) \qquad ...(1)$$

$$\therefore \quad y_1 = \cos (m \sin^{-1} x) \cdot \frac{m}{\sqrt{1 - x^2}}$$

Cross-multiplying, $\sqrt{1-x^2}\, y_1 = m \cos(m \sin^{-1} x)$.

Squaring both sides,

$$(1-x^2)y_1^2 = m^2 \cos^2(m \sin^{-1} x)$$

$$= m^2[1 - \sin^2(m \sin^{-1} x)] \qquad | \because \cos^2\theta = 1 - \sin^2\theta$$

$$= m^2(1-y^2) \qquad \text{[By (1)]}$$

Again differentiating both sides w.r.t. x,

$$(1-x^2)\, 2y_1 y_2 - 2xy_1^2 = -2m^2 yy_1$$

Dividing every term by $2y_1$,

$$(1-x^2)y_2 - xy_1 = -m^2 y$$

or $$(1-x^2)y_2 - xy_1 + m^2 y = 0.$$

Now differentiating every term n times by Leibnitz Theorem, we have

$$[y_{n+2}(1-x^2) + {}^nC_1\, y_{n+1}(-2x) + {}^nC_2\, y_n(-2)] - [y_{n+1} \cdot x + {}^nC_1\, y_n] + m^2 y_n = 0$$

or $$(1-x^2)y_{n+2} - 2nxy_{n+1} - n(n-1)y_n - xy_{n+1} - ny_n + m^2 y_n = 0$$

$\Rightarrow$ $$(1-x^2)y_{n+2} - (2n+1)\, xy_{n+1} - (n^2 - m^2)\, y_n = 0$$

Example 6. *If $y = a \cos(\log x) + b \sin(\log x)$, prove that*

(i) $x^2y_2 + xy_1 + y = 0$ and

(ii) $x^2y_{n+2} + (2n+1)xy_{n+1} + (n^2+1)y_n = 0$.

Sol. (*i*) $$y = a \cos(\log x) + b \sin(\log x) \qquad ...(1)$$

$\therefore$ $$y_1 = -a \sin(\log x) \cdot \frac{1}{x} + b \cos(\log x) \cdot \frac{1}{x}$$

$\Rightarrow$ $$xy_1 = -a \sin(\log x) + b \cos(\log x)$$

Again, $$xy_2 + y_1 = -a \cos(\log x) \cdot \frac{1}{x} - b \sin(\log x) \cdot \frac{1}{x}$$

$\Rightarrow$ $$x^2y_2 + xy_1 = -y$$

$\Rightarrow$ $$x^2y_2 + xy_1 + y = 0 \qquad ...(2)$$

(*ii*) Differentiating equation (2) n times by Leibnitz Theorem, we get

$$[y_{n+2} \cdot (x^2) + {}^nC_1\, y_{n+1} \cdot (2x) + {}^nC_2\, y_n \cdot (2)] + [y_{n+1} \cdot (x) + {}^nC_1 \cdot y_n (1)] + y_n = 0$$

$\Rightarrow$ $$x^2\, y_{n+2} + y_{n+1}(2nx + x) + y_n[n(n-1) + n + 1] = 0$$

$\Rightarrow$ $$x^2 y_{n+2} + (2n+1)\, xy_{n+1} + (n^2+1)\, y_n = 0$$

Example 7. *If $y = (x^2 - 1)^n$, prove that $(x^2-1)\, y_{n+2} + 2xy_{n+1} - n(n+1)\, y_n = 0$*

[U.K.T.U. 2010]

Hence, if $P_n = \dfrac{d^n}{dx^n}(x^2-1)^n$, show that $\dfrac{d}{dx}\left\{(1-x^2)\dfrac{dP_n}{dx}\right\} + n(n+1)\, P_n = 0$.

Sol. $$y = (x^2-1)^n \qquad ...(1)$$

$\therefore$ $$y_1 = n(x^2-1)^{n-1} \cdot 2x$$

$\Rightarrow$ $$(x^2-1)\, y_1 = 2nxy \qquad ...(2) \quad | \text{ Using (1)}$$

Differentiating again, we get

$$(x^2 - 1) y_2 + 2xy_1 = 2n (xy_1 + y)$$

$\Rightarrow$ $(x^2 - 1) y_2 + 2(1 - n) xy_1 - 2ny = 0$

Differentiating n times by Leibnitz theorem, we get

$$[y_{n+2} (x^2 - 1) + {}^nC_1 y_{n+1}(2x) + {}^nC_2 y_n (2)] + 2(1 - n) [y_{n+1} (x) + {}^nC_1 y_n(1)] - 2ny_n = 0$$

$$\Rightarrow \quad (x^2 - 1) y_{n+2} + (2nx + 2x - 2nx) y_{n+1} + (n^2 - n + 2n - 2n^2 - 2n) y_n = 0$$

$$\Rightarrow \quad (x^2 - 1) y_{n+2} + 2xy_{n+1} - n(n + 1)y_n = 0 \quad \text{...(3)}$$

Again, $$P_n = \frac{d^n}{dx^n} (x^2 - 1)^n = y_n \quad \text{...(4)}$$

$$\therefore \quad \frac{d}{dx}\left\{(1-x^2)\frac{dP_n}{dx}\right\} = \frac{d}{dx} \{ (1 - x^2) y_{n+1}\}$$

$$= (1 - x^2) y_{n+2} - 2x y_{n+1} = - \{(x^2 - 1) y_{n+2} + 2xy_{n+1}\}$$

$$= - n(n + 1) y_n \quad \text{| Using (3)}$$

$$= - n(n + 1) P_n \quad \text{| Using (4)}$$

Hence, $$\frac{d}{dx}\left\{(1-x^2)\frac{dP_n}{dx}\right\} + n (n + 1) P_n = 0.$$

Example 8. *If $y^{1/m} + y^{-1/m} = 2x$, prove that*

$$(x^2 - 1) y_{n+2} + (2n + 1) xy_{n+1} + (n^2 - m^2)y_n = 0.$$

[U.P.T.U. 2015; U.K.T.U. 2011]

Sol. $y^{1/m} + y^{-1/m} = 2x$

Put $y^{1/m} = z$

$$\therefore \quad z + \frac{1}{z} = 2x \quad \text{or} \quad z^2 - 2xz + 1 = 0$$

$$\therefore \quad z = \frac{2x \pm 2\sqrt{x^2 - 1}}{2} = x \pm \sqrt{x^2 - 1}$$

i.e., $$y^{1/m} = x \pm \sqrt{x^2 - 1} \quad \text{or} \quad y = (x \pm \sqrt{x^2 - 1})^m$$

If $y = (x + \sqrt{x^2 - 1})^m$, then

$$y_1 = m(x + \sqrt{x^2 - 1})^{m-1} \left(1 + \frac{1}{2} . \frac{1}{\sqrt{x^2 - 1}} . 2x\right)$$

$$= m(x + \sqrt{x^2 - 1})^{m-1} \left(\frac{x + \sqrt{x^2 - 1}}{\sqrt{x^2 - 1}}\right) = \frac{m(x + \sqrt{x^2 - 1})^m}{\sqrt{x^2 - 1}} = \frac{my}{\sqrt{x^2 - 1}}$$

And, similarly, if $y = (x - \sqrt{x^2 - 1})^m$,

$$y_1 = \frac{- my}{\sqrt{x^2 - 1}}.$$

In either case, squaring, we get

$$y_1^2 = \frac{m^2 y^2}{x^2 - 1} \quad \text{or} \quad (x^2 - 1)\, y_1^2 = m^2 y^2.$$

Differentiating again,

$$(x^2 - 1) \,.\, 2y_1 y_2 + y_1^2 \,.\, 2x = m^2 \,.\, 2yy_1$$

Dividing both sides by $2y_1$, we have

$$(x^2 - 1)y_2 + xy_1 - m^2 y = 0$$

Differentiating n times by Leibnitz theorem, we get

$$[y_{n+2}(x^2 - 1) + n \,.\, y_{n+1} \,.\, 2x + \frac{n(n-1)}{2!} \,.\, y_n \,.\, 2] + [y_{n+1} \,.\, x + ny_n \,.\, 1] - m^2 y_n = 0$$

or $$(x^2 - 1)\, y_{n+2} + (2n + 1)\, xy_{n+1} + (n^2 - n + n - m^2) y_n = 0$$

or $$(x^2 - 1) y_{n+2} + (2n + 1)\, xy_{n+1} + (n^2 - m^2) y_n = 0.$$

Example 9. (*i*) *If* $\cos^{-1}\left(\frac{y}{b}\right) = \log\left(\frac{x}{m}\right)^m$, *prove that*

$$x^2 y_{n+2} + (2n + 1)\, xy_{n+1} + (n^2 + m^2)\, y_n = 0.$$

(*ii*) *If* $x = \cosh\left(\frac{1}{m}\log y\right)$, *prove that* $(x^2 - 1)\, y_{n+2} + (2n + 1)\, xy_{n+1} + (n^2 - m^2)\, y_n = 0.$

Sol. (*i*) $$\cos^{-1}\left(\frac{y}{b}\right) = m \log\left(\frac{x}{m}\right)$$

$$\Rightarrow \quad y = b \cos\left\{m \log\left(\frac{x}{m}\right)\right\} \quad \text{...(1)}$$

$$\therefore \quad y_1 = -b \sin\left\{m \log\left(\frac{x}{m}\right)\right\} \,.\, m \frac{1}{(x/m)} \,.\, \frac{1}{m}$$

$$\Rightarrow \quad x\, y_1 = -bm \sin\left\{m \log\left(\frac{x}{m}\right)\right\}$$

Again differentiating,

$$xy_2 + y_1 = -bm \cos\left\{m \log\left(\frac{x}{m}\right)\right\} \,.\, m \,.\, \frac{1}{(x/m)} \,.\, \frac{1}{m}$$

$$\therefore \quad x^2 y_2 + xy_1 = -m^2 b \cos\left\{m \log\left(\frac{x}{m}\right)\right\} = -m^2 y$$

$$\Rightarrow \quad x^2 y_2 + xy_1 + m^2 y = 0 \quad \text{...(2)}$$

Differentiating (2) n times by Leibnitz theorem,

$$[y_{n+2}(x^2) + {}^nC_1\, y_{n+1}\,(2x) + {}^nC_2 \,.\, y_n\,(2)] + [y_{n+1}(x) + {}^nC_1\, y_n\,(1)] + m^2 y_n = 0$$

$$\Rightarrow \quad x^2 y_{n+2} + (2n + 1)\, xy_{n+1} + (n^2 + m^2)\, y_n = 0$$

(*ii*) $$x = \cosh\left(\frac{1}{m}\log y\right)$$

$$\Rightarrow \quad \cosh^{-1} x = \frac{1}{m}\log y$$

$\Rightarrow \qquad y = e^{m \cosh^{-1} x}$ ...(1)

$\therefore \qquad y_1 = e^{m \cosh^{-1} x} . m . \dfrac{1}{\sqrt{x^2 - 1}}$

$\Rightarrow \qquad (\sqrt{x^2 - 1})\, y_1 = me^{m \cosh^{-1} x} = my$ | Using (1)

Squaring both sides,

$$(x^2 - 1)\, y_1^2 = m^2 y^2$$

Differentiating, we get

$$(x^2 - 1)\, 2y_1 y_2 + y_1^2\, (2x) = 2m^2 y y_1$$

$\Rightarrow \qquad (x^2 - 1)y_2 + xy_1 - m^2 y = 0$ ...(2)

Differentiating (2) n times by Leibnitz theorem, we get

$$[y_{n+2}\,(x^2 - 1) + {}^nC_1 y_{n+1}(2x) + {}^nC_2\, y_n\, (2)] + [y_{n+1}\,(x) + {}^nC_1\, y_n(1)] - m^2 y_n = 0$$

$\Rightarrow \qquad (x^2 - 1)\, y_{n+2} + (2n + 1)\, xy_{n+1} + (n^2 - m^2) y_n = 0.$

Example 10. *(i) If* $y = (1 - x)^{-\alpha} e^{-\alpha x}$, *prove that* $(1 - x)\, y_{n+1} - (n + \alpha x) y_n - n\alpha y_{n-1} = 0.$

(U.K.T.U. 2011)

(ii) If $y = \log\,(x + \sqrt{1 + x^2}\,)$, *prove that* $(1 + x^2)\, y_{n+2} + (2n + 1)\, xy_{n+1} + n^2 y_n = 0.$

Sol. *(i)* $\qquad y = (1 - x)^{-\alpha}\, e^{-\alpha x}$ ...(1)

$$y = (1 - x)^{-\alpha}\, (-\alpha e^{-\alpha x}) + e^{-\alpha x}\, (-\alpha)\, (1 - x)^{-\alpha - 1}\, (-1)$$

$$= e^{-\alpha x}\, (1 - x)^{-\alpha} \left(-\alpha + \frac{\alpha}{1 - x}\right)$$

$\Rightarrow \qquad (1 - x)\, y_1 = \alpha xy$ ...(2)

Differentiating (2) n times by Leibnitz theorem, we get

$$y_{n+1}\,(1 - x) + {}^nC_1\, y_n\, (-1) = \alpha\{y_n(x) + {}^nC_1\, y_{n-1}(1)\}$$

$\therefore \qquad (1 - x)y_{n+1} + (-n - ax)y_n - n\alpha y_{n-1} = 0$

$\Rightarrow \qquad (1 - x)\, y_{n+1} - (n + \alpha x)y_n - n\alpha\, y_{n-1} = 0$

(ii) $\qquad y = \log\,(x + \sqrt{1 + x^2}\,)$ ...(1)

$$\therefore \qquad y_1 = \frac{1}{x + \sqrt{1 + x^2}} \left(1 + \frac{x}{\sqrt{1 + x^2}}\right) = \frac{1}{\sqrt{1 + x^2}}$$

$\Rightarrow \qquad \sqrt{1 + x^2}\;\; y_1 = 1$

or $\qquad (1 + x^2)\, y_1^2 = 1$ ...(2)

Differentiating again, we get

$$(1 + x^2)\, 2y_1 y_2 + y_1^2\, (2x) = 0$$

$\Rightarrow \qquad (1 + x^2)y_2 + xy_1 = 0$ ...(3)

Differentiating (3) n times by Leibnitz theorem, we get

$$\{y_{n+2}\,(1 + x^2) + {}^nC_1 y_{n+1}(2x) + {}^nC_2\, y_n\,(2)\} + \{y_{n+1}\,(x) + {}^nC_1\, y_n\,(1)\} = 0$$

$\Rightarrow \qquad (1 + x^2)\, y_{n+2} + (2n + 1)\, xy_{n+1} + n^2 y_n = 0.$

Example 11. *(i) If* $x = \tan(\log y)$, *prove that*

$$(1 + x^2)\, y_{n+1} + (2nx - 1)\, y_n + n(n-1)\, y_{n-1} = 0$$

and $$(1 + x^2)y_{n+2} + [2(n+1)\,x - 1]y_{n+1} + n(n+1)\, y_n = 0 \qquad \text{(U.P.T.U. 2014)}$$

(ii) If $y = \sin \log_e (x^2 + 2x + 1)$, *prove that*

$$(1 + x)^2\, y_{n+2} + (2n+1)(1+x)\, y_{n+1} + (n^2 + 4)\, y_n = 0. \qquad \text{(A.K.T.U. 2019)}$$

Sol. (*i*)
$$x = \tan(\log y)$$

$$\Rightarrow \quad y = e^{\tan^{-1} x} \qquad \text{...(1)}$$

$$\therefore \quad y_1 = e^{\tan^{-1} x} \cdot \frac{1}{1+x^2}$$

$$\Rightarrow \quad (1 + x^2)\, y_1 = y \qquad \text{...(2) | Using (1)}$$

Differentiating (2) n times by Leibnitz theorem, we get

$$y_{n+1}(1 + x^2) + {}^nC_1\, y_n\,(2x) + {}^nC_2\, y_{n-1}\,(2) = y_n$$

$$\Rightarrow \quad (1 + x^2)\, y_{n+1} + (2nx - 1)\, y_n + n(n-1)\, y_{n-1} = 0 \qquad \text{...(3)}$$

Differentiating eqn. (3) again w.r.t. x, we get

$$y_{n+2}(1 + x^2) + 2xy_{n+1} + (2nx - 1)\, y_{n+1} + 2ny_n + n(n-1)\, y_n = 0$$

$$\Rightarrow \quad (1 + x^2)y_{n+2} + [2(n+1)\,x - 1]\, y_{n+1} + n(n+1)\, y_n = 0 \qquad \text{...(4)}$$

Equations (3) and (4) give the required results.

(*ii*)
$$y = \sin \log_e (x^2 + 2x + 1) \qquad \text{...(1)}$$

$$\therefore \quad y_1 = \cos \log_e (x^2 + 2x + 1) \cdot \frac{1}{x^2 + 2x + 1}\,(2x + 2)$$

$$= \cos \log_e (x^2 + 2x + 1) \cdot \frac{2}{x+1}$$

$$\Rightarrow \quad (x + 1)\, y_1 = 2 \cos \log_e (x^2 + 2x + 1)$$

Squaring both sides,

$$(x + 1)^2\, y_1^2 = 4 \cos^2 \log_e (x^2 + 2x + 1) = 4(1 - y^2) \qquad \text{...(2)}$$

Differentiating (2) again w.r.t. x, we get

$$(x + 1)^2 \cdot 2y_1y_2 + y_1^2 \cdot 2(x + 1) = 4(-2yy_1)$$

$$\Rightarrow \quad (x + 1)^2\, y_2 + (x + 1)y_1 + 4y = 0 \qquad \text{...(3)}$$

Differentiating (3) n times by Leibnitz theorem, we get

$$[y_{n+2}(x + 1)^2 + {}^nC_1\, y_{n+1} \cdot 2(x+1) + {}^nC_2\, y_n \cdot 2] + [y_{n+1}(x + 1) + {}^nC_1\, y_n\,(1)] + 4y_n = 0$$

$$\Rightarrow \quad (x + 1)^2\, y_{n+2} + (2nx + 2n + x + 1)\, y_{n+1} + (n^2 - n + n + 4)\, y_n = 0$$

$$\Rightarrow \quad (1 + x)^2\, y_{n+2} + (2n+1)(1+x)\, y_{n+1} + (n^2 + 4)\, y_n = 0.$$

Example 12. *If* $y = x^n \log x$, *prove that*

(i) $y_{n+1} = \dfrac{n!}{x}$ *(ii)* $x^2y_{p+2} + (2p - 2n + 1)\, xy_{p+1} + (p - n)^2\, y_p = 0.$

Sol. (*i*)
$$y = x^n \log x \qquad \text{...(1)}$$

$$\therefore \quad y_1 = x^n \left(\frac{1}{x}\right) + nx^{n-1} \log x$$

$$\Rightarrow \quad xy_1 = x^n + nx^n \log x = x^n + ny \qquad \text{...(2) | Using (1)}$$

Differentiating (2) n times, we get

$$y_{n+1} \cdot (x) + {}^nC_1\, y_n\,(1) = n\,! + ny_n$$

$$\Rightarrow \quad y_{n+1}\,(x) = n\,!$$

or $$y_{n+1} = \frac{n\,!}{x}$$

(ii) $$y_1 = x^n \cdot \frac{1}{x} + nx^{n-1} \log x \qquad \text{| Diff. } y \text{ w.r.t. } x$$

$$\Rightarrow \quad xy_1 = x^n + nx^n \log x \qquad \text{| Multiplying by } x$$

$$\Rightarrow \quad xy_1 = x^n + ny \qquad \ldots(1)$$

Differentiating equation (1) w.r.t. x, we get

$$xy_2 + y_1 = nx^{n-1} + ny_1$$

$$xy_2 + (1-n)\,y_1 = nx^{n-1}$$

$$x^2y_2 + (1-n)\,xy_1 = nx^n = n\,(xy_1 - ny) \qquad \text{| Using (1)}$$

$$\Rightarrow \quad x^2y_2 + (1-2n)\,xy_1 + n^2y = 0 \qquad \ldots(2)$$

Differentiating equation (2), p times by Leibnitz theorem,

$$[y_{p+2}\,(x^2) + {}^pC_1\, y_{p+1}\,(2x) + {}^pC_2 y_p (2)] + (1-2n)\,[y_{p+1}\,(x) + {}^pC_1\, y_p(1)] + n^2y_p = 0$$

$$\Rightarrow \quad x^2 y_{p+2} + y_{p+1}\{2px + (1-2n)\,x\} + y_p\{p\,(p-1) + (1-2n)\,p + n^2\} = 0$$

$$\Rightarrow \quad x^2 y_{p+2} + (2p - 2n + 1)\,xy_{p+1} + (p-n)^2\, y_p = 0.$$

TEST YOUR KNOWLEDGE

Apply Leibnitz Theorem to find y_n in the following cases (1–2):

1. (i) $x^3 e^{ax}$ (ii) $x^{n-1} \log x$. *(A.K.T.U. 2010, 2012, 2018)*
2. (i) $x^3 \cos x$ (ii) $x^2 e^x \cos x$.
3. Find y_3 when $y = \sqrt{1+x^2}\,.\,\sin x$ by Leibnitz theorem.
4. If $y = x^2 \sin x$, prove that $\dfrac{d^n y}{dx^n} = (x^2 - n^2 + n) \sin\left(x + \dfrac{n\pi}{2}\right) - 2nx \cos\left(x + \dfrac{n\pi}{2}\right)$. *(U.P.T.U. 2014)*
5. If $y = \sin^{-1} x$, prove that $(1 - x^2)y_{n+2} - (2n+1)\,xy_{n+1} - n^2y_n = 0$.
6. (a) If $y = e^{m \sin^{-1} x}$, prove that:
   (i) $(1 - x^2)\, y_2 - xy_1 = m^2y$ *(A.K.T.U. 2016)* (ii) $(1 - x^2)\, y_{n+2} - (2n+1)\, xy_{n+1} - (n^2 + m^2)y_n = 0$.
   (b) If $x = \sin\left(\dfrac{\log y}{a}\right)$, prove that $(1 - x^2)\, y_{n+2} - (2n+1)\, xy_{n+1} - (n^2 + a^2)\, y_n = 0$. *(U.K.T.U. 2012)*
7. If $y = \dfrac{\sin^{-1} x}{\sqrt{1-x^2}}$, prove that $(1 - x^2)\, y_{n+2} - (2n+3)\, x\, y_{n+1} - (n+1)^2\, y_n = 0$.
8. If $y = e^{\tan^{-1} x}$, prove that
   (i) $(1 + x^2)y_2 + (2x - 1)\, y_1 = 0$.
   (ii) $(1 + x^2)\, y_{n+2} + [(2n+2)x - 1]y_{n+1} + n(n+1)y_n = 0$. *(A.K.T.U. 2018, 2021)*
9. If $y = \cos(\log x)$, prove that $x^2y_{n+2} + (2n+1)xy_{n+1} + (n^2+1)y_n = 0$.
10. If $y = x \cos(\log x)$, prove that $x^2\, y_{n+2} + (2n-1)\, xy_{n+1} + (n^2 - 2n + 2)y_n = 0$.

11. If $y = (\sin^{-1} x)^2$, prove that

(i) $(1 - x^2)\, y_2 - xy_1 - 2 = 0$ (ii) $(1 - x^2)y_{n+2} - (2n + 1)\, xy_{n+1} - n^2 y_n = 0.$

[A.K.T.U. 2014, 2018]

12. If $y = (\tan^{-1} x)^2$, prove that $(x^2 + 1)^2\, y_2 + 2x(1 + x^2)\, y_1 = 2$

13. If $\cos^{-1}\left(\dfrac{y}{b}\right) = \log\left(\dfrac{x}{n}\right)^n$, prove that $x^2\, y_{n+2} + (2n + 1)\, xy_{n+1} + 2n^2 y_n = 0$

14. If $y = \cos(m \sin^{-1} x)$, prove that $(1 - x^2)y_{n+2} - (2n + 1)\, xy_{n+1} - (n^2 - m^2)y_n = 0$

15. If $y = \sin(m \sinh^{-1} x)$, prove that $(1 + x^2)\, y_{n+2} + (2n + 1)\, xy_{n+1} + (n^2 + m^2)y_n = 0$

16. If $y = (\sinh^{-1} x)^2$, prove that $(1 + x^2)\, y_{n+2} + (2n + 1)xy_{n+1} + n^2 y_n = 0.$ (U.P.T.U. 2015)

17. If $y = \sqrt{1 - x^2}\, \sin^{-1} x$, prove that

(i) $y_3(1 - x^2) - 3y_2 x + 2 = 0$; (ii) $y_{n+3}(1 - x^2) - (2n + 3)\, xy_{n+2} - n(n + 2)\, y_{n+1} = 0.$

18. If $y = A(x + \sqrt{x^2 - 1})^n + B(x - \sqrt{x^2 - 1})^n$, prove that

(a) $(x^2 - 1)\, y_2 + xy_1 - n^2 y = 0$ (b) $(x^2 - 1)\, y_{n+2} + (2n + 1)\, xy_{n+1} = 0.$

19. If $y = \dfrac{\log x}{x}$, prove that $y_n = \dfrac{(-1)^n\, n!}{x^{n+1}}\left[\log x - 1 - \dfrac{1}{2} - \dfrac{1}{3} - \ldots - \dfrac{1}{n}\right].$

20. If $x + y = 1$, prove that $\dfrac{d^n}{dx^n}(x^n y^n) = n!\,[y^n - ({}^nC_1)^2 y^{n-1} . x + ({}^nC_2)^2\, y^{n-2} . x^2 - \ldots + (-1)^n x^n].$

21. If $x \sin\phi + y\cos\phi = a$ and $x\cos\phi - y\sin\phi = b$, show that $\dfrac{d^m x}{d\phi^m} . \dfrac{d^n y}{d\phi^n} - \dfrac{d^m y}{d\phi^m} . \dfrac{d^n x}{d\phi^n} =$ constant.

22. If $y = A(x + \sqrt{x^2 + a^2})^n + B(x + \sqrt{x^2 + a^2})^{-n}$, prove that

$(x^2 + a^2)y_{m+2} + (2m + 1)\, xy_{m+1} + (m^2 - n^2)\, y_m = 0.$

23. Evaluate $D^n[\log(x^2 - 2x\cos\alpha + 1)]$ hence or otherwise find $D^n\left[\tan^{-1}\left(\dfrac{x\sin\alpha}{1 - x\cos\alpha}\right)\right].$

24. Prove that $1 + \dfrac{n^2}{1^2} + \dfrac{n^2(n-1)^2}{1^2 \cdot 2^2} + \dfrac{n^2(n-1)^2(n-2)^2}{1^2 \cdot 2^2 \cdot 3^2} + \ldots = \dfrac{2n!}{(n!)^2}.$

[**Hint.** Find $D^n(x^n . x^n)$ and $D^n(x^{2n})$ then equate two values.]

25. If $y\sqrt{x^2 - 1} = \log(x + \sqrt{x^2 - 1})$, prove that $(x^2 - 1)\, y_{n+1} + (2n + 1)\, xy_n + n^2 y_{n-1} = 0.$

26. If $y = \left(\dfrac{1 + x}{1 - x}\right)^{1/2}$, prove that

$(1 - x^2)y_n - [2(n - 1)x + 1]y_{n-1} - (n - 1)(n - 2)y_{n-2} = 0.$ (M.T.U. 2012)

27. If $y = \tan^{-1}\left(\dfrac{a + x}{a - x}\right)$, prove that

$(a^2 + x^2)y_{n+2} + 2(n + 1)xy_{n+1} + n(n + 1)y_n = 0$ (G.B.T.U. 2012)

28. If $I_n = \dfrac{d^n}{dx^n}(x^n \log x)$, then prove that $I_n = nI_{n-1} + (n - 1)!$ (A.K.T.U. 2017)

and hence show that

$\dfrac{I_n}{n!} = \log x + 1 + \dfrac{1}{2} + \dfrac{1}{3} + \ldots + \dfrac{1}{n}.$

Answers

1. (i) $e^{ax} \cdot a^{n-3}\,[a^3x^3 + 3n \cdot a^2x^2 + 3n(n-1)\,ax + n(n-1)(n-2)]$ (ii) $\dfrac{n-1!}{x}$.

2. (i) $x^3 \cos\left(x + \dfrac{n}{2}\pi\right) + 3nx^2 \cos\left(x + \dfrac{n-1}{2}\pi\right) + 3n(n-1)\,x \cos\left(x + \dfrac{n-2}{2}\pi\right)$
$$+\, n(n-1)(n-2)\cos\left(x + \frac{n-3}{2}\pi\right)$$

(ii) $e^x\left[2^{n/2}\,x^2 \cos\left(x + \dfrac{n\pi}{4}\right) + 2^{\frac{n+1}{2}}\, nx \cos\left(x + \overline{n-1}\,\dfrac{\pi}{4}\right) + 2^{\frac{n-2}{2}}\, n(n-1)\cos\left(x + \overline{n-2}\,\dfrac{\pi}{4}\right)\right]$

3. $y_3 = -\cos x\sqrt{1+x^2} - \dfrac{3x \sin x}{\sqrt{1+x^2}} + \dfrac{3\cos x}{(1+x^2)^{3/2}} - \dfrac{3x \sin x}{(1+x^2)^{5/2}}$

23. $\dfrac{(-1)^{n-1}(n-1)!}{(x^2 - 2x\cos\alpha + 1)^{n/2}}\left[2\cos\left(n\tan^{-1}\dfrac{\sin\alpha}{x-\cos\alpha}\right)\right]$; $\dfrac{(-1)^{n-1}(n-1)!}{(x^2 - 2x\cos\alpha + 1)^{n/2}}\sin\left(n\tan^{-1}\dfrac{\sin\alpha}{x-\cos\alpha}\right)$.

2.5. DETERMINATION OF n^{TH} DERIVATIVE OF A FUNCTION AT $x = 0$

Sometimes it is required to find the nth derivative of a function for $x = 0$.

The **working rule** to find $(y_n)_{x=0}$ is being given below:

1. Put the given function equal to y.

2. Find $y_1 = \dfrac{dy}{dx}$.

Then (*i*) Take LCM (if possible).

(*ii*) Square both sides if square roots are there.

(*iii*) Try to get y in RHS (if possible).

3. Again differentiate both sides w.r.t. x to get an equation in y_2, y_1, y.

4. Differentiate both sides n times w.r.t. x by Leibnitz Theorem.

Leibnitz Theorem is

$$(uv)_n = u_n \cdot v + {}^nC_1 \cdot u_{n-1} \cdot v_1 + {}^nC_2 \cdot u_{n-2} \cdot v_2 + \ldots + {}^nC_n\, uv_n.$$

5. Put $x = 0$ in equations of steps 1, 2, 3, 4.

6. Put $n = 1, 2, 3, 4$ in last equation of step 5.

7. Discuss the two cases when n is even and when n is odd.

ILLUSTRATIVE EXAMPLES

Example 1. *If* $y = e^{m\cos^{-1}x}$, *show that*

$$(1-x^2)\,y_{n+2} - (2n+1)\,xy_{n+1} - (n^2+m^2)\,y_n = 0 \text{ and calculate } y_n(0).$$

(A.K.T.U. 2016, 2020)

Sol. Here, $$y = e^{m\cos^{-1}x} \quad \ldots(1)$$

$$\therefore \quad y_1 = e^{m\cos^{-1}x} \cdot \frac{-m}{\sqrt{1-x^2}} = -\frac{my}{\sqrt{1-x^2}} \quad \ldots(2)$$

Squaring both sides, we have

$$(1 - x^2)\, y_1^2 = m^2 y^2$$

Differentiating w.r.t. x, we have

$$(1 - x^2)\, 2y_1 y_2 - 2xy_1^2 = 2m^2 yy_1$$

Dividing by $2y_1$, we get

$$(1 - x^2)\, y_2 - xy_1 = m^2 y \qquad ...(3)$$

Differentiating n times by Leibnitz Theorem, we get

$$(1 - x^2)y_{n+2} - 2nxy_{n+1} - n(n-1)y_n - xy_{n+1} - ny_n = m^2 y_n$$

i.e.,

$$(1 - x^2)y_{n+2} - (2n + 1)xy_{n+1} - (n^2 + m^2)y_n = 0 \qquad ...(4)$$

By putting $x = 0$, in (1), (2), (3) and (4), we get

$$y(0) = e^{m\,.\,\pi/2}$$

$$y_1(0) = -\,m\,.\,e^{m\,.\,\pi/2}$$

$$y_2(0) = m^2\,.\,y(0) = m^2\,.\,e^{m\,.\,\pi/2}$$

$$y_{n+2}(0) = (n^2 + m^2)y_n(0) \qquad ...(5)$$

Putting $n = 1, 2, 3, 4, ...$ in (5), we have

$$y_3(0) = (1^2 + m^2)y_1(0) = -\,m(1^2 + m^2)e^{m\,.\,\pi/2}$$

$$y_4(0) = (2^2 + m^2)y_2(0) = m^2(2^2 + m^2)e^{m\,.\,\pi/2}$$

$$y_5(0) = (3^2 + m^2)y_3(0) = -\,m(1^2 + m^2)(3^2 + m^2)e^{m\,.\,\pi/2}$$

$$y_6(0) = (4^2 + m^2)y_4(0) = m^2(2^2 + m^2)(4^2 + m^2)e^{m\,.\,\pi/2} \text{ and so on.}$$

In general,

$$y_n(0) = \begin{cases} -\,m\,.\,e^{m\,.\,\pi/2}\,(1^2 + m^2)(3^2 + m^2)\ldots[(n-2)^2 + m^2], & \text{when } n \text{ is odd.} \\ m^2\,.\,e^{m\,.\,\pi/2}\,(2^2 + m^2)(4^2 + m^2)\ldots[(n-2)^2 + m^2], & \text{when } n \text{ is even.} \end{cases}$$

Example 2. *If $y = (\sin^{-1} x)^2$, prove that*

(i) $(1 - x^2)\dfrac{d^2y}{dx^2} - x\dfrac{dy}{dx} - 2 = 0$ *(ii)* $(1 - x^2)\, y_{n+2} - (2n + 1)\, xy_{n+1} - n^2 y_n = 0.$

Also, find the value of nth derivative of y for $x = 0$. [A.K.T.U. 2018, 2014]

Sol.

$$y = (\sin^{-1} x)^2$$

Differentiating both sides w.r.t. x,

$$y_1 = 2(\sin^{-1} x)\,.\,\frac{1}{\sqrt{1 - x^2}} \qquad ...(1)$$

Squaring both sides, $(1 - x^2)y_1^2 = 4\,(\sin^{-1} x)^2 = 4y.$

Differentiating again, $2(1 - x^2)\, y_1 y_2 - 2xy_1^2 = 4y_1$

Dividing both sides by $2y_1$, $(1 - x^2)\, y_2 - xy_1 - 2 = 0 \qquad ...(2)$

Differentiating n times by Leibnitz Theorem,

$$(1 - x^2)y_{n+2} + {}^nC_1\, y_{n+1}(-2x) + {}^nC_2\, y_n(-2) - xy_{n+1} - {}^nC_1\, y_n = 0$$

or

$$(1 - x^2)y_{n+2} - (2n + 1)xy_{n+1} - n^2 y_n = 0 \qquad ...(3)$$

Put $x = 0$ in (1), (2) and (3), then

$$y_1(0) = 0 \quad \text{and} \quad y_2(0) = 2,$$

and

$$y_{(n+2)}(0) = n^2 y_n(0) \qquad ...(4)$$

Putting $n = 1, 2, 3, 4, \ldots$ in (4), we get

$$y_3(0) = 1^2 \cdot y_1(0) = 0$$
$$y_4(0) = 2^2 \cdot y_2(0) = 2 \cdot 2^2$$
$$y_5(0) = 3^2 \cdot y_3(0) = 0$$
$$y_6(0) = 4^2 \cdot y_4(0) = 2 \cdot 2^2 \cdot 4^2$$
$$y_7(0) = 5^2 \cdot y_5(0) = 0$$
$$y_8(0) = 6^2 \cdot y_6(0) = 2 \cdot 2^2 \cdot 4^2 \cdot 6^2.$$

....

....

In general,

When n is odd, $y_n(0) = 0.$

When n is even, $y_n(0) = 2 \cdot 2^2 \cdot 4^2 \cdot 6^2 \ldots (n-2)^2.$

Example 3. *If $y = \tan^{-1} x$, prove that $(1 + x^2)y_{n+1} + 2nxy_n + n(n-1)\, y_{n-1} = 0.$ Hence determine the values of all the derivatives of y with respect to x when $x = 0$.*

Sol. Here $$y = \tan^{-1} x \quad \ldots(1)$$

$\therefore$ $$y_1 = \frac{1}{1+x^2} \quad \ldots(2)$$

or $$y_1(1 + x^2) = 1$$

Differentiating n times by Leibnitz Theorem, we have

$$y_{n+1}(1 + x^2) + ny_n \cdot 2x + \frac{n(n-1)}{2} y_{n-1} \cdot 2 = 0$$

or $$(1 + x^2)\, y_{n+1} + 2nxy_n + n(n-1)\, y_{n-1} = 0 \quad \ldots(3)$$

Putting $x = 0$ in (1), (2) and (3), we have

$$y(0) = 0,\ y_1(0) = 1$$

and $$y_{n+1}(0) = -\,n(n-1)y_{n-1}(0) \quad \ldots(4)$$

Putting $n = 1, 2, 3, 4, \ldots$ in (4), we get

$$y_2(0) = -1 \cdot (0) \cdot y(0) = 0$$
$$y_3(0) = -2 \cdot (1) \cdot y_1(0) = -2 = (-1)^1\, 2\,!$$
$$y_4(0) = -3 \cdot (2) \cdot y_2(0) = 0$$
$$y_5(0) = -4 \cdot (3) \cdot y_3(0) = -4 \cdot (3) \cdot (-2) = (-1)^2\, 4\,!$$

....

....

In general,

When n is even, $y_n(0) = 0.$

When n is odd, $y_n(0) = (-1)^{\frac{n-1}{2}} (n-1)\,!.$

Example 4. *If $y = \sin(a \sin^{-1} x)$, find $y_n(0)$.* [A.K.T.U. 2018, 2019, 2021]

Sol. $$y = \sin(a \sin^{-1} x) \quad \ldots(1)$$

$\therefore$ $$y_1 = \cos(a \sin^{-1} x) \cdot \frac{a}{\sqrt{1-x^2}} \quad \ldots(2)$$

Cross-multiplying,

$$\sqrt{1-x^2}\, y_1 = a \cos (a \sin^{-1} x)$$

Squaring both sides,

$$(1-x^2)\, y_1^2 = a^2 \cos^2 (a \sin^{-1} x)$$
$$= a^2 [1 - \sin^2 (a \sin^{-1} x)] \qquad | \because \cos^2 x = 1 - \sin^2 x$$
$$= a^2 (1 - y^2) \qquad | \text{ By (1)}$$

Again differentiating both sides w.r.t. x,

$$(1-x^2)\, 2y_1 y_2 - 2xy_1^2 = -2a^2 y y_1$$

$$\Rightarrow \quad (1-x^2)\, y_2 - xy_1 + a^2 y = 0 \qquad \text{...(3)} \quad | \text{ Cancelling } 2y_1$$

Now, differentiating (3) n times by Leibnitz theorem,

$$[y_{n+2}(1-x^2) + {}^nC_1 y_{n+1}(-2x) + {}^nC_2 y_n(-2)] - [y_{n+1}(x) + {}^nC_1 y_n(1)] + a^2 y_n = 0$$

$$\Rightarrow \quad (1-x^2) y_{n+2} - (2n+1)\, xy_{n+1} - (n^2 - a^2) y_n = 0 \qquad \text{...(4)}$$

By putting $x = 0$ in equations (1), (2), (3), and (4), we get

$$y(0) = 0$$
$$y_1(0) = a$$
$$y_2(0) = -a^2 y(0) = 0$$
$$y_{n+2}(0) = (n^2 - a^2)\, y_n(0) \qquad \text{...(5)}$$

Putting $n = 1, 2, 3, 4, \ldots$ in (5), we get

$$y_3(0) = (1^2 - a^2)\, y_1(0) = (1^2 - a^2)\, a$$
$$y_4(0) = (2^2 - a^2)\, y_2(0) = 0$$
$$y_5(0) = (3^2 - a^2)(1^2 - a^2)\, a$$
$$y_6(0) = (4^2 - a^2)(2^2 - a^2)\, y_2(0) = 0$$

....

....

In general,

$$y_n(0) = \begin{cases} 0, & \text{when } n \text{ is even} \\ [(n-2)^2 - a^2][(n-4)^2 - a^2] \ldots (3^2 - a^2)(1^2 - a^2)\, a, & \text{when } n \text{ is odd} \end{cases}$$

TEST YOUR KNOWLEDGE

1. If $y = x^2 \cdot \exp(2x)$, determine $(y_n)_0$. *(M.T.U. 2013)*
2. If $y = \sin^{-1} x$, prove that $(1-x^2)\, y_{n+2} - (2n+1)\, xy_{n+1} - n^2 y_n = 0$
   Also find the value of y_n when $x = 0$.
3. If $y = \dfrac{\sin^{-1} x}{\sqrt{1-x^2}}$, prove that $y_n = (n-1)^2\, y_{n-2}$ for $x = 0$.
4. (*i*) Find $y_n(0)$, when $y = \log(x + \sqrt{1+x^2})$.

   (*ii*) If $f(x) = \tan x$, then prove that $f^{(n)}(0) - {}^nC_2 f^{(n-2)}(0) + {}^nC_4 f^{(n-4)}(0) - \ldots = \sin \dfrac{n\pi}{2}$

   [**Hint.** $f(x) \cos x = \sin x$.]

5. (*i*) If $y = [\log(x + \sqrt{1+x^2})]^2$, find all the derivatives of y w.r.t. x, when $x = 0$.
(*ii*) If $y = (\sinh^{-1} x)^2$, prove that $(1 + x^2)\, y_{n+2} + (2n + 1)\, xy_{n+1} + n^2 y_n = 0$.
Hence find y_n when $x = 0$.

6. If $y = e^{a \sin^{-1} x}$, prove that $(1 - x^2)\, y_{n+2} - (2n + 1)\, xy_{n+1} - (n^2 + a^2)\, y_n = 0$.

Deduce that $\underset{x \to 0}{\text{Lt}} \dfrac{y_{n+2}}{y_n} = n^2 + a^2$. Hence find $y_n(0)$.

7. If $y = [x + \sqrt{1+x^2}]^m$, find $y_n(0)$. (*A.K.T.U. 2013, 2018, 2022*)

8. Prove that the value when $x = 0$ of $D^n(\tan^{-1} x)$ is 0, $(n - 1)!$ or $-(n - 1)!$ according as n is of the form $2p$, $4p + 1$ or $4p + 3$ respectively.

9. If $y = \log(x + \sqrt{x^2 + a^2})$, prove that $(a^2 + x^2)\, y_2 + xy_1 = 0$.

Differentiate this differential equation n times and prove that $\lim\limits_{x \to 0} \dfrac{y_{n+2}}{y_n} = -\dfrac{n^2}{a^2}$.

10. If $\log y = \tan^{-1} x$, show that $(1 + x^2)\, y_{n+2} + \{2(n + 1)\, x - 1\}\, y_{n+1} + n(n + 1)\, y_n = 0$ and hence find y_3, y_4 and y_5 at $x = 0$.

Answers

1. $(y_n)_0 = n(n-1)\, 2^{n-2}$
2. When n is even, $y_n(0) = 0$
   When n is odd, $y_n(0) = 1^2 \cdot 3^2 \cdot 5^2 \ldots (n-2)^2$
4. (*i*) When n is even, $y_n(0) = 0$
   When n is odd, $y_n(0) = (-1)^{\frac{n-1}{2}}\, 1^2 \cdot 3^2 \cdot 5^2 \ldots (n-2)^2$
5. (*i*), (*ii*) When n is even, $y_n(0) = (-1)^{\frac{n}{2}-1}\, 2 \cdot 2^2 \cdot 4^2 \cdot 6^2 \ldots (n-2)^2$
   When n is odd, $y_n(0) = 0$
6. When n is even, $y_n(0) = a^2(2^2 + a^2)(4^2 + a^2) \ldots [(n-2)^2 + a^2]$
   When n is odd, $y_n(0) = a(1^2 + a^2)(3^2 + a^2) \ldots [(n-2)^2 + a^2]$
7. When n is even, $y_n(0) = m^2(m^2 - 2^2)(m^2 - 4^2) \ldots [m^2 - (n-2)^2]$
   When n is odd, $y_n(0) = m(m^2 - 1^2)(m^2 - 3^2) \ldots [m^2 - (n-2)^2]$
10. $(y_3)_0 = -1$, $(y_4)_0 = -7$, $(y_5)_0 = 5$.

ASYMPTOTES

2.6. BRANCHES OF A CURVE

Consider the equation of the ellipse $\dfrac{x^2}{a^2} + \dfrac{y^2}{b^2} = 1$.

Solving for y, we get

$$y = b\sqrt{1 - \frac{x^2}{a^2}} \quad \ldots(1) \qquad \text{or} \qquad y = -b\sqrt{1 - \frac{x^2}{a^2}} \quad \ldots(2)$$

These two equations (1) and (2) represent explicity the two branches of the ellipse which in ordinary language we call the upper and lower half of the ellipse. It is to be noticed that the ellipse lies inside the rectangle whose sides are $x = \pm a$ and $y = \pm b$. Thus we see that both the branches of the ellipse lie wholly within a *finite part* of the *x-y* plane and we say, therefore, that both branches of the ellipse are **finite.**

Now let us consider the rectangular hyperbola $x^2 - y^2 = a^2$.

Solving for *y*, we get $y = \sqrt{x^2 - a^2}$ or $y = -\sqrt{x^2 - a^2}$.

If $x \to \pm\infty$, *y* also tends to $\pm\infty$. In this case both the branches *extend to infinity* and are said to be the **infinite branches** of the rectangular hyperbola.

Again, let the curve $x^2y^2 = x^2 - y^2$, be considered. Solving for *y*, we have

$$y = \frac{x}{\sqrt{x^2 + 1}} \quad \text{or} \quad y = -\frac{x}{\sqrt{x^2 + 1}}.$$

Now as $x \to \infty$, $y \to 1$ and as $x \to -\infty$, $y \to -1$, along the first branch. Also in the case of second branch, as $x \to \infty$, $y \to -1$ and as $x \to -\infty$, $y \to 1$. Here both the branches are infinite, and *x* is capable of taking arbitrarily large values whereas *y* remains finite.

We are already familiar with the symbols $x \to \pm\infty$ and $y \to \pm\infty$. But what does $P \to \infty$ stand for, P being a point on an infinite branch of curve. We give the following definition:

Definition. *A point P(x, y) on an infinite branch of a curve is said to tend to infinity along the curve if either x or y or both tend to* $+\infty$ *or* $-\infty$ *as P travels along the branch of the curve.*

2.7. ASYMPTOTE

Definition 1. *A straight line, at a finite distance from the origin is said to be a* **(rectilinear) asymptote** *to an infinite branch of a curve, if the perpendicular distance of a point P on that branch from the straight line tends to zero, as P tends to infinity along the branch.*

Definition 2. A **(rectilinear) asymptote** *to an infinite branch of a curve is the limiting position of the tangent whose point of contact tends to infinity along the branch, but which itself remains at a finite distance from the origin.*

Definition 3. If a straight line cuts a curve in two points at an infinite distance from the origin and yet is not itself wholly at infinity, it is called an asymptote to the curve.

Note 1. In the discussion which follows, we drop the word *rectilinear* and use simply *asymptote* to mean rectilinear asymptote, unless otherwise stated.

Note 2. The asymptotes may be parallel to either *x*-axis or the *y*-axis and accordingly they are called *horizontal* and *vertical* asymptotes. If an asymptote is not parallel to *y*-axis; it is called an *oblique asymptote.*

2.8. ASYMPTOTES PARALLEL TO AXES OF COORDINATES

(i) To find the asymptote parallel to y-axis, of the curve $y = f(x)$.

Let the line $x = k$ ...(1)

parallel to *y*-axis be an asymptote to the curve $y = f(x)$.

Then it is required to determine the value of *k*.

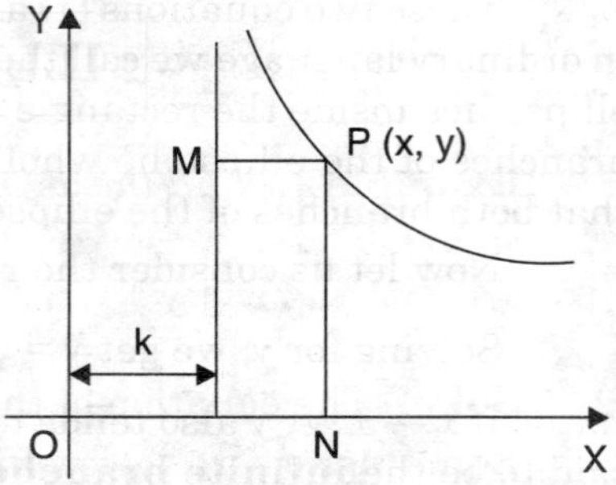

Let PM be the distance of a point P(x, y) on the curve from the equation of line (1), then

$$PM = (x - k)$$

Now by definition of asymptote, if line (i) is an asymptote to the curve, then PM $\to 0$, as P $\to \infty$.

$\therefore$ As P $\to \infty$, PM $= (x - k) \to 0$ or $x \to k$.

If P(x, y) tends to infinity as $x \to k$, only y co-ordinate tends to infinity (*i.e.*, $+\infty$ or $-\infty$), and this gives

$$\lim_{y \to \infty} x = k \quad i.e., \quad x \to k \text{ as } y \to \infty \qquad ...(2)$$

Hence to find the asymptotes parallel to y-axis, we find from the given equation, the definite values $k_1, k_2, k_3, ...$ *to which x tends, as* $y \to +\infty$ *or* $-\infty$*. Then* $x = k_1, x = k_2, x = k_3, ...$ *are the asymptotes parallel to y-axis.*

(ii) Asymptotes parallel to x-axis. Proceeding as above, we arrive at the following method of finding asymptotes parallel to x-axis. *From the given equation, find the definite value* $d_1, d_2, d_3, ...$ *to which y tends as* $x \to +\infty$ *or* $-\infty$ *; then* $y = d_1, y = d_2, y = d_3, ...$ *are asymptotes parallel to the x-axis.*

Asymptotes parallel to co-ordinate axes for algebraic curve $f(x, y) = 0$.

Case (i) Asymptotes parallel to y-axis.

Now let the equation of the algebraic curve $f(x, y) = 0$, after arranging in descending powers of y, be

$$y^n \phi_0(x) + y^{n-1} \phi_1(x) + y^{n-2}\phi_2(x) + ... + \phi_n(x) = 0 \qquad ...(3)$$

where $\phi_0(x), \phi_1(x), \phi_2(x), ..., \phi_n(x)$, are polynomials in x.

Dividing equation (3) throughout by y^n ($\because$ $y \to \infty$), we get

$$\phi_0(x) + \frac{1}{y}\phi_1(x) + \frac{1}{y^2}\phi_2(x) + ... + \frac{1}{y^n}\phi_n(x) = 0$$

Taking limit as $y \to \infty$ and $x \to k$ [From (2)]

we get $\phi_0(k) = 0$ ...(4)

Eliminating k from eqns. (1) and (4), combined equation of asymptotes parallel to y-axis is $\phi_0(x) = 0$.

Thus we arrive at the following *working* rule for finding asymptotes || to y-axis.

Rule. To find asymptotes parallel to y-axis, equate to zero the co-efficient of the highest power of y, present in the given equation of the curve. *Resolve it now into real linear factors.*

If the co-efficient of the highest power of y is either a *constant* or *not resolvable into real linear factors;* then there are no asymptotes parallel to y-axis.

Case (ii) Asymptotes parallel to x-axis for the curve $f(x, y) = 0$.

By interchanging y and x in Case (i), we arrive at the following rule:

Rule. To find asymptotes parallel to x-axis, equate to zero the co-efficient of the highest power of x, present in given equation of the curve. Resolve LHS into real linear factors.

If the coefficient of the highest power of x is either a *constant* or *not resolvable into real linear factors,* there are *no* asymptotes parallel to x-axis.

ILLUSTRATIVE EXAMPLES

Example 1. *Find the asymptotes, parallel to the axes for the curve:*

(*i*) $x^2y^2 = a^2(x^2 + y^2)$ (*ii*) $\frac{a^2}{x^2} - \frac{b^2}{y^2} = 1.$

Sol. (*i*) The equation of the curve is $x^2y^2 - a^2(x^2 + y^2) = 0$.

Equating to zero the coefficient of x^2, the highest power of x, we get

$$y^2 - a^2 = 0 \quad \text{or} \quad y = \pm a,$$

which are the asymptotes parallel to x-axis.

Again equating to zero the coefficient of y^2, the highest power of y, we get

$$x^2 - a^2 = 0 \quad \text{or} \quad x = \pm a,$$

which are the asymptotes parallel to y-axis.

(*ii*) The equation of the curve is $\frac{a^2}{x^2} - \frac{b^2}{y^2} = 1$

or $$a^2y^2 - b^2x^2 = x^2y^2 \quad \text{or} \quad x^2y^2 + b^2x^2 - a^2y^2 = 0$$

Equating to zero the coefficient of x^2, the highest power of x, we get

$$y^2 + b^2 = 0 \quad \text{or} \quad y = \pm ib,$$

which gives imaginary values of y, and therefore there is no asymptote parallel to x-axis.

Again equating to zero the coefficient of y^2, the highest power of y, we get

$$x^2 - a^2 = 0 \quad \text{or} \quad x = \pm a,$$

which are the asymptotes parallel to y-axis.

Example 2. *Find the asymptotes parallel to y-axis of the curve* $\frac{a^2}{x} + \frac{b^2}{y} = 1.$

Sol. The given curve is $a^2y + b^2x = xy$.

Equating the coefficient of highest power of y to zero, we get

$$a^2 - x = 0 \qquad | \text{ Coefficient of } y$$

$$\Rightarrow \quad x = a^2$$

which is the required asymptote parallel to y-axis of the given curve.

Example 3. *Find the asymptotes of the curve* $xy^3 + x^3y = a^4$, *which are parallel to the axes.*

Sol. Here, the coefficient of higher power of x is y. Hence, the equation of the asymptote parallel to x-axis is $y = 0$. Also the coefficient of highest power of y is x, hence the equation of the asymptote parallel to y-axis is $x = 0$.

TEST YOUR KNOWLEDGE

Find the asymptotes, parallel to axes, of the following curves:

1. $x^2y^2 - a^2(x^2 + y^2) - a^3(x + y) + a^4 = 0.$
2. $\frac{a^2}{x^2} + \frac{b^2}{y^2} = 1.$
3. $xy^3 - x^3 = a(x^2 + y^2).$
4. $x^2y^2 - y^2 - 2 = 0.$
5. $y^3 - xy^2 = x^2 + 1.$
6. $y = x(x - 2)(x - 3).$

Answers

1. $y = \pm a, x = \pm a$	**2.** $y = \pm b$; $x = \pm a$	**3.** $x = 0$
4. $y = 0, x = \pm 1$	**5.** No.	**6.** No.

2.9. OBLIQUE ASYMPTOTES

If $y = mx + c$ is an oblique asymptote to any curve $f(x, y) = 0$; then

$$m = \underset{x \to \infty}{\text{Lt}} \frac{y}{x} \text{ and } c = \underset{x \to \infty}{\text{Lt}} (y - mx).$$

Note. The above article requires m and c to be finite. In particular m may be zero and therefore the asymptotes parallel to x-axis can also be found by using this article.

ILLUSTRATIVE EXAMPLES

Example 1. *Find the asymptotes of* $\frac{x^2}{a^2} - \frac{y^2}{b^2} = 1.$

Sol.
$$\frac{x^2}{a^2} - \frac{y^2}{b^2} = 1$$

$$\Rightarrow \quad \frac{1}{a^2} - \frac{1}{b^2}\frac{y^2}{x^2} = \frac{1}{x^2}$$

Take limit as $x \to \infty$,

$$\frac{1}{a^2} - \frac{1}{b^2} \lim_{x \to \infty} \left(\frac{y}{x}\right)^2 = 0$$

or
$$\lim_{x \to \infty} \left(\frac{y}{x}\right)^2 = \frac{b^2}{a^2} \quad \textit{i.e.,} \quad \lim_{x \to \infty} \frac{y}{x} = \pm \frac{b}{a}$$

$\therefore$ Values of m are $\pm \frac{b}{a}$.

Now,
$$c = \lim_{x \to \infty} \left(y - x\frac{dy}{dx}\right) = \lim_{x \to \infty} \left(y - x \cdot \frac{b^2 x}{a^2 y}\right) = \lim_{x \to \infty} \frac{b^2}{y}\left(\frac{y^2}{b^2} - \frac{x^2}{a^2}\right)$$

$$= \lim_{x \to \infty} \left(\frac{-b^2}{y}\right) = 0 \qquad \left|\ \because \underset{x \to \infty}{\text{Lt}}\ y = \pm \frac{b}{a} x\right.$$

Hence the asymptotes are $y = \pm \frac{b}{a} x$.

2.10. ASYMPTOTES OF ALGEBRAIC CURVES

Working rule. Suppose $y = mx + c$ is an asymptote of the curve.

Step 1. Put $mx + c$ for y in the equation of the curve and arrange it in descending powers of x.

Step 2. Equate the coefficients of two highest degree terms to zero.

Step 3. Solve these equations to find the values of m and c.

Step 4. For every pair of values of m and c, there is one asymptote $y = mx + c$ of the curve.

Example 2. *Find the asymptotes of $y^3 - x^2 y + 2y^2 + 4y + 1 = 0$.*

Sol. Putting $y = mx + c$ in the equation of the curve, we get

$$(mx + c)^3 - x^2 (mx + c) + 2 (mx + c)^2 + 4 (mx + c) + 1 = 0$$

or
$$(m^3 - m) x^3 + (3m^2 c - c + 2m^2) x^2 + \text{... etc.} = 0$$

Equating to zero the coefficients of x^3 and x^2, we have

$$m^3 - m = 0 \qquad \text{...(1)}$$

and
$$3m^2 c - c + 2m^2 = 0 \qquad \text{...(2)}$$

From eqn. (2), $m = 0, 1, -1$

From eqn. (3), $c = \dfrac{2m^2}{1-3m^2}$

If $m = 0$, we get $c = 0$;

If $m = 1$, we get $c = -1$;

If $m = -1$, we get $c = -1$

$\therefore$ Three asymptotes are, $y = 0$; $y = x - 1$ and $y = -x - 1$.

Example 3. *Find the asymptotes of the curve*

$$x^3 + 2x^2 y - xy^2 - 2y^3 + 4y^2 + 2xy + y - 1 = 0.$$

Sol. Put $y = mx + c$ in the given equation, we get

$$x^3 + 2x^2 (mx + c) - x (mx + c)^2 - 2 (mx + c)^3 + 4 (mx + c)^2 + 2x(mx + c) + mx + c - 1 = 0$$

$$\Rightarrow \quad (1 + 2m - m^2 - 2m^3) x^3 + 2 (c - mc - 3m^2c + 2m^2 + m) x^2 + \text{... etc.} = 0$$

Equating the coefficients of x^2 and x^3 to zero, we get

$$1 + 2m - m^2 - 2m^3 = 0 \qquad \text{...(1)}$$

and
$$c - mc - 3m^2 c + 2m^2 + m = 0 \qquad \text{...(2)}$$

From eqn. (1), $2m^3 + m^2 - 2m - 1 = 0$

$\Rightarrow \quad m = -1, 1, -1/2$

From eqn. (2), $c = -\left(\dfrac{2m^2 + m}{1 - m - 3m^2}\right)$

If $m = -1$, we have $c = 1$

If $m = 1$, we have $c = 1$

If $m = -1/2$, we have $c = 0$

Hence the asymptotes are $y = -x + 1$, $y = x + 1$ and $y = -\dfrac{1}{2}x$.

2.11. OBLIQUE ASYMPTOTES OF THE GENERAL RATIONAL ALGEBRAIC EQUATION

Let the general algebraic equation of nth degree be

$$(a_0y^n + a_1y^{n-1}x + a_2y^{n-2}x^2 + \cdots + a_{n-1}yx^{n-1} + a_nx^n) + (b_1y^{n-1} + b_2y^{n-2} \cdot x + \cdots + b_{n-1}yx^{n-2} + b_nx^{n-1}) + \cdots + (l_{n-1}y + l_nx) + k_n = 0 \quad \ldots(1)$$

which can be written in the form

$$x^n \phi_n\left(\frac{y}{x}\right) + x^{n-1}\phi_{n-1}\left(\frac{y}{x}\right) + \cdots + x\phi_1\left(\frac{y}{x}\right) + \phi_0\left(\frac{y}{x}\right) = 0 \quad \ldots(2)$$

where $\phi_r\left(\frac{y}{x}\right)$ is a polynomial in $\frac{y}{x}$ of degree r.

Let the straight line $\quad y = mx + c \quad \ldots(3)$

be an asymptote of curve (2), where m and c are finite;

then $$m = \underset{x \to \infty}{\text{Lt}} \frac{y}{x}; \; c = \underset{x \to \infty}{\text{Lt}} (y - mx)$$

To find m, divide both sides of (2) by x^n, we get

$$\phi_n\left(\frac{y}{x}\right) + \frac{1}{x} \cdot \phi_{n-1}\left(\frac{y}{x}\right) + \frac{1}{x^2}\phi_{n-2}\left(\frac{y}{x}\right) + \cdots + \cdots = 0$$

Proceeding to limit as $x \to \infty$, so that $\underset{x \to \infty}{\text{Lt}} \frac{y}{x} = m$, we get

$$\phi_n(m) = 0 \quad \ldots(4)$$

which gives the slopes of the asymptotes.

To find c, corresponding to a value m.

Let $y - mx = p$, so that as $x \to \infty, p \to c$ $\quad \left[\because \; c = \underset{x \to \infty}{\text{Lt}} (y - mx)\right]$

Then $\frac{y}{x} = m + \frac{p}{x}$. Putting this value for $\frac{y}{x}$ in (2), we get

$$x^n\phi_n\left(m + \frac{p}{x}\right) + x^{n-1}\phi_{n-1}\left(m + \frac{p}{x}\right) + \ldots + x\phi_1\left(m + \frac{p}{x}\right) + \phi_0\left(m + \frac{p}{x}\right) = 0$$

Now expanding each term by Taylor's theorem, we have

$$x^n\left[\phi_n(m) + \frac{p}{x}\phi'_n(m) + \frac{p^2}{2!\,x^2}\phi''_n(m) + \ldots\right] + x^{n-1}\left[\phi_{n-1}(m) + \frac{p}{x}\phi'_{n-1}(m) + \ldots\right]$$
$$+ x^{n-2}\left[\phi_{n-2}(m) + \frac{p}{x} \cdot \phi'_{n-2}(m) + \ldots\right] + \ldots = 0 \quad \ldots(5)$$

But $\phi_n(m) = 0$, from (4). Putting $\phi_n(m) = 0$ and arranging the remaining terms in (5) is descending powers of x, we get

$$x^{n-1}[p\phi'_n(m) + \phi_{n-1}(m)] + x^{n-2}\left[\frac{p^2}{2!} \cdot \phi''_n(m) + p \cdot \phi'_{n-1}(m) + \phi_{n-2}(m)\right] + \ldots = 0 \quad \ldots(6)$$

Dividing throughout by x^{n-1}, we have

$$p\phi'_n(m) + \phi_{n-1}(m) + \frac{1}{x}\left[\frac{p^2}{2!}\phi''_n(m) + p\,.\,\phi'_{n-1}(m) + \phi_{n-2}(m)\right] + \ldots = 0$$

Now proceeding to limits as $x \to \infty$, so that $p \to c$, we have

$$\mathbf{c\phi'_n(m) + \phi_{n-1}(m) = 0} \qquad \ldots(7)$$

$\therefore$ If $\phi'_n(m) \neq 0$, which will be case, if equation $\phi_n(m) = 0$, has no repeated roots, then we have

$$c = -\frac{\phi_{n-1}(m)}{\phi'_n(m)}$$

$\therefore$ If $m_1, m_2, m_3, \ldots$ be the *non-repeated* roots of $\phi_n(m) = 0$, and $c_1, c_2, c_3, \ldots$, are the corresponding values of c determined from (7), then $y = m_1 x + c_1$, $y = m_2 x + c_2$, $y = m_3 x + c_3 \ldots$ are the asymptotes.

Exceptional Case. When $\phi_n'(m) = 0$, but $\phi_{n-1}(m) \neq 0$, the finite value of c cannot be determined from (7), and there is *no asymptote* in this case.

Case of parallel asymptotes. If $\phi_n'(m) = 0$ and $\phi_{n-1}(m) = 0$, which is generally there when two roots of $\phi_n(m) = 0$ *i.e.*, two values of m given by $\phi_n(m) = 0$ are equal, then equation (7) becomes an identity and in this case, the equation (6) reduces to

$$x^{n-2}\left[\frac{p^2}{2!}\phi''_n(m) + p\phi'_{n-1}(m) + \phi_{n-2}(m)\right] + x^{n-3}[\ldots] + \ldots = 0$$

Dividing throughout by x^{n-2}, we get

$$\frac{p^2}{2!}\phi''_n(m) + p\phi'_{n-1}(m) + \phi_{n-2}(m) + \frac{1}{x}[\ldots] + \ldots = 0$$

Now proceeding to limits as $x \to \infty$, so that $p \to c$, we have

$$\mathbf{\frac{c^2}{2!}\phi''_n(m) + c\phi'_{n-1}(m) + \phi_{n-2}(m) = 0}$$

which gives two values say c', c'' for c, provided $\phi_n''(m) \neq 0$.

Thus we have two asymptotes $y = mx + c'$; $y = mx + c''$, corresponding to a given slope m. These are evidently *parallel*.

2.11.1. Working Rule

(i) Find the polynomial $\phi_n(m)$, *which can be obtained by putting* $x = 1$, $y = m$ *in the highest degree terms of the given equation of the curve.*

(ii) Put $\phi_n(m)$ *equal to zero, solve for m, and let the roots be* $m_1, m_2, m_3, \ldots$

(iii) Find the polynomial $\phi_{n-1}(m)$ *by putting* $x = 1$, $y = m$ *in next lower degree terms of the given equation. Similarly, polynomial* $\phi_{n-2}(m)$ *can be found out by putting* $x = 1$, $y = m$ *in the next lower degree terms of the equation, and so on.*

(iv) Find the values $c_1, c_2, c_3, \ldots$ *corresponding to values* $m_1, m_2, m_3, \ldots$ *by using the relation*

$$c = -\frac{\phi_{n-1}(m)}{\phi'_n(m)}. \qquad [\textit{Provided } \phi_n'(m) \neq 0]$$

(v) The required asymptotes are $y = m_1 x + c_1$, $y = m_2 x + c_2$, $y = m_3 x + c_3 \ldots$

(vi) If $\phi_n'(m) = 0$ for some value of m, but $\phi_{n-1}(m) \neq 0$, then corresponding to that value of m there is no asymptote.

(vii) If $\phi_n'(m) = 0 = \phi_{n-1}(m)$, for some value of m (i.e., two roots of $\phi_n(m) = 0$ are equal), then the values of c are found from the equation

$$\frac{c^2}{2!}\phi_n''(m) + c \cdot \phi'_{n-1}(m) + \phi_{n-2}(m) = 0.$$

This gives two values of c, and therefore there are two parallel asymptotes corresponding to this value of m.

(viii) If the three roots of $\phi_n(m) = 0$ are equal; [i.e., $\phi_n''(m) = 0$, $\phi'_{n-1}(m) = 0$, $\phi_{n-2}(m) = 0$], then values of c corresponding to that value of m are obtained from the equation

$$\frac{c^3}{3!}\phi_n'''(m) + \frac{c^2}{2!}\phi''_{n-1}(m) + c\phi'_{n-2}(m) + \phi_{n-3}(m) = 0.$$

Note. When it is required to find the asymptotes of a curve, it is advisable to first find the asymptotes parallel to axes, if there be any, and *then* search for the oblique asymptotes.

Example 4. *Find all the asymptotes of the following curves:*

(i) $x^3 + 2x^2y - xy^2 - 2y^3 + 3xy + 3y^2 + x + 1 = 0.$

(ii) $x^3 + 4x^2y + 4xy^2 + 5x^2 + 15xy + 10y^2 - 2y + 1 = 0.$

Sol. (*i*) The equation of the curve is

$$x^3 + 2x^2y - xy^2 - 2y^3 + 3xy + 3y^2 + x + 1 = 0 \qquad ...(1)$$

Since the coefficients of x^3 and y^3, the highest degree terms in x and y, are constant, therefore, there are no asymptotes parallel to x-axis or y-axis. *Now to find oblique asymptotes;* putting $x = 1$, $y = m$ in the third and second degree terms in (1), we get

$$\phi_3(m) = 1 + 2m - m^2 - 2m^3 \text{ and } \phi_2(m) = 3m + 3m^2$$

The slopes of asymptotes are the roots of $\phi_n(m) = 0$ *i.e.*, $\phi_3(m) = 0$,

i.e., of $1 + 2m - m^2 - 2m^3 = 0$ or $2m^3 + m^2 - 2m - 1 = 0$.

$m = 1$ is a root of this equation by inspection. Using synthetic division

1	2	1	–2	–1
		2	3	1
	2	3	1	0

$\therefore$ The reduced equation is $2m^2 + 3m + 1 = 0$

Solving for m, we get $\quad m = -1, \dfrac{-1}{2}$

$$\therefore \qquad m = -1, 1, -\frac{1}{2}$$

Also, $\quad \phi_3'(m) = 2 - 2m - 6m^2 = -(6m^2 + 2m - 2)$

Now, c is given by $\quad c\phi_3'(m) + \phi_2(m) = 0 \qquad | \; c\phi_n'(m) + \phi_{n-1}(m) = 0$

$$\therefore \qquad c = -\frac{\phi_2(m)}{\phi_3'(m)} = \frac{3m + 3m^2}{6m^2 + 2m - 2}$$

When $m = -1$, $\quad c = \dfrac{-3 + 3}{6 - 2 - 2} = 0$

When $m = 1$, $\quad c = \dfrac{3+3}{6+2-2} = 1$

and when $m = -\dfrac{1}{2}$, $\quad c = \dfrac{-\dfrac{3}{2}+\dfrac{3}{4}}{\dfrac{3}{2}-1-2} = \dfrac{1}{2}$.

Putting these values of m and c in $y = mx + c$, the corresponding asymptotes are

$$y = -x + 0,\ y = x + 1 \text{ and } y = -\frac{1}{2}x + \frac{1}{2}$$

or
$$x + y = 0,\ x - y + 1 = 0 \text{ and } x + 2y - 1 = 0.$$

(*ii*) The equation of the curve is

$$x^3 + 4x^2y + 4xy^2 + 5x^2 + 15xy + 10y^2 - 2y + 1 = 0 \qquad ...(1)$$

Since the coefficient of x^3, the highest degree term in x, is constant, $\therefore$ there is no asymptote parallel to x-axis. Again equating to zero the coefficient of y^2, the highest degree term in y, the asymptote parallel to y-axis is given by the equation

$$4x + 10 = 0 \text{ or } 2x + 5 = 0$$

Now to find oblique asymptotes, putting $x = 1$, $y = m$ in the third and second degree terms in the equation of the curve, we get

$$\phi_3(m) = 1 + 4m + 4m^2,$$

and
$$\phi_2(m) = 5 + 15m + 10m^2$$

Slopes of asymptotes are given by the roots of equation $\phi_3(m) = 0$

i.e., by roots of $1 + 4m + 4m^2 = 0$ or $(1 + 2m)^2 = 0$ $\therefore$ $m = -\frac{1}{2}, -\frac{1}{2}$ (Equal)

$\therefore$ c is now given by the equation

$$\frac{c^2}{2!}\phi_n''(m) + c\phi'_{n-1}(m) + \phi_{n-2}(m) = 0$$

$$\frac{1}{2!}c^2\phi_3''(m) + c\phi_2'(m) + \phi_1(m) = 0 \qquad ...(2)$$

Now, $\quad \phi_3'(m) = 4 + 8m,\ \phi_3''(m) = 8,\ \phi_2'(m) = 15 + 20m$ and $\phi_1(m) = -2m$

Equation (2) becomes $\quad \dfrac{1}{2!}c^2 \cdot 8 + c(15 + 20m) - 2m = 0$

For $m = -\dfrac{1}{2}$, it becomes $\quad 4c^2 + c(15 - 10) + 1 = 0$

or
$$4c^2 + 5c + 1 = 0 \quad \text{or} \quad (4c + 1)(c + 1) = 0$$

$$\therefore \quad c = -1,\ c = -\frac{1}{4}$$

Hence the corresponding asymptotes are

$$y = -\frac{1}{2}x - 1 \text{ and } y = -\frac{1}{2}x - \frac{1}{4}$$

or
$$x + 2y + 2 = 0 \text{ and } 2x + 4y + 1 = 0.$$

Example 5. *Show that the parabola $y^2 - 4ax = 0$ has no asymptotes.*

Sol. Since the coefficient of y^2, the highest degree term in y, in the equation of the parabola is a constant, therefore, there are no asymptotes parallel to y-axis. Again, since the coefficient of x, the highest degree term in x is also a constant, $\therefore$ there is no asymptote parallel to x-axis.

For *finding oblique asymptotes* putting $x = 1$, $y = m$ in the second degree and first degree terms in the equation of the parabola, we get

$$\phi_2(m) = m^2 \text{ and } \phi_1(m) = -4a.$$

Slopes of asymptotes are given by the roots of equation $\phi(m) = 0$

i.e., by the equation $m^2 = 0$ or $m = 0, 0$.

But asymptotes (if any) corresponding to $m = 0$ are parallel to x-axis. But it has been proved above that there are no asymptotes parallel to x-axis.

Hence the curve has no oblique asymptote.

Since the parabola has neither oblique asymptotes nor asymptotes parallel to the axes, $\therefore$ it has no asymptotes.

Example 6. *Find asymptotes of the curve $(x - y)^2 (x + 2y - 1) = 3x + y - 7$.*

Sol. The equation of the curve is

$$(x - y)^2 (x + 2y - 1) = 3x + y - 7$$

or $$(x - y)^2 (x + 2y) - (x - y)^2 - 3x - y + 7 = 0 \textbf{ (Note)} \qquad ...(1)$$

Since the coefficients of x^3 and y^3, the highest degree term in x and y are constants, therefore there are no asymptotes parallel to x-axis or y-axis.

Now to find oblique asymptotes; putting $x = 1$, $y = m$ in the third, second and first degree terms in (1), we get

$$\phi_3(m) = (1 - m)^2 (1 + 2m),$$
$$= 2m^3 - 3m^2 + 1$$
$$\phi_2(m) = -(1 - m)^2$$
$$\phi_1(m) = -3 - m$$

$$\phi_3'(m) = 6m^2 - 6m,$$
$$\phi_3''(m) = 12m - 6$$
$$\phi_2'(m) = 2(1 - m)$$

Slopes of the asymptotes are given by $\phi_3(m) = 0$

i.e., by $$(1 - m)^2 (1 + 2m) = 0$$

$\therefore$ $$m = 1, 1, -\frac{1}{2}$$

c is given by the equation $c\phi_3'(m) + \phi_2(m) = 0$

or $$c(6m^2 - 6m) - (1 - m)^2 = 0 \qquad ...(2)$$

For $m = -\frac{1}{2}$, we have

$$c\left(\frac{6}{4} + \frac{6}{2}\right) - \left(1 + \frac{1}{2}\right)^2 = 0$$

or $$\frac{9c}{2} - \frac{9}{4} = 0 \quad \therefore \quad c = \frac{1}{2}$$

$\therefore$ Equation of the asymptote is $y = mx + c$

or $$y = -\frac{1}{2}x + \frac{1}{2} \quad \text{or} \quad \mathbf{2y + x - 1 = 0}$$

For **equal** values $m = 1$, values of c are given by

$$\frac{c^2}{2!}\phi_3''(m) + c\phi_2'(m) + \phi_1(m) = 0$$

or $$\frac{c^2}{2!} \cdot (12m - 6) + 2c(1 - m) + (-3 - m) = 0$$

For $m = 1$, we have $3c^2 - 4 = 0$

or $$c^2 = \frac{4}{3} \qquad \therefore \quad c = \pm \frac{2}{\sqrt{3}}$$

$\therefore$ Equation of asymptote is $y = mx + c$

or $$y = x \pm \frac{2}{\sqrt{3}}. \qquad (\because \quad m = 1)$$

Example 7. *Find the asymptotes of*

$$(x-y)^2 (x^2+y^2) - 10(x-y)x^2 + 12y^2 + 2x + y = 0.$$

Sol. The equation of the curve is

$$(x-y)^2 (x^2+y^2) - 10(x-y)x^2 + 12y^2 + 2x + y = 0 \qquad ...(1)$$

Since coefficients of x^4 and y^4; the highest degree terms in x and y are constants, therefore there are no asymptotes parallel to x-axis or y-axis.

Now *to find oblique asymptotes* putting $x = 1$, $y = m$ in the fourth, third and second degree terms in (1), we get

$$\phi_4(m) = (1-m)^2 (1+m^2)$$
$$= m^4 - 2m^3 + 2m^2 - 2m + 1 \qquad \phi_4'(m) = 4m^3 - 6m^2 + 4m - 2$$
$$\phi_3(m) = -10(1-m) \qquad \phi_4''(m) = 12m^2 - 12m + 4$$
$$\phi_2(m) = 12m^2 \qquad \phi_3'(m) = 10$$

Slopes of the asymptotes are given by $\phi_4(m) = 0$

i.e., by $(1-m)^2 (1+m^2) = 0.$

$\therefore$ $m = 1, 1$ ($\because$ $m^2 + 1 = 0$ gives only imaginary values of m)

For equal values of $m = 1$, values of c are given by

$$\frac{c^2}{2!} \phi_4''(m) + c\phi_3'(m) + \phi_2(m) = 0$$

or by $$c^2 (6m^2 - 6m + 2) + 10c + 12m^2 = 0$$

For $m = 1$, we have $$2c^2 + 10c + 12 = 0$$

or $$c^2 + 5c + 6 = 0 \text{ i.e., } (c+2)(c+3) = 0$$

$\therefore$ $$c = -2, \; c = -3$$

$\therefore$ Asymptotes are $y = mx + c$ *i.e.*, $y = x - 2$ and $y = x - 3$.

2.12. MISCELLANEOUS METHODS OF FINDING ASYMPTOTES OF ALGEBRAIC CURVES

Method I. Let the equation of the curve be of degree n.

Case I. When $y - m_1 x$ is a non-repeated factor of nth degree terms in the equation of curve

Let the curve be $(y - m_1 x) F_{n-1} + P_{n-1} = 0$ ...(1)

where F_{n-1} contains terms of degree $(n-1)$ only and P_{n-1} contains terms of degree not higher than $(n-1)$.

Clearly, m_1 is a root of the eqn. $\phi_n(m) = 0$ so there may be an asymptote $y - m_1 x = c_1$ if a finite value of c_1 can be found out.

$$y - m_1 x = -\frac{P_{n-1}}{F_{n-1}}$$

$$\therefore \qquad c_1 = \lim_{x \to \infty} (y - m_1 x) = \lim_{x \to \infty} -\left(\frac{P_{n-1}}{F_{n-1}}\right)$$

where the limit is to be evaluated by using $\lim\limits_{x \to \infty} \dfrac{y}{x} = m_1$.

Thus the required asymptote corresponding to the factor $y - m_1 x$ is

$$y - m_1 x = \lim_{\substack{x \to \infty \\ \frac{y}{x} \to m_1}} -\left(\frac{P_{n-1}}{F_{n-1}}\right)$$

If there are a number of non-repeated linear factors nth degree terms of the given equation, we can proceed similarly.

Corollary. If we have a factor $ax + by + c$ instead of $y - m_1 x$ in above case, then

$$(ax + by + c)\, F_{n-1} + P_{n-1} = 0$$

Thus asymptote corresponding to the factor $ax + by + c$ is given by

$$ax + by + c + \lim_{\substack{x \to \infty \\ \frac{y}{x} \to -\frac{a}{b}}} \left(\frac{P_{n-1}}{F_{n-1}}\right) = 0$$

If a factor is of the form (*e.g.* $x^2 + xy + y^2$), since it cannot be factorised into real linear factors, therefore there is no asymptote corresponding to it.

Case II. When $(y - m_1 x)^2$ is a factor of nth degree term while $y - m_1 x$ is not of $(n - 1)^{\text{th}}$ one.

Equation of the curve is $(y - m_1 x)^2 F_{n-2} + P_{n-1} = 0$

where F_{n-2} contains term of degree $n - 2$ and

P_{n-1} contains other terms none of which is of degree higher than $(n - 1)$.

$$\lim_{\substack{x \to \infty \\ \frac{y}{x} \to m_1}} (y - m_1 x)^2 = \lim_{x \to \infty} \left(\frac{P_{n-1}}{F_{n-2}}\right) = \infty \quad \text{or} \quad -\infty$$

There is no asymptote parallel to $y - m_1 x = 0$.

Case III. Let the equation be of the form

$$(y - m_1 x)^2 F_{n-2} + (y - m_1 x)\, G_{n-2} + P_{n-2} = 0$$

where F_{n-2} and G_{n-2} contain terms of degree $n - 2$ and P_{n-2} contains terms of degree $n - 2$ at the most.

Dividing by F_{n-2}, $\quad (y - m_1 x)^2 + (y - m_1 x) \dfrac{G_{n-2}}{F_{n-2}} + \dfrac{P_{n-2}}{F_{n-2}} = 0$

Take limit as $x \to \infty$ and $\dfrac{y}{x} \to m_1$, we have the equation.

$$(y - m_1 x)^2 + A(y - m_1 x) + B = 0$$

which on solving for $y - m_1 x$ gives us two parallel asymptotes;

$$y - m_1 x = c_1 \quad \text{and} \quad y - m_1 x = c_2$$

If the equation of the curve is of the form

$$(ax + by + c)^2 + (ax + by + c)\ \frac{G_{n-2}}{F_{n-2}} + \frac{P_{n-2}}{F_{n-2}} = 0$$

then two parallel asymptotes are,

$$(ax + by + c)^2 + (ax + by + c)\ \lim_{\substack{x\to\infty\\ \frac{y}{x}\to\frac{-a}{b}}} \frac{G_{n-2}}{F_{n-2}} + \lim_{\substack{x\to\infty\\ \frac{y}{x}\to\frac{-a}{b}}} \frac{P_{n-2}}{F_{n-2}} = 0.$$

Example 8. *Find the asymptotes of the curve*

$$4x^3 + 2x^2 - 3xy^2 - y^3 - 1 - xy - y^2 = 0.$$

Sol. The given equation of the curve may be written as

$$4x^3 - 4xy^2 + xy^2 - y^3 + 2x^2 - 2xy + xy - y^2 - 1 = 0$$

or $$(4x^2 + 4xy + y^2)(x - y) = 2xy + y^2 - 2x^2 - xy + 1.$$

$\Rightarrow$ $$(y + 2x)^2 (x - y) + (y + 2x)(x - y) - 1 = 0$$

Asymptotes corresponding to $y + 2x$ are given by

$$(y + 2x)^2 + (y + 2x) - \lim_{\substack{x\to\infty\\ \frac{y}{x}\to -2}} \frac{1}{(y+2x)^2} = 0$$

$\Rightarrow$ $$(y + 2x)^2 + (y + 2x) = 0$$

$\Rightarrow$ $$(y + 2x)(y + 2x + 1) = 0$$

or $$y + 2x = 0,\ y + 2x + 1 = 0 \qquad \text{...(1)}$$

Asymptote corresponding to $x - y$ is given by

$$(x - y) + \lim_{\substack{x\to\infty\\ \frac{y}{x}\to 1}} \left(\frac{1}{y + 2x}\right) - \lim_{\substack{x\to\infty\\ \frac{y}{x}\to 1}} \left\{\frac{1}{(y+2x)^2}\right\} = 0$$

$\Rightarrow$ $$x - y = 0$$

Hence required asymptotes are $x - y = 0$, $y + 2x = 0$ and $y + 2x + 1 = 0$.

Example 9. *Find the asymptotes of the curve*

$$y^4 - 2xy^3 + 2x^3 y - x^4 - 3x^3 + 3x^2 y + 3xy^2 - 3y^3 - 2x^2 + 2y^2 - 1 = 0.$$

Sol. Since the coefficients of y^4 and x^4, the highest degree terms in y and x, are constants, therefore there are no asymptotes parallel to y-axis or x-axis.

The given equation may be written as

$$(y - x)^3 (y + x) - 3(y - x)^2 (y + x) + 2(y - x)(y + x) - 1 = 0$$

Asymptote corresponding to $y + x$ is

$$(y + x) - 3 \lim_{\substack{x\to\infty\\ \frac{y}{x}\to -1}} \left(\frac{y + x}{y - x}\right) + 2 \lim_{\substack{x\to\infty\\ \frac{y}{x}\to -1}} \left\{\frac{y + x}{(y - x)^2}\right\} - \lim_{\substack{x\to\infty\\ \frac{y}{x}\to -1}} \left\{\frac{1}{(y - x)^3}\right\} = 0$$

$\Rightarrow$ $$(y + x) - 3 \lim_{\substack{x\to\infty\\ \frac{y}{x}\to -1}} \left(\frac{\frac{y}{x} + 1}{\frac{y}{x} - 1}\right) + 2 \lim_{\substack{x\to\infty\\ \frac{y}{x}\to -1}} \frac{\left(\frac{y}{x} + 1\right)}{x\left(\frac{y}{x} - 1\right)^2} - \lim_{\substack{x\to\infty\\ \frac{y}{x}\to -1}} \frac{1}{(y - x)^3} = 0$$

$\Rightarrow \quad y + x = 0 \quad ...(1) \quad$ | Since all other terms vanish

Again, asymptotes corresponding to $(y - x)$ is

$$(y-x)^3 - 3(y-x)^2 + 2(y-x) - \lim_{\substack{x \to \infty \\ \frac{y}{x} \to 1}} \left(\frac{1}{y+x}\right) = 0$$

$$\Rightarrow \quad (y-x)^3 - 3(y-x)^2 + 2(y-x) = 0$$

$$\Rightarrow \quad y - x = 0,\ y - x = 1,\ y - x = 2 \quad ...(2)$$

Hence the required asymptotes are, $y + x = 0$, $y - x = 0$, $y - x = 1$ and $y - x = 2$.

Example 10. *Find the asymptotes of the curve* $(x^2 - y^2)^2 - 4y^2 + y = 0$.

Sol. There is no asymptote parallel to coordinate axes.

The given equation is, $(x + y)^2 (x - y)^2 = 4y^2 - y$

Asymptotes corresponding to $(x + y)^2$ are

$$(x+y)^2 = \lim_{\substack{x \to \infty \\ \frac{y}{x} \to -1}} \left\{\frac{4y^2 - y}{(x-y)^2}\right\} = \lim_{\substack{x \to \infty \\ \frac{y}{x} \to -1}} \left\{\frac{4\left(\frac{y}{x}\right)^2 - \frac{y}{x^2}}{\left(1 - \frac{y}{x}\right)^2}\right\} = 1$$

$\therefore$ Asymptotes parallel to $y + x = 0$ are $y + x = \pm 1$.

Again, asymptotes corresponding to $(x - y)^2$ are

$$(x-y)^2 = \lim_{\substack{x \to \infty \\ \frac{y}{x} \to 1}} \left\{\frac{4\left(\frac{y}{x}\right)^2 - \frac{y}{x^2}}{\left(1 + \frac{y}{x}\right)^2}\right\} = 1$$

$$\therefore \quad x - y = \pm 1$$

$\therefore$ Required asymptotes are $x + y = \pm 1$, $x - y = \pm 1$.

Example 11. *Find all the asymptotes of the curve* $xy^2 + x^2y + xy + y^2 + 3x = 0$.

Sol. Asymptote parallel to y-axis is $x + 1 = 0$

Asymptote parallel to x-axis is $y = 0$

Also, the curve is $xy(x + y) + y(x + y) + 3x = 0$

Asymptote corresponding to $x + y$ is

$$(x+y) + \lim_{\substack{x \to \infty \\ \frac{y}{x} \to -1}} \left(\frac{x+y}{x}\right) + \lim_{\substack{x \to \infty \\ \frac{y}{x} \to -1}} 3 \cdot \frac{1}{y} = 0$$

$$\Rightarrow \quad x + y = 0$$

$\therefore$ Required asymptotes are $y = 0$, $x + 1 = 0$ and $x + y = 0$.

Method II. Let $f(x, y) = 0$ be a curve. Collect the highest degree terms in it and factorise them into real linear factors. From I method, we know that there is one asymptote corresponding to each linear factor parallel to it. Now, put each such linear factor equal to k. Substitute either x (or y) in terms of y and k (or x and k) from it. Equate the coeff. of highest power of resulting equation to zero and find value of k. This method is repeated for other factors.

Example 12. *Find the asymptotes of the curve*

$$x^3 - 2x^2 y + xy^2 + x^2 - xy + 2 = 0.$$

Sol. The equation of curve is

$$x(x - y)^2 + x(x - y) + 2 = 0 \quad ...(1)$$

Let $x = k$ be an asymptote of the curve.

From eqn. (1), $k(k - y)^2 + k(k - y) + 2 = 0$

Equating the higher power coefficient of y, to zero, we get $k = 0$

$\therefore$ Asymptote parallel to $x = 0$ is $x = 0$ itself.

Again, put $x - y = k_1$ in eqn. (1), we get

$$xk_1^2 + xk_1 + 2 = 0 \quad \Rightarrow \quad x(k_1^2 + k_1) + 2 = 0$$

Equating the coefficient of highest power of x to zero, we get

$$k_1^2 + k_1 = 0 \quad \Rightarrow \quad k_1 = 0, -1$$

Hence the asymptotes parallel to $x - y = 0$ are $x - y = 0$ itself and $x - y = -1$.

Hence all asymptotes are $x = 0$, $x = y$, $x = y - 1$.

TEST YOUR KNOWLEDGE

Find all the asymptotes of the following curves:

1. $x^2y^2 = a^2(x^2 + y^2)$
2. $x^3 + y^3 = 3axy$ (*U.P.T.U. 2014*)
3. $xy^2 = 4a^2(2a - x)$
4. (*a*) $x^3 + 3xy^2 - x^2y - 3y^3 + x^2 - 2xy + 3y^2 + 4x + 7 = 0$
   (*b*) $x^3 - 6x^2y + 11xy^2 - 6y^3 + x + y + 1 = 0$
5. $y^3 - x^2y - 2xy^2 + 2x^3 - 7xy + 3y^2 + 2x^2 + 2x + 2y + 1 = 0$ (*M.T.U. 2012*)
6. $x^3 + 4x^2y + 5xy^2 + 2y^3 + 2x^2 + 4xy + 2y^2 - x - 9y + 1 = 0$
7. $y^3 - 5xy^2 + 8x^2y - 4x^3 - 3y^2 + 9xy - 6x^2 + 2y - 2x + 1 = 0$
8. (*a*) $4x^3 - 3xy^2 - y^3 + 2x^2 - xy - y^2 - 1 = 0$ (*b*) $x^3 + 2x^2y + xy^2 - x^2 - xy + 2 = 0$
9. (*a*) $x(y - x)^2 = x(y - x) + 2$ (*b*) $4x^2(y - x) + y(y - 2)(x - y) = 4x + 4y - 7$
10. $x^2(x - y)^2 + a^2(x^2 - y^2) = a^2xy$
11. $(x - 1)(x - 2)(x + y) + x^2 + x + 1 = 0$
12. $x^3 + 3x^2y - 4y^3 - x + y + 3 = 0$
13. $x^3 - x^2y - xy^2 + y^3 + 2x^2 - 4y^2 + 2xy + x + y + 1 = 0$
14. $x^2y^2(x^2 - y^2)^2 = (x^2 + y^2)^3$
15. $(x + y)^2(x^2 + xy + y^2) = a^2x^2 + a^3(y - x)$
16. $(x - y)^2(x - 2y)(x - 3y) - 2a(x^3 - y^3) - 2a^2(x - 2y)(x + y) = 0$
17. **Prove that the asymptotes of the curve $x^2y^2 = a^2(x^2 + y^2)$ are the sides of a square.**
18. **Show that the asymptotes of $x^2y^2 - a^2(x^2 + y^2) - a^3(x + y) + a^4 = 0$ form a square, through two of whose vertices the curve passes.**
19. **Find rectangular asymptotes of the curve $y = \dfrac{x}{x^2 - 1}$.**
    [**Hint.** Rectangular Asymptotes $\Rightarrow$ Asymptotes at right angles to each other.]
20. **Find the asymptotes of the curve $(x + y + 1)(x^2 + y^2 - xy) - 3xy + x^2 + y^2 + 2x - 3y + 5 = 0$.**
21. **Find the asymptotes of the curve $(x + y + 1)^2(x^2 + y^2 - xy) + 3xy - 7x^2 - 2y^2 - 7x + 8 = 0$.**
    [**Hint.** The above equation can be written as
    $$(x + y)^2(x^2 + y^2 - xy) + 2(x + y)(x^2 + y^2 - xy) - 6x^2 - y^2 + 2xy - 7x + 8 = 0.]$$
22. **Find the asymptotes of the curve $(x^2 - y^2)(y^2 - 4x^2) - 6x^3 + 5x^2y + 3xy^2 - 2y^3 - x^2 + 3xy - 1 = 0$.**
23. **Find the asymptotes of $y^2 = \dfrac{x^4}{a^2 - x^2}$.**

24. (*i*) Find the asymptotes of $(x - y - 1)^2 (x^2 + y^2 + 2) + 6(x - y - 1)(xy + 7) - 8x^2 - 2x - 1 = 0$.

(*ii*) Find the real asymptote of $x^3 + 3x^2y + 3xy^2 + 2y^3 = x^2 + y^2 + x$.

25. Find the asymptotes of the curve $y^2 (x^2 - a^2) = x^2 (x^2 - 4a^2)$.

Answers

1. $x = \pm a, y = \pm a$ **2.** $x + y + a = 0$ **3.** $x = 0$

4. (*a*) $2x - 2y + 1 = 0$ (*b*) $x - y = 0, x - 2y = 0, x - 3y = 0$

5. $y = -x - 2, y = x - 1, y = 2x$ **6.** $x + 2y + 2 = 0, x + y \pm 2\sqrt{2} = 0$

7. $x - y = 0, 2x - y + 1 = 0, 2x - y + 2 = 0$

8. (*a*) $2x + y = 0, 2x + y + 1 = 0, x - y = 0$ (*b*) $x = 0, x + y = 0, x + y = 1$

9. (*a*) $x = 0, y = x, y = x + 1$ (*b*) $x - y = 0, 2x - y + 1 = 0$ and $2x + y - 1 = 0$

10. $x = \pm a, y = x \pm a$ **11.** $x = 1, x = 2, x + y + 1 = 0$

12. $y - x = 0, 2y + x - 1 = 1$ and $2y + x + 1 = 0$.

13. $x + y - 1 = 0, 2x - 2y + (3 + \sqrt{5}) = 0, 2x - 2y + (3 - \sqrt{5}) = 0$

14. $x \pm y = \pm \sqrt{2}, x = \pm 1, y = \pm 1$ **15.** $x + y \pm a = 0$

16. $x = y + a, x = y + 2a, 2y = x + 14a, 3y = x - 13a$. **19.** $x = \pm 1, y = 0$

20. $3x + 3y + 8 = 0$ **21.** $x + y - 1 = 0$ and $x + y + 3 = 0$

22. $y = x, y = 2x, y + x + 1 = 0, y + 2x + 1 = 0$ **23.** $x = \pm a$

24. (*i*) $x - y - 2 = 0, x - y + 3 = 0$ (*ii*) $x + 2y = \frac{5}{3}$ **25.** $x = \pm a, y = \pm x$.

2.13. TOTAL NUMBER OF ASYMPTOTES

To show that a curve of degree n can never have more than n asymptotes.

Consider the equation of algebraic curve of nth degree of the form

$$x^n \phi_n\left(\frac{y}{x}\right) + x^{n-1} \phi_{n-1}\left(\frac{y}{x}\right) + x^{n-2} \phi_{n-2}\left(\frac{y}{x}\right) + \ldots = 0,$$

where $\phi_r\left(\frac{y}{x}\right)$ is a polynomial in $\frac{y}{x}$ of degree r.

The slopes of oblique asymptotes are given by $\phi_n(m) = 0$, which is an equation of degree n and thus has *at the most* n real roots.

Since for each value of m, we have, in general, one value of c determined by the equation $c\phi_n'(m) + \phi_{n-1}(m) = 0$, $\therefore$ it follows that a *curve of degree n has, in general, at the most n asymptotes.* In case the curve has one or more asymptotes parallel to y-axis, then the degree of equation $\phi_n(m) = 0$, is smaller than n by at least the same number.

Further when equation determining c is an equation of second degree ; than $\phi_n'(m) = 0$ and $\therefore$ it follows that $\phi_n(m) = 0$, has two equal roots. Thus, there are two values of c corresponding to two equal roots and there would be at the most $(n - 2)$ other asymptotes corresponding to the remaining roots. Hence it follows that a curve of degree n can never have more than n asymptotes.

2.14. ASYMPTOTES BY INSPECTION

If equation of a curve of degree n, is of the form

$$F_n + F_{n-2} = 0$$

where F_n *is of degree n [i.e., contains terms of degree n and may also contain lower degree terms], and* F_{n-2} *is of degree (n – 2) at the most, then every linear factor of* F_n *equated to zero,* *will be an asymptote, provided no two linear factors of F_n are either coincident or differ by a constant.*

Let $(ax + by + c)$ be a non-repeated factor of F_n and let

$$F_n = (ax + by + c)\,F_{n-1}, \text{ where } F_{n-1} \text{ is of degree } (n-1).$$

The asymptote parallel to $ax + by + c = 0$, is given by

$$ax + by + c + \underset{\substack{x \to \infty \\ \frac{y}{x} \to -\frac{a}{b}}}{\text{Lt}} \frac{F_{n-2}}{F_{n-1}} = 0$$

Now to find the limit $\frac{F_{n-2}}{F_{n-1}}$, the numerator as well as denominator is divided by x^{n-1} and it would be seen that $\frac{1}{x}$ appears as a factor

$\therefore$ Lt. $\frac{F_{n-2}}{F_{n-1}} \to 0$, as $x \to \infty$.

Hence $ax + by + c = 0$, is an asymptote.

Cor. If the equation of the curve is of the form

$$U_1 \cdot U_2 \cdot U_3 \ldots U_n + F_{n-2} = 0$$

where $U_1, U_2, U_3, \ldots, U_n$ are linear factor, then the curve has $U_1 = 0, U_2 = 0, U_3 = 0, \ldots, U_n = 0$ as asymptotes provided no two of the factors are coincident or differ by a constant.

Note. Conversely, a curve having straight lines $U_1 = 0, U_2 = 0, \ldots, U_n = 0$, as asymptotes, must have an equation of the form $U_1 \cdot U_2 \cdot U_3 \ldots U_n + F_{n-2} = 0$.

ILLUSTRATIVE EXAMPLES

Example 1. *(a) Find the asymptotes of $\frac{x^2}{a^2} - \frac{y^2}{b^2} = 1$.*

(b) Find asymptotes of the curve $xy(x^2 - y^2) + 2x^2 + 2y^2 + 1 = 0$.

Sol. (*a*) The equation of the curve is $\left(\frac{x^2}{a^2} - \frac{y^2}{b^2}\right) - 1 = 0$

This is of the form $F_n + F_{n-2} = 0$ $[n = 2]$, where F_2 can be split up into non-repeated linear factors.

Thus the asymptotes are given by $F_2 = 0$

or $$\frac{x^2}{a^2} - \frac{y^2}{b^2} = 0 \quad \text{or} \quad \left(\frac{x}{a} + \frac{y}{b}\right)\left(\frac{x}{a} - \frac{y}{b}\right) = 0$$

$$\therefore \quad \frac{x}{a} + \frac{y}{b} = 0 \quad \text{and} \quad \frac{x}{a} - \frac{y}{b} = 0,$$

are the required asymptotes.

(*b*) The equation of the curve is $xy(x^2 - y^2) + 2x^2 + 2y^2 + 1 = 0$

This is of the form $F_n + F_{n-2} = 0$ $(n = 4)$. The linear factors of $F_4 = xy(x^2 - y^2)$ are x, y, $(x - y)$ and $(x + y)$.

Since none of them is repeated, the four asymptotes of the curve are

$$x = 0,\ y = 0,\ x - y = 0 \quad \text{and} \quad x + y = 0.$$

Example 2. *Find the asymptotes of the curves*

(*i*) $2x^3 + 3x^2y - 2xy^2 - 3y^3 + 6x + y + 3 = 0$

(*ii*) $(x - y)(x + y)(x + 2y - 1) = 3x + 4y + 5.$

Sol. (*i*) The given curve is

$$(2x + 3y)(x - y)(x + y) + 6x + y + 3 = 0$$

It is of the form $F_3 + F_1 = 0$ and no two straight lines obtained from $F_3 = 0$ are parallel. Hence by inspection, the asymptotes are

$$x - y = 0,\ x + y = 0 \text{ and } 2x + 3y = 0.$$

(*ii*) The equation of the curve is of the form $F_3 + F_1 = 0$ and no two straight lines obtained from $F_3 = 0$ are parallel. Hence by inspection, the asymptotes are

$$x - y = 0,\ x + y = 0, \text{ and } x + 2y - 1 = 0.$$

2.15. INTERSECTION OF A CURVE WITH ITS ASYMPTOTES

To prove that any asymptote of an algebraic curve of the nth degree cuts the curve in $(n - 2)$ points.

Let $\quad y = mx + c \quad$ ...(1)

be an asymptote of the algebraic curve

$$x^n\phi_n\left(\frac{y}{x}\right) + x^{n-1}\,\phi_{n-1}\left(\frac{y}{x}\right) + x^{n-2}\,\phi_{n-2}\left(\frac{y}{x}\right) + \ldots = 0 \quad \text{...(2)}$$

To find points of intersections of asymptote (1) and curve (2) let us solve them for x and y.

$\therefore$ Putting value of y from (1) and (2) $\left(i.e., \text{ Putting } \frac{y}{x} = m + \frac{c}{x}\right)$

the abscissae of the points of intersection are the roots of the equation

$$x^n\phi_n\left(m + \frac{c}{x}\right) + x^{n-1}\phi_{n-1}\left(m + \frac{c}{x}\right) + x^{n-2}\phi_{n-2}\left(m + \frac{c}{x}\right) + \ldots = 0$$

Expanding by Taylor's theorem, we have

$$x^n\left[\phi_m(m) + \frac{c}{x}\,\phi_n'(m) + \frac{1}{2!}\frac{c^2}{x^2}\,.\,\phi_n''(m) + \ldots\right]$$

$$+ x^{n-1}\left[\phi_{n-1}(m) + \frac{c}{x}\phi'_{n-1}(m) + \ldots\right] + x^{n-2}\,[\phi_{n-2}(m) + \ldots] + \ldots 0$$

Or arranging the terms according to descending power of x, we get

$$x^n\phi_n(m) + x^{n-1}\,[c\,.\,\phi_n'(m) + \phi_{n-1}(m)] + x^{n-2}\left[\frac{c^2}{2!}\phi_n''(m) + c\phi'_{n-1}(m) + \phi_{n-2}(m)\right] + \ldots = 0$$

...(3)

Since (1) is an asymptote of (2), $\therefore$ m and c are given by

$$\phi_n(m) = 0 \quad \text{and} \quad c\phi_n'(m) + \phi_{n-1}(m) = 0$$

Substituting these values in (*iii*), it reduces to

$$x^{n-2}\left[\frac{c^2}{2!}\phi_n''(m) + c\phi'_{n-1}(m) + \phi_{n-2}(m)\right] + x^{n-3}\,[\ldots] + \ldots = 0$$

which is an equation of $(n - 2)$th degree and determines $(n - 2)$ values of x.

Hence (1) cuts (2) in $(n - 2)$ points.

Cor. 1. *If a curve of nth degree has n asymptotes, then they cut the curve in n(n – 2) points.*

Cor. 2. *If the equation of a curve of nth degree can be put in the form $F_n + F_{n-2} = 0$, where F_{n-2} is of degree (n – 2) at the most and F_n consists of n non-repeated linear factors, then the n(n – 2) points of intersection of the curve and its asymptotes lie on the curve $F_{n-2} = 0$.*

Proof. The joint equation of *n* asymptotes of curve is $F_n = 0$, and the *n*(*n* – 2) point of intersection of the curve and its asymptote satisfy the two equations $F_n + F_{n-2} = 0$ and $F_n = 0$, simultaneously.

∴ They also satisfy the equation.

$$(F_n + F_{n-2}) - F_n = 0 \quad i.e., \quad F_{n-2} = 0$$

Hence the result.

For example:

(*i*) The asymptotes of a *cubic* curve, cut the curve in 3(3 – 2) = 3 points which lie on a curve of degree 3 – 2 = 1 *i.e.*, on a *straight line.*

(*ii*) The asymptotes of a *biquadratic or quartic curve,* cut the curve in 4(4 – 2) = 8 points which lie on a curve of degree 4 – 2 = 2, *i.e.*, on a *conic.*

Example 3. *Show that the points of intersection of the curve*

$$2y^3 - 2x^2y - 4xy^2 + 4x^3 - 14xy + 6y^2 + 4x^2 + 6y + 1 = 0,$$

and its asymptotes lie on the straight line $8x + 2y + 1 = 0$.

Sol. The equation of the curve is

$$2y^3 - 2x^2y - 4xy^2 + 4x^3 - 14xy + 6y^2 + 4x^2 + 6y + 1 = 0 \qquad ...(1)$$

Since the coefficients of x^3 and y^3, the highest degree terms in *x* and *y* are constants, ∴ the curve has no asymptote parallel either to *x*-axis or *y*-axis.

Now to find oblique asymptotes, putting $x = 1$, $y = m$ in the third and second degree terms in (1), we get

$$\phi_3(m) = 2m^3 - 2m - 4m^2 + 4 \text{ and } \phi_2(m) = -14m + 6m^2 + 4.$$

The slopes of asymptotes are the roots of $\phi_3(m) = 0$

i.e., of $\quad 2m^3 - 2m - 4m^2 + 4 = 0$ or $(2m - 4)(m^2 - 1) = 0$

∴ $m = 2, 1, -1.$

Also $\quad \phi_3'(m) = 6m^2 - 2 - 8m = 2(3m^2 - 4m - 1)$

Now *c* is given by, $c\phi_3'(m) + \phi_2(m) = 0$

$$\therefore \quad c = -\frac{\phi_2(m)}{\phi_3'(m)} = \frac{-14m + 6m^2 + 4}{2(3m^2 - 4m - 1)} = \frac{7m - 3m^2 - 2}{3m^2 - 4m - 1}$$

When $m = 2$, $\quad c = \dfrac{14 - 12 - 2}{12 - 8 - 1} = 0$

When $m = 1$, $\quad c = \dfrac{7 - 3 - 2}{3 - 4 - 1} = -1$

and when $m = -1$, $\quad c = \dfrac{-7 - 3 - 2}{3 + 4 - 1} = -2.$

Putting values of *m* and *c* in $y = mx + c$, the corresponding asymptotes are

$$y = 2x + 0, \; y = x - 1 \text{ and } y = -x - 2$$

Making RHS of each equation as zero,

or $$2x - y = 0,\ x - y - 1 = 0 \text{ and } x + y + 2 = 0.$$

These three asymptotes will cut the curve again in 3(3 – 2) = 3 points. The joint equation of the asymptotes is

$$(2x - y)(x - y - 1)(x + y + 2) = 0$$

or $$2x^3 - 2xy^2 - x^2y + y^3 - 7xy + 3y^2 + 2x^2 + 2y - 4x = 0$$

Multiplying by 2 throughout,

$$2y^3 - 2x^2y - 4xy^2 + 4x^3 - 14xy + 6y^2 + 4x^2 + 4y - 8x = 0 \quad ...(2)$$

Now equation of the given curve is

$$2y^3 - 2x^2y - 4xy^2 + 4x^3 - 14xy + 6y^2 + 4x^2 + 6y + 1 = 0 \quad ...(3)$$

Hence the 3 points of intersection of asymptotes (2) and curve (3) also lie on the curve (2) – (3)

i.e., on $$8x + 2y + 1 = 0$$

which is the required straight line.

Example 4. *Find the equation of the cubic which has the same asymptotes as the curve $x^3 - 6x^2y + 11xy^2 - 6y^3 + 4x + 5y + 7 = 0$, and which passes through the points (0, 0), (0, 2) and (2, 0).*

Sol. The equation of given curve is

$$x^3 - 6x^2y + 11xy^2 - 6y^3 + 4x + 5y + 7 = 0$$

or $$(x - y)(x - 2y)(x - 3y) + (4x + 5y + 7) = 0$$

which is of the form $F_n + F_{n-2} = 0$; and F_n consists of non-repeated linear factors.

∴ By the method of inspection, the asymptotes are given by

$$F_n = 0 \text{ or } (x - y)(x - 2y)(x - 3y) = 0$$

or asymptotes are $(x - y) = 0,\ x - 2y = 0$ and $x - 3y = 0$

The joint equation of the asymptotes is

$$F_3 \equiv (x - y)(x - 2y)(x - 3y) = 0.$$

Since the equation of any curve having these asymptotes is of the form $F_3 + F_1 = 0$, where F_1 is of first degree in x and y.

∴ The equation of required cubic may be written as

$$F_3 + (ax + by + c) = 0$$

or $$(x - y)(x - 2y)(x - 3y) + ax + by + c = 0 \quad ...(1)$$

where $ax + by + c$, is the general linear expression.

Since curve (1) passes through (0, 0), ∴ $c = 0$

Again curve (1) passes through (2, 0), ∴ $8 + 2a + c = 0$

and curve (1) passes through (0, 2), ∴ $-48 + 2b + c = 0$

which give, $c = 0$, $a = -4$ and $b = 24$.

Putting these values in (1), the required cubic is

$$(x - y)(x - 2y)(x - 3y) - 4x + 24y = 0$$

or $$x^3 - 6x^2y + 11x^2y - 6y^3 - 4x + 24y = 0.$$

Example 5. *Find the equation of the quartic curve which has $x = 0$, $y = 0$, $y = x$ and $y = -x$, for asymptotes and which passes through (a, b) and which cuts the asymptotes in eight points that lie on the circle $x^2 + y^2 = a^2$.*

Sol. The equations of the asymptotes are

$$x = 0,\ y = 0,\ y - x = 0 \text{ and } y + x = 0$$

The joint equation of the asymptotes is

$$xy(y - x)(y + x) = 0 \text{ or } xy(y^2 - x^2) = 0 \quad ...(1)$$

The equation of given circle is $x^2 + y^2 - a^2 = 0$ ...(2)

The equation of any quartic curve whose asymptotes are given by (*i*) and whose points of intersection with its asymptotes lie on (2) is given by

$$xy(y^2 - x^2) + \lambda(x^2 + y^2 - a^2) = 0 \quad ...(3)$$

where λ is a constant.

This passes through the point (a, b).

$$\therefore \quad ab(b^2 - a^2) + \lambda(a^2 + b^2 - a^2) = 0 \quad \text{or} \quad \lambda = \frac{a(a^2 - b^2)}{b}$$

Putting this value of λ in (3), the equation of required curve is

$$xy(y^2 - x^2) + \frac{a(a^2 - b^2)}{b}(x^2 + y^2 - a^2) = 0$$

or $$bxy(y^2 - x^2) + a(a^2 - b^2)(x^2 + y^2 - a^2) = 0.$$

TEST YOUR KNOWLEDGE

1. Find the asymptotes of the curve $x^2y - xy^2 + xy + y^2 + x - y = 0$, and show that they cut the curve in three points on the straight line $x + y = 0$.
2. (*i*) Show that the asymptotes of the cubic, $x^3 - 2y^3 + xy(2x - y) + y(x - y) + 1 = 0$, cut the curve in three points which again lie on the straight line $x - y + 1 = 0$.

   (*ii*) Find all the asymptotes of the curve $3x^3 + 2x^2y - 7xy^2 + 2y^3 - 14xy + 7y^2 + 4x + 5y = 0$. Show that the asymptotes meet the curve again in three points which lie on a straight line. Find the equation of this line.
3. (*i*) Find the asymptotes of the curve $x(x^2 - y^2) + y(3y - x) = 0$ and prove that the three points where these asymptotes cut the curve lies on $7x - 3y + 6 = 0$.

   (*ii*) Find the asymptotes of the curve $(x + a)\, y^2 = (y + b)\, x^2$ and show that the points of intersection of the curve and its asymptotes lie on a straight line.
4. Show that the points of intersection of the curve $4x^3 - 2x^2y - 4xy^2 + 2y^3 + 6y^2 - 14xy + 4x^2 + 6y + 1 = 0$ and its asymptotes lie on $8x + 2y + 1 = 0$.
5. (*i*) Show that the asymptotes of the curve $(x^2 - y^2)(y^2 - 4x^2) + 6x^3 - 5x^2y - 3xy^2 + 2y^3 - x^2 + 3xy - 1 = 0$ cut the curve again in eight points which lie on a circle of radius unity.

   (*ii*) Show that the asymptotes of the curve $(x^2 - 4y^2)(x^2 - 9y^2) + 5x^2y - 5xy^2 - 30y^3 + xy + 7y^2 - 1 = 0$ cut the curve again in eight points which lie on a circle of radius unity.
6. Find the asymptotes of the curve $xy(x^2 - y^2) + x^2 + y^2 = a^2$, and show that the eight points of intersection of the curve with its asymptotes lie on a circle whose centre is at the origin.
7. Show that the eight points of intersection of curve $x^4 - 5x^2y^2 + 4y^4 + x^2 - y^2 + x + y + 1 = 0$ and its asymptotes lie on a rectangular hyperbola.
8. Show that four asymptotes of the curve $xy(x^2 - y^2) + 25y^2 + 9x^2 - 144 = 0$ cut it again in eight points on an ellipse whose eccentricity is $\frac{4}{5}$.
9. Show that the asymptotes of the curve $4(x^4 + y^4) + 17x^2y^2 - 4x(4y^2 - x^2) + 2(x^2 - 2) = 0$ cut the curve in eight points which lie on the ellipse $x^2 + 4y^2 = 4$.
10. (*i*) Find the equation of the hyperbola having $x + y - 1 = 0$, and $x - y + 2 = 0$, as its asymptotes, and passing through the origin.

    (*ii*) Find the cubic curve whose asymptotes are $x + a = 0$, $y - a = 0$, $x + y + a = 0$ and which touches the x-axis at the origin and passes through the point $(-2a, -2a)$.

11. Find the equation of the cubic, which has the same asymptotes as the curve $x^3 - 6x^2y + 11xy^2 - 6y^3 + x + y + 1 = 0$ and which passes through the points (0, 0), (1, 0) and (0, 1).
12. Find the cubic which has the same asymptotes as the curve $x^3 - 6x^2y + 11xy^2 - 6y^3 + x + y + 1 = 0$ and which touches the axis of y at the origin and goes through the point (3, 2).
[**Hint.** Tangent at the origin is obtained by equating the lowest degree terms to zero.]

Answers

1. $y = 0, x - 1 = 0$ and $x - y + 2 = 0$
2. (*ii*) $6(y - x) + 7 = 0, 2(y - 3x) + 3 = 0, 3(2y + x) + 5 = 0$; $106y - 381x + 105 = 0$
3. (*i*) $x = 3, y = x + 1, x + y + 2 = 0$ (*ii*) $b^2(x + a) = a^2(y + b)$
6. $x = 0, y = 0, x \pm y = 0$
10. (*i*) $(x + y - 1)(x - y + 2) + 2 = 0$ (*ii*) $xy(x + y) + a(y^2 - x^2 + xy) - 6a^2y = 0$
11. $x^3 - 6x^2y + 11xy^2 - 6y^3 - x + 6y = 0$
12. $x^3 - 6x^2y + 11xy^2 - 6y^3 - x = 0$.

2.16. ASYMPTOTES IN POLAR CO-ORDINATES

Theorem 1. *The polar equation of any line is $p = r\cos(\theta - \alpha)$, where p is the length of the perpendicular from the pole to the line and α, is the angle which this perpendicular makes with x-axis.*

Theorem 2. *The length of the perpendicular from a point $P(r_1, \theta_1)$ on the line $p = r\cos(\theta - \alpha)$ is $\pm\{p - r_1\cos(\theta_1 - \alpha)\}$.*

Working rule for finding the perpendicular distance of a point from the line $p = r\cos(\theta - \alpha)$.

1. Make R.H.S. of equation of the line as zero.
2. Substitute the co-ordinates of the point in LHS to get perpendicular distance.

2.17. POLAR EQUATION OF AN ASYMPTOTE TO THE CURVE $r = f(\theta)$ IS

$\boldsymbol{p = r\sin(\theta_1 - \theta)}$, **where** $\boldsymbol{p = \underset{\theta \to \theta_1}{\text{Lt}} -\dfrac{d\theta}{du}}\left(u = \dfrac{1}{r}\right)$ *and θ_1 is a root of the equation obtained by putting $u = 0$.*

2.17.1. Working Rule for Finding Polar Asymptotes

Step I. *Put $r = \dfrac{1}{u}$ in the given equation. Also change all T-ratios, if any, into sin θ and cos θ.*

Step II. *Find the limit of θ as $u \to 0$. Let θ_1 be this limit or one of the limits if more than one such limits exist.*

Step III. *Determine* $p = \underset{\substack{\theta \to \theta_1 \\ u \to 0}}{\text{Lt}}\left(-\dfrac{d\theta}{du}\right)$ *for value or values of θ obtained in Step II.*

Step IV. *Putting the values of p and θ_1 in the equation $p = r\sin(\theta_1 - \theta)$, we get the corresponding asymptote.*

Note. If $\underset{\theta \to \theta_1}{\text{Lt}}\left(-\dfrac{d\theta}{du}\right)$ does not tend to a finite limit ; then there is no asymptote corresponding to the value $\theta = \theta_1$.

ILLUSTRATIVE EXAMPLES

Example 1. *Find the asymptotes of the hyperbolic spiral $r\,\theta = a$.*

Sol. The equation of the curve is $r\theta = a$.

Put $r = \dfrac{1}{u}$ $\therefore$ $\dfrac{\theta}{u} = a$ $\therefore$ $u = \dfrac{\theta}{a}$.

As $u \to 0$ $\dfrac{\theta}{a} \to 0$ *i.e.,* $\theta \to 0$

$$\therefore \quad \theta_1 = 0$$

Since $u = \dfrac{\theta}{a}$, therefore $\dfrac{du}{d\theta} = \dfrac{1}{a}$

$$p = \underset{\theta \to \theta_1}{\text{Lt}} \left(-\frac{d\theta}{du}\right) = \underset{\theta \to 0}{\text{Lt}} \; -a = -a.$$

$\therefore$ Equation of the asymptote is

$$p = r \sin(\theta_1 - \theta)$$

or $$-a = r \sin(-\theta) = -r \sin \theta$$

i.e., $$r \sin \theta = a.$$

Example 2. *Find the asymptotes of the curve $r \cos \theta = a \sin^2 \theta$.*

Sol. The equation of the curve is $r \cos \theta = a \sin^2 \theta$

Put $r = \dfrac{1}{u}$ $\therefore$ $\dfrac{1}{u} \cos \theta = a \sin^2 \theta$

$$\therefore \quad u = \frac{\cos \theta}{a \sin^2 \theta}$$

As $u \to 0$, $\dfrac{\cos \theta}{a \sin^2 \theta} \to 0$ *i.e.,* $\cos \theta \to 0$

$$\therefore \quad \theta \to (2n+1)\frac{\pi}{2} \quad \textit{i.e.,} \quad \theta_1 = (2n+1)\frac{\pi}{2}$$

Since $$u = \frac{\cos \theta}{a \sin^2 \theta}$$

$$\therefore \quad \frac{du}{d\theta} = \frac{1}{a}\left[\frac{\sin^2 \theta(-\sin \theta) - \cos \theta \,.\, 2 \sin \theta \cos \theta}{\sin^4 \theta}\right]$$

$$= \frac{-\sin \theta(\sin^2 \theta + 2\cos^2 \theta)}{a \sin^4 \theta} = \frac{-(\sin^2 \theta + 2 \cos^2 \theta)}{a \sin^3 \theta}$$

$$p = \underset{\theta \to \theta_1}{\text{Lt}} \left(-\frac{d\theta}{du}\right) = \underset{\theta \to (2n+1)\pi/2}{\text{Lt}} \frac{a \sin^3 \theta}{\sin^2 \theta + 2 \cos^2 \theta}$$

or $$p = a \frac{((-1)^n)^3}{((-1)^n)^2 + 0} = a(-1)^n$$

$\therefore$ Equation of the asymptote is

$$p = r \sin(\theta_1 - \theta)$$

or $$a(-1)^n = r \sin\left[(2n+1)\frac{\pi}{2} - \theta\right] = r \sin\left(n\pi + \frac{\pi}{2} - \theta\right)$$

or $$a(-1)^n = (-1)^n\, r \sin\left(\frac{\pi}{2} - \theta\right)$$

or $$a = r \cos \theta.$$

Example 3. *Prove that the curve* $r = \dfrac{a}{1 - \cos \theta}$ *has no asymptotes.*

Sol. The equation of the curve is $r = \dfrac{a}{1 - \cos \theta}$

Put $$r = \frac{1}{u} \quad \therefore \quad \frac{1}{u} = \frac{a}{1 - \cos \theta}$$

$$\therefore \quad u = \frac{1 - \cos \theta}{a}$$

As $u \to 0$; $\dfrac{1 - \cos \theta}{a} \to 0$ *i.e.,* $\cos \theta \to 1$

or $$\cos \theta \to \cos 0$$

$$\therefore \quad \theta \to 2n\pi \pm 0 = 2n\pi$$

$\therefore$ $\theta_1 = 2n\pi$, where n is any integer

Since $$u = \frac{1 - \cos \theta}{a} \Rightarrow \frac{du}{d\theta} = \frac{\sin \theta}{a}$$

$$\therefore \quad \underset{\theta \to \theta_1}{\text{Lt}} \left(-\frac{d\theta}{du}\right) = \underset{\theta \to 2n\pi}{\text{Lt}} -\frac{a}{\sin \theta} \to -\frac{a}{\sin 2n\pi} \to -\infty$$

$\therefore$ $\underset{\theta \to \theta_1}{\text{Lt}} -\dfrac{d\theta}{du}$ does not exist.

$\therefore$ The curve $r = \dfrac{a}{1 - \cos \theta}$ has no asymptote.

TEST YOUR KNOWLEDGE

Find the asymptotes of the following curves:

1. (*i*) $r = \dfrac{a\theta}{\theta - 1}$ (*ii*) $r^2\,\theta = a^2$ (*iii*) $r(\theta^2 - \pi^2) = 2a\theta$
2. (*i*) $r = \dfrac{2a}{1 - 2\cos \theta}$ (*ii*) $r = \dfrac{2a}{1 + 2\cos \theta}$ (*iii*) $r \cos \theta = a \sin \theta$ or $r = a \tan \theta$
3. $r \sin n\theta = a$
4. (*i*) $r \cos 2\theta = a \sin 3\theta$ (*ii*) $r \sin \theta = 2 \cos 2\theta$ (*iii*) $r \cos \theta = a \cos 2\theta$.
5. (*i*) $r^n\, n\theta = a^n$, where n is a positive integer > 1. (*ii*) $r = \dfrac{3a \sin \theta \cos \theta}{\sin^3 \theta + \cos^3 \theta}$
6. (*i*) $r = 4(\sec \theta + \tan \theta)$ (*ii*) $r = a \operatorname{cosec} \theta + b$ (*iii*) $r^2 = a^2(\sec^2 \theta + \operatorname{cosec}^2 \theta)$
   (*iv*) $r = a + b \cot n\theta$ (*v*) $r = a \operatorname{cosec} \theta + b \cot \theta$.

7. $r(1 - e^{\theta}) = a.$
8. (i) $r = a \log \theta$ (ii) $r \log \theta = a.$
9. (i) $r(\pi + \theta) = ae^{\theta}.$ (ii) $r\theta \cos \theta = a \cos 2\theta.$
10. Find asymptotes of $r^n f_n(\theta) + r^{n-1} f_{n-1}(\theta) + \cdots + f_0(\theta) = 0.$
11. Show that all the asymptotes of the curve $r \tan n\theta = a$ touch the circle $r = \dfrac{a}{n}$.

Answers

1. (i) $r \sin(\theta - 1) = a$ (ii) $\theta = 0$ (iii) $r \sin \theta = -a$
2. (i) $r \sin\left(\dfrac{\pi}{3} \pm \theta\right) = \dfrac{-2a}{\sqrt{3}}$ (ii) $r \sin\left(\dfrac{2\pi}{3} \pm \theta\right) = \dfrac{2a}{\sqrt{3}}$ (iii) $r \cos \theta = (-1)^n a$
3. $r \sin\left(\theta - \dfrac{m\pi}{n}\right) = a\,\dfrac{(-1)^m}{n}$, where m is any integer
4. (i) $\dfrac{a}{2} \sin\left(\dfrac{n\pi}{2} + \dfrac{3\pi}{4}\right) = r \sin\left[(2n+1)\dfrac{\pi}{4} - \theta\right]$ (ii) $r \sin \theta = 2$
   (iii) $r \cos \theta + a = 0$
5. (i) $\theta = \dfrac{m\pi}{n}$, where m is any integer (ii) $\sqrt{2}r \sin\left(\dfrac{\pi}{4} + \theta\right) + a = 0.$
6. (i) $4[1 + (-1)^n] = (-1)^n r \cos \theta$ (ii) $r \sin \theta = a$
   (iii) $\pm a(-1)^n = r \sin\left(\dfrac{n\pi}{2} - \theta\right)$ (iv) $r \sin\left(\theta - \dfrac{m\pi}{n}\right) = \dfrac{b}{n}$, where m is an integer
   (v) $a(-1)^n + b = (-1)^n r \sin \theta$
7. $r \sin \theta + a = 0$
8. (i) $\theta = 0$ (ii) $a = r \sin(\theta - 1)$
9. (i) $r \sin \theta\, e^{\pi} + a = 0$ (ii) $r \sin \theta = a$, $(2n + 1)\pi r \cos \theta + 2a = 0$, where n is any *integer*.
10. $\dfrac{f_{n-1}(\theta_1)}{f_n'(\theta_1)} = r \sin(\theta_1 - \theta)$, where θ_1 is any root of the equation $f_n(\theta) = 0.$

2.18. CIRCULAR ASYMPTOTES

If the equation to a curve be $r = f(\theta)$ and if $\lim\limits_{\theta \to \infty} f(\theta) = a$, then the circle $r = a$ is called a circular asymptote of the curve $r = f(\theta)$. Evidently, the curve approaches the circle as $\theta \to \infty$.

Example 1. *Find the circular asymptote of the curve* $r = \dfrac{\theta^2 - 1}{\theta^2 + 1}$.

Sol.
$$f(\theta) = \frac{\theta^2 - 1}{\theta^2 + 1} = \frac{1 - \dfrac{1}{\theta^2}}{1 + \dfrac{1}{\theta^2}}$$

$$\therefore \quad \underset{\theta \to \infty}{\text{Lt}}\, f(\theta) = 1$$

Hence the circular asymptote is $r = 1$.

Example 2. *Prove that the circular asymptote of the curve $r(e^\theta - 1) = a(e^\theta + 1)$ is $r = a$.*

Sol. The curve is, $$r = a\left(\frac{e^\theta + 1}{e^\theta - 1}\right) = a\left(\frac{1 + e^{-\theta}}{1 - e^{-\theta}}\right)$$

Taking limit as $\theta \to \infty$, we get $f(\theta) = a$.

$\therefore$ Circular asymptote is $r = a$.

2.19. ASYMPTOTES BY EXPANSION

To show that $y = mx + c$ is an asymptote of the curve of which the equation can be expressed in the form

$$y = mx + c + \frac{A}{x} + \frac{B}{x^2} + \frac{C}{x^3} + \cdots \qquad ...(1)$$

Dividing both sides of (1) by x, we get

$$\frac{y}{x} = m + \frac{c}{x} + \frac{A}{x^2} + \frac{B}{x^3} + \frac{C}{x^4} + \ldots$$

so that when $x \to \infty$, we have $\text{Lt}\ \dfrac{y}{x} = m$ ...(2)

Again from (1), we have $y - mx = c + \dfrac{A}{x} + \dfrac{B}{x^2} + \dfrac{C}{x^3} + \ldots$

so that we have $\underset{x \to \infty}{\text{Lt}} (y - mx) = c.$ ...(3)

Hence from (2) and (3), it follows that $y = mx + c$, is an asymptotes of curve (1).

2.20. SINGULAR POINT

A point on a curve at which the curve exhibits an extraordinary or unusual behaviour is called a **singular point.**

We are already familiar with one type of singular points namely points of maxima and minima. Here two main types of the singular points (*i*) *multiple points* and (*ii*) *points of inflexion* shall be considered.

2.21. MULTIPLE POINTS

A point on a curve through which more than one branch of the curve passes, is called a multiple point of the curve. In particular a point on a curve is called a double point if two branches of the curve pass through it, a triple point if three branches pass through it and so on. In general, if r branches of the curve pass through a point, that point is called a *multiple point of rth order*. In general, a curve has r tangents (real, coincident or imaginary), at a multiple point of rth order, one for each of the r branches. Thus a curve has two tangents at a double point, one for each branch, and these tangents may be real and different, real and coincident or imaginary.

2.22. TYPES OF DOUBLE POINTS

There are three kinds of double points.

(*i*) Node. A double point on a curve through which pass *two real branches of the curve* and the tangents at which are *real and distinct,* is called a **node.** Thus, point P in the figure (*i*) is a node.

(*ii*) Cusp. A double point on a curve through which pass two real branches of the curve and the tangents at which are *real and coincident* is called a **cusp.** Thus, the point Q in the figures (*ii*) and (*iii*) is a cusp.

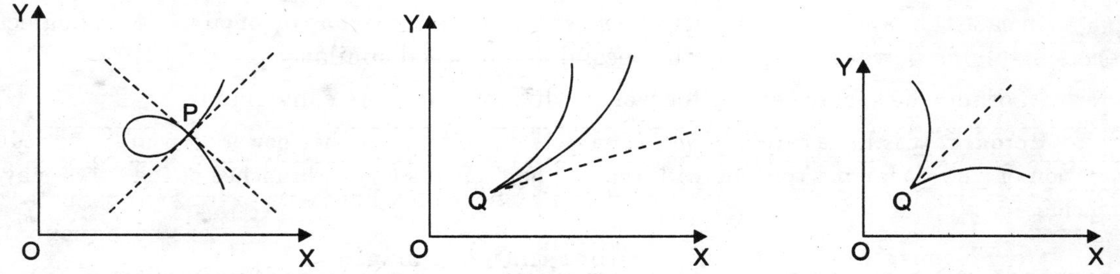

Fig. (*i*) Fig. (*ii*) Fig. (*iii*)

(*iii*) Conjugate point or Isolated point. If there are no real points on the curve in the neighbourhood of a point R on the curve, then R is called a **Conjugate point** *or* **an isolated point.** For example, (2, 0) is a conjugate point on the curve $y^2 = (x-2)^2\,(x-5)$ because (2, 0) satisfies the equation of the curve, but for all value of x very close to 2, y^2 is negative and hence y is imaginary *i.e.*, there is no point on the curve in the neighbourhood of (2, 0).

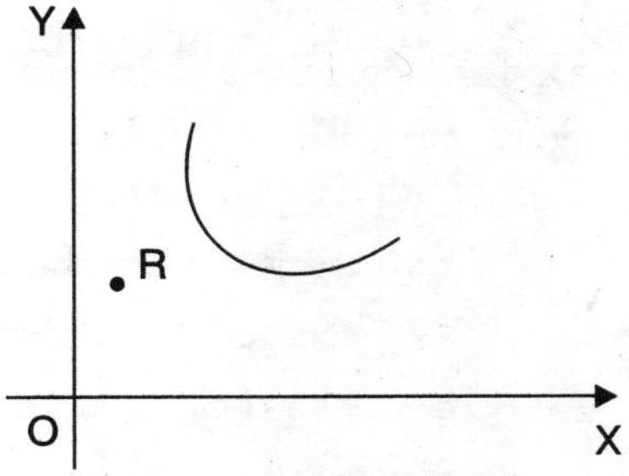

Remark. Since there is no other point in the neighbourhood of conjugate point R, the *tangents at R are generally imaginary* but in some of the cases the *tangents at R may be real.* In such cases the nature of the curve in the neighbourhood of the point under consideration, has to be examined. If the branches of the curve are imaginary, then the point is a conjugate point.

2.23. TANGENTS AT THE ORIGIN TO AN ALGEBRAIC CURVE

The equation to the tangent or tangents at the origin is obtained by equating to zero the terms of lowest degree in the equation of the curve.

2.24. WORKING RULE FOR FINDING THE NATURE OF ORIGIN IF IT IS A DOUBLE POINT

(*a*) (*i*) Write the equation of tangents at the origin by equating to zero the lowest degree terms in the equation of the curve. If origin is a double point, then we shall get two tangents which may be real or imaginary.

(*ii*) If the *two tangents are imaginary,* then *origin, is a conjugate point i.e., an isolated point.*

(*iii*) If the two tangents are *real and distinct*, then *origin is either a node* or *a conjugate point*. To be definite, nature of the curve in the neighbourhood of the origin is to be examined. If the curve has *two real branches passing through the origin, then it is a node otherwise* it is a *conjugate point*.

Remark. The tangents may be real at a conjugate point.

(*b*) *Method for testing the nature of the curve at the origin.*

If the tangents at the origin are $y^2 = 0$, solve the equation for y, neglecting the third and higher powers of y. If for small values of x, the values of y are real, the branches of curve through the origin are real otherwise they are imaginary.

In case the tangents at the origin are $x^2 = 0$, solve the equation of curve for x, neglecting terms involving powers of x above the second and proceed similarly.

In other cases, solve either for y or x, whichever is more convenient.

Remark. It is to be carefully *noted* that while neglecting higher powers of y or x, the reduced equation of curve, may not coincide with that of the tangents or two branches of the curve may not coincide.

Cor. Nature of double points other than the origin

If the point (h, k) is a double point of the curve known in advance, shift the origin to the point (h, k) (*i.e.*, put $x = X + h$ and $y = Y + k$). The double point is the origin now and its nature can be studied by proceeding as in the above article.

Example. *Find the nature of the origin for the curve* $y^2(a^2 + x^2) = x^2 (a^2 - x^2)$.

Sol. The equation of the curve is $x^4 + x^2y^2 + a^2y^2 - a^2x^2 = 0$.

Equating to zero the lowest degree terms, the tangents at the origin are

$$a^2y^2 - a^2x^2 = 0 \text{ or } y^2 - x^2 = 0 \text{ } i.e., \text{ } y = \pm x.$$

Because there are two tangents at the origin, therefore origin is a double point.

Since the two tangents are real and distinct,

∴ the origin is either a node or a conjugate point.

Solving the equation of the curve for y, we get

$$y = \pm x\sqrt{\frac{a^2 - x^2}{a^2 + x^2}}$$

Since for small values of $x \neq 0$, $(a^2 - x^2)$ is positive,

∴ y is real in the neighbourhood of the origin. Hence origin is a node.

2.25. CONDITIONS FOR THE EXISTENCE OF A MULTIPLE POINT ON A CURVE, *AND* THE CLASSIFICATION OF DOUBLE POINTS

The necessary and sufficient conditions for any point $P(x, y)$ on the curve $f(x, y) = 0$ to be a multiple point are

$$\frac{\partial f}{\partial x} = 0 \quad \text{and} \quad \frac{\partial f}{\partial y} = 0 \qquad \ldots(1)$$

∴ *Those of the simultaneous solutions of equations (1) which satisfy equation $f(x, y) = 0$ determine double points or multiple points on the curve.*

Classification of double points

A double point will be a node, cusp or a conjugate point, according as

$$\left(\frac{\partial^2 f}{\partial x \partial y}\right)^2 > \; = \; \text{or} \; < \frac{\partial^2 f}{\partial x^2} \cdot \frac{\partial^2 f}{\partial y^2} \qquad \ldots(2)$$

Note 1. The condition (2) is *not a sure test* for the point (x, y) to be a node, a cusp or a conjugate point. This is, in fact, the condition for two tangents at the double point to be real and distinct, real and coincident or imaginary. This result may, *sometimes,* lead to an incorrect result. To be absolutely correct, we have to examine the nature of curve in the immediate neighbourhood of the double point (x, y).

Note 2. If $\frac{\partial^2 f}{\partial y^2} = \frac{\partial^2 f}{\partial x \partial y} = \frac{\partial^2 f}{\partial x^2} = 0$, at the point P$(x, y)$, then P will be a multiple point of order higher than two.

Note 3. Equation $\frac{\partial^2 f}{\partial y^2}\left(\frac{dy}{dx}\right)^2 + 2\frac{\partial^2 f}{\partial x \partial y}\frac{dy}{dx} + \frac{\partial^2 f}{\partial x^2} = 0$ gives us the slopes of the tangents at the double point (x, y).

2.26. WORKING RULE FOR FINDING THE POSITION *AND* NATURE OF DOUBLE POINTS ON A CURVE

1. Write the equation of the curve in form $f(x, y) = 0$ *i.e.,* make the R.H.S. as zero.

2. Find five partial derivatives $\frac{\partial f}{\partial x}, \frac{\partial f}{\partial y}, \frac{\partial^2 f}{\partial x^2}, \frac{\partial^2 f}{\partial y^2}, \frac{\partial^2 f}{\partial x \partial y}$.

3. Solve the equations $\frac{\partial f}{\partial x} = 0$ and $\frac{\partial f}{\partial y} = 0$.

All the pairs of values of x, y obtained by solving the above two equations are the **possible** double points. Out of these, only those which satisfy the equation of the curve are the double points.

4. At each double point, calculate $D = \left(\frac{\partial^2 f}{\partial x \partial y}\right)^2 - \frac{\partial^2 f}{\partial x^2}\frac{\partial^2 f}{\partial y^2}$.

(*a*) If D is +ve, double point is a node or conjugate point.

(*b*) If D is 0, double point is a cusp or conjugate point.

(*c*) If D is –ve, double point is a conjugate point.

5. For cases (*a*), (*b*) of step 4, when D is +ve or 0 ; test the nature of the curve in the immediate neighbourhood of the double point (x, y).

2.27. KINDS OF CUSPS

We know that at a cusp two branches of a curve have a common tangent and therefore a common normal. A cusp is called a **single cusp** or a **double cusp** according as the *two branches* of the curve lie entirely *on one side* or *on both sides of the common normal.*

Species of Cusps. A cusp, single or double, is said to be *of first species* or *of second species*, according as the *two branches* of the curve, *lie on opposite or same side of the common tangent.*

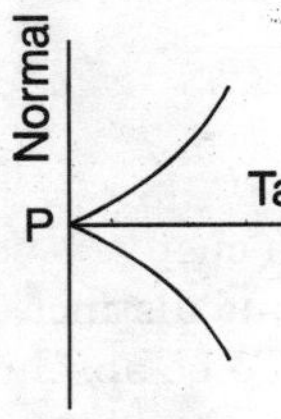

Fig. (*i*)
Single cusp of first species

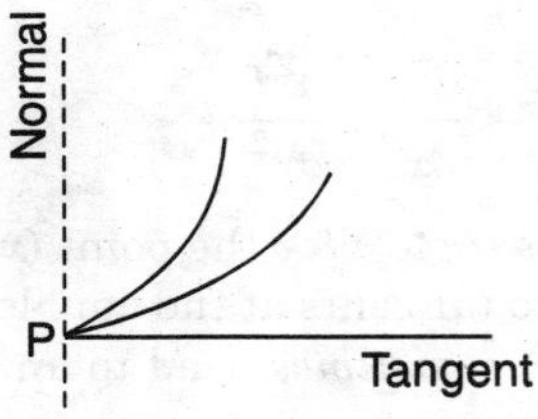

Fig. (*ii*)
Single cusp of second species

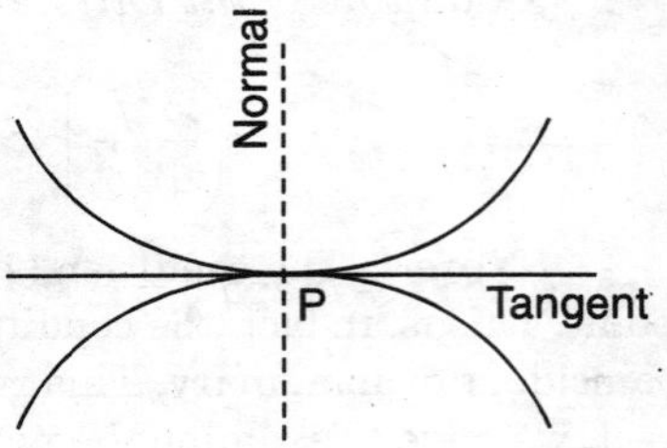

Fig. (*iii*)
Double cusp of first species

Point of oscul-inflexion. A double cusp *with the change of species on the two sides* of the *common normal* is called a *point of oscul-inflexion* [See Fig. (*v*)].

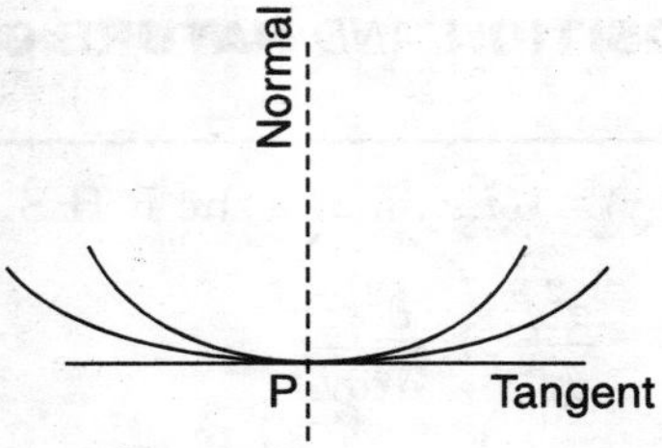

Fig. (*iv*)
Double cusp of second species

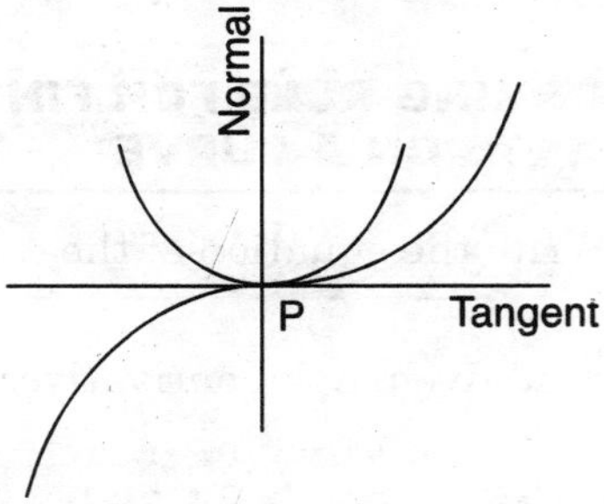

Fig. (*v*)
Point of oscul-inflexion

2.28. WORKING RULE FOR DETERMINING THE NATURE OF A CUSP AT THE ORIGIN

Case 1. *When the cuspidal tangents are $y^2 = 0$, i.e., x-axis.*

Solve the equation of the curve for y, neglecting terms containing powers of y higher than two.

(*i*) If the *roots are real for only one sign of x*, the cusp is a *single cusp.*

(*ii*) If the *roots are real for both signs of x*, the cusp is a *double cusp.*

(*iii*) If the *roots are opposite in sign*, then the cusp is of the *first species.*

(*iv*) If the *roots are of the same sign*, then cusp is of the *second species.*

Case 2. *When the cuspidal tangents are $x^2 = 0$, i.e., y-axis.*

Solve the equation of curve for x, neglecting terms containing powers of x higher than two and distinguish between different cases that arise on similar lines, as in case 1.

Case 3. *When the cuspidal tangents are of the form* $(ax + by)^2 = 0$.

Let p be the length of perpendicular from any point (x, y) of the curve on the tangent line $ax + by = 0$.

Then $$p = \frac{ax + by}{\sqrt{a^2 + b^2}}$$

or p is proportional to $ax + by$.

Let us put $p = ax + by$ and then eliminate y between this equation and the equation of the curve, we shall get a relation between p and x.

Solving this relation for p (neglecting p^3 and higher powers of p), we shall be able to decide the nature of the cusp from the real or non-real values of p and the signs of p (like those of y in case (1) for small values of x, positive and negative.

Note. If it is more convenient to eliminate x (instead of y) between $p = ax + by$, and the equation of the curve, then a relation between p and y, found by eliminating x, also serves the same purpose, when small values of y are considered.

Corollary. The *nature of the cusp at a point other than origin*, can be found by transferring the origin to that point and proceeding as above.

2.29. CONCAVITY *AND* CONVEXITY

Let P be a point on curve $y = f(x)$, where the tangent PT is not parallel to y-axis. Then the curve is said to be

(*i*) *Concave upwards or convex downwards* at P, if in the immediate neighbourhood of P, the curve lies above the tangent PT on both sides, [as in Fig. (1)]; and

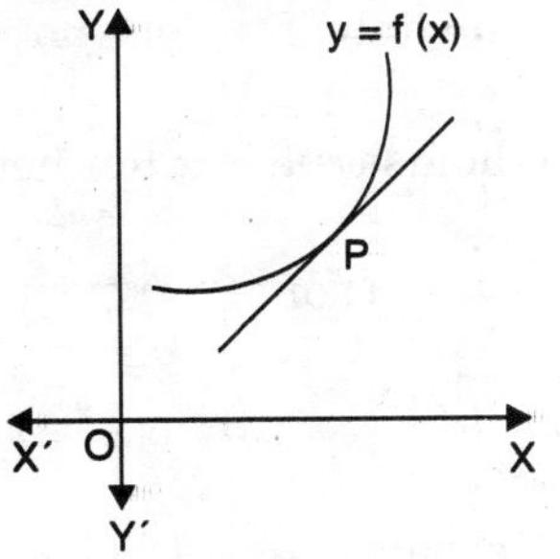

Fig. 1

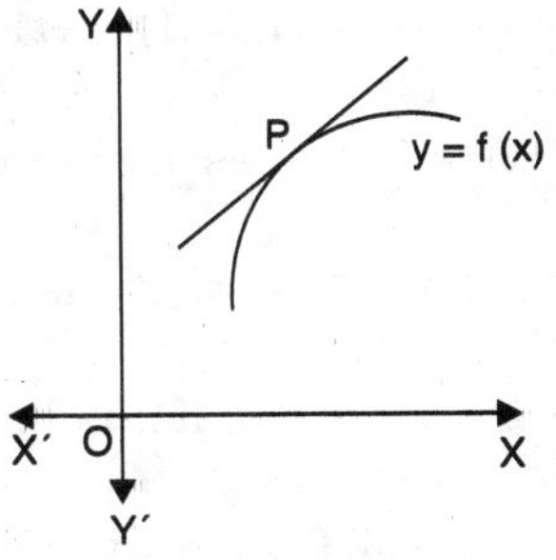

Fig. 2

(*ii*) *Concave downwards or convex upwards* at P, if in the immediate neighbourhood of P, the curve lies below the tangent PT on both sides [as in Fig. (2)].

Note. When the tangents at P are parallel to y-axis, we may define *concavity to the right or convexity to the left, and concavity to the left or convexity to the right* (as at P_1 and P_2 respectively in the adjoining figure) on similar lines.

2.30. CRITERIA FOR CONCAVITY *AND* CONVEXITY

At a point P on the curve $y = f(x)$, the curve is

(*i*) *Concave upwards (or in the positive direction of y-axis) if* $f''(x)$ *or* $\frac{d^2y}{dx^2} > 0$, *and*

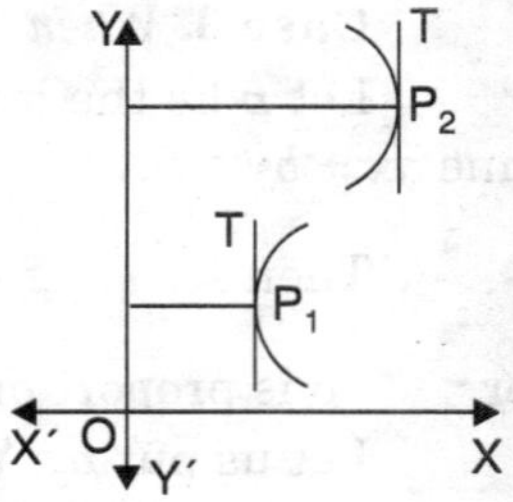

(ii) Concave downwards (or in the negative direction of y-axis) if $f''(x)$ *or* $\dfrac{d^2y}{dx^2}$ *is* < 0.

In general, the curve $y = f(x)$, *is concave upwards or downwards point* $P(x, y)$ *according as* $f^n(x)$ *is positive or negative and n is even.*

ILLUSTRATIVE EXAMPLES

Example 1. *Show that the curve* $y = e^x$ *is everywhere concave upwards.*

Sol. We have $y = e^x$ $\quad\therefore\quad \dfrac{dy}{dx} = e^x$ and $\dfrac{d^2y}{dx^2} = e^x$

Since $\dfrac{d^2y}{dx^2}$ is always positive, $\therefore$ the curve is concave upwards everywhere.

Example 2. *Find the intervals of x for which the graph of the function* $f(x) = (x^2 + 4x + 5)\, e^{-x}$ *is (i) concave upwards (ii) concave downwards.*

Sol. $\qquad f(x) = (x^2 + 4x + 5)\, e^{-x}$

$\therefore \qquad f'(x) = (x^2 + 4x + 5)\,.\, e^{-x}(-1) + e^{-x}(2x + 4) = -(x^2 - 2x + 1)\, e^{-x}$

and $\qquad f''(x) = -(x^2 + 2x + 1)\,.\, e^{-x}(-1) - e^{-x}(2x + 2) = (x - 1)(x + 1)\, e^{-x}.$

Since $e^{-x} > 0$ for each x, +ve or –ve, hence the sign of $f''(x)$ depends upon the sign of $(x - 1)(x + 1)$.

Now $f''(x) > 0$ implies $(x - 1)(x + 1) > 0$, which is true only if x does not lie between -1 and 1 *i.e.*, if $x < -1$ or $x > 1$.

Further $f''(x) < 0$ implies $(x - 1)(x + 1) < 0$, which holds true if x lies between -1 and 1 *i.e.*, if $-1 < x < 1$.

Thus, if $x < -1$ or $x > 1$ *i.e.*, if x is in the interval $(-\infty, -1)$ or $(1, \infty)$, $f''(x) > 0$ so that the graph of f is concave upwards.

Again, if $-1 < x < 1$ *i.e.*, if x is in the open interval $(-1, 1)$, then $f''(x) < 0$ so that the graph of f will be concave downwards.

Hence the graph of f is concave upwards in the intervals $(-\infty, -1)$, $(1, \infty)$ and is concave downwards in the interval $(-1, 1)$.

2.31. POINT OF INFLEXION

Definition. *A point on a curve at which the curve changes from concavity to convexity or vice versa is called a* **point of inflexion.**

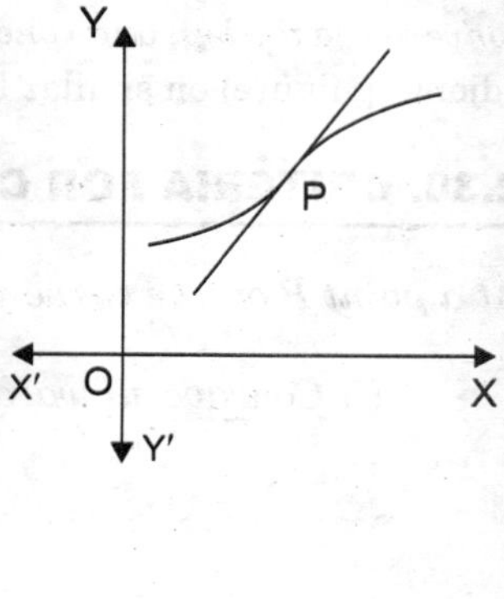

Since a change from concavity to convexity or *vice versa* at a point is possible only if the curve crosses the tangent at that point. Therefore, we may also define *the point of inflexion* as *a point on the curve at which the curve crosses the tangent.*

Remark. It is to be noted that the curve at a point of inflexion exhibits an extra-ordinary or unusual behaviour, for the curve does not ordinarily cross the tangent, as it does at a point of inflexion. Hence a point of inflexion on a curve is also a *singular point.*

2.32. CRITERIA FOR A POINT OF INFLEXION

By definition at a point of inflexion, the curve changes from concavity to convexity or *vice versa*.

$\therefore \quad \frac{d^2y}{dx^2}$ is positive on one side of the point of inflexion P and negative on other side of P. Also if $\frac{d^2y}{dx^2}$ is continuous at P, it can pass from positive values to negative values or *vice versa,* only by passing through zero value of P.

Hence *a point P is a point of inflexion,* if

(i) $\frac{d^2y}{dx^2} = 0$ *at this point, and*

(ii) $\frac{d^2y}{dx^2}$ *changes sign in passing through this point i.e.,* $\frac{d^3y}{dx^3} \neq 0$, *at this point.*

Note. If $\frac{dy}{dx}$ becomes infinite, then $\frac{d^2y}{dx^2}$ does not exist and the points of inflexion are found by considering $\frac{d^2x}{dy^2} = 0$ and its change of sign.

2.33. CURVE TRACING

Let us consider the problem of tracing curves *i.e.*, of finding approximate shape of curves from their cartesian, polar or parametric equations without having to plot a large number of points on them. We shall consider mainly those curves whose equations can be solved for y, x or r. In the following article, we shall explain the various steps which are helpful in tracing a curve. But these steps and (particularly their order) is by no means rigid and some steps can be omitted and their order can be varied to suit the particular problem.

2.34. PROCEDURE FOR TRACING CURVES HAVING CARTESIAN EQUATIONS

I. Symmetry

(i) The curve is symmetrical about the **x-axis**, if the equation of the curve remains unchanged when **y is replaced by – y** *i.e.*, if the equation of the curve contains only **even powers of y.**

e.g., $y^2 = x^2 + 4$ is symmetrical about x-axis, whereas $y^2 + yx = x^2$ is not symmetrical about x-axis.

(ii) The curve is symmetrical about the **y-axis** if the equation of the curve does not change when **x is replaced by – x** *i.e.*, if the curve contains **only even powers of x.**

e.g., $x^2 = 4ay$ is symmetrical about y-axis, whereas $y = x^3 - 3ax^2$ is not symmetrical about y-axis.

(iii) The curve is symmetrical about the **Line y = x** if the equation of the curve remains unchanged when **x is changed to y** and **y is changed to x.**

e.g., $x^3 + y^3 = 3axy$ is symmetrical about the line $y = x$, whereas $x^5 + y^5 = 5ax^2y$ is not symmetrical about the line $y = x$.

(*iv*) The curve is symmetrical about the **Line $y = -x$** if the equation of the curve remains unchanged when **x is changed to $-y$** and **y is changed to $-x$.**

e.g., $x^4 + y^4 = 4a^2xy$ is symmetrical about the line $y = -x$, whereas $x^3 + y^3 = 3axy$ is not symmetrical about this line.

(*v*) The curve is symmetrical in **opposite quadrants** if the equation of the curve remains unchanged when **x is changed to $-x$** and **y is changed to $-y$.**

e.g., $x^5 + y^5 = 5ax^2y$ is symmetrical in opposite quadrants, whereas $x^3 + y^3 = 3axy$ is not symmetrical in opposite quadrants.

II. Origin

(*a*) Find whether the curve passes through the origin. If the constant term is missing from the equation of the algebraic curve, then it passes through the origin.

If the algebraic curve passes through the origin, then write down the equation of the tangents at the origin **by equating the lowest degree terms to 0.**

If the origin is a *double point* (*i.e.*, there are two tangents at the origin); then find its nature whether a *node* (if the tangents are *real, distinct*) or a *cusp* (if the tangents are *real, coincident*) or a *conjugate* (or *isolated*) point if the tangents are *imaginary*. If a *cusp*, find its type.

A cusp is called a **single cusp** or a **double cusp** according as the two branches of the curve lie entirely *on one side* or *on both sides of the common normal.*

A cusp *single* or *double* is said to be of *first kind* or *second kind* according as the two branches of the curve, *lie on opposite or same side of the common tangent.*

(*b*) **Position of curve w.r.t. tangent at the origin**

Find $y_c - y_t$ (or $y_c^2 - y_t^2$ if square roots are there), where y_c and y_t are respectively the ordinates of a point on the curve and a point on the tangent at the origin for the same value of x.

Then discuss the two cases namely $x > 0$ and $x < 0$.

If $(y_c - y_t)$ is + ve *i.e.*, $y_c > y_t$; then the curve lies above the tangent and if $(y_c - y_t)$ is –ve *i.e.*, $y_c < y_t$; then the curve lies below the tangent.

III. Asymptotes

Find all the asymptotes of the curve. The curve will not go beyond its asymptotes.

(*i*) Position of the curve w.r.t. asymptotes || to x-axis.

Let the asymptote parallel to x-axis be $y = a$.

Find the value of $y - a$ from the equation of the curve.

(Generally, $(y - a)$ will occur as a factor in the equation of the curve.)

Now in the R.H.S. of this value of $y - a$, put $y = a$.

Then discuss the two cases when x is > 0 (Near ∞) and $x < 0$ (Near $-\infty$).

If $(y - a)$ is +ve, the curve lies above the asymptote $y = a$

and if $(y - a)$ is –ve, the curve lies below the asymptote $y = a$.

(*ii*) Position of the curve w.r.t. asymptotes || to y-axis.

Let the asymptote parallel to y-axis be $x = a$.

Find the value of $x - a$ from the equation of the curve.

(Generally $(x - a)$ will occur as a factor in the equation of the curve).

Now in the R.H.S. of this value of $x - a$, put $x = a$.

Then discuss the two cases when $y > 0$ and $y < 0$.

(It may be noted that $y > 0$ in First and Second Quadrants and $y < 0$ in Third and Fourth Quadrants.)

If $(x - a)$ is +ve, the curve lies to the right of the asymptote $x = a$ and if $(x - a)$ is –ve, the curve lies to the left of the asymptote $x = a$.

(*iii*) Position of the curve w.r.t. oblique asymptotes

Write the equation of the curve or one of its branches is in the form

$$y = mx + c + \frac{A}{x} + \frac{B}{x^2} + \ldots,$$

then $y = mx + c$ is the asymptote to the curve.

$\therefore\quad y_c - y_a = \frac{A}{x} + \frac{B}{x^2} + \ldots$ where y_c and y_a stand for the ordinates of a point on the curve and a point on the asymptote (both having same abscissa). Now discuss the two cases namely $x > 0$ and $x < 0$.

If $(y_c - y_a)$ is positive, curve lies above the asymptote.

If $(y_c - y_a)$ is negative, curve lies below the asymptote.

IV. Points of Intersections. (These can be obtained by solving the two equations).

Find intersections of the curve (*i*) with the x-axis ($y = 0$); (*ii*) with the y-axis ($x = 0$)

(*iii*) with the line $y = x$ particularly if the curve is symmetrical about it.

(*iv*) With the line $y = -x$ particularly if the curve is symmetrical about it.

(*v*) With the asymptotes (if necessary).

V. Region. Find regions in the four quadrants to which the curve is limited. This is usually done by solving for y or for x (or for each of the two) separately and considering both positive and negative values of x (or y). Values of x (or y) which make y (or x) imaginary are to be rejected.

Again, such values which make the left and right members of an equation opposite in sign are to be rejected.

Note. If $x^2 \le a^2$; then $-a \le x \le a$. If $x^2 \ge a^2$; then either $x \ge a$ or $x \le -a$.

If $(x - \alpha)(x - \beta) \le 0$; then x lies between α and β.

If $(x - \alpha)(x - \beta) \ge 0$ then x does not lies between α and β.

VI. Special points (or singularities)

(*a*) **Find** $\frac{dy}{dx}$ $\left(\frac{dy}{dx}\right.$ represents the slope of the tangent to the curve at the point $(x, y)\left.\right)$

Find the points where

(*i*) $\frac{dy}{dx} = 0$ *i.e.*, tangents are || to x-axis

(*ii*) $\frac{dy}{dx} = \infty$ *i.e.*, tangent are || to y-axis

(*iii*) $\frac{dy}{dx}$ is positive *i.e.*, the function is increasing *i.e.*, the curve is rising.

(*iv*) $\frac{dy}{dx}$ is negative *i.e.*, the curve is falling.

Find $\dfrac{d^2y}{dx^2}$ (If not tedious) and find the points where

(*i*) $\dfrac{d^2y}{dx^2}$ is +ve *i.e.*, the curve is concave upwards.

(*ii*) $\dfrac{d^2y}{dx^2}$ is –ve *i.e.*, the curve is concave downwards.

(*iii*) Also find **points of Inflexion.**

At a point of Inflexion, curve crosses the tangent. At a point of inflexion $\dfrac{d^2y}{dx^2} = 0$ and $\dfrac{d^3y}{dx^3} \neq 0$ or $\dfrac{d^2y}{dx^2}$ changes sign while passing through it.

Remark. The steps of Art. 2.34 can be remembered as:

R—SOAP

where S ≡ Symmetry, O ≡ Origin, A ≡ Asymptotes ; P ≡ Points and R ≡ Region,

Note. Sometimes inconvenient steps may be omitted without any disadvantages.

ILLUSTRATIVE EXAMPLES

Example 1. *Trace the curve* $y^2(2a - x) = x^3$ **(Cissoid).** (A.K.T.U. 2016)

Sol. The equation of the curve is $y^2(2a - x) = x^3$ ...(1)

1. Symmetry. Since equation (1) contains only even powers of y, the curve is symmetrical about x-axis.

2. Origin. The curve passes through the origin. The tangents at the origin are given by $y^2 = 0$ *i.e.*, $y = 0$, $y = 0$. Since the two tangents are real and coincident, ∴ origin is a cusp.

3. Asymptotes. Equating to zero, the co-efficient of y^2, the highest degree term in y, the asymptote ∥ to y-axis is $x - 2a = 0$ *i.e.*, $x = 2a$. There is no other asymptote of the curve.

We can discuss the position of the curve w.r.t. asymptote $x = 2a$.

4. Points of intersection with axes. The curve meets x-axis and y-axis at the origin only.

5. Region. From (1), $y = x\sqrt{\dfrac{x}{2a - x}}$

when $x < 0$, y is imaginary.

∴ No portion of the curve lies to the left of the line $x = 0$ *i.e.*, y-axis.

When $0 < x < 2a$, y is Real.

When $x > 2a$, y is Imaginary.

∴ No portion of the curve lies to the right of the line $x = 2a$.

6. Special points

From (1), $$y = \frac{x^{3/2}}{\sqrt{2a - x}} \qquad ...(2)$$

$$\therefore \qquad \frac{dy}{dx} = \frac{\sqrt{x}\,(3a - x)}{(2a - x)^{3/2}}$$

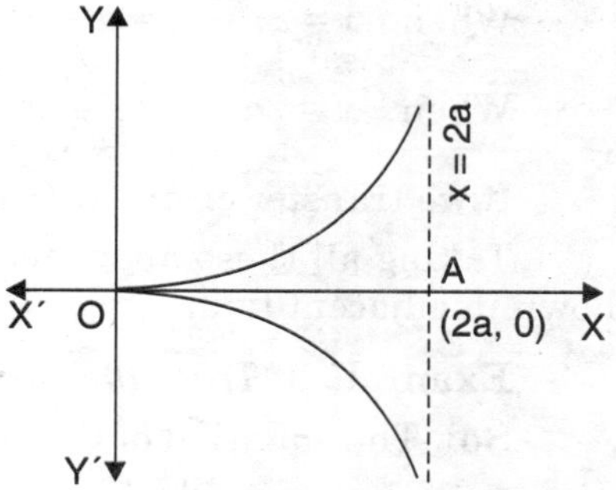

Now, $\frac{dy}{dx} = 0$ when $\sqrt{x}\,(3a - x) = 0$

or $x = 0, \quad x = 3a$

Rejecting $x = 3a$, because when $x = 3a$, from (2), y is imaginary.

When $x = 0, y = 0$ $\therefore$ Tangent at (0, 0) *i.e.*, at origin is || to x-axis *i.e.*, the tangent at origin is x-axis.

When $0 < x < 2a$, $\frac{dy}{dx}$ is positive.

$\therefore$ For positive values of y, y is an increasing function of x *i.e.*, the curve rises for values of x between 0 and $2a$.

Thus, the approximate shape of curve is as shown in the above figure.

Example 2. *Trace the curve:* $y^2 (a + x) = x^2 (3a - x)$. (A.K.T.U. 2018)

Sol. The equation of the curve is

$$y^2 (a + x) = x^2 (3a - x) \qquad ...(1)$$

1. Symmetry. Since equation (1) contains only even powers of y, the curve is symmetrical about x-axis.

2. Origin. The curve passes through the origin (0, 0). The tangents at the origin are given by

$$y^2 = 3x^2$$

$$\Rightarrow \qquad y = \pm\sqrt{3}\,x$$

which represent two non-coincident straight lines.

Hence we may expect a node at the origin.

3. Asymptotes. Equating to zero, the co-efficient of y^2, the highest degree term in y, the asymptote parallel to y-axis is

$$a + x = 0$$

$$\Rightarrow \qquad x = -a$$

4. Points of intersection with axes. The curve meets x-axis at (0, 0) and $(3a, 0)$ while it meets y-axis only at (0, 0).

5. Region. Equation (1) can be rewritten as

$$y = x\sqrt{\frac{3a - x}{a + x}}$$

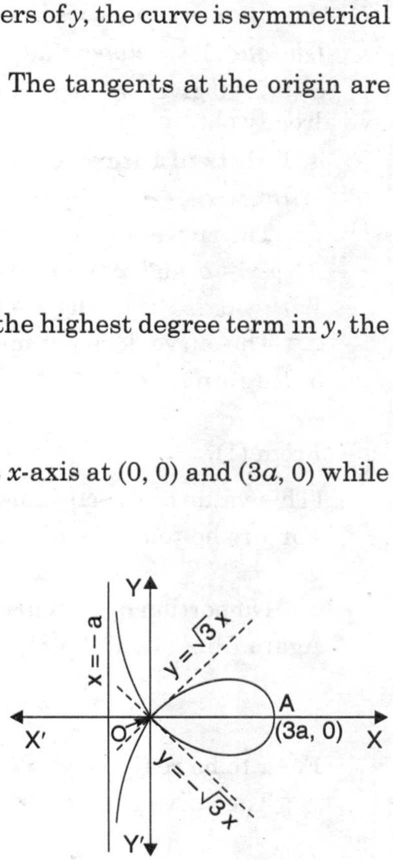

If $x = 0$, then $y = 0$.

When x is positive and small, y is real

Also $\frac{3a - x}{a + x} < \frac{3a}{a}$ *i.e.*, < 3

$$\therefore \qquad y < \sqrt{3}\,x$$

Hence curve lies below tangent $y = \sqrt{3}\,x$.

At $x = 3a, \quad y = 0$

When $x > 3a$, y is imaginary

When $x = a, \quad y = a$

When $x = 2a, \quad y = \dfrac{2a}{\sqrt{3}} = 1.2\,a$ (nearly)

If we transfer origin at $(3a, 0)$, then by equation (1), $x = 0$ will be tangent.

Taking all these above points in consideration, the approximate shape of the curve is shown in adjacent figure.

Example 3. *Trace the curve:* $x^2y^2 = x^2 - a^2$.

Sol. The equation of the curve is $x^2y^2 = x^2 - a^2$ ...(1)

1. Symmetry. Since equation (1) contains only even powers of y and only even powers of x, therefore, the curve is symmetrical about both the axes. Also the curve is symmetrical in opposite quadrants.

2. Origin. The curve does not pass through the origin because (0, 0) does not satisfy the given equation.

3. Asymptotes. *Asymptotes || to x-axis* are given by $y^2 - 1 = 0 \quad \therefore \quad y = 1$ and $y = -1$ are the asymptotes.

Asymptotes || to y-axis is $x = 0$.

Oblique Asymptotes. $\phi_4(m) = m^2$.

$\phi_4(m) = 0$ gives $m = 0$ and this value of m gives asymptotes parallel to x-axis (which we have already obtained).

4. Points of intersection

With x-axis. Putting $y = 0$ in (1), $x^2 = a^2 \quad \therefore \quad x = \pm a$.

$\therefore$ The curve meets the x-axis in the points $(a, 0)$ and $(-a, 0)$.

Also $x = a$ and $x = -a$ are the tangents at the points $(a, 0)$ and $(-a, 0)$ respectively.

With y-axis. Putting $x = 0$ in (1), we get $0 = -a^2$, which is impossible.

$\therefore$ The curve does not meet the y-axis.

5. Region

From (1), $$y = \frac{\sqrt{x^2 - a^2}}{x} \quad ...(2)$$

(This value of y corresponds to the part of the curve in first two quadrants.)

For y to be real, $x^2 - a^2$ must be positive *i.e.*, $x^2 > a^2$.

$\therefore \quad x > a \quad$ or $\quad x < -a$.

$\therefore$ No portion of the curve lies between the lines $x = a$ and $x = -a$ ...(3)

Again from (1), $x^2(1 - y^2) = a^2$

$$\therefore \quad x = \frac{a}{\sqrt{1 - y^2}}.$$

For x to be real, $1 - y^2 > 0$ *i.e.*, $y^2 < 1$. $\therefore \quad -1 < y < 1$.

$\therefore$ The curve lies entirely between the lines $y = -1$ and $y = 1$.

6. Special points

From (2), $\dfrac{dy}{dx} = \dfrac{a^2}{x^2\sqrt{x^2 - a^2}}$

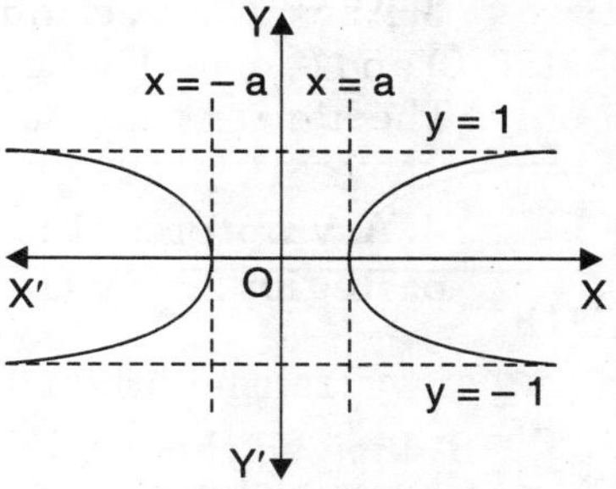

$\dfrac{dy}{dx}$ never becomes zero. $\dfrac{dy}{dx}$ is always positive.

$\dfrac{dy}{dx} \to \infty$ when $x^2 - a^2 = 0$ *i.e.,* $x = \pm a$

For $x = \pm a$, $y = 0$ from (1)

$\therefore$ Tangents at the points $(a, 0)$ and $(-a, 0)$ are parallel to y-axis.

Thus, the shape of the curve is as shown in the adjoining figure.

Example 4. *Trace the following curve:*

$$xy^2 = a^2(a - x).$$ **[Witch of Agnesi]**

Sol. The equation of the curve is

$$xy^2 = a^2(a - x) \qquad ...(1)$$

1. Symmetry. Since (1) contains only even powers of y, so the curve is symmetrical about x-axis.

2. Origin. The curve does not pass through the origin.

3. Points of intersection with axes. The curve meets X-axis at the point A(a, 0). The curve does not meet Y-axis. Shifting the origin to A(a, 0), (1) transforms to

$$(X + a)Y^2 = a^2[a - (X + a)]$$

$$\Rightarrow \qquad (X + a)Y^2 + a^2X = 0$$

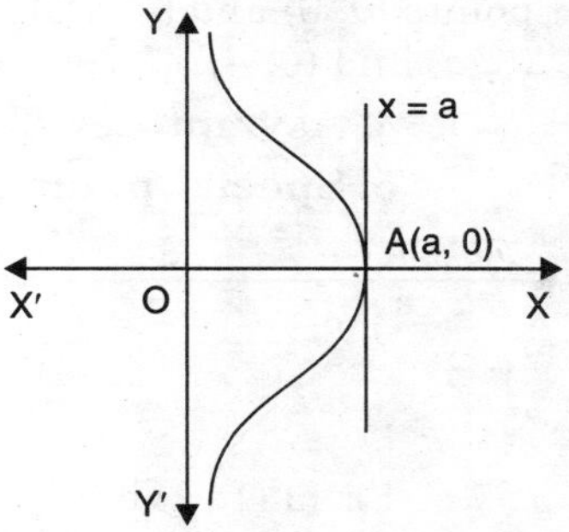

Equating to zero, the lowest degree terms, the tangent at the new origin is X = 0 *i.e.,* new y-axis.

Hence, at the new origin, *i.e.,* at A(a, 0), the tangent is ⊥ to x-axis.

4. Asymptotes. The only asymptotes is $x = 0$.

5. Region. From (1),

$$y = a\sqrt{\frac{a - x}{x}}$$

when $x < 0$ or when $x > a$, y is imaginary. Thus, no portion of the curve lies to the left of y-axis and to the right of $x = a$.

When $x = a$, $y = 0$. As x decreases from a to 0, y increases from 0 to ∞.

Hence, the shape of the curve is as shown in the above figure.

Example 5. *Trace the following curve :*

$$y^2(a - x) = x^2(a + x).$$ **(Strophoid)**

Sol. The given curve is

$$y^2(a - x) = x^2(a + x) \qquad ...(1)$$

1. Symmetry. Since powers of y are even, therefore the curve is symmetrical about x-axis.

2. Origin. The curve passes through the origin. The tangents at the origin are $y = \pm x$. Since the tangents are real and distinct, therefore origin is a node.

3. Axes intersection. The curve intersects x-axis at (0, 0) and $(-a, 0)$. The curve intersects y-axis at (0, 0) only. The tangent at $(-a, 0)$ can be found to be $x + a = 0$.

4. Asymptotes. The only asymptote is $x = a$.

5. Region. From (1),

$$y = x\sqrt{\frac{a+x}{a-x}}.$$

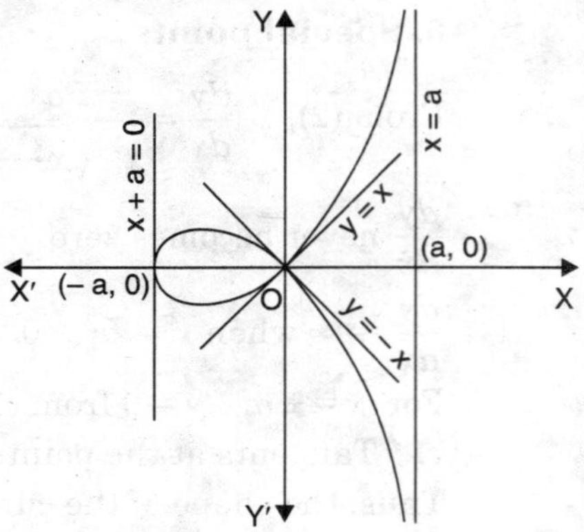

For $x > a$ or $x < -a$, y is imaginary. Hence, the curve does not exist for $x > a$ or $x < -a$.

The shape of the curve is as shown in the figure.

Example 6. *Trace the following curve:*

$$x^{2/3} + y^{2/3} = a^{2/3}.$$ **(Astroid)**

Sol. The equation of the curve is

$$x^{2/3} + y^{2/3} = a^{2/3}. \quad ...(1)$$

1. Symmetry. The curve is symmetrical about both the axes. The curve is symmetrical about the lines $y = x$ and $y = -x$.

2. Origin. The curve does not pass through the origin.

3. Axes intersection. The curve meets X-axis at the points $(a, 0)$ and $(-a, 0)$. The curve meets Y-axis at the points $(0, a)$ and $(0, -a)$.

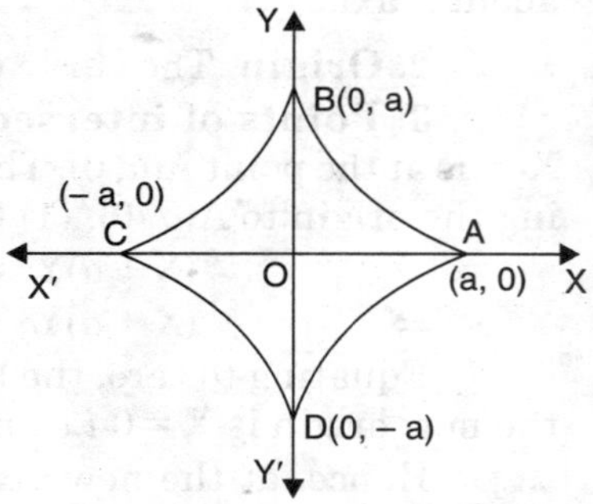

4. Asymptotes. The curve has no asymptotes.

5. Special points. From (1),

$$\frac{dy}{dx} = -\frac{y^{1/3}}{x^{1/3}}$$

$$\therefore \quad \frac{dy}{dx} = 0, \text{ when } y = 0.$$

From (1), When $y = 0$, $x = \pm a$

Hence, the tangents to the curve are parallel to x-axis at the points $(\pm a, 0)$.

Again, $\frac{dy}{dx} \to 0$, when $x = 0$.

From (1), When $x = 0$, $y = \pm a$

Hence, the tangents to the curve are parallel to y-axis at the points $(0, \pm a)$.

6. Region. From (1),

$$y^{2/3} = (a^{2/3} - x^{2/3})$$

If $\quad |x| > a$, $y^{2/3}$ is –ve

$\therefore$ y^2 is –ve

$\therefore$ y is imaginary.

Hence, the curve does not lie beyond the lines $x = \pm a$.

Similarly, the curve does not lie beyond the lines $y = \pm a$.

Again when $x = 0$, $y = a$. As x increases from 0 to a, y decreases from a to 0.

Hence, the shape of the curve is as shown in the figure.

Example 7. *Trace the following curve and write its asymptotes:*

$x^3 + y^3 = 3axy.$ **(Folium of Descartes)**

Sol. The equation of the curve is

$$x^3 + y^3 = 3axy \quad ...(1)$$

1. Symmetry. The curve is symmetrical about the line $y = x$.

2. Origin. The curve passes through the origin. The tangents at origin are $x = 0$, $y = 0$. These tangents being real and distinct, origin is a node.

3. Axes intersection. The curve meets the axes only at the origin (0, 0).

4. Asymptotes. The only asymptote is

$$x + y + a = 0.$$

5. Special points. The curve (1) meets the line of symmetry $y = x$, where

$$2x^3 = 3ax^2$$

$$\Rightarrow \quad x^2(2x - 3a) = 0$$

$$\Rightarrow \quad x = 0, \frac{3a}{2}$$

$$\therefore \quad y = 0, \frac{3a}{2}$$

$\therefore$ The two points of intersection are (0, 0) and $\left(\frac{3a}{2}, \frac{3a}{2}\right)$.

Slope of the tangent at A $\left(\frac{3a}{2}, \frac{3a}{2}\right)$

Differentiating (1) w.r.t. x, we get

$$3x^2 + 3y^2 \frac{dy}{dx} = 3a\left(x\frac{dy}{dx} + y\right)$$

$$\Rightarrow \quad \frac{dy}{dx} = \frac{ay - x^2}{y^2 - ax}$$

At A $\left(\frac{3a}{2}, \frac{3a}{2}\right)$, $\quad \frac{dy}{dx} = -1$

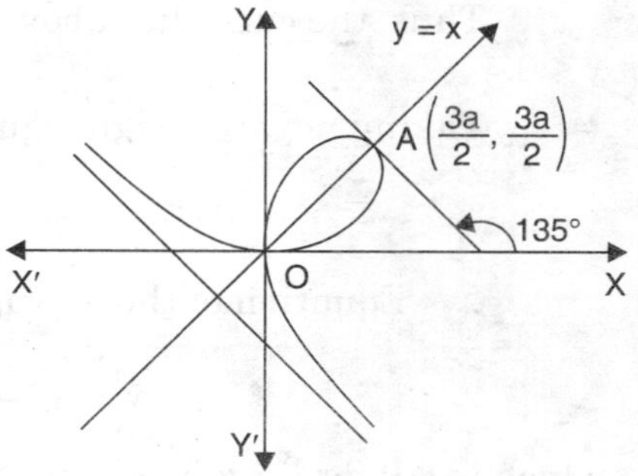

$$\Rightarrow \quad \tan\psi = -1 \quad \Rightarrow \quad \psi = 135°$$

Hence, the tangent at A to the curve makes an angle of 135° with the axis of X.

6. Region. x and y both cannot be negative, because that will make L.H.S. of (1) negative and RHS of (1) positive.

$\Rightarrow$ No portion of the curve lies in the third quadrant.

$\therefore$ The shape of one curve is as shown in the figure.

Example 8. *Trace the curve $y^2(x - a) = x^2 (x + a)$.*

Sol. The equation of the curve is $y^2(x - a) = x^2(x + a)$. ...(1)

1. Symmetry. The curve is symmetrical about the x-axis only.

2. Origin. The curve passes through the origin and the tangents at the origin are given by $x^2 + y^2 = 0$ or $y = \pm ix$, which are imaginary.

$\therefore$ Origin is an isolated point.

3. Asymptotes. There is no asymptote || to x-axis.

Equating to zero, the co-efficient of y^2, the asymptotes parallel to y-axis is $x - a = 0$.

Let us find oblique asymptotes.

From (1), $$y^2 = \frac{x^2(x+a)}{x-a} = \frac{x^2 \cdot x\left(1+\frac{a}{x}\right)}{x\left(1-\frac{a}{x}\right)} = x^2\left(1+\frac{a}{x}\right)\left(1-\frac{a}{x}\right)^{-1}$$

$\therefore$ $$y = \pm x\left(1+\frac{a}{x}\right)^{1/2}\left(1-\frac{a}{x}\right)^{-1/2}$$

or $$y = \pm x\left[1+\frac{1}{2}\frac{a}{x}-\frac{1}{8}\cdot\frac{a^2}{x^2}+\ldots\right]\left[1+\frac{a}{2x}+\frac{3}{8}\frac{a^2}{x^2}+\ldots\right]$$

or $$y = \pm x\left[1+\frac{a}{x}+\frac{a^2}{2x^2}+\ldots\right] = \pm\left[x+a+\frac{a^2}{2x^2}+\ldots\right]$$

Taking +ve sign, equation of one branch is

$$y = x + a + \frac{a^2}{2x} + \ldots$$

$\therefore$ Equation of the asymptote is $y = x + a$

$\therefore$ $$y_c - y_a = \frac{a^2}{2x} + \ldots$$

which is > 0 for large $x > 0$ and < 0 for $x < 0$.

Thus, the curve lies above the asymptote $y = x + a$ for $x > 0$ and below the asymptote for $x < 0$.

Taking negative sign, equation of another branch is

$$y = -x - a - \frac{a^2}{2x} - \ldots$$

$\therefore$ Equation of the asymptote is $y = -x - a$.

$\therefore$ $$y_c - y_a = \frac{-a^2}{2x} \ldots$$

which is < 0 for large $x > 0$ and > 0 for $x < 0$.

$\therefore$ The curve lies below the asymptote $y = -x - a$ for $x > 0$ and above the asymptote for $x < 0$.

4. Points of intersection

The curve cuts the x-axis, where putting $y = 0$, we get

$$x^2(x+a) = 0 \quad \textit{i.e.,} \quad x = 0 \text{ or } -a.$$

Thus, the curve passes through (0, 0) and $(-a, 0)$.

The curve meets y-axis, where putting $x = 0$, we get $-ay^2 = 0$

$\therefore$ $$y = 0 \quad \textit{i.e.,} \quad \text{at } (0, 0)$$

5. Region. From (1), $y = x\sqrt{\dfrac{x+a}{x-a}}$

When $x < -a$, y is Real.

When $-a < x < a$, y is imaginary. Hence no portion of the curve lies between the lines $x = \pm a$.

When $x > a$, y is Real.

6. Special points

From (1), $$y = x\sqrt{\frac{x+a}{x-a}} \qquad \text{...(2)}$$

$$\therefore \quad \frac{dy}{dx} = \frac{x^2 - ax - a^2}{(x+a)^{1/2}\,(x-a)^{3/2}}$$

Now, $\dfrac{dy}{dx} = 0$, when

$x^3 - ax - a^2 = 0$ or, when $x = \frac{1}{2}(1 \pm \sqrt{5})a$.

When $x = \left(\dfrac{1-\sqrt{5}}{2}\right) a$, from (2), y is imaginary.

So the tangent is $\parallel$ to x-axis when $x = \dfrac{1+\sqrt{5}}{2}\, a$.

Thus, the approximate shape of the curve is as shown in the above figure.

Example 9. *Trace the curve:* $y = x^3 - 3ax^2$.

Sol. The curve is $\qquad y = x^3 - 3ax^2 \qquad$ ...(1)

1. Symmetry. No symmetry.

2. Origin. The curve passes through the origin.

The equation of tangent at the origin is $y = 0$ *i.e.*, x-axis.

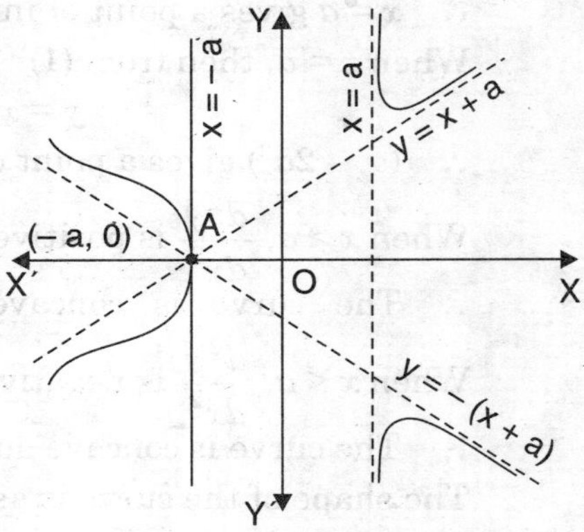

3. Asymptotes. No asymptotes.

4. Points of intersection

Intersection with x-axis. Putting $y = 0$ in (1), we get

$$x^3 - 3ax^2 = 0 \quad \text{or} \quad x^2(x - 3a) = 0$$

$$\therefore \qquad x = 0 \quad \text{or} \quad x = 3a$$

$\therefore$ Intersections with x-axis are (0, 0) and ($3a$, 0).

5. Region

From (1), $\qquad y = x^2(x - 3a)$

When $x < 0$, y is –ve

$\therefore$ No portion of the curve lies in the second quadrant.

When $0 < x < 3a$, y is –ve.

$\therefore$ The curve lies in fourth quadrant for values of x s.t. $0 < x < 3a$.

When $x > 3a$, y is +ve.

$\therefore$ The curve lies in the first quadrant for values of $x > 3a$.

6. Special points

From (1), $$\frac{dy}{dx} = 3x^2 - 6ax = 3x(x - 2a) \qquad \text{...(2)}$$

$$\frac{dy}{dx} = 0 \text{ gives } x = 0 \text{ and } x = 2a.$$

When $x = 0$, then from (1) $y = 0$ $\therefore$ Tangent at (0, 0) is x-axis.

when $x = 2a$, from (1) $y = 8a^3 - 12a^3 = -4a^3$

$\therefore$ Tangent at $(2a, -4a^3)$ is parallel to x-axis.

When $x < 0$, then from (2), $\dfrac{dy}{dx}$ is +ve.

$\therefore$ y increases as x increases *i.e.*, the curve is rising.

When $0 < x < 2a$, $\dfrac{dy}{dx}$ is –ve; $\therefore$ The curve falls in this portion.

When $x > 2a$, $\dfrac{dy}{dx}$ is +ve. $\therefore$ Again y increases as x increases. $(\forall\, x > 2a)$

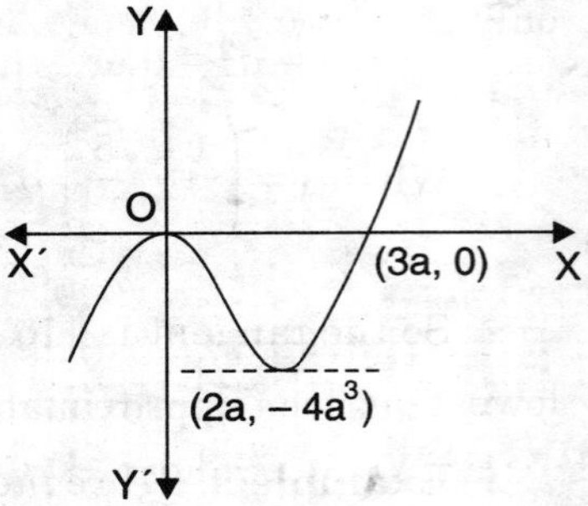

From (2), $$\frac{d^2y}{dx^2} = 6x - 6a = 6(x - a)$$

$$\frac{d^2y}{dx^2} = 0 \text{ gives } x = a, \quad \frac{d^3y}{dx^3} = 6.$$

At $x = a$, $$\frac{d^3y}{dx^3} = 6 \neq 0$$

$\therefore$ $x = a$ gives a point of inflexion.

When $x = a$, then from (1)

$$y = x^3 - 3ax^2 = a^3 - 3a^3 = -2a^3$$

$\therefore$ $(a, -2a^3)$ gives a point of inflexion.

When $x > a$, $\dfrac{d^2y}{dx^2}$ is positive.

$\therefore$ The curve is concave upwards for values of $x > a$.

When $x < a$, $\dfrac{d^2y}{dx^2}$ is negative.

$\therefore$ The curve is concave downwards for values of $x < a$.

The shape of the curve is as shown in the above figure.

Example 10. *Sketch the graph* $y = 1 - \dfrac{1}{1 + x^2}$.

Sol. The equation of given curve is

$$y = 1 - \frac{1}{1 + x^2} \qquad \ldots(1)$$

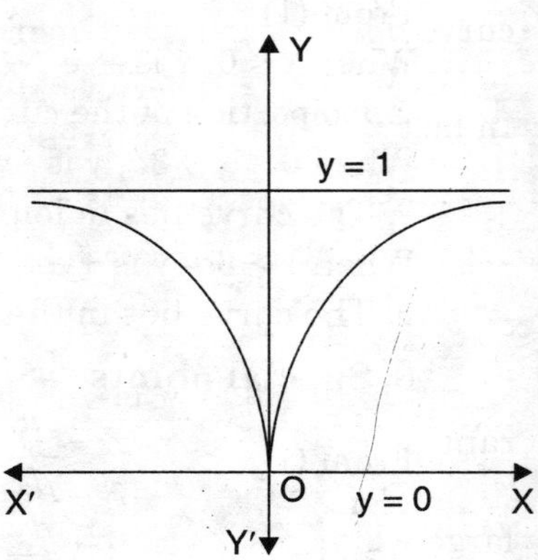

1. Symmetry. The curve is symmetrical about y-axis.

2. Origin. The curve passes through the origin. Equation (1) can be written as

$$x^2y + y = x^2$$

Tangents at the origin are given by

$$y = 0$$

3. Asymptotes. Asymptote parallel to x-axis is

$$y - 1 = 0$$

$$\Rightarrow \qquad y = 1$$

4. Points of intersection. Curve meets the axes only at the origin (0, 0)

5. Special points.

x :	-2	-1	0	1	2
y :	4/5	1/2	0	1/2	4/5

Hence the approximate shape of the given curve is shown in the figure.

Example 11. *Trace the curve* $y^2(x + 3a) = x(x - a)(x - 2a)$.

Sol. 1. Symmetry. The curve is symmetrical about x-axis.

2. Origin. The curve passes through the origin, where the tangent is $x = 0$.

3. Asymptote and Region. The equation of curve is already solved for y, considering only the (+)ve values of y, we find that

(*a*) When $x = 0$, $y = 0$; when $0 < x < a$, y is real ;

when $x = a$, $y = 0$ again $\left\{\text{when } x = \frac{a}{2}, y = \frac{a}{3} \text{ (app.)}\right\}$ and when $a < x < 2a$, y is imaginary.

when $x = 2a$, $y = 0$ and when $x > 2a$, y is real.

Tangent at $(2a, 0)$ is parallel to y-axis obtained by shifting origin to it and then equating lowest degree terms to zero.

When $x = 3a$, $y = a$ and when $x = 4a$, $y = 1.8a$.

When $x \to \infty$, $y \to \infty$ *i.e.* when x, y are very large,

$$\frac{y^2}{x^2} = \frac{\left(1 - \frac{a}{x}\right)\left(1 - \frac{2a}{x}\right)}{\left(1 + \frac{3a}{x}\right)} = 1 - \frac{6a}{x} + \frac{20a^2}{x^2} + \ldots$$

$$\Rightarrow \qquad \frac{y}{x} = \pm\left(1 - \frac{3a}{x} + \frac{11a^2}{2x^2} + \ldots\right)$$

or $$y = x - 3a + \frac{11a^2}{2x} + \ldots \qquad \ldots(1) \mid \text{Leaving } (-)\text{ve values of } y/x$$

$y = x - 3a$ is an asymptote and the curve lies above this asymptote in the I quadrant.

(*b*) When x is negative and numerically less than $3a$, y is imaginary. Therefore the curve does not exist between $x = -3a$ and $x = 0$.

When $x = -3a$, the expression for y becomes meaningless but $x \to -3a$ from left, $y^2 \to \infty$. In fact $x = -3a$ is an asymptote.

When $x \to -\infty$, $y^2 \to \infty$.

For large values of x and y, taking (–)ve values of y/x from (1), we get

$$y = -x + 3a - \frac{11a^2}{2x} + \ldots$$

$\therefore$ $y = 3a - x$ is an asymptote and the curve lies above the asymptote in the II quadrant.

$\therefore$ y is very large in the neighbourhood of $x = -3a$ in II quadrant and is again very large when x is (–)ve and numerically large.

Considering all these points, approximate shape of the curve is given below.

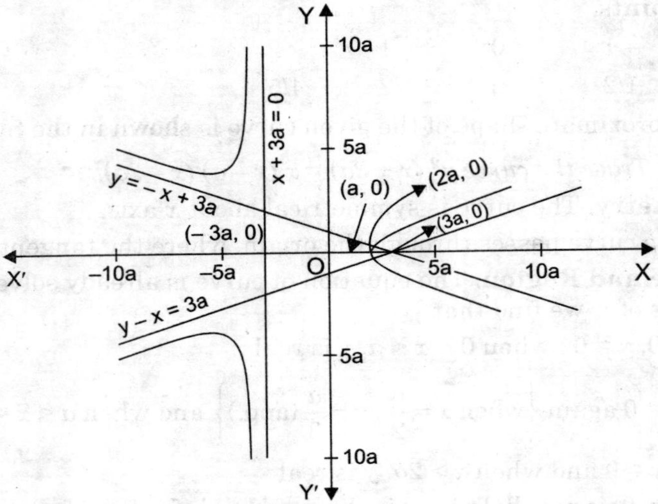

Example 12. *Trace the curve:* $x^{1/2} + y^{1/2} = a^{1/2}$.

Sol. The equation of the curve is

$$x^{1/2} + y^{1/2} = a^{1/2} \quad \ldots(1)$$

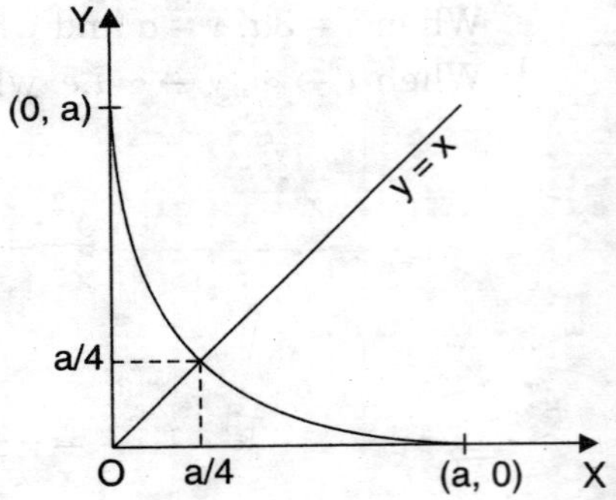

1. Symmetry. The curve is symmetrical about the line $y = x$.

2. Origin. The curve does not pass through origin.

3. Asymptotes. No asymptotes parallel to coordinate axes.

4. Points of intersections. When $x = 0$, $y = a$

When $y = 0$, $x = a$

Hence the curve meets the coordinate axes at $(0, a)$ and $(a, 0)$.

5. Special points.

x:	a	$a/4$	$a/2$	$3a/4$	0
y:	0	$a/4$	$\left(\frac{\sqrt{2}-1}{\sqrt{2}}\right)^2 a$	$\left(\frac{2-\sqrt{3}}{2}\right)^2 a$	a

The approximate shape of the given curve is shown in the above figure.

TEST YOUR KNOWLEDGE

Trace the following curves:

1. $y^2 = x^3$ (semi-cubical parabola)
2. $a^2y^2 = x^2(a^2 - x^2)$
3. $a^2y^2 = x^3(2a - x)$
4. $a^4y^2 = a^2x^4 - x^6$
5. $y^2(a^2 + x^2) = x^2(a^2 - x^2)$ or $x^2(x^2 + y^2) = a^2(x^2 - y^2)$
6. (*i*) $9ay^2 = x(x - 3a)^2$ (*ii*) $4ay^2 = x(x - 2a)^2$ (*U.P.T.U. 2014*)
7. $y^2(x^2 + y^2) + a^2(x^2 - y^2) = 0$ or $x^2(y^2 + a^2) + y^2(y^2 - a^2) = 0$.

8. $y^2 = (x-2)^2(x-5)$

9. $a^2x^2 = y^3(2a - y)$

10. $y^2 = (x-1)(x-2)(x-3)$

11. $xy^2 = 4a^2(2a - x)$

12. (a) $x^2y^2 = (a+y)^2(a^2 - y^2)$ (b) $x^2y^2 = (a^2 + y^2)(a^2 - y^2)$

13. $y(x^2 + 4a^2) = 8a^3$

14. $y^2 x = a^2(x - a)$

15. $y^2(x^2 - 1) = x$

16. (a) $y = \dfrac{x^2+1}{x^2-1}$ (b) $x^2y^2 = x^2 + 1$

17. $x^2y^2 = a^2(y^2 - x^2)$

18. $y^2(a+x) = x^2(a-x)$ or $x(x^2+y^2) = a(x^2 - y^2)$

19. $x^6 + y^6 = a^2x^2y^2$

20. $y = x^3$ (cubical parabola)

21. $y = x^3 - x$ *(U.P.T.U. 2014)*

22. $x^3 y - x - 1 = 0$

23. $(x^2 + a^2)\, y = a^2x$

24. $(x^2 - a^2)(y^2 - b^2) = a^2b^2$

25. $y^3 = x^3 + ax^2$

26. $x^3 + y^3 = a^2x$

27. (i) $y = (x-2)(x+1)^2$ (ii) $\left(\dfrac{x}{a}\right)^{2/3} + \left(\dfrac{y}{b}\right)^{2/3} = 1.$ (Hypocycloid)

Answers

1.

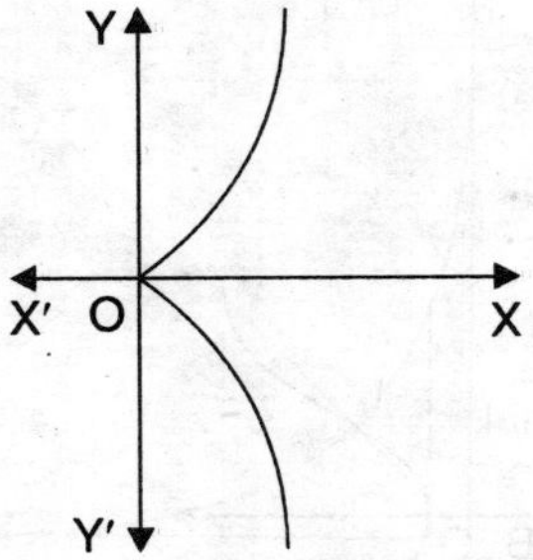

2.

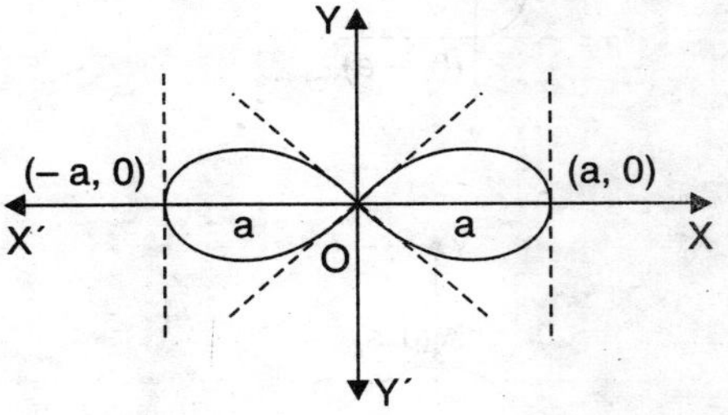

3.

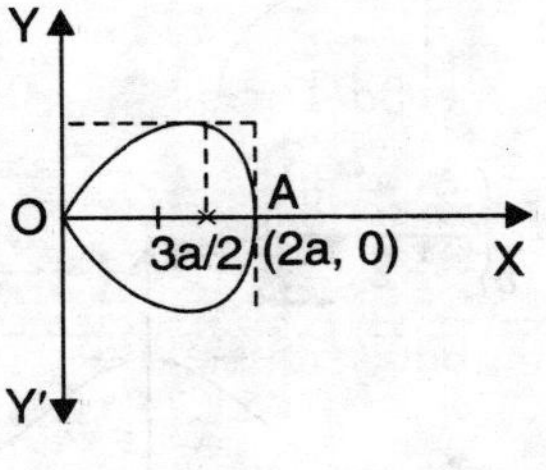

4.

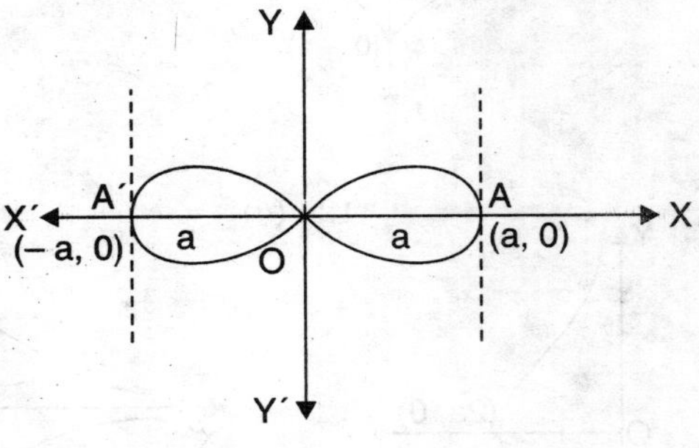

5.

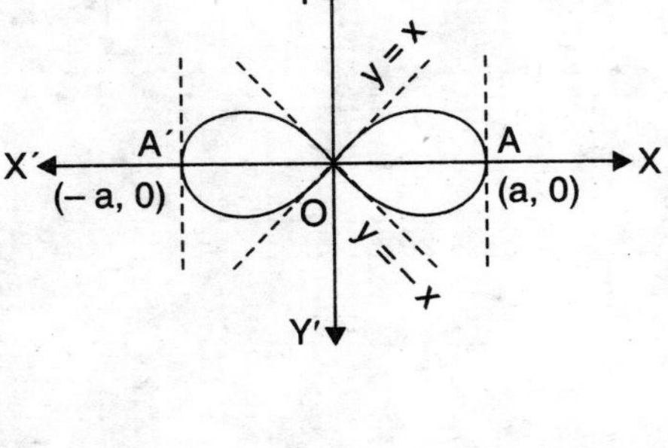

6. (*i*)

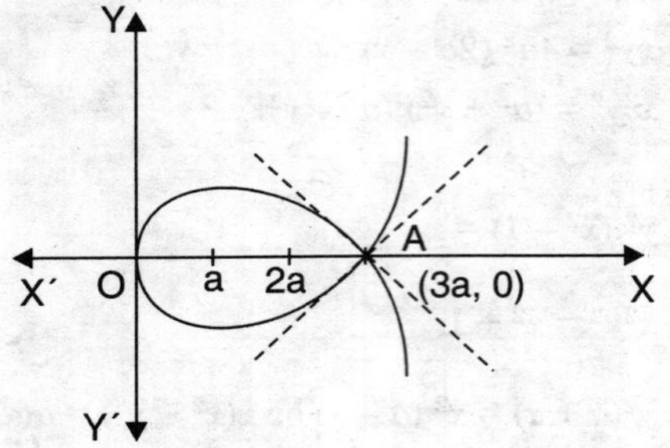

(*ii*)

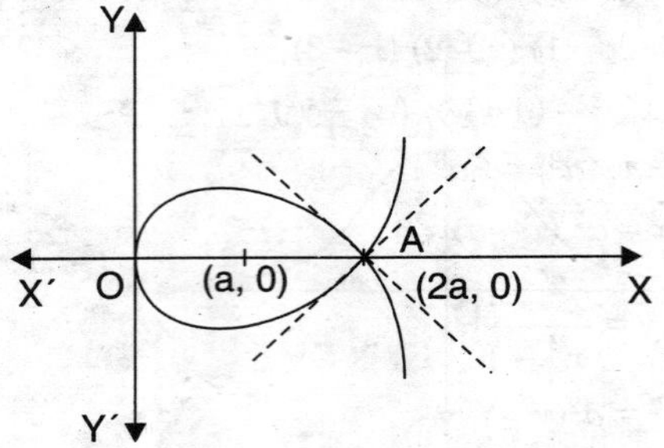

7.

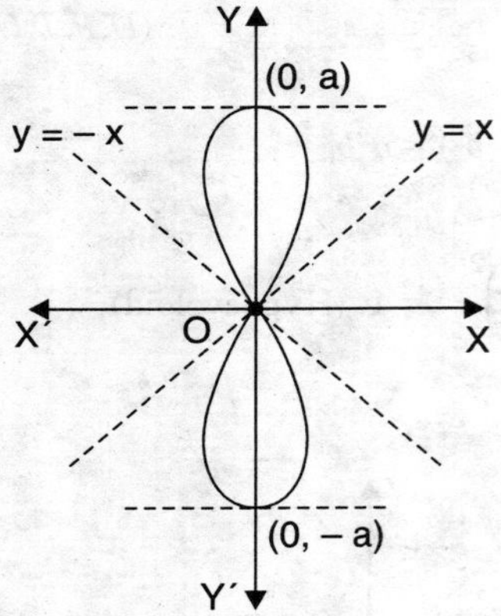

8.

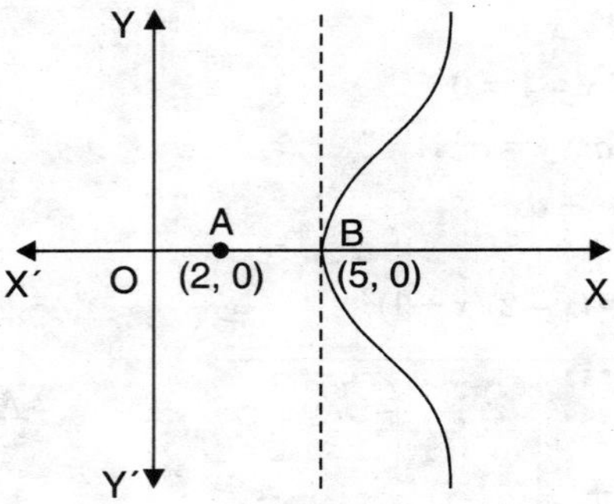

9.

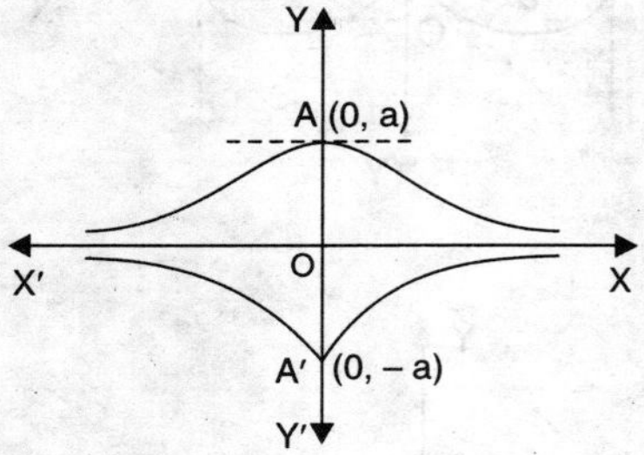

10.

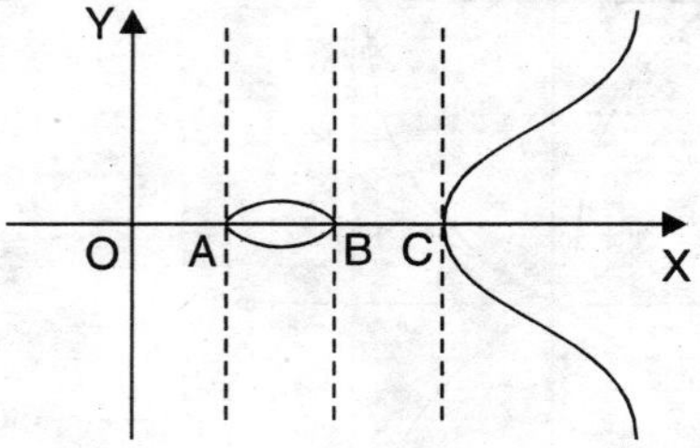

11.

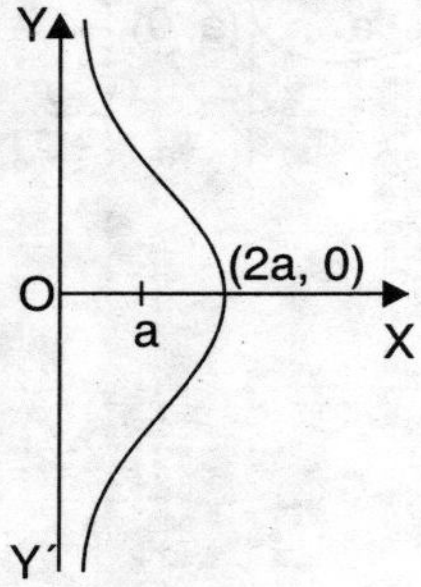

12 . (*a*)

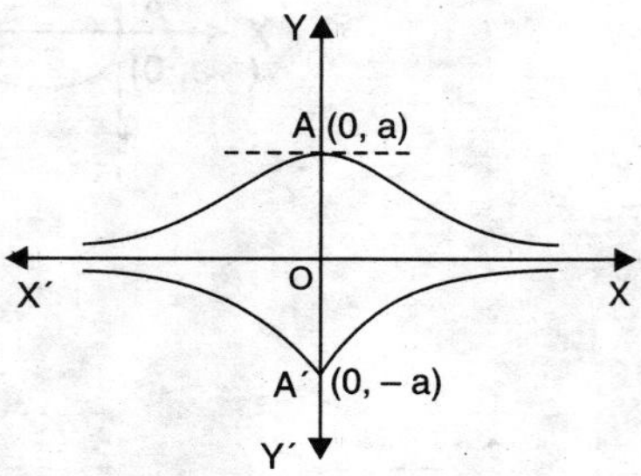

(*b*)

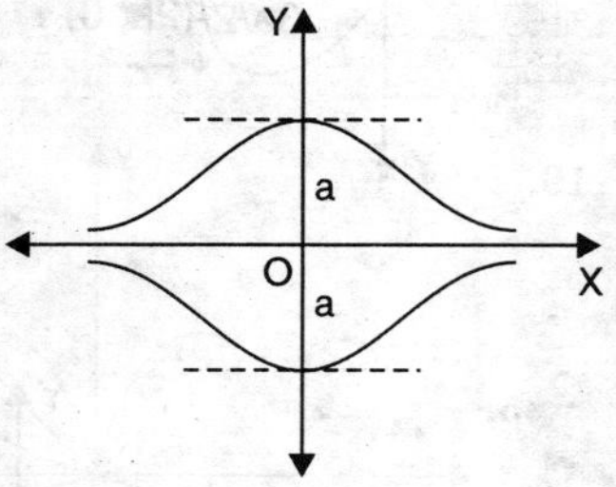

13.

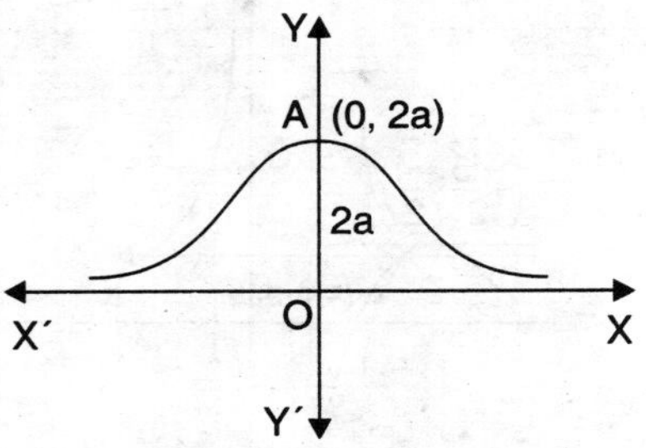

14.

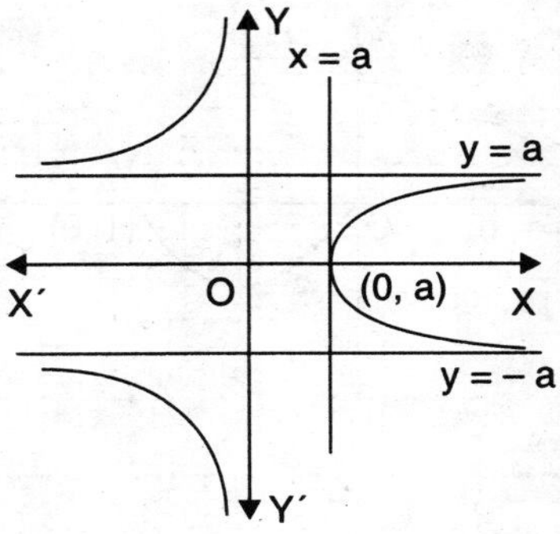

15.

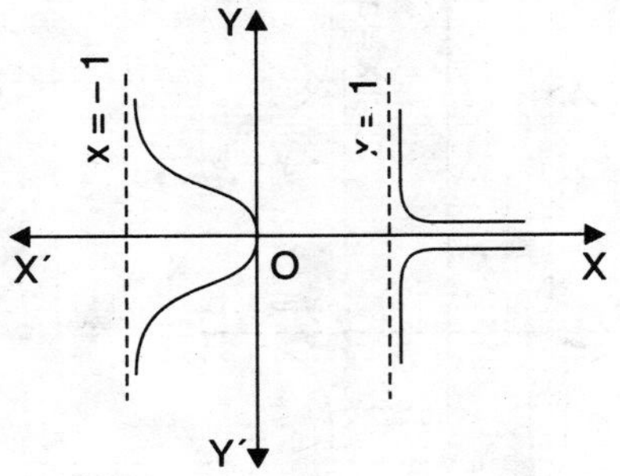

16. (a)

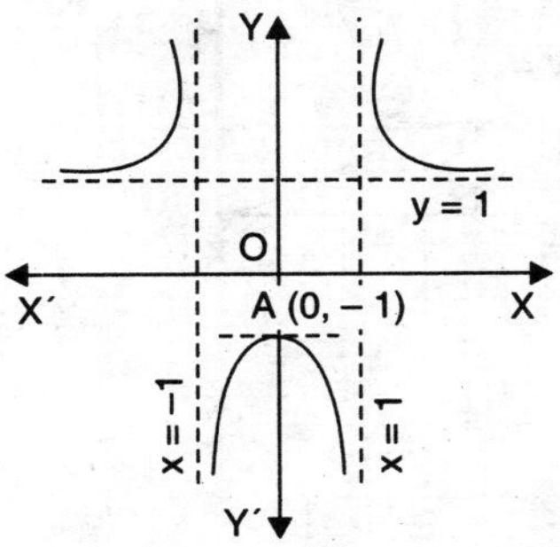

(b)

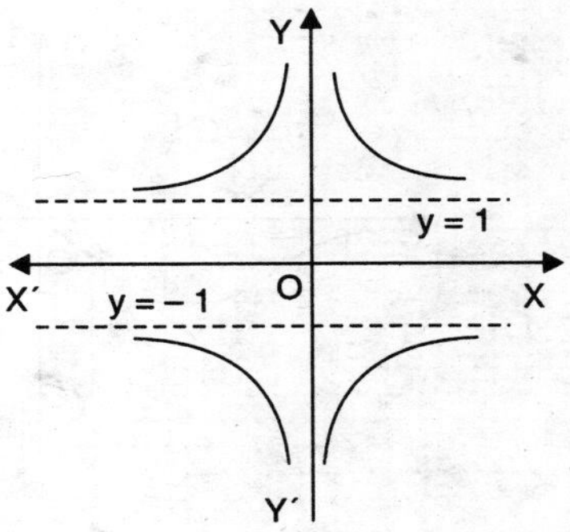

17.

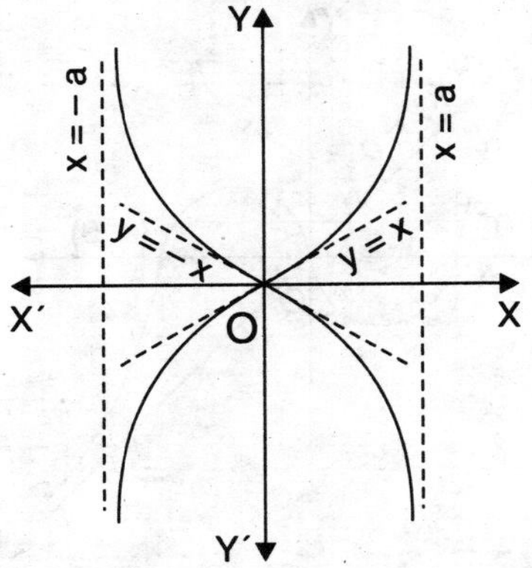

18.

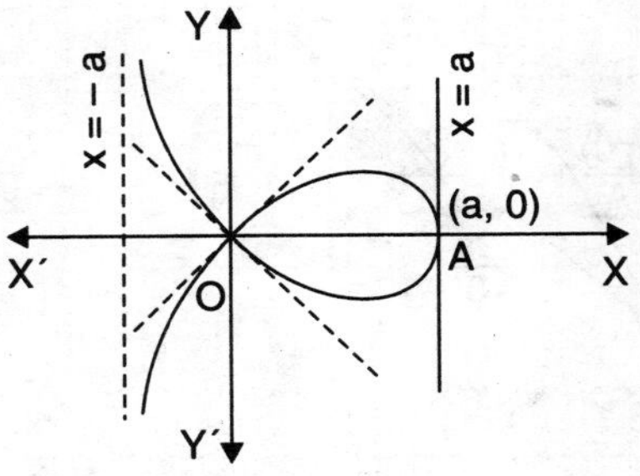

19.

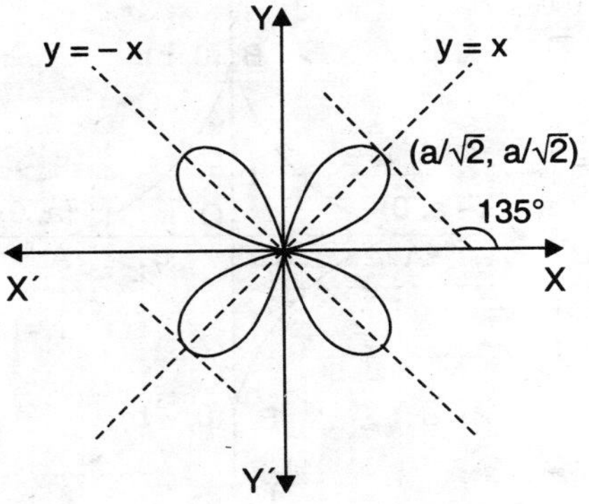

20.

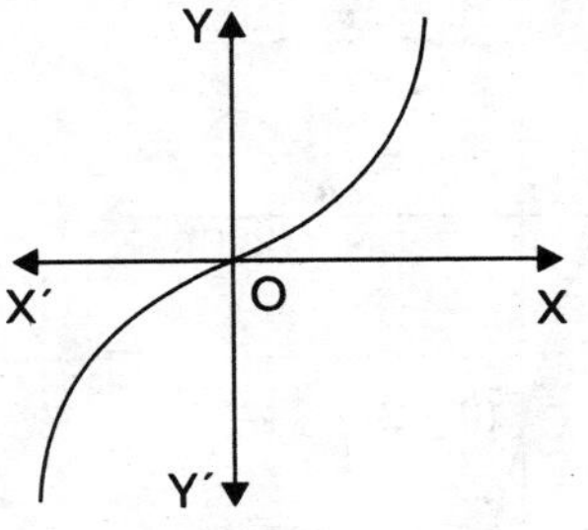

21.

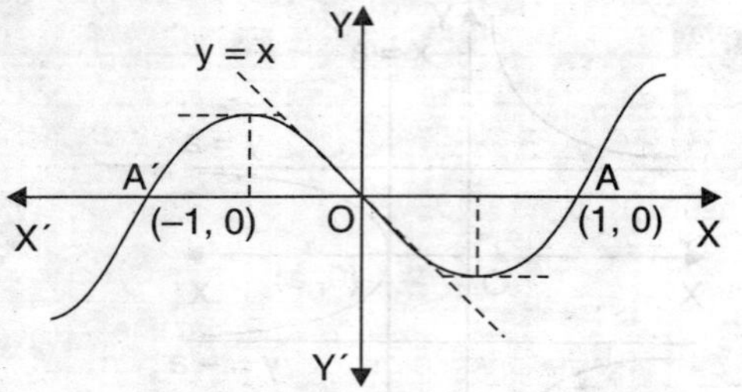

22.

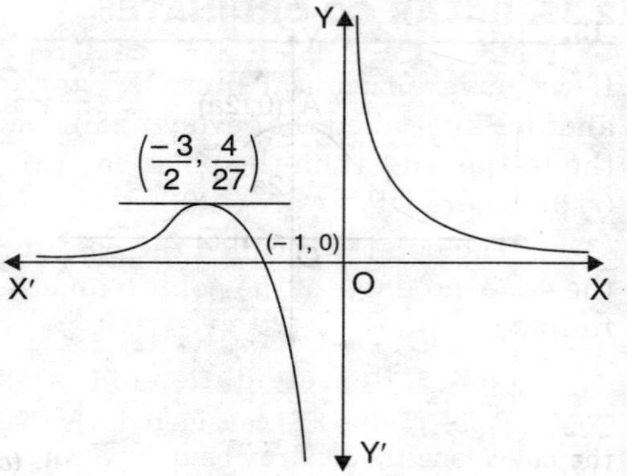

23.

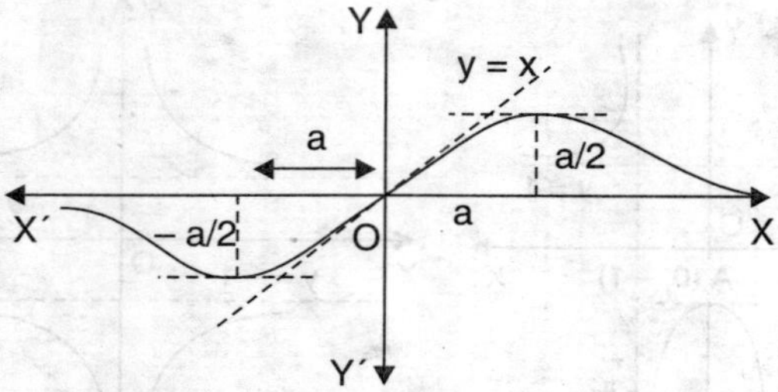

24.

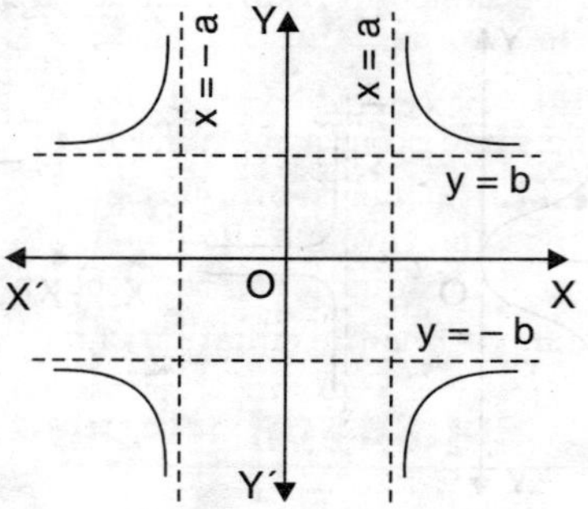

25.

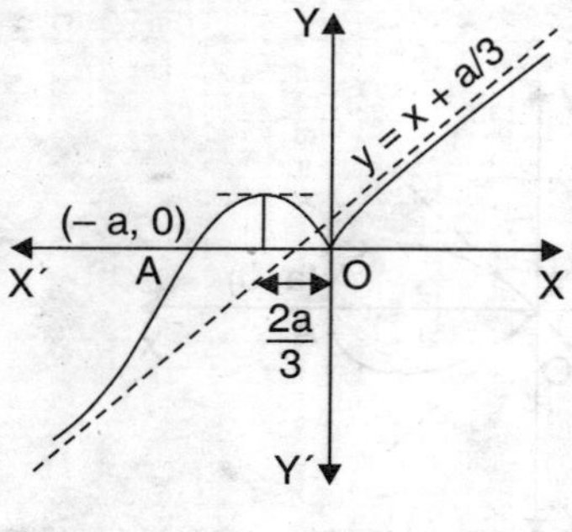

26.

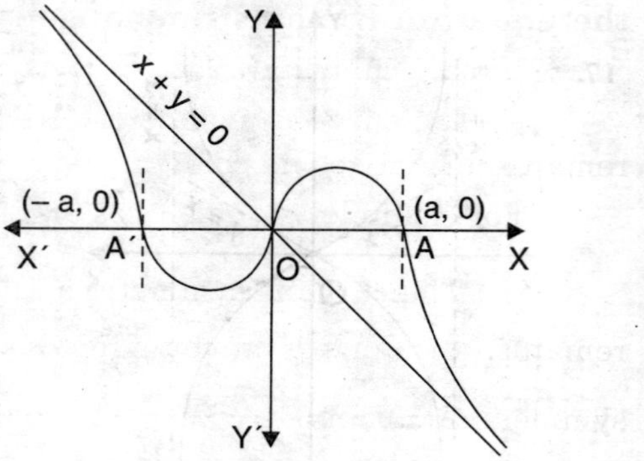

27. (*i*)

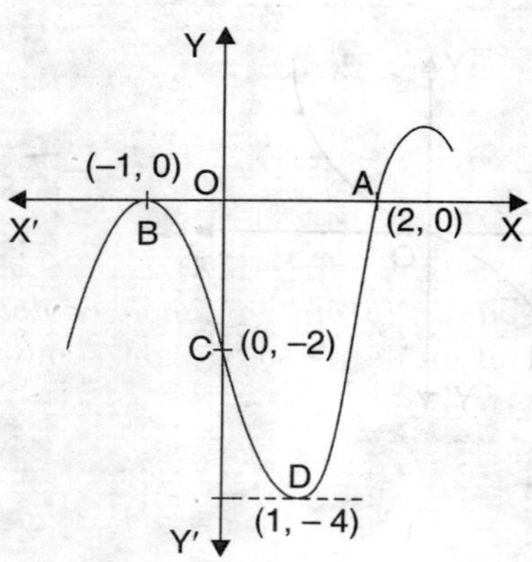

(*ii*)

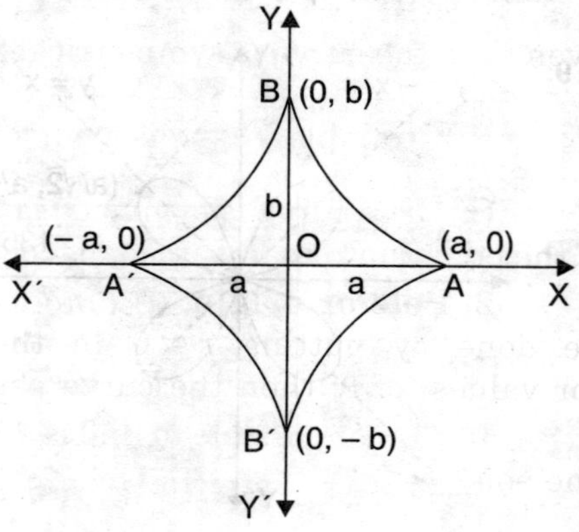

2.35. POLAR CO-ORDINATES

If we have horizontal line OX called the *initial line* and another line called the revolving line, makes an angle θ with the initial line, then the polar co-ordinates of a point P are (r, θ), where OP = r arc (r, θ).

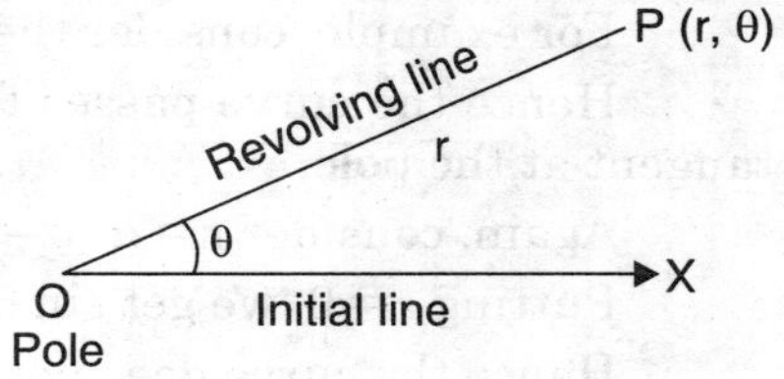

The point O is called the *pole* and the angle θ is called the *vectorial angle* of the point P and the length r is called its *radius vector*.

Note 1. Representation of $(-r, \theta)$, when $r > 0$.

Polar radius r is considered positive if it is measured from the pole along the half ray bounding the vectorial angle.

Let P be the point (r, θ). Produce OP backwards and cut off OP′ = OP. Then P′ is $(-r, \theta)$.

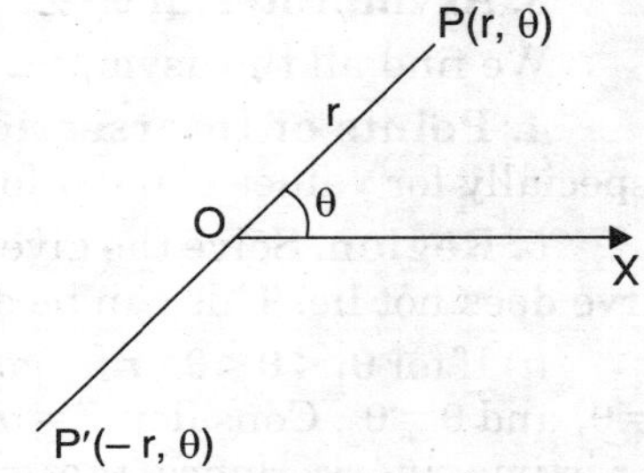

Remark. For $r \geq 0$, $(-r, \theta)$ is the same point as $(r, \theta + \pi)$ or (r, θ) is the same point as $(-r, \theta + \pi)$.

Note 2. In polar co-ordinates $\theta = \alpha$ is the equation of the *half ray* OP passing through the pole and making an angle α with the initial line. The equation of the other half ray OP′ is $\theta = \pi + \alpha$.

Note 3. $r = a$ is the equation of a circle having centre at the pole and radius a.

2.36. PROCEDURE FOR TRACING CURVES HAVING POLAR EQUATIONS

1. Symmetry. A curve is symmetrical about the ray $\theta = \alpha$ if on changing θ to $2\alpha - \theta$, the equation remains unchanged.

In particular:

(*i*) The curve is symmetrical about the initial line $\theta = 0$ (*i.e.*, x-axis) if the equation remains unchanged on changing θ to $-\theta$.

For example, $r = a(1 + \cos\theta)$, $r = a\cos 3\theta$ are symmetrical about the initial line.

(*ii*) The curve is symmetrical about the half ray $\theta = \frac{\pi}{2}$ (*i.e.*, y-axis) if the equation remains unchanged on changing θ to $(\pi - \theta)$ *e.g.*, $r = a(1 + \sin\theta)$ and $r = a\sin 3\theta$ are symmetrical about $\theta = \frac{\pi}{2}$.

If the equation of the curve remains unchanged *when θ is changed to $-\theta$ and r to $-r$*, even then the curve is symmetrical about the half ray $\theta = \frac{\pi}{2}$.

For example, $r\theta = a$ is symmetrical about $\theta = \frac{\pi}{2}$.

(*iii*) The curve is symmetrical about the **pole** (*i.e.*, symmetrical in opposite quadrants) if the equation remains unchanged on changing r to $-r$ **or** θ to $\pi + \theta$.

2. Pole or origin. (*i*) *Find whether the curve passes through the pole or not.* This can be done by putting $r = 0$ in the equation and if on doing so, we get some real value (or values) of θ; then the curve passes through the pole.

If it is not possible to find real value of θ for which $r = 0$, the curve does not pass through the pole.

(*ii*) *Find the tangents at the pole.* **Putting $r = 0$, the real values of θ give the tangents at the pole.**

For example, consider the curve $r = a\,(1 - \cos\theta)$. Putting $r = 0$, we get $\cos\theta = 1 \therefore \theta = 0$.

Hence the curve passes through the pole and the line $\theta = 0$ *i.e.*, the initial line is the tangent at the pole.

Again, consider $r = a\,(3 - \sin\theta)$.

Putting $r = 0$, we get $\sin\theta = 3$, which does not give any real value of θ [$\because$ $|\sin\theta| \le 1$].

Hence the curve does not pass through the pole.

(*iii*) Find the points, where the curve cuts the initial line and the line $\theta = \pi/2$.

3. Asymptotes. If $\theta \to \theta_1$ (some finite value) when $r \to \infty$, then there is an asymptote. We find all the asymptotes of the curve.

4. Points of Intersection. Find some points on the curve for convenient values of θ. (especially for values of $\theta = \alpha$ for which the curve is symmetrical).

5. Region. Solve the given equation for r or θ (if possible). Find the regions in which the curve does not lie. This can be done as :

(*i*) If for $\theta_1 < \theta < \theta_2$, r is imaginary, then there is no portion of the curve between the lines $\theta = \theta_1$ and $\theta = \theta_2$. Consider $r^2 = a^2 \cos 2\theta$. For $\pi/4 < \theta < 3\pi/4$, $\cos 2\theta$ is negative. $\therefore$ r^2 is –ve and so r is imaginary. Hence the curve does not lie between the lines $\theta = \pi/4$ and $\theta = 3\pi/4$.

(*ii*) If the greatest and least numerical values of r be respectively a and b, the curve lies entirely within the circle $r = a$ and entirely outside the circle $r = b$.

(*iii*) Trace the variations of r when θ varies in the intervals $(0, \infty)$ and $(-\infty, 0)$ making the values of θ for which $r = 0$ or attains maximum and minimum values. Plot the points so obtained.

Note. When r is a periodic function of θ, the negative values of θ need not be considered. We may consider values form $\theta = 0$ to those values of θ, where the values begin to repeat.

6. Value of ϕ. Find $\tan\phi = r\dfrac{d\theta}{dr}$ and ϕ. (ϕ is the angle between the tangent and radius vector). Find the points, where ϕ is 0 or $\dfrac{\pi}{2}$.

Again, r increases as θ increases if $\dfrac{dr}{d\theta}$ is positive and r decreases as θ increases if $\dfrac{dr}{d\theta}$ is negative.

Note. Conversion into cartesians. Transform the equation into cartesians, when tracing of the curve becomes easy on transformation to cartesian system of co-ordinates.

ILLUSTRATIVE EXAMPLES

Example 1. *Trace the curve $r = a(1 - \cos\theta)$.* **[Cardioid]**

Sol. The equation of the curve is $r = a(1 - \cos\theta)$ ...(1)

1. Symmetry. The curve is symmetrical about the initial line because the equation of curve remains unaltered when θ is changed to $-\theta$.

2. Pole or origin. (*i*) When $\theta = 0$, $r = 0$ hence the curve passes through the pole and the tangent at the pole is the line $\theta = 0$, *i.e.*, the initial line.

(*ii*) The curve meets the initial line $\theta = 0$ at (0, 0) and the lines $\theta = \pi/2$ and π in the points $(a, \pi/2)$ and $(2a, \pi)$ respectively.

3. Asymptotes. Since for any finite value of θ, r does not tend to infinity, $\therefore$ the curve has no asymptote.

4. Special points. The corresponding values of θ and r are given below:

θ :	0	$\pi/3$	$\pi/2$	$2\pi/3$	π
r :	0	$a/2$	a	$3a/2$	$2a$.

[We need not trace the curve for values of θ from π to 2π as the curve is symmetrical about the initial line.]

5. Region. (*i*) $\because$ $|\cos\theta| \le 1$ $\therefore$ From (1), $r \le 2a$.

$\therefore$ The curve lies entirely within the circle $r = 2a$.

(*ii*) When θ increases from 0 to π, r remains positive and increases from 0 to $2a$.

When θ increases from π to 2π, r is positive and decreases from $2a$ to 0.

6. Value of ϕ.

From (1), $\dfrac{dr}{d\theta} = a\sin\theta$.

$$\therefore \quad \tan\phi = r\frac{d\theta}{dr} = \frac{r}{\left(\dfrac{dr}{d\theta}\right)} = \frac{a(1-\cos\theta)}{a\sin\theta} = \tan\frac{\theta}{2}.$$

$$\therefore \quad \phi = \frac{\theta}{2}.$$

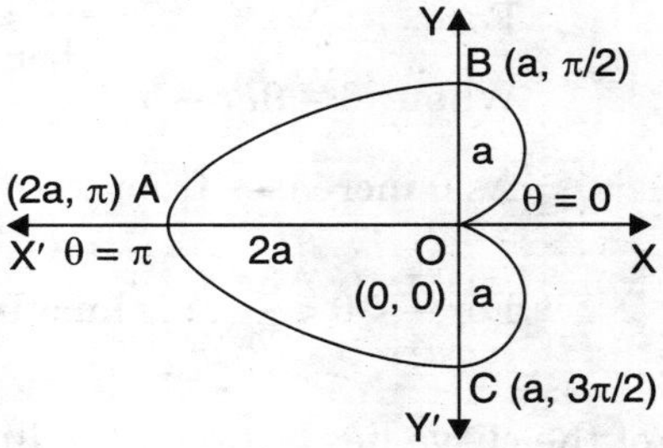

Now, $\phi = 0$ when $\theta = 0$ and $\phi = \pi/2$, when $\theta = \pi$.

$\therefore$ At $(2a, \pi)$ the tangent is perpendicular to the line $\theta = \pi$.

The shape of the curve is as shown in the figure.

Note. We will not consider the values of θ after 2π, because r is periodic with period 2π and therefore we do not get any new point on the curve.

Example 2. *Trace the curve:*

$$r^2 = a^2\cos 2\theta. \quad (\textit{Bernoulli's Lemniscate}) \qquad \text{(A.K.T.U. 2012, 2015, 2020)}$$

or $$(x^2 + y^2)^2 = a^2(x^2 - y^2).$$

Sol. The equation of the curve is $r^2 = a^2\cos 2\theta$ ...(1)

1. Symmetry. The curve is symmetrical about the initial line, the line $\theta = \dfrac{\pi}{2}$ and pole.

2. Origin or Pole. Put $r = 0$ in (1), we get

$$\cos 2\theta = 0 = \cos\left(\pm\frac{\pi}{2}\right)$$

$$\Rightarrow \quad \theta = \pm\frac{\pi}{4}, \text{ which are real.}$$

Hence, the curve passes through the pole and the tangents at the pole are $\theta = \pm\dfrac{\pi}{4}$.

Also, when $\theta = 0$, $r = \pm a$

Hence, the curve meets the initial at $(\pm a, 0)$.

3. Asymptotes. The curve has no asymptote.

4. Value of ϕ. $\phi = \frac{\pi}{2} + 2\theta$

when $\theta = 0, \phi = \frac{\pi}{2}$

For $\theta = 0, r = \pm a$

Hence, at the points $(a, 0)$ and $(-a, 0)$, the tangents are perpendicular to the initial line.

5. Special Points and Region

From (1), $r = a\sqrt{\cos 2\theta}$ | Taking +ve root as the curve is symmetrical about the pole

$$\Rightarrow \quad \frac{dr}{d\theta} = \frac{-a \sin 2\theta}{\sqrt{\cos 2\theta}}$$

For $0 < \theta < \frac{\pi}{4}$, $\frac{dr}{d\theta}$ is –ve $\Rightarrow$ r decreases in this range

For $\frac{3\pi}{4} < \theta < \pi$, $\frac{dr}{d\theta}$ is +ve $\Rightarrow$ r increases in this range.

When $\theta = 0, r = a$

As θ increases from 0 to $\frac{\pi}{4}$. r decreases from a to 0.

For $\frac{\pi}{4} < \theta < \frac{3\pi}{4}$, r is imaginary. Hence, no portion of the curve lies between the lines $\theta = \frac{\pi}{4}$ and $\theta = \frac{3\pi}{4}$.

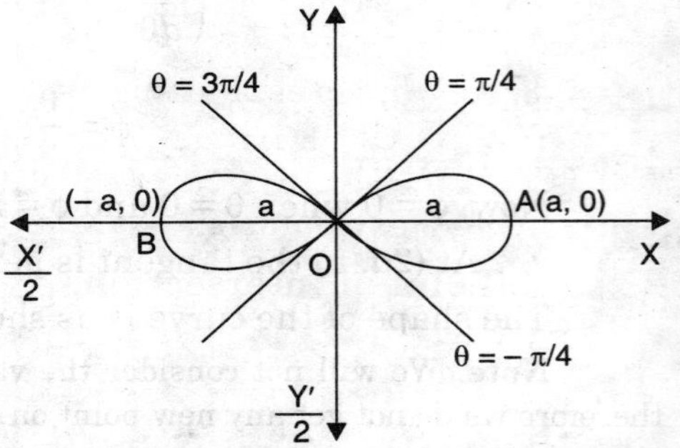

Again, as θ increases from $\frac{3\pi}{4}$ to π, r is +ve and increases from 0 to a. Thus, we can trace the part of the curve about the initial line. The part of the curve below the initial line can be traced by symmetry. Hence, the shape of the curve is as shown in the figure.

Example 3. *Trace the equiangular spiral $r = ae^{m\theta}$.* [*a*, *m* are (+)ve]

Sol. The equation of the curve is

$$r = ae^{m\theta} \quad \ldots(1)$$

1. Symmetry. The curve has no symmetry.

2. Pole. $\because$ r is not zero for any finite real value of θ.

$\therefore$ The curve does not pass through the pole.

3. Value of ϕ. ϕ is constant.

4. Region. (*i*) r is always positive but there is no limit to its values.

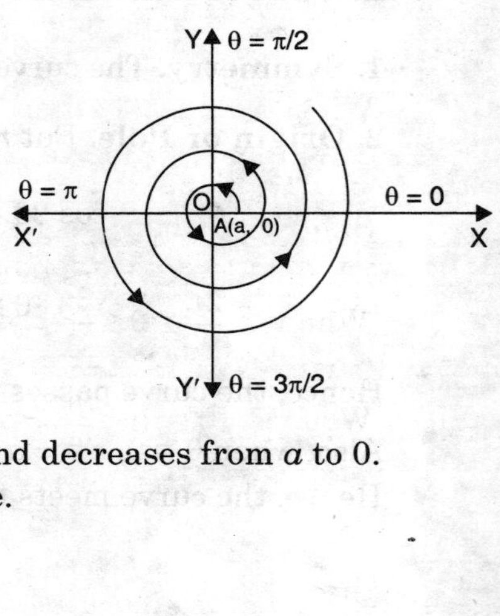

(*ii*) When $\theta = 0, r = a$

$\therefore$ $(a, 0)$ lies on the curve.

(*iii*) As θ increases from 0 to ∞, r remains +ve and increases from a to ∞.

As θ decreases from 0 to $-\infty$, r still remains positive and decreases from a to 0.

Hence, the shape of the curve is as shown in the figure.

Example 4. *Trace the curve $r = a \sin 3\theta$.* **(Three leaved rose)**

Sol. The equation of the curve is $r = a \sin 3\theta$. ...(1)

1. Symmetry.

From (1), $r = a \sin 3\theta = a \sin (\pi - 3\theta) = a \sin (3\pi - 3\theta)$ etc.

or $r = a \sin 3\theta = a \sin 3\left(\frac{\pi}{3} - \theta\right) = a \sin 3 (\pi - \theta)$ etc.

$\therefore$ (1) remains unchanged when θ is changed to $\frac{\pi}{3} - \theta$, or $\pi - \theta$.

$\therefore$ The curve is symmetrical about the rays

$$\theta = \frac{\pi}{6}, \frac{\pi}{2}\left(=\frac{3\pi}{6}\right), \frac{5\pi}{6}, \frac{7\pi}{6}, \frac{9\pi}{6}, \frac{11\pi}{6}.$$

2. Pole or origin. Putting $r = 0$, we get $\sin 3\theta = 0$

$\therefore$ $3\theta = n\pi$, where n is zero or any other integer.

$\therefore$ $\theta = 0, \frac{\pi}{3}, \frac{2\pi}{3}, \frac{3\pi}{3}, \frac{4\pi}{3}, \frac{5\pi}{3}, \frac{6\pi}{3}.$

$\therefore$ The curve passes through the pole and the tangents at the pole are

$$\theta = 0, \theta = \frac{\pi}{3}, \theta = \frac{2\pi}{3}, \theta = \frac{3\pi}{3}, \theta = \frac{4\pi}{3}, \theta = \frac{5\pi}{3}, \theta = \frac{6\pi}{3}.$$

3. Asymptotes. Putting $r = \frac{1}{u}$; from (1), $u = \frac{1}{a \sin 3\theta}$

$u = 0$ does not give any finite value of θ.

$\therefore$ No asymptotes.

4. Point of Intersection.

When $\theta = 0, \frac{\pi}{6}, \frac{3\pi}{6}, \frac{5\pi}{6}, \frac{7\pi}{6}, \frac{9\pi}{6}, \frac{11\pi}{6}$ (Rays of symmetry)

From (1), $r = 0, a, -a, a, -a, a, -a$.

5. Region. (*i*) From (1), $r \le a$ numerically [$\because$ $\sin 3\theta \le 1$ numerically]

$\therefore$ The curve lies entirely within the circle $r = a$.

(*ii*) *Region w.r.t. tangents at the pole.* Tangents at the pole are

$$\theta = 0, \theta = \frac{\pi}{3}, \theta = \frac{2\pi}{3}, \theta = \frac{3\pi}{3}, \theta = \frac{4\pi}{3}, \theta = \frac{5\pi}{3}$$

When $0 \le \theta \le \pi/3$; then $0 \le 3\theta \le \pi$ *i.e.,* 3θ lies in first or second quadrant.

$\therefore$ From (1), r is +ve [$\because$ $\sin \theta$ is +ve in I or II quadrant]

Similarly, when $\frac{\pi}{3} \le \theta \le \frac{2\pi}{3}$, r is –ve.

When $\frac{2\pi}{3} \le \theta \le \frac{3\pi}{3}$, r is +ve; When $\frac{3\pi}{3} \le \theta \le \frac{4\pi}{3}$, r is –ve.

When $\frac{4\pi}{3} \le \theta \le \frac{5\pi}{3}$, r is +ve; When $\frac{5\pi}{3} \le \theta \le \frac{6\pi}{3}$, r is –ve.

6. Value of ϕ. From (1), $r = a \sin 3\theta$

$$\therefore \quad \frac{dr}{d\theta} = 3a \cos 3\theta$$

$$\therefore \quad \tan \phi = \frac{r}{\left(\frac{dr}{d\theta}\right)} = \frac{1}{3} \tan 3\theta$$

$$\tan \phi = \infty \left(i.e., \quad \phi = \frac{\pi}{2} \right)$$

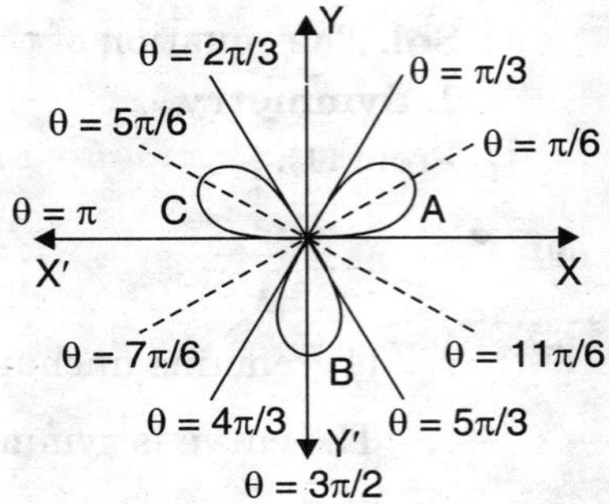

for $$\theta = \frac{\pi}{6}, \frac{3\pi}{6}, \frac{5\pi}{6}, \frac{7\pi}{6}, \frac{9\pi}{6}, \frac{11\pi}{6}$$

$\therefore$ Tangent is perpendicular to the radius vector at the points

$$\theta = \frac{\pi}{6}, \frac{3\pi}{6}, \frac{5\pi}{6}, \frac{7\pi}{6}, \frac{9\pi}{6}, \frac{11\pi}{6}.$$

Thus, the shape of the curve is as shown in the figure.

Remark. We have not considered values of θ outside $[0, 2\pi]$ because $r = a \sin 3\theta$ is periodic.

Note. The curve r = a sin nθ or r = a cos nθ consists of n or 2n loops according as n is odd or even.

Example 5. *Trace the curve $r^2 \cos \theta = a^2 \sin 3\theta$.*

Sol. The equation of the curve is $r^2 \cos \theta = a^2 \sin 3\theta$ ...(1)

1. Symmetry. On changing r to $-r$, equation (1) remains unchanged.

$\therefore$ The curve is symmetrical about the pole, *i.e.*, the curve is symmetrical in opposite quadrants.

2. Origin or Pole.

Putting $r = 0$ in (1), we get $\sin 3\theta = 0$

$\therefore$ $3\theta = n\pi$ when n is zero or any other integer.

$$\therefore \quad \theta = 0, \frac{\pi}{3}, \frac{2\pi}{3}, \frac{3\pi}{3}, \frac{4\pi}{3}, \frac{5\pi}{3}, \frac{6\pi}{3}$$

$\therefore$ The curve passes through the pole and the tangents at the pole are

$$\theta = 0, \theta = \frac{\pi}{3}, \theta = \frac{2\pi}{3}, \theta = \frac{3\pi}{3}, \theta = \frac{4\pi}{3}, \theta = \frac{5\pi}{3}, \theta = \frac{6\pi}{3}.$$

3. Asymptotes. Putting $r = \frac{1}{u}$, from (1),

$$u^2 = \frac{\cos \theta}{a^2 \sin 3\theta} \qquad \text{...(2)}$$

Let $u \to 0$. Therefore, $\cos \theta \to 0$, *i.e.*, $\theta \to \frac{\pi}{2}, \frac{3\pi}{2}$

Diff. (2) w.r.t. θ, $$2u \frac{du}{d\theta} = \frac{1}{a^2} \left[\frac{-\sin 3\theta \sin \theta - \cos \theta\, (3 \cos 3\theta)}{\sin^2 3\theta} \right]$$

$$\therefore \quad \frac{du}{d\theta} = -\frac{(\sin 3\theta \sin \theta + 3 \cos \theta \cos 3\theta)}{2a^2 u \sin^2 3\theta}$$

For $\theta_1 = \frac{\pi}{2}$, $\quad p = \underset{\substack{\theta \to \theta_1 \\ u \to 0}}{\text{Lt}} \left(-\frac{d\theta}{du}\right) = \underset{\substack{\theta \to \pi/2 \\ u \to 0}}{\text{Lt}} \frac{2ua^2 \sin^2 3\theta}{\sin 3\theta \sin \theta + 3 \cos \theta \cos 3\theta} = 0$

∴ The asymptote is $p = r \sin (\theta_1 - \theta)$

or $\quad 0 = r \sin \left(\frac{\pi}{2} - \theta\right)$ *i.e.,* $r \cos \theta = 0$ or $x = 0$ $\quad [\because \; x = r \cos \theta]$

i.e., y-axis is an asymptote to the curve.

Again, for $\theta_1 = \frac{3\pi}{2}$, y-axis is an asymptote.

4. Points of Intersection. When $\theta = 0$; From (1), $r = 0$ (Also see step 2).

When $\quad \theta = \frac{\pi}{6}, r^2 = \frac{2a}{\sqrt{3}} \quad \therefore \quad r = \pm \sqrt{\frac{2a}{\sqrt{3}}}$

∴ The two points are $\left(\pm \sqrt{\frac{2a}{\sqrt{3}}}, \frac{\pi}{6}\right)$, when $\theta = \frac{\pi}{3}$, $r = 0$ again.

5. Region w.r.t. tangents at the pole.

From (1), $\quad r^2 = \frac{a^2 \sin 3\theta}{\cos \theta}$

when $\quad 0 \le \theta \le \frac{\pi}{3}$, r^2 is +ve $\therefore$ r is real.

when $\quad \frac{\pi}{3} \le \theta \le \frac{\pi}{2}$, r^2 is –ve $\therefore$ r is imaginary.

∴ No portion of the curve lies between the two half-rays $\theta = \frac{\pi}{3}$ and $\theta = \frac{\pi}{2}$.

when $\quad \frac{\pi}{2} \le \theta \le \frac{2\pi}{3}$ r^2 is +ve $\therefore$ r is real

when $\quad \frac{2\pi}{3} \le \theta \le \pi$ r^2 is –ve

∴ r is imaginary.

∴ No portion of the curve lies between the half rays

$$\theta = \frac{2\pi}{3} \text{ and } \theta = \pi.$$

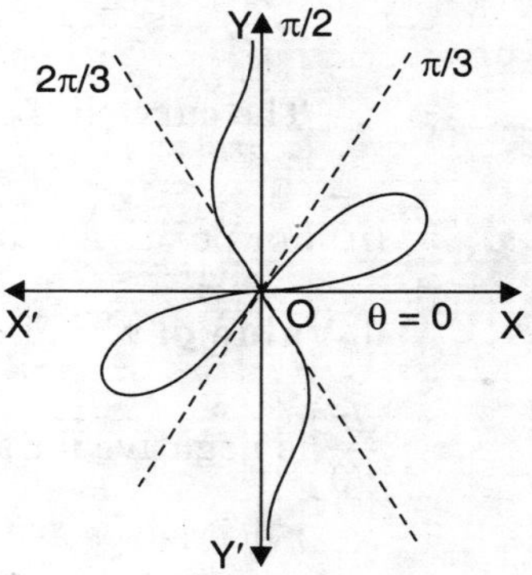

Hence the shape of the curve is as shown.

Note. We have not discussed the values of θ between π and 2π because the curve is symmetrical in opposite quadrants.

Example 6. *Trace the curve reciprocal spiral rθ = a.*

Sol. The equation of the curve is $r\theta = a$ ...(1)

Firstly, we consider positive values of θ, *i.e.,* values of θ in $(0, \infty)$.

1. Symmetry. Equation (1) remains unchanged, where θ is changed to – θ and r to – r, therefore the curve is symmetrical about the line $\theta = \frac{\pi}{2}$ *i.e.,* y-axis.

2. Origin. Since r does not become zero for any real finite value of θ, therefore the curve does not pass through the pole.

3. Asymptotes. Equation (1) is $r\theta = a$.

Putting $r = \frac{1}{u}$, we have $\frac{\theta}{u} = a, \quad u = \frac{\theta}{a}$

$u \to 0$ gives $\theta \to 0$ $\quad \therefore \quad \theta_1 = 0$

$$\frac{du}{d\theta} = \frac{1}{a}$$

$$p = \underset{\substack{\theta \to \theta_1 \\ u \to 0}}{\text{Lt}} \left(-\frac{d\theta}{du} \right) = \underset{\theta \to 0}{\text{Lt}} \ (-a) = -a$$

Putting the values of p and θ_1 in $p = r \sin (\theta_1 - \theta)$, the equation of the asymptote is

$$-a = r \sin (-\theta) \text{ or } r \sin \theta = a$$

or $\quad y = a \qquad (\because \quad r \sin \theta = y)$

4. Points of intersection.

when $\quad \theta = 0, \frac{\pi}{6}, \frac{\pi}{3}, \frac{\pi}{2}, \frac{2\pi}{3}, \pi, \frac{3\pi}{2}, 2\pi, \ldots$

$\therefore$ From (1) ; $\quad r = \infty, \frac{6a}{\pi}, \frac{3a}{\pi}, \frac{2a}{\pi}, \frac{3a}{2\pi}, \frac{a}{\pi}, \frac{2a}{3\pi}, \frac{a}{2\pi}, \ldots$

5. Region. (*i*) From (1), $r = \frac{a}{\theta}$

$\therefore \quad r \sin \theta = \frac{a}{\theta} \sin \theta = a \frac{\sin \theta}{\theta} < a \qquad \left[\because \quad \frac{\sin \theta}{\theta} < 1 \right]$

or $\quad y < a \qquad (\because \quad y = r \sin \theta)$

$\therefore$ The curve lies entirely below the asymptote

$$r \sin \theta = a \quad \text{or} \quad y = a.$$

(*ii*) For positive values of θ, r is positive.

6. Value of ϕ. From (1), $\quad r = \frac{a}{\theta} \quad \therefore \quad \frac{dr}{d\theta} = -\frac{a}{\theta^2}$

$\frac{dr}{d\theta}$ is negative for all values of θ.

$\therefore$ r decreases as θ increases.

As θ increases from 0 to ∞, r is +ve and decreases from ∞ to 0.

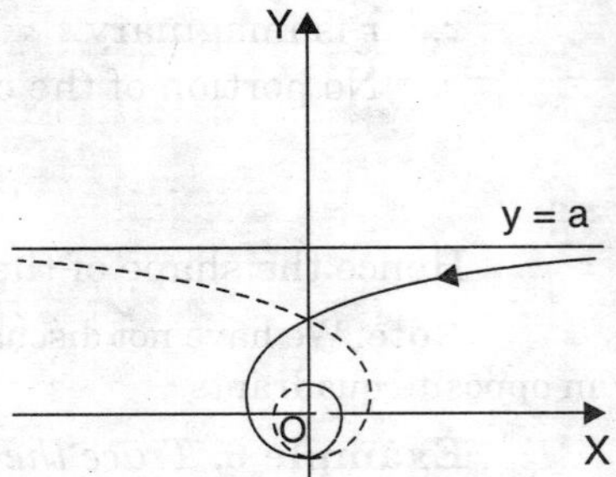

$$\tan \phi = \frac{r}{\left(\frac{dr}{d\theta}\right)} = \frac{a}{\theta} \left(-\frac{\theta^2}{a} \right) = -\theta.$$

Hence for positive values of θ, the shape of the curve as shown by thick lines in the figure.

The part of the curve for –ve values of θ (or here r also because from $r = \frac{a}{\theta}$, r is negative when θ is negative) is obtained by turning the curve for positive value of θ through two right angles. This branch is shown in the figure by the dotted line.

Example 7. *Trace the four leaved rose $r = a \sin 2\theta$.*

Sol. The equation of the given curve is

$$r = a \sin 2\theta \qquad ...(1)$$

1. Symmetry. The curve is symmetrical about the line $\theta = \frac{\pi}{4}$ and the line $\theta = \frac{\pi}{2}$.

2. Pole. Put $r = 0$ in (1), we get

$$\sin 2\theta = 0$$

$$\Rightarrow \quad 2\theta = n\pi \quad \Rightarrow \quad \theta = \frac{n\pi}{2}, \text{ where } n = 0, \pm 1, \pm 2, \pm 3, ...$$

$$\Rightarrow \quad \theta = 0, \pm \frac{\pi}{2}, \pm \pi, \pm \frac{3\pi}{2}, ...$$

Hence, the curve passes through the pole and the tangents at the pole are $\theta = 0, \theta = \frac{\pi}{2}$, because the other values of θ give the same tangents.

3. Asymptotes. The curve has no asymptotes.

4. Special points and region.

(*i*) $\because \quad |\sin 2\theta| \le 1$

$\therefore \quad$ From (1), $|r| \le a$

Hence, the curve lies entirely within the circle $r = a$.

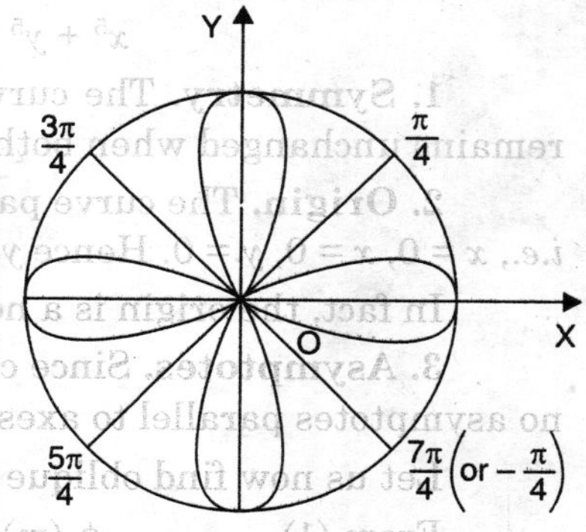

(*ii*) From (1),

$$\frac{dr}{d\theta} = 2a \cos 2\theta \quad \text{and} \quad \frac{d^2r}{d\theta^2} = -4a \sin 2\theta$$

For r to be maximum or minimum, $\frac{dr}{d\theta} = 0$

$$\Rightarrow \quad 2a \cos 2\theta = 0$$

$$\Rightarrow \quad \cos 2\theta = 0$$

$$\Rightarrow \quad 2\theta = (2n + 1)\frac{\pi}{2}, \text{ where } n \in I$$

$$\Rightarrow \quad \theta = (2n + 1)\frac{\pi}{4}, \text{ where } n \in I$$

$$\Rightarrow \quad \theta = \frac{\pi}{4}, \frac{3\pi}{4}, \frac{5\pi}{4}, ...$$

For r to be maximum, $\frac{d^2r}{d\theta^2}$ should be –ve and we find that $\frac{d^2r}{d\theta^2}$ is –ve for $\theta = \frac{\pi}{4}, \frac{5\pi}{4}, ...$ and the maximum value of r is

$$r_{\text{max.}} = a \sin\left(2 \cdot \frac{\pi}{4}\right) = a \sin \frac{\pi}{2} = a$$

The corresponding values of r and θ are given below:

θ :	0	$\frac{\pi}{4}$	$\frac{\pi}{2}$	$\frac{3\pi}{4}$	π
r :	0	a	0	$-a$	0

As θ increases from 0 to $\frac{\pi}{4}$, r is +ve and increases from 0 to a. As θ increases from $\frac{\pi}{4}$ to $\frac{\pi}{2}$, r is +ve and decreases from a to 0. Thus, we get a loop between the lines $\theta = 0$ and $\theta = \frac{\pi}{2}$.

Similarly, we get a loop between the lines $\theta = \frac{\pi}{2}$ and $\theta = \pi$ but in opposite direction ($\because$ r is –ve).

Thus, we trace the two loops I and II.

Since the curve is symmetrical about the line $\theta = \frac{\pi}{2}$, the other two loops are traced by symmetry. Thus, in all we get four equal loops.

Hence, the shape of the curve is as shown in the figure.

Example 8. *Trace the curve $x^5 + y^5 = 5a^2x^2y$.*

Sol. The equation of the curve is

$$x^5 + y^5 = 5a^2x^2y \qquad ...(1)$$

1. Symmetry. The curve is symmetrical in opposite quadrants because equation (1) remains unchanged when both x and y are changed to $-x$ and $-y$ respectively.

2. Origin. The curve passes through the origin. The tangents at the origin are $x^2y = 0$ *i.e.*, $x = 0$, $x = 0$, $y = 0$. Hence y-axis is a cuspidal tangent.

In fact, the origin is a node as well as a cusp.

3. Asymptotes. Since co-efficients of x^5 and y^5 are constants, therefore, the curve has no asymptotes parallel to axes.

Let us now find oblique asymptotes:

From (1), $\phi_5(m) = 1 + m^5$, $\phi_4(m) = 0$

Putting $\phi_5(m) = 0$, we have $m^5 + 1 = 0$ or $m^5 = -1$ $\therefore$ $m = -1$.

To find c, $c\phi_5'(m) + \phi_4(m) = 0$

or $c(5m^4) = 0$ or $c = 0$.

$\therefore$ Equation of the asymptote is $y = mx + c$ or $y = -x$.

Position of the curve w.r.t. the asymptote. In the second quadrant, x is negative and y is positive;

$\therefore$ RHS of (1) is +ve.

$\therefore$ LHS of (1) namely $y^5 + x^5$ must also be +ve.

Hence, y should be numerically > x.

Thus, the curve lies above the asymptote $y = -x$.

Because of symmetry in opposite quadrants, the curve approaches the other end (in the fourth quadrant) from below:

4. Points of Intersection.

(*i*) The curve meets the coordinate axes only at (0, 0).

(*ii*) *Intersections with the line $y = x$.* Putting $y = x$ in (1), we have

$$2x^5 = 5a^2x^3.$$

$\therefore \qquad x = 0 \quad \text{or} \quad x = \pm\sqrt{\frac{5}{2}}\,.\,a$

$\therefore \qquad y = 0 \quad \text{and} \quad y = \pm\sqrt{\frac{5}{2}}\,.\,a$

$\therefore$ Point of intersection are (0, 0) ; $\left(\pm\sqrt{\frac{5}{2}}\,a, \pm\sqrt{\frac{5}{2}}\,a\right)$.

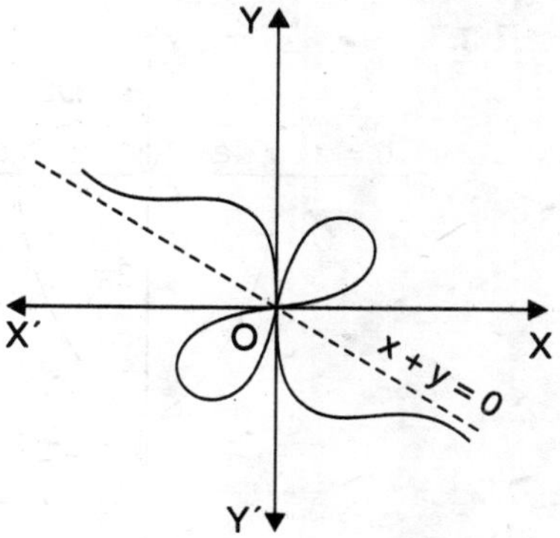

(iii) *Intersections with the asymptote* $y = -x$. Solving (1) and $y = -x$, we again have (0, 0) as their intersection.

5. Region. Transforming to polars [by putting $x = r\cos\theta$ and $y = r\sin\theta$ in (1), we have]

$$r^2 = \frac{5a^2\cos^2\theta\sin\theta}{\cos^5\theta + \sin^5\theta}.$$

For values of θ between $\frac{3\pi}{4}$ and π, r^2 is negative and so r is Imaginary.

Thus, no portion of the curve lies between the lines $\theta = \frac{3\pi}{4}$ and $\theta = \pi$. Thus, the shape of the curve is as shown in the figure.

TEST YOUR KNOWLEDGE

Trace the following curves:

1. $r = a(1 + \cos\theta)$ **[Cardioid]** (*U.P.T.U. 2015*)
2. $r = a(1 + \sin\theta)$
3. $r = a(1 - \sin\theta)$
4. $r = 2a\cos\theta$ **[Circle]**
5. $r = a + b\cos\theta\,;\ a > b$ **[Limacon]**
6. $\frac{2a}{r} = 1 + \cos\theta$ **[Parabola]**
7. $r^2\cos 2\theta = a^2$
8. $r = a\cos 2\theta$
9. $r = a\cos 3\theta$
10. $r = a\cos 5\theta$
11. $r = \frac{a\theta^2}{1 + \theta^2}$
12. $r = ae^{\theta\cot\alpha}$
13. $x^5 + y^5 = 5ax^2y^2$
14. $r = a\,(1 + \sec\theta)$
15. $x^4 + y^4 = 4axy^2$
16. $x^4 + y^4 = 4a^2\,xy$
17. $y^4 - x^4 + xy = 0$
18. $r = a\theta$ **[Spiral of Archimedes]**

Answers

1.

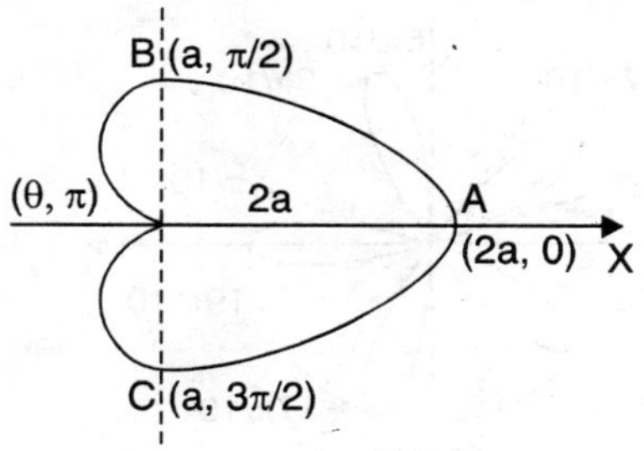

2.

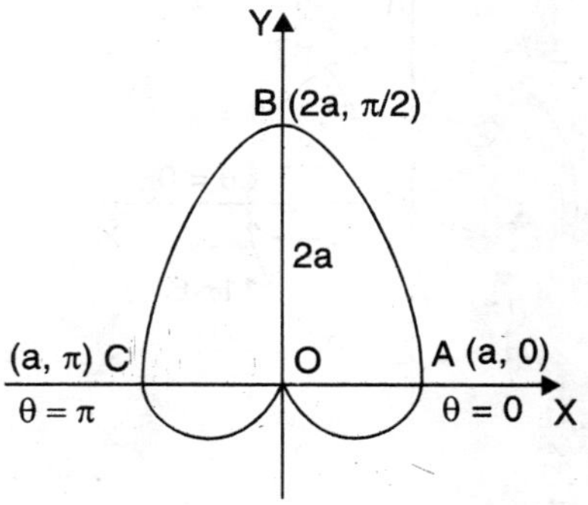

3.

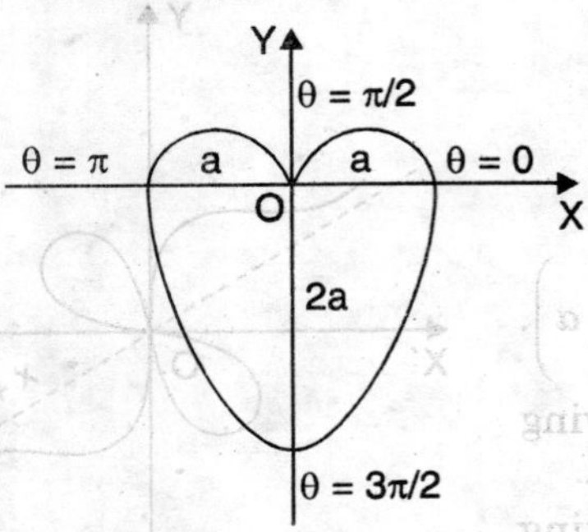
Y
θ = π/2
θ = π
a
a
θ = 0
O
X
2a
θ = 3π/2

4.

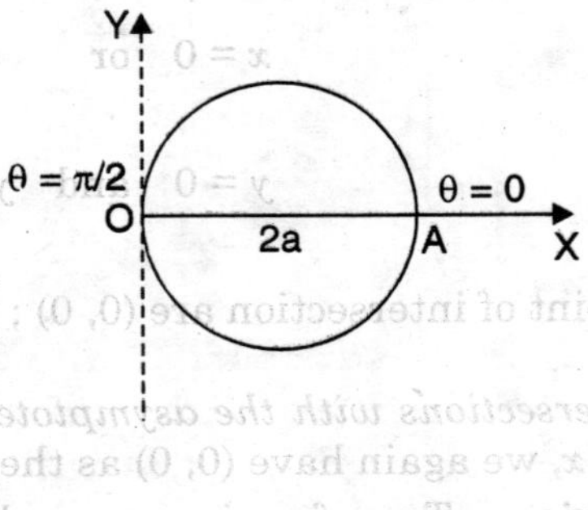
Y
θ = π/2
θ = 0
O
2a
A
X

5.

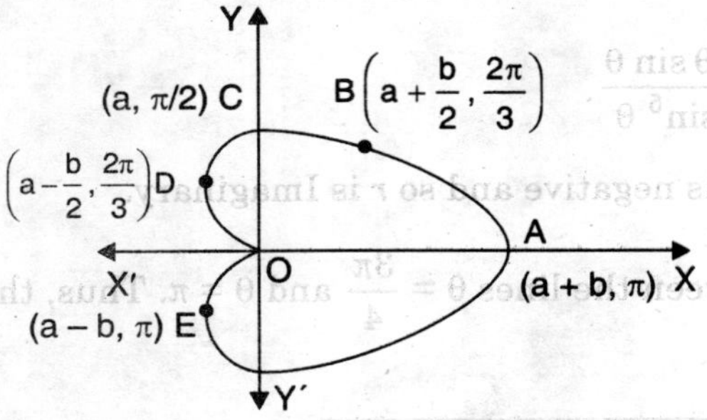
Y
(a, π/2) C
B (a + b/2, 2π/3)
(a − b/2, 2π/3) D
A
X′
O
(a + b, π)
X
(a − b, π) E
Y′

6.

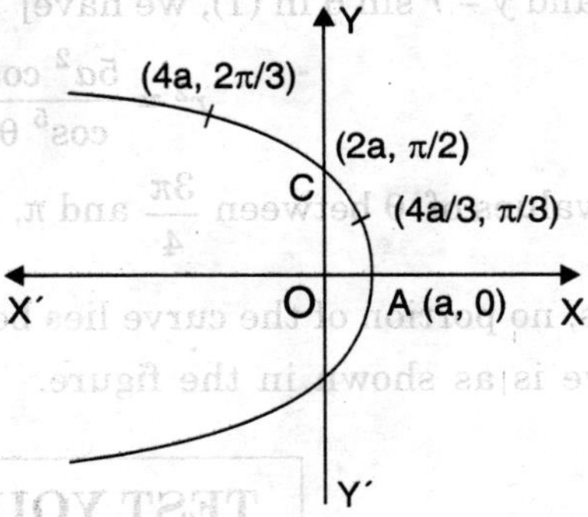
Y
(4a, 2π/3)
(2a, π/2)
C
(4a/3, π/3)
X′
O
A (a, 0)
X
Y′

7.

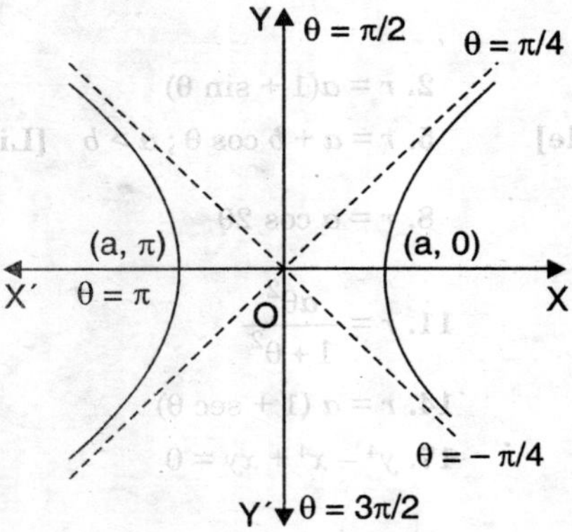
Y
θ = π/2
θ = π/4
(a, π)
(a, 0)
X′
θ = π
O
X
θ = − π/4
Y′
θ = 3π/2

8.

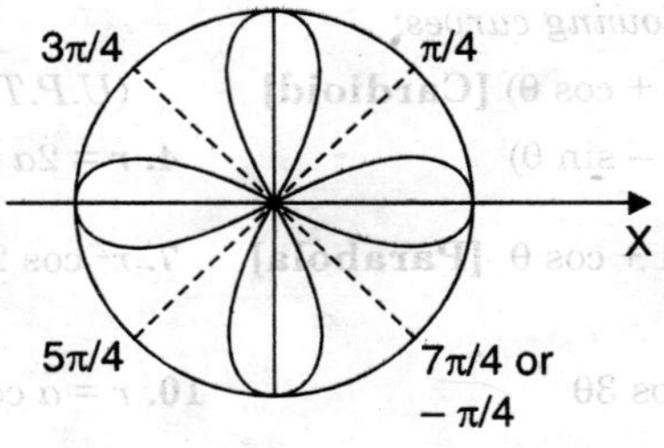
3π/4
π/4
X
5π/4
7π/4 or
− π/4

9.

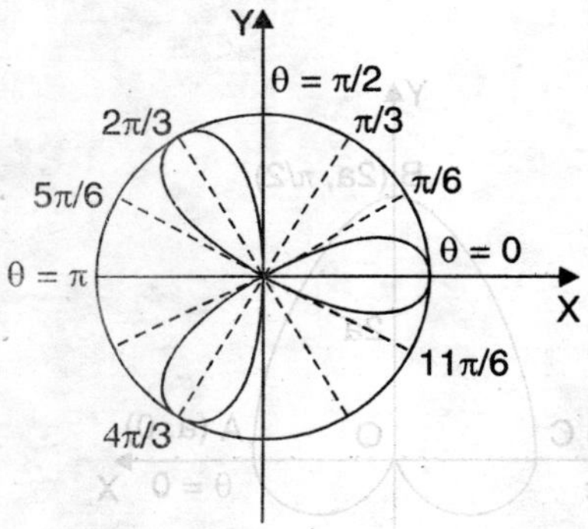
Y
θ = π/2
2π/3
π/3
5π/6
π/6
θ = π
θ = 0
X
11π/6
4π/3

10.

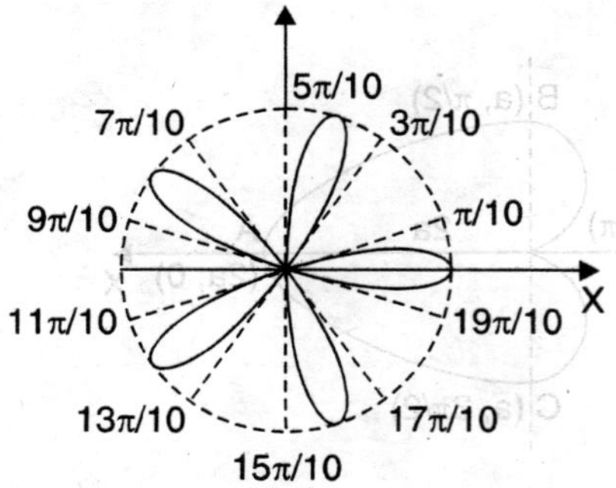
5π/10
7π/10
3π/10
9π/10
π/10
11π/10
19π/10
X
13π/10
17π/10
15π/10

11.

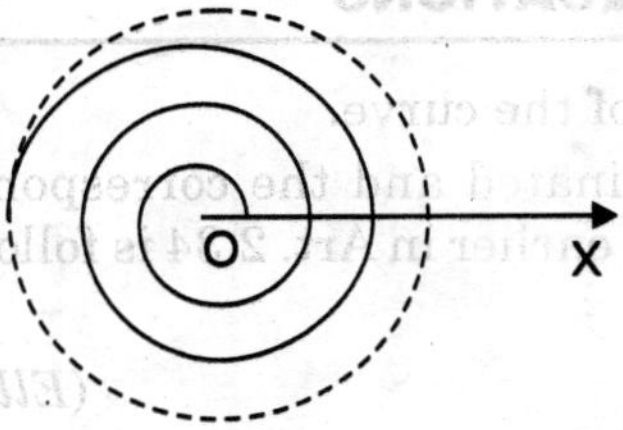

Circular Asymptote $r = a$ is shown dotted. The part of the curve for negative values is its reflection in the initial line.

12.

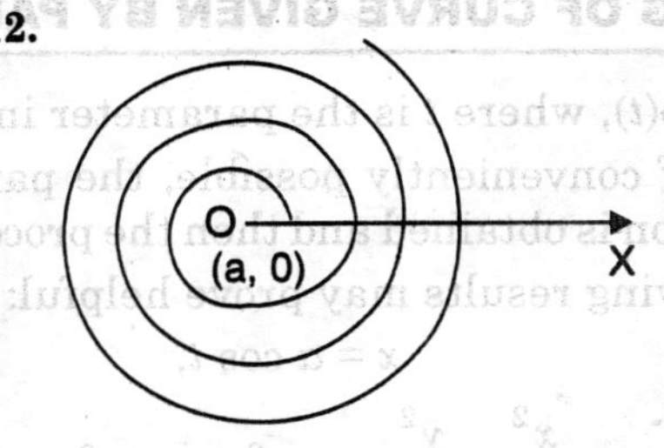

13.

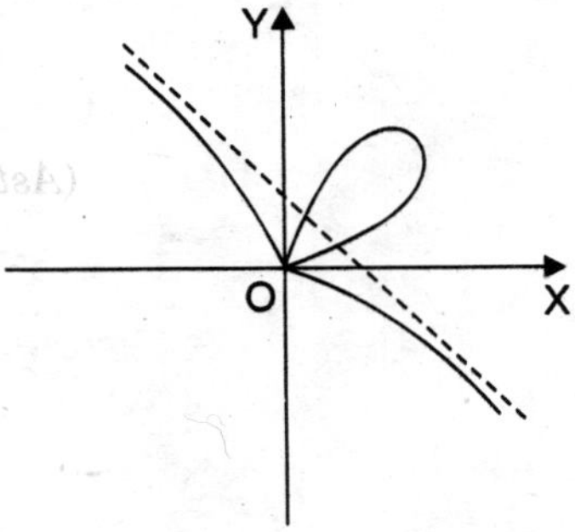

14.

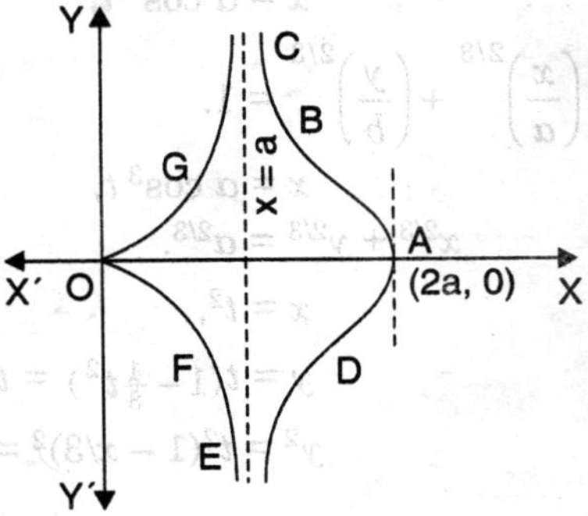

15.

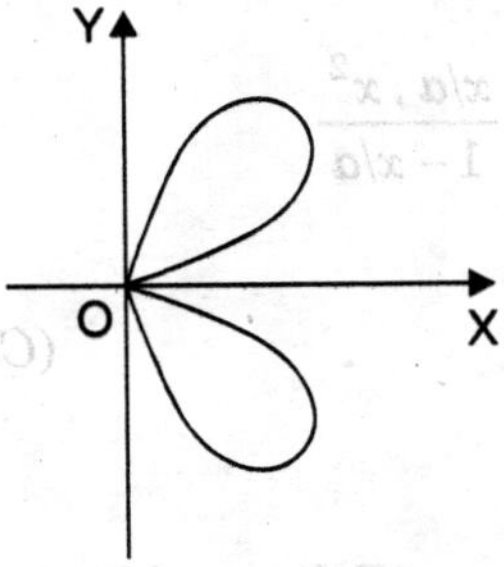

16.

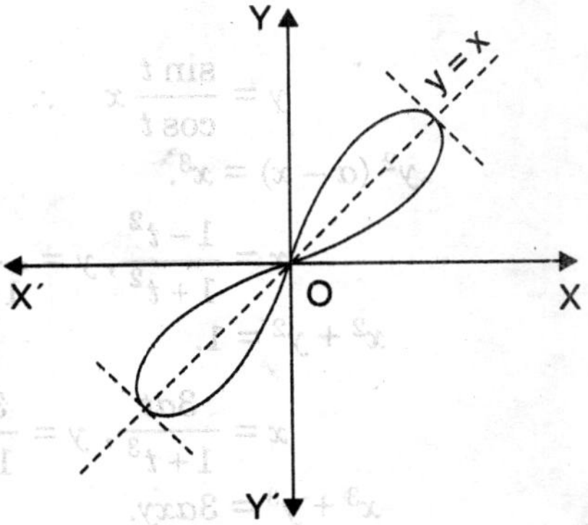

17.

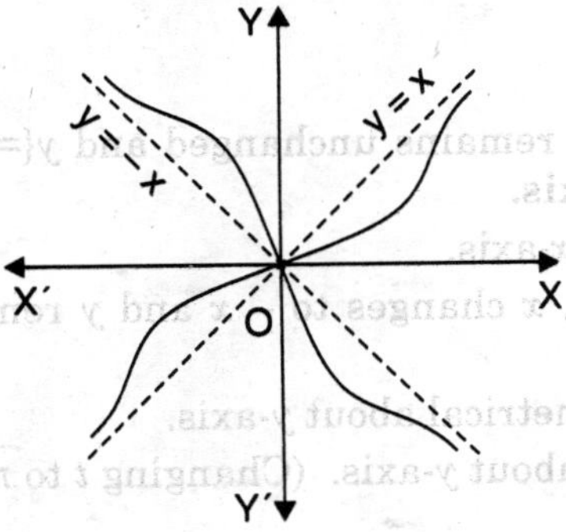

18.

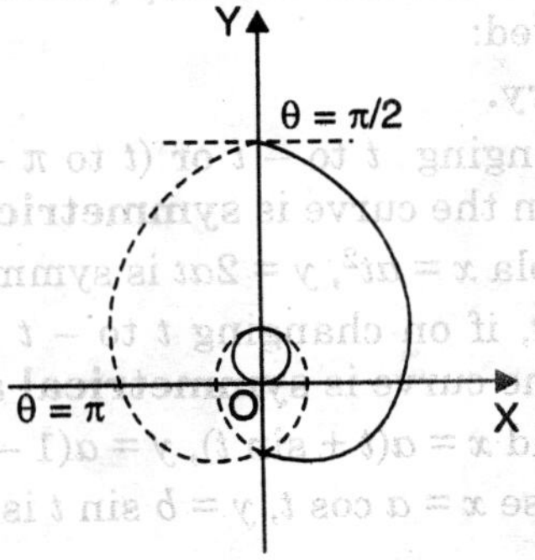

2.37. TRACING OF CURVE GIVEN BY PARAMETRIC EQUATIONS

Let $x = f(t)$, $y = \phi(t)$, where t is the parameter in the equations of the curve.

Case I. If conveniently possible, the parameter is eliminated and the corresponding cartesian equation is obtained and then the procedure explained earlier in Art. 2.34 is followed.

The following results may prove helpful:

1. $x = a \cos t, \qquad y = b \sin t.$ *(Ellipse)*

$\therefore \quad \dfrac{x^2}{a^2} + \dfrac{y^2}{b^2} = \cos^2 t + \sin^2 t = 1.$

2. $x = a \cos t, \qquad y = a \sin t.$ *(Circle)*

$\therefore \quad x^2 + y^2 = a^2 (\cos^2 t + \sin^2 t) = a^2$

3. $x = a \cos^3 t, \qquad y = b \sin^3 t$ *(Hypo-cycloid)*

$\therefore \quad \left(\dfrac{x}{a}\right)^{2/3} + \left(\dfrac{y}{b}\right)^{2/3} = 1.$

4. $x = a \cos^3 t, \qquad y = a \sin^3 t$ *(Astroid)*

$x^{2/3} + y^{2/3} = a^{2/3}.$

5. $x = t^2, \qquad y = t - \frac{1}{3} t^3$

$\therefore \quad y = t(1 - \frac{1}{3} t^2) = t(1 - x/3)$

$\therefore \quad y^2 = t^2(1 - x/3)^2 = x(1 - x/3)^2.$

6. $x = a \sin^2 t, \qquad y = a \dfrac{\sin^3 t}{\cos t}$ *(Cissoid)*

$\therefore \quad y = \dfrac{\sin t}{\cos t} x \quad \therefore \quad y^2 = \dfrac{\sin^2 t}{1 - \sin^2 t} . x^2 = \dfrac{x/a . x^2}{1 - x/a}$

or $\quad y^2 (a - x) = x^3.$

7. $x = \dfrac{1 - t^2}{1 + t^2}, \; y = \dfrac{2t}{1 + t^2}$ *(Circle)*

$\therefore \quad x^2 + y^2 = 1.$

8. $x = \dfrac{3at}{1 + t^3}, \; y = \dfrac{3at^2}{1 + t^3}.$ *(Folium of Descarte's)*

$x^3 + y^3 = 3axy.$

Case II. If it is not conveniently possible to eliminate the parameter, then the following procedure is followed:

1. Symmetry.

(*i*) If on changing t to $-t$ or (t to $\pi - t$); x ($= f(t)$) remains unchanged and $y\{= \phi(t)\}$ changes to $-y$; then the curve is **symmetrical about x-axis.**

e.g., The parabola $x = at^2$, $y = 2at$ is symmetrical about x-axis.

(*ii*) Similarly, if on changing t to $-t$ (or t to $\pi - t$), x changes to $-x$ and y remains unchanged, then the curve is **symmetrical about y-axis.**

e.g., (1) the cycloid $x = a(t + \sin t)$, $y = a(1 - \cos t)$ is symmetrical about y-axis.

(2) The ellipse $x = a \cos t$, $y = b \sin t$ is symmetrical about y-axis. (Changing t to $\pi - t$)

(*iii*) If on changing t to $-t$; x changes to $-x$ and y to $-y$, the curve is symmetrical in *opposite quadrants.*

e.g., the rectangular hyperbola $x = ct$, $y = \frac{c}{t}$ is symmetrical in opposite quadrants.

2. Origin. If on putting $x = 0$ (or $y = 0$) a real value of t can be found out, which makes $y = 0$ (or $x = 0$), then the curve passes through the origin.

or Put both x and y equal to zero and find values of t. If there is any common value of t, then the curve passes through the origin.

3. Asymptotes. Find asymptotes if any.

4. Points of Intersection. Find the points of intersection of the curve with coordinate axes.

5. Region. If easily possible, find the greatest and least values of x and y and therefore, the lines parallel to the axes between which the curve lies or does not lie.

For Example. If the curve is $x = a \cos \theta$, $y = b \sin \theta$; then the greatest value of x is a and the least value is $-a$. ($\because$ greatest and least values of $\cos \theta$ are 1 and -1)

$\therefore$ The curve lies entirely between the lines $x = \pm a$.

Similarly, the curve lies entirely between the lines $y = \pm b$.

6. Special Points. Find $\frac{dx}{dt}$ and $\frac{dy}{dt}$ and then consider the signs of $\frac{dx}{dt}$ and $\frac{dy}{dt}$ in suitable intervals of t.

Find $\frac{dy}{dx}\left(=\frac{dy/dt}{dx/dt}\right)$. Give to the parameter t certain values and find the corresponding values of x, y and the slope of the tangent namely $\frac{dy}{dx}$. Plot these points (whose cartesian co-ordinates are known to us). Find those points on the curve for which $\frac{dy}{dx} = 0$ or ∞. Also find $\frac{d^2y}{dx^2}$ and discuss for concavity and points of inflexion.

ILLUSTRATIVE EXAMPLES

Example 1. *Trace the cycloid $x = a(\theta + \sin \theta)$, $y = a(1 + \cos \theta)$.*

Sol. The parametric equations of the cycloid are

$$x = a(\theta + \sin \theta),\ y = a(1 + \cos \theta) \quad ...(1)$$

Let us trace this curve firstly for values of θ in the interval $[-\pi, \pi]$.

1. Symmetry. On changing θ to $-\theta$ in (1), changes to $-x$ and y remains unchanged.

$\therefore$ The curve is symmetrical about y-axis.

2. Origin. Putting $y = 0$, $a(1 + \cos \theta) = 0$

$$\therefore \quad \cos \theta = -1 \text{ or } \theta = \pi$$

For $\theta = \pi$, $x = a(\theta + \sin \theta) = a\pi \neq 0$

$\therefore$ The curve does not pass through the origin.

3. Asymptotes. The curve has no asymptotes.

4. Points of Intersection

(*i*) *Intersections with x-axis.* Putting $y = 0$ in (1), we get $\theta = \pi$, which gives $x = a\pi$.

$\therefore$ Intersection with x-axis is $(a\pi, 0)$.

(*ii*) *Intersection with y-axis.* Putting $x = 0$ in (1), we have $\theta + \sin\theta = 0$, which is satisfied by only $\theta = 0$ and for $\theta = 0$, $y = a(1 + \cos\theta) = 2a$.

$\therefore$ Intersection with y-axis is $(0, 2a)$.

5. Region. We know that $-1 \le \cos\theta \le 1$

$$\therefore \quad 1 - 1 \le 1 + \cos\theta \le 1 + 1$$

or $$0 \le (1 + \cos\theta) \le 2$$

$$\therefore \quad 0 \le a(1 + \cos\theta) \le 2a \text{ or } 0 \le y \le 2a$$

$\therefore$ Curve lies entirely between the lines $y = 0$ and $y = 2a$.

6. Special points. From (1),

$$\frac{dx}{d\theta} = a(1 + \cos\theta) \text{ and } \frac{dy}{d\theta} = -a\sin\theta$$

$$\because \quad \frac{dy}{dx} = \frac{dy/d\theta}{dx/d\theta} = -\frac{a\sin\theta}{a(1+\cos\theta)} = -\tan\frac{\theta}{2} \quad \ldots(2)$$

Corresponding values of x, y and dy/dx for different values of θ are given below:

	$\theta = -\pi$	$-\pi/2$	0	$\pi/2$	π
From (1),	$x = -a\pi$	$-a\left(\frac{\pi}{2}+1\right)$	0	$a\left(\frac{\pi}{2}+1\right)$	$a\pi$
	$y = 0$	a	$2a$	a	0
From (2),	$\frac{dy}{dx} = \infty$	1	0	-1	$-\infty$

$\therefore$ Points on the curve are $(-a\pi, 0)$, $\left[-a\left(\frac{\pi}{2}+1\right), a\right]$, $[0, 2a]$, $\left[a\left(\frac{\pi}{2}+1\right), a\right]$, $[a\pi, 0]$.

Tangents at the points $[-a\pi, 0]$ and $[a\pi, 0)$ are parallel to y-axis. ($\because$ $dy/dx = \infty$ or $-\infty$)

Tangent at the point $(0, 2a)$ is $\parallel$ to x-axis ($\because$ $dy/dx = 0$)

From (2), $$\frac{dy}{dx} = -\tan\frac{\theta}{2}$$

Again diff. w.r.t. x, $$\frac{d^2y}{dx^2} = -\left(\sec^2\frac{\theta}{2}\right)\frac{1}{2}\frac{d\theta}{dx}$$

$$\frac{d^2y}{dx^2} = \frac{-1}{2}\sec^2\frac{\theta}{2}\cdot\frac{1}{2a}\sec^2\frac{\theta}{2} \qquad \left|\because \frac{d\theta}{dx} = \frac{1}{2a}\sec^2\frac{\theta}{2}\right.$$

or $$\frac{d^2y}{dx^2} = \frac{-1}{4a}\sec^4\frac{\theta}{2}$$

$\therefore$ $\frac{d^2y}{dx^2}$ is negative for all values of θ.

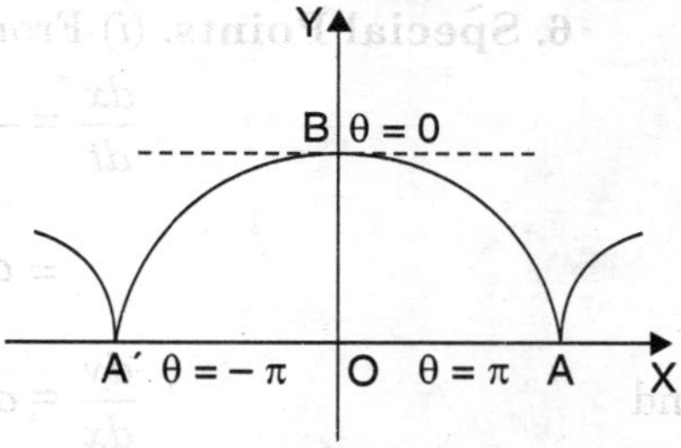

∴ The curve is concave downwards.

For values of θ > π or < – π, the same types of branches of the curve will be obtained.

The curve consists of congruent **arches** on both sides of *y*-axis which extend to infinity.

Hence the shape of the curve is as shown in the figure.

Note. B is called **vertex** of the cycloid and Line A′A *i.e.,* the line joining the two end points is called **base** of the cycloid and *a* is called the radius of the generating circle.

Remark. We have taken θ from – π to π to get one **arch** of the cycloid or one complete cycloid. We will again take θ from – π to π for the cycloid

$$x = a\,(\theta + \sin\theta),\ y = a\,(1 - \cos\theta)$$

i.e., **We take θ in [– π, π] for the cycloids**

$$\boldsymbol{x = a\,(\theta + \sin\theta),\ y = a\,(1 \pm \cos\theta)}$$

We will take θ in [0, 2π] for the cycloids

$$\boldsymbol{x = a\,(\theta - \sin\theta),\ y = a\,(1 \pm \cos\theta)}$$

(∵ For these two cycloids, for values of θ < 0 or θ > 2π, the same types of branches of the curve will be obtained.)

Example 2. *Trace the curve*

$$x = a\left[\cos t + \frac{1}{2}\log\tan^2\frac{t}{2}\right],\ y = a\sin t. \qquad \textbf{(Tractrix)}$$

Sol. The equations of the curve are

$$x = a\cos t + \frac{a}{2}\log\tan^2\frac{t}{2},\ y = a\sin t \qquad \ldots(1)$$

1. Symmetry. (*i*) On changing *t* to – *t* in (1), *x* remains unchanged and *y* changes to – *y*.

∴ The curve is symmetrical about *x*-axis.

(*ii*) On changing *t* to π – *t* in (1), *y* is unchanged and *x* changes to – *x*.

∴ The curve is symmetrical about *y*-axis.

2. Origin. Putting *y* = 0, we get sin *t* = 0 or *t* = 0 and then

$$x = a + \frac{a}{2}\log 0 = -\infty.$$

∴ The curve does not pass through the origin.

3. Asymptotes. When *t* = 0, *x* → – ∞ and *y* = 0.

Thus *y* = 0, *i.e., x*-axis is an asymptote to the curve.

4. Points of Intersection. The curve does not meet *x*-axis.

The curve meets *y*-axis (putting *x* = 0), where

$$x = a\cos t + \frac{1}{2}a\log\tan^2\frac{t}{2} = 0, \text{ which is satisfied by } t = \pm\frac{\pi}{2}.$$

and then $y = a\sin\left(\pm\frac{\pi}{2}\right) = \pm a$. Thus, the curve meets *y*-axis in the points (0, ± *a*).

5. Region. We know that – 1 ≤ sin *t* ≤ 1

∴ $-a \le a\sin t \le a$ or $-a \le y \le a$

Thus, the curve lies entirely between the lines *y* = ± *a*.

6. Special Points. (*i*) From (1),

$$\frac{dx}{dt} = -a\sin t + \frac{1}{2}a \cdot \frac{1}{\tan^2 t/2} \cdot 2\tan\frac{t}{2} \cdot \sec^2\frac{t}{2} \cdot \frac{1}{2}$$

$$= a\left(-\sin t + \frac{1}{2\sin t/2\cos t/2}\right) = a\left(-\sin t + \frac{1}{\sin t}\right) = \frac{a\cos^2 t}{\sin t}$$

and $$\frac{dy}{dx} = a\cos t$$

$$\therefore \quad \frac{dy}{dx} = \frac{\frac{dy}{dt}}{\frac{dx}{dt}} = \frac{a\cos t}{\frac{a\cos^2 t}{\sin t}} = \tan t \qquad \ldots(2)$$

$$\frac{dy}{dx} = \infty, \text{ when } t = \pm\frac{\pi}{2} \text{ and then } x = 0,\ y = \pm a.$$

Thus, at the points $(0, \pm a)$ the tangent is || to y-axis.

i.e., y-axis itself is a tangent at the points $(0, \pm a)$.

$\frac{dy}{dx} = 0$, when $t = 0$ and, then $x \to \infty$, $y = 0$, showing that $y = 0$ or x-axis is an asymptote as proved earlier.

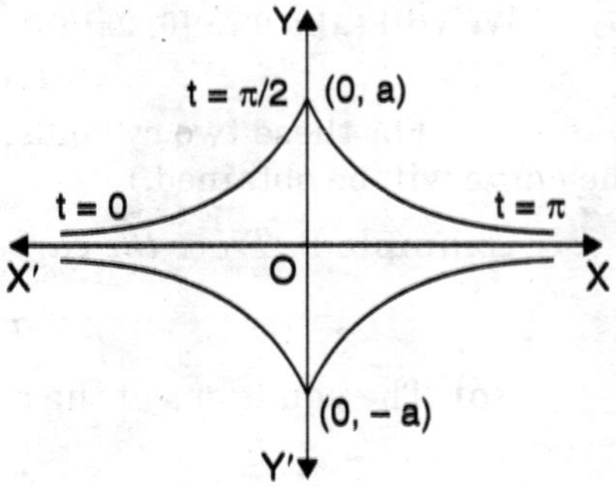

From (2), $$\frac{dy}{dx} = \tan t$$

Diff. w.r.t. x, $$\frac{d^2y}{dx^2} = \sec^2 t\,\frac{dt}{dx}$$

$$= \sec^2 t \cdot \frac{\sin t}{a\cos^2 t} = \frac{\sin t}{a\cos^4 t}$$

$\therefore$ $\frac{d^2y}{dx^2}$ is +ve for values of t in $[0, \pi]$.

$\therefore$ The curve is concave upwards for values of t in $[0, \pi]$.

The shape of the curve is as shown in the figure.

Example 3. *Trace the curve* $x = \frac{a(t+t^3)}{1+t^4}$, $y = \frac{a(t-t^3)}{1+t^4}$.

Sol. We have, $$x = \frac{at\,(1+t^2)}{1+t^4} \quad \ldots(1) \quad \text{and} \quad y = \frac{at\,(1-t^2)}{1+t^4} \qquad \ldots(2)$$

Let us eliminate t from equations (1) and (2) to form Cartesian Equation.

Dividing eqn. (1) by eqn. (2), we have

$$\frac{x}{y} = \frac{1+t^2}{1-t^2} \Rightarrow t^2 = \frac{x-y}{x+y} \qquad \ldots(3)$$

From (1), we have

$$x^2(1+t^4)^2 = a^2t^2(1+t^2)^2 \qquad | \text{ Squaring}$$

Putting the value of t^2 from (3),

$$x^2\left[1+\frac{(x-y)^2}{(x+y)^2}\right]^2 = \frac{a^2\,(x-y)}{x+y}\left(1+\frac{x-y}{x+y}\right)^2$$

or $$x^2\,\frac{[(x+y)^2+(x-y)^2]^2}{(x+y)^4} = \frac{a^2\,(x-y)}{(x+y)}\left[\frac{2x}{(x+y)}\right]^2$$

or $$(x^2+y^2)^2 = a^2(x^2-y^2) \qquad \ldots(4)$$

1. Symmetry. Curve (4) is symmetrical both about x-axis and y-axis.

2. Origin. Curve (4) passes through the origin.

Tangents at the origin are $a^2(x^2-y^2)=0$ (Equating lowest degree terms to zero)

But $a \neq 0$ $\therefore$ $x^2-y^2=0$ or $y^2=x^2$ $\therefore$ $y=\pm x$.

3. Asymptotes. The curve has no asymptotes.

4. Points of Intersection.

***(i)* Intersections with x-axis.**

Putting $y=0$ in eqn. (4), $x^4=a^2x^2$ or $x^2\,(x^2-a^2)=0$

$\therefore$ Either $x=0$ or $x^2=a^2$ *i.e.,* $x=\pm a$

$\therefore$ Intersections of the curve (4) with x-axis are $(-a, 0)$ $(0, 0)$ and $(a, 0)$.

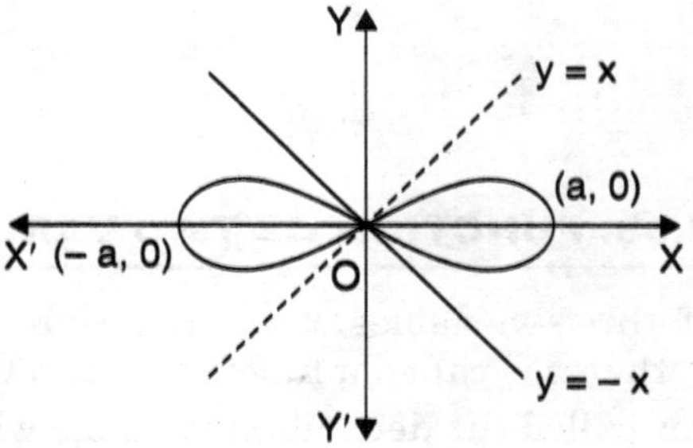

***(ii)* Intersections with y-axis.**

Putting $x=0$ in eqn. (4), $y^4=-a^2y^2$

or $y^4+a^2y^2=0$ or $y^2\,(y^2+a^2)=0$

$\therefore$ $y=0$ or $y=\pm ia$

$\therefore$ The only point of intersection of curve (4) with y-axis is (0, 0).

Thus the shape of the curve is as shown in the figure.

TEST YOUR KNOWLEDGE

Trace the following curves:

1. $x=a(\theta+\sin\theta)$, $y=a(1-\cos\theta)$
2. $x=a(\theta-\sin\theta)$, $y=a(1+\cos\theta)$
3. $x=a(\theta-\sin\theta)$, $y=a(1-\cos\theta)$ (*A.K.T.U. 2017*)
4. $x=t^2$, $y=t-\frac{1}{3}t^3$
5. $x=\dfrac{a\,(1-t^2)}{1+t^2}$, $y=\dfrac{at\,(1-t^2)}{1+t^2}$
6. $x=a\,[\cos\theta-\log(1+\cos\theta)]$, $y=a\sin\theta$

Answers

1.

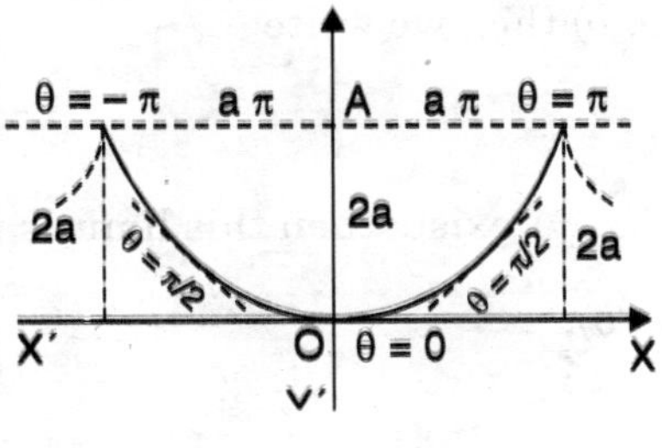

2.

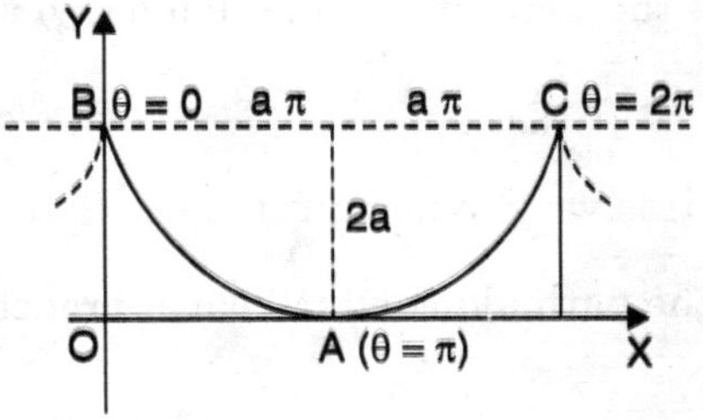

3.

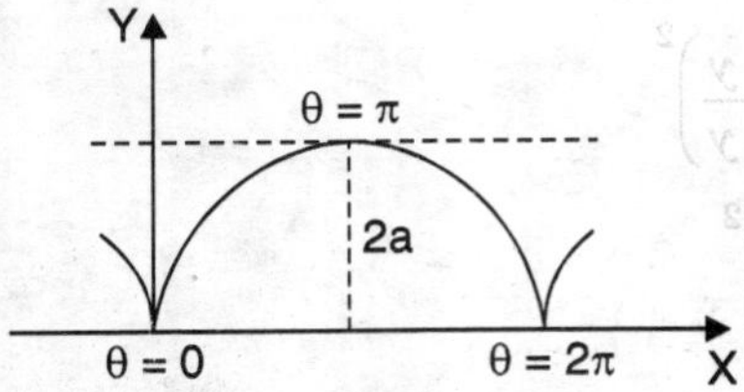

4.

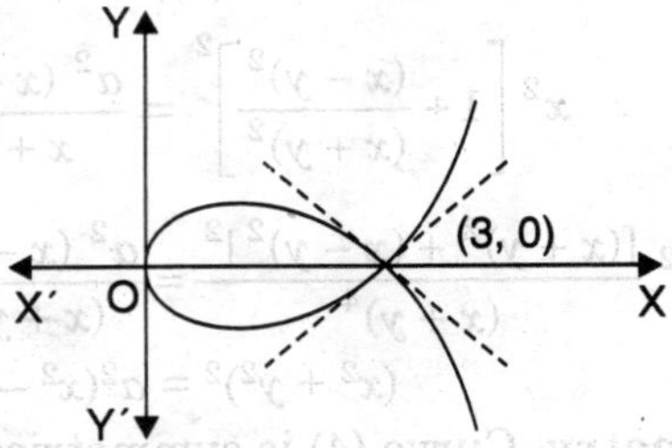

5.

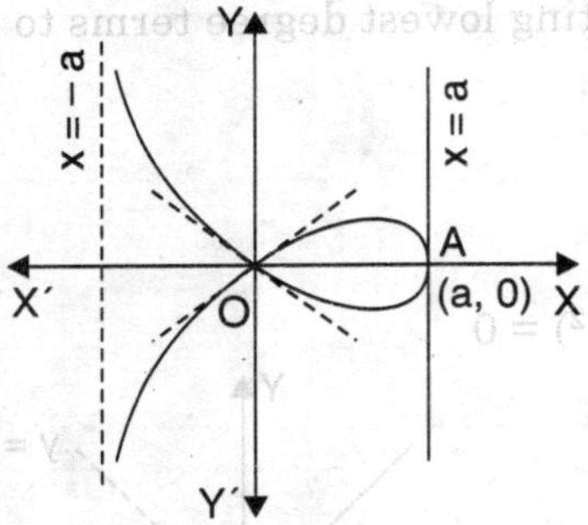

6.

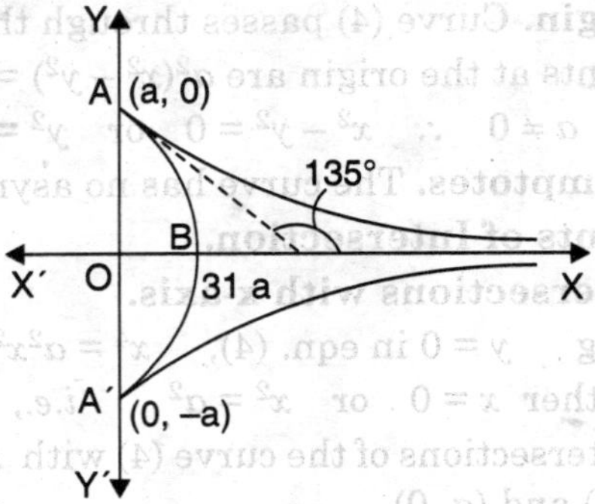

2.38. FUNCTION OF TWO VARIABLES

If three variables x, y, z are so related that the value of z depends upon the values of x and y, then z is called a function of two variables x and y and this is denoted by $z = f(x, y)$.

z is called the dependent variable while x and y are called independent variables.

For example, the area of a triangle is determined when its base and altitude are known. Thus, area of a triangle is a function of two variables, base and altitude.

In a similar way, a function of more than two variables can be defined.

Geometrically, Let $z = f(x, y)$ be a function of two independent variables x and y defined for all pairs of values of x and y which belong to an area A of the xy-plane. Then to each point (x, y) of this area corresponds a value of z given by the relation $z = f(x, y)$. Representing all these values (x, y, z) by points in space, we get a surface.

Hence the function $z = f(x, y)$ represents a surface.

2.39. LIMIT

Let $f(x, y)$ be a function of two variables then a number $L \in R$ is called the limit of the function at the point $(a, b) \in R^2$ if for given $\varepsilon > 0$, there exists positive number δ such that

$$|f(x, y) - L| < \varepsilon, \quad 0 < |x - a| < \delta, \quad 0 < |y - b| < \delta$$

If L is the limit of the function $f(x, y)$ at the point (a, b) then we write

$$\operatorname*{Lim}_{(x, y) \to (a, b)} f(x, y) = L$$

Here it is very important to note that if $\operatorname*{Lim}_{(x, y) \to (a, b)} f(x, y)$ exists then this limit is independent of the path along which we approach the point (a, b).

2.40. THEOREMS

1. If $\lim\limits_{(x, y) \to (a, b)} f(x, y)$ exists then it is unique.
2. The limit of a sum (or difference) of two functions of two variables is equal to the sum (or difference) of their limits.
3. The limit of a product of two functions of two variables is equal to the product of their limits.
4. The limit of product of a constant and a function of two variables is equal to the product of the constant and the limit of the function.
5. The limit of a quotient of two functions of two variables is equal to the quotient of their limits provided the limit of denominator is not equal to zero.
6. If a function $f(x, y)$ has limit L (finite or infinite) as $(x, y) \to (a, b)$ then $f(x, y)$ tends to L as $(x, y) \to (a, b)$ along any path.
7. If a function $f(x, y)$ has distinct limit as $(x, y) \to (a, b)$ along two distinct paths then $\lim\limits_{(x, y) \to (a, b)} f(x, y)$ does not exist.

2.41. CONTINUITY

A function $f(x, y)$ is said to be continuous at a point (a, b) if, for any arbitrarily chosen positive number ε, however small, we can find a corresponding number δ such that $| f(x, y) - f(a, b) | < \varepsilon$ for every point (x, y) within the circle with its centre at (a, b) and radius δ. *i.e.*, whenever $0 < (x - a)^2 + (y - b)^2 < \delta^2$.

Alternatively, $f(x, y)$ is said to be continuous at (a, b) if $\underset{\substack{x \to a \\ y \to b}}{\text{Lt}} f(x, y) = f(a, b)$ irrespective of the path along with $x \to a, y \to b$.

It should not be assumed that the path along which the point (x, y) tends to (a, b) is immaterial, because $\underset{x \to a}{\text{Lt}} \{ \underset{y \to b}{\text{Lt}} f(x, y)\}$ is not always equal to $\underset{y \to b}{\text{Lt}} \{ \underset{x \to a}{\text{Lt}} f(x, y)\}$.

However, *usually*, the limit is the same irrespective of the path of approach.

A function which is not continuous at any point is said to be discontinuous at the point. A function $f(x, y)$ is said to be continuous in a domain D if it is continuous at every point of the domain D. The sum, difference, product and quotient of two continuous functions of two variables is a continuous function.

ILLUSTRATIVE EXAMPLES

Example 1. *Let $A = \{(x, y) : 0 < x < 1, 0 < y < 1, x, y \in R\}$. Let $f : A \to R$ be defined by $f(x, y) = x + y$. Show that*

$$\lim_{(x, y) \to \left(0, \frac{1}{2}\right)} f(x, y) = \frac{1}{2}$$

Sol. $\left| f(x, y) - \frac{1}{2} \right| = \left| x + y - \frac{1}{2} \right|$

$$= \left| (x-0) + \left(y - \frac{1}{2} \right) \right|$$

$$\leq |x - 0| + \left| y - \frac{1}{2} \right| \qquad ...(1)$$

Let $\varepsilon > 0$, choose $\delta = \frac{\varepsilon}{2}$ then for $|x - 0| < \delta$ and $\left| y - \frac{1}{2} \right| < \delta$, we have from (1),

$$\left| f(x, y) - \frac{1}{2} \right| < \varepsilon, \quad |x - 0| < \delta \quad \text{and} \quad \left| y - \frac{1}{2} \right| < \delta$$

Hence, $\displaystyle \lim_{(x, y) \to \left(0, \frac{1}{2}\right)} f(x, y) = \frac{1}{2}$.

Example 2. *Using the $\delta - \varepsilon$ approach, show that* $\displaystyle \lim_{(x, y) \to (2, 1)} 3x + 4y = 10.$

Sol. $|(3x + 4y) - 10| = |3(x - 2) + 4(y - 1)|$

$$\leq 3|x - 2| + 4|y - 1|$$

If we take $|x - 2| < \delta$ and $|y - 1| < \delta$, we get

$|f(x, y) - 10| < 7\delta < \varepsilon$ which is satisfied when $\delta < \frac{\varepsilon}{7}$.

Hence, $\displaystyle \lim_{(x, y) \to (2, 1)} f(x, y) = 10.$

Example 3. *Show that the following limits:*

(*i*) $\displaystyle \lim_{(x, y) \to (0, 0)} \frac{xy}{x^2 + y^2}$ (*ii*) $\displaystyle \lim_{(x, y) \to (0, 0)} \frac{xy^3}{x^2 + y^6}$

do not exist.

Sol. The limit does not exist if it is not finite or if it depends on a particular path.

(*i*) Choose the path $y = mx$. As $(x, y) \to (0, 0)$, we get $x \to 0$.

Therefore, $$\lim_{(x, y) \to (0, 0)} \frac{xy}{x^2 + y^2} = \lim_{x \to 0} \frac{mx^2}{(1 + m^2) x^2} = \frac{m}{1 + m^2}$$

which depends on m. For different values of m, we obtain different limits. Hence the limit does not exist.

(*ii*) Choose the path $x = my^3$. As $(x, y) \to (0, 0)$, we get $y \to 0$

Therefore, $$\lim_{(x, y) \to (0, 0)} \frac{xy^3}{x^2 + y^6} = \lim_{y \to 0} \frac{my^6}{(m^2 + 1) y^6} = \frac{m}{m^2 + 1}$$

which depends on m. Hence the limit does not exist.

Example 4. *Show that the function $f(x, y) = x - y$ is continuous for all $(x, y) \in R^2$.*

Sol. Let $(a, b) \in R^2$ then $f(a, b) = a - b$

$$\therefore \quad |f(x, y) - f(a, b)| = |(x - y) - (a - b)|$$
$$= |(x - a) + (b - y)|$$
$$\leq |x - a| + |y - b| \qquad [\because \; |x| = |-x|] \quad ...(1)$$

Let $\varepsilon > 0$. Choose $\delta = \frac{\varepsilon}{2}$ then for $|x - a| < \delta$ and $|y - b| < \delta$, we have from (1)

$$|f(x, y) - f(a, b)| < \frac{\varepsilon}{2} + \frac{\varepsilon}{2} = \varepsilon$$

Hence the function $f(x, y) = x - y$ is continuous for all $(a, b) \in R^2$. But (a, b) is an arbitrary element of R^2 so $f(x, y) = x - y$ is continuous for all $(x, y) \in R^2$.

Example 5. *Show that the function* $f(x, y) = \begin{cases} \dfrac{x - y}{x + y}, & (x, y) \neq (0, 0) \\ 0, & (x, y) = (0, 0) \end{cases}$ *is discontinuous at the point (0, 0).*

Sol. Choose the path $y = mx$.

As $(x, y) \to (0, 0)$, we get $x \to 0$. Therefore,

$$\lim_{(x, y) \to (0, 0)} \frac{x - y}{x + y} = \lim_{x \to 0} \frac{(1 - m)\,x}{(1 + m)\,x} = \frac{1 - m}{1 + m}$$

which depends on m. Since the limit does not exist, the function is not continuous at (0, 0).

2.42. PARTIAL DERIVATIVES OF FIRST ORDER

Let $z = f(x, y)$ be a function of two independent variables x and y. If y is kept constant and x alone is allowed to vary, then z becomes a function of x only. The derivative of z, with respect to x, treating y as constant, is called partial derivative of z w.r.t. x and is denoted by $\frac{\partial z}{\partial x}$ or $\frac{\partial f}{\partial x}$ or f_x.

Thus,
$$\frac{\partial z}{\partial x} = \underset{h \to 0}{\text{Lt}} \frac{f(x + h, y) - f(x, y)}{h}$$

Similarly, the derivative of z, with respect to y, treating x as constant, is called partial derivative of z w.r.t. y and is denoted by $\frac{\partial z}{\partial y}$ or $\frac{\partial f}{\partial y}$ or f_y

Thus,
$$\frac{\partial z}{\partial y} = \underset{k \to 0}{\text{Lt}} \frac{f(x, y + k) - f(x, y)}{k}$$

$\frac{\partial z}{\partial x}$ and $\frac{\partial z}{\partial y}$ are called **first order partial derivatives of z.**

[**In general,** if z is a function of two or more independent variables, then the **partial derivative of z** w.r.t. **any one** of the independent variables is the **ordinary derivative of z** w.r.t. that variable, **treating all other variables as constant.**]

Geometrically, Let $z = f(x, y)$ be a function of two variables x and y. Then by Art. 2.38, it represents a surface S. If $y = k$, a constant, then $y = k$ represents a plane parallel to the zx-plane.

$\therefore$ $z = f(x, y)$ and $y = k$ represent a plane curve C which is the section of S by $y = k$.

$\dfrac{\partial z}{\partial x}$ represents the slope of tangent to C at (x, k, z).

Thus, $\dfrac{\partial z}{\partial x}$ gives the slope of the tangent drawn to the curve of intersection of the surface $z = f(x, y)$ and a plane parallel to zx-plane.

Similarly, $\dfrac{\partial z}{\partial y}$ gives the slope of the tangent drawn to the curve of intersection of the surface $z = f(x, y)$ and a plane parallel to yz-plane.

2.43. PARTIAL DERIVATIVES OF HIGHER ORDER

Since the first order partial derivatives $\dfrac{\partial z}{\partial x}$ and $\dfrac{\partial z}{\partial y}$ are themselves functions of x and y, they can be further differentiated partially w.r.t. x as well as y. These are called second order partial derivatives of z. The usual notations for these second order partial derivatives are:

$$\frac{\partial}{\partial x}\left(\frac{\partial z}{\partial x}\right) = \frac{\partial^2 z}{\partial x^2} \quad \text{or} \quad f_{xx}\,; \qquad \frac{\partial}{\partial y}\left(\frac{\partial z}{\partial y}\right) = \frac{\partial^2 z}{\partial y^2} \quad \text{or} \quad f_{yy}$$

$$\frac{\partial}{\partial x}\left(\frac{\partial z}{\partial y}\right) = \frac{\partial^2 z}{\partial x \partial y} \quad \text{or} \quad f_{xy}\,; \qquad \frac{\partial}{\partial y}\left(\frac{\partial z}{\partial x}\right) = \frac{\partial^2 z}{\partial y \partial x} \quad \text{or} \quad f_{yx}$$

In general, $$\frac{\partial^2 z}{\partial x \partial y} = \frac{\partial^2 z}{\partial y \partial x} \quad \text{or} \quad f_{xy} = f_{yx}.$$

Note 1. If $z = f(x)$, a function of single independent variable x, we get $\dfrac{dz}{dx}$.

If $z = f(x, y)$, a function of two independent variables x and y, we get $\dfrac{\partial z}{\partial x}$ and $\dfrac{\partial z}{\partial y}$.

Similarly, for a function of more than two independent variables $x_1, x_2, \ldots, x_n$, we get

$$\frac{\partial z}{\partial x_1}, \frac{\partial z}{\partial x_2}, \ldots, \frac{\partial z}{\partial x_n}.$$

Note 2. (*i*) If $z = u + v$, where $u = f(x, y)$, $v = \phi(x, y)$, then z is a function of x and y.

$$\therefore \quad \frac{\partial z}{\partial x} = \frac{\partial u}{\partial x} + \frac{\partial v}{\partial x}\,; \quad \frac{\partial z}{\partial y} = \frac{\partial u}{\partial y} + \frac{\partial v}{\partial y}.$$

(*ii*) If $z = uv$, where $u = f(x, y)$, $v = \phi(x, y)$ then

$$\frac{\partial z}{\partial x} = \frac{\partial}{\partial x}(uv) = u\frac{\partial v}{\partial x} + v\frac{\partial u}{\partial x}, \quad \frac{\partial z}{\partial y} = \frac{\partial}{\partial y}(uv) = u\frac{\partial v}{\partial y} + v\frac{\partial u}{\partial y}.$$

(*iii*) If $z = \dfrac{u}{v}$, where $u = f(x, y)$, $v = \phi(x, y)$, then

$$\frac{\partial z}{\partial x} = \frac{\partial}{\partial x}\left(\frac{u}{v}\right) = \frac{v\dfrac{\partial u}{\partial x} - u\dfrac{\partial v}{\partial x}}{v^2}, \qquad \frac{\partial z}{\partial y} = \frac{\partial}{\partial y}\left(\frac{u}{v}\right) = \frac{v\dfrac{\partial u}{\partial y} - u\dfrac{\partial v}{\partial y}}{v^2}$$

Example 6. *Find the first order partial derivatives of the following:*

(i) $u = \tan^{-1} \dfrac{x^2 + y^2}{x + y}$ *(ii)* $u = \cos^{-1} \left(\dfrac{x}{y}\right)$.

Sol. *(i)* $u = \tan^{-1} \dfrac{x^2 + y^2}{x + y}$

Differentiating u partially w.r.t. x, we get

$$\frac{\partial u}{\partial x} = \frac{1}{1 + \left(\dfrac{x^2 + y^2}{x + y}\right)^2} \cdot \frac{\partial}{\partial x}\left(\frac{x^2 + y^2}{x + y}\right)$$

$$= \frac{(x + y)^2}{(x + y)^2 + (x^2 + y^2)^2} \cdot \frac{(x + y)\dfrac{\partial}{\partial x}(x^2 + y^2) - (x^2 + y^2)\dfrac{\partial}{\partial x}(x + y)}{(x + y)^2}$$

$$= \frac{(x + y) \,.\, 2x - (x^2 + y^2) \,.\, 1}{(x + y)^2 + (x^2 + y^2)^2}$$

$$\Rightarrow \quad \frac{\partial u}{\partial x} = \frac{x^2 + 2xy - y^2}{(x + y)^2 + (x^2 + y^2)^2} \quad ...(1)$$

Since u is symmetrical w.r.t. x and y, $\dfrac{\partial u}{\partial y} = \dfrac{y^2 + 2xy - x^2}{(x + y)^2 + (x^2 + y^2)^2}$.

(ii) $u = \cos^{-1}\left(\dfrac{x}{y}\right)$

$$\therefore \quad \frac{\partial u}{\partial x} = \frac{-1}{\sqrt{1 - \left(\dfrac{x}{y}\right)^2}} \cdot \frac{\partial}{\partial x}\left(\frac{x}{y}\right) = \frac{-y}{\sqrt{y^2 - x^2}} \cdot \frac{1}{y} = \frac{-1}{\sqrt{y^2 - x^2}}$$

and

$$\frac{\partial u}{\partial y} = \frac{-1}{\sqrt{1 - \left(\dfrac{x}{y}\right)^2}} \cdot \frac{\partial}{\partial y}\left(\frac{x}{y}\right) = \frac{-y}{\sqrt{y^2 - x^2}}\left(-\frac{x}{y^2}\right) = \frac{x}{y\sqrt{y^2 - x^2}}.$$

Example 7. *If* $z(x + y) = x^2 + y^2$, *show that* $\left(\dfrac{\partial z}{\partial x} - \dfrac{\partial z}{\partial y}\right)^2 = 4\left(1 - \dfrac{\partial z}{\partial x} - \dfrac{\partial z}{\partial y}\right)$.

Sol. $z = \dfrac{x^2 + y^2}{x + y}$

$$\therefore \quad \frac{\partial z}{\partial x} = \frac{(x + y) \,.\, 2x - (x^2 + y^2) \,.\, 1}{(x + y)^2} = \frac{x^2 + 2xy - y^2}{(x + y)^2}$$

Similarly, $\dfrac{\partial z}{\partial y} = \dfrac{y^2 + 2xy - x^2}{(x + y)^2}$ | $\because$ z is symmetrical w.r.t. x and y

Now, $$\left(\frac{\partial z}{\partial x}-\frac{\partial z}{\partial y}\right)^2=\left[\frac{2x^2-2y^2}{(x+y)^2}\right]^2=\frac{4(x+y)^2\,(x-y)^2}{(x+y)^4}=\frac{4(x-y)^2}{(x+y)^2}$$

$$4\left(1-\frac{\partial z}{\partial x}-\frac{\partial z}{\partial y}\right)=4\left[1-\frac{x^2+2xy-y^2}{(x+y)^2}-\frac{y^2+2xy-x^2}{(x+y)^2}\right]$$

$$=4\left[\frac{x^2+2xy+y^2-x^2-2xy+y^2-y^2-2xy+x^2}{(x+y)^2}\right]$$

$$=\frac{4(x^2-2xy+y^2)}{(x+y)^2}=\frac{4(x-y)^2}{(x+y)^2}$$

$\therefore$ $$\left(\frac{\partial z}{\partial x}-\frac{\partial z}{\partial y}\right)^2=4\left(1-\frac{\partial z}{\partial x}-\frac{\partial z}{\partial y}\right).$$

Example 8. *If* $z = e^{ax+by} f(ax-by)$, *show that* $b\dfrac{\partial z}{\partial x}+a\dfrac{\partial z}{\partial y}=2abz.$

Sol. We have, $$z = e^{ax+by}\,.\,f(ax-by) \qquad ...(1)$$

Differentiating (1) w.r.t. x partially, we get

$$\frac{\partial z}{\partial x}=e^{ax+by}\,.\,f'(ax-by)\,.\,a+f(ax-by)\,.\,e^{ax+by}\,.\,a$$

$\therefore$ $$b\frac{\partial z}{\partial x}=ab\,e^{ax+by}\,[f'(ax-by)+f(ax-by)] \qquad ...(2)$$

Again differentiating (1) w.r.t. y, partially, we get

$$\frac{\partial z}{\partial y}=e^{ax+by}\,.\,f'(ax-by)\,.\,(-b)+f(ax-by)\,.\,e^{ax+by}\,.\,b$$

$$a\frac{\partial z}{\partial y}=ab\,e^{ax+by}\,[-f'(ax-by)+f(ax-by)] \qquad ...(3)$$

Adding (2) and (3), we get

$$b\frac{\partial z}{\partial x}+a\frac{\partial z}{\partial y}=2abz.$$

Example 9. *If* $u(x, y, z) = \log(\tan x + \tan y + \tan z)$, *show that*

$$\sin 2x\frac{\partial u}{\partial x}+\sin 2y\frac{\partial u}{\partial y}+\sin 2z\frac{\partial u}{\partial z}=2. \qquad \text{[G.B.T.U. 2012]}$$

Sol. Differentiating u partially w.r.t. x, we get

$$\frac{\partial u}{\partial x}=\frac{1}{\tan x+\tan y+\tan z}\,(\sec^2 x) \qquad ...(1)$$

Differentiating u partially w.r.t. y, we get

$$\frac{\partial u}{\partial y}=\frac{1}{\tan x+\tan y+\tan z}\,(\sec^2 y) \qquad ...(2)$$

Differentiating u partially w.r.t. z, we get

$$\frac{\partial u}{\partial z}=\frac{1}{\tan x+\tan y+\tan z}\,(\sec^2 z) \qquad ...(3)$$

Now, $\text{LHS} = \sin 2x \dfrac{\partial u}{\partial x} + \sin 2y \dfrac{\partial u}{\partial y} + \sin 2z \dfrac{\partial u}{\partial z}$

$$= \frac{\sin 2x \sec^2 x + \sin 2y \sec^2 y + \sin 2z \sec^2 z}{\tan x + \tan y + \tan z} \qquad | \text{ Using (1), (2) and (3)}$$

$$= \frac{2\tan x + 2\tan y + 2\tan z}{\tan x + \tan y + \tan z} = 2 = \text{RHS}.$$

Example 10. *If* $\theta = t^n e^{-\frac{r^2}{4t}}$, *find the value of n which will make* $\dfrac{1}{r^2}\dfrac{\partial}{\partial r}\left(r^2 \dfrac{\partial \theta}{\partial r}\right) = \dfrac{\partial \theta}{\partial t}$.

Sol. $\theta = t^n e^{-\frac{r^2}{4t}}$

$$\frac{\partial \theta}{\partial r} = t^n . e^{-\frac{r^2}{4t}} . \left(-\frac{2r}{4t}\right) = -\frac{1}{2} r t^{n-1} e^{-\frac{r^2}{4t}}$$

$\therefore$
$$r^2 \frac{\partial \theta}{\partial r} = -\frac{1}{2} r^3 . t^{n-1} e^{-\frac{r^2}{4t}}$$

$$\frac{\partial}{\partial r}\left(r^2 \frac{\partial \theta}{\partial r}\right) = -\frac{1}{2} t^{n-1}\left[3r^2 e^{-\frac{r^2}{4t}} + r^3 e^{-\frac{r^2}{4t}}\left(-\frac{2r}{4t}\right)\right] = -\frac{1}{2} t^{n-1} r^2 e^{-\frac{r^2}{4t}}\left[3 - \frac{r^2}{2t}\right]$$

$\therefore$
$$\frac{1}{r^2}\frac{\partial}{\partial r}\left(r^2 \frac{\partial \theta}{\partial r}\right) = \frac{1}{2} t^{n-1} e^{-\frac{r^2}{4t}}\left(\frac{r^2}{2t} - 3\right)$$

Also,
$$\frac{\partial \theta}{\partial t} = n t^{n-1} e^{-\frac{r^2}{4t}} + t^n e^{-\frac{r^2}{4t}} . \left(\frac{r^2}{4t^2}\right) = t^{n-1} e^{-\frac{r^2}{4t}}\left(n + \frac{r^2}{4t}\right)$$

Since $\dfrac{1}{r^2}\dfrac{\partial}{\partial r}\left(r^2 \dfrac{\partial \theta}{\partial r}\right) = \dfrac{\partial \theta}{\partial t}$ [Given]

$\therefore$
$$\frac{1}{2} t^{n-1} e^{-\frac{r^2}{4t}}\left(\frac{r^2}{2t} - 3\right) = t^{n-1} e^{-\frac{r^2}{4t}}\left(n + \frac{r^2}{4t}\right).$$

$\Rightarrow$
$$\frac{r^2}{4t} - \frac{3}{2} = n + \frac{r^2}{4t}$$

$\therefore$
$$n = -\frac{3}{2}.$$

Example 11. *If* $u = \log(x^3 + y^3 + z^3 - 3xyz)$, *show that*

(*i*) $\left(\dfrac{\partial}{\partial x} + \dfrac{\partial}{\partial y} + \dfrac{\partial}{\partial z}\right)^2 u = -\dfrac{9}{(x+y+z)^2}$ (U.K.T.U. 2011)

(*ii*) $\dfrac{\partial^2 u}{\partial x^2} + \dfrac{\partial^2 u}{\partial y^2} + \dfrac{\partial^2 u}{\partial z^2} + 2\dfrac{\partial^2 u}{\partial y \partial z} + 2\dfrac{\partial^2 u}{\partial z \partial x} + 2\dfrac{\partial^2 u}{\partial x \partial y} = \dfrac{-9}{(x+y+z)^2}$.

Sol. (*i*) $u = \log (x^3 + y^3 + z^3 - 3xyz)$

$$\frac{\partial u}{\partial x} = \frac{3x^2 - 3yz}{x^3 + y^3 + z^3 - 3xyz}\,;\ \frac{\partial u}{\partial y} = \frac{3y^2 - 3zx}{x^3 + y^3 + z^3 - 3xyz}\,;\ \frac{\partial u}{\partial z} = \frac{3z^2 - 3xy}{x^3 + y^3 + z^3 - 3xyz}$$

Adding, $\dfrac{\partial u}{\partial x} + \dfrac{\partial u}{\partial y} + \dfrac{\partial u}{\partial z} = \dfrac{3(x^2 + y^2 + z^2 - xy - yz - zx)}{x^3 + y^3 + z^3 - 3xyz} = \dfrac{3}{x + y + z}$

$[\because\ x^3 + y^3 + z^3 - 3xyz = (x + y + z)(x^2 + y^2 + z^2 - xy - yz - zx)]$

Now, $\left(\dfrac{\partial}{\partial x} + \dfrac{\partial}{\partial y} + \dfrac{\partial}{\partial z}\right)^2 u = \left(\dfrac{\partial}{\partial x} + \dfrac{\partial}{\partial y} + \dfrac{\partial}{\partial z}\right)\left(\dfrac{\partial}{\partial x} + \dfrac{\partial}{\partial y} + \dfrac{\partial}{\partial z}\right) u$

$$= \left(\frac{\partial}{\partial x} + \frac{\partial}{\partial y} + \frac{\partial}{\partial z}\right)\left(\frac{\partial u}{\partial x} + \frac{\partial u}{\partial y} + \frac{\partial u}{\partial z}\right) = \left(\frac{\partial}{\partial x} + \frac{\partial}{\partial y} + \frac{\partial}{\partial z}\right)\left(\frac{3}{x + y + z}\right)$$

$$= -\frac{3}{(x + y + z)^2} - \frac{3}{(x + y + z)^2} - \frac{3}{(x + y + z)^2} = -\frac{9}{(x + y + z)^2} \quad ...(1)$$

(*ii*) $\left(\dfrac{\partial}{\partial x} + \dfrac{\partial}{\partial y} + \dfrac{\partial}{\partial z}\right)^2 u = \left(\dfrac{\partial}{\partial x} + \dfrac{\partial}{\partial y} + \dfrac{\partial}{\partial z}\right)\left(\dfrac{\partial}{\partial x} + \dfrac{\partial}{\partial y} + \dfrac{\partial}{\partial z}\right) u = \left(\dfrac{\partial}{\partial x} + \dfrac{\partial}{\partial y} + \dfrac{\partial}{\partial z}\right)\left(\dfrac{\partial u}{\partial x} + \dfrac{\partial u}{\partial y} + \dfrac{\partial u}{\partial z}\right)$

$$= \frac{\partial}{\partial x}\left(\frac{\partial u}{\partial x} + \frac{\partial u}{\partial y} + \frac{\partial u}{\partial z}\right) + \frac{\partial}{\partial y}\left(\frac{\partial u}{\partial x} + \frac{\partial u}{\partial y} + \frac{\partial u}{\partial z}\right) + \frac{\partial}{\partial z}\left(\frac{\partial u}{\partial x} + \frac{\partial u}{\partial y} + \frac{\partial u}{\partial z}\right)$$

$$= \frac{\partial^2 u}{\partial x^2} + \frac{\partial^2 u}{\partial x \partial y} + \frac{\partial^2 u}{\partial x \partial z} + \frac{\partial^2 u}{\partial y \partial x} + \frac{\partial^2 u}{\partial y^2} + \frac{\partial^2 u}{\partial y \partial z} + \frac{\partial^2 u}{\partial z \partial x} + \frac{\partial^2 u}{\partial z \partial y} + \frac{\partial^2 u}{\partial z^2}$$

$$= \frac{\partial^2 u}{\partial x^2} + \frac{\partial^2 u}{\partial y^2} + \frac{\partial^2 u}{\partial z^2} + 2\frac{\partial^2 u}{\partial y \partial z} + 2\frac{\partial^2 u}{\partial z \partial x} + 2\frac{\partial^2 u}{\partial x \partial y}$$

$$\left[\because\ \frac{\partial^2 u}{\partial y \partial x} = \frac{\partial^2 u}{\partial x \partial y}, \frac{\partial^2 u}{\partial z \partial y} = \frac{\partial^2 u}{\partial y \partial z}, \frac{\partial^2 u}{\partial x \partial z} = \frac{\partial^2 u}{\partial z \partial x}\right]$$

$$\therefore \quad \frac{\partial^2 u}{\partial x^2} + \frac{\partial^2 u}{\partial y^2} + \frac{\partial^2 u}{\partial z^2} + 2\frac{\partial^2 u}{\partial y \partial z} + 2\frac{\partial^2 u}{\partial z \partial x} + 2\frac{\partial^2 u}{\partial x \partial y} = \left(\frac{\partial}{\partial x} + \frac{\partial}{\partial y} + \frac{\partial}{\partial z}\right)^2 u = \frac{-9}{(x + y + z)^2}$$

[From (1)]

Example 12. *If* $x^x y^y z^z = c$, *show that at* $x = y = z$, $\dfrac{\partial^2 z}{\partial x \partial y} = -(x \log ex)^{-1}$.

[M.T.U. (C.O.) 2013]

Sol. $x^x y^y z^z = c$ defines z as a function of x and y.

Taking log, $x \log x + y \log y + z \log z = \log c$

Differentiating partially w.r.t. y, we have

$$y \cdot \frac{1}{y} + 1 \cdot \log y + z \cdot \frac{1}{z}\frac{\partial z}{\partial y} + 1 \cdot \log z \cdot \frac{\partial z}{\partial y} = 0$$

or $\quad 1 + \log y + (1 + \log z)\dfrac{\partial z}{\partial y} = 0 \quad ...(1)$

or $\quad \dfrac{\partial z}{\partial y} = -\dfrac{1 + \log y}{1 + \log z}$

Similarly, $\quad \dfrac{\partial z}{\partial x} = -\dfrac{1 + \log x}{1 + \log z} \quad ...(2)$

Differentiating (1) partially w.r.t. x, we have

$$\left(\frac{1}{z}\frac{\partial z}{\partial x}\right)\frac{\partial z}{\partial y} + (1 + \log z)\,\frac{\partial^2 z}{\partial x \partial y} = 0$$

or $$\frac{\partial^2 z}{\partial x \partial y} = -\frac{1}{z(1+\log z)}\,\frac{\partial z}{\partial x}\cdot\frac{\partial z}{\partial y} \qquad \ldots(3)$$

When $x = y = z$

From (2), $$\frac{\partial z}{\partial y} = -1,\ \frac{\partial z}{\partial x} = -1$$

From (3), $$\frac{\partial^2 z}{\partial x \partial y} = -\frac{1}{x(1+\log x)}\,(-1)(-1)$$

$$= -\frac{1}{x(\log e + \log x)} = -\frac{1}{x(\log ex)} = -(x \log ex)^{-1}.$$

Example 13. *If $u = f(r)$, where $r^2 = x^2 + y^2$, prove that* $\dfrac{\partial^2 u}{\partial x^2} + \dfrac{\partial^2 u}{\partial y^2} = f''(r) + \dfrac{1}{r}\,f'(r)$. (A.K.T.U. 2016)

Sol. $$r^2 = x^2 + y^2 \qquad \ldots(1)$$

Differentiating partially w.r.t. x, we get

$$2r\,\frac{\partial r}{\partial x} = 2x \quad \text{or} \quad \frac{\partial r}{\partial x} = \frac{x}{r}$$

Similarly, $$\frac{\partial r}{\partial y} = \frac{y}{r}$$

Now, $$u = f(r)$$

$\therefore$ $$\frac{\partial u}{\partial x} = f'(r)\,.\,\frac{\partial r}{\partial x} = \frac{x}{r}\,f'(r)$$

Differentiating again w.r.t. x, we get

$$\frac{\partial^2 u}{\partial x^2} = \frac{1}{r}\,f'(r) + x\,.\left(-\frac{1}{r^2}\frac{\partial r}{\partial x}\right)f'(r) + \frac{x}{r}\,f''(r)\,.\,\frac{\partial r}{\partial x}$$

$$\left[\because\ \frac{\partial}{\partial x}(uvw) = vw\,\frac{\partial}{\partial x}(u) + uw\,\frac{\partial}{\partial x}(v) + uv\,\frac{\partial}{\partial x}(w)\right]$$

$$= \frac{1}{r}\,f'(r) - \frac{x}{r^2}\cdot\frac{x}{r}\,f'(r) + \frac{x}{r}\,.\,f''(r)\,.\,\frac{x}{r} = \frac{1}{r}\,f'(r) - \frac{x^2}{r^3}\,f'(r) + \frac{x^2}{r^2}\,f''(r)$$

$$= \frac{r^2 - x^2}{r^3}\,f'(r) + \frac{x^2}{r^2}\,f''(r) = \frac{y^2}{r^3}\,f'(r) + \frac{x^2}{r^2}\,f''(r) \qquad |\ \text{Using (1)}$$

Similarly, $$\frac{\partial^2 u}{\partial y^2} = \frac{x^2}{r^3}\,f'(r) + \frac{y^2}{r^2}\,f''(r)$$

$\therefore$ $$\frac{\partial^2 u}{\partial x^2} + \frac{\partial^2 u}{\partial y^2} = \frac{x^2 + y^2}{r^3}\,f'(r) + \frac{x^2 + y^2}{r^2}\,f''(r) = \frac{r^2}{r^3}\,f'(r) + \frac{r^2}{r^2}\,f''(r) = f''(r) + \frac{1}{r}\,f'(r).$$

TEST YOUR KNOWLEDGE

1. Find the first order partial derivatives of the following functions:
   (i) $u = y^x$ (ii) $u = \log (x^2 + y^2)$.
2. If $u = x^2 + y^2 + z^2$, show that $xu_x + yu_y + zu_z = 2u$.
3. If $z = \log (x^2 + xy + y^2)$, show that $x \frac{\partial z}{\partial x} + y \frac{\partial z}{\partial y} = 2$.
4. If $u = x^2y + y^2z + z^2x$, show that $u_x + u_y + u_z = (x + y + z)^2$.
5. If $u = \frac{y}{z} + \frac{z}{x} + \frac{x}{y}$, show that $x \frac{\partial u}{\partial x} + y \frac{\partial u}{\partial y} + z \frac{\partial u}{\partial z} = 0$.
6. If $f(x, y) = x^3y - xy^3$, find $\left[\frac{1}{\frac{\partial f}{\partial x}} + \frac{1}{\frac{\partial f}{\partial y}} \right]_{\substack{x=1 \\ y=2}}$.
7. If $u = x^y$, show that $\frac{\partial^3 u}{\partial x^2 \partial y} = \frac{\partial^3 u}{\partial x \partial y \partial x}$.
8. If $f(x, y, z) = \begin{vmatrix} x^2 & y^2 & z^2 \\ x & y & z \\ 1 & 1 & 1 \end{vmatrix}$, show that $f_x + f_y + f_z = 0$.
9. Verify $\frac{\partial^2 u}{\partial x \partial y} = \frac{\partial^2 u}{\partial y \partial x}$ for the following functions:
   (i) $u = ax^2 + 2hxy + by^2$ (ii) $u = \tan^{-1}\left(\frac{x}{y}\right)$ (iii) $u = \log\left(\frac{x^2 + y^2}{xy}\right)$.
10. If $z = \log (e^x + e^y)$, show that $rt - s^2 = 0$, where $r = \frac{\partial^2 z}{\partial x^2}$, $t = \frac{\partial^2 z}{\partial y^2}$, $s = \frac{\partial^2 z}{\partial x \partial y}$. *(U.P.T.U. 2015)*
11. If $u = \tan^{-1} \frac{xy}{\sqrt{1 + x^2 + y^2}}$, show that $\frac{\partial^2 u}{\partial x \partial y} = (1 + x^2 + y^2)^{-3/2}$.
12. If $u = e^{xyz}$, prove that $\frac{\partial^3 u}{\partial x \partial y \partial z} = (1 + 3xyz + x^2y^2z^2)u$.
13. If $u = \log (x^2 + y^2) + \tan^{-1} \frac{y}{x}$, show that $\frac{\partial^2 u}{\partial x^2} + \frac{\partial^2 u}{\partial y^2} = 0$.
14. (i) If $z = f(x + ct) + \phi (x - ct)$, show that $\frac{\partial^2 z}{\partial t^2} = c^2 \frac{\partial^2 z}{\partial x^2}$. *(G.B.T.U. 2011)*

    (ii) If $z = \tan (y + ax) - (y - ax)^{3/2}$, show that $\frac{\partial^2 z}{\partial x^2} = a^2 \frac{\partial^2 z}{\partial y^2}$.
15. Show that $\frac{\partial^2 z}{\partial x^2} - 2 \frac{\partial^2 z}{\partial x \partial y} + \frac{\partial^2 z}{\partial y^2} = 0$, where $z = xf(x + y) + yg(x + y)$.

16. Find the value of n so that the equation $V = r^n (3\cos^2\theta - 1)$ satisfies the relation

$$\frac{\partial}{\partial r}\left(r^2 \frac{\partial V}{\partial r}\right) + \frac{1}{\sin\theta}\frac{\partial}{\partial\theta}\left(\sin\theta\frac{\partial V}{\partial\theta}\right) = 0.$$ *(U.P.T.U. 2014)*

17. If $e^{-\frac{z}{x^2-y^2}} = x - y$ then show that $y\frac{\partial z}{\partial x} + x\frac{\partial z}{\partial y} = x^2 - y^2$. *(A.K.T.U. 2017)*

18. If $u = (x^2 + y^2 + z^2)^{-1/2}$, show that *(A.K.T.U. 2016)*

(i) $x\frac{\partial u}{\partial x} + y\frac{\partial u}{\partial y} + z\frac{\partial u}{\partial z} = -u$ (ii) $\frac{\partial^2 u}{\partial x^2} + \frac{\partial^2 u}{\partial y^2} + \frac{\partial^2 u}{\partial z^2} = 0.$

19. If $x^2 = au + bv$, $y^2 = au - bv$, prove that $\left(\frac{\partial u}{\partial x}\right)_y \cdot \left(\frac{\partial x}{\partial u}\right)_v = \frac{1}{2} = \left(\frac{\partial v}{\partial y}\right)_x \cdot \left(\frac{\partial y}{\partial v}\right)_u$. *(A.K.T.U. 2018)*

20. If $u = \log(x^2 + y^2 + z^2)$, show that $x\frac{\partial^2 u}{\partial y\partial z} = y\frac{\partial^2 u}{\partial z\partial x} = z\frac{\partial^2 u}{\partial x\partial y}$.

21. If $u = \log\sqrt{x^2 + y^2 + z^2}$, show that $(x^2 + y^2 + z^2)\left(\frac{\partial^2 u}{\partial x^2} + \frac{\partial^2 u}{\partial y^2} + \frac{\partial^2 u}{\partial z^2}\right) = 1.$ *(U.K.T.U. 2011)*

22. If $u = x^2\tan^{-1}\frac{y}{x} - y^2\tan^{-1}\frac{x}{y}$, find the value of $\frac{\partial^2 u}{\partial x\partial y}$. *(A.K.T.U. 2018)*

23. If $x = r\cos\theta$, $y = r\sin\theta$, prove that

(i) $\frac{\partial^2 r}{\partial x^2}\cdot\frac{\partial^2 r}{\partial y^2} = \left(\frac{\partial^2 r}{\partial x\partial y}\right)^2$ (ii) $\frac{\partial^2 r}{\partial x^2} + \frac{\partial^2 r}{\partial y^2} = \frac{1}{r}\left[\left(\frac{\partial r}{\partial x}\right)^2 + \left(\frac{\partial r}{\partial y}\right)^2\right]$.

24. If $u = \sqrt{x^2 + y^2 + z^2}$, show that

(i) $\left(\frac{\partial u}{\partial x}\right)^2 + \left(\frac{\partial u}{\partial y}\right)^2 + \left(\frac{\partial u}{\partial z}\right)^2 = 1$ (ii) $\frac{\partial^2 u}{\partial x^2} + \frac{\partial^2 u}{\partial y^2} + \frac{\partial^2 u}{\partial z^2} = \frac{2}{u}$.

25. If $V = f(r)$ and $r^2 = x^2 + y^2 + z^2$, prove that $V_{xx} + V_{yy} + V_{zz} = f''(r) + \frac{2}{r}f'(r)$.

26. Find p and q, if $x = \sqrt{a}(\sin u + \cos v)$, $y = \sqrt{a}(\cos u - \sin v)$, $z = 1 + \sin(u - v)$, where p and q mean $\frac{\partial z}{\partial x}$ and $\frac{\partial z}{\partial y}$ respectively.

$\left[\textbf{Hint.}\ x^2 + y^2 = 2az,\ \therefore\ z = \frac{x^2 + y^2}{2a}.\right]$

27. If $u = f(ax^2 + 2hxy + by^2)$, $v = \phi(ax^2 + 2hxy + by^2)$, show that $\frac{\partial}{\partial y}\left(u\frac{\partial v}{\partial x}\right) = \frac{\partial}{\partial x}\left(u\frac{\partial v}{\partial y}\right)$.

28. If $u = \log(x^3 + y^3 - x^2y - xy^2)$, show that $\frac{\partial^2 u}{\partial x^2} + 2\frac{\partial^2 u}{\partial x\partial y} + \frac{\partial^2 u}{\partial y^2} = -4(x + y)^{-2}$.

[**Hint.** $u = \log\{x^2(x - y) - y^2(x - y)\} = \log(x - y)(x^2 - y^2) = 2\log(x - y) + \log(x + y)$]

29. If $\frac{x^2}{a^2+u}+\frac{y^2}{b^2+u}+\frac{z^2}{c^2+u}=1$, prove that $\left(\frac{\partial u}{\partial x}\right)^2+\left(\frac{\partial u}{\partial y}\right)^2+\left(\frac{\partial u}{\partial z}\right)^2=2\left(x\frac{\partial u}{\partial x}+y\frac{\partial u}{\partial y}+z\frac{\partial u}{\partial z}\right)$.

30. If $x=e^{r\cos\theta}\cos(r\sin\theta)$ and $y=e^{r\cos\theta}\sin(r\sin\theta)$, prove that

$$\frac{\partial x}{\partial r}=\frac{1}{r}\frac{\partial y}{\partial \theta} \quad \text{and} \quad \frac{\partial y}{\partial r}=-\frac{1}{r}\frac{\partial x}{\partial \theta}$$

Hence deduce that

$$\frac{\partial^2 x}{\partial r^2}+\frac{1}{r}\frac{\partial x}{\partial r}+\frac{1}{r^2}\frac{\partial^2 x}{\partial \theta^2}=0.$$

Answers

1. (i) $y^x \log y,\ xy^{x-1}$ (ii) $\frac{2x}{x^2+y^2}, \frac{2y}{x^2+y^2}$ **6.** $-\frac{13}{22}$ **16.** 2, – 3

22. $\frac{x^2-y^2}{x^2+y^2}$ **26.** $p=\frac{x}{a}, q=\frac{y}{a}$.

2.44. HOMOGENEOUS FUNCTIONS

A function $f(x, y)$ is said to be homogeneous of degree (or order) n in the variables x and y if it can be expressed in the form $x^n\,\phi\left(\frac{y}{x}\right)$ or $y^n\,\phi\left(\frac{x}{y}\right)$.

An alternative test for a function $f(x, y)$ to be homogeneous of degree (or order) n is that

$$f(tx, ty)=t^n f(x, y).$$

For example, if $f(x, y)=\frac{x+y}{\sqrt{x}+\sqrt{y}}$, then

$$(i)\ f(x, y)=\frac{x\left(1+\frac{y}{x}\right)}{\sqrt{x}\left(1+\sqrt{\frac{y}{x}}\right)}=x^{1/2}\,\phi\left(\frac{y}{x}\right)$$

$\Rightarrow$ $f(x, y)$ is a homogeneous function of degree $\frac{1}{2}$ in x and y.

$$(ii)\ f(x, y)=\frac{y\left(\frac{x}{y}+1\right)}{\sqrt{y}\left(\sqrt{\frac{x}{y}}+1\right)}=y^{1/2}\,\phi\left(\frac{x}{y}\right)$$

$\Rightarrow$ $f(x, y)$ is a homogeneous function of degree $\frac{1}{2}$ in x and y.

$$(iii)\ f(tx, ty)=\frac{tx+ty}{\sqrt{tx}+\sqrt{ty}}=\frac{t(x+y)}{\sqrt{t}\,(\sqrt{x}+\sqrt{y})}=t^{1/2} f(x, y)$$

$\Rightarrow$ $f(x, y)$ is a homogeneous function of degree $\frac{1}{2}$ in x and y.

Similarly, a function $f(x, y, z)$ is said to be homogeneous of degree (or order) n in the variables x, y, z if $f(x, y, z) = x^n \phi\left(\frac{y}{x}, \frac{z}{x}\right)$ or $y^n \phi\left(\frac{x}{y}, \frac{z}{y}\right)$ or $z^n \phi\left(\frac{x}{z}, \frac{y}{z}\right)$.

Alternative test is $f(tx, ty, tz) = t^n f(x, y, z)$.

2.45. EULER'S THEOREM ON HOMOGENEOUS FUNCTIONS

[*G.B.T.U. 2013; U.K.T.U. 2011*]

If u is a homogeneous function of degree n in x and y, then $x\frac{\partial u}{\partial x} + y\frac{\partial u}{\partial y} = nu$

Since u is a homogeneous function of degree n in x and y, it can be expressed as

$$u = x^n f\left(\frac{y}{x}\right)$$

$$\therefore \quad \frac{\partial u}{\partial x} = nx^{n-1} f\left(\frac{y}{x}\right) + x^n f'\left(\frac{y}{x}\right).\left(-\frac{y}{x^2}\right)$$

$$\Rightarrow \quad x\frac{\partial u}{\partial x} = nx^n f\left(\frac{y}{x}\right) - x^{n-1} yf'\left(\frac{y}{x}\right) \quad \text{...(1)}$$

Also, $$\frac{\partial u}{\partial y} = x^n f'\left(\frac{y}{x}\right).\frac{1}{x} = x^{n-1} f'\left(\frac{y}{x}\right)$$

$$\Rightarrow \quad y\frac{\partial u}{\partial y} = x^{n-1} yf'\left(\frac{y}{x}\right) \quad \text{...(2)}$$

Adding (1) and (2), we get $x\frac{\partial u}{\partial x} + y\frac{\partial u}{\partial y} = nx^n f\left(\frac{y}{x}\right) = nu$.

Note. Euler's theorem can be extended to a homogeneous function of any number of variables. Thus, if u is a homogeneous function of degree n in x, y and z, then $x\frac{\partial u}{\partial x} + y\frac{\partial u}{\partial y} + z\frac{\partial u}{\partial z} = nu$.

2.46. IF *u* IS A HOMOGENEOUS FUNCTION OF DEGREE *n* IN *x* AND *y*, THEN

$$x^2\frac{\partial^2 u}{\partial x^2} + 2xy\frac{\partial^2 u}{\partial x \partial y} + y^2\frac{\partial^2 u}{\partial y^2} = n(n-1)u$$

Since u is a homogeneous function of degree n in x and y

$\therefore$ By Euler's Theorem, we have $x\frac{\partial u}{\partial x} + y\frac{\partial u}{\partial y} = nu$ ...(1)

Differentiating (1) partially w.r.t. x, we have $1.\frac{\partial u}{\partial x} + x\frac{\partial^2 u}{\partial x^2} + y.\frac{\partial^2 u}{\partial x \partial y} = n\frac{\partial u}{\partial x}$ ...(2)

Differentiating (1) partially w.r.t. y, we have $x\frac{\partial^2 u}{\partial y \partial x} + 1.\frac{\partial u}{\partial y} + y.\frac{\partial^2 u}{\partial y^2} = n\frac{\partial u}{\partial y}$

But $$\frac{\partial^2 u}{\partial y \partial x} = \frac{\partial^2 u}{\partial x \partial y}$$

$$\therefore \qquad x \frac{\partial^2 u}{\partial x \partial y} + \frac{\partial u}{\partial y} + y \frac{\partial^2 u}{\partial y^2} = n \frac{\partial u}{\partial y} \qquad ...(3)$$

Multiplying (2) by x, (3) by y and adding

$$x^2 \frac{\partial^2 u}{\partial x^2} + 2xy \frac{\partial^2 u}{\partial x \partial y} + y^2 \frac{\partial^2 u}{\partial y^2} + \left(x \frac{\partial u}{\partial x} + y \frac{\partial u}{\partial y} \right) = n \left(x \frac{\partial u}{\partial x} + y \frac{\partial u}{\partial y} \right)$$

or $$x^2 \frac{\partial^2 u}{\partial x^2} + 2xy \frac{\partial^2 u}{\partial x \partial y} + y^2 \frac{\partial^2 u}{\partial y^2} + nu = n \cdot nu \qquad \text{[Using (1)]}$$

$$\Rightarrow \qquad x^2 \frac{\partial^2 u}{\partial x^2} + 2xy \frac{\partial^2 u}{\partial x \partial y} + y^2 \frac{\partial^2 u}{\partial y^2} = n^2 u - nu = n(n-1)u.$$

2.47. DEDUCTIONS FROM EULER'S THEOREM

If F(u) = V(x, y), where V is a homogeneous function in x and y of degree n, then

(*i*) $$x \frac{\partial u}{\partial x} + y \frac{\partial u}{\partial y} = n \frac{F(u)}{F'(u)}.$$

(*ii*) $$x^2 \frac{\partial^2 u}{\partial x^2} + 2xy \frac{\partial^2 u}{\partial x \partial y} + y^2 \frac{\partial^2 u}{\partial y^2} = \phi(u)\,[\phi'(u) - 1], \quad \textit{where} \quad \phi(u) = n \frac{F(u)}{F'(u)}$$

Proof. (*i*) Since V = F(u) is a homogeneous function of degree n, then by Euler's theorem, we get

$$x \frac{\partial}{\partial x} \{F(u)\} + y \frac{\partial}{\partial y} \{F(u)\} = n\, F(u)$$

or $$x\, F'(u) \frac{\partial u}{\partial x} + y\, F'(u) \frac{\partial u}{\partial y} = n\, F(u)$$

or $$\boxed{x \frac{\partial u}{\partial x} + y \frac{\partial u}{\partial y} = n \frac{F(u)}{F'(u)}}. \qquad ...(1)$$

(*ii*) Let $n \dfrac{F(u)}{F'(u)} = \phi(u)$, then from (1),

$$x \frac{\partial u}{\partial x} + y \frac{\partial u}{\partial y} = \phi(u) \qquad ...(2)$$

Differentiating (2) partially w.r.t. x, we get

$$x \frac{\partial^2 u}{\partial x^2} + \frac{\partial u}{\partial x} + y \frac{\partial^2 u}{\partial x \partial y} = \phi'(u) \frac{\partial u}{\partial x}$$

$$\Rightarrow \qquad x \frac{\partial^2 u}{\partial x^2} + y \frac{\partial^2 u}{\partial x \partial y} = \{\phi'(u) - 1\} \frac{\partial u}{\partial x} \qquad ...(3)$$

Similarly, differentiating (2) partially w.r.t. y, we get

$$y \frac{\partial^2 u}{\partial y^2} + x \frac{\partial^2 u}{\partial y \partial x} = \{\phi'(u) - 1\} \frac{\partial u}{\partial y} \qquad ...(4)$$

Multiplying (3) by x, (4) by y and adding, we get

$$x^2 \frac{\partial^2 u}{\partial x^2} + 2xy \frac{\partial^2 u}{\partial x \partial y} + y^2 \frac{\partial^2 u}{\partial y^2} = \{\phi'(u) - 1\} \left(x \frac{\partial u}{\partial x} + y \frac{\partial u}{\partial y} \right)$$

$$= \{\phi'(u) - 1\} \phi(u). \quad \text{| From (2)}$$

ILLUSTRATIVE EXAMPLES

Example 1. *Verify Euler's theorem for the functions:*

(i) $u = (x^{1/2} + y^{1/2})(x^n + y^n)$ *(ii)* $u = \sin^{-1} \frac{x}{y} + \tan^{-1} \frac{y}{x}$.

Sol. *(i)* $u(x, y) = (x^{1/2} + y^{1/2})(x^n + y^n)$ ...(1)

$$u(tx, ty) = t^{1/2} (x^{1/2} + y^{1/2})\, t^n (x^n + y^n) = t^{n + 1/2} u(x, y)$$

$\Rightarrow$ u is a homogeneous function of degree $(n + \frac{1}{2})$ in x and y

$\therefore$ By Euler's theorem, we should have $x \frac{\partial u}{\partial x} + y \frac{\partial u}{\partial y} = \left(n + \frac{1}{2} \right) u$ ...(2)

From (1),
$$\frac{\partial u}{\partial x} = \frac{1}{2} x^{-1/2} (x^n + y^n) + nx^{n-1}(x^{1/2} + y^{1/2})$$

$$x \frac{\partial u}{\partial x} = \frac{1}{2} x^{1/2} (x^n + y^n) + nx^n(x^{1/2} + y^{1/2})$$

$$\frac{\partial u}{\partial y} = \frac{1}{2} y^{-1/2} (x^n + y^n) + ny^{n-1} (x^{1/2} + y^{1/2})$$

$$y \frac{\partial u}{\partial y} = \frac{1}{2} y^{1/2}(x^n + y^n) + ny^n(x^{1/2} + y^{1/2})$$

$\therefore$
$$x \frac{\partial u}{\partial x} + y \frac{\partial u}{\partial y} = \frac{1}{2} (x^{1/2} + y^{1/2})(x^n + y^n) + n(x^n + y^n)(x^{1/2} + y^{1/2})$$

$$= \frac{1}{2} u + nu = \left(n + \frac{1}{2} \right) u,$$ which is the same as (2). Hence the verification.

(ii) $u(x, y) = \sin^{-1} \frac{x}{y} + \tan^{-1} \frac{y}{x}$ ...(1)

$$u(tx, ty) = \sin^{-1} \frac{tx}{ty} + \tan^{-1} \frac{ty}{tx} = t^0 \left(\sin^{-1} \frac{x}{y} + \tan^{-1} \frac{y}{x} \right) = t^0 u(x, y)$$

$\Rightarrow$ u is a homogeneous function of degree 0 in x and y.

$\therefore$ By Euler's theorem, we should have $x \frac{\partial u}{\partial x} + y \frac{\partial u}{\partial y} = 0 \times u = 0$ ...(2)

From (1),
$$\frac{\partial u}{\partial x} = \frac{1}{\sqrt{1 - \frac{x^2}{y^2}}} \cdot \frac{1}{y} + \frac{1}{1 + \frac{y^2}{x^2}} \left(-\frac{y}{x^2} \right) = \frac{1}{\sqrt{y^2 - x^2}} - \frac{y}{\sqrt{x^2 + y^2}}$$

$$x \frac{\partial u}{\partial x} = \frac{x}{\sqrt{y^2 - x^2}} - \frac{xy}{x^2 + y^2}$$

$$\frac{\partial u}{\partial y} = \frac{1}{\sqrt{1 - \frac{x^2}{y^2}}} \left(-\frac{x}{y^2} \right) + \frac{1}{1 + \frac{y^2}{x^2}} \cdot \frac{1}{x} = -\frac{x}{y\sqrt{y^2 - x^2}} + \frac{x}{x^2 + y^2}$$

$$y\frac{\partial u}{\partial y} = -\frac{x}{\sqrt{y^2 - x^2}} + \frac{xy}{x^2 + y^2}$$

$\therefore \quad x\frac{\partial u}{\partial x} + y\frac{\partial u}{\partial y} = 0$, which is the same as (2). Hence the verification.

Example 2. (*i*) *If* $u = (x^{1/4} + y^{1/4})(x^{1/5} + y^{1/5})$, *apply Euler's theorem to find the value of*

$$x\frac{\partial u}{\partial x} + y\frac{\partial u}{\partial y}.$$ [M.T.U. (SUM) 2011]

(*ii*) *If* $u(x, y) = (\sqrt{x} + \sqrt{y})^5$, *find the value of* $x^2\frac{\partial^2 u}{\partial x^2} + 2xy\frac{\partial^2 u}{\partial x \partial y} + y^2\frac{\partial^2 u}{\partial y^2}$. (M.T.U. 2013)

(*iii*) *If* $z = x^4 y^2 \sin^{-1}\left(\frac{x}{y}\right) + \log x - \log y$, *show that* $x\frac{\partial z}{\partial x} + y\frac{\partial z}{\partial y} = 6x^4y^2 \sin^{-1}\left(\frac{x}{y}\right)$.

Sol. (*i*) $u(x, y) = (x^{1/4} + y^{1/4})(x^{1/5} + y^{1/5})$

$$u(tx, ty) = t^{1/4}(x^{1/4} + y^{1/4})\, t^{1/5}(x^{1/5} + y^{1/5}) = t^{9/20} u(x, y).$$

$\therefore \quad u$ is a homogeneous function in x and y of degree $\frac{9}{20}$.

Hence, By Euler's theorem,

$$x\frac{\partial u}{\partial x} + y\frac{\partial u}{\partial y} = \frac{9}{20}u.$$

(*ii*) $u(x, y) = (\sqrt{x} + \sqrt{y})^5$

$$u(tx, ty) = (\sqrt{tx} + \sqrt{ty})^5 = (\sqrt{t})^5(\sqrt{x} + \sqrt{y})^5 = t^{5/2} u(x, y)$$

$\therefore \quad u(x, y)$ is a homogeneous function in x and y of degree $\frac{5}{2}$.

Hence by using Euler's theorem,

$$x^2\frac{\partial^2 u}{\partial x^2} + 2xy\frac{\partial^2 u}{\partial x \partial y} + y^2\frac{\partial^2 u}{\partial y^2} = n(n-1)u$$ | See Art. 2.46

$$= \frac{5}{2}\left(\frac{5}{2} - 1\right)u = \frac{15}{4}u.$$

(*iii*) Let $u(x, y) = x^4y^2 \sin^{-1}\left(\frac{x}{y}\right)$

and $v(x, y) = \log x - \log y = \log\left(\frac{x}{y}\right)$

Now, $u(tx, ty) = t^6 x^4 y^2 \sin^{-1}\left(\frac{x}{y}\right) = t^6 u(x, y)$

$\therefore \quad u$ is a homogeneous function is x and y of degree 6.

Also, $v(tx, ty) = \log\left(\frac{tx}{ty}\right) = t^0 \log\left(\frac{x}{y}\right) = t^0 v(x, y)$

$\therefore \quad v$ is a homogeneous function in x and y of degree 0.

∴ By Euler's theorem,

$$x\frac{\partial u}{\partial x}+y\frac{\partial u}{\partial y}=6u \qquad ...(1)$$

and

$$x\frac{\partial v}{\partial x}+y\frac{\partial v}{\partial y}=0 \qquad ...(2)$$

Adding (1) and (2), we get

$$x\frac{\partial z}{\partial x}+y\frac{\partial z}{\partial y}=6u=6x^4y^2\sin^{-1}\left(\frac{x}{y}\right). \qquad | \text{ Since } z=u+v$$

Example 3. (*i*) *If* $u=\cos\left(\frac{xy+yz+zx}{x^2+y^2+z^2}\right)$, *prove that* $x\frac{\partial u}{\partial x}+y\frac{\partial u}{\partial y}+z\frac{\partial u}{\partial z}=0$

(*ii*) *Show that* $x\frac{\partial u}{\partial x}+y\frac{\partial u}{\partial y}+z\frac{\partial u}{\partial z}=2\tan u$, *where* $u=\sin^{-1}\left(\frac{x^3+y^3+z^3}{ax+by+cz}\right)$.

(A.K.T.U. 2018)

Sol. (*i*)

$$u(x,y,z)=\cos\left(\frac{xy+yz+zx}{x^2+y^2+z^2}\right)$$

∴

$$u(tx,ty,tz)=\cos\left\{\frac{t^2(xy+yz+zx)}{t^2(x^2+y^2+z^2)}\right\}$$

$$=t^{\circ}\cos\left(\frac{xy+yz+zx}{x^2+y^2+z^2}\right)=t^{\circ}\,u(x,y,z)$$

Hence u is a homogeneous function in x, y and z of degree 0.

Hence by Euler's theorem,

$$x\frac{\partial u}{\partial x}+y\frac{\partial u}{\partial y}+z\frac{\partial u}{\partial z}=0\,.\,u=0$$

(*ii*) u is not a homogeneous function.

Let

$$\sin u=\frac{x^3+y^3+z^3}{ax+by+cz}=f(x,y,z) \qquad ...(1) \quad | \text{ say}$$

Now,

$$f(tx,ty,tz)=\frac{t^3(x^3+y^3+z^3)}{t(ax+by+cz)}=t^2 f(x,y,z)$$

Hence $f(x, y, z)$ is a homogeneous function of degree 2.

∴ By Euler's theorem,

$$x\frac{\partial f}{\partial x}+y\frac{\partial f}{\partial y}+z\frac{\partial f}{\partial z}=2f$$

$$\Rightarrow \quad x\frac{\partial}{\partial x}(\sin u)+y\frac{\partial}{\partial y}(\sin u)+z\frac{\partial}{\partial z}(\sin u)=2\sin u$$

$$\Rightarrow \quad x\cos u\frac{\partial u}{\partial x}+y\cos u\frac{\partial u}{\partial y}+z\cos u\frac{\partial u}{\partial z}=2\sin u$$

$$\Rightarrow \quad x\frac{\partial u}{\partial x}+y\frac{\partial u}{\partial y}+z\frac{\partial u}{\partial z}=2\tan u.$$

Example 4. *If* $u = \tan^{-1} \dfrac{x^3 + y^3}{x - y}$, *prove that*

(*i*) $x \dfrac{\partial u}{\partial x} + y \dfrac{\partial u}{\partial y} = \sin 2u$

(*ii*) $x^2 \dfrac{\partial^2 u}{\partial x^2} + 2xy \dfrac{\partial^2 u}{\partial x \partial y} + y^2 \dfrac{\partial^2 u}{\partial y^2} = \sin 4u - \sin 2u = 2 \cos 3u \sin u.$

Sol. (*i*) Here u is not a homogeneous function.

Let $\quad \tan u = \dfrac{x^3 + y^3}{x - y} = f(x, y) \quad$ ...(1) | say

Now, $\quad f(tx, ty) = \dfrac{t^3 (x^3 + y^3)}{t(x - y)} = t^2 f(x, y)$

Hence $f(x, y) = \tan u$ is a homogeneous function of degree 2.

$\therefore$ By Euler's theorem, we have

$$x \frac{\partial}{\partial x} (\tan u) + y \frac{\partial}{\partial y} (\tan u) = 2 \tan u$$

or $$x \sec^2 u \frac{\partial u}{\partial x} + y \sec^2 u \frac{\partial u}{\partial y} = 2 \tan u$$

or $$x \frac{\partial u}{\partial x} + y \frac{\partial u}{\partial y} = \frac{2 \sin u}{\cos u} . \cos^2 u = 2 \sin u \cos u = \sin 2u.$$

Hence $$x \frac{\partial u}{\partial x} + y \frac{\partial u}{\partial y} = \sin 2u \qquad ...(2)$$

(*ii*) Here $\quad \phi(u) = \sin 2u$

$\therefore \quad \phi'(u) = 2 \cos 2u$

By deduction from Euler's theorem, we know that

$$x^2 \frac{\partial^2 u}{\partial x^2} + 2xy \frac{\partial^2 u}{\partial x \partial y} + y^2 \frac{\partial^2 u}{\partial y^2} = \phi(u) \{\phi'(u) - 1\}$$

$$= \sin 2u (2 \cos 2u - 1) = \sin 4u - \sin 2u = 2 \cos 3u \sin u.$$

Example 5. *If* $u = \sin^{-1} \left(\dfrac{x + 2y + 3z}{\sqrt{x^8 + y^8 + z^8}} \right)$, *show that*

$$x \frac{\partial u}{\partial x} + y \frac{\partial u}{\partial y} + z \frac{\partial u}{\partial z} + 3 \tan u = 0. \qquad \text{(U.K.T.U. 2010)}$$

Sol. Here u is not a homogeneous function. But, we have

$$\sin u = \frac{x + 2y + 3z}{\sqrt{x^8 + y^8 + z^8}} = f(x, y, z) \qquad \text{| say}$$

$$f(tx, ty, tz) = \frac{t(x + 2y + 3z)}{t^4 \sqrt{x^8 + y^8 + z^8}} = t^{-3} f(x, y, z)$$

$\therefore$ $\sin u$ is a homogeneous function of degree -3 in x, y, z.

∴ By Euler's theorem, we have

$$x \frac{\partial}{\partial x}(\sin u) + y \frac{\partial}{\partial y}(\sin u) + z \frac{\partial}{\partial z}(\sin u) = -3 \sin u$$

or
$$x \cos u \frac{\partial u}{\partial x} + y \cos u \frac{\partial u}{\partial y} + z \cos u \frac{\partial u}{\partial z} + 3 \sin u = 0$$

or
$$x \frac{\partial u}{\partial x} + y \frac{\partial u}{\partial y} + z \frac{\partial u}{\partial z} + 3 \tan u = 0.$$

Example 6. *If* $u = \log_e \left(\frac{x^4 + y^4}{x + y} \right)$, *show that* $x \frac{\partial u}{\partial x} + y \frac{\partial u}{\partial y} = 3.$ [U.K.T.U. 2012]

Sol. Here u is not a homogeneous function

$$u = \log_e \left(\frac{x^4 + y^4}{x + y} \right)$$

⇒
$$e^u = \frac{x^4 + y^4}{x + y}$$

which is a homogeneous function of degree 3 in x and y.

∴ By Euler's theorem, we have

$$x \frac{\partial}{\partial x}(e^u) + y \frac{\partial}{\partial y}(e^u) = 3e^u$$

or
$$xe^u \frac{\partial u}{\partial x} + y\, e^u \frac{\partial u}{\partial y} = 3e^u$$

or
$$x \frac{\partial u}{\partial x} + y \frac{\partial u}{\partial y} = 3.$$

Example 7. *If* $u = \sin^{-1} \frac{x + y}{\sqrt{x} + \sqrt{y}}$, *prove that*

(*i*) $x \frac{\partial u}{\partial x} + y \frac{\partial u}{\partial y} = \frac{1}{2} \tan u$ (*ii*) $x^2 \frac{\partial^2 u}{\partial x^2} + 2xy \frac{\partial^2 u}{\partial x \partial y} + y^2 \frac{\partial^2 u}{\partial y^2} = -\frac{\sin u \cos 2u}{4 \cos^3 u}$.

Sol. (*i*) $u = \sin^{-1} \frac{x + y}{\sqrt{x} + \sqrt{y}}$ is not a homogeneous function but

$\sin u = \frac{x + y}{\sqrt{x} + \sqrt{y}}$ is a homogeneous function of degree $\frac{1}{2}$ in x and y.

∴ By Euler's theorem, we have

$$x \frac{\partial}{\partial x}(\sin u) + y \frac{\partial}{\partial y}(\sin u) = \frac{1}{2} \sin u$$

or
$$x \cos u \frac{\partial u}{\partial x} + y \cos u \frac{\partial u}{\partial y} = \frac{1}{2} \sin u$$

or
$$x \frac{\partial u}{\partial x} + y \frac{\partial u}{\partial y} = \frac{1}{2} \tan u \quad ...(1)$$

(*ii*) Here $\phi(u) = \frac{1}{2} \tan u$

∴ $\phi'(u) = \frac{1}{2} \sec^2 u$

By deduction from Euler's theorem, we know that

$$x^2 \frac{\partial^2 u}{\partial x^2} + 2xy \frac{\partial^2 u}{\partial x \partial y} + y^2 \frac{\partial^2 u}{\partial y^2} = \phi(u)\,\{\phi'(u) - 1\}$$

$$= \frac{1}{2} \tan u \left(\frac{1}{2} \sec^2 u - 1 \right) = \frac{1}{2} \frac{\sin u}{\cos u} \left(\frac{1}{2 \cos^2 u} - 1 \right)$$

$$= \frac{1}{2} \frac{\sin u}{\cos u} \left\{ \frac{-(2 \cos^2 u - 1)}{2 \cos^2 u} \right\} = - \frac{\sin u \cos 2u}{4 \cos^3 u}.$$

Example 8. *If* $u = x\phi\left(\frac{y}{x}\right) + \psi\left(\frac{y}{x}\right)$, *show that:*

(*i*) $x \frac{\partial u}{\partial x} + y \frac{\partial u}{\partial y} = x\,\phi\left(\frac{y}{x}\right)$ (*ii*) $x^2 \frac{\partial^2 u}{\partial x^2} + 2xy \frac{\partial^2 u}{\partial x \partial y} + y^2 \frac{\partial^2 u}{\partial y^2} = 0$.

Sol. Let $v = x\,\phi\left(\frac{y}{x}\right)$ and $w = \psi\left(\frac{y}{x}\right) = x^0\,\psi\left(\frac{y}{x}\right)$ so that,

$$u = v + w \qquad \text{...(1)}$$

(*i*) Since v is a homogeneous function of degree $n = 1$ in x, y

$$\therefore \quad x \frac{\partial v}{\partial x} + y \frac{\partial v}{\partial y} = v \qquad \text{...(2)}$$

Since w is a homogeneous function of degree $n = 0$ in x, y

$$\therefore \quad x \frac{\partial w}{\partial x} + y \frac{\partial w}{\partial y} = 0 \qquad \text{...(3)}$$

Adding equations (2) and (3), we get

$$x \frac{\partial}{\partial x} (v + w) + y \frac{\partial}{\partial y} (v + w) = v$$

$$\Rightarrow \quad x \frac{\partial u}{\partial x} + y \frac{\partial u}{\partial y} = x\,\phi\left(\frac{y}{x}\right).$$

(*ii*) Since v is a homogeneous function of degree $n = 1$ in x, y

$$\therefore \quad x^2 \frac{\partial^2 v}{\partial x^2} + 2xy \frac{\partial^2 v}{\partial x \partial y} + y^2 \frac{\partial^2 v}{\partial y^2} = n(n-1)\,v = 1\,(1-1)\,v = 0 \qquad \text{...(4)}$$

Also since w is a homogeneous function of degree $n = 0$ in x, y

$$\therefore \quad x^2 \frac{\partial^2 w}{\partial x^2} + 2xy \frac{\partial^2 w}{\partial x \partial y} + y^2 \frac{\partial^2 w}{\partial y^2} = n(n-1)\,w = 0\,(0-1)\,w = 0 \qquad \text{...(5)}$$

Adding (4) and (5), we get

$$x^2 \frac{\partial^2}{\partial x^2} (v + w) + 2xy \frac{\partial^2}{\partial x \partial y} (v + w) + y^2 \frac{\partial^2}{\partial y^2} (v + w) = 0$$

$$\Rightarrow \quad x^2 \left(\frac{\partial^2 u}{\partial x^2} \right) + 2xy \left(\frac{\partial^2 u}{\partial x \partial y} \right) + y^2 \left(\frac{\partial^2 u}{\partial y^2} \right) = 0$$

TEST YOUR KNOWLEDGE

1. Verify Euler's theorem for the following functions:

(i) $f(x, y) = ax^2 + 2hxy + by^2$ (ii) $u = \dfrac{x}{y} + \dfrac{y}{z} + \dfrac{z}{x}$

(iii) $f(x, y) = \dfrac{x^2(x^2 - y^2)^3}{(x^2 + y^2)^3}$ (iv) $z = \dfrac{x^{1/3} + y^{1/3}}{x^{1/2} + y^{1/2}}$ (A.K.T.U. 2016, 2018)

(v) $u = \log\left(\dfrac{x^2 + y^2}{xy}\right)$ (U.K.T.U. 2011) (vi) $u = x^2 \log\left(\dfrac{y}{x}\right)$

(vii) $u = \dfrac{x\,(x^4 - y^4)}{x^4 + y^4}$ (U.P.T.U. 2014) (viii) $u = \log\left(\dfrac{x^4 + y^4}{x + y}\right)$. (U.P.T.U. 2015)

2. (i) If $u = f\left(\dfrac{y}{x}\right)$, prove that $x\dfrac{\partial u}{\partial x} + y\dfrac{\partial u}{\partial y} = 0$ (G.B.T.U. 2011)

(ii) If $u = xf\left(\dfrac{y}{x}\right)$, prove that $x\dfrac{\partial u}{\partial x} + y\dfrac{\partial u}{\partial y} = u.$

3. (i) If $V = \dfrac{x^3y^3}{x^3 + y^3}$, prove that $x\dfrac{\partial V}{\partial x} + y\dfrac{\partial V}{\partial y} = 3V.$

(ii) If $u = x^3 + y^3 + z^3 + 3xyz$, show that $x\dfrac{\partial u}{\partial x} + y\dfrac{\partial u}{\partial y} + z\dfrac{\partial u}{\partial z} = 3u.$ [G.B.T.U. (C.O.) 2010]

4. If $u(x, y) = \sin^{-1}\dfrac{x}{y} + \tan^{-1}\dfrac{y}{x}$, prove that $x\dfrac{\partial u}{\partial x} + y\dfrac{\partial u}{\partial y} = 0.$

5. (i) If $f(x, y) = \sqrt{x^2 - y^2}\,\sin^{-1}\dfrac{y}{x}$, prove that $x\dfrac{\partial f}{\partial x} + y\dfrac{\partial f}{\partial y} = f(x, y).$

(ii) If $f(x, y) = \sqrt{y^2 - x^2}\,\sin^{-1}\dfrac{x}{y} + \dfrac{x^2 - y^2}{\sqrt{x^2 + y^2}}$, prove that $x\dfrac{\partial f}{\partial x} + y\dfrac{\partial f}{\partial y} - f(x, y) = 0.$

6. If $f(x, y) = \dfrac{1}{x^2} + \dfrac{1}{xy} + \dfrac{\log x - \log y}{x^2 + y^2}$, prove that $x\dfrac{\partial f}{\partial x} + y\dfrac{\partial f}{\partial y} + 2f(x, y) = 0.$

7. If $u = (x^2 + y^2)^{1/3}$, prove that $x^2\dfrac{\partial^2 u}{\partial x^2} + 2xy\dfrac{\partial^2 u}{\partial x\partial y} + y^2\dfrac{\partial^2 u}{\partial y^2} = -\dfrac{2u}{9}$

8. (i) If $z = xy\,f\left(\dfrac{x}{y}\right)$, prove that $x\dfrac{\partial z}{\partial x} + y\dfrac{\partial z}{\partial y} = 2z.$

(ii) If $x = e^u \tan v$, $y = e^u \sec v$; find the value of $\left(x\dfrac{\partial u}{\partial x} + y\dfrac{\partial u}{\partial y}\right).\left(x\dfrac{\partial v}{\partial x} + y\dfrac{\partial v}{\partial y}\right).$

9. (i) If $u = \cos^{-1}\dfrac{x + y}{\sqrt{x} + \sqrt{y}}$, prove that $x\dfrac{\partial u}{\partial x} + y\dfrac{\partial u}{\partial y} + \dfrac{1}{2}\cot u = 0.$ (A.K.T.U. 2019)

(ii) Show that $x\dfrac{\partial u}{\partial x} + y\dfrac{\partial u}{\partial y} = 2u \log u$, where $\log u = \dfrac{x^3 + y^3}{3x + 4y}$. (U.K.T.U. 2011)

(iii) If $u = \log\left[\dfrac{x^2 + y^2}{x + y}\right]$, prove that $x\dfrac{\partial u}{\partial x} + y\dfrac{\partial u}{\partial y} = 1.$

10. (*i*) If $u = \sin^{-1}\left(\dfrac{x^2 + y^2}{x + y}\right)$, prove that $x\dfrac{\partial u}{\partial x} + y\dfrac{\partial u}{\partial y} = \tan u$.

(*ii*) Show that : $xu_x + yu_y + zu_z = -2 \cot u$

where $u = \cos^{-1}\left(\dfrac{x^3 + y^3 + z^3}{ax + by + cz}\right)$. (*U.P.T.U. 2014*)

(*iii*) If $u = \sec^{-1}\left(\dfrac{x^3 - y^3}{x + y}\right)$, prove that $x\dfrac{\partial u}{\partial x} + y\dfrac{\partial u}{\partial y} = 2 \cot u$.

Also evaluate $x^2\dfrac{\partial^2 u}{\partial x^2} + 2xy\dfrac{\partial^2 u}{\partial x \partial y} + y^2\dfrac{\partial^2 u}{\partial y^2}$. (*A.K.T.U. 2018, 2021*)

11. (*a*) If $u = \sin^{-1}\dfrac{\sqrt{x} - \sqrt{y}}{\sqrt{x} + \sqrt{y}}$, prove that $\dfrac{\partial u}{\partial x} = -\dfrac{y}{x}\dfrac{\partial u}{\partial y}$.

(*b*) If $\sin u = \dfrac{x^2 y^2}{x + y}$, prove that $x\dfrac{\partial u}{\partial x} + y\dfrac{\partial u}{\partial y} = 3 \tan u$.

(*c*) Prove that $xu_x + yu_y = \dfrac{5}{2} \tan u$ if $u = \sin^{-1}\left(\dfrac{x^3 + y^3}{\sqrt{x} + \sqrt{y}}\right)$. (*U.P.T.U. 2015*)

12. (*a*) If $u = \log\left(\dfrac{x^5 + y^5 + z^5}{x^2 + y^2 + z^2}\right)$, prove that $x\dfrac{\partial u}{\partial x} + y\dfrac{\partial u}{\partial y} + z\dfrac{\partial u}{\partial z} = 3$.

(*b*) If $u = \dfrac{x^2 y^2 z^2}{x^2 + y^2 + z^2} + \cos\dfrac{xy + yz}{x^2 + y^2 + z^2}$, prove that $x\dfrac{\partial u}{\partial x} + y\dfrac{\partial u}{\partial y} + z\dfrac{\partial u}{\partial z} = \dfrac{4x^2 y^2 z^2}{x^2 + y^2 + z^2}$.

13. If $u = \dfrac{x^2 y^2}{x + y}$, prove that

(*i*) $x\dfrac{\partial^2 u}{\partial x^2} + y\dfrac{\partial^2 u}{\partial x \partial y} = 2\dfrac{\partial u}{\partial x}$ (*ii*) $x\dfrac{\partial^2 u}{\partial x \partial y} + y\dfrac{\partial^2 u}{\partial y^2} = 2\dfrac{\partial u}{\partial y}$.

14. Given $z = x^n f_1\left(\dfrac{y}{x}\right) + y^{-n} f_2\left(\dfrac{x}{y}\right)$, prove that $x^2\dfrac{\partial^2 z}{\partial x^2} + 2xy\dfrac{\partial^2 z}{\partial x \partial y} + y^2\dfrac{\partial^2 z}{\partial y^2} + x\dfrac{\partial z}{\partial x} + y\dfrac{\partial z}{\partial y} = n^2 z$.

15. If $u = \sin^{-1}\left(\dfrac{x^{1/4} + y^{1/4}}{x^{1/6} + y^{1/6}}\right)$ then evaluate $x^2\dfrac{\partial^2 u}{\partial x^2} + 2xy\dfrac{\partial^2 u}{\partial x \partial y} + y^2\dfrac{\partial^2 u}{\partial y^2}$. (*A.K.T.U. 2017*)

16. If $f(x, y)$ and $\phi(x, y)$ are homogeneous functions of x, y of degree m and n respectively and $u = f(x, y) + \phi(x, y)$, prove that

$$f(x, y) = \frac{1}{m(m-n)}\left(x^2\frac{\partial^2 u}{\partial x^2} + 2xy\frac{\partial^2 u}{\partial x \partial y} + y^2\frac{\partial^2 u}{\partial y^2}\right) - \frac{n-1}{m(m-n)}\left(x\frac{\partial u}{\partial x} + y\frac{\partial u}{\partial y}\right).$$

17. If $u = \tan^{-1}\left(\dfrac{y^2}{x}\right)$, prove that $x^2\dfrac{\partial^2 u}{\partial x^2} + 2xy\dfrac{\partial^2 u}{\partial x \partial y} + y^2\dfrac{\partial^2 u}{\partial y^2} = -\sin 2u \sin^2 u$.

18. (*i*) If $u = x^2 \tan^{-1}\left(\dfrac{y}{x}\right) - y^2 \tan^{-1}\left(\dfrac{x}{y}\right)$, evaluate $x^2\dfrac{\partial^2 u}{\partial x^2} + 2xy\dfrac{\partial^2 u}{\partial x \partial y} + y^2\dfrac{\partial^2 u}{\partial y^2}$. (*U.K.T.U. 2010*)

(*ii*) If $u = x \sin^{-1}\left(\dfrac{x}{y}\right) + y \sin^{-1}\left(\dfrac{y}{x}\right)$, evaluate $x^2\dfrac{\partial^2 u}{\partial x^2} + 2xy\dfrac{\partial^2 u}{\partial x \partial y} + y^2\dfrac{\partial^2 u}{\partial y^2}$. (*U.P.T.U. 2015*)

19. (*i*) If $u = \sin^{-1}\left(\dfrac{x^{1/3} + y^{1/3}}{x^{1/2} - y^{1/2}}\right)^{1/2}$, show that $x\dfrac{\partial u}{\partial x} + y\dfrac{\partial u}{\partial y} = -\dfrac{1}{12}\tan u$. (*M.T.U. 2012*)

(*ii*) If $u = \text{cosec}^{-1}\left(\dfrac{x^{1/2} + y^{1/2}}{x^{1/3} + y^{1/3}}\right)^{1/2}$, prove that $x^2\dfrac{\partial^2 u}{\partial x^2} + 2xy\dfrac{\partial^2 u}{\partial x \partial y} + y^2\dfrac{\partial^2 u}{\partial y^2} = \dfrac{\tan u}{144}(13 + \tan^2 u)$.

20. If $V = \log_e \sin\left\{\dfrac{\pi(2x^2 + y^2 + xz)^{1/2}}{2(x^2 + xy + 2yz + z^2)^{1/3}}\right\}$, prove that when $x = 0, y = 1, z = 2$;

$$x\frac{\partial V}{\partial x} + y\frac{\partial V}{\partial y} + z\frac{\partial V}{\partial z} = \frac{\pi}{12}.$$

21. State and prove Euler's theorem for partial differentiation of a homogeneous function $f(x, y)$. [*G.B.T.U. 2013*]

Answers

8. (*ii*) 0 **10.** (*iii*) $-2\cot u\,(2\,\text{cosec}^2 u + 1)$ **15.** $\dfrac{1}{144}\tan u\,(\sec^2 u - 12)$

18. (*i*) $2u$ (*ii*) 0.

2.48. COMPOSITE FUNCTIONS

(*i*) If $u = f(x, y)$, where $x = \phi(t)$, $y = \psi(t)$, then u is called a composite function of (**the single variable**) t and we can find $\dfrac{du}{dt}$.

(*ii*) If $z = f(x, y)$, where $x = \phi(u, v)$, $y = \psi(u, v)$, then z is called a composite function of (**two variables**) u and v so that we can find $\dfrac{\partial z}{\partial u}$ and $\dfrac{\partial z}{\partial v}$.

2.49. DIFFERENTIATION OF COMPOSITE FUNCTIONS

If u is composite function of t, defined by the relations $u = f(x, y)$; $x = \phi(t)$, $y = \psi(t)$, then

$$\boxed{\frac{du}{dt} = \frac{\partial u}{\partial x}\cdot\frac{dx}{dt} + \frac{\partial u}{\partial y}\cdot\frac{dy}{dt}.}$$

Proof. Here, $u = f(x, y)$ ...(1)

Let δt be an increment in t and δx, δy, δu the corresponding increments in x, y and u respectively. Then, we have

$$u + \delta u = f(x + \delta x, y + \delta y) \quad ...(2)$$

Subtracting (1) from (2), we get

$$\delta u = f(x + \delta x, y + \delta y) - f(x, y)$$

$$= f(x + \delta x, y + \delta y) - f(x, y + \delta y) + f(x, y + \delta y) - f(x, y)$$

$$\frac{\delta u}{\delta t} = \frac{f(x + \delta x, y + \delta y) - f(x, y + \delta y)}{\delta t} + \frac{f(x, y + \delta y) - f(x, y)}{\delta t}$$

$$= \frac{f(x + \delta x, y + \delta y) - f(x, y + \delta y)}{\delta x}\cdot\frac{\delta x}{\delta t} + \frac{f(x, y + \delta y) - f(x, y)}{\delta y}\cdot\frac{\delta y}{\delta t} \quad ...(3)$$

As $\delta t \to 0$, δx and δy both $\to 0$, so that

$$\underset{\delta t \to 0}{\text{Lt}} \frac{\delta u}{\delta t} = \frac{du}{dt}, \quad \underset{\delta t \to 0}{\text{Lt}} \frac{\delta x}{\delta t} = \frac{dx}{dt}, \quad \underset{\delta t \to 0}{\text{Lt}} \frac{\delta y}{\delta t} = \frac{dy}{dt}$$

$$\underset{\delta x \to 0}{\text{Lt}} \frac{f(x+\delta x, y+\delta y) - f(x, y+\delta y)}{\delta x} = \frac{\partial f}{\partial x} = \frac{\partial u}{\partial x}$$

and
$$\underset{\delta y \to 0}{\text{Lt}} \frac{f(x, y+\delta y) - f(x, y)}{\delta y} = \frac{\partial f}{\partial y} = \frac{\partial u}{\partial y}$$

$\therefore$ From (1),
$$\frac{du}{dt} = \frac{\partial u}{\partial x} \cdot \frac{dx}{dt} + \frac{\partial u}{\partial y} \cdot \frac{dy}{dt}$$

$\frac{du}{dt}$ is called the **total derivative** of u to distinguish it from the partial derivatives $\frac{\partial u}{\partial x}$ and $\frac{\partial u}{\partial y}$.

Cor. 1. If $u = f(x, y, z)$ and x, y, z are function of t, then u is a composite function of t and

$$\frac{du}{dt} = \frac{\partial u}{\partial x} \cdot \frac{dx}{dt} + \frac{\partial u}{\partial y} \cdot \frac{dy}{dt} + \frac{\partial u}{\partial z} \cdot \frac{dz}{dt}.$$

Cor. 2. If $z = f(x, y)$ and x, y are functions of u and v, then

$$\frac{\partial z}{\partial u} = \frac{\partial z}{\partial x} \cdot \frac{\partial x}{\partial u} + \frac{\partial z}{\partial y} \cdot \frac{\partial y}{\partial u}; \quad \frac{\partial z}{\partial v} = \frac{\partial z}{\partial x} \cdot \frac{\partial x}{\partial v} + \frac{\partial z}{\partial y} \cdot \frac{\partial y}{\partial v}.$$

Cor. 3. If $u = f(x, y)$, where $y = \phi(x)$, then since $x = \psi(x)$, u is a composite function of x.

$$\therefore \quad \frac{du}{dx} = \frac{\partial u}{\partial x} \cdot \frac{dx}{dx} + \frac{\partial u}{\partial y} \cdot \frac{dy}{dx} \quad \Rightarrow \quad \frac{du}{dx} = \frac{\partial u}{\partial x} + \frac{\partial u}{\partial y} \cdot \frac{dy}{dx}.$$

Cor. 4. If we are given an implicit function $f(x, y) = c$, then $u = f(x, y)$, where $u = c$

Using Cor. 3, we have $\frac{du}{dx} = \frac{\partial u}{\partial x} + \frac{\partial u}{\partial y} \cdot \frac{dy}{dx}$

But $\frac{du}{dx} = 0$ $\therefore$ $\frac{\partial u}{\partial x} + \frac{\partial u}{\partial y} \cdot \frac{dy}{dx} = 0$ or $\frac{\partial f}{\partial x} + \frac{\partial f}{\partial y} \cdot \frac{dy}{dx} = 0$

$$\therefore \quad \frac{dy}{dx} = -\frac{\frac{\partial f}{\partial x}}{\frac{\partial f}{\partial y}} = -\frac{f_x}{f_y}$$

Cor. 5. If $f(x, y) = c$, then by Cor. 4, we have $\frac{dy}{dx} = -\frac{f_x}{f_y}$

Differentiating again w.r.t. x, we get

$$\frac{d^2y}{dx^2} = -\left[\frac{f_y \frac{d}{dx}(f_x) - f_x \frac{d}{dx}(f_y)}{f_y^2}\right] = -\frac{f_y\left[\frac{\partial f_x}{\partial x} + \frac{\partial f_x}{\partial y} \cdot \frac{dy}{dx}\right] - f_x\left[\frac{\partial f_y}{\partial x} + \frac{\partial f_y}{\partial y} \cdot \frac{dy}{dx}\right]}{f_y^2}$$

$$= -\frac{f_y\left[f_{xx} - f_{yx}\cdot\frac{f_x}{f_y}\right] - f_x\left[f_{xy} - f_{yy}\cdot\frac{f_x}{f_y}\right]}{f_y^2} = -\left[\frac{f_{xx}f_y^2 - f_xf_yf_{xy} - f_xf_yf_{xy} + f_{yy}f_x^2}{f_y^3}\right]$$

Hence $\dfrac{d^2y}{dx^2} = -\left(\dfrac{f_{xx}f_y^2 - 2f_xf_yf_{xy} + f_{yy}f_x^2}{f_y^3}\right)$.

ILLUSTRATIVE EXAMPLES

Example 1. *(i) If $u = \sin^{-1}(x - y)$, $x = 3t$, $y = 4t^3$, show that* $\dfrac{du}{dt} = \dfrac{3}{\sqrt{1-t^2}}$.

(ii) If $u = x^2 - y^2 + \sin yz$, *where* $y = e^x$ *and* $z = \log x$ *; find* $\dfrac{du}{dx}$.

Sol. (*i*) The given equations define u as a composite function of t.

$$\therefore \quad \frac{du}{dt} = \frac{\partial u}{\partial x}\cdot\frac{dx}{dt} + \frac{\partial u}{\partial y}\cdot\frac{dy}{dt}$$

$$= \frac{1}{\sqrt{1-(x-y)^2}}\cdot 3 + \frac{1}{\sqrt{1-(x-y)^2}}(-1)\cdot 12t^2$$

$$= \frac{3(1-4t^2)}{\sqrt{1-(3t-4t^3)^2}} = \frac{3(1-4t^2)}{\sqrt{1-9t^2+24t^4-16t^6}}$$

$$= \frac{3(1-4t^2)}{\sqrt{(1-t^2)(1-8t^2+16t^4)}} = \frac{3(1-4t^2)}{\sqrt{(1-t^2)(1-4t^2)^2}} = \frac{3}{\sqrt{1-t^2}}.$$

(*ii*)

$$\frac{du}{dx} = \frac{\partial u}{\partial y}\cdot\frac{dy}{dx} + \frac{\partial u}{\partial z}\frac{dz}{dx} + \frac{\partial u}{\partial x}$$

$$= (-2y + z\cos yz)\,e^x + (y\cos yz)\,\frac{1}{x} + 2x$$

$$= \{-2e^x + \log x\cos(e^x\log x)\}\,e^x + \left\{e^x\cos(e^x\log x)\cdot\frac{1}{x}\right\} + 2x$$

$$= 2(x - e^{2x}) + e^x\cos(e^x\log x)\left(\log x + \frac{1}{x}\right).$$

Example 2. *(i) If $u = x\log(xy)$, where $x^3 + y^3 + 3xy = 1$, find* $\dfrac{du}{dx}$

(ii) If $u = f(r, s)$ and $r = x + y$, $s = x - y$; show that $\dfrac{\partial u}{\partial x} + \dfrac{\partial u}{\partial y} = 2\dfrac{\partial u}{\partial r}$.

Sol. (*i*) $\qquad u = x\log xy \qquad$ ...(1)

$$\therefore \quad \frac{\partial u}{\partial x} = x\left(\frac{1}{xy}\cdot y\right) + \log(xy) = 1 + \log xy$$

and

$$\frac{\partial u}{\partial y} = x\left(\frac{1}{xy}\cdot x\right) = \frac{x}{y}$$

Also, $x^3 + y^3 + 3xy = 1$ ...(2)

Differentiating (2) w.r.t. x, we get

$$3x^2 + 3y^2 \frac{dy}{dx} + 3\left(x\frac{dy}{dx} + y\right) = 0$$

$$\Rightarrow \quad \frac{dy}{dx} = -\left(\frac{x^2 + y}{x + y^2}\right) \quad ...(3)$$

We know that $\frac{du}{dx} = \frac{\partial u}{\partial x} + \frac{\partial u}{\partial y}\frac{dy}{dx} = 1 + \log(xy) + \frac{x}{y}\left\{-\left(\frac{x^2+y}{x+y^2}\right)\right\}$

$$= 1 + \log(xy) - \frac{x(x^2+y)}{y(y^2+x)}$$

(*ii*) $$\frac{\partial u}{\partial x} = \frac{\partial u}{\partial r}\cdot\frac{\partial r}{\partial x} + \frac{\partial u}{\partial s}\cdot\frac{\partial s}{\partial x}$$

$$= \frac{\partial u}{\partial r} + \frac{\partial u}{\partial s} \quad ...(1) \quad \Big| \because \frac{\partial r}{\partial x} = \frac{\partial s}{\partial x} = 1$$

$$\frac{\partial u}{\partial y} = \frac{\partial u}{\partial r}\cdot\frac{\partial r}{\partial y} + \frac{\partial u}{\partial s}\cdot\frac{\partial s}{\partial y}$$

$$= \frac{\partial u}{\partial r} - \frac{\partial u}{\partial s} \quad ...(2) \quad \Big| \because \frac{\partial r}{\partial y} = 1, \frac{\partial s}{\partial y} = -1$$

Adding (1) and (2), we get

$$\frac{\partial u}{\partial x} + \frac{\partial u}{\partial y} = 2\frac{\partial u}{\partial r}.$$

Example 3. *If z is a function of x and y, where $x = e^u + e^{-v}$ and $y = e^{-u} - e^v$, show that*

$$\frac{\partial z}{\partial u} - \frac{\partial z}{\partial v} = x\frac{\partial z}{\partial x} - y\frac{\partial z}{\partial y}.$$

Sol. Here z is a composite function of u and v

$$\therefore \quad \frac{\partial z}{\partial u} = \frac{\partial z}{\partial x}\cdot\frac{\partial x}{\partial u} + \frac{\partial z}{\partial y}\cdot\frac{\partial y}{\partial u} = \frac{\partial z}{\partial x}\cdot e^u + \frac{\partial z}{\partial y}(-e^{-u})$$

and $$\frac{\partial z}{\partial v} = \frac{\partial z}{\partial x}\cdot\frac{\partial x}{\partial v} + \frac{\partial z}{\partial y}\cdot\frac{\partial y}{\partial v} = \frac{\partial z}{\partial x}(-e^{-v}) + \frac{\partial z}{\partial y}(-e^v)$$

Subtracting, $$\frac{\partial z}{\partial u} - \frac{\partial z}{\partial v} = (e^u + e^{-v})\frac{\partial z}{\partial x} - (e^{-u} - e^v)\frac{\partial z}{\partial y} = x\frac{\partial z}{\partial x} - y\frac{\partial z}{\partial y}.$$

Example 4. *If $u = f(y - z, z - x, x - y)$, prove that $\frac{\partial u}{\partial x} + \frac{\partial u}{\partial y} + \frac{\partial u}{\partial z} = 0$.* [A.K.T.U. 2021]

Sol. Here $u = f(X, Y, Z)$, where $X = y - z$, $Y = z - x$, $Z = x - y$

$\therefore$ u is a composite function of x, y and z.

$$\frac{\partial u}{\partial x} = \frac{\partial u}{\partial X}\cdot\frac{\partial X}{\partial x} + \frac{\partial u}{\partial Y}\cdot\frac{\partial Y}{\partial x} + \frac{\partial u}{\partial Z}\cdot\frac{\partial Z}{\partial x} = \frac{\partial u}{\partial X}(0) + \frac{\partial u}{\partial Y}(-1) + \frac{\partial u}{\partial Z}(1)$$

$$\frac{\partial u}{\partial y} = \frac{\partial u}{\partial X}\cdot\frac{\partial X}{\partial y} + \frac{\partial u}{\partial Y}\cdot\frac{\partial Y}{\partial y} + \frac{\partial u}{\partial Z}\cdot\frac{\partial Z}{\partial y} = \frac{\partial u}{\partial X}(1) + \frac{\partial u}{\partial Y}(0) + \frac{\partial u}{\partial Z}(-1)$$

$$\frac{\partial u}{\partial z} = \frac{\partial u}{\partial X} \cdot \frac{\partial X}{\partial z} + \frac{\partial u}{\partial Y} \cdot \frac{\partial Y}{\partial z} + \frac{\partial u}{\partial Z} \cdot \frac{\partial Z}{\partial z} = \frac{\partial u}{\partial X}(-1) + \frac{\partial u}{\partial Y}(1) + \frac{\partial u}{\partial Z}(0)$$

Adding, we get $\dfrac{\partial u}{\partial x} + \dfrac{\partial u}{\partial y} + \dfrac{\partial u}{\partial z} = 0.$

Example 5. *If $w = f(x, y)$, $x = r \cos \theta$, $y = r \sin \theta$, show that*

$$\left(\frac{\partial w}{\partial r}\right)^2 + \frac{1}{r^2}\left(\frac{\partial w}{\partial \theta}\right)^2 = \left(\frac{\partial f}{\partial x}\right)^2 + \left(\frac{\partial f}{\partial y}\right)^2.$$

Sol. The given equations define w as a composite function of r and θ.

$$\frac{\partial w}{\partial r} = \frac{\partial w}{\partial x} \cdot \frac{\partial x}{\partial r} + \frac{\partial w}{\partial y} \cdot \frac{\partial y}{\partial r} = \frac{\partial w}{\partial x} \cdot \cos \theta + \frac{\partial w}{\partial y} \cdot \sin \theta$$

or $$\frac{\partial w}{\partial r} = \frac{\partial f}{\partial x} \cos \theta + \frac{\partial f}{\partial y} \sin \theta \qquad \text{...(1) } [\because \ w = f(x, y)]$$

Also, $$\frac{\partial w}{\partial \theta} = \frac{\partial w}{\partial x} \cdot \frac{\partial x}{\partial \theta} + \frac{\partial w}{\partial y} \cdot \frac{\partial y}{\partial \theta} = \frac{\partial w}{\partial x}(-r \sin \theta) + \frac{\partial w}{\partial y}(r \cos \theta)$$

or $$\frac{1}{r}\frac{\partial w}{\partial \theta} = -\frac{\partial f}{\partial x} \sin \theta + \frac{\partial f}{\partial y} \cos \theta \qquad \text{...(2)}$$

Squaring and adding (1) and (2), we get

$$\left(\frac{\partial w}{\partial r}\right)^2 + \frac{1}{r^2}\left(\frac{\partial w}{\partial \theta}\right)^2 = \left(\frac{\partial f}{\partial x}\right)^2 + \left(\frac{\partial f}{\partial y}\right)^2.$$

Example 6. *If $f(x, y) = 0$, $\phi(y, z) = 0$, show that* $\dfrac{\partial f}{\partial y} \cdot \dfrac{\partial \phi}{\partial z} \cdot \dfrac{dz}{dx} = \dfrac{\partial f}{\partial x} \cdot \dfrac{\partial \phi}{\partial y}$.

Sol. $$\frac{dz}{dx} = \frac{dz}{dy} \cdot \frac{dy}{dx} \qquad \text{...(1)}$$

$$f(x, y) = 0 \text{ gives } \frac{dy}{dx} = -\frac{\dfrac{\partial f}{\partial x}}{\dfrac{\partial f}{\partial y}}$$

$$\phi(y, z) = 0 \text{ gives } \frac{dz}{dy} = -\frac{\dfrac{\partial \phi}{\partial y}}{\dfrac{\partial \phi}{\partial z}}$$

$\therefore$ From (1), $$\frac{dz}{dx} = \frac{\dfrac{\partial f}{\partial x} \cdot \dfrac{\partial \phi}{\partial y}}{\dfrac{\partial f}{\partial y} \cdot \dfrac{\partial \phi}{\partial z}}$$

$$\Rightarrow \quad \frac{\partial f}{\partial y} \cdot \frac{\partial \phi}{\partial z} \cdot \frac{dz}{dx} = \frac{\partial f}{\partial x} \cdot \frac{\partial \phi}{\partial y}.$$

Example 7. *If $\phi(x, y, z) = 0$, show that* $\left(\dfrac{\partial y}{\partial z}\right)_x \left(\dfrac{\partial z}{\partial x}\right)_y \left(\dfrac{\partial x}{\partial y}\right)_z = -1.$

Sol. The given relation defines y as a function of x and z. Treating x as constant

$$\left(\frac{\partial y}{\partial z}\right)_x = -\frac{\frac{\partial \phi}{\partial z}}{\frac{\partial \phi}{\partial y}}$$

The given relation defines z as a function of x and y. Treating y as constant

$$\left(\frac{\partial z}{\partial x}\right)_y = -\frac{\frac{\partial \phi}{\partial x}}{\frac{\partial \phi}{\partial z}}$$

Similarly, $$\left(\frac{\partial x}{\partial y}\right)_z = -\frac{\frac{\partial \phi}{\partial y}}{\frac{\partial \phi}{\partial x}}$$

Multiplying, we get the desired result.

Example 8. (*i*) *If* $u = u\left(\frac{y-x}{xy}, \frac{z-x}{xz}\right)$, *show that* $x^2 \frac{\partial u}{\partial x} + y^2 \frac{\partial u}{\partial y} + z^2 \frac{\partial u}{\partial z} = 0.$

(*ii*) *If* $\phi(cx - az, cy - bz) = 0$, *show that* $a \frac{\partial z}{\partial x} + b \frac{\partial z}{\partial y} = c.$

Sol. (*i*) Let $$v = \frac{y-x}{xy} = \frac{1}{x} - \frac{1}{y} \qquad ...(1)$$

and $$w = \frac{z-x}{xz} = \frac{1}{x} - \frac{1}{z} \qquad ...(2)$$

so that $$u = u(v, w) \qquad ...(3)$$

Now, $$\frac{\partial u}{\partial x} = \frac{\partial u}{\partial v} \cdot \frac{\partial v}{\partial x} + \frac{\partial u}{\partial w} \cdot \frac{\partial w}{\partial x} = \frac{\partial u}{\partial v}\left(\frac{-1}{x^2}\right) + \frac{\partial u}{\partial w}\left(\frac{-1}{x^2}\right)$$ | Using (1) and (2)

$\Rightarrow$ $$x^2 \frac{\partial u}{\partial x} = -\frac{\partial u}{\partial v} - \frac{\partial u}{\partial w} \qquad ...(4)$$

Also, $$\frac{\partial u}{\partial y} = \frac{\partial u}{\partial v} \cdot \frac{\partial v}{\partial y} + \frac{\partial u}{\partial w} \cdot \frac{\partial w}{\partial y} = \frac{\partial u}{\partial v}\left(\frac{1}{y^2}\right) + \frac{\partial u}{\partial w}(0)$$ | Using (1) and (2)

$\Rightarrow$ $$y^2 \frac{\partial u}{\partial y} = \frac{\partial u}{\partial v} \qquad ...(5)$$

and $$\frac{\partial u}{\partial z} = \frac{\partial u}{\partial v} \cdot \frac{\partial v}{\partial z} + \frac{\partial u}{\partial w} \cdot \frac{\partial w}{\partial z} = \frac{\partial u}{\partial v}(0) + \frac{\partial u}{\partial w}\left(\frac{1}{z^2}\right)$$

$\Rightarrow$ $$z^2 \frac{\partial u}{\partial z} = \frac{\partial u}{\partial w} \qquad ...(6)$$

Adding (4), (5) and (6), we get

$$x^2 \frac{\partial u}{\partial x} + y^2 \frac{\partial u}{\partial y} + z^2 \frac{\partial u}{\partial z} = 0$$

(*ii*) Let $v = cx - az$ and $w = cy - bz$, then we have

$$\phi(v, w) = 0 \qquad ...(1)$$

$$\therefore \quad \frac{\partial \phi}{\partial x} = \frac{\partial \phi}{\partial v}\cdot\frac{\partial v}{\partial x} + \frac{\partial \phi}{\partial w}\cdot\frac{\partial w}{\partial x}$$

$$0 = \frac{\partial \phi}{\partial v}\left(c - a\frac{\partial z}{\partial x}\right) + \frac{\partial \phi}{\partial w}\left(-b\frac{\partial z}{\partial x}\right)$$

$$0 = c\frac{\partial \phi}{\partial v} + \frac{\partial z}{\partial x}\left(-a\frac{\partial \phi}{\partial v} - b\frac{\partial \phi}{\partial w}\right)$$

$$\Rightarrow \quad a\frac{\partial z}{\partial x} = \frac{ac\dfrac{\partial \phi}{\partial v}}{a\dfrac{\partial \phi}{\partial v} + b\dfrac{\partial \phi}{\partial w}} \qquad ...(2)$$

Also,

$$\frac{\partial \phi}{\partial y} = \frac{\partial \phi}{\partial v}\cdot\frac{\partial v}{\partial y} + \frac{\partial \phi}{\partial w}\cdot\frac{\partial w}{\partial y}$$

$$0 = \frac{\partial \phi}{\partial v}\left(-a\frac{\partial z}{\partial y}\right) + \frac{\partial \phi}{\partial w}\left(c - b\frac{\partial z}{\partial y}\right)$$

$$0 = c\frac{\partial \phi}{\partial w} + \frac{\partial z}{\partial y}\left(-a\frac{\partial \phi}{\partial v} - b\frac{\partial \phi}{\partial w}\right)$$

$$\Rightarrow \quad b\frac{\partial z}{\partial y} = \frac{bc\dfrac{\partial \phi}{\partial w}}{a\dfrac{\partial \phi}{\partial v} + b\dfrac{\partial \phi}{\partial w}} \qquad ...(3)$$

Adding (2) and (3), we get

$$a\frac{\partial z}{\partial x} + b\frac{\partial z}{\partial y} = \frac{c\left(a\dfrac{\partial \phi}{\partial v} + b\dfrac{\partial \phi}{\partial w}\right)}{\left(a\dfrac{\partial \phi}{\partial v} + b\dfrac{\partial \phi}{\partial w}\right)} = c.$$

Example 9. *Prove that* $\dfrac{\partial^2 z}{\partial x^2} + \dfrac{\partial^2 z}{\partial y^2} = \dfrac{\partial^2 z}{\partial u^2} + \dfrac{\partial^2 z}{\partial v^2}$, *where*

$$x = u\cos\alpha - v\sin\alpha,\ y = u\sin\alpha + v\cos\alpha.$$

Sol. Here z is a composite function of u and v

$$\frac{\partial z}{\partial u} = \frac{\partial z}{\partial x}\cdot\frac{\partial x}{\partial u} + \frac{\partial z}{\partial y}\cdot\frac{\partial y}{\partial u} = \cos\alpha\frac{\partial z}{\partial x} + \sin\alpha\frac{\partial z}{\partial y}$$

$$\Rightarrow \quad \frac{\partial}{\partial u} \equiv \cos\alpha\frac{\partial}{\partial x} + \sin\alpha\frac{\partial}{\partial y} \qquad ...(1)$$

Also,

$$\frac{\partial z}{\partial v} = \frac{\partial z}{\partial x}\cdot\frac{\partial x}{\partial v} + \frac{\partial z}{\partial y}\cdot\frac{\partial y}{\partial v} = -\sin\alpha\frac{\partial z}{\partial x} + \cos\alpha\frac{\partial z}{\partial y}$$

$$\Rightarrow \qquad \frac{\partial}{\partial v} \equiv -\sin\alpha \frac{\partial}{\partial x} + \cos\alpha \frac{\partial}{\partial y} \qquad ...(2)$$

Now, we shall make use of the equivalence of operators as given by (1) and (2).

$$\frac{\partial^2 z}{\partial u^2} = \frac{\partial}{\partial u}\left(\frac{\partial z}{\partial u}\right) = \left(\cos\alpha \frac{\partial}{\partial x} + \sin\alpha \frac{\partial}{\partial y}\right)\left(\cos\alpha \frac{\partial z}{\partial x} + \sin\alpha \frac{\partial z}{\partial y}\right)$$

$$= \cos^2\alpha \frac{\partial^2 z}{\partial x^2} + \cos\alpha\sin\alpha \frac{\partial^2 z}{\partial x\partial y} + \sin\alpha\cos\alpha \frac{\partial^2 z}{\partial y\partial x} + \sin^2\alpha \frac{\partial^2 z}{\partial y^2}$$

$$= \cos^2\alpha \frac{\partial^2 z}{\partial x^2} + 2\cos\alpha\sin\alpha \frac{\partial^2 z}{\partial x\partial y} + \sin^2\alpha \frac{\partial^2 z}{\partial y^2} \qquad ...(3)$$

$$\frac{\partial^2 z}{\partial v^2} = \frac{\partial}{\partial v}\left(\frac{\partial z}{\partial v}\right) = \left(-\sin\alpha \frac{\partial}{\partial x} + \cos\alpha \frac{\partial}{\partial y}\right)\left(-\sin\alpha \frac{\partial z}{\partial x} + \cos\alpha \frac{\partial z}{\partial y}\right)$$

$$= \sin^2\alpha \frac{\partial^2 z}{\partial x^2} - \sin\alpha\cos\alpha \frac{\partial^2 z}{\partial x\partial y} - \cos\alpha\sin\alpha \frac{\partial^2 z}{\partial x\partial y} + \cos^2\alpha \frac{\partial^2 z}{\partial y^2}$$

$$= \sin^2\alpha \frac{\partial^2 z}{\partial x^2} - 2\cos\alpha\sin\alpha \frac{\partial^2 z}{\partial x\partial y} + \cos^2\alpha \frac{\partial^2 z}{\partial y^2} \qquad ...(4)$$

Adding (3) and (4), $\dfrac{\partial^2 z}{\partial u^2} + \dfrac{\partial^2 z}{\partial v^2} = \dfrac{\partial^2 z}{\partial x^2} + \dfrac{\partial^2 z}{\partial y^2}$.

TEST YOUR KNOWLEDGE

1. (*i*) If $u = x^3 + y^3$, where $x = a\cos t$, $y = b\sin t$, find $\dfrac{du}{dt}$ and verify the result. (*A.K.T.U. 2020*)

   (*ii*) Find $\dfrac{du}{dt}$ as a total derivative and verify the result by direct substitution if $u = x^2 + y^2 + z^2$ and $x = e^{2t}$, $y = e^{2t}\cos 3t$, $z = e^{2t}\sin 3t$. (*U.P.T.U. 2015*)

2. If $z = u^2 + v^2$, $u = r\cos\theta$, $v = r\sin\theta$, find $\dfrac{\partial z}{\partial r}$ and $\dfrac{\partial z}{\partial \theta}$.

3. If $z = x^2 y$ and $x^2 + xy + y^2 = 1$, show that $\dfrac{dz}{dx} = 2xy - \dfrac{x^2(2x+y)}{x+2y}$.

4. If the curves $f(x, y) = 0$ and $\phi(x, y) = 0$ touch each other, show that at the point of contact
$$\frac{\partial f}{\partial x}\cdot\frac{\partial \phi}{\partial y} - \frac{\partial f}{\partial y}\cdot\frac{\partial \phi}{\partial x} = 0.$$

5. If $x = u + v$, $y = uv$ and z is a function of x, y; show that $u\dfrac{\partial z}{\partial u} + v\dfrac{\partial z}{\partial v} = x\dfrac{\partial z}{\partial x} + 2y\dfrac{\partial z}{\partial y}$.

6. If $u = f(r, s)$, $r = x + at$, $s = y + bt$, show that $\dfrac{\partial u}{\partial t} = a\dfrac{\partial u}{\partial x} + b\dfrac{\partial u}{\partial y}$.

7. (*a*) If $u = f(r, s, t)$ and $r = \dfrac{x}{y}$, $s = \dfrac{y}{z}$, $t = \dfrac{z}{x}$, prove that $x\dfrac{\partial u}{\partial x} + y\dfrac{\partial u}{\partial y} + z\dfrac{\partial u}{\partial z} = 0$. (*A.K.T.U. 2018*)

(b) If $x = u + v + w$, $y = vw + wu + uv$, $z = uvw$ and F is a function of x, y, z, show that

$$u\frac{\partial F}{\partial u} + v\frac{\partial F}{\partial v} + w\frac{\partial F}{\partial w} = x\frac{\partial F}{\partial x} + 2y\frac{\partial F}{\partial y} + 3z\frac{\partial F}{\partial z}.$$

8. (a) If $u = f(2x - 3y, 3y - 4z, 4z - 2x)$, prove that $\frac{1}{2}\frac{\partial u}{\partial x} + \frac{1}{3}\frac{\partial u}{\partial y} + \frac{1}{4}\frac{\partial u}{\partial z} = 0.$ *(A.K.T.U. 2015, 2020)*

(b) If $u = f(x^2 + 2yz, y^2 + 2zx)$, prove that $(y^2 - zx)\frac{\partial u}{\partial x} + (x^2 - yz)\frac{\partial u}{\partial y} + (z^2 - xy)\frac{\partial u}{\partial z} = 0.$

9. If $v = f(r)$; $r^2 = x^2 + y^2 + z^2$, prove that $\frac{\partial^2 v}{\partial x^2} + \frac{\partial^2 v}{\partial y^2} + \frac{\partial^2 v}{\partial z^2} = \frac{\partial^2 v}{\partial r^2} + \frac{2}{r}\frac{\partial v}{\partial r}$.

10. If $w = f(x, y)$, where $x = e^u \cos v$, $y = e^u \sin v$, show that $y\frac{\partial w}{\partial u} + x\frac{\partial w}{\partial v} = e^{2u}\frac{\partial w}{\partial y}$.

11. If $z = xyf\left(\frac{y}{x}\right)$ and z is a constant, show that $\dfrac{f'\left(\frac{y}{x}\right)}{f\left(\frac{y}{x}\right)} = \dfrac{x\left[y + x\frac{dy}{dx}\right]}{y\left[y - x\frac{dy}{dx}\right]}$.

12. If $z = z(u, v)$, $u = x^2 - 2xy - y^2$, $v = y$, show that $(x + y)\frac{\partial z}{\partial x} + (x - y)\frac{\partial z}{\partial y} = 0$ is equivalent to $\frac{\partial z}{\partial v} = 0$.

13. By changing the independent variables x and y to u and v by means of the relations $u = x - ay$, $v = x + ay$, show that $a^2\frac{\partial^2 z}{\partial x^2} - \frac{\partial^2 z}{\partial y^2}$ transforms into $4a^2\frac{\partial^2 z}{\partial u \partial v}$.

14. If z is a function of x and y, and u and v be two other variables such that $u = lx + my$, $v = ly - mx$, show that

$$\frac{\partial^2 z}{\partial x^2} + \frac{\partial^2 z}{\partial y^2} = (l^2 + m^2)\left(\frac{\partial^2 z}{\partial u^2} + \frac{\partial^2 z}{\partial v^2}\right).$$

15. (i) If $z = f(x, y)$, where $x = e^u \cos v$, $y = e^u \sin v$, prove that $\left(\frac{\partial f}{\partial x}\right)^2 + \left(\frac{\partial f}{\partial y}\right)^2 = e^{-2u}\left[\left(\frac{\partial f}{\partial u}\right)^2 + \left(\frac{\partial f}{\partial v}\right)^2\right]$.

(G.B.T.U. 2013)

(ii) If $x = e^r \cos \theta$, $y = e^r \sin \theta$, show that $\frac{\partial^2 u}{\partial x^2} + \frac{\partial^2 u}{\partial y^2} = e^{-2r}\left(\frac{\partial^2 u}{\partial r^2} + \frac{\partial^2 u}{\partial \theta^2}\right)$.

16. Transform the equation $\frac{\partial^2 u}{\partial x^2} + \frac{\partial^2 u}{\partial y^2} = 0$ into polar coordinates.

17. If $w = \sqrt{x^2 + y^2 + z^2}$ and $x = \cos v$, $y = u \sin v$, $z = uv$, then prove that

$$u\frac{\partial w}{\partial u} - v\frac{\partial w}{\partial v} = \frac{u}{\sqrt{1 + v^2}}.$$

(A.K.T.U. 2017)

Answers

1. (i) $-3a^3 \cos^2 t \sin t + 3b^3 \sin^2 t \cos t$ (ii) $2e^{4t}$ **2.** $2r$, 0

16. $\frac{\partial^2 u}{\partial r^2} + \frac{1}{r}\frac{\partial u}{\partial r} + \frac{1}{r^2}\frac{\partial^2 u}{\partial \theta^2} = 0$.

ASSIGNMENT II

1. Find y_n if $y = \dfrac{x^n - 1}{x - 1}$. (M.T.U. 2012)
2. What is the asymptote of the curve $y^2 (2a - x) = x^3$? (M.T.U. 2012)
3. What is the $(2n)^{th}$ derivative of $y = (x^2 - 1)^n$?
4. Find the eighth derivative of $y = 3x^8$.
5. If $y = \sinh x$ then find y_{2n+1}.
6. Find the nth derivative of $y = \sin^3 x$.
7. Find the nth derivative of $y = x^2 \sin x$. (U.P.T.U. 2014)
8. Find the nth derivative of $y = \dfrac{x + 1}{x^2 - 4}$.
9. Find the nth derivative of $x^2 e^{-x}$ at $x = 0$. (A.K.T.U. 2018)
10. Find the nth derivative of $y = e^{\sin^{-1} x}$ at $x = 0$.
11. If $y = e^{a \sin^{-1} x}$, find the value of $(1 - x^2) y_2 - xy_1 - a^2 y$. (A.K.T.U. 2016)
12. About which of the axes, the curve $x^{2/3} + y^{2/3} = a^{2/3}$ is symmetrical? (G.B.T.U. 2011)
13. Write an asymptote of a curve given by
$$x^2y - xy^2 + xy + y^2 + x - y = 0.$$
14. If $y = a \cos (\log x) + b \sin (\log x)$, then find $(y_n)_0$. (G.B.T.U. 2012)
15. If $y = e^{-x} \cos x$, then find the value of $y_4 + 4y$.
16. (i) Find the 8th derivative of $x^2 e^x$. (G.B.T.U. 2013)
    (ii) If $y = x^2 e^x$ then find y_n. (U.P.T.U. 2014)
17. Find all the asymptotes of the curve $xy^2 = 4a^2 (2a - x)$. (U.P.T.U. 2014)
18. Evaluate $\lim\limits_{(x, y) \to (0, 0)} \dfrac{xy}{x^2 + y^2}$.
19. Show that the function $f(x, y) = \begin{cases} x^2 + 4y & \text{when } (x, y) \neq (1, 2) \\ 0 & \text{when } (x, y) = (1, 2) \end{cases}$.
    is not continuous at the point (1, 2).
20. Discuss all the symmetry of the curve $x^2y^2 = x^2 - a^2$. (A.K.T.U. 2021)
21. State Euler's theorem on homogeneous function. (A.K.T.U. 2022)
22. If $u = \cos^{-1}\left(\dfrac{x + y}{\sqrt{x} + \sqrt{y}}\right)$ then find the value of $x\dfrac{\partial u}{\partial x} + y\dfrac{\partial u}{\partial y}$. (A.K.T.U. 2020)
23. Show that the function $f(x, y) = \begin{cases} \dfrac{x - y}{x + y}, & (x, y) \neq (0, 0) \\ 0, & (x, y) = (0, 0) \end{cases}$ is discontinuous at the point (0, 0).
24. Show that the function $f(x, y) = x - y$ is continuous for all $(x, y) \in R^2$.
25. Show that $\lim\limits_{(x, y) \to (0, 0)} \dfrac{xy^3}{x^2 + y^6}$ does not exist.
26. Let $A = \{(x, y);\ 0 < x < 1,\ 0 < y < 1,\ x, y \in R\}$. Let $f: A \to R$ be defined by $f(x, y) = x + y$.
    Show that $\lim\limits_{(x, y) \to \left(0, \frac{1}{2}\right)} f(x, y) = \dfrac{1}{2}$.
27. If $x^2 = au + bv$, $y^2 = au - bv$, then find $\dfrac{\partial u}{\partial x}$. (A.K.T.U. 2018)

28. Find the condition for the contour on $x - y$ plane where the partial derivative of $x^2 + y^2$ with respect to y is equal to the partial derivative of $6y + 4x$ with respect to x. *(A.K.T.U. 2017)*

29. If $z = u^2 + v^2$ and $u = at^2$, $v = 2at$, then find $\dfrac{dz}{dt}$.

30. If $u(x, y) = (\sqrt{x} + \sqrt{y})^5$, find the value of $x^2 \dfrac{\partial^2 u}{\partial x^2} + 2xy \dfrac{\partial^2 u}{\partial x \partial y} + y^2 \dfrac{\partial^2 u}{\partial y^2}$.

31. If $u = x^2yz - 4y^2z^2 + 2xz^3$, then find $x \dfrac{\partial u}{\partial x} + y \dfrac{\partial u}{\partial y} + z \dfrac{\partial u}{\partial z}$.

32. If $u = \log \left(\dfrac{x^2}{y}\right)$, then find $x \dfrac{\partial u}{\partial x} + y \dfrac{\partial u}{\partial y}$.

33. If $u = x^2 + 2xy + y^2 + x + y$, then find $x \dfrac{\partial u}{\partial x} + y \dfrac{\partial u}{\partial y}$.

34. What is the degree of homogeneous function $u(x, y) = \dfrac{x^2(x^2 - y^2)^{1/3}}{(x^2 + y^2)^{2/3}}$?

35. What is the degree of the homogeneous function $u(x, y) = \dfrac{\sqrt{x} - \sqrt{y}}{\sqrt{x} + \sqrt{y}}$?

36. What is the degree of homogeneous function $u(x, y) = x^{1/3} y^{-4/3} \tan^{-1}\left(\dfrac{y}{x}\right)$?

37. If $z = \log (x^3 + y^3 - x^2y - xy^2)$, then find $\dfrac{\partial z}{\partial x} + \dfrac{\partial z}{\partial y}$.

38. If $w = (y - z)(z - x)(x - y)$, then find $\dfrac{\partial w}{\partial x} + \dfrac{\partial w}{\partial y} + \dfrac{\partial w}{\partial z}$.

39. If $x^y = y^x$ then find the value of $\dfrac{dy}{dx}$ at (2, 4).

40. If $V(x, y, z) = \cos 3x \cos 4y \sinh 5z$ then find $V_{xx} + V_{yy} + V_{zz}$.

41. If $u = x^3y^2 \sin^{-1} \dfrac{y}{x}$, then find

(*i*) $x \dfrac{\partial u}{\partial x} + y \dfrac{\partial u}{\partial y}$ *(A.K.T.U. 2019)* (*ii*) $x^2 \dfrac{\partial^2 u}{\partial x^2} + 2xy \dfrac{\partial^2 u}{\partial x \partial y} + y^2 \dfrac{\partial^2 u}{\partial y^2}$

42. If $\log u = \dfrac{x^2y^2}{x + y}$, then find $x \dfrac{\partial u}{\partial x} + y \dfrac{\partial u}{\partial y}$.

43. If $(\sqrt{x} + \sqrt{y}) \cot u = x + y$, then find $4x \dfrac{\partial u}{\partial x} + 4y \dfrac{\partial u}{\partial y}$.

44. If $\cos u = \dfrac{x^5 - 2y^5 + 6z^5}{\sqrt{ax^3 + by^3 + cz^3}}$, then find $x u_x + y u_y + z u_z$.

45. If $u = \sin^{-1} (x^2 + y^2)^{1/5}$, then find the value of $x \dfrac{\partial u}{\partial x} + y \dfrac{\partial u}{\partial y}$. *(U.P.T.U. 2014)*

46. (*i*) If $u = x^2 + 2xy - y \log z$, where $x = s + t^2$, $y = s - t^2$ and $z = 2t$, then find the value of $\dfrac{\partial u}{\partial s}$ at (1, 2, 1).

(*ii*) If $u = x^2 - y^2 + \sin yz$, where $y = e^x$ and $z = \log x$, then find $\dfrac{\partial u}{\partial x}$.

47. (*i*) If $u = xg\left(\frac{y}{x}\right) + h\left(\frac{y}{x}\right)$, then find the value of $x^2 \frac{\partial^2 u}{\partial x^2} + 2xy \frac{\partial^2 u}{\partial x \, \partial y} + y^2 \frac{\partial^2 u}{\partial y^2}$.

(*ii*) If $f(x, y)$ is a homogeneous function in x and y of degree n, then find the value of

$$x^3 \frac{\partial^3 f}{\partial x^3} + 3x^2 y \frac{\partial^3 f}{\partial x^2 \, \partial y} + 3xy^2 \frac{\partial^3 f}{\partial x \, \partial y^2} + y^3 \frac{\partial^3 f}{\partial y^3}.$$

48. If $f(x, y, z, w) = 0$, then find $\frac{\partial x}{\partial y} \times \frac{\partial y}{\partial z} \times \frac{\partial z}{\partial w} \times \frac{\partial w}{\partial x}$. *(A.K.T.U. 2016, 2018)*

49. If $u = e^{x^2 + y^2 + z^2}$, then find $\frac{\partial^3 u}{\partial x \, \partial y \, \partial z}$.

50. (*i*) Evaluate $\lim\limits_{(x, y) \to (0, 0)} \frac{xy}{x^2 + y^2}$.

(*ii*) Show that the function

$$f(x, y) = \begin{cases} x^2 + 4y, & \text{when } (x, y) \neq (1, 2) \\ 0, & \text{when } (x, y) = (1, 2) \end{cases}$$

is not continuous at the point (1, 2).

Answers

1. 0
2. $x = 2a$
3. $2n!$
4. $3(8)!$

5. $\cosh x$
6. $\frac{3}{4} \sin\left(x + \frac{n\pi}{2}\right) - \frac{1}{4} (3)^n \sin\left(3x + \frac{n\pi}{2}\right)$
7. $(n - n^2) \sin \frac{n\pi}{2}$

8. $\frac{(-1)^n \, n!}{4} \left[\frac{3}{(x-2)^{n+1}} + \frac{1}{(x+2)^{n+1}}\right]$
9. $n(n-1)(-1)^n$

10. $y_n(0) = \begin{cases} 1^2 . (1^2 + 2^2)(1^2 + 4^2) \ldots\ldots \{1^2 + (n-2)^2\}, & \text{when } n \text{ is even} \\ 1^2 . (1^2 + 1^2)(1^2 + 3^2) \ldots\ldots \{1^2 + (n-2)^2\}, & \text{when } n \text{ is odd} \end{cases}$

11. 0
12. both the co-ordinate axes
13. $x + y = 0$

14. 0
15. 0
16. (*i*) $e^x(x^2 + 16x + 56)$ (*ii*) $e^x\,[x^2 + 2nx + n(n-1)]$

17. $x = 0$
18. does not exist
19. $4a^2t\,(t^2 + 2)$

20. Symmetrical about x-axis and y-axis
Symmetrical in opposite quadrants

22. $-\frac{1}{2} \cot u$
27. $\frac{x}{a}$
28. $y = 2$
29. $4a^2t(t^2 + 2)$

30. $\frac{15}{4} u$
31. $4u$
32. 1
33. $2x^2 + 4xy + 2y^2 + x + y$

34. 4/3
35. 0
36. -1
37. $\frac{2}{x + y}$

38. 0
39. $4\left(\frac{\log 2 - 1}{2 \log 2 - 1}\right)$
40. 0
41. (*i*) $5u$ (*ii*) $20u$

42. $3u \log u$
43. $-\sin 2u$
44. $-\frac{7}{2} \cot u$
45. $\frac{2}{5} \tan u$

46. (*i*) 8 (*ii*) $2(x - e^{2x}) + e^x \cos(e^x \log x)\left(\log x + \frac{1}{x}\right)$

47. (*i*) 0 (*ii*) $n(n-1)(n-2)f$
48. 1
49. $8xyzu$

50. (*i*) does not exist.

MODULE 3

Differential Calculus – II

EXPANSION OF FUNCTIONS OF ONE VARIABLE

3.1. TAYLOR'S THEOREM

Statement. Let $f(x + h)$ be a function of h(x being independent of h) which can be expanded in powers of h and the expansion be differentiable any number of times, then

$$f(x+h) = f(x) + hf'(x) + \frac{h^2}{2!}f''(x) + \ldots + \frac{h^n}{n!}f^n(x) + \ldots$$

Proof. Let $\quad f(x+h) = B_0 + B_1h + B_2h^2 + B_3h^3 + B_4h^4 + \ldots \qquad \ldots(1)$

where $B_0, B_1, B_2, B_3, B_4, \ldots$ are the functions of x alone which are to be determined.

Now, $\quad \frac{d}{dh}\{f(x+h)\} = \frac{d}{dz}f(z) \cdot \frac{dz}{dh} \quad$ where $z = x + h$

$$= f'(z) \cdot 1 = f'(x+h)$$

By successive differentiation of (1) w.r.t. h, we get

$$f'(x+h) = B_1 + 2B_2h + 3B_3h^2 + 4B_4h^3 + \ldots \qquad \ldots(2)$$

$$f''(x+h) = 2 \cdot 1 \cdot B_2 + 3 \cdot 2 \cdot B_3h + 4 \cdot 3 \cdot B_4h^2 + \ldots \qquad \ldots(3)$$

$$f'''(x+h) = 3 \cdot 2 \cdot 1 \cdot B_3 + 4 \cdot 3 \cdot 2 \cdot B_4h + \ldots \qquad \ldots(4)$$

Put $h = 0$, we get from (1), (2), (3), (4), ...;

$$f(x) = B_0,\ f'(x) = B_1,\ f''(x) = B_2\, 2\,!,\ f'''(x) = 3\,!\ B_3, \ldots$$

Hence, $\quad \boxed{f(x+h) = f(x) + hf'(x) + \frac{h^2}{2!}f''(x) + \frac{h^3}{3!}f'''(x) + \ldots + \frac{h^n}{n!}f^n(x) + \ldots}$

3.1.1. Other Forms of Taylor's Theorem

(*i*) Writing a for x in (1), we get

$$f(a+h) = f(a) + hf'(a) + \frac{h^2}{2!}f''(a) + \frac{h^3}{3!}f'''(a) + \ldots + \frac{h^n}{n!}f^n(a) + \ldots \qquad \ldots(5)$$

(*ii*) Putting $a + h = b$ or $h = b - a$ in (5), we get

$$f(b) = f(a) + (b-a)f'(a) + \frac{(b-a)^2}{2!}f''(a) + \ldots + \frac{(b-a)^n}{n!}f^n(a) + \ldots \qquad \ldots(6)$$

(*iii*) Putting $h = x - a$ in (5), we get

$$f(x) = f(a) + (x-a)f'(a) + \frac{(x-a)^2}{2!}f''(a) + \ldots + \frac{(x-a)^n}{n!}f^n(a) + \ldots \quad \ldots(7)$$

3.1.2. Corollary. Putting $a = 0$ and $h = x$ in (5), we get

$$f(x) = f(0) + x\,f'(0) + \frac{x^2}{2!}f''(0) + \frac{x^3}{3!}f'''(0) + \ldots + \frac{x^n}{n!}f^n(0) + \ldots$$

which is **Maclaurin's theorem.**

Note. If the function $f(x)$ is denoted by y, the expansion may be written in the form

$$y = y(0) + xy_1(0) + \frac{x^2}{2!}y_2(0) + \ldots + \frac{x^n}{n!}y_n(0) + \ldots$$

where $y(0), y_1(0), y_2(0), \ldots, y_n(0)$ etc. denote values of $y_1, y_2, \ldots, y_n$ respectively at $x = 0$.

3.2. FAILURE OF TAYLOR'S *AND* MACLAURIN'S THEOREM

(*a*) **Taylor's theorem** *fails* to expand $f(x + h)$ in an infinite series in the following situations:

(*i*) if any functions $f(x), f'(x), f''(x)$ ... become infinite or does not exist for any value of x in the interval under consideration,

or (*ii*) if $R_n = \frac{h^n}{n!}f^n(x + \theta h)$ does not tend to zero as $n \to \infty$.

(*b*) **Maclaurin's theorem** *fails* to expand $f(x)$ in an infinite series in the following situations:

(*i*) if any of $f(x), f'(x), f''(x)$... becomes infinite or does not exist in the closed interval $[0, x]$,

or (*ii*) if $R_n = \frac{h^n}{n!}(x + \theta h)$ does not tend to zero as $n \to \infty$.

Note. We generally obtain the expansion by *assuming* the possibility of expanding it in an infinite series by assuming that R_n tends to 0 as n tends to infinity.

3.3. APPLICATION OF MACLAURIN'S THEOREM

Working Rule

1st Step. Put the given function equal to $f(x)$.

2nd Step. Differentiate $f(x)$, a number of times and find $f'(x), f''(x), f'''(x)$ ... and so on.

3rd Step. Put $x = 0$ in the results obtained in 2nd Step and find $f(0), f'(0), f''(0), f'''(0)$... and so on.

4th Step. Now substitute the values of $f(0), f'(0), f''(0), f'''(0)$, ... in,

$$f(x) = f(0) + xf'(0) + \frac{x^2}{2!}f''(0) + \frac{x^3}{3!}f'''(0) + \ldots$$

ILLUSTRATIVE EXAMPLES

Example 1. *Expand a^x and e^x in powers of x by Maclaurin's theorem.*

Sol. Let $f(x) = a^x$ $\therefore$ $f(0) = a^0 = 1$

then $f'(x) = a^x \log a$ $\quad f'(0) = \log a$

$f''(x) = a^x (\log a)^2$ $\quad f''(0) = (\log a)^2$

$f'''(x) = a^x (\log a)^3$ $\quad f'''(0) = (\log a)^3$

and so on ....

Now, $$f(x) = f(0) + xf'(0) + \frac{x^2}{2!}f''(0) + \frac{x^3}{3!}f'''(0) + ...$$

$$\therefore \quad \mathbf{a^x = 1 + x \log a + \frac{x^2}{2!} (\log a)^2 + \frac{x^3}{3!} (\log a)^3 + ...} \quad ...(1)$$

Putting $a = e$, we have

$$\mathbf{e^x = 1 + x + \frac{x^2}{2!} + \frac{x^3}{3!} + ...} \quad ...(2)$$

Note. The series (1) and (2) are called exponential series and are convergent for all values of x.

Example 2. *Expand $\log (1 + x)$ in powers of x. Then find series for $\log_e \left(\frac{1+x}{1-x}\right)$ and hence determine the value of $\log_e \left(\frac{11}{9}\right)$ upto five places of decimal.* (M.T.U. 2012)

Sol. Let $f(x) = \log (1 + x)$ $\therefore$ $f(0) = \log 1 = 0$

then $f'(x) = \frac{1}{1+x} = (1 + x)^{-1}$ $\therefore$ $f'(0) = 1$

$f''(x) = (-1)(1 + x)^{-2}$ $\therefore$ $f''(0) = -1$

$f'''(x) = (-1)(-2)(1 + x)^{-3}$ $\therefore$ $f'''(0) = 2$

$f^{iv}(x) = (-1)(-2)(-3)(1 + x)^{-4}$ $\therefore$ $f^{iv}(0) = -2 \times 3 = -6$

and so on.

Putting these values in

$$f(x) = f(0) + xf'(0) + \frac{x^2}{2!}f''(0) + \frac{x^3}{3!}f'''(0) + \frac{x^4}{4!}f^{iv}(0) + ...$$

$$\therefore \quad \log (1 + x) = 0 + x - \frac{x^2}{2} + \frac{x^3}{3} - \frac{x^4}{4} + ...$$

or $$\mathbf{\log (1 + x) = x - \frac{x^2}{2} + \frac{x^3}{3} - \frac{x^4}{4} + ...}$$

Changing x into $-x$, we have

$$\mathbf{\log (1 - x) = -x - \frac{x^2}{2} - \frac{x^3}{3} - \frac{x^4}{4} - ...}$$

Now, $\log (1 + x) - \log (1 - x)$

$$= \left[x - \frac{x^2}{2} + \frac{x^3}{3} - \frac{x^4}{4} + ...\right] - \left[-x - \frac{x^2}{2} - \frac{x^3}{3} - \frac{x^4}{4} - ....\right]$$

$$\therefore \quad \log\left(\frac{1+x}{1-x}\right) = 2\left[x + \frac{x^3}{3} + \frac{x^5}{5} + \frac{x^7}{7} + \ldots\right]$$

Putting $x = \frac{1}{10}$ in the above result, we get

$$\log\left(\frac{11}{9}\right) = 2\left[\frac{1}{10} + \frac{1}{3}\left(\frac{1}{10}\right)^3 + \frac{1}{5}\left(\frac{1}{10}\right)^5 + \frac{1}{7}\left(\frac{1}{10}\right)^7 + \ldots\right] = 0.20067$$

Note. The above series for log $(1 + x)$ is called logarithmic series and is convergent for $|x|$ less than 1 and also for $x = 1$.

Example 3. *Expand* $\sin^{-1} x$ *upto four terms in powers of x.*

Sol. Let $\qquad f(x) = \sin^{-1} x \qquad \therefore f(0) = 0$

then $\qquad f'(x) = \frac{1}{\sqrt{1-x^2}} = (1 - x^2)^{-1/2} \qquad \therefore f'(0) = 1$

Expanding by Binomial theorem,

$$f'(x) = 1 + \left(-\frac{1}{2}\right)(-x^2) + \frac{\left(-\frac{1}{2}\right)\left(-\frac{3}{2}\right)(-x^2)^2}{2!} + \frac{\left(-\frac{1}{2}\right)\left(-\frac{3}{2}\right)\left(-\frac{5}{2}\right)}{3!}(-x^2)^3 + \ldots\ldots$$

$$= 1 + \frac{1}{2}x^2 + \frac{3}{8}x^4 + \frac{15}{18}x^6 + \ldots$$

$$f''(x) = x + \frac{3}{2}x^3 + \frac{15}{8}x^5 + \ldots \qquad \therefore \qquad f''(0) = 0$$

$$f'''(x) = 1 + \frac{9}{2}x^2 + \frac{75}{8}x^4 + \ldots \qquad \therefore \qquad f'''(0) = 1$$

$$f^{iv}(x) = 9x + \frac{75}{2}x^3 + \ldots \qquad f^{iv}(0) = 0$$

$$f^{v}(x) = 9 + \frac{225}{2}x^2 + \ldots \qquad f^{v}(0) = 9$$

$$f^{vi}(x) = 225x + \ldots \qquad f^{vi}(0) = 0$$

and proceeding further, $f^{vii}(0) = 225$, etc.

Putting these values in

$$f(x) = f(0) + xf'(0) + \frac{x^2}{2!}f''(0) + \frac{x^3}{3!}f'''(0) + \ldots$$

$$\therefore \quad \sin^{-1} x = 0 + x + \frac{x^2}{2!}(0) + \frac{x^3}{3!}(1) + \frac{x^4}{4!}(0) + \frac{x^5}{5!}(9) + \frac{x^6}{6!}(0) + \frac{x^7}{7!}(225) + \ldots$$

$$= x + \frac{x^3}{3!} + \frac{9x^5}{5!} + \frac{225x^7}{5!} + \ldots$$

$$= x + \frac{1}{2}\cdot\frac{x^3}{3} + \frac{1\cdot 3}{2\cdot 4}\cdot\frac{x^5}{5} + \frac{1\cdot 3\cdot 5}{2\cdot 4\cdot 6}\cdot\frac{x^7}{7} + \ldots$$

Note. To expand an alone inverse function; find its first derivative. *Expand by Binomial Theorem* and then find other derivatives.

Example 4. *Expand* $tan^{-1} x$.

Sol. Let $f(x) = \tan^{-1} x$, $\therefore$ $f(0) = 0$

then $f'(x) = \dfrac{1}{1+x^2} = (1 + x^2)^{-1}$ $\therefore$ $f'(0) = 1$

$= 1 - x^2 + x^4 - x^6 + \ldots$ [By Binomial Theorem]

$f''(x) = -2x + 4x^3 - 6x^5 + \ldots$ $f''(0) = 0$

$f'''(x) = -2 + 12x^2 - 30x^4 + \ldots$ $f'''(0) = -2$

$f^{iv}(x) = 24x - 120x^3 + \ldots$ $f^{iv}(0) = 0$

$f^{v}(x) = 24 - 360x^2 + \ldots$ $f^{v}(0) = 24$

and so on ....

Putting these values in

$$f(x) = f(0) + xf'(0) + \frac{x^2}{2!} f''(0) + \frac{x^3}{3!} f'''(0) + \ldots$$

$$\therefore \quad \tan^{-1} x = 0 + x + \frac{x^2}{2!}(0) + \frac{x^3}{3!}(-2) + \frac{x^4}{4!}(0) + \frac{x^5}{5!}.(24) + \ldots$$

or

$$\tan^{-1} x = x - \frac{x^3}{3} + \frac{x^5}{5} - \frac{x^7}{7} + \ldots$$

Note 1. The expansion for $\tan^{-1} x$ is valid only if $-1 < x \le 1$.

Note 2. This expansion for $\tan^{-1} x$ known as Gregory's series is very useful in finding the value of π.

Example 5. *Expand tan x by Maclaurin's Theorem as far as* x^5 *and hence find the value of tan 46° 30′ upto four decimal places.*

Sol. Let $f(x) = \tan x$ $\therefore$ $f(0) = 0$

then $f'(x) = \sec^2 x = 1 + \tan^2 x$ $\therefore$ $f'(0) = 1$

$f''(x) = 2 \tan x \sec^2 x = 2 \tan x (1 + \tan^2 x)$

$= 2 \tan x + 2 \tan^3 x$ $\therefore$ $f''(0) = 0$

$f'''(x) = 2 \sec^2 x + 6 \tan^2 x \sec^2 x$

$= 2(1 + \tan^2 x) + 6 \tan^2 x (1 + \tan^2 x)$

$= 2 + 8 \tan^2 x + 6 \tan^4 x$ $\therefore$ $f'''(0) = 2$

$f^{iv}(x) = 16 \tan x \sec^2 x + 24 \tan^3 x \sec^2 x$

$= 8 \sec^2 x (2 \tan x + 3 \tan^3 x)$

$= 8(1 + \tan^2 x)(2 \tan x + 3 \tan^3 x)$

$= 16 \tan x + 40 \tan^3 x + 24 \tan^5 x$ $\therefore$ $f^{iv}(0) = 0$

$f^{v}(x) = 16 \sec^2 x + 120 \tan^2 x \sec^2 x + 120 \tan^4 x \sec^2 x$

$= 8 \sec^2 x (2 + 15 \tan^2 x + 15 \tan^4 x)$ $\therefore$ $f^{v}(0) = 16$

Putting these values in

$$f(x) = f(0) + xf'(0) + \frac{x^2}{2!} f''(0) + \frac{x^3}{3!} f'''(0) + \frac{x^4}{4!} f^{iv}(0) + \frac{x^5}{5!} f^{v}(0) + \ldots$$

$$\therefore \quad \tan x = 0 + x + \frac{x^3}{3!}.2 + \frac{x^5}{5!}.16 \ldots = x + \frac{x^3}{3} + \frac{2}{15}x^5 + \ldots \quad \ldots(1)$$

Deduction. Here $x = 46° \, 30' = \left(46\frac{1}{2}\right)^\circ$ $\quad | \because \; 1° = 60' \;\; \therefore \;\; 30' = (\frac{1}{2})°$

$$= \frac{93}{2} \times \frac{\pi}{180} \text{ Radians} = \frac{31}{120} \times \frac{22}{7} = 0.812$$

Putting $x = 46° \, 30' = 0.812$ in eqn. (1), we have

$$\tan 46° \, 30' = 0.812 + \frac{(0.812)^3}{3} + \frac{2}{15}(0.812)^5 = 1.0374.$$

Example 6. *Expand by Maclaurin's Theorem* $\frac{e^x}{e^x+1}$, *as far as* x^3.

Sol. Let $\quad f(x) = \dfrac{e^x}{e^x+1} \qquad \therefore \quad f(0) = \dfrac{e^0}{e^0+1} = \dfrac{1}{2}$

then $\quad f'(x) = \dfrac{(e^x+1)\,e^x - e^x.\,e^x}{(e^x+1)^2} = \dfrac{e^x}{(e^x+1)^2} \qquad \therefore \quad f'(0) = \dfrac{e^0}{(e^0+1)^2} = \dfrac{1}{2^2}$

$$f''(x) = \frac{(e^x+1)^2 \, e^x - e^x \, 2(e^x+1)e^x}{(e^x+1)^4}$$

$$= \frac{e^{2x} + e^x - 2e^{2x}}{(e^x+1)^3} = \frac{e^x - e^{2x}}{(e^x+1)^3} \qquad \therefore \quad f''(0) = \frac{e^0 - e^0}{(e^0+1)^3} = 0$$

$$f'''(x) = \frac{(e^x+1)^3(e^x - 2e^{2x}) - (e^x - e^{2x}).\,3(e^x+1)^2 e^x}{(e^x+1)^6}$$

$$= \frac{e^x - 4e^{2x} + e^{3x}}{(e^x+1)^4} \qquad \therefore \quad f'''(0) = -\frac{1}{2^3}$$

Putting these values in

$$f(x) = f(0) + xf'(0) + \frac{x^2}{2!} f''(0) + \frac{x^3}{3!} f'''(0) + \ldots$$

$$\therefore \qquad \frac{e^x}{e^x+1} = \frac{1}{2} + \frac{x}{4} - \frac{x^3}{48} + \ldots$$

Example 7. *Prove by Maclaurin's Theorem that*

$$e^{\sin x} = 1 + x + \frac{x^2}{1 \cdot 2} - \frac{3 \cdot x^4}{1 \cdot 2 \cdot 3 \cdot 4} + \ldots$$

Sol. Let $\quad f(x) = e^{\sin x} \qquad \therefore \quad f(0) = e^0 = 1$

$f'(x) = e^{\sin x} \cos x \qquad \therefore \quad f'(0) = e^0 \cos 0 = 1$

$f''(x) = e^{\sin x}(-\sin x) + \cos x \, e^{\sin x} \cos x$

$= e^{\sin x}[\cos^2 x - \sin x] \qquad \therefore \quad f''(0) = e^0[1 - 0] = 1$

$f''' = e^{\sin x}[2\cos x(-\sin x) - \cos x] + e^{\sin x} \cos x \,.\, [\cos^2 x - \sin x]$

$= e^{\sin x} \cos x \,.\, [-2\sin x - 1 + \cos^2 x - \sin x]$

$= -e^{\sin x} \cos x[3\sin x + \sin^2 x] \qquad \therefore \quad f'''(0) = 0$

$f^{iv}(x) = -e^{\sin x} \cos x\,[3\cos x + 2\sin x \cos x] + e^{\sin x} \,.\, \sin x[3\sin x + \sin^2 x]$

$\quad - [3\sin x + \sin^2 x] \cos x \, e^{\sin x} \,.\, \cos x \qquad \therefore \quad f^{iv}(0) = -3$

Putting these value in Maclaurin's Theorem, we get

$$e^{\sin x} = 1 + x \,.\, 1 + \frac{x^2}{2!} \,.\, 1 + \frac{x^3}{3!} \,.\, 0 + \frac{x^4}{4!}\,(-3) + \ldots$$

$$= 1 + x + \frac{x^2}{1 \cdot 2} - \frac{3x^4}{1 \cdot 2 \cdot 3 \cdot 4} + \ldots$$

Example 8. *Expand sin ($e^x - 1$) up to and including the term x^4.*

Sol. Let $f(x) = \sin (e^x - 1)$ $\qquad f(0) = 0$

$\therefore$ $f'(x) = \cos (e^x - 1) \,.\, e^x$ $\qquad f'(0) = 1$

Again, $f''(x) = e^x [-\sin (e^x - 1)]\, e^x + \cos (e^x - 1) \,.\, e^x$

$f''(x) = -e^{2x} \sin (e^x - 1) + e^x \cos (e^x - 1)$ $\qquad f''(0) = 1$

Again

$$f'''(x) = -[e^{2x} \cos (e^x - 1) \,.\, e^x + \sin (e^x - 1) \,.\, e^{2x} \,.\, 2] + e^x [-\sin (e^x - 1) \,.\, e^x] + \cos (e^x - 1) \,.\, e^x$$

or $$f'''(x) = -e^{3x} \cos (e^x - 1) - 3e^{2x} \sin (e^x - 1) + e^x \cos (e^x - 1)$$

Put $x = 0$, $\qquad f'''(0) = -1 - 0 + 1 = 0.$

Again diff. w.r.t. x, $f^{iv}(x) = -[-e^{3x} \sin (e^x - 1) \,.\, e^x + \cos (e^x - 1) \,.\, e^{3x} \,.\, 3]$
$- 3 [e^{2x} \cos (e^x - 1) \,.\, e^x + \sin (e^x - 1) \,.\, e^{2x} \,.\, 2]$
$- e^x \sin (e^x - 1) \,.\, e^x + \cos (e^x - 1) \,.\, e^x.$

Put $x = 0$, $\qquad f^{iv}(0) = -3 - 3(1 + 0) - 0 + 1 = -5$

Putting these values of $f(0), f'(0), f''(0), f'''(0), f^{iv}(0)$ in Maclaurin's expansion

$$\sin (e^x - 1) = 0 + x(1) + \frac{x^2}{2!}\,(1) + \frac{x^3}{3!}\,(0) + \frac{x^4}{4!}\,(-5) + \ldots$$

or $$\sin (e^x - 1) = x + \frac{x^2}{2} - \frac{5x^2}{24} + \ldots$$

Example 9. *Show that* $\log \dfrac{\tan x}{x} = \dfrac{x^2}{3} + \dfrac{7}{90}x^4 + \ldots$

Sol. We know that $\tan x = x + \dfrac{x^3}{3} + \dfrac{2}{15}x^5 + \ldots$

$$\therefore \quad \text{LHS} = \log \left(\frac{x + \dfrac{x^3}{3} + \dfrac{2}{15}x^5 + \ldots}{x} \right) = \log \left[1 + \frac{x^2}{3} + \frac{2}{15}x^4 + \ldots \right]$$

$$= \log (1 + z), \quad \text{where } z = \frac{x^2}{3} + \frac{2}{15}x^4 + \ldots$$

$$= z - \frac{z^2}{2} + \frac{z^3}{3} - \ldots$$

$$= \left(\frac{x^2}{3} + \frac{2}{15}x^4 + \ldots \right) - \frac{1}{2}\left(\frac{x^2}{3} + \frac{2}{15}x^4 + \ldots \right)^2 + \frac{1}{3}\left(\frac{x^2}{3} + \frac{2}{15}x^4 + \ldots \right)^3$$

$$= \frac{x^2}{3} + \frac{2}{15}x^4 + \ldots - \frac{1}{2}\left(\frac{x^4}{9} + \ldots\right) = \frac{x^2}{3} + \frac{2}{15}x^4 - \frac{x^4}{18} - \ldots$$

$$= \frac{x^2}{3} + \frac{7x^4}{90} + \ldots = \text{RHS.}$$

Example 10. *Find the values of a and b such that the expansion of* $\log(1+x) - \dfrac{x(1+ax)}{(1+bx)}$ *in ascending powers of x begins with the term* x^4 *and hence find this term.* (M.T.U. 2013)

Sol. $\log(1+x) - \dfrac{x(1+ax)}{(1+bx)} = \log(1+x) - x(1+ax)(1+bx)^{-1}$

$$= \left(x - \frac{x^2}{2} + \frac{x^3}{3} - \frac{x^4}{4} + \ldots\right) - (x + ax^2)(1 - bx + b^2x^2 - b^3x^3 + \ldots)$$

| Using binomial expansion

$$= \left(x - \frac{x^2}{2} + \frac{x^3}{3} - \frac{x^4}{4} + \ldots\right) - [x + (a-b)x^2 - b(a-b)x^3 + b^2(a-b)x^4 - \ldots]$$

$$= \left(-\frac{1}{2} - a + b\right)x^2 + \left\{\frac{1}{3} + b(a-b)\right\}x^3 - \left\{\frac{1}{4} + b^2(a-b)\right\}x^4 + \ldots \quad \ldots(1)$$

According to the question,

$$-\frac{1}{2} - a + b = 0 \;\Rightarrow\; a - b = -\frac{1}{2} \quad \ldots(2)$$

and $\quad \dfrac{1}{3} + b(a-b) = 0 \;\Rightarrow\; \dfrac{1}{3} - \dfrac{b}{2} = 0 \;\Rightarrow\; b = \dfrac{2}{3}$

$\therefore$ From (2), $\quad a = \dfrac{1}{6}$

$$\text{Required term} = -\left[\frac{1}{4} + b^2(a-b)\right]x^4 = -\left[\frac{1}{4} + \frac{4}{9}\left(-\frac{1}{2}\right)\right]x^4 = -\frac{1}{36}x^4.$$

TEST YOUR KNOWLEDGE

Apply Maclaurin's Theorem to prove (1–3):

1. (*a*) $\log \sec x = \dfrac{x^2}{2} + \dfrac{x^4}{12} + \dfrac{x^6}{45} + \ldots$ (*b*) $\log \cos x = -\dfrac{x^2}{2!} - 2 \cdot \dfrac{x^2}{4!} - 16 \cdot \dfrac{x^6}{6!} - \ldots$
2. (*a*) $\sec x = 1 + \dfrac{x^2}{2!} + \dfrac{5x^4}{4!} + \ldots$ (*b*) $\cos^2 x = 1 - x^2 + \dfrac{x^4}{3} + \ldots$
3. $\tan^{-1}(1+x) = \dfrac{\pi}{4} + \dfrac{x}{2} - \dfrac{x^2}{4} + \dfrac{x^3}{12} + \ldots$
4. (*i*) Show that $e^{ax} \sin bx = bx + abx^2 + \dfrac{b(3a^2 - b^2)}{3!}x^3 + \ldots + \dfrac{(a^2+b^2)^{n/2}}{n!}x^n \sin\left(n \tan^{-1}\dfrac{b}{a}\right) + \ldots$

   (*ii*) Expand $e^{2x} \sin x$ in ascending powers of x up to x^5. (*U.P.T.U. 2014*)

5. (*a*) Show that $\tan^{-1}\frac{2x}{1-x^2}=\sin^{-1}\left(\frac{2x}{1+x^2}\right)=2\left(x-\frac{x^3}{3}+\frac{x^5}{5}-\ldots\right)$. (*U.P.T.U. 2015*)

(*b*) Expand $\sin^{-1}\left(\frac{2x^2}{1+x^4}\right)$ in series of ascending powers of x.

(*c*) Show that $\tan^{-1}\frac{\sqrt{1+x^2}-1}{x}=\frac{x}{2}-\frac{x^3}{6}+\frac{x^5}{10}-\frac{x^7}{14}+\ldots$

6. Use Maclaurin's Theorem to find the expansion in the ascending powers of x of $\log_e(1+e^x)$ up to the term including x^4.

7. Use Maclaurin's Theorem to prove that $\log(1+\sin x)=x-\frac{x^2}{2}+\frac{x^3}{6}-\frac{x^4}{12}+\ldots$

Point out the case of failure, if any.

8. Prove that $e^x\log(1+x)=x+\frac{x^2}{2!}+\frac{2x^3}{3!}+\ldots$

9. (*a*) Apply Maclaurin's Theorem to show that $e^x\cos x=1+x-\frac{2x^3}{3!}-\frac{2^2x^4}{4!}-\frac{2^2x^5}{5!}+\ldots$

(*b*) Prove that $e^x\sec x=1+x+2\,.\,\frac{x^2}{2!}+4\,.\,\frac{x^3}{3!}+\ldots$

10. Expand by Maclaurin's theorem $\tan^2 x$ in ascending powers of x as far as x^4.

11. Expand $\log(1+\sin^2 x)$ in powers of x as far as the term containing x^4.

12. (*a*) By Maclaurin's Theorem expand $\log[1-\log(1-x)]$ in powers of x as far as the term in x^3.

(*b*) Expand by Maclaurin's series $\log[1+\log(1+x)]$ in powers of x as far as term containing x^3.

13. By Maclaurin's Theorem find the first four terms in the expansion of $\log(1+\tan x)$.

14. Show that $\log(1-x+x^2)=-x+\frac{x^2}{2}+\frac{2x^3}{3}-\frac{x^4}{4}-\frac{x^5}{5}-\frac{2x^6}{6}-\frac{x^7}{7}+\frac{x^8}{8}\ldots$

$\left[\textbf{Hint. } \log(1-x+x^2)=\log\left(\frac{1+x^3}{1+x}\right)=\log(1+x^3)-\log(1+x).\right]$

15. If $y^3-6xy-8=0$, prove that, $y=2+x-\frac{1}{2}\cdot\frac{x^3}{3!}+\frac{x^4}{4!}+\ldots$

Answers

4. (*ii*) $x+2x^2+\frac{11}{3!}x^3+x^4+\frac{41}{120}x^5+\ldots+\frac{5^{n/2}}{n!}x^n\sin\left(n\tan^{-1}\frac{1}{2}\right)$

5. (*b*) $2\left[x^2-\frac{1}{3}x^6+\frac{1}{5}x^{10}-\ldots\right]$

6. $\log 2+\frac{x}{2}+\frac{x^2}{8}-\frac{x^4}{192}+\ldots$

7. case of failure if $\sin x=-1$

10. $x^2+\frac{2x^4}{3}+\ldots$

11. $x^2-\frac{5}{6}x^4+\ldots$

12. (*a*) $x+\frac{x^3}{6}+\ldots$ (*b*) $x-x^2+\frac{7}{6}x^3+\ldots$

13. $x-\frac{x^2}{2}+\frac{2x^3}{3}-\frac{7x^4}{12}+\ldots$

3.4. APPLICATION OF TAYLOR'S THEOREM

Working Rule

Step 1. Put the given function equal to $f(x + h)$.

Step 2. Put $h = 0$ and write $f(x)$.

Step 3. Differentiate $f(x)$ a number of times and obtain $f'(x), f''(x), f'''(x)$, etc. ...

Step 4. Now substitute the values of $f(x), f'(x), f''(x)$ ... in

$$f(x+h) = f(x) + hf'(x) + \frac{h^2}{2!} f''(x) + \frac{h^3}{3!} f'''(x) + \ldots$$

ILLUSTRATIVE EXAMPLES

Example 1. *Apply Taylor's Theorem to find the expansion of log sin $(x + h)$.*

Sol. (*i*) Put $f(x + h) = \log \sin (x + h)$

(*ii*) Making $h = 0$, we have $f(x) = \log \sin x$

(*iii*) $\therefore$

$$f'(x) = \frac{\cos x}{\sin x} = \cot x$$

$$f''(x) = -\operatorname{cosec}^2 x$$

$$f'''(x) = -2 \operatorname{cosec} x(-\operatorname{cosec} x \cot x)$$

$$= 2 \operatorname{cosec}^2 x \cot x$$

....

Substituting these values in $f(x+h) = f(x) + hf'(x) + \frac{h^2}{2!} f''(x) + \frac{h^3}{3!} f'''(x) + \ldots$, we get

$$\log \sin (x+h) = \log \sin x + h \cot x - \frac{h^2}{2!} (\operatorname{cosec}^2 x) + \frac{h^3}{3!} (2 \operatorname{cosec}^2 x \cot x) + \ldots$$

Example 2. *Expand $\sin^{-1} (x + h)$ in powers of h.*

Sol. (*i*) Put $f(x + h) = \sin^{-1} (x + h)$

(*ii*) Making $h = 0$, we have $f(x) = \sin^{-1} x$.

(*iii*)

$$f'(x) = \frac{1}{\sqrt{1-x^2}}$$

$$f''(x) = -\frac{1}{2} (1 - x^2)^{-3/2} (-2x) = \frac{x}{(1-x^2)^{3/2}}$$

$$f'''(x) = \frac{(1-x^2)^{3/2} - \frac{3}{2}(1-x^2)^{1/2}(-2x) \cdot x}{(1-x^2)^3}$$

$$= \frac{(1-x^2)^{1/2}(1-x^2+3x^2)}{(1-x^2)^3} = \frac{1+2x^2}{(1-x^2)^{5/2}}$$

....

(*iv*) Now, $f(x+h) = f(x) + hf'(x) + \dfrac{h^2}{2!} f''(x) + \ldots$

$$\therefore \quad \sin^{-1}(x+h) = \sin^{-1} x + \frac{h}{\sqrt{1-x^2}} + \frac{x}{(1-x^2)^{3/2}} \cdot \frac{h^2}{2!} + \frac{1+2x^2}{(1-x^2)^{5/2}} \cdot \frac{h^3}{3!} + \ldots$$

Example 3. (*a*) *Expand sin (x + y) in powers of y and deduce that*

sin (x + y) = sin x cos y + sin y cos x.

(*b*) *Obtain the value of sin 31° correct to four places of decimals.*

Sol. (*a*) (*i*) Put $f(x+y) = \sin(x+y)$

(*ii*) Making $y = 0$, we have $f(x) = \sin x$

(*iii*) $\therefore$ $f'(x) = \cos x$; $f''(x) = -\sin x$,

$f'''(x) = -\cos x$; $f^{iv}(x) = \sin x, \ldots$, and $f^n(x) = \sin(x + n\pi/2)$

Now, $$f(x+y) = f(x) + yf'(x) + \frac{y^2}{2!} f''(x) + \ldots$$

$$\therefore \quad \sin(x+y) = \sin x + y\cos x - \frac{y^2}{2!}\sin x - \frac{y^3}{3!}\cos x + \frac{y^4}{4!}\sin x + \ldots \quad \ldots(1)$$

$$= \sin x \left[1 - \frac{y^2}{2!} + \frac{y^4}{4!} - \ldots\right] + \cos x \left[y - \frac{y^3}{3!} + \frac{y^5}{5!} - \ldots\right]$$

$$= \sin x \cos y + \cos x \sin y.$$

(*b*) Put $x = \dfrac{\pi}{6}$ and $y = 1° = \dfrac{\pi}{180} = .0175$ radians in (1),

Then $$\sin 31° = \sin \frac{\pi}{6} + \frac{\pi}{180}\cos\frac{\pi}{6} - \left(\frac{\pi}{180}\right)^2 \frac{1}{2!}\sin\frac{\pi}{6} + \ldots = 0.5151$$

correct to four places of decimal.

Examples 4. (*i*) *If* $f(x) = x^3 + 8x^2 + 15x - 24$, *calculate the value of* $f\left(\dfrac{11}{10}\right)$ *by the application of Taylor's series.*

(*ii*) $f(x) = x^3 - 2x + 5$, *find the value of f (2.001) with the help of Taylor's Theorem. Find the approximate change in the value of f(x) when x changes from 2 to 2.001.*

Sol. (*i*) By Taylor's Theorem,

$$f(x+h) = f(x) + hf'(x) + \frac{h^2}{2!} f''(x) + \frac{h^3}{3!} f'''(x) + \ldots$$

Put $x = 1$ and $h = \dfrac{1}{10}$ in the above series,

$$f\left(\frac{11}{10}\right) = f\left(1 + \frac{1}{10}\right) = f(1) + \frac{1}{10} f'(1) + \frac{1}{10^2} \cdot \frac{1}{2!} \cdot f''(1) + \frac{1}{3!} \cdot \frac{1}{(10)^3} f'''(1) + \ldots \quad \ldots(1)$$

Now,
$f(x) = x^3 + 8x^2 + 15x - 24 \quad \therefore \quad f(1) = 0$
$f'(x) = 3x^2 + 16x + 15 \quad \therefore \quad f'(1) = 34$
$f''(x) = 6x + 16 \quad \therefore \quad f''(1) = 22$
$f'''(x) = 6 \quad \therefore \quad f'''(1) = 6$
$f^{iv}(x) = 0 \quad \therefore \quad f^{iv}(1) = 0$ and so on.

∴ Substituting values of $f(1), f'(1), f''(1)$ ... etc. in (1), we get

$$f\left(\frac{11}{10}\right) = f\left(1+\frac{1}{10}\right) = 0 + \frac{1}{10}\,34 + \frac{11}{100} + \frac{1}{1000} = 3.511.$$

(*ii*) Here put $x = 2$ and $h = .001$ in Taylor's series,

$$f(2.001) = f(2) + (.001)\, f'(2) + \frac{(.001)^2}{2!} f''(2) + \frac{(.001)^3}{3!} f'''(2) + \ldots \qquad \ldots(2)$$

Now,

$$\begin{aligned} f(x) &= x^3 - 2x + 5 & \therefore\quad f(2) &= 9 \\ f'(x) &= 3x^2 - 2 & \therefore\quad f'(2) &= 10 \\ f''(x) &= 6x & f''(2) &= 12 \\ f'''(x) &= 6 & f'''(2) &= 6 \\ f^{iv}(x) &= 0 & f^{iv}(2) &= 0 \text{ and so on.} \end{aligned}$$

Putting these values in (2),

$$f(2.001) = 9 + (.001)\,10 + \frac{1}{2!}(.001)^2\,(12) + \frac{1}{3!}(.001)^3\,(6) + \ldots = 9.01 \text{ (approx.)}$$

Approximate value of $f(2.001) - f(2) = 9.01 - 9 = .01$.

TEST YOUR KNOWLEDGE

Apply Taylor's Theorem to prove the following expansions (1–4):

1. (*a*) $a^{x+h} = a^x \left[1 + h\log a + \frac{h^2}{2!}(\log a)^2 + \frac{h^3}{3!}(\log a)^3 + \ldots\right]$

(*b*) $e^{x+h} = e^x + he^x + \frac{h^2}{2!}e^x + \ldots$ (*c*) $\frac{1}{x+h} = \frac{1}{x} - \frac{h}{x^2} + \frac{h^2}{x^3} - \frac{h^3}{x^4} + \ldots$

(*d*) $\tan(x+h) = \tan x + h\sec^2 x + h^2\sec^2 x\tan x + \frac{h^3}{3}\sec^2 x(1 + 3\tan^2 x) + \ldots$

2. (*a*) $\cos\left(\frac{\pi}{4} + h\right) = \frac{1}{\sqrt{2}}\left(1 - h - \frac{h^2}{2!} + \frac{h^3}{3!} + \ldots\right)$ (*b*) $\sin\left(\frac{\pi}{4} + \theta\right) = \frac{1}{\sqrt{2}}\left(1 + \theta - \frac{\theta^2}{2!} - \frac{\theta^3}{3!} + \ldots\right)$.

3. (*a*) $\sec^{-1}(x+h) = \sec^{-1} x + \frac{h}{x\sqrt{x^2-1}} - \frac{2x^2-1}{x^2(x^2-1)^{3/2}} \cdot \frac{h^2}{2!} + \ldots$

(*b*) $\tan^{-1}(x+h) = \tan^{-1} x + \frac{h}{(1+x^2)} - \frac{xh^2}{(1+x^2)^2} \ldots$

4. $\frac{f(x+h) + f(x-h)}{2} = f(x) + \frac{h^2}{2!} f''(x) + \frac{h^4}{4!} f^{iv}(x) + \ldots$

5. If $f(x) = 3x^3 - 5x^2 + 7$, find the value of $f\left(\frac{21}{20}\right)$ by Taylor's theorem and find the change in the value of $f(x)$ when x changes from 2 to 2.1.

6. If $f(x) = x^3 + 6x^2 + 9$, find the value of $f\left(\frac{11}{10}\right)$ by Taylor's theorem.

7. Prove that $\log_e(x+h) = \log_e h + \frac{x}{h} - \frac{x^2}{2h^2} + \frac{x^3}{3h^3} - \ldots$

8. Calculate the approximate value of $\sqrt{26}$ to three decimal places by Taylor's Expansion.

9. Prove that $f\left(\dfrac{x^2}{1+x}\right) = f(x) - \dfrac{x}{1+x} f'(x) + \left(\dfrac{x}{1+x}\right)^2 \dfrac{1}{2!} f''(x) - \left(\dfrac{x}{1+x}\right)^3 \dfrac{1}{3!} f'''(x) + \ldots$

[*M.T.U. (C.O.) 2013*]

$\left[\textbf{Hint. } \text{Write } \dfrac{x^2}{1+x} = x - \dfrac{x}{1+x} = x + h, \text{ where } h = -\dfrac{x}{1+x}.\right]$

Answers

5. (*i*) 4.960375, 1.878 **6.** 17.591 **8.** 5.099.

3.5. ANOTHER FORM OF TAYLOR'S SERIES

In this form $f(x)$ is expressed as a series in ascending integral powers of $x - a$.

$f(x) = f(a + x - a)$. Put $x - a = h$. Then

$$f(x) = f(a + h) = f(a) + hf'(a) + \frac{h^2}{2!} f''(a) + \frac{h^3}{3!} f'''(a) + \ldots$$

Now replacing h by $x - a$, we get

$$f(x) = f(a) + (x - a) f'(a) + \frac{(x-a)^2}{2!} f''(a) + \frac{(x-a)^3}{3!} f'''(a) + \ldots$$

ILLUSTRATIVE EXAMPLES

Example 1. *Expand sin x in ascending powers of* $\left(x - \dfrac{\pi}{2}\right)$.

Sol. Here $f(x) = \sin x$

$$f(x) = f\left(\frac{\pi}{2} + x - \frac{\pi}{2}\right) = f(a + h), \text{ where } a = \frac{\pi}{2} \text{ and } h = x - \frac{\pi}{2}$$

i.e.,
$$f(x) = f(a) + hf'(a) + \frac{h^2}{2!} f''(a) + \frac{h^3}{3!} f'''(a) + \frac{h^4}{4!} f^{iv}(a) + \ldots$$

Putting these values of a and h, we get

$$\sin x = f\left(\frac{\pi}{2}\right) + \left(x - \frac{\pi}{2}\right) f'\left(\frac{\pi}{2}\right) + \frac{\left(x - \frac{\pi}{2}\right)^2}{2!} f''\left(\frac{\pi}{2}\right) + \ldots \quad \ldots(1)$$

But
$$f(x) = \sin x \qquad \therefore \qquad f\left(\frac{\pi}{2}\right) = 1$$

$$f'(x) = \cos x \qquad f'\left(\frac{\pi}{2}\right) = 0$$

$$f''(x) = -\sin x \qquad f''\left(\frac{\pi}{2}\right) = -1$$

$$f'''(x) = -\cos x \qquad f'''\left(\frac{\pi}{2}\right) = 0$$

$$f^{iv}(x) = \sin x \qquad f^{iv}\left(\frac{\pi}{2}\right) = 1 \text{ and so on.}$$

Putting these values in (1), we get

$$\sin x = 1 - \frac{\left(x - \frac{\pi}{2}\right)^2}{2!} + \frac{\left(x - \frac{\pi}{2}\right)^4}{4!} - \ldots$$

Example 2. *Obtain the first four terms in the expansion of log sin x in powers of (x – 3).*

Sol. Here, $f(x) = \log \sin x, \qquad a = 3$

Then, $$f(x) = f(3) + (x-3) f'(3) + \frac{(x-3)^2}{2!} f''(3) + \frac{(x-3)^3}{3!} f'''(3) + \ldots \qquad \ldots(1)$$

Now, $f(x) = \log \sin x \qquad \therefore \quad f(3) = \log \sin 3$

Now, $f'(x) = \dfrac{1}{\sin x} (\cos x) = \cot x \qquad \therefore \quad f'(3) = \cot 3$

$f''(x) = -\operatorname{cosec}^2 x = -[1 + \cot^2 x] \qquad \therefore \quad f''(3) = -\operatorname{cosec}^2 3$

$f'''(x) = -[2 \cot x (-\operatorname{cosec}^2 x)]$

$= 2 \cot x \operatorname{cosec}^2 x \qquad \therefore \quad f'''(3) = 2 \cot 3 \operatorname{cosec}^2 3$

Putting these values in (1), we get

$$\log \sin x = \log \sin 3 + (x-3) \cot 3 - \frac{(x-3)^2}{2} \operatorname{cosec}^2 3 + \frac{(x-3)^3}{3} \cot 3 \operatorname{cosec}^2 3 + \ldots$$

Example 3. *Expand log x in powers of (x – 1) by Taylor's theorem and hence find the value of $\log_e$ (1.1).*

Sol. Here $f(x) = \log x, \quad a = 1 \quad \therefore \quad f(1) = 0$

Now, $f'(x) = \dfrac{1}{x}, \qquad f'(1) = 1$

$f''(x) = -\dfrac{1}{x^2}, \qquad f''(1) = -1$

$f'''(x) = \dfrac{2}{x^3}, \qquad f'''(1) = 2$

$f^{(iv)}(x) = \dfrac{-6}{x^4}, \qquad f^{(iv)}(1) = -6$ and so on

We know that,

$$f(x) = f(1) + (x-1) f'(1) + \frac{(x-1)^2}{2!} f''(1) + \frac{(x-1)^3}{3!} f'''(1) + \ldots$$

$$\Rightarrow \quad \log x = 0 + (x-1) \,.\, 1 + \frac{(x-1)^2}{2!}(-1) + \frac{(x-1)^3}{3!} \,.\, 2 + \frac{(x-1)^4}{4!}(-6) + \ldots$$

$$\Rightarrow \quad \log x = (x-1) - \frac{1}{2}(x-1)^2 + \frac{1}{3}(x-1)^3 - \frac{1}{4}(x-1)^4 + \ldots$$

Putting $x = 1.1$, we get

$$\log (1.1) = (1.1 - 1) - \frac{1}{2}(1.1-1)^2 + \frac{1}{3}(1.1-1)^3 - \frac{1}{4}(1.1-1)^4 + \ldots = 0.095305$$

TEST YOUR KNOWLEDGE

1. (*i*) Expand e^x in power of $(x - 2)$. (*ii*) Expand $4x^2 + 7x + 5$ in powers of $x - 3$.
   (*iii*) Expand a^x in powers of $x - a$.
2. Expand log sin x in powers of $(x - 2)$.
3. Use Taylor's Theorem to express the polynomial $2x^3 + 7x^2 + x - 6$ in powers of $(x - 2)$.
4. Expand tan x in powers of $\left(x - \dfrac{\pi}{4}\right)$ up to first four terms.
5. Show that $f(ax) = f(x) + (a - 1)\, xf'(x) + \dfrac{(a-1)^2 x^2}{2!} f''(x) + \ldots$

Answers

1. (*i*) $e^2 \left[1 + (x - 2) + \dfrac{(x-2)^2}{2!} + \ldots\right]$ (*ii*) $62 + 31(x - 3) + 4(x - 3)^2$

   (*iii*) $a^a \left[1 + (x - a)\log a + \dfrac{(x-a)^2}{2!}(\log a)^2 + \dfrac{(x-a)^3}{3!}(\log a)^3 + \ldots\right]$

2. $\log \sin 2 + (x - 2)\cot 2 - \dfrac{(x-2)^2}{2}\operatorname{cosec}^2 2 + \dfrac{(x-2)^3}{3}\cot 2 \operatorname{cosec}^2 2 + \ldots$
3. $40 + 53\,(x - 2) + 19\,(x - 2)^2 + 2(x - 2)^3$.
4. $1 + 2\left(x - \dfrac{\pi}{4}\right) + 2\left(x - \dfrac{\pi}{4}\right)^2 + \dfrac{8}{3}\left(x - \dfrac{\pi}{4}\right)^3 + \ldots$

3.6. METHOD OF DIFFERENTIAL EQUATIONS

Now we shall proceed to expand functions like $(\sin^{-1} x)^2$, $e^{a \sin^{-1} x}$, $\sin (m \sin^{-1} x)$, $\cos (m \sin^{-1} x)$ etc. by the method of differential equations.

Working rule:

1. Put the given function equal to y.

2. Find $y_1 = \dfrac{dy}{dx}$.

Then (*i*) Take LCM (if possible).

(*ii*) Square both sides if square roots are there.

(*iii*) Try to get y in RHS (if possible).

3. Again differentiate both sides w.r.t. x to get an equation in y_2, y_1 and y.

4. Differentiate both sides n times w.r.t. x by Leibnitz Theorem.

5. Put $x = 0$ in equations of steps 1, 2, 3, 4.

6. Put $n = 1, 2, 3, 4$ in last equation of step 5.

7. Discuss the two cases when n is even and when n is odd.

ILLUSTRATIVE EXAMPLES

Example 1. *Prove that:*

$$\sin(m\sin^{-1}x) = mx + \frac{m(1^2-m^2)}{3!}x^3 + \frac{m(1^2-m^2)(3^2-m^2)}{5!}x^5 + \ldots$$
$$+ \ldots + \frac{m(1^2-m^2)(3^2-m^2)\ldots[(2n-3)^2-m^2]x^{2n-1}}{(2n-1)!} + \ldots$$

Sol. Let $y = \sin(m\sin^{-1}x)$ ...(1)

Differentiating w.r.t. x, $y_1 = \cos(m\sin^{-1}x)\dfrac{m}{\sqrt{1-x^2}}$

$\therefore$ $\sqrt{1-x^2}\,y_1 = m\cos(m\sin^{-1}x)$ ...(2)

Squaring both sides

$$(1-x^2)\,y_1^2 = m^2\cos^2(m\sin^{-1}x) = m^2(1-y^2)$$

Again differentiating w.r.t. x,

$$(1-x^2)\,2y_1y_2 + y_1^2(-2x) = m^2(-2yy_1)$$

or $y_2(1-x^2) - xy_1 + m^2y = 0$ ...(3)

Differentiating both sides n times by Leibnitz theorem, we get

$$[y_{n+2}(1-x^2) + {}^nC_1y_{n+1}(-2x) + {}^nC_2y_n(-2)] - [y_{n+1}(x) + {}^nC_1y_n(1)] + m^2y_n = 0$$

$\Rightarrow$ $(1-x^2)\,y_{n+2} - (2n+1)\,xy_{n+1} - (n^2-m^2)\,y_n = 0$ ...(4)

Putting $x = 0$ in (1), (2), (3) and (4), we get

$$y = \sin(0) = 0 = f(0)$$
$$y_1 = m\cos 0 = m = f'(0)$$
$$y_2 = 0 = f''(0)$$
$$y_{n+2} = (n^2-m^2)y_n$$

Putting $n = 1, 2, 3, 4$ in $y_{n+2} = (n^2-m^2)\,y_n$; we get

$$y_3 = (1^2-m^2)y_1 = m(1^2-m^2) = f'''(0)$$
$$y_4 = (2^2-m^2)y_2 = 0 = f^{iv}(0)$$
$$y_5 = (3^2-m^2)y_3 = m(1^2-m^2)(3^2-m^2) = f^{v}(0)$$
$$y_6 = (4^2-m^2)\,y_4 = 0 = f^{vi}(0)$$

$\therefore$ When n is even $y_n = 0$

and when n is odd $y_n = m(1^2-m^2)(3^2-m^2)\ldots[(n-2)^2-m^2]$

In particular, $y_{2n-1} = m(1^2-m^2)(3^2-m^2)\ldots[(2n-3)^2-m^2]$

By Maclaurin's Theorem,

$$f(x) = f(0) + xf(0) + \frac{x^2}{2!}f''(0) + \frac{x^3}{3!}f'''(0) + \frac{x^4}{4!}f^{iv}(0) + \frac{x^5}{5!}f^{v}(0) + \ldots + \frac{x^{2n-1}}{(2n-1)!}f^{2n-1}(0) + \ldots$$

or $$\sin (m \sin^{-1} x) = mx + \frac{m(1^2 - m^2)}{3!}x^3 + \frac{m(1^2 - m^2)(3^2 - m^2)}{5!}x^5 + \ldots$$
$$+ \frac{m(1^2 - m^2)(3^2 - m^2) \ldots [(2n-3)^2 - m^2]}{(2n-1)!}x^{2n-1} + \ldots$$

Example 2. *If* $y = (\sin^{-1} x)^2$, *show that*

(*a*) $(1 - x^2)\, y_2 - xy_1 = 2$ (*b*) $\dfrac{(\sin^{-1}x)^2}{2} = \dfrac{x^2}{2!} + \dfrac{2^2}{4!} . x^4 + \dfrac{2^2 . 4^2}{6!} x^6 + \ldots$

Sol. Let $$y = (\sin^{-1} x)^2 \quad \ldots(1)$$

$\therefore$ $$y_1 = 2 \sin^{-1} x \times \frac{1}{\sqrt{1-x^2}} \quad \ldots(2)$$

i.e., $$\sqrt{1-x^2} \,.\, y_1 = 2\sqrt{y}$$

or $$(1 - x^2)\, y_1^2 = 4y.$$

Differentiating again,
$$(1 - x^2) \,.\, 2y_1\, y_2 + y_1^2\,(-2x) = 4y_1$$

Dividing by $2y_1$, $$(1 - x^2)y_2 - xy_1 = 2 \quad \ldots(3)$$

which proves part (*a*).

Differentiating (3), *n* times by Leibnitz's Theorem,
$$(1 - x^2)\, y_{n+2} + {}^nC_1\, y_{n+1}\,(-2x) + {}^nC_2\, y_n\,(-2) - xy_{n+1} - {}^nC_1\, y_n \,.\, 1 = 0$$

or $$(1 - x)^2\, y_{n+2} - 2nxy_{n+1} - n(n-1)\, y_n - xy_{n+1} - ny_n = 0$$

or $$(1 - x^2)\, y_{n+2} - (2n - 1)\, xy_{n+1} = n^2 y_n \quad \ldots(4)$$

Putting $x = 0$, in (1), (2), (3), (4), we get
$$y = 0, \quad y_1 = 0, \quad y_2 = 2$$
$$(y_{n+2})_0 = n^2 (y_n)_0$$

Putting $n = 1, 2, 3, 4$ in the last equation, we get
$$y_3 = 1^2\, y_1 = 0$$
$$y_4 = 2^2\, y_2 = 2^2 \,.\, 2 = 2.2^2$$
$$y_5 = 3^2\, y_3 = 0$$
$$y_6 = 4^2\, y_4 = 2.2^2 \,.\, 4^2$$
....

$\therefore$ By Maclaurin's Theorem,
$$f(x) = f(0) + xf'(0) + \frac{x^2}{2!} f''(0) + \frac{x^3}{3!} f'''(0) + \ldots$$

$\therefore$ $$(\sin^{-1} x)^2 = 0 + x(0) + \frac{x^2}{2!} \,.\, 2 + \frac{x^3}{3!} \,.\, (0) + \frac{x^4}{4!}(2.2^2) + \frac{x^5}{5!}(0) + \frac{x^6}{6!}(2.2^2 \,.\, 4^2) + \ldots$$

or $$\frac{(\sin^{-1} x)^2}{2} = \frac{x^2}{2!} + \frac{2^2}{4!}x^4 + \frac{2^2 \,.\, 4^2}{6!}x^6 + \ldots$$ which proves part (*b*).

TEST YOUR KNOWLEDGE

1. Show that $e^{a \sin^{-1} x} = 1 + ax + \frac{a^2x^2}{2!} + \frac{a(a^2+1)}{3!}x^3 + \frac{a^2(a^2+2^2)}{4!}x^4 + \frac{a(a^2+1)(a^2+3^2)}{5!}x^5 + ...$

2. If $y = \cos(m \sin^{-1} x)$, prove that $(y_{n+2})_0 = (n^2 - m^2)(y_n)_0$.
Hence evaluate $(y_n)_0$ and expand $y = \cos m(\sin^{-1} x)$.

3. Prove that $(x + \sqrt{1+x^2})^n = 1 + nx + \frac{n^2x^2}{2!} + n\frac{(n^2-1^2)}{3!}x^3 + \frac{n^2(n^2-2^2)}{4!}x^4 + ...$

Answer

2. If n is odd, $y_n = 0$
If n is even, $y_n = -m^2(2^2 - m^2)(4^2 - m^2) \ldots [(n-2)^2 - m^2]$

$$\cos(m \sin^{-1} x) = 1 - \frac{m^2}{2!}x^2 - \frac{m^2(2^2-m^2)}{4!}x^4 - \frac{m^2(2^2-m^2)(4^2-m^2)}{6!}x^6 - ...$$

3.7. METHOD OF DIFFERENTIATION *AND* INTEGRATION

This method of treatment is best indicated by solved examples.

ILLUSTRATIVE EXAMPLES

Example 1. *Expand $\sin^{-1} x$ in ascending powers of x.*

Sol. $$\frac{d}{dx}(\sin^{-1} x) = \frac{1}{\sqrt{1-x^2}} = (1-x^2)^{-1/2}$$

$$= 1 + \left(-\frac{1}{2}\right)(-x^2) + \frac{-\frac{1}{2}\left(-\frac{3}{2}\right)}{2!}(-x^2)^2 + ...$$

$$= 1 + \frac{1}{2}x^2 + \frac{1 \cdot 3}{2 \cdot 4}x^4 + ...$$

Integrating w.r.t. x within the limits 0 and x

$$\sin^{-1} x = x + \frac{1}{2} \cdot \frac{x^3}{3} + \frac{1 \cdot 3}{2 \cdot 4} \cdot \frac{x^5}{5} + ...$$

Example 2. *Prove that $\log(x + \sqrt{1+x^2}) = x - \frac{1 \cdot x^3}{2 \cdot 3} + \frac{1 \cdot 3}{2 \cdot 4} \cdot \frac{x^5}{5} - ...$*

Sol. $$\frac{d}{dx}[\log(x + \sqrt{1+x^2})] = \frac{1}{x+\sqrt{1+x^2}}\left(1 + \frac{x}{\sqrt{1+x^2}}\right) = \frac{1}{\sqrt{1+x^2}} = (1+x^2)^{-1/2}$$

$$= 1 + \left(-\frac{1}{2}\right) x^2 + \frac{\left(-\frac{1}{2}\right)\left(-\frac{1}{2}-1\right)}{2!} (x^2)^2 \qquad \text{(Expanding by Binomial Theorem)}$$

$$= 1 - \frac{1}{2}x^2 + \frac{1.3}{2.4}x^4 - \ldots$$

Integrating w.r.t. x within the limits 0 to x

$$\log\left(x + \sqrt{1+x^2}\right) = x - \frac{1}{2}\cdot\frac{x^3}{3} + \frac{1.3}{2.4}\cdot\frac{x^5}{5} - \ldots$$

TEST YOUR KNOWLEDGE

1. Obtained the expansion in powers of x of

(a) $\sin^{-1}\dfrac{2x}{1+x^2}$ (b) $\tan^{-1}\dfrac{x}{\sqrt{1-x^2}}$ (c) $\tan^{-1}\dfrac{\sqrt{1+x^2}-\sqrt{1-x^2}}{\sqrt{1+x^2}+\sqrt{1-x^2}}$.

(*U.P.T.U. 2015*)

Answers

1. (a) $2\left(x - \dfrac{x^3}{3} + \dfrac{x^5}{5} - \ldots\right)$ (b) $x + \dfrac{1^2.x^3}{3!} + \dfrac{1^2.3^2}{5!}x^5 + \ldots$ (c) $\dfrac{x^2}{2} + \dfrac{x^6}{12} + \ldots$

EXPANSION OF A FUNCTION OF SEVERAL VARIABLES

3.8. TAYLOR'S THEOREM FOR A FUNCTION OF TWO VARIABLES

We know that by Taylor's theorem for a function $f(x)$ of single variable x,

$$f(x+h) = f(x) + hf'(x) + \frac{h^2}{2!}f''(x) + \frac{h^3}{3!}f'''(x) + \ldots$$

Now let $f(x, y)$ be a function of two independent variables x and y. **If y is kept constant,** then by Taylor's theorem for a function of a single variable x, we have

$$f(x+h, \mathbf{y+k}) = f(x, \mathbf{y+k}) + h\frac{\partial}{\partial x}f(x, \mathbf{y+k}) + \frac{h^2}{2!}\cdot\frac{\partial^2}{\partial x^2}f(x, \mathbf{y+k})$$

$$+ \frac{h^3}{3!}\cdot\frac{\partial^3}{\partial x^3}f(x, \mathbf{y+k}) + \ldots \qquad \ldots(1)$$

Now **keeping x constant** and applying Taylor's theorem for a function of a single variable y, we have

$$f(\mathbf{x}, y+k) = f(\mathbf{x}, y) + k\frac{\partial}{\partial y}f(\mathbf{x}, y) + \frac{k^2}{2!}\cdot\frac{\partial^2}{\partial y^2}f(\mathbf{x}, y) + \frac{k^3}{3!}\cdot\frac{\partial^3}{\partial y^3}f(\mathbf{x}, y) + \ldots \qquad \ldots(2)$$

Using (2), we can write (1) as

$$f(x+h, y+k) = \left[f(x,y) + k\frac{\partial}{\partial y}f(x,y) + \frac{k^2}{2!}\frac{\partial^2}{\partial y^2}f(x,y) + \frac{k^3}{3!}\frac{\partial^3}{\partial y^3}f(x,y) + \ldots\right]$$

$$+ h\frac{\partial}{\partial x}\left[f(x,y) + k\frac{\partial}{\partial y}f(x,y) + \frac{k^2}{2!}\frac{\partial^2}{\partial y^2}f(x,y) + \ldots\right]$$

$$+ \frac{h^2}{2!}\frac{\partial^2}{\partial x^2}\left[f(x,y) + k\frac{\partial}{\partial y}f(x,y) + \ldots\right] + \frac{h^3}{3!}\frac{\partial^3}{\partial x^3}[f(x,y) + \ldots] + \ldots$$

$$= \left[f(x,y) + k\frac{\partial f}{\partial y} + \frac{k^2}{2!}\frac{\partial^2 f}{\partial y^2} + \frac{k^3}{3!}\frac{\partial^3 f}{\partial y^3} + \ldots\right] + \left[h\frac{\partial f}{\partial x} + hk\frac{\partial^2 f}{\partial x \partial y} + \frac{hk^2}{2!}\frac{\partial^3 f}{\partial x \partial y^2} + \ldots\right]$$

$$+ \left[\frac{h^2}{2!}\frac{\partial^2 f}{\partial x^2} + \frac{h^2 k}{2!}\frac{\partial^3 f}{\partial x^2 \partial y} + \ldots\right] + \frac{h^3}{3!}\frac{\partial^3 f}{\partial x^3} + \ldots$$

$$= f(x,y) + \left(h\frac{\partial f}{\partial x} + k\frac{\partial f}{\partial y}\right) + \left(\frac{h^2}{2!}\frac{\partial^2 f}{\partial x^2} + hk\frac{\partial^2 f}{\partial x \partial y} + \frac{k^2}{2!}\frac{\partial^2 f}{\partial y^2}\right)$$

$$+ \left(\frac{h^3}{3!}\frac{\partial^3 f}{\partial x^3} + \frac{h^2 k}{2!}\frac{\partial^3 f}{\partial x^2 \partial y} + \frac{hk^2}{2!}\frac{\partial^3 f}{\partial x \partial y^2} + \frac{k^3}{3!}\frac{\partial^3 f}{\partial y^3}\right) + \ldots$$

$$= f(x,y) + \left(h\frac{\partial f}{\partial x} + k\frac{\partial f}{\partial y}\right) + \frac{1}{2!}\left(h^2\frac{\partial^2 f}{\partial x^2} + 2hk\frac{\partial^2 f}{\partial x \partial y} + k^2\frac{\partial^2 f}{\partial y^2}\right)$$

$$+ \frac{1}{3!}\left(h^3\frac{\partial^3 f}{\partial x^3} + 3h^2 k\frac{\partial^3 f}{\partial x^2 \partial y} + 3hk^2\frac{\partial^3 f}{\partial x \partial y^2} + k^3\frac{\partial^3 f}{\partial y^3}\right) + \ldots$$

$$= f(x,y) + \left(h\frac{\partial}{\partial x} + k\frac{\partial}{\partial y}\right)f + \frac{1}{2!}\left(h\frac{\partial}{\partial x} + k\frac{\partial}{\partial y}\right)^2 f + \frac{1}{3!}\left(h\frac{\partial}{\partial x} + k\frac{\partial}{\partial y}\right)^3 f + \ldots$$

Hence,

$$\mathbf{f(x+h, y+k) = f(x,y) + (hf_x + kf_y) + \frac{1}{2!}(h^2 f_{xx} + 2hkf_{xy} + k^2 f_{yy}) + \frac{1}{3!}(h^3 f_{xxx} + 3h^2 kf_{xxy} + 3hk^2 f_{xyy} + k^3 f_{yyy}) + \ldots}$$

Corollary 1. Putting $x = a$ and $y = b$, we have

$$f(a+h, b+k) = f(a,b) + \{hf_x(a,b) + kf_y(a,b)\} + \frac{1}{2!}\{h^2 f_{xx}(a,b) + 2hkf_{xy}(a,b) + k^2 f_{yy}(a,b)\} + \frac{1}{3!}\{h^3 f_{xxx}(a,b) + 3h^2 kf_{xxy}(a,b) + 3hk^2 f_{xyy}(a,b) + k^3 f_{yyy}(a,b)\} + \ldots$$

Corollary 2. In cor. 1, put $h = x - a$ and $k = y - b$, we have

$$f(x,y) = f(a,b) + \{(x-a)f_x(a,b) + (y-b)f_y(a,b)\} + \frac{1}{2!}\{(x-a)^2 f_{xx}(a,b) + 2(x-a)(y-b) f_{xy}(a,b) + (y-b)^2 f_{yy}(a,b)\} + \ldots$$

Corollary 3. Putting $a = 0$ and $b = 0$ in cor. 2, we have

$$f(x, y) = f(0, 0) + \{(xf_x(0, 0) + yf_y(0, 0)\} + \frac{1}{2!}\{x^2 f_{xx}(0, 0) + 2xyf_{xy}(0, 0) + y^2 f_{yy}(0, 0)\} + \ldots$$

This is called Maclaurin's theorem for two variables.

Note. Corollary 3 is used to expand $f(x, y)$ in powers of x and y [or to expand $f(x, y)$ in the neighbourhood of origin (0, 0). Cor. 2 is used to expand $f(x, y)$ in the neighbourhood of (a, b).

ILLUSTRATIVE EXAMPLES

Example 1. *Expand $e^x \sin y$ in powers of x and y as far as terms of the third degree.*

Sol. Here

$$\begin{aligned}
f(x, y) &= e^x \sin y\,; & f(0, 0) &= 0\\
f_x(x, y) &= e^x \sin y, & f_x(0, 0) &= 0\\
f_y(x, y) &= e^x \cos y, & f_y(0, 0) &= 1\\
f_{xx}(x, y) &= e^x \sin y, & f_{xx}(0, 0) &= 0\\
f_{xy}(x, y) &= e^x \cos y, & f_{xy}(0, 0) &= 1\\
f_{yy}(x, y) &= -e^x \sin y, & f_{yy}(0, 0) &= 0\\
f_{xxx}(x, y) &= e^x \sin y, & f_{xxx}(0, 0) &= 0\\
f_{xxy}(x, y) &= e^x \cos y, & f_{xxy}(0, 0) &= 1\\
f_{xyy}(x, y) &= -e^x \sin y, & f_{xyy}(0, 0) &= 0\\
f_{yyy}(x, y) &= -e^x \cos y, & f_{yyy}(0, 0) &= -1
\end{aligned}$$

....

$$\therefore \quad f(x, y) = f(0, 0) + [xf_x(0, 0) + y\, f_y(0, 0)] + \frac{1}{2!}[x^2 f_{xx}(0, 0) + 2xy\, f_{xy}(0, 0) + y^2 f_{yy}(0, 0)]$$

$$+ \frac{1}{3!}[x^3 f_{xxx}(0, 0) + 3x^2y\, f_{xxy}(0, 0) + 3xy^2 f_{xyy}(0, 0) + y^3 f_{yyy}(0, 0)] + \ldots$$

$$\Rightarrow \quad e^x \sin y = 0 + [x.0 + y.1] + \frac{1}{2!}[x^2.0 + 2xy.1 + y^2.0]$$

$$+ \frac{1}{3!}[x^3.0 + 3x^2y.1 + 3xy^2 + 0 + y^3(-1)] + \ldots$$

$$= y + xy + \frac{1}{2}x^2y - \frac{1}{6}y^3 + \ldots$$

Example 2. *Expand $\tan^{-1}\frac{y}{x}$ in the neighbourhood of (1, 1) upto and inclusive of second degree terms. Hence compute f(1.1, 0.9) approximately.* [A.K.T.U. 2014]

Sol. Here $f(x, y) = \tan^{-1}\frac{y}{x}$, $\qquad f(1, 1) = \tan^{-1} 1 = \frac{\pi}{4}$

$$f_x(x, y) = \frac{1}{1+\frac{y^2}{x^2}}\cdot\left(-\frac{y}{x^2}\right) = -\frac{y}{x^2+y^2}, \qquad f_x(1, 1) = -\frac{1}{2}$$

$$f_y(x, y) = \frac{1}{1+\frac{y^2}{x^2}}\cdot\frac{1}{x} = \frac{x}{x^2+y^2}, \qquad f_y(1, 1) = \frac{1}{2}$$

$$f_{xx}(x, y) = -y(-1)(x^2 + y^2)^{-2}.\ 2x = \frac{2xy}{(x^2+y^2)^2}, \qquad f_{xx}(1, 1) = \frac{1}{2}$$

$$f_{xy}(x, y) = \frac{(x^2+y^2).1 - x.2x}{(x^2+y^2)^2} = \frac{y^2 - x^2}{(x^2+y^2)^2}, \qquad f_{xy}(1, 1) = 0$$

$$f_{yy}(x, y) = x(-1)(x^2 + y^2)^{-2}.\ 2y = -\frac{2xy}{(x^2+y^2)^2}, \qquad f_{yy}(1, 1) = -\frac{1}{2}$$

Similarly,

$$f_{xxx}(1, 1) = -\frac{1}{2}, f_{xxy}(1, 1) = -\frac{1}{2}, \quad f_{xyy}(1, 1) = \frac{1}{2}, \quad f_{yyy}(1, 1) = \frac{1}{2} \text{ and so on.}$$

We know that,

$$f(x, y) = f(1, 1) + [(x-1) f_x(1, 1) + (y-1) f_y(1, 1)] + \frac{1}{2!}[(x-1)^2 f_{xx}(1, 1)$$
$$+ 2(x-1)(y-1) f_{xy}(1, 1) + (y-1)^2 f_{yy}(1, 1)]$$
$$+ \frac{1}{3!}[(x-1)^3 f_{xxx}(1, 1) + 3(x-1)^2(y-1) f_{xxy}(1, 1)$$
$$+ 3(x-1)(y-1)^2 f_{xyy}(1, 1) + (y-1)^3 f_{yyy}(1, 1)] + \ldots$$

$$\therefore \quad \tan^{-1}\left(\frac{y}{x}\right) = \frac{\pi}{4} + \left[(x-1).\left(-\frac{1}{2}\right) + (y-1)\frac{1}{2}\right]$$
$$+ \frac{1}{2!}\left[(x-1)^2.\frac{1}{2} + 2(x-1)(y-1).0 + (y-1)^2.\left(-\frac{1}{2}\right)\right]$$
$$- \frac{1}{12}[(x-1)^3 + 3(x-1)^2(y-1) - 3(x-1)(y-1)^2 - (y-1)^3] + \ldots$$
$$= \frac{\pi}{4} - \frac{1}{2}(x-1) + \frac{1}{2}(y-1) + \frac{1}{4}(x-1)^2 - \frac{1}{4}(y-1)^2$$
$$- \frac{1}{12}[(x-1)^3 + 3(x-1)^2(y-1) - 3(x-1)(y-1)^2 - (y-1)^3] + \ldots$$

Now, $$f(1.1, 0.9) = \frac{\pi}{4} - \frac{1}{2}(.1) + \frac{1}{2}(-.1) + \frac{1}{4}(.1)^2 - \frac{1}{4}(-.1)^2$$
$$- \frac{1}{12}[(.1)^3 + 3(.1)^2(-.1) - 3(.1)(-.1)^2 - (-.1)^3] + \ldots$$
$$= \frac{\pi}{4} - \frac{(.2)}{2} - \frac{1}{12}[.001 - .003 - .003 + .001] + \ldots = 0.6857 \text{ (approx.)}$$

Example 3. *Find the first six terms of the expansions of the function $e^x \log(1+y)$ in a Taylor series in the neighbourhood of the point (0, 0).* (U.P.T.U. 2015)

Sol. Here $f(x, y) = e^x \log(1+y)$, $\qquad f(0, 0) = 0$

$f_x(x, y) = e^x \log(1+y)$, $\qquad f_x(0, 0) = 0$

$f_y(x, y) = \dfrac{e^x}{1+y}$, $\qquad f_y(0, 0) = 1$

$$f_{xx}(x, y) = e^x \log (1 + y), \qquad f_{xx}(0, 0) = 0$$

$$f_{xy}(x, y) = \frac{e^x}{1+y}, \qquad f_{xy}(0, 0) = 1$$

$$f_{yy}(x, y) = -\frac{e^x}{(1+y)^2}, \qquad f_{yy}(0, 0) = -1$$

$$f_{xxx}(x, y) = e^x \log (1 + y), \qquad f_{xxx}(0, 0) = 0$$

$$f_{xxy}(x, y) = \frac{e^x}{1+y}, \qquad f_{xxy}(0, 0) = 1$$

$$f_{xyy}(x, y) = -\frac{e^x}{(1+y)^2}, \qquad f_{xyy}(0, 0) = -1$$

$$f_{yyy}(x, y) = \frac{2e^x}{(1+y)^3}, \qquad f_{yyy}(0, 0) = 2$$

.... and so on.

We know that,

$$f(x, y) = f(0, 0) + [xf_x (0, 0) + yf_y(0, 0)] + \frac{1}{2!} [x^2 f_{xx}(0, 0) + 2xyf_{xy} (0, 0) + y^2 f_{yy}(0, 0)]$$

$$+ \frac{1}{3!} [x^3 f_{xxx}(0, 0) + 3x^2y\, f_{xxy}(0, 0) + 3xy^2 f_{xyy}(0, 0) + y^3 f_{yyy}(0, 0)] + ...$$

$$\therefore \quad e^x \log (1 + y) = 0 + [x.0 + y.1] + \tfrac{1}{2} [x^2.0 + 2xy.1 + y^2(-1)]$$

$$+ \frac{1}{6} [x^3.0 + 3x^2y.1 + 3xy^2.(-1) + y^3.2] + ...$$

$$= y + xy - \tfrac{1}{2}y^2 + \tfrac{1}{2}x^2y - \tfrac{1}{2}xy^2 + \tfrac{1}{3}y^3 + ...$$

Example 4. *Expand* $\frac{(x+h)(y+k)}{x+h+y+k}$ *in powers of h, k up to and inclusive of the second degree terms.*

Sol. Here $\qquad f(x + h, y + k) = \dfrac{(x+h)(y+k)}{x+h+y+k}$

Putting $h = k = 0$, we have $f(x, y) = \dfrac{xy}{x+y}$

$$f_x = \frac{(x+y).y - xy.1}{(x+y)^2} = \frac{y^2}{(x+y)^2}; \qquad f_y = \frac{x^2}{(x+y)^2}, \text{ by symmetry}$$

$$f_{xx} = -\frac{2y^2}{(x+y)^3}, f_{yy} = -\frac{2x^2}{(x+y)^3}; \qquad f_{xy} = \frac{(x+y)^2 . 2x - x^2 . 2(x+y)}{(x+y)^4} = \frac{2xy}{(x+y)^3}$$

$$\therefore \quad \frac{(x+h)(y+k)}{x+h+y+k} = f(x+h, y+k) = f(x, y) + [hf_x + kf_y] + \frac{1}{2!} [h^2 f_{xx} + 2hk\, f_{xy} + k^2 f_{yy}] + ...$$

$$= \frac{xy}{x+y} + \left[h \cdot \frac{y^2}{(x+y)^2} + k \cdot \frac{x^2}{(x+y)^2} \right]$$

$$+ \frac{1}{2}\left[h^2 \cdot \frac{-2y^2}{(x+y)^3} + 2hk \frac{2xy}{(x+y)^3} + k^2 \cdot \frac{-2x^2}{(x+y)^3} \right] + \ldots$$

$$= \frac{xy}{x+y} + \frac{y^2}{(x+y)^2} \cdot h + \frac{x^2}{(x+y)^2} \cdot k - \frac{y^2}{(x+y)^3} \cdot h^2$$

$$+ \frac{2xy}{(x+y)^3} \cdot hk - \frac{x^2}{(x+y)^3} \cdot k^2 + \ldots$$

Example 5. *Expand $x^2y + 3y - 2$ in powers of $(x - 1)$ and $(y + 2)$ using Taylor's Theorem.*

Sol. Expansion of $f(x, y)$ in powers of $(x - a)$ and $(y - b)$ is given by

$$f(x, y) = f(a, b) + [(x-a)f_x(a, b) + (y-b)f_y(a, b)] + \frac{1}{2!}[(x-a)^2 f_{xx}(a, b)$$

$$+ 2(x-a)(y-b) f_{xy}(a, b) + (y-b)^2 f_{yy}(a, b)] + \frac{1}{3!}[(x-a)^3 f_{xxx}(a, b)$$

$$+ 3(x-a)^2(y-b)f_{xxy}(a, b) + 3(x-a)(y-b)^2 f_{xyy}(a, b) + (y-b)^3 f_{yyy}(a, b)] + \ldots \quad \ldots(1)$$

Here $f(x, y) = x^2y + 3y - 2$, $a = 1$, $b = -2$; $f(1, -2) = -10$

$f_x = 2xy,$	$f_x(1, -2) = -4;$	$f_y = x^2 + 3,$	$f_y(1, -2) = 4$
$f_{xx} = 2y,$	$f_{xx}(1, -2) = -4;$	$f_{xy} = 2x,$	$f_{xy}(1, -2) = 2$
$f_{yy} = 0,$	$f_{yy}(1, -2) = 0;$	$f_{xxx} = 0,$	$f_{xxx}(1, -2) = 0$
$f_{xxy} = 2,$	$f_{xxy}(1, -2) = 2;$	$f_{xyy} = 0,$	$f_{xyy}(1, -2) = 0$
$f_{yyy} = 0,$	$f_{yyy}(1, -2) = 0$		

All higher order partial derivatives vanish.

$\therefore$ From (1), we have

$$x^2y + 3y - 1 = f(x, y)$$

$$= -10 + [(x-1)(-4) + (y+2)(4)] + \tfrac{1}{2}[(x-1)^2(-4)$$

$$+ 2(x-1)(y+2)(2) + (y+2)^2(0)] + \tfrac{1}{6}[(x-1)^3(0) + 3(x-1)^2(y+2)(2)$$

$$+ 3(x-1)(y+2)^2(0) + (y+2)^3(0)]$$

$$= -10 - 4(x-1) + 4(y+2) - 2(x-1)^2 + 2(x-1)(y+2) + (x-1)^2(y+2).$$

TEST YOUR KNOWLEDGE

1. Show that $e^y \log(1+x) = x + xy - \dfrac{x^2}{2}$ approximately.

   [**Hint.** Find the expansion at (0, 0)]

2. (*i*) Expand $e^x \cos y$ in powers of x and y as far as the terms of third degree.

   (*ii*) Expand $e^{ax} \sin by$ in powers of x and y as far as the terms of third degree. *(G.B.T.U. 2011)*

3. Expand e^{xy} at (1, 1).

4. Expand $(1 + x + y^2)^{1/2}$ at (1, 0).

5. Expand $e^x \cos y$ at $\left(1, \dfrac{\pi}{4}\right)$.

6. Express the function $f(x, y) = x^2 + 3y^2 - 9x - 9y + 26$ as Taylor's series expansion about the point (1, 2). *(A.K.T.U. 2017, 2018)*

7. Expand $(x^2y + \sin y + e^x)$ in powers of $(x - 1)$ and $(y - \pi)$. *(U.P.T.U. 2014)*

8. Expand $\sin xy$ in powers of $(x - 1)$ and $(y - \pi/2)$ up to second degree terms. *[U.K.T.U. 2011]*

9. Expand $f(x, y) = y^x$ about (1, 1) upto second degree terms and hence evaluate $(1.02)^{1.03}$. *(M.T.U. 2013)*

10. Expand x^y in powers of $(x - 1)$ and $(y - 1)$ up to the third degree terms and hence evaluate $(1.1)^{1.02}$. *(A.K.T.U. 2022)*

11. Expand $f(x, y) = e^x \tan^{-1} y$ in powers of $(x - 1)$ and $(y - 1)$ up to two terms of degree 2. *(G.B.T.U. 2013)*

12. Expand $\sin(x + h)(y + k)$ by Taylor's Theorem.

Answers

2. (*i*) $1 + x + \dfrac{1}{2!}(x^2 - y^2) + \dfrac{1}{3!}(x^3 - 3xy^2) + \ldots$ (*ii*) $by + abxy + \dfrac{1}{3!}(3a^2bx^2y - b^3y^3) + \ldots$

3. $e\left\{1 + (x-1) + (y-1) + \dfrac{1}{2!}\left((x-1)^2 + 4(x-1)(y-1) + (y-1)^2\right) + \ldots\right\}$

4. $\sqrt{2}\left\{1 + \dfrac{x-1}{4} - \dfrac{(x-1)^2}{32} + \dfrac{y^2}{4} + \ldots\right\}$

5. $\dfrac{e}{\sqrt{2}}\left\{1 + (x-1) - \left(y - \dfrac{\pi}{4}\right) + \dfrac{(x-1)^2}{2} - (x-1)\left(y - \dfrac{\pi}{4}\right) - \dfrac{1}{2}\left(y - \dfrac{\pi}{4}\right)^2 + \ldots\right\}$

6. $f(x, y) = 12 - 7(x-1) + 3(y-2) + (x-1)^2 + 3(y-2)^2$

7. $\pi + e + (x-1)(2\pi + e) + \dfrac{1}{2}(x-1)^2(2\pi + e) + 2(x-1)(y-\pi) + \ldots$

8. $1 - \dfrac{\pi^2}{8}(x-1)^2 - \dfrac{\pi}{2}(x-1)\left(y - \dfrac{\pi}{2}\right) - \dfrac{1}{2}\left(y - \dfrac{\pi}{2}\right)^2$

9. $y^x = 1 + (y-1) + (x-1)(y-1) + \dfrac{1}{2}(x-1)(y-1)^2 + \ldots$; $(1.02)^{1.03} = 1.020606$

10. $1 + (x-1) + (x-1)(y-1) + \dfrac{1}{2}(x-1)^2(y-1) + \ldots$; 1.1021

11. $e^x \tan^{-1} y = \frac{\pi}{4} e + \frac{e}{2}\left[(x-1)\frac{\pi}{2} + (y-1)\right] + \frac{e}{4}\left[(x-1)^2 \frac{\pi}{2} + 2(x-1)(y-1) - (y-1)^2\right] + \ldots$.

12. $\sin xy + (hy + kx)\cos xy + hk \cos xy - \frac{1}{2}(hy + kx)^2 \sin xy + \ldots$

JACOBIANS

Jacobian is a functional determinant, useful in transformation of variables from cartesian to polar, cylindrical and spherical polar co-ordinates in multiple integrals. Jacobian is named after the German mathematician Carl Gustav Jacob Jacobi (1804–1851) who made significant contributions to mechanics, partial differential equations, astronomy, elliptic functions and the calculus of variations.

3.9. DEFINITIONS

(*i*) If u and v are the functions of two independent variables x and y, then the determinant

$$\begin{vmatrix} \frac{\partial u}{\partial x} & \frac{\partial u}{\partial y} \\ \frac{\partial v}{\partial x} & \frac{\partial v}{\partial y} \end{vmatrix}$$

is called the **Jacobian** or functional determinant of u and v with respect to x and y. It is written as $\frac{\partial(u, v)}{\partial(x, y)}$ or $J(u, v)$.

(*ii*) If u, v and w are the functions of three independent variables x, y and z, then the determinant

$$\begin{vmatrix} \frac{\partial u}{\partial x} & \frac{\partial u}{\partial y} & \frac{\partial u}{\partial z} \\ \frac{\partial v}{\partial x} & \frac{\partial v}{\partial y} & \frac{\partial v}{\partial z} \\ \frac{\partial w}{\partial x} & \frac{\partial w}{\partial y} & \frac{\partial w}{\partial z} \end{vmatrix}$$

is called the Jacobian of u, v and w with respect to x, y and z. It is written as

$$\frac{\partial(u, v, w)}{\partial(x, y, z)} \quad \text{or} \quad J(u, v, w).$$

(*iii*) Similarly if $u_1, u_2, \ldots, u_n$ are the functions of independent variables $x_1, x_2, \ldots, x_n$, then the determinant

$$\begin{vmatrix} \frac{\partial u_1}{\partial x_1} & \frac{\partial u_1}{\partial x_2} & \frac{\partial u_1}{\partial x_3} & \cdots & \frac{\partial u_1}{\partial x_n} \\ \frac{\partial u_2}{\partial x_1} & \frac{\partial u_2}{\partial x_2} & \frac{\partial u_2}{\partial x_3} & \cdots & \frac{\partial u_2}{\partial x_n} \\ \cdots & \cdots & \cdots & \cdots & \cdots \\ \cdots & \cdots & \cdots & \cdots & \cdots \\ \frac{\partial u_n}{\partial x_1} & \frac{\partial u_n}{\partial x_2} & \frac{\partial u_n}{\partial x_3} & \cdots & \frac{\partial u_n}{\partial x_n} \end{vmatrix}$$

is called the Jacobian of $u_1, u_2, ..., u_n$ with respect to variables $x_1, x_2, ..., x_n$. It is written as

$$\frac{\partial(u_1, u_2, \ldots, u_n)}{\partial(x_1, x_2, \ldots, x_n)} \quad \text{or} \quad J(u_1, u_2, \ldots, u_n).$$

3.10. PROPERTIES OF JACOBIANS (Chain Rules)

I. If u, v are functions of r, s where r, s are functions of x, y then,

$$\boxed{\frac{\partial(u, v)}{\partial(x, y)} = \frac{\partial(u, v)}{\partial(r, s)} \times \frac{\partial(r, s)}{\partial(x, y)}}$$

Proof. Since u, v are composite functions of x, y

$$\therefore \left.\begin{aligned} \frac{\partial u}{\partial x} &= \frac{\partial u}{\partial r}\cdot\frac{\partial r}{\partial x} + \frac{\partial u}{\partial s}\cdot\frac{\partial s}{\partial x} = u_r r_x + u_s s_x \\ \frac{\partial u}{\partial y} &= \frac{\partial u}{\partial r}\cdot\frac{\partial r}{\partial y} + \frac{\partial u}{\partial s}\cdot\frac{\partial s}{\partial y} = u_r r_y + u_s s_y \\ \frac{\partial v}{\partial x} &= \frac{\partial v}{\partial r}\cdot\frac{\partial r}{\partial x} + \frac{\partial v}{\partial s}\cdot\frac{\partial s}{\partial x} = v_r r_x + v_s s_x \\ \frac{\partial v}{\partial y} &= \frac{\partial v}{\partial r}\cdot\frac{\partial r}{\partial y} + \frac{\partial v}{\partial s}\cdot\frac{\partial s}{\partial y} = v_r r_y + v_s s_y \end{aligned}\right\} \quad \ldots(1)$$

Now, $$\frac{\partial(u, v)}{\partial(r, s)} \times \frac{\partial(r, s)}{\partial(x, y)} = \begin{vmatrix} u_r & u_s \\ v_r & v_s \end{vmatrix} \cdot \begin{vmatrix} r_x & r_y \\ s_x & s_y \end{vmatrix}$$

$$= \begin{vmatrix} u_r r_x + u_s s_x & u_r r_y + u_s s_y \\ v_r r_x + v_s s_x & v_r r_y + v_s s_y \end{vmatrix} = \begin{vmatrix} \frac{\partial u}{\partial x} & \frac{\partial u}{\partial y} \\ \frac{\partial v}{\partial x} & \frac{\partial v}{\partial y} \end{vmatrix} \quad \text{[Using (1)]}$$

$$= \frac{\partial(u, v)}{\partial(x, y)}.$$

Note. If u_1, u_2, u_3 are functions of y_1, y_2, y_3 and y_1, y_2, y_3 are functions of x_1, x_2, x_3 then

$$\frac{\partial(u_1, u_2, u_3)}{\partial(x_1, x_2, x_3)} = \frac{\partial(u_1, u_2, u_3)}{\partial(y_1, y_2, y_3)} \cdot \frac{\partial(y_1, y_2, y_3)}{\partial(x_1, x_2, x_3)}$$

II. If J_1 is the Jacobian of u, v with respect to x, y and J_2 is the Jacobian of x, y with respect to u, v then

$$J_1 J_2 = 1, \text{ i.e., } \boxed{\frac{\partial(u, v)}{\partial(x, y)} \cdot \frac{\partial(x, y)}{\partial(u, v)} = 1} \qquad (A.K.T.U.\ 2014)$$

Proof. Let $u = u(x, y)$ and $v = v(x, y)$ so that u and v are functions of x, y.

Differentiating partially w.r.t. u and v, we get

$$\left.\begin{aligned}
1 &= \frac{\partial u}{\partial x}\cdot\frac{\partial x}{\partial u}+\frac{\partial u}{\partial y}\cdot\frac{\partial y}{\partial u} = u_x x_u + u_y y_u \\
0 &= \frac{\partial u}{\partial x}\cdot\frac{\partial x}{\partial v}+\frac{\partial u}{\partial y}\cdot\frac{\partial y}{\partial v} = u_x x_v + u_y y_v \\
0 &= \frac{\partial v}{\partial x}\cdot\frac{\partial x}{\partial u}+\frac{\partial v}{\partial y}\cdot\frac{\partial y}{\partial u} = v_x x_u + v_y y_u \\
1 &= \frac{\partial v}{\partial x}\cdot\frac{\partial x}{\partial v}+\frac{\partial v}{\partial y}\cdot\frac{\partial y}{\partial v} = v_x x_v + v_y y_v
\end{aligned}\right\} \quad ...(1)$$

Now, $\dfrac{\partial(u, v)}{\partial(x, y)}\cdot\dfrac{\partial(x, y)}{\partial(u, v)} = \begin{vmatrix} u_x & u_y \\ v_x & v_y \end{vmatrix}\begin{vmatrix} x_u & x_v \\ y_u & y_v \end{vmatrix} = \begin{vmatrix} u_x x_u + u_y y_u & u_x x_v + u_y y_v \\ v_x x_u + v_y y_u & v_x x_v + v_y y_v \end{vmatrix}$

$$= \begin{vmatrix} 1 & 0 \\ 0 & 1 \end{vmatrix} = 1. \qquad \text{[Using (1)]}$$

Note. It can be extended to three variables as

$$\frac{\partial(u, v, w)}{\partial(x, y, z)}\cdot\frac{\partial(x, y, z)}{\partial(u, v, w)} = 1.$$

3.11. THEOREMS ON JACOBIANS

Theorem 1. *If the relations connecting u_i's and x_i's are of the form,*

$$\begin{aligned}
u_1 &= f_1(x_1) \\
u_2 &= f_2(x_1, x_2) \\
&\ \vdots \\
&\ \vdots \\
u_n &= f_n(x_1, x_2, ..., x_n)
\end{aligned}$$

then $$\frac{\partial(u_1, u_2, ..., u_n)}{\partial(x_1, x_2, ..., x_n)} = \frac{\partial u_1}{\partial x_1}\cdot\frac{\partial u_2}{\partial x_2}\cdot\frac{\partial u_3}{\partial x_3}\cdots\frac{\partial u_n}{\partial x_n}.$$

Proof. We know that

$$\frac{\partial(u_1, u_2, ..., u_n)}{\partial(x_1, x_2, ..., x_n)} = \begin{vmatrix}
\frac{\partial u_1}{\partial x_1} & \frac{\partial u_1}{\partial x_2} & \frac{\partial u_1}{\partial x_3} & \cdots & \frac{\partial u_1}{\partial x_n} \\
\frac{\partial u_2}{\partial x_1} & \frac{\partial u_2}{\partial x_2} & \frac{\partial u_2}{\partial x_3} & \cdots & \frac{\partial u_2}{\partial x_n} \\
\cdots & \cdots & \cdots & \cdots & \cdots \\
\cdots & \cdots & \cdots & \cdots & \cdots \\
\frac{\partial u_n}{\partial x_1} & \frac{\partial u_n}{\partial x_2} & \frac{\partial u_n}{\partial x_3} & \cdots & \frac{\partial u_n}{\partial x_n}
\end{vmatrix} \quad ...(1)$$

u_1 is function of x_1 only therefore

$$\frac{\partial u_1}{\partial x_2} = 0, \frac{\partial u_1}{\partial x_3} = 0, \dots, \frac{\partial u_1}{\partial x_n} = 0$$

u_2 is function of x_1 and x_2 hence only $\frac{\partial u_2}{\partial x_1}$ and $\frac{\partial u_2}{\partial x_2}$ will exist

$$\text{the rest } \frac{\partial u_2}{\partial x_3} = 0, \frac{\partial u_2}{\partial x_4} = 0, \dots, \frac{\partial u_2}{\partial x_n} = 0$$

u_3 is function of x_1, x_2 and x_3, therefore $\frac{\partial u_3}{\partial x_1}, \frac{\partial u_3}{\partial x_2}, \frac{\partial u_3}{\partial x_3}$ will exist and the rest will be zero.

u_n is function of x_1, x_2, ..., x_n therefore all $\frac{\partial u_n}{\partial x_1}, \frac{\partial u_n}{\partial x_2}, \dots, \frac{\partial u_n}{\partial x_n}$ will exist.

Putting these values in (1), we get

$$\frac{\partial(u_1, u_2, u_3, \dots, u_n)}{\partial(x_1, x_2, x_3, \dots, x_n)} = \begin{vmatrix} \frac{\partial u_1}{\partial x_1} & 0 & 0 & \dots & 0 \\ \frac{\partial u_2}{\partial x_1} & \frac{\partial u_2}{\partial x_2} & 0 & \dots & 0 \\ \frac{\partial u_3}{\partial x_1} & \frac{\partial u_3}{\partial x_2} & \frac{\partial u_3}{\partial x_3} & \dots & 0 \\ \dots & \dots & \dots & \dots & \dots \\ \dots & \dots & \dots & \dots & \dots \\ \frac{\partial u_n}{\partial x_1} & \frac{\partial u_n}{\partial x_2} & \frac{\partial u_n}{\partial x_3} & \dots & \frac{\partial u_n}{\partial x_n} \end{vmatrix}$$

Expanding the determinant in terms of first row, we get

$$\boxed{\frac{\partial(u_1, u_2, \dots, u_n)}{\partial(x_1, x_2, \dots, x_n)} = \frac{\partial u_1}{\partial x_1} \cdot \frac{\partial u_2}{\partial x_2} \cdot \frac{\partial u_3}{\partial x_3} \cdots \frac{\partial u_n}{\partial x_n}.}$$

Theorem 2. *If functions u, v, w of three independent variables x, y, z are not independent then the Jacobian of u, v, w with respect to x, y, z vanishes.*

Proof. It is given that u, v and w are not independent variables, then there will be a relation $F(u, v, w) = 0$, which will connect these independent variables.

Differentiating this relation with respect to x, y and z, we get

$$\frac{\partial F}{\partial u} \cdot \frac{\partial u}{\partial x} + \frac{\partial F}{\partial v} \cdot \frac{\partial v}{\partial x} + \frac{\partial F}{\partial w} \cdot \frac{\partial w}{\partial x} = 0 \qquad \text{...(1)}$$

$$\frac{\partial F}{\partial u} \cdot \frac{\partial u}{\partial y} + \frac{\partial F}{\partial v} \cdot \frac{\partial v}{\partial y} + \frac{\partial F}{\partial w} \cdot \frac{\partial w}{\partial y} = 0 \qquad \text{...(2)}$$

$$\frac{\partial F}{\partial u} \cdot \frac{\partial u}{\partial z} + \frac{\partial F}{\partial v} \cdot \frac{\partial v}{\partial z} + \frac{\partial F}{\partial w} \cdot \frac{\partial w}{\partial z} = 0 \qquad \text{...(3)}$$

Eliminating $\frac{\partial F}{\partial u}, \frac{\partial F}{\partial v}$ and $\frac{\partial F}{\partial w}$ from (1), (2) and (3), we get

$$\begin{vmatrix} \frac{\partial u}{\partial x} & \frac{\partial v}{\partial x} & \frac{\partial w}{\partial x} \\ \frac{\partial u}{\partial y} & \frac{\partial v}{\partial y} & \frac{\partial w}{\partial y} \\ \frac{\partial u}{\partial z} & \frac{\partial v}{\partial z} & \frac{\partial w}{\partial z} \end{vmatrix} = 0$$

or $$\boxed{\frac{\partial(u, v, w)}{\partial(x, y, z)} = 0.}$$

3.12. JACOBIAN OF IMPLICIT FUNCTIONS

If u_1, u_2 and u_3 are the implicit functions of x_1, x_2, x_3 *i.e.*,

$$F_1(u_1, u_2, u_3, x_1, x_2, x_3) = 0$$
$$F_2(u_1, u_2, u_3, x_1, x_2, x_3) = 0$$
$$F_3(u_1, u_2, u_3, x_1, x_2, x_3) = 0$$

then, $$\boxed{\frac{\partial(u_1, u_2, u_3)}{\partial(x_1, x_2, x_3)} = (-1)^3 \left[\frac{\partial(F_1, F_2, F_3)}{\partial(x_1, x_2, x_3)} \Big/ \frac{\partial(F_1, F_2, F_3)}{\partial(u_1, u_2, u_3)}\right]}$$

Proof. Differentiating F_1, F_2 and F_3 with respect to x_1, x_2 and x_3, we get

$$\frac{\partial F_1}{\partial x_1} + \frac{\partial F_1}{\partial u_1} \cdot \frac{\partial u_1}{\partial x_1} + \frac{\partial F_1}{\partial u_2} \cdot \frac{\partial u_2}{\partial x_1} + \frac{\partial F_1}{\partial u_3} \cdot \frac{\partial u_3}{\partial x_1} = 0 \Rightarrow \sum_{r=1}^{3} \frac{\partial F_1}{\partial u_r} \cdot \frac{\partial u_r}{\partial x_1} = -\frac{\partial F_1}{\partial x_1}$$

$$\frac{\partial F_1}{\partial x_2} + \frac{\partial F_1}{\partial u_1} \cdot \frac{\partial u_1}{\partial x_2} + \frac{\partial F_1}{\partial u_2} \cdot \frac{\partial u_2}{\partial x_2} + \frac{\partial F_1}{\partial u_3} \cdot \frac{\partial u_3}{\partial x_3} = 0 \Rightarrow \sum_{r=1}^{3} \frac{\partial F_1}{\partial u_r} \cdot \frac{\partial u_r}{\partial x_2} = -\frac{\partial F_1}{\partial x_2}$$

$$\frac{\partial F_1}{\partial x_3} + \frac{\partial F_1}{\partial u_1} \cdot \frac{\partial u_1}{\partial x_3} + \frac{\partial F_1}{\partial u_2} \cdot \frac{\partial u_2}{\partial x_3} + \frac{\partial F_1}{\partial u_3} \cdot \frac{\partial u_3}{\partial x_3} = 0 \Rightarrow \sum_{r=1}^{3} \frac{\partial F_1}{\partial u_r} \cdot \frac{\partial u_r}{\partial x_3} = -\frac{\partial F_1}{\partial x_3}.$$

Similarly, $\sum_{r=1}^{3} \frac{\partial F_2}{\partial u_r} \cdot \frac{\partial u_r}{\partial x_1} = -\frac{\partial F_2}{\partial x_1}, \sum_{r=1}^{3} \frac{\partial F_2}{\partial u_r} \cdot \frac{\partial u_r}{\partial x_2} = -\frac{\partial F_2}{\partial x_2}, \sum_{r=1}^{3} \frac{\partial F_2}{\partial u_r} \cdot \frac{\partial u_r}{\partial x_3} = -\frac{\partial F_2}{\partial x_3}$

Also, $\sum_{r=1}^{3} \frac{\partial F_3}{\partial u_r} \cdot \frac{\partial u_r}{\partial x_1} = -\frac{\partial F_3}{\partial x_1}, \sum_{r=1}^{3} \frac{\partial F_3}{\partial u_r} \cdot \frac{\partial u_r}{\partial x_2} = -\frac{\partial F_3}{\partial x_2}$ and $\sum_{r=1}^{3} \frac{\partial F_3}{\partial u_r} \cdot \frac{\partial u_r}{\partial x_3} = -\frac{\partial F_3}{\partial x_3}$

Now, $$\frac{\partial(F_1, F_2, F_3)}{\partial(u_1, u_2, u_3)} \times \frac{\partial(u_1, u_2, u_3)}{\partial(x_1, x_2, x_3)} = \begin{vmatrix} \frac{\partial F_1}{\partial u_1} & \frac{\partial F_1}{\partial u_2} & \frac{\partial F_1}{\partial u_3} \\ \frac{\partial F_2}{\partial u_1} & \frac{\partial F_2}{\partial u_2} & \frac{\partial F_2}{\partial u_3} \\ \frac{\partial F_3}{\partial u_1} & \frac{\partial F_3}{\partial u_2} & \frac{\partial F_3}{\partial u_3} \end{vmatrix} \times \begin{vmatrix} \frac{\partial u_1}{\partial x_1} & \frac{\partial u_1}{\partial x_2} & \frac{\partial u_1}{\partial x_3} \\ \frac{\partial u_2}{\partial x_1} & \frac{\partial u_2}{\partial x_2} & \frac{\partial u_2}{\partial x_3} \\ \frac{\partial u_3}{\partial x_1} & \frac{\partial u_3}{\partial x_2} & \frac{\partial u_3}{\partial x_3} \end{vmatrix}$$

$$= \begin{vmatrix} \Sigma \frac{\partial F_1}{\partial u_r} \cdot \frac{\partial u_r}{\partial x_1} & \Sigma \frac{\partial F_1}{\partial u_r} \cdot \frac{\partial u_r}{\partial x_2} & \Sigma \frac{\partial F_1}{\partial u_r} \cdot \frac{\partial u_r}{\partial x_3} \\ \Sigma \frac{\partial F_2}{\partial u_r} \cdot \frac{\partial u_r}{\partial x_1} & \Sigma \frac{\partial F_2}{\partial u_r} \cdot \frac{\partial u_r}{\partial x_2} & \Sigma \frac{\partial F_2}{\partial u_r} \cdot \frac{\partial u_r}{\partial x_3} \\ \Sigma \frac{\partial F_3}{\partial u_r} \cdot \frac{\partial u_r}{\partial x_1} & \Sigma \frac{\partial F_3}{\partial u_r} \cdot \frac{\partial u_r}{\partial x_2} & \Sigma \frac{\partial F_3}{\partial u_r} \cdot \frac{\partial u_r}{\partial x_3} \end{vmatrix}$$

$$= \begin{vmatrix} -\frac{\partial F_1}{\partial x_1} & -\frac{\partial F_1}{\partial x_2} & -\frac{\partial F_1}{\partial x_3} \\ -\frac{\partial F_2}{\partial x_1} & -\frac{\partial F_2}{\partial x_2} & -\frac{\partial F_2}{\partial x_3} \\ -\frac{\partial F_3}{\partial x_1} & -\frac{\partial F_3}{\partial x_2} & -\frac{\partial F_3}{\partial x_3} \end{vmatrix} \qquad \left| \text{Putting values of each summation} \right.$$

$$= (-1)^3 \begin{vmatrix} \frac{\partial F_1}{\partial x_1} & \frac{\partial F_1}{\partial x_2} & \frac{\partial F_1}{\partial x_3} \\ \frac{\partial F_2}{\partial x_1} & \frac{\partial F_2}{\partial x_2} & \frac{\partial F_2}{\partial x_3} \\ \frac{\partial F_3}{\partial x_1} & \frac{\partial F_3}{\partial x_2} & \frac{\partial F_3}{\partial x_3} \end{vmatrix} = (-1)^3 \frac{\partial(F_1, F_2, F_3)}{\partial(x_1, x_2, x_3)}$$

Hence,
$$\boxed{\frac{\partial(u_1, u_2, u_3)}{\partial(x_1, x_2, x_3)} = (-1)^3 \left[\frac{\partial(F_1, F_2, F_3)}{\partial(x_1, x_2, x_3)} \bigg/ \frac{\partial(F_1, F_2, F_3)}{\partial(u_1, u_2, u_3)} \right]}$$

3.13. FUNCTIONAL RELATIONSHIP [*M.T.U. (SUM) 2011*]

"Let u_1, u_2, u_3, ..., u_n be functions of x_1, x_2, x_3, ..., x_n. Then the necessary condition for the existence of a relation of the form $F(u_1, u_2, ..., u_n) = 0$ is that the Jacobian $\frac{\partial(u_1, u_2, ..., u_n)}{\partial(x_1, x_2, ..., x_n)}$ should vanish identically."

Proof. Let there exist a relation $F(u_1, u_2, ..., u_n) = 0$. ...(1)

We are to show that $\frac{\partial(u_1, u_2, ..., u_n)}{\partial(x_1, x_2, ..., x_n)} = 0.$

Differentiating (1) partially with respect to $x_1, x_2, ..., x_n$, we get

with respect to x_1,

$$\frac{\partial F}{\partial u_1} \cdot \frac{\partial u_1}{\partial x_1} + \frac{\partial F}{\partial u_2} \cdot \frac{\partial u_2}{\partial x_1} + ... + \frac{\partial F}{\partial u_n} \cdot \frac{\partial u_n}{\partial x_1} = 0$$

with respect to x_2,

$$\frac{\partial F}{\partial u_1} \cdot \frac{\partial u_1}{\partial x_2} + \frac{\partial F}{\partial u_2} \cdot \frac{\partial u_2}{\partial x_2} + ... + \frac{\partial F}{\partial u_n} \cdot \frac{\partial u_n}{\partial x_2} = 0$$

... ...

... ...

with respect to x_n,

$$\frac{\partial F}{\partial u_1}\cdot\frac{\partial u_1}{\partial x_n}+\frac{\partial F}{\partial u_2}\cdot\frac{\partial u_2}{\partial x_n}+\ldots+\frac{\partial F}{\partial u_n}\cdot\frac{\partial u_n}{\partial x_n}=0$$

Since, we cannot have at the same time

$$\frac{\partial F}{\partial u_1}=\frac{\partial F}{\partial u_2}=\frac{\partial F}{\partial u_3}=\ldots=\frac{\partial F}{\partial u_n}=0$$

otherwise relation (1) will reduce to a trivial identity, hence eliminating $\frac{\partial F}{\partial u_1}, \frac{\partial F}{\partial u_2}, \ldots, \frac{\partial F}{\partial u_n}$ from these above equations, we get

$$\begin{vmatrix} \frac{\partial u_1}{\partial x_1} & \frac{\partial u_2}{\partial x_1} & \frac{\partial u_3}{\partial x_1} & \cdots & \frac{\partial u_n}{\partial x_1} \\ \frac{\partial u_1}{\partial x_2} & \frac{\partial u_2}{\partial x_2} & \frac{\partial u_3}{\partial x_2} & \cdots & \frac{\partial u_n}{\partial x_2} \\ \frac{\partial u_1}{\partial x_3} & \frac{\partial u_2}{\partial x_3} & \frac{\partial u_3}{\partial x_3} & \cdots & \frac{\partial u_n}{\partial x_3} \\ \cdots & \cdots & \cdots & \cdots & \cdots \\ \cdots & \cdots & \cdots & \cdots & \cdots \\ \frac{\partial u_1}{\partial x_n} & \frac{\partial u_2}{\partial x_n} & \frac{\partial u_3}{\partial x_n} & \cdots & \frac{\partial u_n}{\partial x_n} \end{vmatrix} = 0$$

or

$$\frac{\partial(u_1, u_2, \ldots, u_n)}{\partial(x_1, x_2, \ldots, x_n)} = 0.$$

ILLUSTRATIVE EXAMPLES

Example 1. *If* $x = r\cos\theta$, $y = r\sin\theta$ *find* $\frac{\partial(x, y)}{\partial(r, \theta)}$ *and* $\frac{\partial(r, \theta)}{\partial(x, y)}$. *Also prove that* $JJ' = 1$.

(U.P.T.U. 2014)

Sol.

$$\frac{\partial(x, y)}{\partial(r, \theta)} = \begin{vmatrix} \frac{\partial x}{\partial r} & \frac{\partial x}{\partial \theta} \\ \frac{\partial y}{\partial r} & \frac{\partial y}{\partial \theta} \end{vmatrix} = \begin{vmatrix} \cos\theta & -r\sin\theta \\ \sin\theta & r\cos\theta \end{vmatrix} = r(\cos^2\theta + \sin^2\theta) = r$$

Now,

$$r = \sqrt{x^2 + y^2} \quad \text{and} \quad \theta = \tan^{-1}\frac{y}{x}$$

$$\therefore \quad \frac{\partial(r, \theta)}{\partial(x, y)} = \begin{vmatrix} \frac{\partial r}{\partial x} & \frac{\partial r}{\partial y} \\ \frac{\partial \theta}{\partial x} & \frac{\partial \theta}{\partial y} \end{vmatrix} = \begin{vmatrix} \frac{x}{\sqrt{x^2 + y^2}} & \frac{y}{\sqrt{x^2 + y^2}} \\ \frac{-y}{x^2 + y^2} & \frac{x}{x^2 + y^2} \end{vmatrix} = \frac{1}{\sqrt{x^2 + y^2}} = \frac{1}{r}$$

Hence, $\dfrac{\partial(x, y)}{\partial(r, \theta)} \cdot \dfrac{\partial(r, \theta)}{\partial(x, y)} = r.\dfrac{1}{r} = 1$

$\Rightarrow$ $JJ' = 1.$

Example 2. *If $x = r \sin \theta \cos \phi$, $y = r \sin \theta \sin \phi$, $z = r \cos \theta$, show that*

$$\frac{\partial(x, y, z)}{\partial(r, \theta, \phi)} = r^2 \sin \theta \text{ and find } \frac{\partial(r, \theta, \phi)}{\partial(x, y, z)}.$$

Sol. $$\frac{\partial(x, y, z)}{\partial(r, \theta, \phi)} = \begin{vmatrix} \frac{\partial x}{\partial r} & \frac{\partial x}{\partial \theta} & \frac{\partial x}{\partial \phi} \\ \frac{\partial y}{\partial r} & \frac{\partial y}{\partial \theta} & \frac{\partial y}{\partial \phi} \\ \frac{\partial z}{\partial r} & \frac{\partial z}{\partial \theta} & \frac{\partial z}{\partial \phi} \end{vmatrix} = \begin{vmatrix} \sin \theta \cos \phi & r \cos \theta \cos \phi & -r \sin \theta \sin \phi \\ \sin \theta \sin \phi & r \cos \theta \sin \phi & r \sin \theta \cos \phi \\ \cos \theta & -r \sin \theta & 0 \end{vmatrix}$$

$$= r^2 \sin \theta \begin{vmatrix} \sin \theta \cos \phi & \cos \theta \cos \phi & -\sin \phi \\ \sin \theta \sin \phi & \cos \theta \sin \phi & \cos \phi \\ \cos \theta & -\sin \theta & 0 \end{vmatrix}$$

$$= r^2 \sin \theta \left\{ \cos \theta \begin{vmatrix} \cos \theta \cos \phi & -\sin \phi \\ \cos \theta \sin \phi & \cos \phi \end{vmatrix} + \sin \theta \begin{vmatrix} \sin \theta \cos \phi & -\sin \phi \\ \sin \theta \sin \phi & \cos \phi \end{vmatrix} \right\}$$

$$= r^2 \sin \theta \, [\cos \theta \, (\cos \theta \cos^2 \phi + \cos \theta \sin^2 \phi) + \sin \theta \, (\sin \theta \cos^2 \phi + \sin \theta \sin^2 \phi)]$$

$$= r^2 \sin \theta \, (\cos^2 \theta + \sin^2 \theta) = r^2 \sin \theta$$

$\therefore$ $\dfrac{\partial(r, \theta, \phi)}{\partial(x, y, z)} = \dfrac{1}{r^2 \sin \theta}$ | Using II property

Example 3. *If $x = a \cosh \alpha \cos \beta$, $y = a \sinh \alpha \sin \beta$, then show that*

$$\frac{\partial(x, y)}{\partial(\alpha, \beta)} = \frac{a^2}{2} [\cosh 2\alpha - \cos 2\beta].$$

Sol. $$\frac{\partial(x, y)}{\partial(\alpha, \beta)} = \begin{vmatrix} \frac{\partial x}{\partial \alpha} & \frac{\partial x}{\partial \beta} \\ \frac{\partial y}{\partial \alpha} & \frac{\partial y}{\partial \beta} \end{vmatrix} \quad \ldots(1)$$

Differentiating partially with respect to α and β the equations

$$x = a \cosh \alpha \cos \beta \quad \ldots(2)$$

and $$y = a \sinh \alpha \sin \beta \quad \ldots(3)$$

we get $$\frac{\partial x}{\partial \alpha} = a \sinh \alpha \cos \beta, \quad \frac{\partial x}{\partial \beta} = -a \cosh \alpha \sin \beta$$

$$\frac{\partial y}{\partial \alpha} = a \cosh \alpha \sin \beta \text{ and } \frac{\partial y}{\partial \beta} = a \sinh \alpha \cos \beta$$

Putting in (1)

$$\frac{\partial(x, y)}{\partial(\alpha, \beta)} = \begin{vmatrix} a \sinh \alpha \cos \beta & -a \cosh \alpha \sin \beta \\ a \cosh \alpha \sin \beta & a \sinh \alpha \cos \beta \end{vmatrix}$$

$$= a^2 (\sinh^2 \alpha \cos^2 \beta + \cosh^2 \alpha \sin^2 \beta)$$

$$= a^2 [(\cosh^2 \alpha - 1) \cos^2 \beta + \cosh^2 \alpha (1 - \cos^2 \beta)$$

$$= a^2 [\cosh^2 \alpha - \cos^2 \beta] = \frac{a^2}{2} [\cosh 2\alpha + 1 - 1 - \cos 2\beta]$$

$$= \frac{a^2}{2} [\cosh 2\alpha - \cos 2\beta].$$

Example 4. (*i*) *Calculate the Jacobian* $\frac{\partial(u, v, w)}{\partial(x, y, z)}$ *of the following:*

$$u = x + 2y + z,\ v = x + 2y + 3z,\ w = 2x + 3y + 5z$$

(*ii*) *If* $u = xyz$, $v = xy + yz + zx$, $w = x + y + z$, *then compute the Jacobian* $\frac{\partial(u, v, w)}{\partial(x, y, z)}$.

Sol. (*i*)

$$\frac{\partial(u, v, w)}{\partial(x, y, z)} = \begin{vmatrix} \frac{\partial u}{\partial x} & \frac{\partial u}{\partial y} & \frac{\partial u}{\partial z} \\ \frac{\partial v}{\partial x} & \frac{\partial v}{\partial y} & \frac{\partial v}{\partial z} \\ \frac{\partial w}{\partial x} & \frac{\partial w}{\partial y} & \frac{\partial w}{\partial z} \end{vmatrix} = \begin{vmatrix} 1 & 2 & 1 \\ 1 & 2 & 3 \\ 2 & 3 & 5 \end{vmatrix} = 2$$

(*ii*)

$$\frac{\partial(u, v, w)}{\partial(x, y, z)} = \begin{vmatrix} yz & zx & xy \\ y + z & z + x & x + y \\ 1 & 1 & 1 \end{vmatrix}$$

$$= \begin{vmatrix} yz & z(x - y) & y(x - z) \\ y + z & x - y & x - z \\ 1 & 0 & 0 \end{vmatrix}$$

| By $C_2 \to C_2 - C_1$ and $C_3 \to C_3 - C_1$

$$= \begin{vmatrix} z(x - y) & y(x - z) \\ x - y & x - z \end{vmatrix} = (x - y)(x - z) \begin{vmatrix} z & y \\ 1 & 1 \end{vmatrix}$$

$$= (x - y)(x - z)(z - y) = (x - y)(y - z)(z - x)$$

Example 5. *If* $y_1 = \frac{x_2 x_3}{x_1}$, $y_2 = \frac{x_1 x_3}{x_2}$, $y_3 = \frac{x_1 x_2}{x_3}$, *then show that* $\frac{\partial(y_1, y_2, y_3)}{\partial(x_1, x_2, x_3)} = 4$.

[A.K.T.U. 2018]

Sol. $y_1 = \frac{x_2 x_3}{x_1}$, then $\frac{\partial y_1}{\partial x_1} = -\frac{x_2 x_3}{x_1^2}$, $\frac{\partial y_1}{\partial x_2} = \frac{x_3}{x_1}$, $\frac{\partial y_1}{\partial x_3} = \frac{x_2}{x_1}$

$$y_2 = \frac{x_1x_3}{x_2}, \quad \text{then} \quad \frac{\partial y_2}{\partial x_1} = \frac{x_3}{x_2}, \frac{\partial y_2}{\partial x_2} = \frac{-x_1x_3}{x_2^2}, \frac{\partial y_2}{\partial x_3} = \frac{x_1}{x_2}$$

$$y_3 = \frac{x_1x_2}{x_3}, \quad \text{then} \quad \frac{\partial y_3}{\partial x_1} = \frac{x_2}{x_3}, \frac{\partial y_3}{\partial x_2} = \frac{x_1}{x_3}, \frac{\partial y_3}{\partial x_3} = -\frac{x_1x_2}{x_3^2}.$$

$$\therefore \quad \frac{\partial(y_1, y_2, y_3)}{\partial(x_1, x_2, x_3)} = \begin{vmatrix} \frac{\partial y_1}{\partial x_1} & \frac{\partial y_1}{\partial x_2} & \frac{\partial y_1}{\partial x_3} \\ \frac{\partial y_2}{\partial x_1} & \frac{\partial y_2}{\partial x_2} & \frac{\partial y_2}{\partial x_3} \\ \frac{\partial y_3}{\partial x_1} & \frac{\partial y_3}{\partial x_2} & \frac{\partial y_3}{\partial x_3} \end{vmatrix} = \begin{vmatrix} \frac{-x_2x_3}{x_1^2} & \frac{x_3}{x_1} & \frac{x_2}{x_1} \\ \frac{x_3}{x_2} & \frac{-x_1x_3}{x_2^2} & \frac{x_1}{x_2} \\ \frac{x_2}{x_3} & \frac{x_1}{x_3} & \frac{-x_1x_2}{x_3^2} \end{vmatrix}$$

$$= \frac{1}{x_1^2x_2^2x_3^2} \begin{vmatrix} -x_2x_3 & x_1x_3 & x_1x_2 \\ x_2x_3 & -x_1x_3 & x_1x_2 \\ x_2x_3 & x_1x_3 & -x_1x_2 \end{vmatrix}$$

$$= \frac{x_1^2x_2^2x_3^2}{x_1^2x_2^2x_3^2} \begin{vmatrix} -1 & 1 & 1 \\ 1 & -1 & 1 \\ 1 & 1 & -1 \end{vmatrix}$$

$$= -1(1-1) - 1(-1-1) + 1(1+1) = 0 + 2 + 2 = 4.$$

Example 6. *(i) Verify the chain rule for Jacobians if $x = u$, $y = u \tan v$, $z = w$.*

(ii) If $x = \sqrt{vw}$, $y = \sqrt{wu}$, $z = \sqrt{uv}$ and $u = r \sin \theta \cos \phi$, $v = r \sin \theta \sin \phi$, $w = r \cos \theta$, then calculate the Jacobian $\frac{\partial(x, y, z)}{\partial(r, \theta, \phi)}$. (U.P.T.U. 2015)

Sol. *(i)*

$$J = \frac{\partial(x, y, z)}{\partial(u, v, w)} = \begin{vmatrix} 1 & 0 & 0 \\ \tan v & u \sec^2 v & 0 \\ 0 & 0 & 1 \end{vmatrix} = u \sec^2 v$$

Solving for u, v, w in terms of x, y, z; we have

$$u = x, \; v = \tan^{-1} \frac{y}{x} \quad \text{and} \quad w = z$$

$$\therefore \quad J' = \frac{\partial(u, v, w)}{\partial(x, y, z)} = \begin{vmatrix} 1 & 0 & 0 \\ \frac{-y}{x^2+y^2} & \frac{x}{x^2+y^2} & 0 \\ 0 & 0 & 1 \end{vmatrix}$$

$$= \frac{x}{x^2+y^2} = \frac{1}{x\left[1+\left(\frac{y}{x}\right)^2\right]} = \frac{1}{u \sec^2 v}$$

Therefore, $JJ' = 1$

Hence, chain rule is verified.

(ii)

$$\frac{\partial(x, y, z)}{\partial(u, v, w)} = \begin{vmatrix} \frac{\partial x}{\partial u} & \frac{\partial x}{\partial v} & \frac{\partial x}{\partial w} \\ \frac{\partial y}{\partial u} & \frac{\partial y}{\partial v} & \frac{\partial y}{\partial w} \\ \frac{\partial z}{\partial u} & \frac{\partial z}{\partial v} & \frac{\partial z}{\partial w} \end{vmatrix} = \begin{vmatrix} 0 & \frac{1}{2}\sqrt{\frac{w}{v}} & \frac{1}{2}\sqrt{\frac{v}{w}} \\ \frac{1}{2}\sqrt{\frac{w}{u}} & 0 & \frac{1}{2}\sqrt{\frac{u}{w}} \\ \frac{1}{2}\sqrt{\frac{v}{u}} & \frac{1}{2}\sqrt{\frac{u}{v}} & 0 \end{vmatrix}$$

$$= \frac{1}{8}\sqrt{\frac{w}{v}}\sqrt{\frac{u}{w}}\sqrt{\frac{v}{u}} + \frac{1}{8}\sqrt{\frac{v}{w}}\sqrt{\frac{w}{u}}\sqrt{\frac{u}{v}} = \frac{1}{8} + \frac{1}{8} = \frac{1}{4}$$

$$\frac{\partial(u, v, w)}{\partial(r, \theta, \phi)} = r^2 \sin\theta \qquad \text{| From Ex. 2.}$$

Therefore, by the property of Jacobians, we have

$$\frac{\partial(x, y, z)}{\partial(r, \theta, \phi)} = \frac{\partial(x, y, z)}{\partial(u, v, w)} \cdot \frac{\partial(u, v, w)}{\partial(r, \theta, \phi)} = \frac{1}{4} r^2 \sin\theta.$$

Example 7. *(i) If $y_1 = 1 - x_1$, $y_2 = x_1(1 - x_2)$, $y_3 = x_1 x_2 (1 - x_3)$, ..., $y_n = x_1 x_2 x_3 \ldots x_{n-1}(1 - x_n)$, then show that*

$$\frac{\partial(y_1, y_2, \ldots, y_n)}{\partial(x_1, x_2, \ldots, x_n)} = (-1)^n x_1^{n-1} x_2^{n-2} \ldots x_{n-1}.$$

(ii) If $y_1 = \cos x_1$, $y_2 = \sin x_1 \cos x_2$ and $y_3 = \sin x_1 \sin x_2 \cos x_3$, then show that

$$\frac{\partial(y_1, y_2, y_3)}{\partial(x_1, x_2, x_3)} = -\sin^3 x_1 \sin^2 x_2 \sin x_3.$$

Sol. *(i)* We know that

$$\frac{\partial(y_1, y_2, \ldots, y_n)}{\partial(x_1, x_2, \ldots, x_n)} = \frac{\partial y_1}{\partial x_1} \cdot \frac{\partial y_2}{\partial x_2}, \ldots, \frac{\partial y_n}{\partial x_n}$$

$$= (-1) \cdot (-x_1)(-x_1 x_2) \ldots (-x_1 x_2 \ldots x_{n-1})$$

$$= (-1)^n x_1^{n-1} x_2^{n-2} \ldots x_{n-1}.$$

(ii) $$\frac{\partial(y_1, y_2, y_3)}{\partial(x_1, x_2, x_3)} = \frac{\partial y_1}{\partial x_1} \cdot \frac{\partial y_2}{\partial x_2} \cdot \frac{\partial y_3}{\partial x_3} \qquad \text{| Theorem 1, Art. 3.11}$$

$$= (-\sin x_1)(-\sin x_1 \sin x_2)(-\sin x_1 \sin x_2 \sin x_3)$$

$$= -\sin^3 x_1 \sin^2 x_2 \sin x_3.$$

Example 8. *(i) If $u = xyz$, $v = x^2 + y^2 + z^2$, $w = x + y + z$, find the Jacobian $\frac{\partial(x, y, z)}{\partial(u, v, w)}$.*

(U.K.T.U. 2012)

(ii) If $x + y + z = u$, $y + z = uv$, $z = uvw$, then show that $\frac{\partial(x, y, z)}{\partial(u, v, w)} = u^2 v$. (A.K.T.U. 2016)

Sol. *(i)* $$\frac{\partial(u, v, w)}{\partial(x, y, z)} = \begin{vmatrix} \frac{\partial u}{\partial x} & \frac{\partial u}{\partial y} & \frac{\partial u}{\partial z} \\ \frac{\partial v}{\partial x} & \frac{\partial v}{\partial y} & \frac{\partial v}{\partial z} \\ \frac{\partial w}{\partial x} & \frac{\partial w}{\partial y} & \frac{\partial w}{\partial z} \end{vmatrix} = \begin{vmatrix} yz & zx & xy \\ 2x & 2y & 2z \\ 1 & 1 & 1 \end{vmatrix} = 2 \begin{vmatrix} yz & zx & xy \\ x & y & z \\ 1 & 1 & 1 \end{vmatrix}$$

$$= 2\begin{vmatrix} y(z-x) & x(z-y) & xy \\ x-z & y-z & z \\ 0 & 0 & 1 \end{vmatrix}$$ | Applying $C_1 \to C_1 - C_3$ and $C_2 \to C_2 - C_3$

$$= 2(z-x)(y-z)\begin{vmatrix} y & -x & xy \\ -1 & 1 & z \\ 0 & 0 & 1 \end{vmatrix}$$

$$= 2(y-z)(z-x)(y-x) = -2(x-y)(y-z)(z-x)$$

Since $\dfrac{\partial(u, v, w)}{\partial(x, y, z)} \cdot \dfrac{\partial(x, y, z)}{\partial(u, v, w)} = 1$ | By chain Rule

$$\therefore \quad \frac{\partial(x, y, z)}{\partial(u, v, w)} = \frac{-1}{2(x-y)(y-z)(z-x)}.$$

(*ii*) $x = u - uv = u(1-v)$

$y = uv - uvw = uv(1-w)$

$z = uvw$

$$\therefore \quad \frac{\partial(x, y, z)}{\partial(u, v, w)} = \begin{vmatrix} \frac{\partial x}{\partial u} & \frac{\partial x}{\partial v} & \frac{\partial x}{\partial w} \\ \frac{\partial y}{\partial u} & \frac{\partial y}{\partial v} & \frac{\partial y}{\partial w} \\ \frac{\partial z}{\partial u} & \frac{\partial z}{\partial v} & \frac{\partial z}{\partial w} \end{vmatrix} = \begin{vmatrix} 1-v & -u & 0 \\ v(1-w) & u(1-w) & -uv \\ vw & uw & uv \end{vmatrix}$$

$$= \begin{vmatrix} 1 & 0 & 0 \\ v(1-w) & u(1-w) & -uv \\ vw & uw & uv \end{vmatrix}$$ | Applying $R_1 \to R_1 + (R_2 + R_3)$

$$= u^2v(1-w) + u^2vw = u^2v.$$

Example 9. *If* $u = x(1-r^2)^{-1/2}$, $v = y(1-r^2)^{-1/2}$, $w = z(1-r^2)^{-1/2}$, *where* $r^2 = x^2 + y^2 + z^2$, *then show that* $\dfrac{\partial(u, v, w)}{\partial(x, y, z)} = (1-r^2)^{-5/2}$.

Sol. $$u = \frac{x}{\sqrt{1-x^2-y^2-z^2}} \qquad \text{...(1)} \quad | \text{ Given}$$

Differentiating (1) w.r.t. x partially, we get

$$\frac{\partial u}{\partial x} = \frac{\sqrt{1-x^2-y^2-z^2} - x\left[\frac{1}{2} \cdot \frac{(-2x)}{\sqrt{1-x^2-y^2-z^2}}\right]}{(1-x^2-y^2-z^2)}$$

$$= \frac{1-x^2-y^2-z^2+x^2}{(1-x^2-y^2-z^2)^{3/2}} = \frac{1-y^2-z^2}{(1-x^2-y^2-z^2)^{3/2}}$$

Differentiating (1) partially w.r.t. y, we get

$$\frac{\partial u}{\partial y} = -\frac{x}{2} \cdot \frac{(-2y)}{(1-x^2-y^2-z^2)^{3/2}} = \frac{xy}{(1-x^2-y^2-z^2)^{3/2}}$$

Similarly, $$\frac{\partial u}{\partial z} = \frac{xz}{(1-x^2-y^2-z^2)^{3/2}}$$

In the same way, other partial derivatives can be found.

Now, $$\frac{\partial(u, v, w)}{\partial(x, y, z)} = \frac{1}{(1-x^2-y^2-z^2)^{9/2}} \begin{vmatrix} 1-y^2-z^2 & xy & xz \\ yx & 1-z^2-x^2 & yz \\ zx & zy & 1-x^2-y^2 \end{vmatrix}$$

$$= \frac{xyz}{(1-x^2-y^2-z^2)^{9/2}} \begin{vmatrix} \frac{1-y^2-z^2}{x} & y & z \\ x & \frac{1-z^2-x^2}{y} & z \\ x & y & \frac{1-x^2-y^2}{z} \end{vmatrix}$$

$$= \frac{1}{(1-x^2-y^2-z^2)^{9/2}} \begin{vmatrix} 1-y^2-z^2 & y^2 & z^2 \\ x^2 & 1-z^2-x^2 & z^2 \\ x^2 & y^2 & 1-x^2-y^2 \end{vmatrix}$$

$$= \frac{1}{(1-x^2-y^2-z^2)^{9/2}} \begin{vmatrix} 1-x^2-y^2-z^2 & 0 & -(1-x^2-y^2-z^2) \\ 0 & 1-x^2-y^2-z^2 & -(1-x^2-y^2-z^2) \\ x^2 & y^2 & 1-x^2-y^2 \end{vmatrix}$$

$$= (1-x^2-y^2-z^2)^{-5/2} \begin{vmatrix} 1 & 0 & -1 \\ 0 & 1 & -1 \\ x^2 & y^2 & 1-x^2-y^2 \end{vmatrix}$$

$$= (1-x^2-y^2-z^2)^{-5/2} [(1-x^2-y^2+y^2)+x^2] = (1-r^2)^{-5/2}$$

Example 10. *If* $u^3+v^3+w^3 = x+y+z$, $u^2+v^2+w^2 = x^3+y^3+z^3$ *and* $u+v+w = x^2+y^2+z^2$, *then show that* $\dfrac{\partial(u, v, w)}{\partial(x, y, z)} = \dfrac{(x-y)(y-z)(z-x)}{(u-v)(v-w)(w-u)}$. (A.K.T.U. 2020)

Sol. Let $F_1 \equiv (u^3+v^3+w^3-x-y-z)$,

$F_2 \equiv (u^2+v^2+w^2-x^3-y^3-z^3)$

and $F_3 \equiv (u+v+w-x^2-y^2-z^2)$

$$\frac{\partial(u, v, w)}{\partial(x, y, z)} = (-1)^3 \left[\frac{\dfrac{\partial(F_1, F_2, F_3)}{\partial(x, y, z)}}{\dfrac{\partial(F_1, F_2, F_3)}{\partial(u, v, w)}}\right] \qquad ...(1)$$

Now consider

$$\frac{\partial(F_1, F_2, F_3)}{\partial(x, y, z)} = \begin{vmatrix} \frac{\partial F_1}{\partial x} & \frac{\partial F_1}{\partial y} & \frac{\partial F_1}{\partial z} \\ \frac{\partial F_2}{\partial x} & \frac{\partial F_2}{\partial y} & \frac{\partial F_2}{\partial z} \\ \frac{\partial F_3}{\partial x} & \frac{\partial F_3}{\partial y} & \frac{\partial F_3}{\partial z} \end{vmatrix} = \begin{vmatrix} -1 & -1 & -1 \\ -3x^2 & -3y^2 & -3z^2 \\ -2x & -2y & -2z \end{vmatrix}$$

$$= \begin{vmatrix} 0 & 0 & -1 \\ 3(z^2 - x^2) & 3(z^2 - y^2) & -3z^2 \\ 2(z - x) & 2(z - y) & -2z \end{vmatrix} \quad \left| \begin{array}{l} \text{Applying} \\ C_1 \to C_1 - C_3 \text{ and } C_2 \to C_2 - C_3 \end{array} \right.$$

$$= 6\,(z - x)\,(z - y)\,(-1) \begin{vmatrix} z + x & z + y \\ 1 & 1 \end{vmatrix}$$

$$= -6\,(z - x)\,(z - y)\,(x - y) = 6\,(x - y)\,(y - z)\,(z - x)$$

Now consider

$$\frac{\partial(F_1, F_2, F_3)}{\partial(u, v, w)} = \begin{vmatrix} \frac{\partial F_1}{\partial u} & \frac{\partial F_1}{\partial v} & \frac{\partial F_1}{\partial w} \\ \frac{\partial F_2}{\partial u} & \frac{\partial F_2}{\partial v} & \frac{\partial F_2}{\partial w} \\ \frac{\partial F_3}{\partial u} & \frac{\partial F_3}{\partial v} & \frac{\partial F_3}{\partial w} \end{vmatrix} = \begin{vmatrix} 3u^2 & 3v^2 & 3w^2 \\ 2u & 2v & 2w \\ 1 & 1 & 1 \end{vmatrix}$$

$$= 6 \begin{vmatrix} u^2 - w^2 & v^2 - w^2 & 3w^2 \\ u - w & v - w & 2w \\ 0 & 0 & 1 \end{vmatrix} \quad \left| \begin{array}{l} \text{Applying} \\ C_1 \to C_1 - C_3, C_2 \to C_2 - C_3 \end{array} \right.$$

$$= 6\,(u - w)\,(v - w) \begin{vmatrix} u + w & v + w \\ 1 & 1 \end{vmatrix} = 6\,(u - w)\,(v - w)\,(u - v)$$

$$\therefore \quad \frac{\partial(u, v, w)}{\partial(x, y, z)} = (-1)^3 \frac{6\,(x - y)\,(y - z)\,(z - x)}{6\,(u - v)\,(v - w)\,(u - w)} = \frac{(x - y)\,(y - z)\,(z - x)}{(u - v)\,(v - w)\,(w - u)}.$$

Example 11. (*i*) *If u, v, w are the roots of the cubic* $(\lambda - x)^3 + (\lambda - y)^3 + (\lambda - z)^3 = 0$ *in* λ, *then find*

$$\frac{\partial(u, v, w)}{\partial(x, y, z)} \qquad \text{(A.K.T.U. 2016, 2022)}$$

(*ii*) *If u, v, w are the roots of the equation* $(x - a)^3 + (x - b)^3 + (x - c)^3 = 0$, *then find*

$$\frac{\partial(u, v, w)}{\partial(a, b, c)}. \qquad \text{(A.K.T.U. 2015, 2019)}$$

Sol. (*i*) If u, v, w are the roots of the equation

$$(\lambda - x)^3 + (\lambda - y)^3 + (\lambda - z)^3 = 0$$

i.e., $$3\lambda^3 - 3\lambda^2(x + y + z) + 3\lambda(x^2 + y^2 + z^2) - (x^3 + y^3 + z^3) = 0$$

then, $$S_1 = u + v + w = \frac{3(x + y + z)}{3}$$

$$S_2 = uv + vw + wu = \frac{3(x^2 + y^2 + z^2)}{3}$$

$$S_3 = uvw = \frac{x^3 + y^3 + z^3}{3}$$

Let
$$F_1 \equiv u + v + w - x - y - z = 0$$
$$F_2 \equiv uv + vw + wu - (x^2 + y^2 + z^2) = 0$$
$$F_3 \equiv uvw - \frac{1}{3}(x^3 + y^3 + z^3) = 0$$

$$\therefore \quad \frac{\partial(u, v, w)}{\partial(x, y, z)} = (-1)^3 \left[\frac{\dfrac{\partial(F_1, F_2, F_3)}{\partial(x, y, z)}}{\dfrac{\partial(F_1, F_2, F_3)}{\partial(u, v, w)}} \right] \qquad \ldots(1)$$

Now,
$$\frac{\partial(F_1, F_2, F_3)}{\partial(x, y, z)} = \begin{vmatrix} \frac{\partial F_1}{\partial x} & \frac{\partial F_1}{\partial y} & \frac{\partial F_1}{\partial z} \\ \frac{\partial F_2}{\partial x} & \frac{\partial F_2}{\partial y} & \frac{\partial F_2}{\partial z} \\ \frac{\partial F_3}{\partial x} & \frac{\partial F_3}{\partial y} & \frac{\partial F_3}{\partial z} \end{vmatrix} = \begin{vmatrix} -1 & -1 & -1 \\ -2x & -2y & -2z \\ -x^2 & -y^2 & -z^2 \end{vmatrix}$$

$$= (-2) \begin{vmatrix} 1 & 0 & 0 \\ x & y - x & z - x \\ x^2 & y^2 - x^2 & z^2 - x^2 \end{vmatrix} \qquad \left| \text{By } C_2 \to C_2 - C_1, \; C_3 \to C_3 - C_1 \right.$$

$$= -2(y - x)(z - x)(z + x - y - x)$$
$$= -2(x - y)(y - z)(z - x)$$

Also,
$$\frac{\partial(F_1, F_2, F_3)}{\partial(u, v, w)} = \begin{vmatrix} \frac{\partial F_1}{\partial u} & \frac{\partial F_1}{\partial v} & \frac{\partial F_1}{\partial w} \\ \frac{\partial F_2}{\partial u} & \frac{\partial F_2}{\partial v} & \frac{\partial F_2}{\partial w} \\ \frac{\partial F_3}{\partial u} & \frac{\partial F_3}{\partial v} & \frac{\partial F_3}{\partial w} \end{vmatrix} = \begin{vmatrix} 1 & 1 & 1 \\ v + w & w + u & u + v \\ vw & uw & uv \end{vmatrix}$$

$$= \begin{vmatrix} 1 & 0 & 0 \\ v + w & u - v & u - w \\ vw & w(u - v) & v(u - w) \end{vmatrix} \qquad \left| \text{By } C_2 \to C_2 - C_1, \; C_3 \to C_3 - C_1 \right.$$

$$= \begin{vmatrix} u - v & u - w \\ w(u - v) & v(u - w) \end{vmatrix} = (u - v)(u - w) \begin{vmatrix} 1 & 1 \\ w & v \end{vmatrix}$$

$$= -(u - v)(v - w)(w - u)$$

Hence,
$$\frac{\partial(u, v, w)}{\partial(x, y, z)} = (-1)^3 \left[\frac{-2(x - y)(y - z)(z - x)}{-(u - v)(v - w)(w - u)} \right] \qquad | \text{ Using (1)}$$

$$\Rightarrow \quad J(u, v, w) = -\frac{2(x - y)(y - z)(z - x)}{(u - v)(v - w)(w - u)}.$$

(ii) If u, v, w are the roots of the equation

$$(x - a)^3 + (x - b)^3 + (x - c)^3 = 0$$

i.e.,
$$3x^3 - 3x^2(a + b + c) + 3x(a^2 + b^2 + c^2) - (a^3 + b^3 + c^3) = 0$$

then,
$$S_1 = u + v + w = \frac{3(a + b + c)}{3}$$

$$S_2 = uv + vw + wu = \frac{3(a^2 + b^2 + c^2)}{3}$$

$$S_3 = uvw = \frac{a^3 + b^3 + c^3}{3}$$

Let $F_1 \equiv u + v + w - a - b - c = 0$

$F_2 \equiv uv + vw + wu - a^2 - b^2 - c^2 = 0$

$F_3 \equiv uvw - \frac{1}{3}(a^3 + b^3 + c^3) = 0$

$$\frac{\partial(u, v, w)}{\partial(a, b, c)} = (-1)^3 \left[\frac{\partial(F_1, F_2, F_3)}{\partial(a, b, c)} \bigg/ \frac{\partial(F_1, F_2, F_3)}{\partial(u, v, w)}\right] \qquad ...(1)$$

Now, $$\frac{\partial(F_1, F_2, F_3)}{\partial(a, b, c)} = \begin{vmatrix} -1 & -1 & -1 \\ -2a & -2b & -2c \\ -a^2 & -b^2 & -c^2 \end{vmatrix} = -2(a-b)(b-c)(c-a)$$

Also, $$\frac{\partial(F_1, F_2, F_3)}{\partial(u, v, w)} = \begin{vmatrix} 1 & 1 & 1 \\ v+w & w+u & u+v \\ vw & uw & uv \end{vmatrix} = -(u-v)(v-w)(w-u)$$

Hence, from (1),

$$\frac{\partial(u, v, w)}{\partial(a, b, c)} = (-1)^3 \left[\frac{-2(a-b)(b-c)(c-a)}{-(u-v)(v-w)(w-u)}\right] = -\left[\frac{2(a-b)(b-c)(c-a)}{(u-v)(v-w)(w-u)}\right].$$

Example 12. *Determine functional dependence and find relation between*

(*i*) $u = \frac{x+y}{x-y}$, $v = \frac{xy}{(x-y)^2}$ (*ii*) $u = \frac{x-y}{x+y}$, $v = \frac{x+y}{x}$.

Sol. (*i*) $$\frac{\partial u}{\partial x} = \frac{-2y}{(x-y)^2}, \quad \frac{\partial u}{\partial y} = \frac{2x}{(x-y)^2}$$

$$\frac{\partial v}{\partial x} = \frac{-y(x+y)}{(x-y)^3}, \quad \frac{\partial v}{\partial y} = \frac{x(x+y)}{(x-y)^3}$$

Now, $$\frac{\partial(u, v)}{\partial(x, y)} = \begin{vmatrix} \frac{-2y}{(x-y)^2} & \frac{2x}{(x-y)^2} \\ \frac{-y(x+y)}{(x-y)^3} & \frac{x(x+y)}{(x-y)^3} \end{vmatrix} = \frac{2(x+y)}{(x-y)^5} \begin{vmatrix} -y & x \\ -y & x \end{vmatrix} = 0$$

Hence, the functional relationship exists between u and v.

Now, $$u = \frac{x+y}{x-y} \Rightarrow x = y\left(\frac{u+1}{u-1}\right) \qquad ...(1)$$

We have, $$v = \frac{xy}{(x-y)^2} = \frac{y^2\left(\frac{u+1}{u-1}\right)}{y^2\left(\frac{u+1}{u-1} - 1\right)^2} \qquad | \text{ Using (1)}$$

$$= \left(\frac{u+1}{u-1}\right) \cdot \frac{(u-1)^2}{4} = \frac{u^2-1}{4}$$

$\Rightarrow \qquad u^2 - 4v = 1$

(ii) $$\frac{\partial u}{\partial x} = \frac{2y}{(x+y)^2}, \frac{\partial u}{\partial y} = \frac{-2x}{(x+y)^2}, \frac{\partial v}{\partial x} = \frac{-y}{x^2}, \frac{\partial v}{\partial y} = \frac{1}{x}$$

Now, $$\frac{\partial(u, v)}{\partial(x, y)} = \begin{vmatrix} \frac{2y}{(x+y)^2} & \frac{-2x}{(x+y)^2} \\ \frac{-y}{x^2} & \frac{1}{x} \end{vmatrix} = \frac{2}{x^2(x+y)^2} \begin{vmatrix} y & -x \\ -y & x \end{vmatrix} = 0$$

Hence the functional relationship exists between u and v.

Now, $$u = \frac{x-y}{x+y} \Rightarrow x = y\left(\frac{1+u}{1-u}\right) \qquad ...(1)$$

We have, $$v = \frac{x+y}{x} = 1 + \frac{y}{x} = 1 + \frac{1-u}{1+u} \qquad | \text{ Using (1)}$$

$$v = \frac{2}{1+u}$$

Example 13. *(i) If $u = x + 2y + z$, $v = x - 2y + 3z$ and $w = 2xy - xz + 4yz - 2z^2$, show that they are not independent. Find the relation between u, v and w.* [A.K.T.U. 2017]

(ii) If $u = \frac{x+y}{z}$, $v = \frac{y+z}{x}$, $w = \frac{y(x+y+z)}{xz}$, then show that u, v, w are not independent and find the relation between them. (G.B.T.U. 2010)

Sol. u, v and w will not be independent if $\frac{\partial(u, v, w)}{\partial(x, y, z)} = 0$.

(i) $$\frac{\partial(u, v, w)}{\partial(x, y, z)} = \begin{vmatrix} 1 & 2 & 1 \\ 1 & -2 & 3 \\ 2y - z & 2x + 4z & -x + 4y - 4z \end{vmatrix}$$

$$= \begin{vmatrix} 1 & 0 & 0 \\ 1 & -4 & 2 \\ 2y - z & 2x - 4y + 6z & -x + 2y - 3z \end{vmatrix} \qquad \begin{array}{l} | \text{ By } C_2 \to C_2 - 2C_1, \\ C_3 \to C_3 - C_1 \end{array}$$

$$= -4(-x + 2y - 3z) - 2(2x - 4y + 6z)$$

$$= 4x - 8y + 12z - 4x + 8y - 12z = 0$$

Hence, u, v and w are not independent.

Now $$u + v = 2x + 4z$$

$$u - v = 4y - 2z$$

Multiplying these, we get

$$(u + v)(u - v) = (2x + 4z)(4y - 2z) = 4(2xy + 4yz - zx - 2z^2)$$

$\Rightarrow \qquad u^2 - v^2 = 4w.$

(ii) $$\frac{\partial(u, v, w)}{\partial(x, y, z)} = \begin{vmatrix} \frac{1}{z} & \frac{1}{z} & \frac{-(x+y)}{z^2} \\ \frac{-(y+z)}{x^2} & \frac{1}{x} & \frac{1}{x} \\ \frac{-y^2-yz}{x^2 z} & \frac{x+2y+z}{xz} & \frac{-xy-y^2}{xz^2} \end{vmatrix}$$

$$= \frac{1}{x^4 z^4} \begin{vmatrix} z & z & -(x+y) \\ -(y+z) & x & x \\ -yz(y+z) & xz(x+2y+z) & -xy(x+y) \end{vmatrix}$$

$$= \frac{1}{x^4 z^4} \begin{vmatrix} z & 0 & -(x+y+z) \\ -(y+z) & x+y+z & x+y+z \\ -yz(y+z) & z(x+y)(x+y+z) & y(z-x)(x+y+z) \end{vmatrix}$$

| By $C_2 \to C_2 - C_1$ and $C_3 \to C_3 - C_1$

$$= \frac{(x+y+z)^2}{x^4 z^4} \begin{vmatrix} z & 0 & -1 \\ -(y+z) & 1 & 1 \\ -yz(y+z) & z(x+y) & y(z-x) \end{vmatrix}$$

$$= \frac{(x+y+z)^2}{x^4 z^4} \begin{vmatrix} 0 & 0 & -1 \\ -y & 1 & 1 \\ -yz(y+x) & z(x+y) & y(z-x) \end{vmatrix}$$

| By $C_1 \to C_1 + zC_3$

$$= -\frac{(x+y+z)^2}{x^4 z^4} [-yz(x+y) + yz(x+y)] = 0$$

Hence, u, v and w are not independent.

Now, $$uv = \frac{(x+y)(y+z)}{zx} = \frac{xy + y^2 + yz + zx}{zx} = \frac{y(x+y+z)}{zx} + 1 = w + 1.$$

Example 14. *If $u = \sin^{-1} x + \sin^{-1} y$, $v = x\sqrt{1-y^2} + y\sqrt{1-x^2}$, find $\frac{\partial(u, v)}{\partial(x, y)}$. Is u, v functionally related? If so, find the relationship.*

Sol. $$u = \sin^{-1} x + \sin^{-1} y \qquad \text{...(1)} \quad | \text{ Given}$$

$$\frac{\partial u}{\partial x} = \frac{1}{\sqrt{1-x^2}} \text{ and } \frac{\partial u}{\partial y} = \frac{1}{\sqrt{1-y^2}}$$

Also, $$v = x\sqrt{1-y^2} + y\sqrt{1-x^2} \qquad \text{...(2)} \quad | \text{ Given}$$

$$\frac{\partial v}{\partial x} = \sqrt{1-y^2} - \frac{xy}{\sqrt{1-x^2}}, \quad \frac{\partial v}{\partial y} = \frac{-xy}{\sqrt{1-y^2}} + \sqrt{1-x^2}$$

Now, $$\frac{\partial(u, v)}{\partial(x, y)} = \begin{vmatrix} \frac{\partial u}{\partial x} & \frac{\partial u}{\partial y} \\ \frac{\partial v}{\partial x} & \frac{\partial v}{\partial y} \end{vmatrix} = \begin{vmatrix} \frac{1}{\sqrt{1-x^2}} & \frac{1}{\sqrt{1-y^2}} \\ \sqrt{1-y^2} - \frac{xy}{\sqrt{1-x^2}} & \frac{-xy}{\sqrt{1-y^2}} + \sqrt{1-x^2} \end{vmatrix}$$

$$= \frac{-xy}{\sqrt{(1-x^2)(1-y^2)}} + 1 - 1 + \frac{xy}{\sqrt{(1-x^2)(1-y^2)}} = 0$$

Hence, u and v are not independent. They are functionally related.

We have from (1), $u = \sin^{-1} x + \sin^{-1} y = \sin^{-1} (x\sqrt{1-y^2} + y\sqrt{1-x^2})$

$= \sin^{-1} v$ | From (2)

$\Rightarrow$ $v = \sin u$

Example 15. *If* $u = \dfrac{x+y}{1-xy}$ *and* $v = \tan^{-1} x + \tan^{-1} y$, *find* $\dfrac{\partial(u, v)}{\partial(x, y)}$. *Are u and v functionally related? If so, find the relationship.*

Sol.

$$\frac{\partial(u, v)}{\partial(x, y)} = \begin{vmatrix} \dfrac{\partial u}{\partial x} & \dfrac{\partial u}{\partial y} \\ \dfrac{\partial v}{\partial x} & \dfrac{\partial v}{\partial y} \end{vmatrix} = \begin{vmatrix} \dfrac{1+y^2}{(1-xy)^2} & \dfrac{1+x^2}{(1-xy)^2} \\ \dfrac{1}{1+x^2} & \dfrac{1}{1+y^2} \end{vmatrix} = \frac{1}{(1-xy)^2} - \frac{1}{(1-xy)^2} = 0$$

Hence, u, v are functionally related.

$$\tan^{-1} x + \tan^{-1} y = \tan^{-1}\left(\frac{x+y}{1-xy}\right)$$

$\Rightarrow$ $v = \tan^{-1} u$

Example 16. *Show that* $ax^2 + 2hxy + by^2$ *and* $Ax^2 + 2Hxy + By^2$ *are independent unless*

$$\frac{a}{A} = \frac{b}{B} = \frac{h}{H}.$$

Sol. Let $u = ax^2 + 2hxy + by^2$, $v = Ax^2 + 2Hxy + By^2$

u and v will not be independent if there exists a relationship between them and in that case $\dfrac{\partial(u, v)}{\partial(x, y)}$ should vanish identically.

i.e.,

$$\begin{vmatrix} 2(ax+hy) & 2(hx+by) \\ 2(Ax+Hy) & 2(Hx+By) \end{vmatrix} = 0$$

or, $(ax + hy)(Hx + By) - (hx + by)(Ax + Hy) = 0$

or, $(aH - Ah)x^2 + (aB - Ab)xy + (Bh - bH)y^2 = 0$

Now, the variables x and y are independent and as such the co-efficients of x^2 and y^2 should vanish separately.

$\therefore$ $aH - Ah = 0$ or $\dfrac{a}{A} = \dfrac{h}{H}$

and $Bh - bH = 0$ or $\dfrac{h}{H} = \dfrac{b}{B}$

Hence, $\dfrac{a}{A} = \dfrac{h}{H} = \dfrac{b}{B}$ and these conditions also make the co-efficient of xy zero.

Hence, $\dfrac{a}{A} = \dfrac{b}{B} = \dfrac{h}{H}$ are the required conditions.

TEST YOUR KNOWLEDGE

1. If $x = u(1 + v)$, $y = v(1 + u)$, then find the value of $\dfrac{\partial(x, y)}{\partial(u, v)}$.

2. If $x = uv$, $y = \dfrac{u+v}{u-v}$, find $\dfrac{\partial(u, v)}{\partial(x, y)}$.

3. If $x = r\cos\theta$, $y = r\sin\theta$, $z = z$, find $\dfrac{\partial(x, y, z)}{\partial(r, \theta, z)}$ and $\dfrac{\partial(r, \theta, z)}{\partial(x, y, z)}$. *(A.K.T.U. 2019)*

4. (*i*) If $u = x(1 - y)$, $v = xy$, find $\dfrac{\partial(u, v)}{\partial(x, y)}$. *(A.K.T.U. 2020)*

   (*ii*) If $u = x^2 - y^2$, $v = 2xy$, find $\dfrac{\partial(x, y)}{\partial(u, v)}$.

5. If $u = \dfrac{y - x}{1 + xy}$ and $v = \tan^{-1} y - \tan^{-1} x$, find $\dfrac{\partial(u, v)}{\partial(x, y)}$.

6. If $x = e^v \sec u$, $y = e^v \tan u$, then evaluate $\dfrac{\partial(x, y)}{\partial(u, v)}$. *(A.K.T.U. 2021)*

7. If $y_1 = 1 - x_1$, $y_2 = x_1(1 - x_2)$, $y_3 = x_1x_2(1 - x_3)$, find the value of $\dfrac{\partial(y_1, y_2, y_3)}{\partial(x_1, x_2, x_3)}$.

8. If $u = \dfrac{x}{y - z}$, $v = \dfrac{y}{z - x}$, $w = \dfrac{z}{x - y}$, find $\dfrac{\partial(u, v, w)}{\partial(x, y, z)}$.

9. If $u_1 = x_1 + x_2 + x_3 + x_4$, $u_1u_2 = x_2 + x_3 + x_4$, $u_1u_2u_3 = x_3 + x_4$, $u_1u_2u_3u_4 = x_4$, then show that $\dfrac{\partial(x_1, x_2, x_3, x_4)}{\partial(u_1, u_2, u_3, u_4)} = u_1^3u_2^2u_3$.

10. If $u^3 + v^3 = x + y$, $u^2 + v^2 = x^3 + y^3$, then show that $\dfrac{\partial(u, v)}{\partial(x, y)} = \dfrac{y^2 - x^2}{2uv(u - v)}$.

11. If $u_1 = x_1 + x_2 + x_3$, $u_1^2u_2 = x_2 + x_3$ and $u_1^3\, u_3 = x_3$, find the value of Jacobian $\dfrac{\partial(u_1, u_2, u_3)}{\partial(x_1, x_2, x_3)}$.

12. If $x = v^2 + w^2$, $y = w^2 + u^2$, $z = u^2 + v^2$, then show that $\dfrac{\partial(x, y, z)}{\partial(u, v, w)} \cdot \dfrac{\partial(u, v, w)}{\partial(x, y, z)} = 1$. *(A.K.T.U. 2017)*

13. If $u = x^2 - y^2$, $v = 2xy$ and $x = r\cos\theta$, $y = r\sin\theta$, show that $\dfrac{\partial(u, v)}{\partial(r, \theta)} = 4r^3$.

14. If
$$u^3 + v + w = x + y^2 + z^2$$
$$u + v^3 + w = x^2 + y + z^2$$
$$u + v + w^3 = x^2 + y^2 + z$$
show that $\dfrac{\partial(u, v, w)}{\partial(x, y, z)} = \dfrac{1 - 4(xy + yz + zx) + 16xyz}{2 - 3(u^2 + v^2 + w^2) + 27u^2v^2w^2}$. *(A.K.T.U. 2021)*

15. If λ, μ, ν be the roots of the equation $\dfrac{x}{a+k} + \dfrac{y}{b+k} + \dfrac{z}{c+k} = 1$ in k; prove that

$$\frac{\partial(x, y, z)}{\partial(\lambda, \mu, \nu)} = \frac{-(\nu-\lambda)(\lambda-\mu)(\mu-\nu)}{(a-b)(b-c)(c-a)}.$$

16. Use the Jacobian to prove that the functions $u = x + y - z$, $v = x - y + z$ and $w = x^2 + y^2 + z^2 - 2yz$ are not independent of one another. Find the relation between them.

17. Use the Jacobian to prove that the functions $u = xy + yz + zx$, $v = x^2 + y^2 + z^2$ and $w = x + y + z$ are not independent of one another. Find the relation between them. *(U.K.T.U. 2010)*

18. If $y_1(x_1 - x_2) = 0$, $y_2(x_1^2 + x_1x_2 + x_2^2) = 0$ show that, $\dfrac{\partial(y_1, y_2)}{\partial(x_1, x_2)} = 3y_1y_2\left(\dfrac{x_1 + x_2}{x_1^3 - x_2^3}\right)$.

19. (*i*) Determine functional dependence and find relation between $u = \dfrac{x-y}{x+y}$, $v = \dfrac{xy}{(x+y)^2}$.

(*ii*) Are the functions: $u = \dfrac{x-y}{x+z}$, $v = \dfrac{x+z}{y+z}$ functionally dependent? If so, find the relation between them. *(M.T.U. 2012)*

20. Find the Jacobian of the functions

$$y_1 = (x_1 - x_2)(x_2 + x_3)$$
$$y_2 = (x_1 + x_2)(x_2 - x_3)$$
$$y_3 = x_2(x_1 - x_3)$$

hence show that the functions are not independent. Find the relation between them.

21. Prove that the functions

(*i*) $u = 3x + 2y - z$, $v = x - 2y + z$, $w = x(x + 2y - z)$ are not independent. Find the relation between them.

(*ii*) $u = x + y - z$, $v = x - y + z$, $w = x^2 + y^2 + z^2 - 2yz$ are not independent. Find the relation between them.

22. If $u = x + y + z + t$, $v = x^2 + y^2 + z^2 + t^2$, $w = x^3 + y^3 + z^3 + t^3$, $r = xyz + yzt + zxt + txy$ show that u, v, w, r are not independent and find a relation between them.

23. (*i*) If $u = x + y + z$, $v = x^2 + y^2 + z^2$, $w = x^3 + y^3 + z^3 - 3xyz$, prove that u, v, w are not independent and hence find the relation between them.

(*ii*) Show that $u = y + z$, $v = x + 2z^2$, $w = x - 4yz - 2y^2$ are not independent. Find the relation between them. *(U.P.T.U. 2014)*

24. If $u = x^3 + x^2y + x^2z - z^2(x + y + z)$,

$v = x + z$

$w = x^2 - z^2 + xy - zy$

prove that u, v and w are connected by a functional relation.

25. Show that the functions:

$$u = x + y + z,$$
$$v = x^2 + y^2 + z^2 - 2xy - 2yz - 2zx$$

and, $w = x^3 + y^3 + z^3 - 3xyz$

are functionally related. Find the relation between them. *(A.K.T.U. 2014, 2018)*

Answers

1. $1 + u + v$	**2.** $\frac{(u-v)^2}{4uv}$	**3.** $r, \frac{1}{r}$
4. (*i*) x	(*ii*) $\frac{1}{4(x^2+y^2)}$	**5.** 0
6. $-e^{2v} \sec u$	**7.** $-x_1^2 x_2$	**8.** 0
11. u_1^{-5}	**16.** $(u+v)^2 + (u-v)^2 = 4w$	**17.** $w^2 = v + 2u$
19. (*i*) $4v = 1 - u^2$	(*ii*) Yes, $v = \frac{1}{1-u}$	**20.** $y_1 + y_2 - 2y_3 = 0$
21. (*i*) $u^2 - v^2 = 8w$	(*ii*) $u^2 + v^2 = 2w$	**22.** $u^3 = 3uv - 2w + 6r$
23. (*i*) $2w = u(3v - u^2)$	(*ii*) $2u^2 = v - w$	**24.** $u = wv$
25. $w = \frac{u(u^2+3v)}{4}$.		

3.14. APPROXIMATION OF ERRORS

Let $$z = f(x, y) \quad \text{...(1)}$$

If δx, δy are small increments in x and y respectively and δz, the corresponding increment in z, then

$$z + \delta z = f(x + \delta x, y + \delta y) \quad \text{...(2)}$$

Subtracting (1) from (2), we get

$$\delta z = f(x + \delta x, y + \delta y) - f(x, y) = f(x, y) + \delta x \frac{\partial f}{\partial x} + \delta y \frac{\partial f}{\partial y} + \ldots - f(x, y)$$

$$= \delta x \frac{\partial f}{\partial x} + \delta y \frac{\partial f}{\partial y} \text{ (approximately)} \quad [\text{Neglecting higher powers of } \delta x, \delta y]$$

$$\therefore \quad \delta z = \frac{\partial f}{\partial x} . \delta x + \frac{\partial f}{\partial y} . \delta y \text{ (approximately)}$$

$\Rightarrow$ If δx and δy are small changes (or errors) in x and y respectively, then an approximate change (or error) in z is δz.

Replacing δx, δy, δz by dx, dy, dz respectively, we have

$$\boxed{dz = \frac{\partial f}{\partial x} dx + \frac{\partial f}{\partial y} dy}$$

Note. If δx is the error in x, then **relative error** $= \frac{\delta x}{x}$

and **percentage error** $= \frac{\delta x}{x} \times 100$.

ILLUSTRATIVE EXAMPLES

Example 1. *Find the percentage error in the area of an ellipse when an error of + 1 percent is made in measuring the major and minor axes.* (G.B.T.U. 2011)

Sol. If x and y are semi-major and semi-minor axes of the ellipse,

then its area A is given by $\quad A = \pi xy$

Taking log on both sides, $\log A = \log \pi + \log x + \log y$

Differentiating, we get $\quad \dfrac{\delta A}{A} = 0 + \dfrac{\delta x}{x} + \dfrac{\delta y}{y}$

or
$$\frac{\delta A}{A} \times 100 = \frac{\delta x}{x} \times 100 + \frac{\delta y}{y} \times 100 = 1 + 1 = 2.$$

$\therefore$ Error in the area = 2%.

Example 2. *The period of a simple pendulum is* $T = 2\pi\sqrt{\dfrac{l}{g}}$. *Find the maximum (%) error in T due to the possible errors upto 1% in l and 2.5% in g.* (U.P.T.U. 2014)

Sol.
$$T = 2\pi\sqrt{\frac{l}{g}}$$

Taking log,
$$\log T = \log 2\pi + \frac{1}{2}\log_e l - \frac{1}{2}\log_e g$$

Differentiation yields,
$$\frac{1}{T}\delta T = \frac{1}{2}\frac{\delta l}{l} - \frac{1}{2}\frac{\delta g}{g}$$

$$\Rightarrow \quad \frac{\delta T}{T} \times 100 = \frac{1}{2}\left[\frac{\delta l}{l} \times 100 - \frac{\delta g}{g} \times 100\right] = \frac{1}{2}(1 \pm 2.5)$$

Maximum error in $T = \dfrac{1}{2}(1 + 2.5) = 1.75\%$.

Example 3. *The work that must be done to propel a ship of displacement D for a distance 'S' in time 't' is proportional to* $\dfrac{S^2 D^{2/3}}{t^2}$. *Find approximately the increase of work necessary when the displacement is increased by 1%, the time diminished by 1% and the distance diminished by 2%.*

Sol. Let the work done be W, then $W \propto \dfrac{S^2 D^{2/3}}{t^2}$

$$\Rightarrow \quad W = k \cdot \frac{S^2 D^{2/3}}{t^2}, \text{ where } k \text{ is a constant.}$$

Taking log,
$$\log W = \log k + 2\log S + \tfrac{2}{3}\log D - 2\log t$$

Differentiation yields, $\dfrac{\delta W}{W} = 0 + 2\dfrac{\delta S}{S} + \dfrac{2}{3}\dfrac{\delta D}{D} - 2\dfrac{\delta t}{t}$

or $$\frac{\delta W}{W} \times 100 = 2\frac{\delta S}{S} \times 100 + \frac{2}{3}\frac{\delta D}{D} \times 100 - 2\frac{\delta t}{t} \times 100$$

$$= 2(-2) + \tfrac{2}{3}(1) - 2(-1) = -\frac{4}{3}$$

$\therefore$ Approximate increase of work $= -\dfrac{4}{3}$%.

Example 4. *The diameter and altitude of a can in the shape of a right circular cylinder are measured as 4 cm and 6 cm respectively. The possible error in each measurement is 0.1 cm. Find approximately the maximum possible error in the value computed for the volume and lateral surface.*

Sol. Let the diameter and altitude of the can be denoted by D and H respectively.

Then radius $= \dfrac{D}{2}$

(*i*) The volume V of the can is given by

$$V = \pi r^2 h = \frac{\pi}{4} D^2 H \qquad | = f(D, H)$$

$$\therefore \quad dV = \frac{\partial V}{\partial D} dD + \frac{\partial V}{\partial H} dH = \frac{\pi}{4}[2DH\,dD + D^2 dH]$$

$$= \frac{\pi}{4}[2 \times 4 \times 6 \times 0.1 + 4^2 \times 0.1] = \frac{\pi}{4}(6.4) = 1.6\,\pi \text{ cu. cm.}$$

(*ii*) The lateral surface S of the can is given by

$$S = 2\pi r h = \pi DH \qquad | = f(D, H)$$

$$\therefore \quad dS = \frac{\partial S}{\partial D} dD + \frac{\partial S}{\partial H} dH$$

$$= \pi[H\,dD + D\,dH] = \pi[6 \times 0.1 + 4 \times 0.1] = \pi \text{ sq. cm.}$$

Example 5. *If the base radius and height of a cone are measured as 4 and 8 inches with a possible error of 0.04 and 0.08 inches respectively, calculate the percentage (%) error in calculating volume of the cone.* (G.B.T.U. 2012)

Sol. Volume $$V = \frac{1}{3}\pi r^2 h \qquad ...(1)$$

Taking log, $$\log V = \log\frac{1}{3} + \log \pi + 2\log r + \log h$$

Differentiating, we get

$$\frac{\delta V}{V} = 2\frac{\delta r}{r} + \frac{\delta h}{h}$$

$$\Rightarrow \quad \frac{\delta V}{V} = 2\left(\frac{.04}{4}\right) + \left(\frac{.08}{8}\right) = .03$$

$\therefore$ Percentage (%) error in volume = .03 × 100 = 3%.

Example 6. *A balloon is in the form of right circular cylinder of radius 1.5 m and length 4 m and is surmounted by hemispherical ends. If the radius is increased by 0.01 m and length by 0.05 m, find the percentage change in the volume of balloon.* (A.K.T.U. 2014, 2018)

Sol. Here, $r = 1.5$ m, $h = 4$ m.

Let V be the volume of the balloon then,

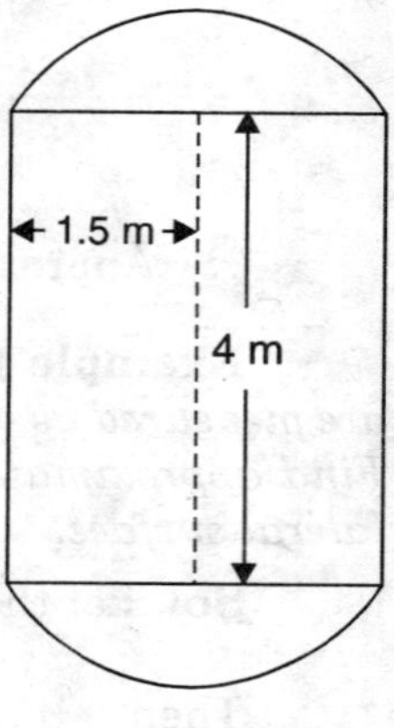

$$V = \pi r^2 h + \frac{2}{3}\pi r^3 + \frac{2}{3}\pi r^3 = \pi r^2 h + \frac{4}{3}\pi r^3$$

Now,
$$\delta V = \pi \,.\, 2r\,\delta r \,.\, h + \pi r^2\,\delta h + \frac{4}{3}\pi \,.\, 3r^2\,\delta r$$

$$= \pi r\,[2h\,\delta r + r\delta h + 4r\,\delta r]$$

$$\frac{\delta V}{V} = \frac{\pi r[2(h+2r)\,\delta r + r\delta h]}{\pi r^2 h + \frac{4}{3}\pi r^3} = \frac{2(h+2r)\,\delta r + r\delta h}{rh + \frac{4}{3}r^2}$$

$$= \frac{2(4+3)(.01) + (1.5)(.05)}{(1.5\times 4) + \frac{4}{3}(1.5)^2} = \frac{0.215}{9}$$

$$\therefore \qquad \frac{\delta V}{V}\times 100 = \frac{0.215}{9}\times 100 = 2.389\%$$

Hence the percentage change in the volume of the balloon is 2.389.

Example 7. (*i*) *What error in the common logarithm of a number will be produced by an error of 1% in the number?* [A.K.T.U. 2018]

(*ii*) *Prove that the relative error of a quotient does not exceed the sum of the relative errors of dividend and the divisor.*

Sol. (*i*) Consider x as any number and

Let
$$y = \log_{10} x$$

Then,
$$\delta y = \frac{1}{x}\log_{10} e\,(\delta x)$$

$$= \frac{\delta x}{x}\log_{10} e = \left(\frac{\delta x}{x}\times 100\right)\left(\frac{1}{100}\log_{10} e\right)$$

$$= \frac{1}{100}\log_{10} e \qquad \left|\ \because\ \frac{\delta x}{x}\times 100 = 1\right.$$

$$\Rightarrow \qquad \delta y = \frac{.43429}{100} = .0043429.$$

(*ii*) Let
$$z = \frac{x}{y} \qquad ...(1)$$

where z is quotient, x is dividend and y is divisor. Taking log on both sides of (1),

$$\log z = \log x - \log y$$

Differentiating, we get

$$\frac{\delta z}{z} = \frac{\delta x}{x} - \frac{\delta y}{y}$$

$$\left|\frac{\delta z}{z}\right| \le \left|\frac{\delta x}{x}\right| + \left|\frac{\delta y}{y}\right|$$

Hence the result.

Example 8. (*i*) *If the sides and angles of a triangle ABC vary in such a way that its circum-radius remains constant, prove that* $\frac{da}{\cos A} + \frac{db}{\cos B} + \frac{dc}{\cos C} = 0$.

(*ii*) *The side a and the opposite angle A of a ΔABC remain constant. Show that when the other sides and angles are slightly varied,*

$$\frac{\delta b}{\cos B} + \frac{\delta c}{\cos C} = 0$$

Sol. (*i*) The circum-radius R of a ΔABC is given by

$$R = \frac{a}{2 \sin A} = \frac{b}{2 \sin B} = \frac{c}{2 \sin C}$$

Now, $a = 2R \sin A$ (R is constant)

Taking differentials, $da = 2R \cos A \, dA$ or $\frac{da}{\cos A} = 2R \, dA$

Similarly, $\frac{db}{\cos B} = 2R \, dB$; $\frac{dc}{\cos C} = 2R \, dC$

Adding $\frac{da}{\cos A} + \frac{db}{\cos B} + \frac{dc}{\cos C} = 2R(dA + dB + dC)$ ...(1)

Also, $A + B + C = \pi$

Taking differentials, $dA + dB + dC = 0$

∴ From (1), $\frac{da}{\cos A} + \frac{db}{\cos B} + \frac{dc}{\cos C} = 0.$

(*ii*) $a = b \cos C + c \cos B$

$$\delta a = -b \sin C \, \delta C + \cos C \, \delta b - c \sin B \, \delta B + \cos B \, \delta c$$

$$= -(\delta C + \delta B) + \cos C \, \delta b + \cos B \, \delta c$$

$\left| \because \frac{b}{\sin B} = \frac{c}{\sin C} \right.$ by sine formula

$$= \delta A + \cos C \, \delta b + \cos B \, \delta c$$

| ∵ $A + B + C = \pi$
∴ $\delta B + \delta C = -\delta A$

⇒ $0 = \cos C \, \delta b + \cos B \, \delta c$ | ∵ $\delta a = 0$ and $\delta A = 0$

⇒ $\frac{\delta b}{\cos B} + \frac{\delta c}{\cos C} = 0$

Example 9. *The angles of a triangle are calculated from the sides a, b, c. If small changes δa, δb and δc are made in the sides, find δA, δB and δC approximately, where Δ is the area of the triangle and A, B, C are angles opposite to sides a, b, c respectively. Also show that $\delta A + \delta B + \delta C = 0$.* (G.B.T.U. 2012)

Sol. We know that, $a^2 = b^2 + c^2 - 2bc \cos A$ ...(1)

Differentiating eqn. (1), we get

$$2a\,\delta a = 2b\,\delta b + 2c\,\delta c - 2b\,\delta c \cos A - 2c\,\delta b \cos A + 2bc \sin A\,\delta A \quad \text{(approx.)}$$

or $$bc \sin A\,\delta A = a\delta a - (b - c \cos A)\,\delta b - (c - b \cos A)\,\delta c$$

or $$2\Delta\delta A = a\delta a - (a \cos C + c \cos A - c \cos A)\,\delta b - (a \cos B + b \cos A - b \cos A)\,\delta c$$

or $$\delta A = \frac{a}{2\Delta}(\delta a - \delta b \cos C - \delta c \cos B) \quad ...(2)$$

Similarly, we get $\delta B = \dfrac{b}{2\Delta}(\delta b - \delta c \cos A - \delta a \cos C)$ ...(3)

(By symmetry)

and $$\delta C = \frac{c}{2\Delta}(\delta c - \delta a \cos B - \delta b \cos A) \quad ...(4)$$

Adding eqns. (2), (3) and (4), we get

$$\delta A + \delta B + \delta C = \frac{1}{2\Delta}[(a - b \cos C - c \cos B)\,\delta a + (b - c \cos A - a \cos C)\,\delta b + (c - a \cos B - b \cos A)\,\delta c]$$

$$= \frac{1}{2\Delta}[(a - a)\,\delta a + (b - b)\,\delta b + (c - c)\,\delta c]$$

$$= \frac{1}{2\Delta}(0 + 0 + 0) = 0.$$

Example 10. *In the manufacture of closed rectangular boxes with specified sides a, b, c ($a \neq b \neq c$), small changes of A%, B%, C% occurred in a, b, and c respectively from box to box from the specified dimension. However the volume and surface area of all boxes were according to specification, show that $\dfrac{A}{a(b-c)} = \dfrac{B}{b(c-a)} = \dfrac{C}{c(a-b)}$.*

Sol. Volume $\quad V = abc$

$\therefore \quad \delta V = bc\,\delta a + ca\,\delta b + ab\,\delta c$

$$0 = bc\,\delta a + ca\,\delta b + ab\,\delta c$$

$$0 = \frac{\delta a}{a} + \frac{\delta b}{b} + \frac{\delta c}{c}$$

$$0 = \left(\frac{\delta a}{a} \times 100\right) + \left(\frac{\delta b}{b} \times 100\right) + \left(\frac{\delta c}{c} \times 100\right)$$

$\Rightarrow \quad A + B + C = 0$ ...(1)

Surface area $\quad S = 2(ab + bc + ca)$

$\therefore \quad \dfrac{1}{2}\delta S = (b + c)\,\delta a + (c + a)\,\delta b + (a + b)\,\delta c$

$$0 = \left(\frac{b+c}{bc}\right)\frac{\delta a}{a} + \left(\frac{c+a}{ac}\right)\frac{\delta b}{b} + \left(\frac{a+b}{ab}\right)\frac{\delta c}{c}$$

$$0 = \left(\frac{b+c}{bc}\right)\left(\frac{\delta a}{a} \times 100\right) + \left(\frac{c+a}{ac}\right)\left(\frac{\delta b}{b} \times 100\right) + \left(\frac{a+b}{ab}\right)\left(\frac{\delta c}{c} \times 100\right)$$

$$\Rightarrow \quad \left(\frac{b+c}{bc}\right)A + \left(\frac{c+a}{ac}\right)B + \left(\frac{a+b}{ab}\right)C = 0 \quad ...(2)$$

Cross-multiplication in (1) and (2) yields

$$\frac{A}{\left(\frac{a+b}{ab}\right) - \left(\frac{a+c}{ac}\right)} = \frac{B}{\left(\frac{b+c}{bc}\right) - \left(\frac{a+b}{ab}\right)} = \frac{C}{\left(\frac{c+a}{ca}\right) - \left(\frac{b+c}{bc}\right)}$$

$$\frac{A}{a(b-c)} = \frac{B}{b(c-a)} = \frac{C}{c(a-b)}.$$

Example 11. (*i*) *In estimating the number of bricks in a pile which is measured to be (5 m × 10 m × 5 m), the count of bricks is taken as 100 bricks per m³. Find the error in the cost when the tape is stretched 2% beyond its standard length. The cost of bricks is ₹ 2000 per thousand bricks.*

(*ii*) *In estimating the cost of a pile of bricks measured as 6 m × 50 m × 4 m, the tape is stretched 1% beyond the standard length. If the count is 12 bricks in 1 m³ and bricks cost ₹ 100 per 1,000, find the approximate error in the cost.* (U.K.T.U. 2010)

Sol. (*i*) Let x, y and z m be the length, breadth and height of the pile so that its volume $V = xyz$

Taking log, we get

$$\log V = \log x + \log y + \log z$$

Differentiating, we get

$$\frac{\delta V}{V} = \frac{\delta x}{x} + \frac{\delta y}{y} + \frac{\delta z}{z}$$

$$\Rightarrow \quad \frac{\delta V}{V} \times 100 = \left(\frac{\delta x}{x} \times 100\right) + \left(\frac{\delta y}{y} \times 100\right) + \left(\frac{\delta z}{z} \times 100\right) = 2 + 2 + 2 = 6$$

$$\therefore \quad \delta V = \frac{6}{100} V = \frac{6}{100}(250) = 15 \text{ m}^3$$

| Since $V = 5 \times 10 \times 5 = 250 \text{ m}^3$

$\therefore$ Number of bricks in $\delta V = 15 \times 100 = 1500$

Thus, error in the cost $= 1500 \times \dfrac{2000}{1000} =$ ₹ 3000

Hence, ₹ 3000 is the required error in the cost.

(*ii*) Let x, y and z m be the length, breadth and height of the pile so that its volume

$$V = xyz$$

Taking log, we get

$$\log V = \log x + \log y + \log z$$

Differentiating, we get

$$\frac{\delta V}{V} = \frac{\delta x}{x} + \frac{\delta y}{y} + \frac{\delta z}{z}$$

$$\Rightarrow \quad \frac{\delta V}{V} \times 100 = \left(\frac{\delta x}{x} \times 100\right) + \left(\frac{\delta y}{y} \times 100\right) + \left(\frac{\delta z}{z} \times 100\right) = 1 + 1 + 1 = 3$$

$$\Rightarrow \quad \delta V = \frac{3}{100}(V) = \frac{3}{100}(1200) = 36 \text{ m}^3 \qquad \left| \begin{aligned} \text{Since } V &= 6 \times 50 \times 4 \\ &= 1200 \text{ m}^3 \end{aligned}\right.$$

$\therefore$ Number of bricks in $\delta V = 36 \times 12 = 432$

Thus, error in cost $= 432 \times \dfrac{100}{1000} = ₹\, 43.20.$

Example 12. *Find the possible percentage error in computing the parallel resistance r of three resistances r_1, r_2, r_3 from the formula $\frac{1}{r} = \frac{1}{r_1} + \frac{1}{r_2} + \frac{1}{r_3}$ if r_1, r_2, r_3 are each in error by + 1.2%.* (U.K.T.U. 2011)

Sol. We have
$$\frac{1}{r} = \frac{1}{r_1} + \frac{1}{r_2} + \frac{1}{r_3} \qquad ...(1)$$

Differentiating, we get

$$-\frac{1}{r^2}\delta r = -\frac{1}{r_1^2}\delta r_1 - \frac{1}{r_2^2}\delta r_2 - \frac{1}{r_3^2}\delta r_3$$

$$\Rightarrow \quad \frac{1}{r}\left(\frac{\delta r}{r} \times 100\right) = \frac{1}{r_1}\left(\frac{\delta r_1}{r_1} \times 100\right) + \frac{1}{r_2}\left(\frac{\delta r_2}{r_2} \times 100\right) + \frac{1}{r_3}\left(\frac{\delta r_3}{r_3} \times 100\right)$$

$$= \frac{1}{r_1}(1.2) + \frac{1}{r_2}(1.2) + \frac{1}{r_3}(1.2)$$

$$= 1.2\left(\frac{1}{r_1} + \frac{1}{r_2} + \frac{1}{r_3}\right) = 1.2\left(\frac{1}{r}\right) \qquad | \text{ From (1)}$$

$$\Rightarrow \quad \frac{\delta r}{r} \times 100 = 1.2$$

Hence, % error in $r = 1.2$.

Example 13. (*i*) *If Δ is the area of a triangle, prove that the error in Δ resulting from a small error in c is given by*

$$\delta\Delta = \frac{\Delta}{4}\left[\frac{1}{s} + \frac{1}{s-a} + \frac{1}{s-b} - \frac{1}{s-c}\right]\delta c.$$

(ii) Two sides a, b of a triangle and included angle C are measured. Show that the error δc in the computed length of third side c due to a small error in the angle C is given by

$$\delta c = a \sin B.\ \delta C.$$

Sol. (*i*) We know that

$$\Delta = \sqrt{s(s-a)(s-b)(s-c)}$$

$$\Rightarrow \quad \log \Delta = \frac{1}{2}\,[\log s + \log (s-a) + \log (s-b) + \log (s-c)]$$

Differentiating w.r.t. c, we get

$$\frac{1}{\Delta}\frac{\delta\Delta}{\delta c} = \frac{1}{2}\left[\frac{1}{s}\frac{\delta s}{\delta c} + \frac{1}{(s-a)}\frac{\delta(s-a)}{\delta c} + \frac{1}{(s-b)}\frac{\delta(s-b)}{\delta c} + \frac{1}{(s-c)}\frac{\delta(s-c)}{\delta c}\right] \quad \text{...(1)}$$

Now, $\quad s = \frac{1}{2}(a+b+c) \;\Rightarrow\; \frac{\delta s}{\delta c} = \frac{1}{2}$

Also, $\quad s - c = \frac{1}{2}(a+b-c) \;\Rightarrow\; \frac{\delta(s-c)}{\delta c} = -\frac{1}{2}$

Similarly, $\quad s - a = \frac{1}{2}(b+c-a) \;\Rightarrow\; \frac{\delta(s-a)}{\delta c} = \frac{1}{2}$

and $\quad s - b = \frac{1}{2}(a+c-b) \;\Rightarrow\; \frac{\delta(s-b)}{\delta c} = \frac{1}{2}$

From (1),

$$\frac{1}{\Delta}\frac{\delta\Delta}{\delta c} = \frac{1}{2}\left[\frac{1}{s}\left(\frac{1}{2}\right) + \frac{1}{s-a}\left(\frac{1}{2}\right) + \frac{1}{s-b}\left(\frac{1}{2}\right) + \frac{1}{s-c}\left(\frac{-1}{2}\right)\right]$$

$$\Rightarrow \quad \delta\Delta = \frac{\Delta}{4}\left[\frac{1}{s} + \frac{1}{s-a} + \frac{1}{s-b} - \frac{1}{s-c}\right]\delta c.$$

(*ii*) We know that,

$$\cos C = \frac{a^2 + b^2 - c^2}{2\,ab} \quad \text{...(1)} \quad | \text{ By cosine formula}$$

$$\Rightarrow \quad c^2 = a^2 + b^2 - 2\,ab \cos C \quad \text{...(2)}$$

Differentiating (2), we get

$$2c\,\delta c = -\,2\,ab(-\sin C\,\delta C)$$

$$\Rightarrow \quad c\,\delta c = ab \sin C\,\delta C \quad \text{...(3)}$$

By sine formula, we know that

$$\frac{b}{\sin B} = \frac{c}{\sin C}$$

$$\Rightarrow \quad b \sin C = c \sin B$$

$\therefore$ From (3), $c\,\delta c = ac \sin B\,\delta C$

$$\Rightarrow \quad \delta c = a \sin B.\ \delta C.$$

Example 14. *In determining the specific gravity by the formula* $S = \frac{A}{A-w}$ *where A is the weight in air and w is the weight in water ; A can be read within 0.01 gram and w within 0.02 gram. Find approximately the maximum error in S if the readings are A = 1.1 gram, w = 0.6 gram. Find also the maximum relative error.*

Sol. $$S = \frac{A}{A - w}$$

Differentiation yields,

$$\delta S = \frac{(A - w)\,\delta A - A(\delta A - \delta w)}{(A - w)^2} \quad ...(1)$$

Maximum error in S can be obtained if we take $\delta A = -\,.01$ and $\delta w = .02$

$\therefore$ From (1), $$\delta S = \frac{(1.1 - .6)(-\,.01) - (1.1)(-\,.01 - .02)}{(1.1 - .6)^2} = 0.112$$

Maximum relative error in

$$S = \frac{(\delta S)_{max}}{S} = \frac{0.112}{\left(\dfrac{1.1}{1.1 - .6}\right)} = .05091.$$

Example 15. *If the kinetic energy T is given by* $T = \frac{1}{2} mv^2$, *find approximate the change in T as the mass m changes from 49 to 49.5 and the velocity v changes from 1600 to 1590.*

Sol. $$T = \frac{1}{2} mv^2$$

$\therefore$ $$\delta T = \frac{1}{2} \{(\delta m)\, v^2 + m(2v\, \delta v)\} = \frac{1}{2} v^2\, \delta m + mv\, \delta v \quad ...(1)$$

It is given that m changes from 49 to 49.5

$\therefore$ $$\delta m = 0.5 \quad ...(2)$$

Also, v changes from 1600 to 1590

$\therefore$ $$\delta v = -\,10 \quad ...(3)$$

From (1), $$\delta T = \frac{1}{2} (1600)^2 (0.5) + 49(1600)(-\,10) = -\,144000$$

Thus, T decreases by 144000 units.

Example 16. *Compute an approximate value of* $(1.04)^{3.01}$. (U.P.T.U. 2014)

Sol. Let $f(x, y) = x^y$

then, $$\frac{\partial f}{\partial x} = yx^{y-1}, \quad \frac{\partial f}{\partial y} = x^y \log x$$

Here, let $x = 1$, $dx = 0.04$, $y = 3$, $dy = 0.01$

Now, $$df = \frac{\partial f}{\partial x} dx + \frac{\partial f}{\partial y} dy$$ | By total differentiation

$$= yx^{y-1}\, dx + x^y \log x\, dy = 3(1)^{3-1}\,(.04) + (1)^3\,(\log 1)(0.01) = 0.12$$

$\therefore$ $$(1.04)^{3.01} = f(1, 3) + df = (1)^3 + 0.12 = 1.12.$$

Example 17. *Find approximate value of* $[(0.98)^2 + (2.01)^2 + (1.94)^2]^{1/2}$.

(U.P.T.U. 2015)

Sol. Let $f(x, y, z) = (x^2 + y^2 + z^2)^{1/2}$ ...(1)

Taking $x = 1, y = 2$ and $z = 2$ so that

$$dx = -.02,\ dy = .01 \text{ and } dz = -0.06$$

From (1), $\frac{\partial f}{\partial x} = x(x^2 + y^2 + z^2)^{-1/2}, \frac{\partial f}{\partial y} = y(x^2 + y^2 + z^2)^{-1/2}, \frac{\partial f}{\partial z} = z(x^2 + y^2 + z^2)^{-1/2}$

Now, $df = \frac{\partial f}{\partial x} dx + \frac{\partial f}{\partial y} dy + \frac{\partial f}{\partial z} dz$ | By total differentiation

$$= (x^2 + y^2 + z^2)^{-1/2} (x\,dx + y\,dy + z\,dz)$$

$$= \frac{1}{3}(-.02 + .02 - .12) = -.04$$

$\therefore \quad [(0.98)^2 + (2.01)^2 + (1.94)^2]^{1/2} = f(1, 2, 2) + df = 3 + (-.04) = 2.96.$

Example 18. (*i*) *The focal length of a mirror is given by the formula* $\frac{1}{v} - \frac{1}{u} = \frac{1}{f}$.

If equal errors δ *are made in the determination of u and v, show that the relative error* $\frac{\delta f}{f}$ *in the focal length is given by* $\delta\left(\frac{1}{u} + \frac{1}{v}\right)$.

(*ii*) *The area of a* ΔABC *is determined from the side a and the two angles B and C. If there are small errors in values of B and C, show that the resulting error in the calculated value of the area will be* $\frac{1}{2}(b^2\Delta C + c^2\Delta B)$.

Sol. (*i*) We have, $\frac{1}{v} - \frac{1}{u} = \frac{1}{f}$ ...(1)

Differentiating, we get

$$-\frac{1}{v^2}\delta v + \frac{1}{u^2}\delta u = -\frac{1}{f^2}\delta f$$

$$\Rightarrow \quad -\frac{1}{v^2}\delta + \frac{1}{u^2}\delta = -\frac{1}{f}\left(\frac{\delta f}{f}\right) \quad | \because \delta v = \delta u = \delta \text{ (given)}$$

$$\Rightarrow \quad -\delta\left(\frac{1}{v} - \frac{1}{u}\right)\left(\frac{1}{v} + \frac{1}{u}\right) = -\frac{1}{f}\left(\frac{\delta f}{f}\right)$$

$$\Rightarrow \quad -\delta\left(\frac{1}{f}\right)\left(\frac{1}{v} + \frac{1}{u}\right) = -\frac{1}{f}\left(\frac{\delta f}{f}\right)$$

$$\Rightarrow \quad \frac{\delta f}{f} = \delta\left(\frac{1}{u} + \frac{1}{v}\right).$$

(*ii*) Let S be the area of ΔABC. Then,

$$S = \frac{1}{2} ab \sin C \quad ...(1)$$

and $$\frac{a}{\sin A} = \frac{b}{\sin B} = \frac{c}{\sin C} \quad ...(2)$$

$\therefore$ From (1) and (2),

$$S = \frac{1}{2} a^2 \frac{\sin B \sin C}{\sin A} \quad ...(3)$$

Taking log, $\log S = \log\left(\frac{a^2}{2}\right) + \log \sin B + \log \sin C - \log \sin A$

Differentiating, we get

$$\frac{\Delta S}{S} = \frac{\cos B}{\sin B}\Delta B + \frac{\cos C}{\sin C}\Delta C - \frac{\cos A}{\sin A}\Delta A$$

$$= \frac{\cos B}{\sin B}\Delta B + \frac{\cos C}{\sin C}\Delta C - \frac{\cos A}{\sin A}(-\Delta B - \Delta C) \quad | \because \Delta A + \Delta B + \Delta C = 0$$

$$= \left(\frac{\cos B}{\sin B} + \frac{\cos A}{\sin A}\right)\Delta B + \left(\frac{\cos C}{\sin C} + \frac{\cos A}{\sin A}\right)\Delta C$$

$$= \frac{\sin (A + B)}{\sin A \sin B}\Delta B + \frac{\sin (A + C)}{\sin A \sin C}\Delta C$$

$$= \frac{\sin C}{\sin A \sin B}\Delta B + \frac{\sin B}{\sin A \sin C}\Delta C \quad ...(4)$$

Since, $S = \frac{1}{2} b^2 \frac{\sin A \sin C}{\sin B} = \frac{1}{2} c^2 \frac{\sin A \sin B}{\sin C}$ | From (3)

$\therefore$ From (4), $\frac{\Delta S}{S} = \frac{1}{2}\frac{c^2}{S}\Delta B + \frac{1}{2}\frac{b^2}{S}\Delta C$

$\Rightarrow$ $\Delta S = \frac{1}{2}(b^2 \Delta C + c^2 \Delta B).$

Example 19. *The height h and the semi-vertical angle α of a cone are measured and from them, A, the total surface area of the cone, including the base, is calculated. If h and α are in error by small quantities δh and δα respectively, find the corresponding error in the area.*

Show further that, if $\alpha = \frac{\pi}{6}$, *an error of +1% in h will be approximately compensated by an error of –0.33° in α.*

Sol. Base radius $r = h \tan \alpha$

Slant height $l = h \sec \alpha$

Total Area $= \pi r^2 + \pi rl = \pi r (r + l)$

$= \pi h \tan \alpha (h \tan \alpha + h \sec \alpha)$

$A = \pi h^2 (\tan^2 \alpha + \sec \alpha \tan \alpha)$

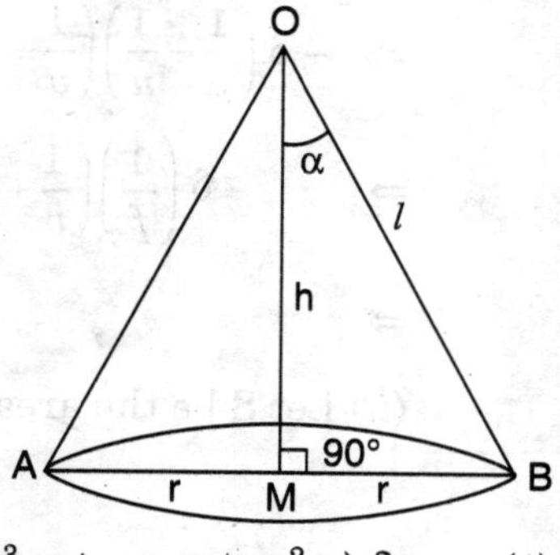

$\therefore$ $\delta A = \frac{\partial A}{\partial h}\delta h + \frac{\partial A}{\partial \alpha}\delta\alpha$

$\Rightarrow$ $\delta A = 2\pi h (\tan^2 \alpha + \sec \alpha \tan \alpha)\, \delta h$

$$+ \pi h^2 (2 \tan \alpha \sec^2 \alpha + \sec^3 \alpha + \sec \alpha \tan^2 \alpha)\, \delta\alpha \quad ...(1)$$

which gives the error in A.

Setting $\alpha = \frac{\pi}{6}$, $\delta h = \frac{h}{100}$ in (1), we get

$$\delta A = 2\pi h\left(\frac{1}{3}+\frac{2}{3}\right)\frac{h}{100} + \pi h^2\left[\frac{2}{\sqrt{3}}\left(\frac{4}{3}\right)+\frac{8}{3\sqrt{3}}+\frac{2}{3\sqrt{3}}\right]\delta\alpha$$

$$= \frac{\pi h^2}{50} + 2\sqrt{3}\,\pi h^2\,\delta\alpha$$

Since the error in h is to be compensated by the error in α,

$\therefore$ $\delta A = 0$

$\Rightarrow$ $\delta\alpha = -\frac{1}{100\sqrt{3}}$ radians $= -\frac{57.3°}{173.2} = -0.33°$ | $\because$ 1 radian = 57.3° nearly

TEST YOUR KNOWLEDGE

1. (*i*) Find the percentage error in the area of a rectangle when an error of +1 percent is made in measuring its length and breadth.

(*ii*) Find the percentage error in measuring the volume of a rectangular box when the error of 1% is made in measuring each side. (*A.K.T.U. 2017*)

2. (*i*) The period T of a simple pendulum is $T = 2\pi\sqrt{\frac{l}{g}}$. Find the maximum percentage error in T due to possible errors of 2% in l and 5% in g.

(*ii*) The time 'T' of a complete oscillation of a simple pendulum of length 'L' is governed by the equation $T = 2\pi\sqrt{\frac{L}{g}}$, g is constant. Find the approximate error in the calculated value of T corresponding to an error of 2% in the value of L.

3. (*i*) The power 'P' required to propel a steamer of length 'l' at a speed 'u' is given by $P = \lambda u^3 l^3$ where λ is constant. If u is increased by 3% and l is decreased by 1%, find the corresponding increase in 'P'. (*U.K.T.U. 2012*)

(*ii*) If the H.P. required to propel a steamer varies as the cube of the velocity and the square of length, prove that a 3% increase in velocity and 4% increase in length will require an increase of about 17% in H.P.

4. The indicated horsepower I of an engine is calculated from the formula $I = \frac{PLAN}{33000}$, where $A = \frac{\pi}{4}d^2$. Assuming that errors of r percent may have been made in measuring P, L, N and d, find the greatest possible error in I.

5. A ballon in the form of right circular cylinder of radius 1.5 m and length 4 m is surmounted by hemispherical ends. If the radius is increased by 0.01 m; find the percentage change in the volume of the balloon. (*A.K.T.U. 2018*)

6. The diameter and height of a right circular cylinder are found by measurement to be 8.0 cm and 12.5 cm respectively with possible errors of 0.05 in each measurement. Find the maximum possible approximate error in the computed volume.

7. The diameter and altitude of a can in the shape of a right circular cylinder are measured as 40 cm and 64 cm respectively. The possible error in each measurement is ± 5%. Find approximately the maximum possible percentage errors in the computed value for the volume and the lateral surface.

8. The voltage V across a resistor is measured with an error h and the resistance R is measured with an error k. Show that the error in calculating the power $W = \frac{V^2}{R}$ generated in the resistor is $\frac{V}{R^2}(2Rh - Vk)$.

9. The torsional rigidity of a length of a wire is obtained from the formula $N = \frac{8\pi I l}{t^2 r^4}$. If l is decreased by 2%, r increased by 2%, t increased by 1.5%, show that the value of N is diminished by 13% approximately.

10. (*i*) Find the possible percentage error in computing the parallel resistance r of two resistances r_1 and r_2 from the formula $\frac{1}{r} = \frac{1}{r_1} + \frac{1}{r_2}$ where the error in both r_1 and r_2 is +2% each.

(*U.K.T.U. 2011*)

(*ii*) Find the possible percentage error in computing the resultant r of three resistances r_1, r_2 and r_3 if they are connected in series. It is given that error in r_1, r_2, r_3 each is 2%.

11. The deflection at the centre of a rod of length l and diameter d supported at its ends and loaded at the centre with a weight w varies as wl^3d^{-4}. What is the increase in the deflection corresponding to p% increase in w, q% decrease in l and r% increase in d?

12. The rate of flow Q of water per second over the sharp-edged notch of length l, the height of the general level of the water above the bottom of the notch being h, is given by the formula,

$$Q = C\left(l - \frac{h}{5}\right) h^{3/2}, \text{ where C is a constant.}$$

Show that for small error δh in the measurement of h, the error δQ in Q is:

$$\frac{1}{2} C(3l - h)\, h^{1/2} \,.\, \delta h.$$

(*M.T.U. 2013*)

13. (*i*) In estimating the cost of a pile of bricks measured as 2 m × 15 m × 1.2 m, the tape is stretched 1% beyond the standard length. If the count is 450 bricks to 1 m³ and bricks cost ₹ 130 per thousand, find the approximate error in the cost.

(*ii*) In estimating the cost of a pile of bricks measured as 8 by 50 by 5 ft., the tape is stretched 1% beyond the standard length. If the count is 12 bricks to 1 ft.³ and bricks cost ₹ 1400 per thousand, find the approximate error in the cost.

14. The two sides of a triangle are measured as 50 cm and 70 cm and the angle between them is 30°. If there are possible errors of 0.5% in the measurement of the sides and 0.5 degree in that of the angle, find the maximum approximate percentage error in measuring the area of the triangle.

$\left[\textbf{Hint. } 1° = \frac{\pi}{180} \text{ radian.}\right]$ (*G.B.T.U. 2013*)

15. Compute an approximate value of $[(3.82)^2 + 2(2.1)^3]^{1/5}$. (*U.P.T.U. 2014*)

16. If $f(x, y) = x^2y^{1/10}$, compute the value of f when $x = 1.99$ and $y = 3.01$.

17. (*i*) Evaluate log $[(1.01)^{1/3} + (0.99)^{1/4} - 1]$ approximately.

(*ii*) Evaluate log $[(1.03)^{1/3} + (0.98)^{1/4} - 1]$ approximately.

18. (*i*) Calculate the error in R if E = RI and possible errors in E and I are 20% and 10% respectively.

(*ii*) The resistance 'R' of a circuit was found by formula $I = \frac{E}{R}$. If there is an error of 0.1 amp. in reading I, 0.5 volts in E, find the corresponding possible percentage error in R when readings are I = 15 amp. and E = 100 volts.

(*iii*) Show that the relative error in c due to a given error in θ is minimum when θ = 45° if $c = k \tan \theta$.

19. (*i*) Find the percentage error in calculated volume of a right circular cone whose altitude is same as the base radius and is measured as 5 cm with a possible error of 0.02 cm.

(*ii*) Considering area of a circular ring as an increment of area of a circle, find approximately area of a ring whose inner and outer radii are 3 cm and 3.02 cm respectively.

(*iii*) If the radius of sphere is measured as 5 cm with a possible error of 0.2 cm, find approximately the greatest possible error and percentage error in the compound value of the volume.

20. Calculate the percentage increase in the pressure p corresponding to a reduction of $\frac{1}{2}$% in the volume V if the p and V are related as $pV^{1.4} = c$ where c is a constant.

Answers

1. (*i*) 2%	(*ii*) 3%	**2.** (*i*) 3.5%	(*ii*) 1%
3. (*i*) 6%	**4.** 5*r*%	**5.** 1.55%	**6.** 3.3 π cu. cm.
7. ± 15%, ± 10%	**10.** (*i*) 2%	(*ii*) 2%	**11.** (*p* – 3*q* – 4*r*)%
13. (*i*) ₹ 63.18	(*ii*) 1008	**14.** 2.5115%	**15.** 2.012
16. 4.4213	**17.** (*i*) 0.00083	(*ii*) .005	
18. (*i*) 10%	(*ii*) – 16.66%		
19. (*i*) 1.2%	(*ii*) 0.12 π cm^2	(*iii*) ±20 π, ±12%	**20.** 0.7%.

3.15. MAXIMA *AND* MINIMA OF FUNCTIONS OF TWO VARIABLES

A function $f(x, y)$ is said to have a **maximum value at $x = a, y = b$ if**

$f(a, b) > f(a + h, b + k)$, for small and independent values of h and k, positive or negative.

A function $f(x, y)$ is said to have a **minimum value at $x = a, y = b$** if

$f(a, b) < f(a + h, b + k)$, for small and independent values of h and k, positive or negative.

Thus, $f(x, y)$ has a maximum or minimum value at a point (a, b) according as

$$\Delta f = f(a + h, b + k) - f(a, b) < \text{ or } > 0.$$

Geometrically, the surface $z = f(x, y)$ has a maximum at the point (a, b) if $f(a, b)$ is the greatest in the small neighbourhood of the point (a, b). As such, the surface descends in all directions at this point. The point of maxima may well be compared with the highest point of a dome. The surface has a minimum at the point (a, b) if $f(a, b)$ is the smallest in the small neighbourhood of the point (a, b). As such, the surface ascends in all directions at this point. The point of minima may well be compared with the lowest point of a bowl. However, sometimes, at the point (a, b), the tangent plane is horizontal and the surface descends in certain directions and ascends in other directions. Such a point is called a saddle point.

A maximum or a minimum value of a function is called its **extreme value.**

3.16. CONDITIONS FOR $f(x, y)$ TO BE MAXIMUM *OR* MINIMUM

By Taylor's theorem, we have

$$\Delta f = f(a + h, b + k) - f(a, b)$$

$$= [hf_x(a, b) + kf_y(a, b)] + \frac{1}{2!}[h^2 f_{xx}(a, b) + 2hkf_{xy}(a, b) + k^2 f_{yy}(a, b)] + \ldots \quad \ldots(1)$$

For small values of h and k, the second and higher order terms are still smaller and may be neglected. Thus, sign of Δf = sign of $[hf_x(a, b) + kf_y(a, b)]$.

Taking $h = 0$, the sign of Δf changes with the sign of k. Similarly taking $k = 0$, the sign of Δf changes with the sign of h. Since Δf changes sign with h and k, $f(x, y)$ cannot have a maximum or a minimum value at (a, b) unless $f_x(a, b) = 0 = f_y(a, b)$.

Hence the necessary conditions for (x, y) to have a maximum or a minimum value at (a, b) are

$$f_x(a, b) = 0, \; f_y(a, b) = 0.$$

If these conditions are satisfied, then for small values of h and k, we have, from (1),

$$\Delta f = \frac{1}{2!}[h^2 f_{xx}(a, b) + 2hkf_{xy}(a, b) + k^2 f_{yy}(a, b)] + \ldots$$

$$= \frac{1}{2!}[h^2 r + 2hks + k^2 t] + \ldots \text{ where } r = f_{xx}(a, b), \quad s = f_{xy}(a, b), \quad t = f_{yy}(a, b)$$

or
$$\Delta f = \frac{1}{2r}[h^2 r^2 + 2hkrs + k^2 rt] + \ldots = \frac{1}{2r}[(h^2 r^2 + 2hkrs + k^2 s^2) + k^2 rt - k^2 s^2)] + \ldots$$

$$= \frac{1}{2r}[(hr + ks)^2 + k^2(rt - s^2)] + \ldots \quad \ldots(2)$$

Now $(hr + ks)^2$ is always positive and $k^2(rt - s^2)$ will be positive if $rt - s^2 > 0$. In this case, Δf will have the same sign as that of r for all values of h and k.

Hence if $rt - s^2 > 0$, then $f(x, y)$ has a maximum or a minimum at (a, b) according as $r < 0$ or $r > 0$.

If $rt - s^2 < 0$, then Δf changes sign with h and k. Hence there is neither a maximum nor a minimum value at (a, b). The point (a, b) is a saddle point in this case.

If $rt - s^2 = 0$, then from (2), we have the quadratic expression $rh^2 + 2hks + tk^2 = \frac{1}{r}(rh + ks)^2$ and therefore of the same sign as r. Hence in this case, the maximum or minimum value of the function will depend on the sign of r.

If $rt - s^2 = 0$ and $\frac{h}{k} = -\frac{s}{r} = \alpha$ (say), then second degree terms of RHS of (2) vanish and we shall have to consider higher degree terms on the RHS of (2). Now the third degree terms must collectively vanish when $\frac{h}{k} = \alpha$ otherwise by changing the sign of both h and k, the sign of the expression Δf can be changed. And in this case, the fourth degree terms must collectively be of the same sign as r and t when $\frac{h}{k} = \alpha$.

Note. The point (a, b) is called a *stationary point* if $f_x(a, b) = 0$, $f_y(a, b) = 0$. The value $f(a, b)$ is called a *stationary value*. Thus every extreme value is a stationary value but the converse may not be true.

3.17. RULE TO FIND THE EXTREME VALUES OF A FUNCTION $z = f(x, y)$

(*i*) Find $\dfrac{\partial z}{\partial x}$ and $\dfrac{\partial z}{\partial y}$.

(*ii*) Solve $\dfrac{\partial z}{\partial x} = 0$ and $\dfrac{\partial z}{\partial y} = 0$ simultaneously.

Let (a, b); (c, d) ... be the solutions of these equations.

(*iii*) For each solution in step (*ii*), find $r = \dfrac{\partial^2 f}{\partial x^2}, s = \dfrac{\partial^2 f}{\partial x \partial y}, t = \dfrac{\partial^2 f}{\partial y^2}$.

(*iv*) (*a*) If $rt - s^2 > 0$ and $r < 0$ for a particular solution (a, b) of step (*ii*), then z has a maximum value at (a, b).

(*b*) If $rt - s^2 > 0$ and $r > 0$ for a particular solution (a, b) of step (*ii*), then z has a minimum value at (a, b).

(*c*) If $rt - s^2 < 0$ for a particular solution (a, b) of step (*ii*), then z has no extreme value at (a, b).

3.18. CONDITIONS FOR $f(x, y, z)$ TO BE MAXIMUM *OR* MINIMUM

Working Rule

Let the function be $u = f(x, y, z)$.

(*i*) Find $\dfrac{\partial u}{\partial x}, \dfrac{\partial u}{\partial y}, \dfrac{\partial u}{\partial z}$ and equate them to zero.

(*ii*) Find $\dfrac{\partial^2 u}{\partial x^2}, \dfrac{\partial^2 u}{\partial y^2}, \dfrac{\partial^2 u}{\partial z^2}, \dfrac{\partial^2 u}{\partial y \partial z}, \dfrac{\partial^2 u}{\partial z \partial x}, \dfrac{\partial^2 u}{\partial x \partial y}$. They are denoted by A, B, C, F, G, H respectively.

(*iii*) Also find $AB - H^2 = \begin{vmatrix} A & H \\ H & B \end{vmatrix} = D_1$ (say) and find $\begin{vmatrix} A & H & G \\ H & B & F \\ G & F & C \end{vmatrix} = D_2$ (say).

The given function u will have a minimum if $A > 0, D_1 > 0, D_2 > 0$

and will have a maximum if $A < 0, D_1 > 0, D_2 < 0$.

(*iv*) If these above conditions are not satisfied, we have neither maximum nor minimum.

ILLUSTRATIVE EXAMPLES

Example 1. *Find the extreme values of function* $x^3 + y^3 - 3axy$. [G.B.T.U. 2012]

Sol. Here $f(x, y) = x^3 + y^3 - 3axy$

$$f_x = 3x^2 - 3ay,\ f_y = 3y^2 - 3ax,\ r = f_{xx} = 6x,\ s = f_{xy} = -3a,\ t = f_{yy} = 6y$$

Now, $f_x = 0$ and $f_y = 0$

$\Rightarrow$ $x^2 - ay = 0$ ...(1) and $y^2 - ax = 0$ ...(2)

From (1), $y = \dfrac{x^2}{a}$

$\therefore$ From (2), $\dfrac{x^4}{a^2} - ax = 0$ or $x(x^3 - a^3) = 0$ or $x = 0, a$

when $x = 0, y = 0$; when $x = a, y = a$.

$\therefore$ There are two stationary points (0, 0) and (a, a).

Now, $rt - s^2 = 36xy - 9a^2$

At (0, 0) $rt - s^2 = -9a^2 < 0$

$\Rightarrow$ There is no extreme value at (0, 0).

At (a, a) $rt - s^2 = 36a^2 - 9a^2 = 27a^2 > 0$

$\Rightarrow$ $f(x, y)$ has extreme value at (a, a)

Now, $r = 6a$

If $a > 0$, $r > 0$ so that $f(x, y)$ has a **minimum** value at (a, a).

Minimum value $= a^3 + a^3 - 3a^3 = -a^3$.

If $a < 0$, $r < 0$ so that $f(x, y)$ has a **maximum** value at (a, a).

Maximum value $= -a^3 - a^3 + 3a^3 = a^3$.

Example 2. *Test the function $f(x, y) = x^3y^2(6 - x - y)$ for maxima and minima for points not at the origin.*

Sol. Here $f(x, y) = x^3y^2 (6 - x - y) = 6x^3y^2 - x^4y^2 - x^3y^3$

$$\therefore \quad f_x = 18x^2y^2 - 4x^3y^2 - 3x^2y^3$$

$$f_y = 12x^3y - 2x^4y - 3x^3y^2$$

$$r = f_{xx} = 36xy^2 - 12x^2y^2 - 6xy^3 = 6xy^2(6 - 2x - y)$$

$$s = f_{xy} = 36x^2y - 8x^3y - 9x^2y^2 = x^2y(36 - 8x - 9y)$$

$$t = f_{yy} = 12x^3 - 2x^4 - 6x^3y = x^3(12 - 2x - 6y)$$

Now, $f_x = 0$

$$\Rightarrow \quad x^2y^2(18 - 4x - 3y) = 0 \qquad \text{...(1)}$$

and $f_y = 0$

$$\Rightarrow \quad x^3y(12 - 2x - 3y) = 0 \qquad \text{...(2)}$$

From (1) and (2),

$$4x + 3y = 18$$

$$2x + 3y = 12 \quad \text{and} \quad x = 0 = y$$

Solving, we get $x = 3$, $y = 2$ and $x = 0 = y$. Leaving $x = 0 = y$, we get $x = 3$, $y = 2$.

Hence, (3, 2) is the only stationary point under consideration.

Now, $rt - s^2 = 6x^4y^2(6 - 2x - y)(12 - 2x - 6y) - x^4y^2(36 - 8x - 9y)^2$

At (3, 2) $rt - s^2 = (+)$ ve (> 0)

Also, $r = 6(3)(4)(6 - 6 - 4) = (-)$ ve (< 0)

$\therefore$ $f(x, y)$ has a maximum value at (3, 2).

Example 3. *Examine for extreme values:*

(*i*) $x^2 + y^2 + 6x + 12$ (G.B.T.U. 2012) (*ii*) $x^3 + y^3 - 63(x + y) + 12xy$. [U.K.T.U. 2011]

Sol. (*i*) Let $f(x, y) = x^2 + y^2 + 6x + 12$ ...(1)

$$\therefore \quad f_x = 2x + 6, f_y = 2y, f_{xx} = 2, f_{yy} = 2, f_{xy} = 0$$

For maxima and minima,

$$f_x = 0 \quad \text{and} \quad f_y = 0$$

$$\Rightarrow \quad 2x + 6 = 0 \quad \text{and} \quad 2y = 0$$

$$\Rightarrow \quad x = -3 \quad \text{and} \quad y = 0$$

Hence, (– 3, 0) is the stationary point.

At(– 3, 0), $rt - s^2 = 4(> 0)$

Also, $r = 2(> 0)$

Hence, $f(x, y)$ is minimum when $x = -3$ and $y = 0$.

Minimum value $= (-3)^2 + 6(-3) + 12 = 3$.

(*ii*) Let $f(x, y) = x^3 + y^3 - 63(x + y) + 12xy$ ...(1)

$\therefore$ $f_x = 3x^2 - 63 + 12y,\quad f_y = 3y^2 - 63 + 12x$

$f_{xx} = 6x,\quad f_{yy} = 6y,\quad f_{xy} = 12$

For extremum, $f_x = 0$ and $f_y = 0$

$\Rightarrow$ $3x^2 - 63 + 12y = 0$ and $3y^2 - 63 + 12x = 0$

$\Rightarrow$ $x^2 + 4y = 21$ ...(2)

and $y^2 + 4x = 21$ ...(3)

Solving (2) and (3), we get

$x^2 - y^2 - 4(x - y) = 0$

$\Rightarrow$ $(x - y)(x + y - 4) = 0$

$\Rightarrow$ $x = y$ and $x + y = 4$ ...(4)

If $x = y$, from (2), $x^2 + 4x - 21 = 0$

$\Rightarrow$ $(x + 7)(x - 3) = 0$

$\Rightarrow$ $x = -7, 3$

$\therefore$ $y = -7, 3$

Hence, stationary points are (– 7, – 7), (3, 3)

If $x + y = 4$, from (2),

$x^2 + 4(4 - x) = 21$

$\Rightarrow$ $x^2 - 4x - 5 = 0$

$\Rightarrow$ $x = -1, 5$

$\therefore$ $y = 5, -1$

Hence, other stationary points are (– 1, 5), (5, – 1).

At (– 7, – 7),

$rt - s^2 = 36(-7)(-7) - 144 = (+)\text{ ve } (> 0)$

Also, $r = 6(-7) = -42(< 0)$

Hence, $f(x, y)$ is maximum at (– 7, – 7).

Maximum value $= (-7)^3 + (-7)^3 - 63(-14) + 12(-7)(-7) = 784$

At (3, 3),

$rt - s^2 = 36(3)(3) - 144 = 180(> 0)$

Also, $r = 6(3) = 18(> 0)$

Hence, $f(x, y)$ is minimum at (3, 3).

Minimum value $= (3)^3 + (3)^3 - 63(3 + 3) + 12(9) = -216$

At (– 1, 5),

$rt - s^2 = 36(-1)(5) - 144 = -324(< 0)$

Hence, $f(x, y)$ has neither maximum nor minimum at (– 1, 5). (– 1, 5) is a saddle point.

At (5, – 1),

$rt - s^2 = 36(5)(-1) - 144 = -324(< 0)$

Hence, $f(x, y)$ has neither maximum nor minimum also at (5, – 1). (5, – 1) is a saddle point.

Example 4. *Examine for minimum and maximum values: sin x + sin y + sin (x + y).*

Sol. Here $f(x, y) = \sin x + \sin y + \sin (x + y)$

$$f_x = \cos x + \cos (x + y)$$
$$f_y = \cos y + \cos (x + y)$$
$$r = f_{xx} = -\sin x - \sin (x + y)$$
$$s = f_{xy} = -\sin (x + y)$$
$$t = f_{yy} = -\sin y - \sin (x + y)$$

Now, $f_x = 0$ and $f_y = 0$

$\Rightarrow$ $\cos x + \cos (x + y) = 0$ ...(1) and $\cos y + \cos (x + y) = 0$ ...(2)

Subtracting equation (2) from (1),

$$\cos x - \cos y = 0 \text{ or } \cos x = \cos y \quad \therefore \quad x = y$$

From (1), $\cos x + \cos 2x = 0$

or $$\cos 2x = -\cos x = \cos (\pi - x)$$

or $$2x = \pi - x \quad \therefore \quad x = \frac{\pi}{3}$$

$\therefore$ $x = y = \dfrac{\pi}{3}$ is a stationary point.

At $\left(\frac{\pi}{3}, \frac{\pi}{3}\right)$, $$r = -\frac{\sqrt{3}}{2} - \frac{\sqrt{3}}{2} = -\sqrt{3},\ s = \frac{\sqrt{3}}{2},\ t = -\frac{\sqrt{3}}{2} - \frac{\sqrt{3}}{2} = -\sqrt{3}$$

$\therefore$ $$rt - s^2 = 3 - \frac{3}{4} = \frac{9}{4} > 0$$

Also $r < 0$

$\therefore$ $f(x, y)$ has a maximum value at $\left(\frac{\pi}{3}, \frac{\pi}{3}\right)$.

$$\text{Maximum value} = f\left(\frac{\pi}{3}, \frac{\pi}{3}\right) = \sin \frac{\pi}{3} + \sin \frac{\pi}{3} + \sin \frac{2\pi}{3} = \frac{\sqrt{3}}{2} + \frac{\sqrt{3}}{2} + \frac{\sqrt{3}}{2} = \frac{3\sqrt{3}}{2}.$$

Example 5. *In a plane triangle ABC, find the maximum value of cos A cos B cos C.*

(G.B.T.U. 2010)

Sol. $\cos A \cos B \cos C = \cos A \cos B \cos [\pi - (A + B)]$ $\quad | \because A + B + C = \pi$

$$= -\cos A \cos B \cos (A + B) = f(A, B)$$

$$\frac{\partial f}{\partial A} = -\cos B[-\sin A \cos (A + B) - \cos A \sin (A + B)]$$
$$= \cos B \sin [A + (A + B)] = \cos B \sin (2A + B)$$
$$\frac{\partial f}{\partial B} = -\cos A[-\sin B \cos (A + B) - \cos B \sin (A + B)]$$
$$= \cos A \sin [B + (A + B)] = \cos A \sin (A + 2B)$$
$$r = 2 \cos B \cos (2A + B)$$
$$s = -\sin A \sin (A + 2B) + \cos A \cos (A + 2B)$$
$$= \cos [A + (A + 2B)] = \cos (2A + 2B)$$
$$t = 2 \cos A \cos (A + 2B)$$

Now, $\frac{\partial f}{\partial A} = 0$ and $\frac{\partial f}{\partial B} = 0$

$\Rightarrow$ $\cos B \sin(2A + B) = 0$ ...(1)

$\cos A \sin(A + 2B) = 0$ ...(2)

If $\cos B = 0$, then $B = \frac{\pi}{2}$

From (2), $\cos A \sin(A + \pi) = 0$ or $\cos A(-\sin A) = 0$

$\Rightarrow$ either $\cos A = 0$, *i.e.*, $A = \frac{\pi}{2}$ which is not possible ($\because$ $A + B + C = \pi \Rightarrow C = 0$)

or $\sin A = 0$, *i.e.*, $A = 0$ or π which is not possible.

$\therefore$ $\cos B \neq 0$. Similarly $\cos A \neq 0$

$\therefore$ From (1), $\sin(2A + B) = 0$ or $2A + B = \pi$

From (2), $\sin(A + 2B) = 0$ or $A + 2B = \pi$

Solving these equations, $A = B = \frac{\pi}{3}$

At $A = B = \frac{\pi}{3}$,

$$r = 2\cos\frac{\pi}{3}\cos\pi = -1,\ s = \cos\frac{4\pi}{3} = -\frac{1}{2},\quad t = 2\cos\frac{\pi}{3}\cos\pi = -1$$

$rt - s^2 = 1 - \frac{1}{4} = \frac{3}{4}$ (> 0). Also $r = -1$ (< 0)

$\Rightarrow$ $f(A, B)$ is maximum at $A = B = \frac{\pi}{3}$

$$\text{Maximum value} = f\left(\frac{\pi}{3}, \frac{\pi}{3}\right) = -\cos\frac{\pi}{3}\cos\frac{\pi}{3}\cos\frac{2\pi}{3} = -\left(\frac{1}{2}\right)\left(\frac{1}{2}\right)\left(-\frac{1}{2}\right) = \frac{1}{8}.$$

Example 6. *A rectangular box, open at the top, is to have a given capacity. Find the dimensions of the box requiring least material for its construction.*

Sol. Let x, y and z be the length, breadth and height respectively. Let V be the given capacity and S, the surface.

V is given $\Rightarrow$ V is constant

$$V = xyz \quad \text{or} \quad z = \frac{V}{xy} \quad ...(1)$$

$$S = xy + 2xz + 2yz = xy + \frac{2V}{y} + \frac{2V}{x} = f(x, y) \quad \text{[Using (1)]}$$

$$f_x = y - \frac{2V}{x^2},\ f_y = x - \frac{2V}{y^2}$$

$$r = f_{xx} = \frac{4V}{x^3},\ s = f_{xy} = 1,\ t = f_{yy} = \frac{4V}{y^3}$$

Now, $f_x = 0$ and $f_y = 0$

$\Rightarrow$ $y - \frac{2V}{x^2} = 0$ ...(1) $x - \frac{2V}{y^2} = 0$ ...(2)

From (1), $y = \frac{2V}{x^2}$

$\therefore$ From (2), $\quad x - 2V.\dfrac{x^4}{4V^2} = 0$

or $\quad x\left(1 - \dfrac{x^3}{2V}\right) = 0$ or $x = (2V)^{1/3}$ $\quad | \because x \neq 0$

and $\quad y = \dfrac{2V}{x^2} = \dfrac{2V}{(2V)^{2/3}} = (2V)^{1/3}$

$\therefore$ $x = y = (2V)^{1/3}$ is a stationary point.

At this point, $\quad r = \dfrac{4V}{2V} = 2 > 0,\ s = 1,\ t = \dfrac{4V}{2V} = 2$

so that $\quad rt - s^2 = 4 - 1 = 3 > 0$ and $r > 0$

$\Rightarrow$ S is minimum when $x = y = (2V)^{1/3}$

Also, $\quad z = \dfrac{V}{xy} = \dfrac{V}{(2V)^{2/3}} = \dfrac{(2V)^{1/3}}{2} = \dfrac{y}{2}$

Example 7. *Find the shortest distance between the lines*

$$\frac{x-3}{1} = \frac{y-5}{-2} = \frac{z-7}{1} \quad \textit{and} \quad \frac{x+1}{7} = \frac{y+1}{-6} = \frac{z+1}{1}.$$

Sol. Any point P on the first line is $(\lambda + 3, 5 - 2\lambda, 7 + \lambda)$. Similarly, any point Q on the second line is $(7\mu - 1, -1 - 6\mu, \mu - 1)$

Distance between these two points is

$$PQ = \sqrt{(\lambda + 3 - 7\mu + 1)^2 + (5 - 2\lambda + 1 + 6\mu)^2 + (7 + \lambda - \mu + 1)^2}$$

$$= \sqrt{6\lambda^2 + 86\mu^2 - 40\lambda\mu + 116}$$

If the distance is max. or min. so will be the square of the distance.

Let $\quad u = (PQ)^2 = 6\lambda^2 + 86\mu^2 - 40\lambda\mu + 116 \quad$ ...(1)

$$\frac{\partial u}{\partial \lambda} = 12\lambda - 40\mu, \quad \frac{\partial u}{\partial \mu} = 172\mu - 40\lambda$$

For stationary points,

$$\frac{\partial u}{\partial \lambda} = 0 \Rightarrow 12\lambda - 40\mu = 0 \quad \text{...(2)}$$

and $$\frac{\partial u}{\partial \mu} = 0 \Rightarrow 172\mu - 40\lambda = 0 \quad \text{...(3)}$$

Solving (2) and (3), we get $\lambda = 0, \mu = 0$ $\therefore$ (0, 0) is stationary point

Now, $\quad \dfrac{\partial^2 u}{\partial \lambda^2} = 12,\ \dfrac{\partial^2 u}{\partial \mu^2} = 172,\ \dfrac{\partial^2 u}{\partial \lambda\, \partial \mu} = -40$

$$rt - s^2 = (12)(172) - (-40)^2 > 0$$

Also, $r > 0$

Hence, u is minimum at (0, 0). Shortest distance is given by $PQ = \sqrt{116} = 2\sqrt{29}$.

Example 8. (*i*) *Given* $f(x, y, z) = \dfrac{5xyz}{x + 2y + 4z}$. *Find the values of x, y, z for which f(x, y, z) is a maximum, subject to the condition xyz = 8.*

(*ii*) *Examine for extreme values:* $f(x, y, z) = x^2 + y^2 + z^2 - xy + x - 2z$.

Sol. (*i*)

$$u = \frac{5xyz}{x + 2y + 4z} \qquad ...(1)$$

$$xyz = 8 \qquad ...(2)$$

From (1) and (2),

$$u = \frac{40}{x + 2y + 4z}$$

$\therefore$

$$du = -\frac{40}{(x + 2y + 4z)^2}(dx + 2dy + 4dz)$$

For max. and min. of u, $du = 0$

$\Rightarrow$

$$dx + 2dy + 4dz = 0 \qquad ...(3)$$

From (2),

$$\log x + \log y + \log z = \log 8$$

Differentiation gives, $\dfrac{1}{x}dx + \dfrac{1}{y}dy + \dfrac{1}{z}dz = 0 \qquad ...(4)$

From (3) and (4), after introducing λ such that

$$dx\left(1 + \frac{\lambda}{x}\right) + dy\left(2 + \frac{\lambda}{y}\right) + dz\left(4 + \frac{\lambda}{z}\right) = 0$$

$\Rightarrow$

$$1 + \frac{\lambda}{x} = 0 \qquad \Rightarrow \quad x = -\lambda$$

$$2 + \frac{\lambda}{y} = 0 \qquad \Rightarrow \quad y = -\frac{\lambda}{2}$$

$$4 + \frac{\lambda}{z} = 0 \qquad \Rightarrow \quad z = -\frac{\lambda}{4}$$

From (2),

$$-\frac{\lambda^3}{8} = 8 \qquad \Rightarrow \quad \lambda = -4$$

$\therefore$ u is stationary at the point $x = 4, y = 2, z = 1$.

Now,

$$u = \frac{40}{x + 2y + 4z}$$

$\therefore$

$$\frac{\partial u}{\partial x} = \frac{-40}{(x + 2y + 4z)^2} \cdot \left(1 + 4\frac{\partial z}{\partial x}\right)$$

From (2), $\log x + \log y + \log z = \log 8$

Differentiating w.r.t. x partially, we get

$$\frac{1}{x} + \frac{1}{z}\frac{\partial z}{\partial x} = 0 \qquad \Rightarrow \quad \frac{\partial z}{\partial x} = \frac{-z}{x}$$

$$\therefore \quad \frac{\partial z}{\partial x} = \frac{-40}{(x+2y+4z)^2}\cdot\left(1-\frac{4z}{x}\right)$$

$$\therefore \quad \frac{\partial^2 u}{\partial x^2} = \frac{80}{(x+2y+4z)^3}\left(1+4\frac{\partial z}{\partial x}\right)\left(1-\frac{4z}{x}\right)-\frac{40}{(x+2y+4z)^2}\left(\frac{4z}{x^2}-\frac{4}{x}\cdot\frac{\partial z}{\partial x}\right)$$

$$= \frac{80}{(x+2y+4z)^3}\left(1-\frac{4z}{x}\right)^2-\frac{40}{(x+2y+4z)^2}\left(\frac{8z}{x^2}\right)$$

At the stationary point,

$$r = \frac{80}{(12)^3}(1-1)^2-\frac{40}{144}\left(\frac{1}{4}+\frac{1}{4}\right) < 0$$

$\therefore$ u is maximum at the point (4, 2, 1).

(*ii*) We have,

$$u = x^2+y^2+z^2-xy+x-2z \qquad ...(1)$$

$$\frac{\partial u}{\partial x} = 2x-y+1,\ \frac{\partial u}{\partial y} = 2y-x,\ \frac{\partial u}{\partial z} = 2z-2$$

For extreme values,

$$\frac{\partial u}{\partial x} = 0,\quad \frac{\partial u}{\partial y} = 0 \quad\text{and}\quad \frac{\partial u}{\partial z} = 0$$

$$\therefore \quad 2x-y+1=0,\ 2y-x=0 \text{ and } 2z-2=0$$

Solving above equations, we get

$$x = -\frac{2}{3},\ y = -\frac{1}{3},\ z = 1$$

Hence, $\left(-\frac{2}{3},-\frac{1}{3},1\right)$ is the stationary point.

Now,
$$A = \frac{\partial^2 u}{\partial x^2} = 2(>0),\quad B = \frac{\partial^2 u}{\partial y^2} = 2,\quad C = \frac{\partial^2 u}{\partial z^2} = 2$$

$$F = \frac{\partial^2 u}{\partial y\,\partial z} = 0,\quad G = \frac{\partial^2 u}{\partial z\,\partial x} = 0,\quad H = \frac{\partial^2 u}{\partial x\,\partial y} = -1$$

$$\therefore \quad D_1 = AB - H^2 = 4-1 = 3(>0)$$

$$D_2 = \begin{vmatrix} A & H & G \\ H & B & F \\ G & F & C \end{vmatrix} = \begin{vmatrix} 2 & -1 & 0 \\ -1 & 2 & 0 \\ 0 & 0 & 2 \end{vmatrix} = 6(>0)$$

Since $A > 0$, $D_1 > 0$ and $D_2 > 0$,

hence the given function u will have a minimum at $\left(-\frac{2}{3},-\frac{1}{3},1\right)$

and minimum value $= \left(-\frac{2}{3}\right)^2+\left(-\frac{1}{3}\right)^2+1-\frac{2}{9}-\frac{2}{3}-2 = -\frac{4}{3}$.

3.19. LAGRANGE'S METHOD OF UNDETERMINED MULTIPLIERS

To find the maximum or minimum values of a function of three (or more) variables, when the variables are not independent but are connected by some given relation, we try to convert the given function to the one, having least number of independent variables with the help of the given relation.

When this procedure is not practicable, we use Lagrange's method.

Let $f(x, y, z)$ be a function of x, y, z which is to be examined for maximum or minimum value.

Let the variables x, y, z be connected by the relation $\phi(x, y, z) = 0$ ...(1)

For $f(x, y, z)$ to have a maximum or minimum value, the necessary condition is

$$\frac{\partial f}{\partial x} = 0, \frac{\partial f}{\partial y} = 0, \frac{\partial f}{\partial z} = 0$$

$$\therefore \quad \frac{\partial f}{\partial x}\, dx + \frac{\partial f}{\partial y}\, dy + \frac{\partial f}{\partial z}\, dz = 0 \qquad ...(2)$$

Also, from (1), taking differentials, we get $\frac{\partial \phi}{\partial x}\, dx + \frac{\partial \phi}{\partial y}\, dy + \frac{\partial \phi}{\partial z}\, dz = 0$ ...(3)

Multiplying (3) by a parameter λ and adding to (2), we get

$$\left(\frac{\partial f}{\partial x} + \lambda \frac{\partial \phi}{\partial x}\right) dx + \left(\frac{\partial f}{\partial y} + \lambda \frac{\partial \phi}{\partial y}\right) dy + \left(\frac{\partial f}{\partial z} + \lambda \frac{\partial \phi}{\partial z}\right) dz = 0$$

This equation will hold good if $\frac{\partial f}{\partial x} + \lambda \frac{\partial \phi}{\partial x} = 0, \frac{\partial f}{\partial y} + \lambda \frac{\partial \phi}{\partial y} = 0, \frac{\partial f}{\partial z} + \lambda \frac{\partial \phi}{\partial z} = 0.$

These equations together with equation (1), give the values of x, y, z and λ for a maximum or minimum.

Lagrange's method does not enable us to find whether there is a maximum or minimum. This fact is determined from the physical considerations of the problem.

Note. The above equations can be easily obtained by considering Lagrange's function

$$F(x, y, z) = f(x, y, z) + \lambda\phi(x, y, z)$$

and considering the stationary values of $F(x, y, z)$. For stationary values of $F(x, y, z)$, $dF = 0$.

$$\Rightarrow \quad \left(\frac{\partial f}{\partial x} + \lambda \frac{\partial \phi}{\partial x}\right) dx + \left(\frac{\partial f}{\partial y} + \lambda \frac{\partial \phi}{\partial y}\right) dy + \left(\frac{\partial f}{\partial z} + \lambda \frac{\partial \phi}{\partial z}\right) dz = 0$$

$$\Rightarrow \quad \frac{\partial f}{\partial x} + \lambda \frac{\partial \phi}{\partial x} = 0, \frac{\partial f}{\partial y} + \lambda \frac{\partial \phi}{\partial y} = 0, \frac{\partial f}{\partial z} + \lambda \frac{\partial \phi}{\partial z} = 0.$$

Example 9. *A wire of length b is cut into two parts which are bent in the form of a square and circle respectively. Find the least value of the sum of the areas so found using Lagrange's method of multipliers.*

Sol. Let x and y be two parts into which the given wire is cut so that

$$x + y = b \qquad ...(1)$$

Let the piece of wire of length x be bent into a square so that each side $= \frac{x}{4}$

$$\therefore \quad \text{Area of square} = \frac{x^2}{16} \quad \text{...(2)}$$

Similarly, the piece of wire of length y is bent into a circle with perimeter y.

$$\therefore \quad \text{Area of circle} = \pi\left(\frac{y}{2\pi}\right)^2 = \frac{y^2}{4\pi} \quad \text{...(3)}$$

Consider Lagrange's function

$$F(x, y) = \left(\frac{x^2}{16} + \frac{y^2}{4\pi}\right) + \lambda(x + y - b) \quad \text{...(4)}$$

For stationary values,

$$dF = 0$$

$$\Rightarrow \quad \frac{x}{8} + \lambda = 0 \quad \text{...(5)}$$

and $$\frac{y}{2\pi} + \lambda = 0 \quad \text{...(6)}$$

$$\Rightarrow \quad x = -8\lambda \quad \text{and} \quad y = -2\pi\lambda$$

Substituting these values in (1), we get

$$-8\lambda - 2\pi\lambda = b$$

$$\lambda = \frac{-b}{8 + 2\pi} \quad \text{...(7)}$$

$$\therefore \quad x = -8\lambda = \frac{8b}{8 + 2\pi} \quad \text{and} \quad y = -2\pi\lambda = \frac{2\pi b}{8 + 2\pi}$$

The least value of the sum of areas

$$= \frac{x^2}{16} + \frac{y^2}{4\pi} = \frac{1}{16} \cdot \frac{64b^2}{(8 + 2\pi)^2} + \frac{1}{4\pi} \cdot \frac{4\pi^2 b^2}{(8 + 2\pi)^2}$$

$$= \frac{b^2(\pi + 4)}{4(\pi + 4)^2} = \frac{b^2}{4(\pi + 4)}.$$

Example 10. *The sum of three positive numbers is constant. Prove that their product is maximum when they are equal.*

Sol. Let x, y, z be the three positive numbers.

$$x + y + z = k \quad \text{(say)} \quad \text{...(1)}$$

Let $$u = xyz \quad \text{...(2)}$$

Consider Lagrange's function

$$F(x, y, z) = xyz + \lambda(x + y + z - k)$$

For stationary values,

$$dF = 0$$

$$\Rightarrow \quad (yz + \lambda)\, dx + (zx + \lambda)\, dy + (xy + \lambda)\, dz = 0$$

$$\Rightarrow \quad yz + \lambda = 0 \quad \text{...(3)}$$

$$zx + \lambda = 0 \quad \text{...(4)}$$

$$xy + \lambda = 0 \quad \text{...(5)}$$

Multiplying (3), (4) and (5) by x, y and z respectively, and adding, we get

$$3xyz + \lambda(x + y + z) = 0$$

$$\Rightarrow \quad 3u + \lambda k = 0$$

$$\Rightarrow \quad \lambda = -\frac{3u}{k}$$

From (3), (4) and (5),

$$yz - \frac{3u}{k} = 0$$

$$zx - \frac{3u}{k} = 0$$

and

$$xy - \frac{3u}{k} = 0$$

$$\Rightarrow \quad xyz - \frac{3ux}{k} = 0 \quad \Rightarrow \quad u - \frac{3ux}{k} = 0$$

$$xyz - \frac{3uy}{k} = 0 \quad \Rightarrow \quad u - \frac{3uy}{k} = 0$$

$$xyz - \frac{3uz}{k} = 0 \quad \Rightarrow \quad u - \frac{3uz}{k} = 0$$

$$\Rightarrow \quad x = \frac{k}{3},\ y = \frac{k}{3},\ z = \frac{k}{3}.$$

Hence, $\left(\frac{k}{3}, \frac{k}{3}, \frac{k}{3}\right)$ is the stationary point. Now to find whether u is maximum or minimum.

Let z be a function of x and y.

$$u = xy(k - x - y) = kxy - x^2y - xy^2$$

$$\frac{\partial u}{\partial x} = ky - 2xy - y^2 \quad \text{and} \quad \frac{\partial u}{\partial y} = kx - x^2 - 2xy$$

Now,

$$r = \frac{\partial^2 u}{\partial x^2} = -2y, \quad s = \frac{\partial^2 u}{\partial x \partial y} = k - 2x - 2y, \quad t = \frac{\partial^2 u}{\partial y^2} = -2x$$

$$\therefore \quad rt - s^2 = 4xy - (k - 2x - 2y)^2 = 4\frac{k}{3}.\frac{k}{3} - \left(k - \frac{2k}{3} - \frac{2k}{3}\right)^2 = \frac{4k^2}{9} - \frac{k^2}{9} = \frac{k^2}{3} > 0$$

Also,

$$r = -2y = -2\frac{k}{3} < 0$$

Hence, u is maximum at $x = y = z = \frac{k}{3}$ and maximum value $= \frac{k}{3}.\frac{k}{3}.\frac{k}{3} = \frac{k^3}{27}$.

Example 11. *Show that the rectangular solid of maximum volume that can be inscribed in a given sphere is a cube.*

Sol. Let $2x, 2y, 2z$ be the length, breadth and height of the rectangular solid and let R be the radius of the sphere.

Then, $x^2 + y^2 + z^2 = R^2$ ...(1)

Volume $V = 8xyz$ ...(2)

Consider Lagrange's function

$$F(x, y, z) = 8xyz + \lambda(x^2 + y^2 + z^2 - R^2)$$

For stationary values,

$$dF = 0$$

$$\Rightarrow \{8yz + \lambda(2x)\}\, dx + \{8xz + \lambda(2y)\}\, dy + \{8xy + \lambda(2z)\}\, dz = 0$$

$$\Rightarrow \quad 8yz + 2\lambda x = 0 \quad ...(3)$$

$$8zx + 2\lambda y = 0 \quad ...(4)$$

$$8xy + 2\lambda z = 0 \quad ...(5)$$

From (3), $2\lambda x^2 = -8xyz$

From (4), $2\lambda y^2 = -8xyz$

From (5), $2\lambda z^2 = -8xyz$

$\therefore$ $2\lambda x^2 = 2\lambda y^2 = 2\lambda z^2$

or $x^2 = y^2 = z^2$ or $x = y = z.$

Hence, rectangular solid is a cube.

Example 12. *Find the volume of the largest rectangular parallelopiped that can be inscribed in the ellipsoid* $\dfrac{x^2}{a^2} + \dfrac{y^2}{b^2} + \dfrac{z^2}{c^2} = 1.$ (A.K.T.U. 2020)

Sol. Let (x, y, z) be a vertex of the parallelopiped then it lies on the given ellipsoid

$$\therefore \quad \phi(x, y, z) \equiv \frac{x^2}{a^2} + \frac{y^2}{b^2} + \frac{z^2}{c^2} - 1 = 0 \quad ...(1)$$

Let $2x, 2y, 2z$ be the length, breadth and height of the rectangular parallelopiped inscribed in the ellipsoid.

$\therefore$ Volume $V = 8xyz$ ...(2)

Consider Lagrange's function

$$F(x, y, z) = 8xyz + \lambda\left(\frac{x^2}{a^2} + \frac{y^2}{b^2} + \frac{z^2}{c^2} - 1\right)$$

For stationary values,

$$dF = 0$$

$$\Rightarrow \left\{8yz + \lambda\left(\frac{2x}{a^2}\right)\right\} dx + \left\{8zx + \lambda\left(\frac{2y}{b^2}\right)\right\} dy + \left\{8xy + \lambda\left(\frac{2z}{c^2}\right)\right\} dz = 0$$

$$\Rightarrow \quad 8yz + \lambda\left(\frac{2x}{a^2}\right) = 0 \quad ...(3)$$

$$8zx + \lambda\left(\frac{2y}{b^2}\right) = 0 \qquad ...(4)$$

$$8xy + \lambda\left(\frac{2z}{c^2}\right) = 0 \qquad ...(5)$$

Multiplying (3), (4), (5) by x, y, z respectively and adding, we get

$$24xyz + 2\lambda\left(\frac{x^2}{a^2} + \frac{y^2}{b^2} + \frac{z^2}{c^2}\right) = 0$$

$\Rightarrow$ $$24xyz + 2\lambda = 0 \Rightarrow \lambda = -12xyz$$

From (3), $$8yz - 12xyz\left(\frac{2x}{a^2}\right) = 0$$

$\Rightarrow$ $$1 - \frac{3x^2}{a^2} = 0 \Rightarrow x = \frac{a}{\sqrt{3}}$$

Similarly from (4) and (5), we get $y = \frac{b}{\sqrt{3}}$, $z = \frac{c}{\sqrt{3}}$.

$\therefore$ Volume of the largest rectangular parallelopiped $= 8xyz = \frac{8abc}{3\sqrt{3}}$.

Example 13. *The temperature T at any point (x, y, z) in space is $T = 400\, xyz^2$. Find the highest temperature at the surface of a unit sphere $x^2 + y^2 + z^2 = 1$.*

Sol. Consider Lagrange's function

$$F(x, y, z) = 400\, xyz^2 + \lambda(x^2 + y^2 + z^2 - 1) \qquad ...(1)$$

For stationary values,

$$dF = 0$$

$\Rightarrow$ $[400yz^2 + \lambda(2x)]\, dx + [400\, xz^2 + \lambda(2y)]\, dy + [800\, xyz + \lambda(2z)]\, dz = 0$

$\Rightarrow$ $$400yz^2 + 2\,\lambda x = 0 \qquad ...(2)$$

$$400xz^2 + 2\,\lambda y = 0 \qquad ...(3)$$

$$800xyz + 2\,\lambda z = 0 \qquad ...(4)$$

Multiplying (2) by x, (3) by y and (4) by z and adding, we get

$$1600xyz^2 + 2\,\lambda(x^2 + y^2 + z^2) = 0$$

$\Rightarrow$ $$\lambda = -800xyz^2 \qquad ...(5) \qquad |\because\ x^2 + y^2 + z^2 = 1$$

From (2), $400yz^2 - 1600x^2yz^2 = 0$ | Using (5)

$\Rightarrow$ $$x = \pm\frac{1}{2}$$

Similarly, $$y = \pm\frac{1}{2}$$

From (4) $800xyz - 1600xyz^3 = 0$

$\Rightarrow$ $$1 - 2z^2 = 0$$

$\Rightarrow$ $$z = \pm\frac{1}{\sqrt{2}}$$

Putting values of x, y, z in T, we get

$$T = 400\left(\frac{1}{2}\right)\left(\frac{1}{2}\right)\left(\frac{1}{2}\right) = 50.$$

Example 14. *Find the maximum and minimum distances of the point (3, 4, 12) from the sphere* $x^2 + y^2 + z^2 = 1$. (U.K.T.U. 2010)

Sol. Let (x, y, z) be any point on the sphere.

Distance of the point A(3, 4, 12) from (x, y, z) is given by $\sqrt{(x-3)^2 + (y-4)^2 + (z-12)^2}$.

If the distance is maximum or minimum, so will be the square of the distance.

Let $$f(x, y, z) = (x-3)^2 + (y-4)^2 + (z-12)^2 \quad ...(1)$$

subject to the condition that $\phi(x, y, z) \equiv x^2 + y^2 + z^2 - 1 = 0 \quad ...(2)$

Consider Lagrange's function

$$F(x, y, z) = f(x, y, z) + \lambda\phi(x, y, z)$$
$$= (x-3)^2 + (y-4)^2 + (z-12)^2 + \lambda(x^2 + y^2 + z^2 - 1)$$

For stationary values, $dF = 0$.

$$\Rightarrow \quad [2(x-3) + 2\lambda x]dx + [2(y-4) + 2\lambda y]dy + [2(z-12) + 2\lambda z]dz = 0$$

$$\Rightarrow \quad 2(x-3) + 2\lambda x = 0 \quad ...(3)$$
$$2(y-4) + 2\lambda y = 0 \quad ...(4)$$
$$2(z-12) + 2\lambda z = 0 \quad ...(5)$$

Multiplying (3) by x, (4) by y, (5) by z and adding, we get

$$2(x^2 + y^2 + z^2) - 6x - 8y - 24z + 2\lambda(x^2 + y^2 + z^2) = 0$$

or $$2 - 6x - 8y - 24z + 2\lambda = 0 \quad | \text{ Using (2)}$$

or $$3x + 4y + 12z = 1 + \lambda \quad ...(6)$$

From (3), (4) and (5), $x = \frac{3}{1+\lambda}, y = \frac{4}{1+\lambda}, z = \frac{12}{1+\lambda}$.

Putting these values of x, y, z in (6), we have

$$\frac{9}{1+\lambda} + \frac{16}{1+\lambda} + \frac{144}{1+\lambda} = 1 + \lambda$$

or $$(1+\lambda)^2 = 169 \text{ or } 1 + \lambda = \pm 13$$

$\therefore$ $$\lambda = 12 \text{ or } -14$$

When $\lambda = 12$, $$x = \frac{3}{13}, y = \frac{4}{13}, z = \frac{12}{13}$$

When $\lambda = -14$, $$x = -\frac{3}{13}, y = -\frac{4}{13}, z = -\frac{12}{13}$$

Thus, we get two points $P\left(\frac{3}{13}, \frac{4}{13}, \frac{12}{13}\right)$ and $Q\left(-\frac{3}{13}, -\frac{4}{13}, -\frac{12}{13}\right)$

on the sphere which are at a *maximum or minimum* distance from the given point A.

Now, $$AP = \sqrt{\left(3 - \frac{3}{13}\right)^2 + \left(4 - \frac{4}{13}\right)^2 + \left(12 - \frac{12}{13}\right)^2} = 12$$

$$AQ = \sqrt{\left(3+\frac{3}{13}\right)^2+\left(4+\frac{4}{13}\right)^2+\left(12+\frac{12}{13}\right)^2} = 14$$

$\therefore$ P $\left(\frac{3}{13}, \frac{4}{13}, \frac{12}{13}\right)$ is at a minimum distance from A and the minimum distance = 12.

Q $\left(-\frac{3}{13}, -\frac{4}{13}, -\frac{12}{13}\right)$ is at a maximum distance from A and the maximum distance = 14.

Example 15. *(i) Find the minimum distance from the point (1, 2, 0) to the cone $z^2 = x^2 + y^2$.*
(ii) Find a point on the paraboloid $z = x^2 + y^2$ nearest to the point (3, – 6, 4).

Sol. *(i)* Let (x, y, z) be any point on the cone then distance from the point (1, 2, 0) is

$$D = \sqrt{(x-1)^2+(y-2)^2+z^2}$$

If the distance is maximum or minimum, so will be the square of the distance.

Let $\quad u = D^2 = (x-1)^2 + (y-2)^2 + z^2 \quad$ ...(1)

Also, $\quad \phi \equiv x^2 + y^2 - z^2 = 0 \quad$ ...(2)

Consider Lagrange's function

$$F(x, y, z) = u + \lambda\phi = (x-1)^2 + (y-2)^2 + z^2 + \lambda(x^2 + y^2 - z^2) \quad ...(3)$$

For stationary values,

$$dF = 0$$

$\Rightarrow \quad \{2(x-1) + \lambda(2x)\}\, dx + \{2(y-2) + \lambda(2y)\}\, dy + \{2z + \lambda(-2z)\}\, dz = 0$

$\Rightarrow \quad 2(x-1) + \lambda(2x) = 0 \quad$ ...(4)

$2(y-2) + \lambda(2y) = 0 \quad$ ...(5)

$2z + \lambda(-2z) = 0 \quad$ ...(6)

$\Rightarrow \quad x = \frac{1}{1+\lambda},\ y = \frac{2}{1+\lambda},\ \lambda = 1$

$\therefore \quad x = \frac{1}{2},\ y = 1$

Putting the values of x and y in eqn. (2), we get

$$\frac{1}{4} + 1 - z^2 = 0 \Rightarrow z^2 = \frac{5}{4} \Rightarrow z = \pm\frac{\sqrt{5}}{2}$$

Hence, minimum distance from the point (1, 2, 0) is

$$u = \left(\frac{1}{2}-1\right)^2 + (1-2)^2 + \frac{5}{4} = \frac{5}{2} = D^2$$

$\therefore \quad D = \sqrt{\frac{5}{2}}$

(ii) Let (x, y, z) be any point on paraboloid nearest to the point (3, – 6, 4).

$\therefore \quad D = \sqrt{(x-3)^2+(y+6)^2+(z-4)^2}$

Let $\quad u = D^2 = (x-3)^2 + (y+6)^2 + (z-4)^2 \quad$ ...(1)

Also, $\quad \phi \equiv x^2 + y^2 - z = 0 \quad$ ...(2)

Consider Lagrange's function

$$F(x, y, z) = (x-3)^2 + (y+6)^2 + (z-4)^2 + \lambda(x^2 + y^2 - z) \quad ...(3)$$

For stationary values,

$$dF = 0$$

$$\therefore \quad 2(x-3) + \lambda(2x) = 0 \quad ...(4)$$

$$2(y+6) + \lambda(2y) = 0 \quad ...(5)$$

$$2(z-4) + \lambda(-1) = 0 \quad ...(6)$$

From (4), (5) and (6),

$$x = \frac{3}{1+\lambda},\ y = \frac{-6}{1+\lambda},\ z = \frac{\lambda+8}{2}$$

Putting values of x, y, z in (2), we get

$$\left(\frac{3}{1+\lambda}\right)^2 + \left(\frac{-6}{1+\lambda}\right)^2 = \frac{\lambda+8}{2}$$

$$\Rightarrow \lambda^3 + 10\lambda^2 + 17\lambda - 82 = 0$$

$$\Rightarrow \quad \lambda = 2$$

Hence, $\quad x = 1, \quad y = -2, z = 5$

Example 16. *Find the maximum and minimum distances from the origin to the curve*

$$x^2 + 4xy + 6y^2 = 140.$$

(M.T.U. 2012)

Sol. Let (x, y) be any point on the curve. Distance of the point A(0, 0) from (x, y) is given by $\sqrt{(x-0)^2 + (y-0)^2}$.

If the distance is maximum or minimum, so will be the square of the distance.

Let $\quad f(x, y) = x^2 + y^2 \quad ...(1)$

subject to the condition

$$\phi(x, y) \equiv x^2 + 4xy + 6y^2 - 140 = 0 \quad ...(2)$$

Consider Lagrange's function

$$F(x, y) = f(x, y) + \lambda\phi(x, y) = x^2 + y^2 + \lambda(x^2 + 4xy + 6y^2 - 140)$$

For stationary values,

$$dF = 0$$

$$\Rightarrow \quad [2x + \lambda(2x + 4y)]\,dx + [2y + \lambda(4x + 12y)]\,dy = 0$$

$$\Rightarrow \quad 2x + \lambda(2x + 4y) = 0 \quad ...(3)$$

$$2y + \lambda(4x + 12y) = 0 \quad ...(4)$$

Multiplying (3) by x and (4) by y and adding,

$$2x^2 + 2y^2 + \lambda(2x^2 + 8xy + 12y^2) = 0$$

$$\Rightarrow \quad \lambda = -\left(\frac{x^2+y^2}{140}\right) = -\frac{f}{140} \quad \mid \because x^2 + 4xy + 6y^2 = 140$$

From (3) and (4),

$$2x - \frac{f}{70}(x + 2y) = 0$$

and

$$2y - \frac{f}{35}(x + 3y) = 0$$

$$\Rightarrow \quad (140 - f)x - 2fy = 0 \quad \text{...(5)}$$

and
$$-fx + (70 - 3f)y = 0 \quad \text{...(6)}$$

Solving (5) and (6), we get

$$f^2 - 490f + 9800 = 0$$

$$\Rightarrow \quad f = \frac{490 \pm \sqrt{(490)^2 - 4(9800)}}{2} = 245 \pm 35\sqrt{41} = 469.1093,\ 20.8906$$

Hence, maximum distance $= \sqrt{469.1093} = 21.6589$

minimum distance $= \sqrt{20.8906} = 4.5706.$

Example 17. *Find the minimum value of $x^2 + y^2 + z^2$, given that $ax + by + cz = p$.*

(U.K.T.U. 2012)

Sol. Let
$$u = x^2 + y^2 + z^2 \quad \text{...(1)}$$

where
$$\phi(x, y, z) = ax + by + cz - p = 0 \quad \text{...(2)}$$

Consider Lagrange's function, $F(x, y, z) = (x^2 + y^2 + z^2) + \lambda(ax + by + cz - p)$

For stationary values, $dF = 0$

$$\Rightarrow \quad (2x + \lambda a)dx + (2y + \lambda b)dy + (2z + \lambda c)dz = 0$$

$$\Rightarrow \quad 2x + \lambda a = 0 \quad \text{...(3)}$$

$$2y + \lambda b = 0 \quad \text{...(4)}$$

$$2z + \lambda c = 0 \quad \text{...(5)}$$

Multiplying (3) by x, (4) by y, (5) by z and adding, we get

$$2(x^2 + y^2 + z^2) + \lambda(ax + by + cz) = 0$$

or
$$2u + \lambda p = 0 \quad | \text{ Using (1) and (2)}$$

$$\therefore \quad \lambda = -\frac{2u}{p}$$

From (3), (4) and (5),
$$x = \frac{au}{p},\ y = \frac{bu}{p},\ z = \frac{cu}{p}$$

$\therefore$ From (1),
$$u = \frac{(a^2 + b^2 + c^2)u^2}{p^2}$$

or
$$u = \frac{p^2}{a^2 + b^2 + c^2}.$$

This is the **maximum or minimum** value of u. Now u is the square of the distance of any point $P(x, y, z)$ on the plane (2) from the origin. Also, the length of perpendicular from O on the plane is $\dfrac{p}{\sqrt{a^2 + b^2 + c^2}}$. Clearly, OP is least when P coincides with M, the foot of the perpendicular from O on the plane.

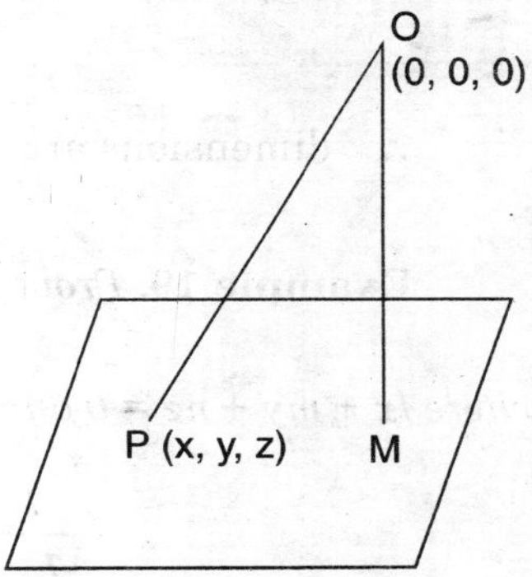

Hence, the minimum value of $u = \dfrac{p^2}{a^2 + b^2 + c^2}.$

Example 18. *Find the dimensions of a rectangular box of maximum capacity whose surface area is given when*

(*i*) *box is open at the top* (A.K.T.U. 2016)

(*ii*) *box is closed.* (A.K.T.U. 2016)

Sol. Let x, y, z be the dimensions of the rectangular box so that its volume is

$$V = xyz \quad ...(1)$$

Total surface area of the box is

$$S = nxy + 2yz + 2zx = \text{given constant} \quad ...(2)$$

Here, $n = 1$ if the box is open at the top and $n = 2$ if the box is closed on all sides.

Consider Lagrange's function

$$F(x, y, z) \equiv xyz + \lambda (nxy + 2yz + 2zx - S) \quad ...(3)$$

For stationary values,

$$dF = 0$$

$$\Rightarrow \quad \{yz + \lambda (ny + 2z)\} dx + \{zx + \lambda(nx + 2z)\} dy + \{xy + \lambda (2y + 2x)\} = 0$$

$$\Rightarrow \quad yz + \lambda (ny + 2z) = 0 \quad ...(4)$$

$$zx + \lambda (nx + 2z) = 0 \quad ...(5)$$

$$xy + \lambda (2y + 2x) = 0 \quad ...(6)$$

Multiplying eqns. (4), (5) and (6) by x, y and z respectively, we get

$$xyz + \lambda (nxy + 2zx) = 0 \quad ...(7)$$

$$xyz + \lambda (nxy + 2zy) = 0 \quad ...(8)$$

$$xyz + \lambda (2yz + 2xz) = 0 \quad ...(9)$$

Equating, we get

$$nx = ny = 2z$$

(*i*) When box is open at the top, $n = 1$

$$x = y = 2z$$

From eqn. (2),

$$S = x^2 + x^2 + x^2 = 3x^2$$

$$\therefore \quad \text{dimensions are,} \quad x = \sqrt{\frac{S}{3}} = y, \qquad z = \frac{1}{2}\sqrt{\frac{S}{3}}$$

(*ii*) When box is closed, $n = 2$

$$x = y = z$$

from eqn. (2),

$$S = 2x^2 + 2x^2 + 2x^2 = 6x^2$$

$$\therefore \quad \text{dimensions are,} \quad x = \sqrt{\frac{S}{6}} = y = z$$

Example 19. *Prove that the stationary values of* $u = \frac{x^2}{a^4} + \frac{y^2}{b^4} + \frac{z^2}{c^4}$ *where* $lx + my + nz = 0$ *and* $\frac{x^2}{a^2} + \frac{y^2}{b^2} + \frac{z^2}{c^2} = 1$ *are the roots of the equation*

$$\frac{l^2 a^4}{1 - a^2 u} + \frac{m^2 b^4}{1 - b^2 u} + \frac{n^2 c^4}{1 - c^2 u} = 0.$$

Sol. Consider Lagrange's function,

$$F(x, y, z) = \left(\frac{x^2}{a^4} + \frac{y^2}{b^4} + \frac{z^2}{c^4}\right) + \lambda\,(lx + my + nz) + \mu\left(\frac{x^2}{a^2} + \frac{y^2}{b^2} + \frac{z^2}{c^2} - 1\right)$$

For stationary values, $dF = 0$

$$\Rightarrow \quad \left(\frac{2x}{a^4} + \lambda l + \frac{2\mu x}{a^2}\right) dx + \left(\frac{2y}{b^4} + \lambda m + \frac{2\mu y}{b^2}\right) dy + \left(\frac{2z}{c^4} + \lambda n + \frac{2\mu z}{c^2}\right) dz = 0$$

$$\Rightarrow \quad \frac{2x}{a^4} + \lambda l + \frac{2\mu x}{a^2} = 0 \qquad \text{...(1)}$$

$$\frac{2y}{b^4} + \lambda m + \frac{2\mu y}{b^2} = 0 \qquad \text{...(2)}$$

$$\frac{2z}{c^4} + \lambda n + \frac{2\mu z}{c^2} = 0 \qquad \text{...(3)}$$

Multiplying (1), (2), (3) by x, y, z respectively and adding, we get

$$2\left(\frac{x^2}{a^4} + \frac{y^2}{b^4} + \frac{z^2}{c^4}\right) + \lambda\,(lx + my + nz) + 2\mu\left(\frac{x^2}{a^2} + \frac{y^2}{b^2} + \frac{z^2}{c^2}\right) = 0$$

$$\Rightarrow \quad 2u + \lambda(0) + 2\mu\,(1) = 0 \qquad |\text{ from given relations}$$

$$\Rightarrow \quad \mu = -u$$

$\therefore$ Equation (1) becomes $\dfrac{2x}{a^4} + \lambda l - \dfrac{2ux}{a^2} = 0$ or $\dfrac{2x}{a^4}(1 - a^2u) = -\lambda l$ or $x = -\dfrac{\lambda l a^4}{2(1 - a^2u)}$

Similarly, $\qquad y = -\dfrac{\lambda m b^4}{2(1 - b^2u)}, \; z = -\dfrac{\lambda n c^4}{2(1 - c^2u)}$

To eliminate λ between them, multiply these values of x, y, z by l, m, n respectively and add. Then

$$lx + my + nz = -\frac{\lambda}{2}\left[\frac{l^2a^4}{1 - a^2u} + \frac{m^2b^4}{1 - b^2u} + \frac{n^2c^4}{1 - c^2u}\right]$$

Since $lx + my + nz = 0$, we have $\dfrac{l^2a^4}{1 - a^2u} + \dfrac{m^2b^4}{1 - b^2u} + \dfrac{n^2c^4}{1 - c^2u} = 0$

which is a quadratic in u and gives two stationary values of u.

Example 20. (*i*) *Use Lagrange's method of undetermined multipliers to find the minimum value of* $x^2 + y^2 + z^2$ *subject to the conditions* $x + y + z = 1$, $xyz + 1 = 0$.

(*ii*) *If x and y satisfy the relation* $ax^2 + by^2 = ab$, *prove that the extreme values of the function* $u = x^2 + xy + y^2$ *are given by the roots of the equation* $4(u - a)(u - b) = ab$.

Sol. (*i*) Here $\qquad u = x^2 + y^2 + z^2$

subject to $x + y + z = 1$ and $xyz + 1 = 0$.

For a minimum u,

$$du = 2x\,dx + 2y\,dy + 2z\,dz = 0$$

$$\Rightarrow \quad x\,dx + y\,dy + z\,dz = 0 \qquad \text{...(1)}$$

Also, $$dx + dy + dz = 0 \qquad \text{...(2)}$$

and $$yz\,dx + zx\,dy + xy\,dz = 0 \qquad \text{...(3)}$$

Multiplying (1) by 1, (2) by λ and (3) by μ and adding, we get

$$x + \lambda + \mu\, yz = 0 \quad \ldots(4)$$

$$y + \lambda + \mu\, zx = 0 \quad \ldots(5)$$

$$z + \lambda + \mu\, xy = 0 \quad \ldots(6)$$

Subtracting (5) from (4), we get

$$x - y + \mu\, z(y - x) = 0$$

$$\Rightarrow \quad (x - y)(1 - \mu z) = 0$$

$$\Rightarrow \quad x = y \quad \text{or} \quad \mu z = 1$$

Similarly, $\quad y = z \quad \text{or} \quad \mu x = 1$ | From (5) and (6)

From these, we choose the solution,

$$x = y \quad \text{and} \quad \mu = \frac{1}{x} \quad \ldots(7)$$

Multiplying (4) and (5) by x and y respectively and subtracting (5) from (4), we get

$$x^2 - y^2 + \lambda\,(x - y) = 0$$

$$\Rightarrow \quad (x - y)(x + y + \lambda) = 0$$

$$\Rightarrow \quad x = y \quad \text{or} \quad x + y + \lambda = 0$$

Similarly, $\quad y = z \quad \text{or} \quad y + z + \lambda = 0 \quad \ldots(8)$ | From (5) and (6)

From (4), (7) and (8), $\quad x - (y + z) + \dfrac{1}{x}(yz) = 0$

$$\Rightarrow \quad x - (1 - x) + \frac{1}{x}\left(-\frac{1}{x}\right) = 0$$

$$\Rightarrow \quad 2x^3 - x^2 - 1 = 0$$

$$\Rightarrow \quad (1 - x)(2x^2 + x + 1) = 0$$

Second factor gives imaginary roots. A real solution is

$$x = 1,\ y = x = 1,\ z = \frac{xyz}{xy} = -1$$

Hence, the minimum $u = 1 + 1 + (-1)^2 = 3$.

Since u can increase indefinitely for numerically large x, y, z therefore the above value is a minimum.

(*ii*) Here $\quad u = x^2 + xy + y^2$

Consider Lagrange's function

$$F(x, y) = x^2 + xy + y^2 + \lambda\left(\frac{x^2}{b} + \frac{y^2}{a} - 1\right). \quad \ldots(1)$$

For stationary values,

$$dF = 0$$

$$\Rightarrow \quad \left[(2x + y) + \lambda\left(\frac{2x}{b}\right)\right]dx + \left[x + 2y + \lambda\left(\frac{2y}{a}\right)\right]dy = 0$$

$$\Rightarrow \quad 2x + y + \frac{2\lambda x}{b} = 0 \quad \ldots(2)$$

$$x + 2y + \frac{2\lambda y}{a} = 0 \quad \ldots(3)$$

Multiplying (2) by x and (3) by y and adding, we get

$$2(x^2 + xy + y^2) + 2\lambda\left(\frac{x^2}{b} + \frac{y^2}{a}\right) = 0$$

$$\Rightarrow \qquad 2u + 2\lambda(1) = 0 \qquad \left|\ \because \frac{x^2}{b} + \frac{y^2}{a} = 1\right.$$

$$\Rightarrow \qquad \lambda = -u \qquad ...(4)$$

From (2), $$2x + y - \frac{2ux}{b} = 0 \qquad |\text{ Using (4)}$$

$$\Rightarrow \qquad 2\left(1 - \frac{u}{b}\right)x + y = 0 \qquad ...(5)$$

From (3), $$x + 2y - \frac{2uy}{a} = 0 \qquad |\text{ Using (4)}$$

$$\Rightarrow \qquad x + 2\left(1 - \frac{u}{a}\right)y = 0 \qquad ...(6)$$

From (5), $$\frac{x}{y} = \frac{-1}{2\left(1 - \frac{u}{b}\right)}$$

From (6), $$\frac{x}{y} = -2\left(1 - \frac{u}{a}\right)$$

Equating values of $\frac{x}{y}$, we get

$$\frac{-1}{2\left(1 - \frac{u}{b}\right)} = -2\left(1 - \frac{u}{a}\right)$$

$$\Rightarrow \qquad 4\left(\frac{u}{a} - 1\right)\left(\frac{u}{b} - 1\right) = 1 \Rightarrow 4(u - a)(u - b) = ab$$

Example 21. *Find the maxima and minima of $u = x^2 + y^2 + z^2$ subject to the conditions $ax^2 + by^2 + cz^2 = 1$ and $lx + my + nz = 0$. Interpret the result geometrically.*

Sol. Let us define a function F such that

$$F = x^2 + y^2 + z^2 + \lambda_1(ax^2 + by^2 + cz^2 - 1) + \lambda_2(lx + my + nz) \qquad ...(1)$$

For maxima and minima of F, we have

$$\frac{\partial F}{\partial x} = 2x + 2ax\lambda_1 + l\lambda_2 = 0 \qquad ...(2)$$

$$\frac{\partial F}{\partial y} = 2y + 2by\lambda_1 + m\lambda_2 = 0 \qquad ...(3)$$

$$\frac{\partial F}{\partial z} = 2z + 2cz\lambda_1 + n\lambda_2 = 0 \qquad ...(4)$$

Multiplying eqns. (2), (3), (4) by x, y, z respectively and adding, we get

$$2(x^2 + y^2 + z^2) + 2\lambda_1(ax^2 + by^2 + cz^2) + \lambda_2(lx + my + nz) = 0$$

$$\Rightarrow \qquad 2u + 2\lambda_1 = 0 \Rightarrow \lambda_1 = -u$$

$\therefore$ From (2), $2x - 2axu + l\lambda_2 = 0$

$$\Rightarrow \qquad x = \frac{l\lambda_2}{2(au-1)}$$

Similarly, $$y = \frac{m\lambda_2}{2(bu-1)}, \quad z = \frac{n\lambda_2}{2(cu-1)}$$

Substituting these values of x, y, z in $lx + my + nz = 0$, we get

$$l \cdot \frac{l\lambda_2}{2(au-1)} + m \cdot \frac{m\lambda_2}{2(bu-1)} + n \cdot \frac{n\lambda_2}{2(cu-1)} = 0$$

$$\Rightarrow \qquad \frac{l^2}{au-1} + \frac{m^2}{bu-1} + \frac{n^2}{cu-1} = 0$$

which gives the reqd. maximum and minimum values of u.

Geometrical interpretation. **$x^2 + y^2 + z^2$ is the square of the distance of any point (x, y, z) from the origin (0, 0, 0). Hence in this problem, we have found the maximum and minimum values of the square of distance of the origin from the point of intersection of the central conicoid $ax^2 + by^2 + cz^2 = 1$ by the central plane $lx + my + nz = 0$.**

TEST YOUR KNOWLEDGE

1. Examine for extreme values:
   (*i*) $x^3 + y^3 - 3xy$ (*ii*) $3x^2 - y^2 + x^3$
   (*iii*) $x^2y^2 - 5x^2 - 8xy - 5y^2$ (*iv*) $x^3 + 3xy^2 - 15x^2 - 15y^2 + 72x$
   (*v*) $x^2 - xy + y^2 + 3x - 2y + 1$ (*vi*) $xy + \frac{a^3}{x} + \frac{a^3}{y}$
   (*vii*) $\sin x \sin y \sin (x + y)$ (*viii*) $\cos x + \cos y + \cos (x + y)$
   (*ix*) $x^3y^2(1 - x - y)$ (*x*) $2(x - y)^2 - x^4 - y^4$.
2. (*i*) Divide 24 into three parts such that the continued product of the first, square of the second and the cube of the third may be maximum. (*A.K.T.U. 2021*)
   (*ii*) Divide a number, say 120 into three parts so that the sum of their product taken two at a time shall be maximum.
   (*iii*) Divide a number into three parts such that the product of first, square of the second and cube of third is maximum. (*A.K.T.U. 2017*)
3. (*i*) Find the maximum and minimum values of $ax + by$ when $xy = c^2$.
   (*ii*) Find the minimum value of $x^2 + y^2$ when $ax + by = c$.
4. (*i*) Find the stationary points of the function $z = x^3y^2(12 - x - y)$ satisfying the condition $x > 0$, $y > 0$ and examine their nature.
   (*ii*) Locate the stationary points of $x^4 + y^4 - 2x^2 + 4xy - 2y^2$ and determine their nature. (*M.T.U. 2013*)
   [**Hint.** At (0, 0), third order partial derivatives vanish while fourth order partial derivatives are positive and rest all zero.]
5. Find the minimum values of $x^2 + y^2 + z^2$ when
   (*i*) $x + y + z = 3a$ (*ii*) $xyz = a^3$.
6. Find the minimum value of the function $x + y + z$ subject to the condition $\frac{a}{x} + \frac{b}{y} + \frac{c}{z} = 1$.
7. Find the minimum value of $ax + by + cz$ subject to the condition $\frac{a^2}{x} + \frac{b^2}{y} + \frac{c^2}{z} = 1$.

8. Find the minimum value of the function $x^2 + y^2 + z^2$ subject to the condition
(i) $ax + by + cz = a + b + c$ (ii) $xy + yz + zx = 3a^2$.
9. (i) Find the points on the surface $z^2 = xy + 1$ nearest to the origin.
(ii) Find the shortest distance from origin to the surface $xyz^2 = 2$.
10. (i) Find the shortest distance from the point (1, 2, 2) to the sphere $x^2 + y^2 + z^2 = 36$.
(ii) Use Lagrange's method to determine the minimum distance from the origin to the plane $3x + 2y + z = 12$.
11. Show that the maximum and minimum values of $r^2 = x^2 + y^2$ where $ax^2 + 2hxy + by^2 = 1$ are given by the roots of the quadratic equation

$$\left(a - \frac{1}{r^2}\right)\left(b - \frac{1}{r^2}\right) = h^2.$$

12. Find the extreme value of $x^2 + y^2 + z^2$ subject to the condition $xy + yz + zx = p$.
13. (i) Find the shortest and the longest distances from the point (1, 2, – 1) to the sphere

$$x^2 + y^2 + z^2 = 24.$$ *(A.K.T.U. 2015, 2018, 2019)*

(ii) Find the maximum and minimum distances from the origin to the curve

$$5x^2 + 6xy + 5y^2 - 8 = 0.$$

14. Find the maxima and minima of $u = x^2 + y^2 + z^2$ where $ax^2 + by^2 + cz^2 = 1$
15. The temperature 'T' at any point (x, y, z) in space is $T(x, y, z) = Kxyz^2$ where K is constant. Find the highest temperature on the surface of the sphere $x^2 + y^2 + z^2 = a^2$.
16. If $u = a^3x^2 + b^3y^2 + c^3z^2$, where $x^{-1} + y^{-1} + z^{-1} = 1$, show that the stationary value of u is given by

$$x = \frac{\Sigma a}{a}, y = \frac{\Sigma a}{b}, z = \frac{\Sigma a}{c}.$$

17. (i) Find the maximum value of $x^m y^n z^p$, when $x + y + z = a$.
(ii) Find the maximum value of $x^p y^q z^r$ when the variables x, y, z are subject to the condition

$$ax + by + cz = p + q + r.$$

18. (i) A rectangular box, which is open at the top, has a capacity of 32 c.c. Determine, using Lagrange's method of multipliers, the dimensions of the box such that the least material is required for the construction of the box. *(A.K.T.U. 2014, 2015, 2022)*
(ii) A rectangular box, which is open at the top, has a capacity of 256 cubic feet. Determine the dimensions of the box such that the least material is required for the construction of the box. Use Lagrange's method of multipliers to obtain the solution. *(U.K.T.U. 2011)*
19. If $u = ax^2 + by^2 + cz^2$ where $x^2 + y^2 + z^2 = 1$ and $lx + my + nz = 0$, prove that stationary values of u satisfy the equation $\frac{l^2}{a-u} + \frac{m^2}{b-u} + \frac{n^2}{c-u} = 0.$ *(U.K.T.U. 2011)*
20. Find a point in the plane $x + 2y + 3z = 13$ nearest to the point (1, 1, 1) using the method of Lagrange's multipliers.

Answers

1. (i) Min. value = – 1 at (1, 1) (ii) Max. value = 4 at (– 2, 0)
(iii) Max. value = 0 at (0, 0)
(iv) Max. value = 112 at (4, 0) ; Min. value = 108 at (6, 0)
(v) Min. value $= -\frac{4}{3}$ at $\left(-\frac{4}{3}, \frac{1}{3}\right)$ (vi) Min. value = $3a^2$ at (a, a)
(vii) Max. value $= \frac{3\sqrt{3}}{8}$ at $\left(\frac{\pi}{3}, \frac{\pi}{3}\right)$; Min. value $= -\frac{3\sqrt{3}}{8}$ at $\left(\frac{2\pi}{3}, \frac{2\pi}{3}\right)$
(viii) Min. value $= -\frac{3}{2}$ at $\left(\frac{2\pi}{3}, \frac{2\pi}{3}\right)$ (ix) Max. value $= \frac{1}{432}$ at $\left(\frac{1}{2}, \frac{1}{3}\right)$
(x) Max. value = 8 at $(\sqrt{2}, -\sqrt{2})$ and $(-\sqrt{2}, \sqrt{2})$.

2. (*i*) 4, 8, 12 (*ii*) 40, 40, 40

(*iii*) $\frac{a}{6}, \frac{a}{3}, \frac{a}{2}$ if number is '*a*'

3. (*i*) Min. value $= 2c\sqrt{ab}$, Max. value $= -2c\sqrt{ab}$ (*ii*) $\frac{c^2}{a^2 + b^2}$

4. (*i*) Maximum at (6, 4)

(*ii*) Max. at (0, 0), max. value = 0 ; Min. at $(\sqrt{2}, -\sqrt{2})$, min. value = – 8

Min. at $(-\sqrt{2}, \sqrt{2})$, min. value = – 8

5. (*i*) $3a^2$ (*ii*) $3a^2$

6. $(\sqrt{a} + \sqrt{b} + \sqrt{c})^2$ at $\{\sqrt{a}(\sqrt{a} + \sqrt{b} + \sqrt{c}), \sqrt{b}(\sqrt{a} + \sqrt{b} + \sqrt{c}), \sqrt{c}(\sqrt{a} + \sqrt{b} + \sqrt{c})\}$

7. $(a^{3/2} + b^{3/2} + c^{3/2})^2$

8. (*i*) $\frac{(a + b + c)^2}{a^2 + b^2 + c^2}$ (*ii*) $3a^2$ at (a, a, a) and $(-a, -a, -a)$

9. (*i*) (0, 0, ± 1) (*ii*) 2

10. (*i*) 3 (*ii*) 3.2071

12. p

13. (*i*) $\sqrt{6}, 3\sqrt{6}$ (*ii*) 4, 1

14. Maximum and minimum values of u are the roots of $\left(u - \frac{1}{a}\right)\left(u - \frac{1}{b}\right)\left(u - \frac{1}{c}\right) = 0$

15. $\frac{ka^4}{8}$

17. (*i*) $\frac{m^m n^n p^p \, a^{m+n+p}}{(m + n + p)^{m+n+p}}$ (*ii*) $\left(\frac{p}{a}\right)^p \left(\frac{q}{b}\right)^q \left(\frac{r}{c}\right)^r$

18. (*i*) $x = 4, y = 4, z = 2$ (*ii*) $x = 8, y = 8, z = 4$

20. $\left(\frac{3}{2}, 2, \frac{5}{2}\right)$.

ASSIGNMENT III

1. Determine the point(s) where the function $u = x^2 + y^2 + 6x + 12$ has a maximum or minimum. (*G.B.T.U. 2012*)
2. Find the stationary points of $f(x, y) = 5x^2 + 10y^2 + 12xy - 4x - 6y + 1$. (*U.P.T.U. 2013, 2014*)
3. What is the maximum value of function $f(x, y) = 1 - x^2 - y^2$?
4. Find the Taylor's series expansion of $f(x, y) = x^3 + xy^2$ about point (2, 1). (*A.K.T.U. 2018*)
5. If $A = F_{xx}(a, b)$, $B = F_{xy}(a, b)$, $C = F_{yy}(a, b)$, then what is the condition for the function $F(x, y)$ to be maximum at (a, b)?
6. At what point, $u = x^2 + 2y^2 + 3z^2 - 2xy - 2yz - 2$ is maximum?
7. What is the condition for two functions $u(x, y)$ and $v(x, y)$ to be functionally dependent?
8. If δx is the error in x, then find the relative error in x.
9. The formula $V = kr^4$ says that the volume V of fluid flowing through a small pipe or tube in a unit of time at a fixed pressure is a constant times the fourth power of the tube's radius r. How will a 10% increase in r affect V?
10. State the Taylor's theorem for two variables. (*A.K.T.U. 2019*)

11. (*i*) If $u = lx + my$, $v = mx - ly$, then find $\frac{\partial (x, y)}{\partial (u, v)}$ (*U.P.T.U. 2014*)

(*ii*) If $x = uv$, $y = \frac{u+v}{u-v}$, then find $\frac{\partial (x, y)}{\partial (u, v)}$. (*U.P.T.U. 2014*)

(*iii*) If $x = r\cos\theta$ and $y = r\sin\theta$, then find $\frac{\partial(x, y)}{\partial(r, \theta)}$.

(*iv*) If $x = u(1 + v)$, $y = v(1 + u)$, then find the Jacobian of u, v with respect to x, y. (*A.K.T.U. 2018*)

(*v*) If $u = e^x \sin y$ and $v = e^x \cos y$, evaluate: $\frac{\partial(u, v)}{\partial(x, y)}$. (*G.B.T.U. 2012*)

(*vi*) Calculate $\frac{\partial(u, v)}{\partial(x, y)}$ for $x = e^u \cos v$ and $y = e^u \sin v$. (*U.P.T.U. 2015*)

(*vii*) If $u = x \sin y$ and $v = y \sin x$, then find $\frac{\partial(u, v)}{\partial(x, y)}$.

(*viii*) Find the Jacobian $\frac{\partial(u, v)}{\partial(x, y)}$ for the functions $u = e^x \sin y$, $v = x + \log \sin y$. (*U.P.T.U. 2009*)

12. What is the relation between $J = \frac{\partial(u, v)}{\partial(x, y)}$ and $J^* = \frac{\partial(x, y)}{\partial(u, v)}$. (*U.P.T.U. 2014*)

13. If $u = x + \frac{y^2}{x}$ and $v = \frac{y^2}{x}$ then find $J^* = \frac{\partial(x, y)}{\partial(u, v)}$.

14. If $u = \frac{yz}{x}$, $v = \frac{zx}{y}$, $w = \frac{xy}{z}$, then find $\frac{\partial(u, v, w)}{\partial(x, y, z)}$.

15. If $u = e^x \sin y$, $v = x \log \sin y$, then find the Jacobian $\frac{\partial(u, v)}{\partial(x, y)}$. (*G.B.T.U. 2010*)

16. If $u = \frac{2yz}{x}$, $v = \frac{3zx}{y}$, $w = \frac{4xy}{z}$, then find $\frac{\partial(u, v, w)}{\partial(x, y, z)}$.

17. If $u = xyz$, $v = xy + yz + zx$, $w = x + y + z$, then find $\frac{\partial(u, v, w)}{\partial(x, y, z)}$.

18. If $u = 3x + 2y - z$, $v = x - y + z$, $w = x + 2y - z$, then find $\frac{\partial(u, v, w)}{\partial(x, y, z)}$.

19. If $x = u^2 v$, $y = uv^2$ and $u = x^2 - y^2$, $v = xy$, then find the Jacobian $\frac{\partial(x, y)}{\partial(u, v)}$.

20. What is the functional relation between $u = \frac{x}{y}$ and $v = \frac{x+y}{x-y}$.

21. What is the functional relation between

(*i*) $u = \frac{x^2 - y^2}{x^2 + y^2}$, $v = \frac{2xy}{x^2 + y^2}$ (*ii*) $u = \frac{x - y}{x + a}$, $v = \frac{x + a}{y + a}$ (where a is constant)

(*iii*) $u = \frac{x}{y - z}$, $v = \frac{y}{z - x}$, $w = \frac{z}{x - y}$.

22. What is the functional relation between $u = \frac{3x^2}{2(y + z)}$, $v = \frac{2(y + z)}{3(x - y)^2}$ and $w = \frac{x - y}{x}$?

23. What is the minimum value of $x^2 + y^2$?

24. Find the maximum value of the function $f(x, y, z) = z - 2x^2 - 2y^2$ where $3xy - z + 7 = 0$. (*A.K.T.U. 2017*)

25. Find the Jacobian $\dfrac{\partial(u, v)}{\partial(x, y)}$ if

(i) $u = 3x + 5y$, $v = 4x - 3y$ (ii) $u = x + y$, $v = \dfrac{y}{x + y}$

(iii) $u = x + y$, $v = xy$ (iv) $u = \dfrac{x + y}{1 - xy}$, $v = \tan^{-1} x + \tan^{-1} y$.

26. Find approximately:

(i) $(2.98)^3$ (ii) $\sqrt{4.05}$ (iii) $\dfrac{1}{2.1}$ (iv) $(83.7)^{1/4}$.

27. What is the shortest distance from the point (1, 2, – 1) to the sphere $x^2 + y^2 + z^2 = 24$? *(A.K.T.U. 2018)*

28. Which point is the plane $3x - 4y + 5z = 50$ exists nearest to the origin?

29. Compute the approximate value of $(1.04)^{3.01}$. *(U.P.T.U. 2014)*

30. Find the Taylor's series expansion of $f(x) = x^5 + 2x^4 + x - x^2 + 1$ about the point $x = -1$.

31. If $f(x) = f(0) + kf_1(0) + \dfrac{k^2}{2!} f_2(\theta k)$, $0 < \theta < 1$, then find the value of θ when $k = 1$ and $f(x) = (1 - x)^{5/2}$. *(G.B.T.U. 2010)*

32. If $pv^2 = k$ and the relative errors in p and v are respectively 0.05 and 0.025, show that the error in k is 10%. *(A.K.T.U. 2016)*

33. Determine the functional relation between:

(i) $u = x^2 e^{-y} \cosh z$, $v = x^2 e^{-y} \sinh z$, $w = 3x^4 e^{-2y}$

(ii) $u = 4x^2 + 9y^2 + 16z^2$, $v = 2x + 3y + 4z$, $w = 12xy + 16xz + 24yz$

(iii) $u = x + y + z$, $v = x^3 + y^3 + z^3 - 3xyz$, $w = x^2 + y^2 + z^2 - xy - yz - zx$

(iv) $u = x + y + z$, $v = x^2 + y^2 + z^2$, $w = x^3 + y^3 + z^3 - 3xyz$.

34. Find the stationary point of $f(x, y) = x^3 + y^3 + 3axy$, $a > 0$. *(A.K.T.U. 2019)*

35. Find the critical points of the function $f(x, y) = x^3 + y^3 - 3axy$. *(A.K.T.U. 2022)*

Answers

1. (–3, 0), Minimum, Min. value = 3 **2.** $\left(\dfrac{1}{7}, \dfrac{3}{14}\right)$ **3.** 1

4. $f(x, y) = 10 + [13(x - 2) + 4(y - 1)] + 2[3(x - 2)^2 + (y - 1)^2 + (x - 2)(y - 1)] + [(x - 2)^3 + (x - 2)(y - 1)^2] + \ldots$

5. $F_x = 0 = F_y$, $AC > B^2$ and $A < 0$ **6.** (1, 1)

7. their Jacobian is zero. **8.** $\dfrac{\delta x}{x}$ **9.** 40% increase

11. (i) – 1 (ii) $\dfrac{4uv}{(u - v)^2}$ (iii) r (iv) $\dfrac{1}{1 + u + v}$

(v) $-e^{2x}$ (vi) $\dfrac{1}{x^2 + y^2}$ or e^{-2u} (vii) $\sin x \sin y - xy \cos x \cos y$

(viii) 0

12. $JJ^* = 1$ **13.** $\dfrac{x}{2y}$ or $\dfrac{u - v}{2\sqrt{v(u - v)}}$ **14.** 4

15. $e^x \cos y\,(x - \log \sin y)$ **16.** 96 **17.** $(x - y)(y - z)(z - x)$

18. -2 **19.** $6x^2y^2(x^2+y^2)(x^2-y^2)^2$ **20.** $v = \dfrac{u+1}{u-1}$

21. (*i*) $u^2 + v^2 = 1$ (*ii*) $v = \dfrac{1}{1-u}$ (*iii*) $vw + wu + uv + 1 = 0$

22. $uvw^2 = 1$ **23.** 0 **24.** 7

25. (*i*) -29 (*ii*) $\dfrac{1}{x+y}$ (*iii*) $x - y$ (*iv*) 0

26. (*i*) 26.46 (*ii*) 2.13 (*iii*) 0.475 (*iv*) 3.025

27. $\sqrt{6}$ **28.** $(3, -4, 5)$ **29.** 1.12.

30. $f(x) = (x+1)^2 + 2(x+1)^3 - 3(x+1)^4 + (x+1)^5$ **31.** 0.36

33. (*i*) $3u^2 - 3v^2 = w$ (*ii*) $v^2 = u + w$ (*iii*) $uw = v$ (*iv*) $2w = u(3v - v^2)$

34. $(0, 0)$; $(-a, -a)$ **35.** $(0, 0)$; (a, a)

MODULE 4

Multiple Integration

Multiple integral is a natural extension of a definite integral to a function of two variables (double integral) or three variables (triple integral) or more variables. Double and triple integrals are useful in finding area, volume, mass, centroid and moments of inertia of plane and solid regions.

4.1. DOUBLE INTEGRALS

The definite integral $\int_a^b f(x)\, dx$ is defined as the limits of the sum $f(x_1)\delta x_1 + f(x_2)\delta x_2 + \ldots + f(x_n)\delta x_n$ when $n \to \infty$ and each of the lengths $\delta x_1, \delta x_2, \ldots, \delta x_n$ tends to zero. Here $\delta x_1, \delta x_2, \ldots, \delta x_n$ are n sub-intervals into which the range $b-a$ has been divided and $x_1, x_2, \ldots, x_n$ are values of x lying respectively in the first, second, ..., n^{th} sub-interval.

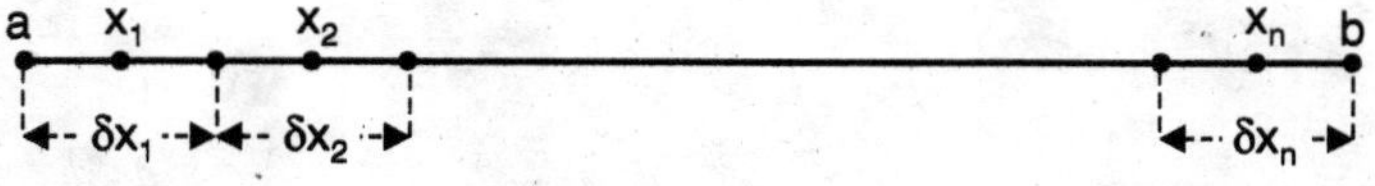

A double integral is its counterpart in two dimensions. Let a single-valued and bounded function $f(x, y)$ of two independent variables x, y be defined in a closed region R of the xy-plane. Divide the region R into sub-regions by drawing lines parallel to co-ordinate axes. Number the rectangles which lie entirely inside the region R, from 1 to n. Let (x_r, y_r) be any point inside the rth rectangle whose area is δA_r.

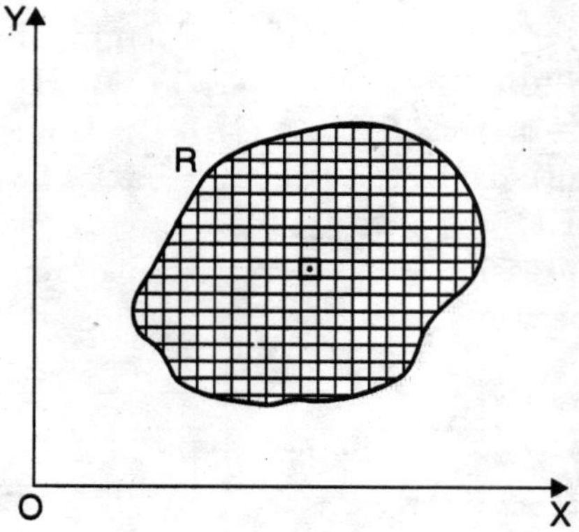

Consider the sum

$$f(x_1, y_1)\delta A_1 + f(x_2, y_2)\delta A_2 + \ldots + f(x_n, y_n)\delta A_n$$

$$= \sum_{r=1}^{n} f(x_r, y_r)\delta A_r \qquad \ldots(1)$$

Let the number of these sub-regions increase indefinitely, such that the largest linear dimension (*i.e.*, diagonal) of δA_r approaches zero. The limit of the sum (1), if it exists, irrespective of the mode of sub-division, is called the *double integral* of $f(x, y)$ over the region R and is denoted by $\iint_R f(x, y)\, dA$

In other words, $$\underset{\substack{n \to \infty \\ \delta A_r \to 0}}{\text{Lt}} \sum_{r=1}^{n} f(x_r, y_r)\, \delta A_r = \iint_R f(x, y)\, dA$$

which is also expressed as $$\iint_R f(x, y)\, dx\, dy \quad \text{or} \quad \iint_R f(x, y)\, dy\, dx$$

4.2. EVALUATION OF DOUBLE INTEGRALS

The methods of evaluating the double integrals depend upon the nature of the curves bounding the region R. Let the region R be bounded by the curves $x = x_1$, $x = x_2$ and $y = y_1$, $y = y_2$.

(*i*) **When x_1, x_2 are functions of y and y_1, y_2 are constants.** Let AB and CD be the curves $x_1 = \phi_1(y)$ and $x_2 = \phi_2(y)$.

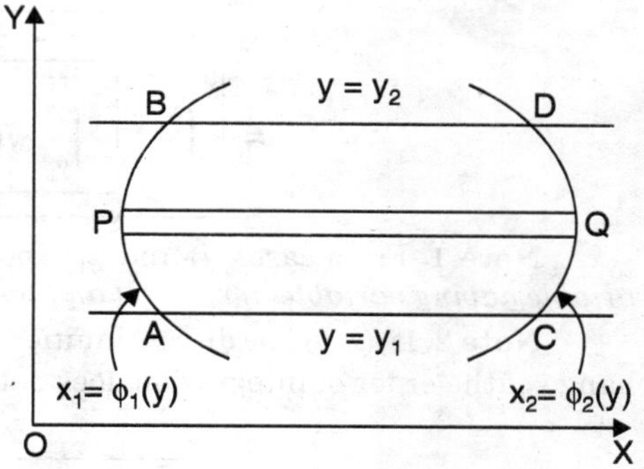

Take a horizontal strip PQ of width δy. Here the double integral is evaluated first w.r.t. x (treating y as a constant). The resulting expression which is a function of y is integrated w.r.t. y between the limits $y = y_1$ and $y = y_2$. Thus,

$$\iint_R f(x, y)\, dx\, dy = \boxed{\int_{y_1}^{y_2} \boxed{\int_{x_1 = \phi_1(y)}^{x_2 = \phi_2(y)} f(x, y)\, dx}\; dy}$$

the integration being carried from the inner to the outer rectangle. Geometrically, the integral in the inner rectangle indicates that the integration is performed along the horizontal strip PQ (keeping y constant) while the outer rectangle corresponds to the sliding of the strip PQ from AC to BD thus covering the entire region ABDC of integration.

(*ii*) **When y_1, y_2 are functions of x and x_1, x_2 are constants.** Let AB and CD be the curves $y_1 = \phi_1(x)$ and $y_2 = \phi_2(x)$. Take a vertical strip PQ of width δx. Here the double integral is evaluated first w.r.t. y (treating x as constant). The resulting expression which is a function of x is integrated w.r.t. x between the limits $x = x_1$ and $x = x_2$. Thus,

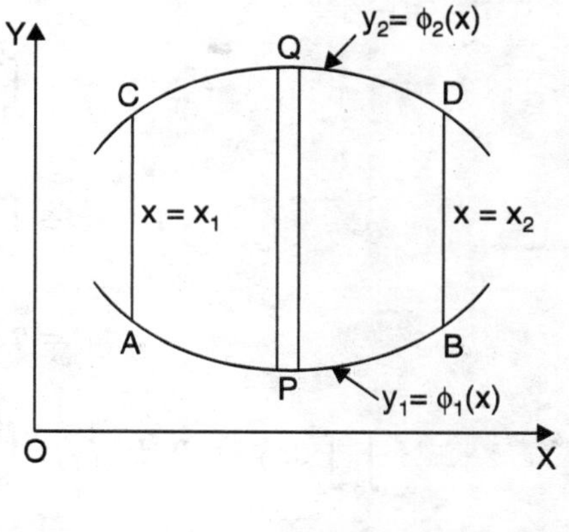

$$\iint_R f(x, y)\, dx\, dy = \boxed{\int_{x_1}^{x_2} \boxed{\int_{y_1 = \phi_1(x)}^{y_2 = \phi_2(x)} f(x, y)\, dy}\; dx}$$

the integration being carried from the inner to the outer rectangle. Geometrically, the integral in the inner rectangle indicates that the integration is performed along the vertical strip PQ (keeping x constant) while the outer rectangle corresponds to the sliding of the strip PQ from AC to BD thus covering the entire region ABDC of integration.

(iii) **When x_1, x_2, y_1, y_2 are constants.** Here the region of integration R is the rectangle ABDC. It is immaterial whether we integrate first along the horizontal strip PQ and then slide it from AB to CD ; or we integrate first along the vertical strip P′Q′ and then slide it from AC to BD. Thus the order of integration is immaterial, provided the limits of integration are changed accordingly.

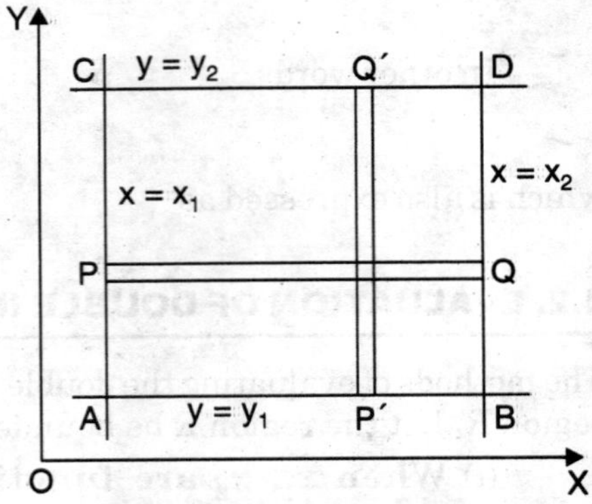

$$\iint_R f(x, y)\, dx\, dy = \int_{y_1}^{y_2} \left[\int_{x_1}^{x_2} f(x, y)\, dx \right] dy$$

$$= \int_{x_1}^{x_2} \left[\int_{y_1}^{y_2} f(x, y)\, dy \right] dx$$

Note 1. From cases *(i)* and *(ii)* above, we observe that *integration is to be performed w.r.t. that variable having variable limits first and then w.r.t. the variable with constant limits.*

Note 2. If $f(x, y)$ has discontinuities within or on the boundary of the region of integration, then the change of the order of integration does not result into the same integrals.

ILLUSTRATIVE EXAMPLES

Example 1. *Prove that* $\int_1^2 \int_3^4 (xy + e^y)\, dy\, dx = \int_3^4 \int_1^2 (xy + e^y)\, dx\, dy.$

Sol. $\int_1^2 \int_3^4 (xy + e^y)\, dy\, dx = \int_1^2 \left[\int_3^4 (xy + e^y)\, dy \right] dx$

$$= \int_1^2 \left[\frac{xy^2}{2} + e^y \right]_3^4 dx = \int_1^2 \left(8x + e^4 - \frac{9}{2}x - e^3 \right) dx$$

$$= \int_1^2 \left(\frac{7}{2}x + e^4 - e^3 \right) dx = \left[\frac{7x^2}{4} + (e^4 - e^3)x \right]_1^2$$

$$= 7 + 2(e^4 - e^3) - \frac{7}{4} - (e^4 - e^3) = \frac{21}{4} + e^4 - e^3$$

$$\int_3^4 \int_3^4 (xy + e^y)\, dx\, dy = \int_3^4 \left[\int_1^2 (xy + e^y)\, dx \right] dy = \int_3^4 \left[\frac{yx^2}{2} + xe^y \right]_1^2 dy$$

$$= \int_3^4 \left(2y + 2e^y - \frac{y}{2} - e^y \right) dy = \int_3^4 \left(\frac{3y}{2} + e^y \right) dy$$

$$= \left[\frac{3y^2}{4} + e^y \right]_3^4 = 12 + e^4 - \frac{27}{4} - e^3 = \frac{21}{4} + e^4 - e^3$$

Hence the result.

Example 2. *Evaluate* $\int_0^1 \int_0^1 \frac{dx\,dy}{\sqrt{(1-x^2)(1-y^2)}}$. (A.K.T.U. 2016]

Sol. $$\int_0^1\int_0^1 \frac{dx\,dy}{\sqrt{(1-x^2)(1-y^2)}} = \int_0^1\left[\int_0^1 \frac{dx}{\sqrt{(1-x^2)(1-y^2)}}\right]dy = \int_0^1 \frac{1}{\sqrt{1-y^2}}\left[\sin^{-1}x\right]_0^1 dy$$

$$= \int_0^1 \frac{1}{\sqrt{1-y^2}}\cdot\frac{\pi}{2}\,dy = \frac{\pi}{2}\left[\sin^{-1}y\right]_0^1 = \frac{\pi}{2}\cdot\frac{\pi}{2} = \frac{\pi^2}{4}.$$

Example 3. *Show that* $\int_0^1 dx \int_0^1 \frac{x-y}{(x+y)^3}\,dy \neq \int_0^1 dy \int_0^1 \frac{x-y}{(x+y)^3}\,dx.$

Sol. $$\text{LHS} = \int_0^1 dx \int_0^1 \frac{2x-(x+y)}{(x+y)^3}\,dy = \int_0^1 dx \int_0^1 \left[\frac{2x}{(x+y)^3} - \frac{1}{(x+y)^2}\right]dy$$

$$= \int_0^1 \left[2x\cdot\frac{(x+y)^{-2}}{-2} - \frac{(x+y)^{-1}}{-1}\right]_0^1 dx = \int_0^1 \left[\frac{-x}{(x+y)^2} + \frac{1}{x+y}\right]_0^1 dx$$

$$= \int_0^1 \left[\frac{-x}{(x+1)^2} + \frac{1}{x+1} + \frac{1}{x} - \frac{1}{x}\right]dx = \int_0^1 \frac{-x+x+1}{(x+1)^2}\,dx = \int_0^1 \frac{1}{(x+1)^2}\,dx$$

$$= \left[-\frac{1}{x+1}\right]_0^1 = -\frac{1}{2} + 1 = \frac{1}{2}$$

$$\text{RHS} = \int_0^1 dy \int_0^1 \frac{(x+y)-2y}{(x+y)^3}\,dx = \int_0^1 dy \int_0^1 \left[\frac{1}{(x+y)^2} - \frac{2y}{(x+y)^3}\right]dx$$

$$= \int_0^1 \left[\frac{(x+y)^{-1}}{-1} - 2y\cdot\frac{(x+y)^{-2}}{-2}\right]_0^1 dy = \int_0^1 \left[-\frac{1}{x+y} + \frac{y}{(x+y)^2}\right]_0^1 dy$$

$$= \int_0^1 \left[-\frac{1}{1+y} + \frac{y}{(1+y)^2} + \frac{1}{y} - \frac{1}{y}\right]dy = \int_0^1 \frac{-1-y+y}{(1+y)^2}\,dy = -\int_0^1 \frac{1}{(1+y)^2}\,dy$$

$$= \left[\frac{1}{1+y}\right]_0^1 = \frac{1}{2} - 1 = -\frac{1}{2}$$

∴ The two integrals are not equal.

Example 4. *Evaluate* $\int_0^1 \int_0^{\sqrt{1+x^2}} \frac{dy\,dx}{1+x^2+y^2}$.

Sol. $$I = \int_0^1 \left[\int_0^{\sqrt{1+x^2}} \frac{1}{(1+x^2)+y^2}\,dy\right]dx$$

$$= \int_0^1 \frac{1}{\sqrt{1+x^2}} \left[\tan^{-1} \frac{y}{\sqrt{1+x^2}}\right]_0^{\sqrt{1+x^2}} dx$$

$$= \int_0^1 \frac{1}{\sqrt{1+x^2}} [\tan^{-1} 1 - \tan^{-1} 0]\, dx = \frac{\pi}{4} \int_0^1 \frac{dx}{\sqrt{1+x^2}}$$

$$= \frac{\pi}{4} \left[\log (x + \sqrt{1+x^2})\right]_0^1 = \frac{\pi}{4} [\log (1 + \sqrt{2}) - \log 1] = \frac{\pi}{4} \log (\sqrt{2} + 1).$$

Example 5. *Evaluate* $\int_0^a \int_0^{\sqrt{a^2-y^2}} \sqrt{a^2 - x^2 - y^2}\, dx\, dy.$

Sol. $$I = \int_0^a \left[\int_0^{\sqrt{a^2-y^2}} \sqrt{(a^2 - y^2) - x^2}\, dx\right] dy$$

$$= \int_0^a \left[\frac{x\sqrt{a^2 - y^2 - x^2}}{2} + \frac{a^2 - y^2}{2} \sin^{-1} \frac{x}{\sqrt{a^2 - y^2}}\right]_0^{\sqrt{a^2-y^2}} dy$$

$$= \int_0^a \frac{a^2 - y^2}{2} (\sin^{-1} 1)\, dy = \frac{\pi}{4} \int_0^a (a^2 - y^2)\, dy$$

$$= \frac{\pi}{4} \left[a^2 y - \frac{y^3}{3}\right]_0^a = \frac{\pi}{4} \left[a^3 - \frac{a^3}{3}\right] = \frac{\pi a^3}{6}.$$

Example 6. *Evaluate* $\iint e^{2x + 3y}\, dx\, dy$ *over the triangle bounded by* $x = 0$, $y = 0$ *and* $x + y = 1$.

Sol. The region R of integration is the triangle OAB. Here x varies from 0 to 1 and y varies from x-axis upto the line $x + y = 1$ *i.e.*, from 0 to $1 - x$.

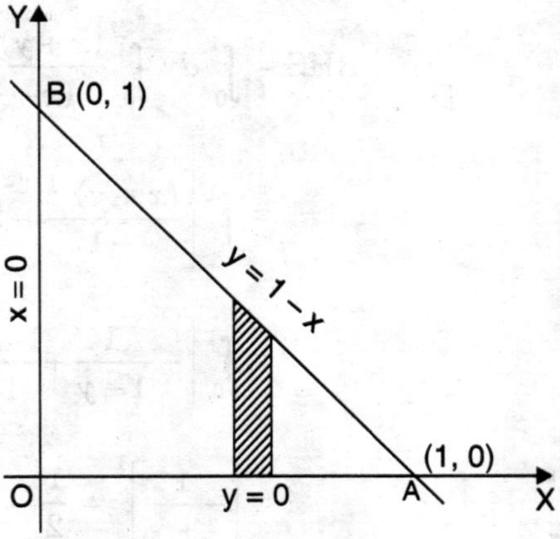

∴ The region R can be expressed as

$0 \le x \le 1,\ 0 \le y \le 1 - x$

∴ $\iint_R e^{2x + 3y}\, dx\, dy$

$$= \int_0^1 \int_0^{1-x} e^{2x + 3y}\, dy\, dx$$

$$= \int_0^1 \left[\frac{1}{3} e^{2x + 3y}\right]_0^{1-x} dx$$

$$= \frac{1}{3} \int_0^1 (e^{3-x} - e^{2x})\, dx$$

$$= \frac{1}{3}\left[- e^{3-x} - \frac{1}{2} e^{2x}\right]_0^1 = - \frac{1}{3} \left[\left(e^2 + \frac{1}{2} e^2\right) - \left(e^3 + \frac{1}{2}\right)\right]$$

$$= -\frac{1}{3}\left[-e^2(e-1)+\frac{1}{2}(e^2-1)\right] = \frac{1}{6}\ (e-1)(2e^2-e-1)$$

$$= \frac{1}{6}\ (e-1)^2(2e+1).$$

Example 7. *Evaluate* $\iint_R y\ dx\ dy$, *where R is the region bounded by the parabolas* $y^2 = 4x$ *and* $x^2 = 4y$.

Sol. Solving $y^2 = 4x$ and $x^2 = 4y$, we have

$$\left(\frac{x^2}{4}\right)^2 = 4x \quad \text{or} \quad x(x^3-64) = 0$$

$\therefore$ $x = 0, 4$

When $x = 4$, $y = 4$

$\therefore$ Co-ordinates of A are (4, 4)

The region R can be expressed as

$$0 \le x \le 4,\ \frac{x^2}{4} \le y \le 2\sqrt{x}$$

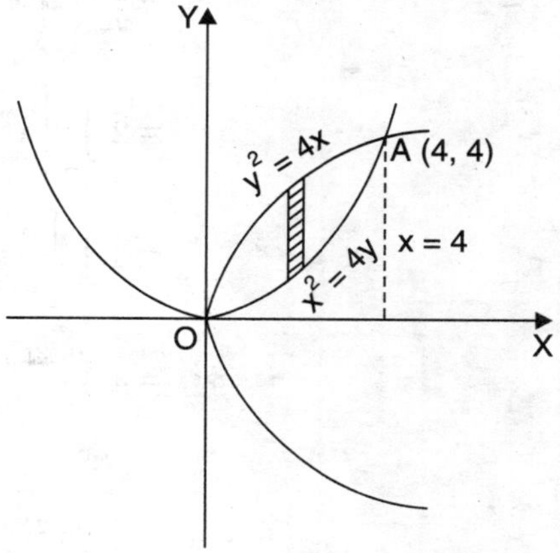

$$\therefore \quad \iint_R y\ dx\ dy = \int_0^4 \int_{x^2/4}^{2\sqrt{x}} y\ dy\ dx$$

$$= \int_0^4 \frac{1}{2}\Big[y^2\Big]_{x^2/4}^{2\sqrt{x}} dx = \frac{1}{2}\int_0^4 \left(4x - \frac{x^4}{16}\right) dx$$

$$= \frac{1}{2}\left[2x^2 - \frac{x^5}{80}\right]_0^4 = \frac{1}{2}\left[32 - \frac{1024}{80}\right] = \frac{48}{5}.$$

Example 8. *Evaluate* $\iint (x+y)^2\ dx\ dy$ *over the area bounded by the ellipse* $\frac{x^2}{a^2} + \frac{y^2}{b^2} = 1$.

(U.K.T.U. 2011)

Sol. For the ellipse

$$\frac{y}{b} = \pm\sqrt{1-\frac{x^2}{a^2}} \quad \text{or} \quad y = \pm\frac{b}{a}\sqrt{a^2-x^2}$$

$\therefore$ The region of integration R can be expressed as

$$-a \le x \le a,\ -\frac{b}{a}\sqrt{a^2-x^2} \le y \le \frac{b}{a}\sqrt{a^2-x^2}$$

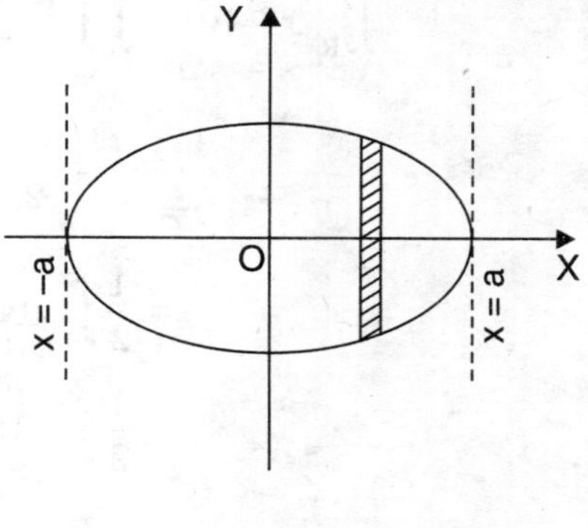

$$\therefore \quad \iint_R (x+y)^2\ dx\ dy = \iint_R (x^2+y^2+2xy)\ dx\ dy$$

$$= \int_{-a}^{a}\int_{(-b/a)\sqrt{a^2-x^2}}^{(b/a)\sqrt{a^2-x^2}} (x^2+y^2+2xy)\ dy\ dx$$

$$= \int_{-a}^{a} \int_{(-b/a)\sqrt{a^2-x^2}}^{(b/a)\sqrt{a^2-x^2}} (x^2+y^2)\, dy\, dx + \int_{-a}^{a} \int_{(-b/a)\sqrt{a^2-x^2}}^{(b/a)\sqrt{a^2-x^2}} 2xy\, dy\, dx$$

$$= \int_{-a}^{a} \int_{0}^{(b/a)\sqrt{a^2-x^2}} 2(x^2+y^2)\, dy\, dx + 0$$

[Since $(x^2 + y^2)$ is an even function of y and $2xy$ is an odd function of y]

$$= \int_{-a}^{a} \left[2\left(x^2 y + \frac{y^3}{3} \right) \right]_0^{(b/a)\sqrt{a^2-x^2}} dx$$

$$= 2\int_{-a}^{a} \left[x^2 \cdot \frac{b}{a}\sqrt{a^2-x^2} + \frac{1}{3}\cdot\frac{b^3}{a^3}(a^2-x^2)^{3/2} \right] dx$$

$$= 4\int_{0}^{a} \left[\frac{b}{a}x^2\sqrt{a^2-x^2} + \frac{b^3}{3a^3}(a^2-x^2)^{3/2} \right] dx$$

$$= 4\int_{0}^{\pi/2} \left(\frac{b}{a}\cdot a^2 \sin^2\theta \cdot a\cos\theta + \frac{b^3}{3a^3}\cdot a^3\cos^3\theta \right) \times a\cos\theta\, d\theta$$

on putting $x = a\sin\theta$ and $dx = a\cos\theta\, d\theta$

$$= 4\int_{0}^{\pi/2} \left(a^3 b \sin^2\theta\cos^2\theta + \frac{ab^3}{3}\cos^4\theta \right) d\theta$$

$$= 4\left[a^3 b \cdot \frac{1.1}{4.2}\cdot\frac{\pi}{2} + \frac{ab^3}{3}\cdot\frac{3.1}{4.2}\cdot\frac{\pi}{2} \right] = \frac{\pi}{4}(a^3 b + ab^3) = \frac{\pi}{4}ab(a^2+b^2).$$

Example 9. *Evaluate* $\iint y\, dx\, dy$ *over the part of the plane bounded by the line* $y = x$ *and the parabola* $y = 4x - x^2$.

Sol. The line $y = x$ meets the parabola $y = 4x - x^2$ in two distinct points O(0, 0) and A(3, 3).

The region of integration R can be expressed as $0 \le x \le 3,\ x \le y \le 4x - x^2$.

$$\therefore \quad \iint_R y\, dx\, dy = \int_0^3 \int_x^{4x-x^2} y\, dy\, dx$$

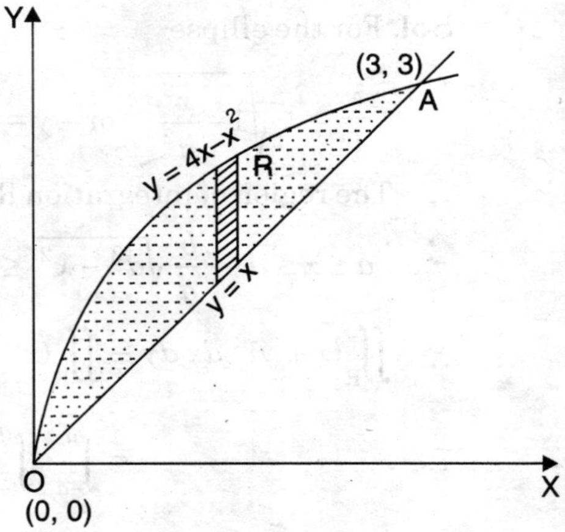

$$= \int_0^3 \left(\frac{y^2}{2} \right)_x^{4x-x^2} dx$$

$$= \frac{1}{2}\int_0^3 [(4x-x^2)^2 - x^2]\, dx$$

$$= \frac{1}{2}\int_0^3 (15x^2 + x^4 - 8x^3)\, dx$$

$$= \frac{1}{2}\left(5x^3 + \frac{x^5}{5} - 2x^4 \right)_0^3 = \frac{54}{5}.$$

Example 10. *When the region R of integration is the triangle given by y = 0, y = x and x = 1, show that* $\iint_R \sqrt{4x^2 - y^2}\, dx\, dy = \frac{1}{3}\left(\frac{\pi}{3} + \frac{\sqrt{3}}{2}\right)$.

Sol. The region of integration R can be expressed as $0 \le x \le 1,\ 0 \le y \le x$.

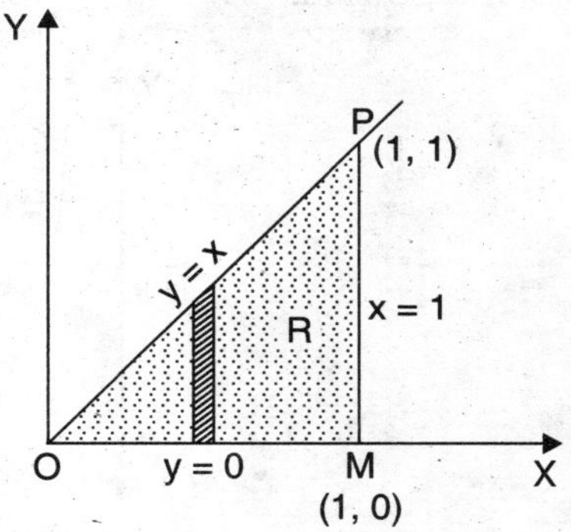

$$\therefore \quad \iint_R \sqrt{4x^2 - y^2}\, dx\, dy = \int_0^1 \int_0^x \sqrt{4x^2 - y^2}\, dy\, dx$$

$$= \int_0^1 \left[\frac{1}{2} y\sqrt{4x^2 - y^2} + \frac{1}{2} . 4x^2 \sin^{-1}\frac{y}{2x}\right]_0^x dx$$

$$= \int_0^1 \frac{1}{2}\left(\sqrt{3}x^2 + 4x^2 . \frac{\pi}{6}\right) dx$$

$$= \frac{1}{2}\left(\sqrt{3}\frac{x^3}{3} + \frac{2\pi}{3} . \frac{x^3}{3}\right)_0^1 = \frac{1}{2}\left(\sqrt{3} . \frac{1}{3} + \frac{2\pi}{3} . \frac{1}{3}\right)$$

$$= \frac{1}{3}\left(\frac{\pi}{3} + \frac{\sqrt{3}}{2}\right).$$

Example 11. *Evaluate* $\iint_S \sqrt{xy - y^2}\, dx\, dy$, *where S is a triangle with vertices (0, 0), (10, 1) and (1, 1).*

Sol. Let OAB be the triangle whose vertices are (0, 0), (10, 1) and (1, 1) as shown in the figure.

The equation of the line joining O(0, 0) and A(1, 1) is

$$y - 0 = \frac{1-0}{1-0}(x - 0) \quad \Rightarrow \quad y = x$$

The equation of the line joining O(0, 0) and B(10, 1) is

$$y - 0 = \frac{1-0}{10-0}(x - 0) \quad \Rightarrow \quad x = 10y$$

Hence the region of integration can be expressed as $y \le x \le 10y,\ 0 \le y \le 1$

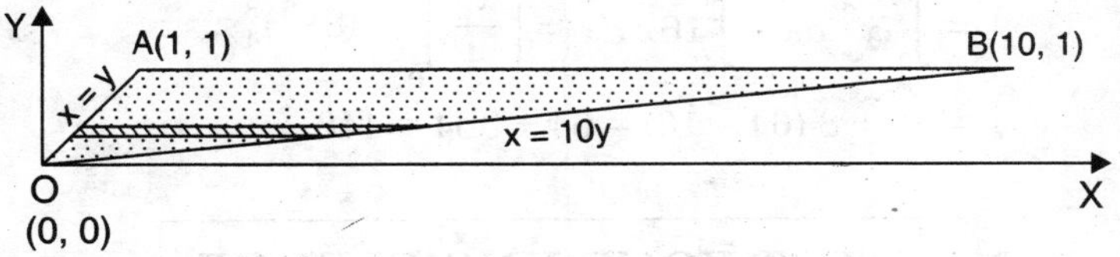

$$\therefore \quad \iint_S \sqrt{xy - y^2}\, dx\, dy = \int_0^1 \int_y^{10y} \sqrt{xy - y^2}\, dx\, dy$$

$$= \int_0^1 \left[\frac{2}{3}\frac{(xy - y^2)^{3/2}}{y}\right]_y^{10y} dy = \int_0^1 \frac{2}{3y}(9y^2)^{3/2}\, dy$$

$$= 18\int_0^1 y^2\, dy = 18\left(\frac{y^3}{3}\right)_0^1 = 18\left(\frac{1}{3}\right) = 6.$$

Example 12. *Let D be the region in the first quadrant bounded by the curves xy = 16, x = y, y = 0 and x = 8. Sketch the region of integration of the following integral* $\iint_D x^2\, dx\, dy$ *and evaluate it by expressing it as an appropriate repeated integral.*

Sol. The straight line $x = y$ intersects the hyperbola $xy = 16$ at A(4, 4). The line $x = 8$ intersects the hyperbola $xy = 16$ at B(8, 2). The shaded portion in the following figure is the region of integration.

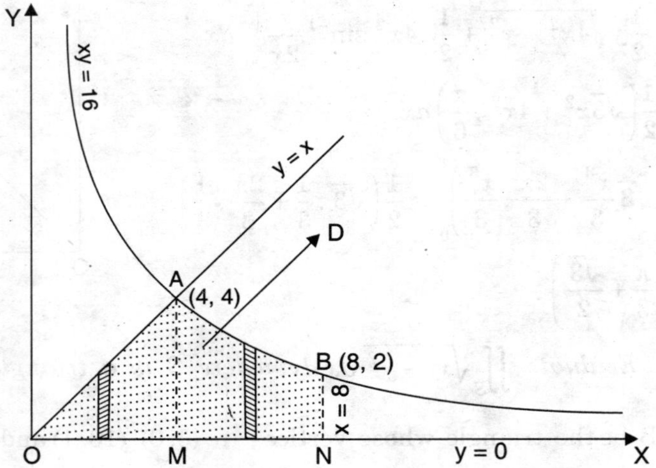

To evaluate the given integral, we divide the area OABNO into two parts by AM as shown in the figure.

Then, $$\iint_D x^2\, dx\, dy = \int_{x=0}^{x=4}\int_{y=0}^{y=x} x^2\, dy\, dx + \int_{x=4}^{x=8}\int_{y=0}^{y=\frac{16}{x}} x^2\, dy\, dx$$

$$= \int_0^4 x^2\, dx \int_0^x dy + \int_4^8 x^2\, dx \int_0^{\frac{16}{x}} dy = \int_0^4 x^2 (y)_0^x\, dx + \int_4^8 x^2 \left(y\right)_0^{16/x} dx$$

$$= \int_0^4 x^3\, dx + \int_4^8 16x\, dx = \left(\frac{x^4}{4}\right)_0^4 + (8x^2)_4^8$$

$$= 64 + 8\,(64 - 16) = 64 + 384 = 448.$$

TEST YOUR KNOWLEDGE

Evaluate the following integrals (1–10):

1. $\int_0^3\int_0^1 (x^2 + 3y^2)\, dy\, dx.$

2. $\int_0^3\int_1^2 xy\,(1 + x + y)\, dy\, dx.$

3. $\int_1^a\int_1^b \frac{dy\, dx}{xy}.$

4. $\int_1^2\int_0^x \frac{dy\, dx}{x^2 + y^2}.$

5. $\int_0^1 dx \int_0^x e^{y/x}\, dy$.

6. $\int_0^1 \int_{x^2}^{x} (x^2 + 3y + 2)\, dy\, dx$.

7. $\int_0^1 \int_0^{x^2} e^{y/x}\, dy\, dx$. (*A.K.T.U. 2021, 2019*)

8. $\int_0^1 \int_y^{y^2+1} x^2 y\, dx\, dy$.

9. $\int_0^1 \int_{y^2}^{y} (1 + xy^2)\, dx\, dy$

10. $\int_0^1 \int_x^{\sqrt{x}} (x^2 + y^2)\, dy\, dx$.

11. Evaluate $\iint (x^2 + y^2)\, dx\, dy$ over the region in the positive quadrant for which $x + y \le 1$.

12. Evaluate $\iint x^2y^2\, dx\, dy$ over the circle $x^2 + y^2 = 1$.

13. Evaluate $\iint xy\, dx\, dy$ over the positive quadrant of the circle $x^2 + y^2 = a^2$.

14. (*i*) Compute the value of $\iint_R y\, dx\, dy$, where R is the region in the first quadrant bounded by the ellipse $\frac{x^2}{a^2} + \frac{y^2}{b^2} = 1$.

(*ii*) Evaluate $\iint_R \left(1 - \frac{x^2}{a^2} + \frac{y^2}{b^2}\right) dx\, dy$ over the first quadrant of the ellipse $\frac{x^2}{a^2} + \frac{y^2}{b^2} = 0$.
(*G.B.T.U. 2010*)

15. Evaluate $\iint xy\, dx\, dy$ over the region in the positive quadrant for which $x + y \le 1$.

16. Evaluate $\iint xy\,(x + y)\, dx\, dy$ over the area between $y = x^2$ and $y = x$. (*U.K.T.U. 2011, 2012*)

17. Evaluate $\iint_A xy\, dx\, dy$, where A is the domain bounded by x-axis, ordinate $x = 2a$ and the curve $x^2 = 4ay$. [*M.T.U. 2012*]

18. Evaluate $\iint (x^2 + y^2)\, dx\, dy$ throughout the area enclosed by the curves $y = 4x$, $x + y = 3$, $y = 0$ and $y = 2$.

Answers

1. 12	2. $30\frac{3}{4}$	3. $\log a \log b$	4. $\frac{\pi}{4} \log 2$
5. $\frac{1}{2}(e - 1)$	6. $\frac{7}{12}$	7. $\frac{1}{2}$	8. $\frac{67}{120}$
9. $\frac{41}{210}$	10. $\frac{3}{35}$	11. $\frac{1}{6}$	12. $\frac{\pi}{24}$
13. $\frac{a^4}{8}$	14. (*i*) $\frac{ab^2}{3}$ (*ii*) $\frac{\pi ab}{4}$	15. $\frac{1}{24}$	16. $\frac{3}{56}$
17. $\frac{a^4}{3}$	18. $\frac{463}{48}$.		

4.3. EVALUATION OF DOUBLE INTEGRALS IN POLAR COORDINATES

To evaluate $\int_{\theta_1}^{\theta_2} \int_{r_1}^{r_2} f(r, \theta)\, dr\, d\theta$ over the region bounded by the straight lines $\theta = \theta_1$, $\theta = \theta_2$ and the curves $r = r_1$, $r = r_2$, we first integrate w.r.t. r between the limits $r = r_1$ and $r = r_2$ (treating θ as a constant). The resulting expression is then integrated w.r.t. θ between the limits $\theta = \theta_1$ and $\theta = \theta_2$.

Geometrically, AB and CD are the curves $r = f_1(\theta)$ and $r = f_2(\theta)$ bounded by the lines $\theta = \theta_1$ and $\theta = \theta_2$ so that ACDB is the region of integration. PQ is a wedge of angular thickness $\delta\theta$. Then $\int_{r=r_1}^{r=r_2} f(r, \theta)\, dr$ indicates that the integration is performed along PQ (*i.e.*, r varies, θ is constant) and the integration w.r.t. θ means rotation of this strip PQ from AC to BD. dr indicates that the integration is performed along PQ (*i.e.*, r varies, θ is constant) and the integration w.r.t. θ means rotation of this strip PQ from AC to BD.

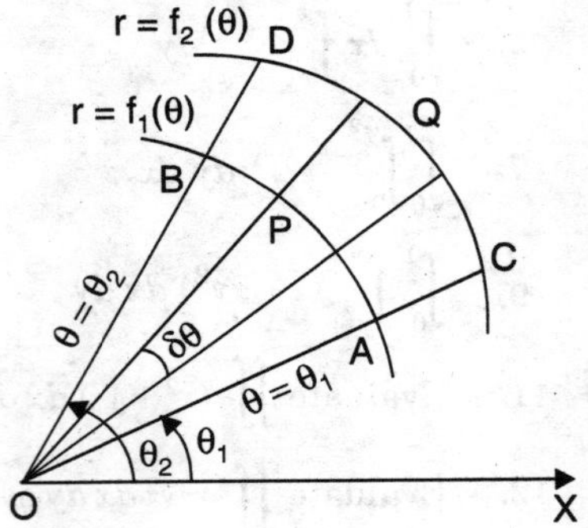

ILLUSTRATIVE EXAMPLES

Example 1. *Evaluate* $\int_0^{\pi/2}\left[\int_0^{a\cos\theta} r\sqrt{a^2 - r^2}\, dr\right] d\theta.$

Sol.
$$I = \int_0^{\pi/2}\left[\int_0^{a\cos\theta} -\frac{1}{2}(a^2 - r^2)^{1/2}(-2r)\, dr\right] d\theta$$

$$= \int_0^{\pi/2}\left[-\frac{1}{2}\cdot\frac{(a^2 - r^2)^{3/2}}{3/2}\right]_0^{a\cos\theta} d\theta$$

$$= -\frac{1}{3}\int_0^{\pi/2}(a^3\sin^3\theta - a^3)\, d\theta = -\frac{a^3}{3}\left[\frac{2}{3} - \frac{\pi}{2}\right] = \frac{a^3}{18}(3\pi - 4).$$

Example 2. *Evaluate* $\int_0^{\pi}\int_0^{a(1-\cos\theta)} r^2 \sin\theta\, dr\, d\theta.$

Sol.
$$I = \int_0^{\pi}\left(\frac{r^3}{3}\right)_0^{a(1-\cos\theta)} \sin\theta\, d\theta = \frac{a^3}{3}\int_0^{\pi}\sin\theta(1-\cos\theta)^3\, d\theta$$

$$= \frac{a^3}{3}\int_0^2 t^3\, dt = \frac{a^3}{3}\left(\frac{t^4}{4}\right)_0^2 = \frac{4}{3}a^3.$$

Put $1 - \cos\theta = t$
$\therefore$ $\sin\theta\, d\theta = dt$

Example 3. *Evaluate* $\iint r^3\, dr\, d\theta$, *over the area bounded between the circles* $r = 2\cos\theta$ *and* $r = 4\cos\theta$.

Sol. The region of integration R is shown shaded. Here r varies from $2\cos\theta$ to $4\cos\theta$ while θ varies from $-\frac{\pi}{2}$ to $\frac{\pi}{2}$.

$$\therefore \quad \iint_R r^3\, dr\, d\theta = \int_{-\pi/2}^{\pi/2}\int_{2\cos\theta}^{4\cos\theta} r^3 dr\, d\theta$$

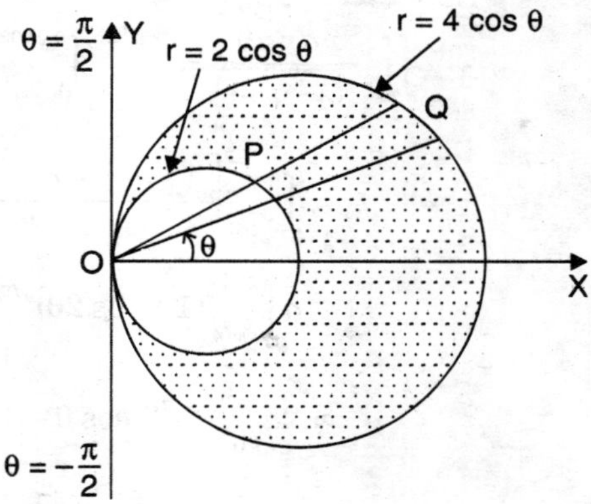

$$= \int_{-\pi/2}^{\pi/2} \left[\frac{r^4}{4}\right]_{2\cos\theta}^{4\cos\theta} d\theta$$

$$= \int_{-\pi/2}^{\pi/2} \frac{1}{4}(256\cos^4\theta - 16\cos^4\theta)\, d\theta$$

$$= 60\int_{-\pi/2}^{\pi/2} \cos^4\theta\, d\theta = 120\int_0^{\pi/2} \cos^4\theta\, d\theta$$

$$= 120 \times \frac{3\times 1}{4\times 2}\cdot\frac{\pi}{2} = \frac{45}{2}\pi.$$

Example 4. *Evaluate* $\iint r\sin\theta\, dr\, d\theta$ *over the area of the cardioid* $r = a(1+\cos\theta)$ *above the initial line.*

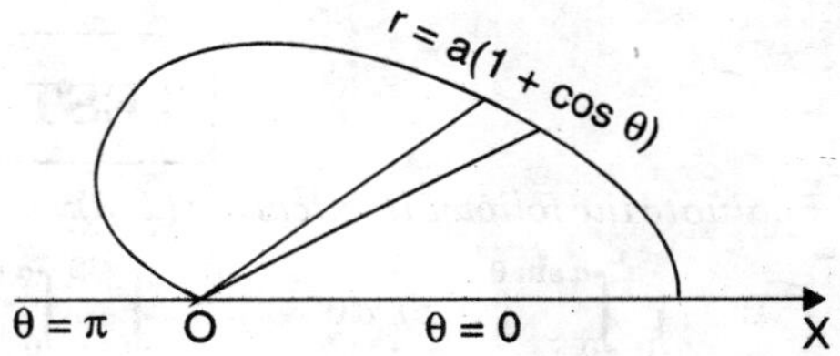

Sol. The region of integration R is covered by radial strips whose ends are $r = 0$ and $r = a(1+\cos\theta)$, the strips starting from $\theta = 0$ and ending at $\theta = \pi$.

$$\therefore \quad \iint_R r\sin\theta\, dr\, d\theta = \int_0^{\pi}\int_0^{a(1+\cos\theta)} r\sin\theta\, dr\, d\theta\cdot$$

$$= \int_0^{\pi} \sin\theta \left[\frac{r^2}{2}\right]_0^{a(1+\cos\theta)} d\theta = \frac{1}{2}\int_0^{\pi} \sin\theta\, .\, a^2(1+\cos\theta)^2\, d\theta$$

$$= \frac{a^2}{2}\int_0^2 t^2 dt \qquad \text{| Putting } 1+\cos\theta = t \quad \therefore \quad \sin\theta\, d\theta = -dt$$

$$= \frac{a^2}{2}\left(\frac{t^3}{3}\right)_0^2 = \frac{4}{3}a^2$$

Example 5. *Evaluate* $\iint \frac{r\, dr\, d\theta}{\sqrt{a^2+r^2}}$ *over one loop of the lemniscate* $r^2 = a^2\cos 2\theta$.

Sol. The region of integration R is covered by radial strips whose ends are $r = 0$ and $r = a\sqrt{\cos 2\theta}$, the strips starting from $\theta = -\frac{\pi}{4}$ and ending at $\theta = \frac{\pi}{4}$.

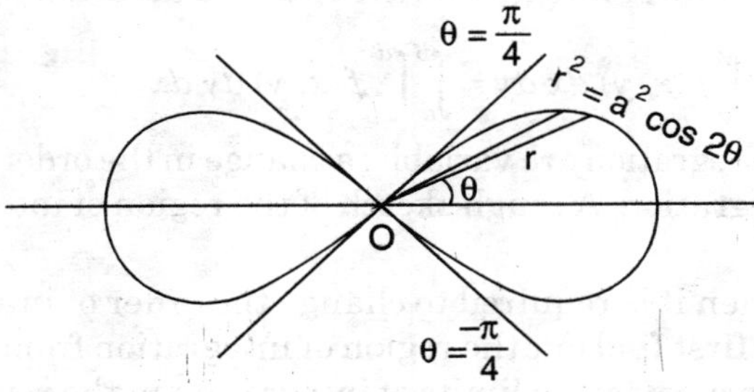

$$\therefore \quad \iint_R \frac{r\,dr\,d\theta}{\sqrt{a^2+r^2}} = \int_{-\pi/4}^{\pi/4}\int_0^{a\sqrt{\cos 2\theta}} \tfrac{1}{2}(a^2+r^2)^{-1/2}\,.\,2r\,dr\,d\theta$$

$$= \int_{-\pi/4}^{\pi/4}\left[\frac{1}{2}\cdot\frac{(a^2+r^2)^{1/2}}{1/2}\right]_0^{a\sqrt{\cos 2\theta}} d\theta = \int_{-\pi/4}^{\pi/4}[(a^2+a^2\cos 2\theta)^{1/2} - a]\,d\theta$$

$$= a\int_{-\pi/4}^{\pi/4}[(1+\cos 2\theta)^{1/2} - 1]\,d\theta = a\int_{-\pi/4}^{\pi/4}(\sqrt{2}\cos\theta - 1)\,d\theta$$

$$= 2a\int_0^{\pi/4}(\sqrt{2}\cos\theta - 1)\,d\theta = 2a\left[\sqrt{2}\sin\theta - \theta\right]_0^{\pi/4}$$

$$= 2a\left[\sqrt{2}\,.\,\frac{1}{\sqrt{2}} - \frac{\pi}{4}\right] = 2a\left(1-\frac{\pi}{4}\right).$$

TEST YOUR KNOWLEDGE

Evaluate the following integrals (1–3):

1. $\int_0^{\pi}\int_0^{a\sin\theta} r\,dr\,d\theta$ 2. $\int_0^{\pi/2}\int_0^{a\cos\theta} r\sin\theta\,dr\,d\theta$ 3. $\int_0^{\pi}\int_0^{a(1+\cos\theta)} r^2\cos\theta\,dr\,d\theta$
4. Evaluate $\iint_A r^2\,dr\,d\theta$, where A is the area between the circles $r = a\cos\theta$ and $r = 2a\cos\theta$.
5. Show that $\iint_R r^2\sin\theta\,dr\,d\theta = \frac{2a^3}{3}$, where R is the region bounded by the semi-circle $r = 2a\cos\theta$, above the initial line.
6. Evaluate $\iint r^3\,dr\,d\theta$ over the area included between the circles $r = 2\sin\theta$ and $r = 4\sin\theta$.

Answers

1. $\frac{\pi a^2}{4}$ 2. $\frac{a^2}{6}$ 3. $\frac{5}{8}\pi a^3$
4. $\frac{28a^3}{9}$ 6. $\frac{45\pi}{2}$.

4.4. CHANGE OF ORDER OF INTEGRATION

In a double integral, if the limits of integration are constant, then the order of integration is immaterial, provided the limits of integration are changed accordingly. Thus,

$$\int_c^d\int_a^b f(x, y)\,dx\,dy = \int_a^b\int_c^d f(x, y)\,dy\,dx.$$

But if the limits of integration are variable, a change in the order of integration necessitates change in the limits of integration. A rough sketch of the region of integration helps in fixing the new limits of integration.

Working Rule. When it is required to change the order of integration in an integral for which limits are given, we first find out the region of integration from the limits. When region R of integration is known, we next assign limits of integration in the reverse order from geometric considerations.

ILLUSTRATIVE EXAMPLES

Example 1. *Evaluate the following integral by changing the order of integration*

$$\int_0^1 \int_{e^x}^{e} \frac{dy\,dx}{\log y}.$$

Sol. The given limits show that the region of integration is bounded by the curves

$$y = e^x,\ y = e,\ x = 0,\ x = 1$$

It is the area shaded in the diagram. It can also be considered as bounded by

$$x = 0,\ x = \log y,\ y = 1,\ y = e$$

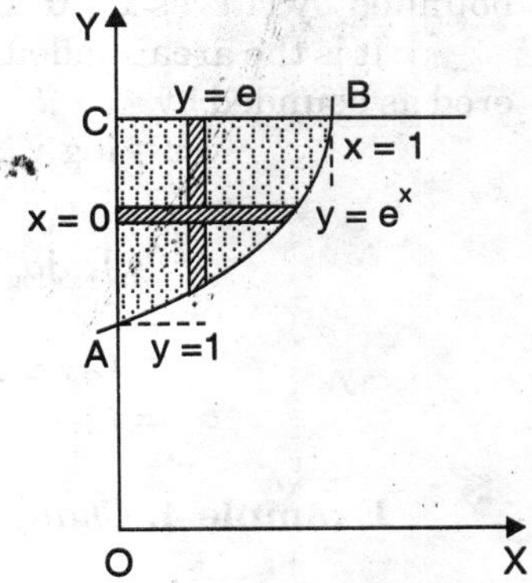

Hence, $\displaystyle \int_0^1 \int_{e^x}^{e} \frac{dy\,dx}{\log y} = \int_1^e \int_0^{\log y} \frac{dx\,dy}{\log y}$

$$= \int_1^e \left(\frac{x}{\log y} \right)_0^{\log y} dy$$

$$= \int_1^e 1.\,dy = \left(y \right)_1^e = e - 1.$$

Example 2. *Change the order of integration in* $\displaystyle \int_0^a \int_y^a \frac{x\,dx\,dy}{x^2 + y^2}$ *and hence evaluate the same.*

Sol. From the limits of integration, it is clear that the region of integration is bounded by $x = y$, $x = a$, $y = 0$ and $y = a$. Thus the region of integration is the ΔOAB and is divided into horizontal strips. For changing the order of integration, we divide the region of integration into vertical strips. The new limits of integration become: y varies from 0 to x and x varies from 0 to a.

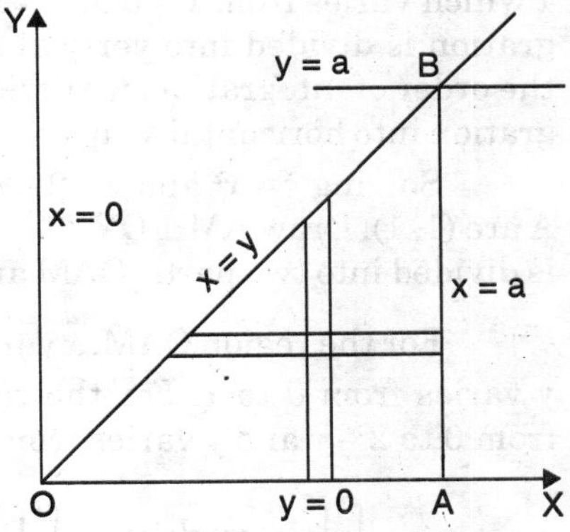

$$\int_0^a \int_y^a \frac{x\,dx\,dy}{x^2 + y^2} = \int_0^a \int_0^x \frac{x\,dy\,dx}{x^2 + y^2}$$

$$= \int_0^a x . \left[\frac{1}{x} \tan^{-1} \frac{y}{x} \right]_0^x dx$$

$$= \int_0^a \frac{\pi}{4}\,dx = \frac{\pi a}{4}.$$

Example 3. (*i*) *Evaluate by changing the order of integration:*

$$\int_0^1 \int_{2y}^2 e^{x^2}\,dx\,dy.$$

(*ii*) *Evaluate* $\displaystyle \int_0^2 \int_1^{e^x} dy\,dx$ *by changing the order of integration.*

Sol. (*i*) The given limits show that the region of integration is bounded by the lines:

$$x = 2y,\ x = 2,\ y = 0,\ y = 1$$

We can also consider it as bounded by $y = 0,\ y = x/2,\ x = 0,\ x = 2$

$$\therefore \quad \int_0^1 \int_{2y}^2 e^{x^2}\, dx\, dy = \int_0^2 \int_0^{x/2} e^{x^2}\, dy\, dx$$

$$= \int_0^2 e^{x^2} (y)_0^{x/2}\, dx = \frac{1}{2}\int_0^2 x\, e^{x^2}\, dx = \frac{1}{4}\left(e^{x^2}\right)_0^2 = \frac{(e^4 - 1)}{4}.$$

(*ii*) The given limits show that the region of integration is bounded by curves $x = 0$, $x = 2$, $y = 1$ and $y = e^x$.

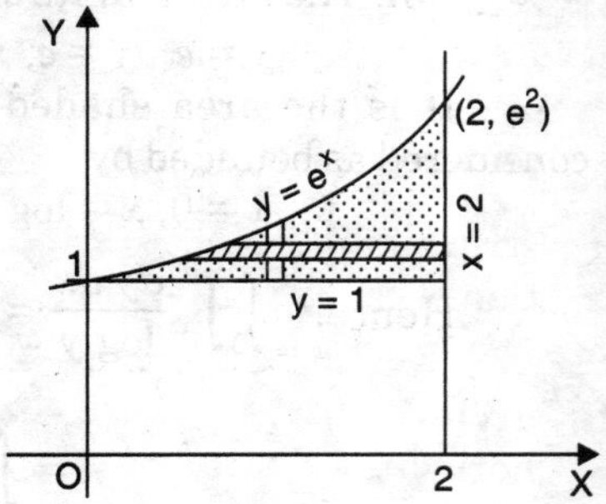

It is the area shaded in the diagram. It can also be considered as bounded by

$$x = \log y,\ x = 2,\ y = 1 \text{ and } y = e^2$$

$$\therefore \quad I = \int_1^{e^2} \int_{\log y}^2 dx\, dy = \int_1^{e^2} (2 - \log y)\, dy$$

$$= (2y - y \log y + y)_1^{e^2} = (3y - y \log y)_1^{e^2}$$
$$= (3e^2 - 2e^2) - 3 = e^2 - 3.$$

Example 4. *Change the order of integration in* $I = \int_0^1 \int_{x^2}^{2-x} xy\, dy\, dx$ *and hence evaluate the same.* (A.K.T.U. 2014, 2015, 2016, 2017, 2020, 2022)

Sol. From the limits of integration, it is clear that we have to integrate first with respect to y which varies from $y = x^2$ to $y = 2 - x$ and then with respect to x which varies from $x = 0$ to $x = 1$. The region of integration is divided into vertical strips. For changing the order of integration, we divide the region of integration into horizontal strips.

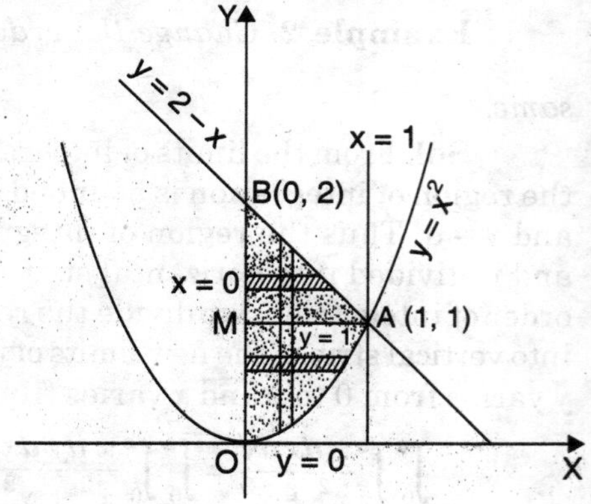

Solving $y = x^2$ and $y = 2 - x$, the co-ordinates of A are (1, 1). Draw AM ⊥ OY. The region of integration is divided into two parts, OAM and MAB.

For the region OAM, x varies from 0 to $\sqrt{y}$ and y varies from 0 to 1. For the region MAB, x varies from 0 to $2 - y$ and y varies from 1 to 2.

$$\therefore \quad \int_0^1 \int_{x^2}^{2-x} xy\, dx\, dy = \int_0^1 \int_0^{\sqrt{y}} xy\, dx\, dy + \int_1^2 \int_0^{2-y} xy\, dx\, dy$$

$$= \int_0^1 y \cdot \left[\frac{x^2}{2}\right]_0^{\sqrt{y}} dy + \int_1^2 y \cdot \left[\frac{x^2}{2}\right]_0^{2-y} dy = \frac{1}{2}\int_0^1 y^2 dy + \frac{1}{2}\int_1^2 y(2-y)^2\, dy$$

$$= \frac{1}{2}\left[\frac{y^3}{3}\right]_0^1 + \frac{1}{2}\int_1^2 (4y - 4y^2 + y^3)\, dy = \frac{1}{6} + \frac{1}{2}\left[2y^2 - \frac{4y^3}{3} + \frac{y^4}{4}\right]_1^2$$

$$= \frac{1}{6} + \frac{1}{2}\left[\left(8 - \frac{32}{3} + 4\right) - \left(2 - \frac{4}{3} + \frac{1}{4}\right)\right] = \frac{3}{8}.$$

Example 5. *Change the order of integration in the following integral and evaluate:*

$$\int_0^{4a} \int_{x^2/4a}^{2\sqrt{ax}} dy\, dx.$$

Sol. From the limits of integration, it is clear that we have to integrate first w.r.t. y which varies from $y = \frac{x^2}{4a}$ to $y = 2\sqrt{ax}$ and then w.r.t. x which varies from $x = 0$ to $x = 4a$. Thus integration is first performed along the vertical strip PQ which extends from a point P on the parabola $y = \frac{x^2}{4a}$ (*i.e.*, $x^2 = 4ay$) to the point Q on the parabola $y = 2\sqrt{ax}$ (*i.e.*, $y^2 = 4ax$). Then the strip slides from O to A $(4a, 4a)$, the point of intersection of the two parabolas.

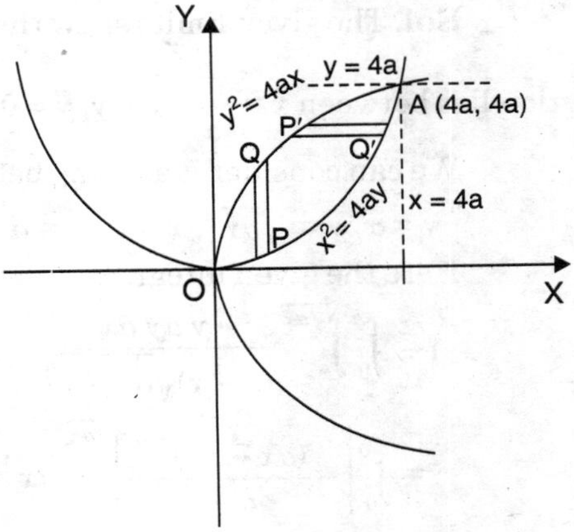

For changing the order of integration, we divide the region of integration OPAQO into horizontal strips P′Q′ which extend from P′ on the parabola $y^2 = 4ax$ *i.e.*, $x = \frac{y^2}{4a}$ to Q′ on the parabola $x^2 = 4ay$ *i.e.*, $x = 2\sqrt{ay}$. Then this strip slides from O to A$(4a, 4a)$, *i.e.*, y varies from 0 to $4a$.

$$\therefore \quad \int_0^{4a} \int_{x^2/4a}^{2\sqrt{ax}} dy\, dx = \int_0^{4a} \int_{y^2/4a}^{2\sqrt{ay}} dx\, dy$$

$$= \int_0^{4a} \left(2\sqrt{ay} - \frac{y^2}{4a}\right) dy$$

$$= \left[2\sqrt{a}\,.\frac{y^{3/2}}{3/2} - \frac{y^3}{12a}\right]_0^{4a} = \frac{4}{3}\sqrt{a}\,.(4a)^{3/2} - \frac{64a^3}{12a}$$

$$= \frac{32a^2}{3} - \frac{16a^2}{3} = \frac{16a^2}{3}.$$

Example 6. *Change the order of integration in the integral* $\int_{-a}^{a} \int_0^{\sqrt{a^2 - y^2}} f(x, y)\, dx\, dy$.

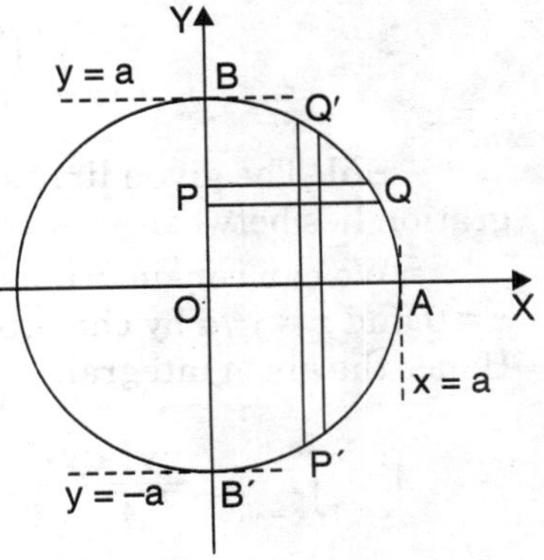

Sol. From the limits of integration, it is clear that we have to integrate first w.r.t. x which varies from $x = 0$ to $x = \sqrt{a^2 - y^2}$ and then w.r.t. y which varies from $y = -a$ to $y = a$. Thus integration is first performed along the horizontal strip PQ which extends from a point P on $x = 0$ (*i.e.*, y-axis) to the point Q on the circle $x = \sqrt{a^2 - y^2}$ (*i.e.*, $x^2 + y^2 = a^2$). Then the strip slides from B′ to B.

For changing the order of integration, we divide the region of integration B′AQBPB′ into vertical strips P′Q′ which extend from P′ on the circle $y = -\sqrt{a^2 - x^2}$ to Q′ on the circle $y = \sqrt{a^2 - x^2}$. Then this strip slides from y-axis ($x = 0$) to A, where $x = a$.

$$\therefore \quad \int_{-a}^{a}\int_{0}^{\sqrt{a^2-y^2}} f(x,y)\,dx\,dy = \int_{0}^{a}\int_{-\sqrt{a^2-x^2}}^{\sqrt{a^2-x^2}} f(x,y)\,dy\,dx.$$

Example 7. *By changing the order of integration, evaluate* $\displaystyle\int_0^a \int_{y^2/a}^{y} \frac{y}{(a-x)\sqrt{ax-y^2}}\,dx\,dy$.

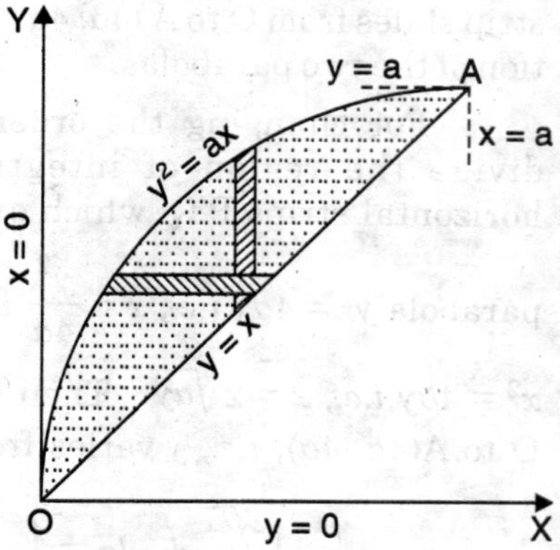

Sol. The given limits show that the area of integration lies between $x = \dfrac{y^2}{a}$, $x = y$, $y = 0$, $y = a$.

We can consider it as lying between

$y = x$, $y = \sqrt{ax}$, $x = 0$, $x = a$

Here the given integral,

$$I = \int_0^a \int_x^{\sqrt{ax}} \frac{y\,dy\,dx}{(a-x)\sqrt{ax-y^2}}$$

$$= \int_0^a \left[-\frac{(ax-y^2)^{1/2}}{a-x} \right]_x^{\sqrt{ax}} dx$$

$$= \int_0^a \frac{(ax-x^2)^{1/2}}{(a-x)}\,dx = \int_0^a \left(\frac{x}{a-x}\right)^{1/2} dx \qquad \Big|\ \text{Put } x = a\sin^2\theta \quad \therefore \quad dx = 2a\sin\theta\cos\theta\,d\theta$$

$$= \int_0^{\pi/2} \left(\frac{a\sin^2\theta}{a\cos^2\theta}\right)^{1/2} .\, 2a\sin\theta\cos\theta\,d\theta$$

$$= \int_0^{\pi/2} 2a\sin^2\theta\,d\theta = 2a\,.\,\frac{1}{2}\,.\,\frac{\pi}{2} = \frac{\pi a}{2}.$$

Example 8. *Change the order of integration and hence evaluate*

$$\int_0^a \int_{\sqrt{ax}}^{a} \frac{y^2\,dy\,dx}{\sqrt{y^4 - a^2x^2}}.$$

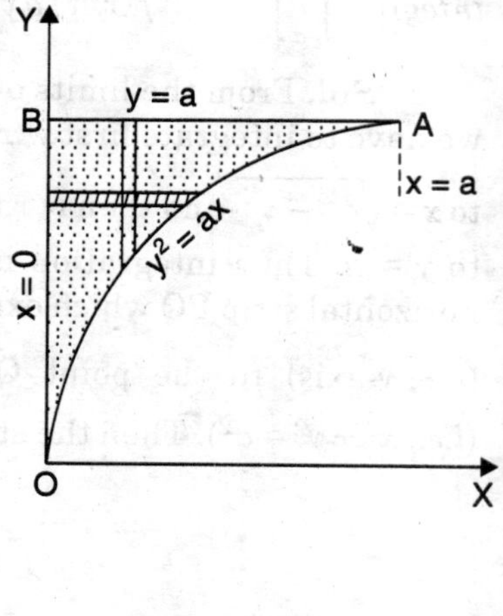

Sol. The given limits show that the area of integration lies between $y^2 = ax$, $y = a$, $x = 0$ and $x = a$.

We can consider it as lying between $y = 0$, $y = a$, $x = 0$ and $x = y^2/a$ by changing the order of integration. Hence the given integral,

$$\int_{x=0}^{a}\int_{y=\sqrt{ax}}^{a} \frac{y^2\,dy\,dx}{\sqrt{y^4-a^2x^2}} = \int_{y=0}^{a}\int_{x=0}^{\frac{y^2}{a}} \frac{y^2\,dx\,dy}{\sqrt{y^4-a^2x^2}}$$

$$= \frac{1}{a}\int_0^a \int_0^{\frac{y^2}{a}} \frac{y^2\,dx\,dy}{\sqrt{\left(\frac{y^2}{a}\right)^2 - x^2}} = \frac{1}{a}\int_0^a y^2 \left[\sin^{-1}\left(\frac{ax}{y^2}\right)\right]_0^{y^2/a} dy$$

$$= \frac{1}{a}\int_0^a y^2\,[\sin^{-1}(1) - \sin^{-1}(0)]\,dy$$

$$= \frac{\pi}{2a}\int_0^a y^2\,dy = \frac{\pi}{2a}\left(\frac{y^3}{3}\right)_0^a = \frac{\pi}{6a}(a^3) = \frac{\pi a^2}{6}.$$

Example 9. *By changing the order of integration of* $\int_0^\infty \int_0^\infty e^{-xy} \sin px\,dx\,dy$, *show that*

$$\int_0^\infty \frac{\sin px}{x}\,dx = \frac{\pi}{2}.$$

Sol. Let $\quad I = \int_0^\infty \int_0^\infty e^{-xy} \sin px\,dx\,dy$

$$= \int_0^\infty \sin px \left\{\int_0^\infty e^{-xy}\,dy\right\} dx = \int_0^\infty \sin px \left(\frac{e^{-xy}}{-x}\right)_0^\infty dx = \int_0^\infty \frac{\sin px}{x}\,dx \quad \text{...(1)}$$

Again, $\quad I = \int_0^\infty \int_0^\infty e^{-xy} \sin px\,dx\,dy$

$$= \int_0^\infty \left[\int_0^\infty e^{-xy} \sin px\,dx\right] dy = \int_0^\infty \left[\frac{-e^{-xy}}{p^2 + y^2}(p \cos px + y \sin px)\right]_0^\infty dy$$

$$= \int_0^\infty \frac{p}{p^2 + y^2}\,dy = \left[\tan^{-1}\left(\frac{y}{p}\right)\right]_0^\infty = \frac{\pi}{2} \quad \text{...(2)}$$

Hence from (1) and (2),

$$\int_0^\infty \frac{\sin px}{x}\,dx = \frac{\pi}{2}.$$

Example 10. *Show that* $\int_0^{\pi/2} \int_0^{\pi/2} \sin x \sin^{-1}(\sin x \sin y)\,dx\,dy = \frac{\pi^2}{4} - \frac{\pi}{2}.$

Sol. We have,

$$I = \int_0^{\pi/2} \sin x\,dx \int_0^{\pi/2} \sin^{-1}(\sin x \sin y)\,dy$$

Put $\quad \sin x \sin y = \sin z \quad$ so that $\sin x \cos y\,dy = \cos z\,dz$

$$\therefore \quad I = \int_0^{\pi/2} dx \int_0^x \frac{z \cos z}{\cos y}\,dz$$

$$= \int_0^{\pi/2} dx \int_0^x \frac{z \cos z}{\sqrt{1 - \frac{\sin^2 z}{\sin^2 x}}} dz$$

$$= \int_0^{\pi/2} dx \int_0^x \frac{z \sin x \cos z}{\sqrt{\sin^2 x - \sin^2 z}} dz$$

$$= \int_0^{\pi/2} dx \int_0^x \frac{z \sin x \cos z}{\sqrt{\cos^2 z - \cos^2 x}} dz$$

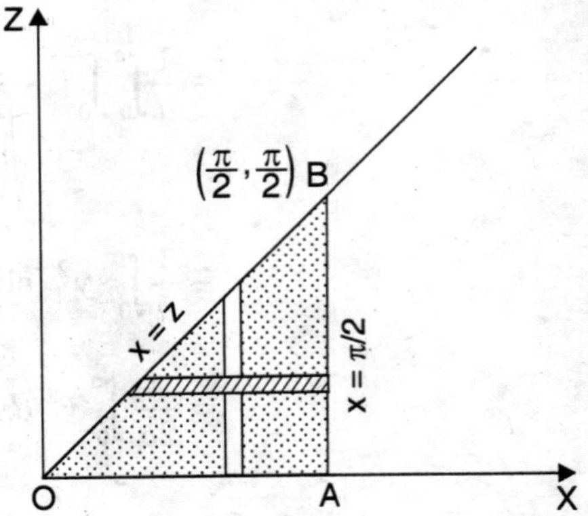

Changing the order of integration, we get

$$I = \int_{z=0}^{\pi/2} \int_{x=z}^{\pi/2} \frac{z \cos z \sin x \, dz \, dx}{\sqrt{\cos^2 z - \cos^2 x}}$$

$$= -\int_0^{\pi/2} z \cos z \, dz \int_z^{\pi/2} \frac{d(\cos x)}{\sqrt{\cos^2 z - \cos^2 x}} = -\int_0^{\pi/2} z \cos z \, dz \left[\sin^{-1} \frac{\cos x}{\cos z}\right]_z^{\pi/2}$$

$$= \int_0^{\pi/2} z \cos z \, . \frac{\pi}{2} dz = \frac{\pi}{2}\left(\frac{\pi}{2} - 1\right) = \frac{\pi^2}{4} - \frac{\pi}{2}.$$

Example 11. *Evaluate the following integral by changing the order of integration:*

$$\int_0^\infty \int_x^\infty \frac{e^{-y}}{y} dy \, dx$$

(A.K.T.U. 2021)

Sol. The given limits show that the area of integration lies between $y = x$, $y = \infty$, $x = 0$ and $x = \infty$.

We can consider it as lying between $x = 0$, $x = y$, $y = 0$ and $y = \infty$ by changing the order of integration.

Hence the given integral,

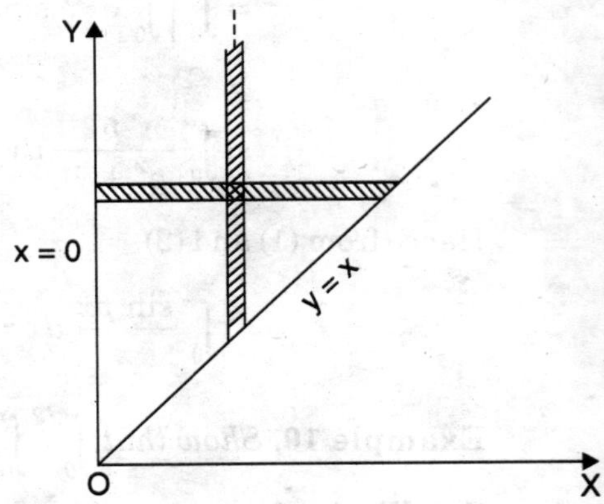

$$\int_0^\infty \int_x^\infty \frac{e^{-y}}{y} dy \, dx = \int_0^\infty \int_0^y \frac{e^{-y}}{y} dx \, dy$$

$$= \int_0^\infty \frac{e^{-y}}{y} \Big(x \Big)_0^y dy = \int_0^\infty e^{-y} dy$$

$$= \Big(-e^{-y} \Big)_0^\infty = 1 - 0 = 1.$$

Example 12. *Evaluate* $\int_0^a \int_0^x \frac{f'(y) \, dy \, dx}{[(a-x)(x-y)]^{1/2}}$.

Sol. Here the limits are $x = 0$, $x = a$; $y = 0$, $y = x$

Evidently the region of integration is OABO.

By changing the order of integration,

$$I = \int_0^a \int_y^a \frac{f'(y)\,dx\,dy}{\sqrt{(a-x)(x-y)}}$$

$$= \int_0^a f'(y)\,dy \int_y^a \frac{dx}{\sqrt{(a-x)(x-y)}} \quad \ldots(1)$$

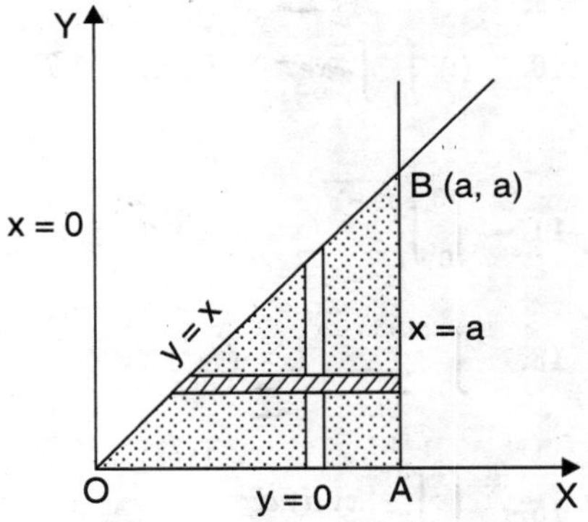

Put $x = a\cos^2\theta + y\sin^2\theta$

$\therefore \quad a - x = a\sin^2\theta - y\sin^2\theta = (a-y)\sin^2\theta$

$\Rightarrow \quad -dx = 2(a-y)\sin\theta\cos\theta\,d\theta$, keeping y constant.

Also, $\quad x - y = a\cos^2\theta + y(\sin^2\theta - 1) = (a-y)\cos^2\theta$

When $\quad x = y$ then

$x - y = 0$

i.e., $\quad (a-y)\cos^2\theta = 0 \Rightarrow \theta = \dfrac{\pi}{2}$

When $\quad x = a$

$\Rightarrow \quad a - x = 0$

i.e., $\quad (a-y)\sin^2\theta = 0$ or $\sin\theta = 0$ or $\theta = 0$

Thus $\quad x = a \Rightarrow \theta = 0$

and $\quad x = y \Rightarrow \theta = \dfrac{\pi}{2}$

$$\therefore \quad \int_y^a \frac{dx}{\sqrt{(a-x)(x-y)}} = \int_{\pi/2}^0 \frac{-2(a-y)\sin\theta\cos\theta\,d\theta}{(a-y)\sin\theta\cos\theta} = 2\int_0^{\pi/2} d\theta = 2\left(\frac{\pi}{2}\right) = \pi.$$

Now, $$I = \int_0^a \pi f'(y)\,dy = \pi\Big[f(y)\Big]_0^a = \pi[f(a) - f(0)]$$

TEST YOUR KNOWLEDGE

Evaluate the following integrals (1–21) by changing the order of integration:

1. $\displaystyle\int_0^1\int_0^{\sqrt{1-x^2}} y^2\,dy\,dx$

2. $\displaystyle\int_0^1\int_{4y}^4 e^{x^2}\,dx\,dy$

3. $\displaystyle\int_0^4\int_y^4 \frac{x\,dx\,dy}{x^2+y^2}$

4. $\displaystyle\int_0^{a/\sqrt{2}}\int_y^{\sqrt{a^2-y^2}} x\,dx\,dy$

5. $\displaystyle\int_0^1\int_x^{\sqrt{2-x^2}} \frac{x\,dy\,dx}{\sqrt{x^2+y^2}}$ *(U.K.T.U. 2012)*

6. $\displaystyle\int_0^b\int_0^{(a/b)\sqrt{b^2-y^2}} xy\,dx\,dy$

7. (*i*) $\displaystyle\int_0^a\int_{x^2/a}^{2a-x} xy\,dy\,dx$ *(U.K.T.U. 2010)* (*ii*) $\displaystyle\int_0^1\int_{y^2}^{2-y} xy\,dx\,dy$ *(U.P.T.U. 2014)*

8. $\displaystyle\int_{y=0}^1\int_y^1 x^2\cos(x^2 - xy)\,dx\,dy$

9. $\displaystyle\int_0^1\int_x^1 \sin y^2\,dy\,dx$ *(U.P.T.U. 2015)*

10. (i) $\int_0^{\infty}\int_0^{x} xe^{-\frac{x^2}{y}}\, dy\, dx$ (M.T.U. 2013) (ii) $\int_0^{\infty}\int_0^{y} ye^{-\frac{y^2}{x}}\, dx\, dy$ (G.B.T.U. 2013)

11. $\int_0^3\int_1^{\sqrt{4-y}} (x+y)\, dx\, dy$

12. $\int_0^a\int_{x/a}^{\sqrt{x/a}} (x^2+y^2)\, dy\, dx$

13. $\int_0^{2a}\int_{\frac{x^2}{4a}}^{3a-x} (x^2+y^2)\, dy\, dx$ (U.K.T.U. 2011)

14. $\int_1^4\int_1^{\sqrt{x}} (x+y^2)\, dy\, dx$

15. $\int_0^2\int_{\frac{x^2}{4}}^{3-x} xy\, dy\, dx$ (A.K.T.U. 2018, 2019)

16. $\int_0^a\int_0^{\frac{bx}{a}} x\, dy\, dx$

17. $\int_0^a\int_{a-\sqrt{a^2-y^2}}^{a+\sqrt{a^2-y^2}} xy\, dx\, dy$

18. $\int_0^1\int_0^{\sqrt{1-x^2}} \frac{e^y}{(e^y+1)\sqrt{1-x^2-y^2}}\, dy\, dx$ (M.T.U. 2012)

19. (i) $\int_0^a\int_0^x \frac{\sec^2 y}{\sqrt{(a-x)(a-y)}}\, dy\, dx$ (ii) $\int_0^3\int_0^{6/x} x^2\, dy\, dx$ (U.P.T.U. 2015)

20. $\int_0^1\int_1^{\sqrt{2-y^2}} \frac{y}{\sqrt{(2-x^2)(1-x^2y^2)}}\, dx\, dy$

21. $\int_0^{\infty} dx \int_0^1 \frac{dy}{1+yx^2}$

Change the order of integration in (22–27):

22. (i) $\int_0^a\int_{\frac{x^2}{a}}^{2a-x} f(x,y)\, dy\, dx$ (ii) $\int_0^1\int_{x^2}^{2-x} f(x,y)\, dy\, dx$ (A.K.T.U. 2018)

23. $\int_0^{2a} dx \int_0^{x^2/4a} (x+y)^3\, dy$

24. $\int_0^a\int_{mx}^{lx} V(x,y)\, dy\, dx$

25. $\int_0^a\int_0^{\frac{b}{b+x}} f(x,y)\, dy\, dx$

26. $\int_0^{2a}\int_{\sqrt{2ax-x^2}}^{\sqrt{2ax}} V(x,y)\, dy\, dx$

27. $\int_0^a\int_{\sqrt{ax-x^2}}^{\sqrt{ax}} f(x,y)\, dy\, dx$

28. Changing the order of integration in the double integral $I = \int_0^8\int_{x/4}^2 f(x,y)\, dy\, dx$ leads to the value $I = \int_r^s\int_p^q f(x,y)\, dx\, dy$. What is the value of q? (A.K.T.U. 2017)

29. Reverse the order of integration in the following integral:

$$I = \int_{\pi/2}^{5\pi/2}\int_{\sin x}^{1} f(x,y)\, dy\, dx$$

30. If $n \geq 0$, show that $\int_a^b\int_a^y (y-x)^n f(x)\, dx\, dy = \frac{1}{n+1}\int_a^b (b-x)^{n+1} f(x)\, dx$

Answers

1. $\frac{\pi}{16}$
2. $\frac{1}{8}(e^{16}-1)$
3. π
4. $\frac{a^3\sqrt{2}}{6}$
5. $1-\frac{1}{\sqrt{2}}$
6. $\frac{1}{8}a^2b^2$
7. (i) $\frac{3a^4}{8}$ (ii) 3/8
8. $\frac{1}{2}(1-\cos 1)$
9. $\frac{1}{2}(1-\cos 1)$
10. (i) $\frac{1}{2}$ (ii) $\frac{1}{2}$
11. $\frac{241}{60}$
12. $\frac{a^3}{28}+\frac{a}{20}$
13. $\frac{314}{35}a^4$
14. $\frac{241}{30}$
15. $\frac{8}{3}$
16. $\frac{1}{3}a^2b$
17. $\frac{2}{3}a^4$
18. $\frac{\pi}{2}\log\left(\frac{e+1}{2}\right)$
19. (i) $2\tan a$ (ii) 27
20. $\frac{\pi}{4}$
21. π
22. (i) $\int_0^a\int_0^{\sqrt{ay}} f(x, y)\, dx\, dy + \int_a^{2a}\int_0^{2a-y} f(x, y)\, dx\, dy$

    (ii) $\int_0^1\int_0^{\sqrt{y}} f(x, y)\, dx\, dy + \int_1^2\int_0^{2-y} f(x, y)\, dx\, dy$
23. $\int_0^a dy \int_{\sqrt{4ay}}^{2a} (x+y)^3\, dx$
24. $\int_0^{ma}\int_{y/l}^{y/m} V(x, y)\, dx\, dy + \int_{ma}^{la}\int_{y/l}^{a} V(x, y)\, dx\, dy$
25. $\int_0^{\frac{b}{a+b}}\int_0^a f(x, y)\, dx\, dy + \int_{\frac{b}{a+b}}^1\int_0^{\frac{b(1-y)}{y}} f(x, y)\, dx\, dy$
26. (i) $\int_0^a\int_{\frac{y^2}{2a}}^{a-\sqrt{a^2-y^2}} V(x, y)\, dx\, dy + \int_0^a\int_{a+\sqrt{a^2-y^2}}^{2a} V(x, y)\, dx\, dy + \int_a^{2a}\int_{\frac{y^2}{2a}}^{2a} V(x, y)\, dx\, dy$
27. $\int_0^a\int_{\frac{y^2}{a}}^{\frac{a}{2}-\sqrt{\frac{a^2}{4}-y^2}} f(y, x)\, dx\, dy + \int_0^{a/2}\int_{\frac{a}{2}+\sqrt{\frac{a^2}{4}-y^2}}^{a} f(x, y)\, dx\, dy + \int_{a/2}^{a}\int_{\frac{y}{a}}^{a} f(x, y)\, dx\, dy$
28. $4y$
29. $\int_{-1}^1\int_{\pi-\sin^{-1}y}^{2\pi+\sin^{-1}y} f(x, y)\, dx\, dy$

4.5. TRIPLE INTEGRALS

Consider a function $f(x, y, z)$ which is continuous at every point of a finite region V of three dimensional space. Divide the region V into n sub-regions of respective volumes $\delta V_1, \delta V_2, ..., \delta V_n$. Let (x_r, y_r, z_r) be an arbitrary point in the rth sub-region.

Consider the sum $\sum_{r=1}^{n} f(x_r, y_r, z_r)\, \delta V_r$.

The limit of this sum as $n \to \infty$ and $\delta V_r \to 0$, if it exists, is called the *triple integral* of $f(x, y, z)$ over the region V and is denoted by $\iiint_V f(x, y, z)\, dV$.

For purpose of evaluation, it can be expressed as the repeated integral

$$\int_{z_1}^{z_2} \int_{y_1}^{y_2} \int_{x_1}^{x_2} f(x, y, z)\, dx\, dy\, dz, \qquad ...(1)$$

the order of integration depending upon the limits.

Let x_1, x_2 be function of y, z ; y_1, y_2 be function of z and z_1, z_2 be constants, *i.e.*,

Let $x_1 = f_1(y, z)$, $x_2 = f_2(y, z)$, $y_1 = \phi_1(z)$, $y_2 = \phi_2(z)$ and $z_1 = a$, $z_2 = b$.

Then the integral (*i*) is evaluated as follows:

$$\int_{z_1 = a}^{z_2 = b} \left[\int_{y_1 = \phi_1(z)}^{y_2 = \phi_2(z)} \left[\int_{x_1 = f_1(y, z)}^{x_2 = f_2(y, z)} f(x, y, z)\, dx \right] dy \right] dz$$

First $f(x, y, z)$ is integrated w.r.t. x (keeping y and z constant) between the limits x_1 and x_2. The resulting expression, which is a function of y and z is then integrated w.r.t. y (keeping z constant) between the limits y_1 and y_2. The resulting expression, which is a function of z only is then integrated w.r.t. z between the limits z_1 and z_2. The order of integration is from the innermost rectangle to the outermost rectangle.

Limits involving two variables are kept innermost, then the limits involving one variable and finally the constant limits.

If x_1, x_2 ; y_1, y_2 and z_1, z_2 are all constants, then the order of integration is immaterial, provided the limits are changed accordingly. Thus

$$\int_{z_1}^{z_2} \int_{y_1}^{y_2} \int_{x_1}^{x_2} f(x, y, z)\, dx\, dy\, dz = \int_{x_1}^{x_2} \int_{z_1}^{z_2} \int_{y_1}^{y_2} f(x, y, z)\, dy\, dz\, dx = \int_{x_1}^{x_2} \int_{y_1}^{y_2} \int_{z_1}^{z_2} f(x, y, z)\, dz\, dy\, dx$$

ILLUSTRATIVE EXAMPLES

Example 1. *Evaluate* $\displaystyle\int_0^1 \int_0^{\sqrt{1-x^2}} \int_0^{\sqrt{1-x^2-y^2}} \frac{1}{\sqrt{1-x^2-y^2-z^2}}\, dz\, dy\, dx$.

Sol. Let $\displaystyle I = \int_0^1 \int_0^{\sqrt{1-x^2}} \int_0^{\sqrt{1-x^2-y^2}} \frac{1}{\sqrt{(1-x^2-y^2)-z^2}}\, dz\, dy\, dx$

$$= \int_0^1 \int_0^{\sqrt{1-x^2}} \left[\sin^{-1} \frac{z}{\sqrt{1-x^2-y^2}} \right]_0^{\sqrt{1-x^2-y^2}} dy\, dx$$

$$= \int_0^1 \int_0^{\sqrt{1-x^2}} (\sin^{-1} 1 - \sin^{-1} 0)\, dy\, dx$$

$$= \int_0^1 \int_0^{\sqrt{1-x^2}} \frac{\pi}{2}\, dy\, dx = \frac{\pi}{2} \int_0^1 \sqrt{1-x^2}\, dx$$

$$= \frac{\pi}{2}\left[\frac{x\sqrt{1-x^2}}{2} + \frac{1}{2}\sin^{-1} x\right]_0^1 = \frac{\pi}{4}[\sin^{-1} 1] = \frac{\pi}{4}\cdot\frac{\pi}{2} = \frac{\pi^2}{8}.$$

Example 2. *Evaluate* $\int_0^4 \int_0^{2\sqrt{z}} \int_0^{\sqrt{4z-x^2}} dy\, dx\, dz.$

Sol. $$I = \int_{z=0}^4 \int_{x=0}^{2\sqrt{z}} (y)_0^{\sqrt{4z-x^2}}\, dx\, dz = \int_0^4 \int_0^{2\sqrt{z}} \sqrt{4z-x^2}\, dx\, dz$$

$$= \int_0^4 \left[\frac{1}{2} x \sqrt{4z-x^2} + \frac{1}{2}.\ 4z \sin^{-1}\left(\frac{x}{\sqrt{4z}}\right)\right]_0^{2\sqrt{z}} dz$$

$$= \int_0^4 2z \sin^{-1}(1)\, dz = \pi \int_0^4 z\, dz = \pi\left(\frac{z^2}{2}\right)_0^4 = 8\pi.$$

TEST YOUR KNOWLEDGE

Evaluate the following integrals (1–12):

1. $\int_0^1 \int_0^1 \int_0^1 e^{x+y+z}\, dx\, dy\, dz$

2. $\int_0^a \int_0^a \int_0^a (yz + zx + xy)\, dx\, dy\, dz$

3. $\int_{-1}^1 \int_0^z \int_{x-z}^{x+z} (x + y + z)\, dy\, dx\, dz$

4. $\int_1^3 \int_{1/x}^1 \int_0^{\sqrt{xy}} xyz\, dz\, dy\, dx$

5. $\int_0^1 \int_{y^2}^1 \int_0^{1-x} x\, dz\, dx\, dy$

6. $\int_0^2 \int_0^x \int_0^{x+y} e^x (y + 2z)\, dz\, dy\, dx$

7. $\int_0^a \int_0^{a-x} \int_0^{a-x-y} (x + y + z)\, dz\, dy\, dx$

8. $\int_{-c}^c \int_{-b}^b \int_{-a}^a (x^2 + y^2 + z^2)\, dx\, dy\, dz$

9. $\int_0^1 \int_0^{\sqrt{1-x^2}} \int_0^{\sqrt{1-x^2-y^2}} xyz\, dz\, dy\, dx$ (A.K.T.U. 2017)

10. $\int_0^{\log 2} \int_0^x \int_0^{x+\log y} e^{x+y+z}\, dz\, dy\, dx$

11. $\int_0^{\pi/2} d\theta \int_0^{a\sin\theta} dr \int_0^{(a^2-r^2)/a} r dz$

12. $\int_0^{\pi/2} \int_0^{a\cos\theta} \int_0^{\sqrt{a^2-r^2}} r dz\, dr\, d\theta$

13. Evaluate $\iiint_R (x - 2y + z)\, dzdydx,$ where R is the region determined by $0 \le x \le 1, 0 \le y \le x^2,$ $0 \le z \le x + y.$

14. Evaluate $\iiint_R (x + y + z)\, dxdydz$ where R : $0 \le x \le 1,\ 1 \le y \le 2,\ 2 \le z \le 3.$ [*A.K.T.U. 2016, 2018*]

15. Evaluate $\iiint_S x\, dV$ where S is the region bounded by the surfaces $y = x^2$, $y = x + 2$, $4z = x^2 + y^2$ and $z = x + 3.$

Answers

1. $(e-1)^3$	**2.** $\frac{3}{4}a^5$	**3.** 0	**4.** $\frac{13}{9}-\frac{1}{6}\log 3$
5. $\frac{4}{35}$	**6.** $19\left(\frac{e^2}{3}+1\right)$	**7.** $\frac{a^4}{8}$	**8.** $\frac{8}{3}abc\,(a^2+b^2+c^2)$
9. $\frac{1}{48}$	**10.** $\frac{8}{3}\log 2-\frac{19}{9}$	**11.** $\frac{5a^3\pi}{64}$	**12.** $\frac{a^3}{3}\left(\frac{\pi}{2}-\frac{2}{3}\right)$
13. $\frac{8}{35}$	**14.** $\frac{9}{2}$	**15.** $\frac{837}{160}$	

4.6. CHANGE OF VARIABLES

Quite often, the evaluation of a double or triple integral is greatly simplified by a suitable change of variables.

Let the variables x, y in the double integral $\iint_R f(x, y)\, dx\, dy$ be changed to u, v by means of the relations $x = \phi(u, v)$, $y = \psi(u, v)$, then the double integral is transformed into

$$\iint_{R'} f\{\phi(u, v), \psi(u, v)\}\, |\,J\,|\, du\, dv \text{ where } J = \frac{\partial(x, y)}{\partial(u, v)} = \begin{vmatrix} \frac{\partial x}{\partial u} & \frac{\partial x}{\partial v} \\ \frac{\partial y}{\partial u} & \frac{\partial y}{\partial v} \end{vmatrix} \text{ is the Jacobian of transformation}$$

from (x, y) to (u, v) co-ordinates and R′ is the region in the uv-plane which corresponds to the region R in the xy-plane.

(*i*) *To change cartesian co-ordinates (x, y) to polar co-ordinates (r, θ).*

Here we have $x = r\cos\theta$, $y = r\sin\theta$ so that $x^2 + y^2 = r^2$

$$J = \begin{vmatrix} \frac{\partial x}{\partial r} & \frac{\partial x}{\partial \theta} \\ \frac{\partial y}{\partial r} & \frac{\partial y}{\partial \theta} \end{vmatrix} = \begin{vmatrix} \cos\theta & -r\sin\theta \\ \sin\theta & r\cos\theta \end{vmatrix} = r(\cos^2\theta + \sin^2\theta) = r$$

$$\therefore \quad \boxed{\iint_R f(x, y)\, dx\, dy = \iint_{R'} f(r\cos\theta, r\sin\theta)\, r\, dr\, d\theta}$$

i.e., replace x by $r\cos\theta$, y by $r\sin\theta$ and $dx\,dy$ by $r dr d\theta$.

(*ii*) *To change cartesian co-ordinates (x, y, z) to spherical polar co-ordinates (r, θ, ϕ).*

Here, we have $x = r\sin\theta\cos\phi$

$$y = r\sin\theta\sin\phi$$

$$z = r\cos\theta$$

so that $\quad x^2 + y^2 + z^2 = r^2$

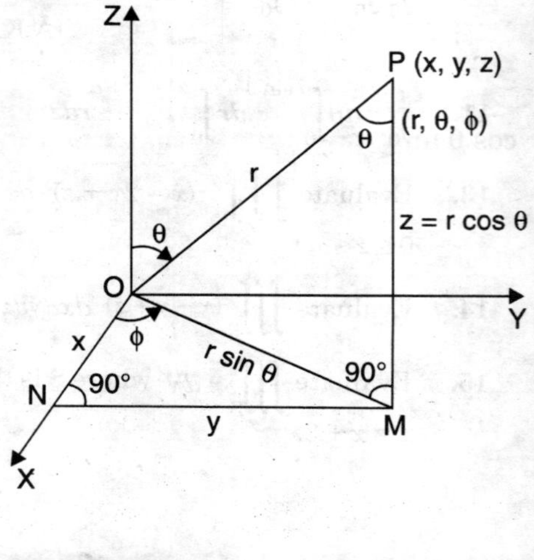

$$J = \begin{vmatrix} \frac{\partial x}{\partial r} & \frac{\partial x}{\partial \theta} & \frac{\partial x}{\partial \phi} \\ \frac{\partial y}{\partial r} & \frac{\partial y}{\partial \theta} & \frac{\partial y}{\partial \phi} \\ \frac{\partial z}{\partial r} & \frac{\partial z}{\partial \theta} & \frac{\partial z}{\partial \phi} \end{vmatrix} = \begin{vmatrix} \sin\theta\cos\phi & r\cos\theta\cos\phi & -r\sin\theta\sin\phi \\ \sin\theta\sin\phi & r\cos\theta\sin\phi & r\sin\theta\cos\phi \\ \cos\theta & -r\sin\theta & 0 \end{vmatrix} = r^2\sin\theta$$

$$\therefore \quad \iiint_V f(x, y, z)\, dx\, dy\, dz = \iiint_{V'} f(r\sin\theta\cos\phi, r\sin\theta\sin\phi, r\cos\theta)\, r^2 \sin\theta\, dr\, d\theta\, d\phi$$

(iii) To change cartesian co-ordinates (x, y, z) to cylindrical co-ordinates (r, φ, z).

Here we have $x = r\cos\phi$

$y = r\sin\phi$

$z = z$

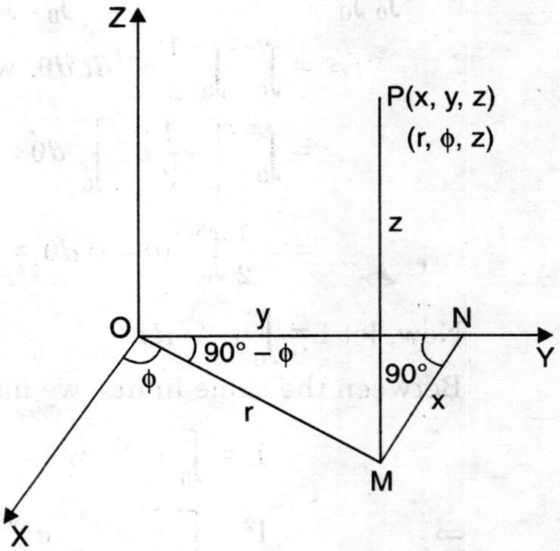

$$J = \begin{vmatrix} \frac{\partial x}{\partial r} & \frac{\partial x}{\partial \phi} & \frac{\partial x}{\partial z} \\ \frac{\partial y}{\partial r} & \frac{\partial y}{\partial \phi} & \frac{\partial y}{\partial z} \\ \frac{\partial z}{\partial r} & \frac{\partial z}{\partial \phi} & \frac{\partial z}{\partial z} \end{vmatrix}$$

$$= \begin{vmatrix} \cos\phi & -r\sin\phi & 0 \\ \sin\phi & r\cos\phi & 0 \\ 0 & 0 & 1 \end{vmatrix}$$

$$= r(\cos^2\phi + \sin^2\phi) = r$$

$$\therefore \quad \iiint_V f(x, y, z)\, dx\, dy\, dz = \iiint_{V'} f(r\cos\phi, r\sin\phi, z)\, r\, dr\, d\phi\, dz.$$

ILLUSTRATIVE EXAMPLES

Example 1. *Evaluate the following by changing into polar coordinates:*

$$\int_0^a \int_0^{\sqrt{a^2 - y^2}} y^2 \sqrt{x^2 + y^2}\, dx\, dy$$

Sol. Changing to polar coordinates by putting $x = r\cos\theta$ and $y = r\sin\theta$ so that $x^2 + y^2 = 1$, we get

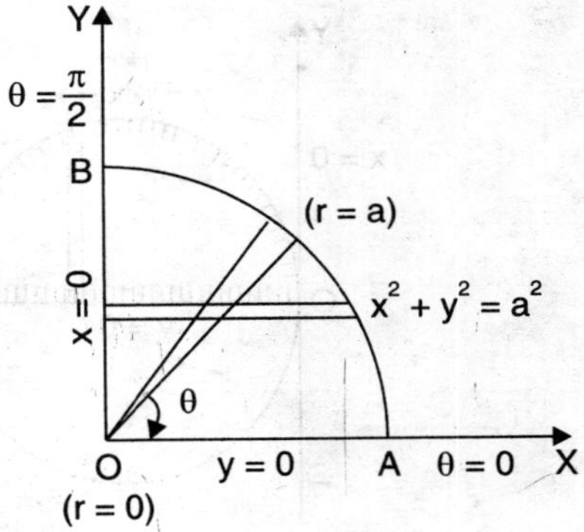

$$I = \int_0^{\pi/2} \int_0^a r^2 \sin^2\theta \,.\, r \,.\, r\, dr\, d\theta$$

$$= \int_0^{\pi/2} \left(\frac{r^5}{5}\right)_0^a \sin^2\theta\, d\theta$$

$$= \frac{a^5}{5} \cdot \frac{1}{2} \cdot \frac{\pi}{2} = \frac{\pi a^5}{20}$$

Example 2. *Change into polar coordinates and evaluate* $\int_0^\infty \int_0^\infty e^{-(x^2+y^2)}\, dy\, dx$.

Hence show that $\int_0^\infty e^{-x^2}\, dx = \frac{\sqrt{\pi}}{2}$. (A.K.T.U. 2015, 2019)

Sol. For the region of integration in cartesian co-ordinates, y varies from 0 to ∞ and x also varies from 0 to ∞. Thus the region of integration is the plane XOY. Changing to polar co-ordinates by putting $x = r\cos\theta$, $y = r\sin\theta$ so that $x^2 + y^2 = r^2$; for the region of integration r varies from 0 to ∞ and θ varies from 0 to $\frac{\pi}{2}$.

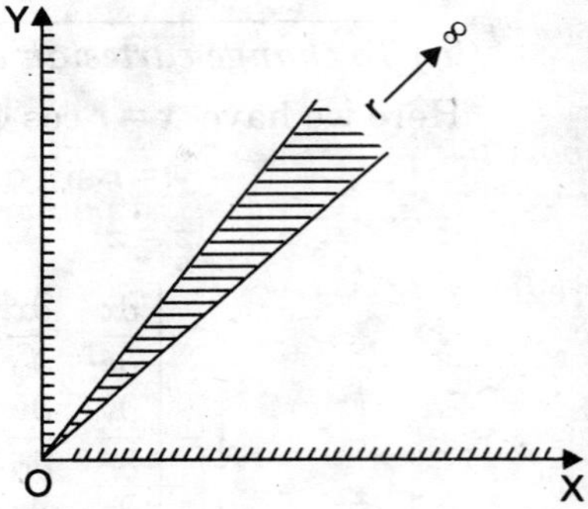

$$\therefore \quad \int_0^\infty \int_0^\infty e^{-(x^2+y^2)}\, dy\, dx = \int_0^{\pi/2} \int_0^\infty e^{-r^2} . r\, dr\, d\theta$$

$$= \int_0^{\pi/2} \int_0^\infty \frac{1}{2} e^{-t} dt\, d\theta, \text{ where } t = r^2$$

$$= \int_0^{\pi/2} \left[-\frac{1}{2} e^{-t} \right]_0^\infty d\theta$$

$$= -\frac{1}{2} \int_0^{\pi/2} (0 - 1)\, d\theta = \frac{1}{2} \left[\theta \right]_0^{\pi/2} = \frac{\pi}{4}$$

Now, let $I = \int_0^\infty e^{-x^2}\, dx$

Between the same limits, we have

$$I = \int_0^\infty e^{-y^2}\, dy$$

$$\Rightarrow \quad I^2 = \int_0^\infty \int_0^\infty e^{-x^2} . e^{-y^2} dx\, dy = \int_0^\infty \int_0^\infty e^{-(x^2+y^2)}\, dx\, dy = \frac{\pi}{4} \qquad | \text{ Proved already}$$

$$\Rightarrow \quad I = \frac{\sqrt{\pi}}{2}.$$

Example 3. *Evaluate* $\int_0^2 \int_0^{\sqrt{2x-x^2}} \frac{x\, dy\, dx}{\sqrt{x^2+y^2}}$ *by changing to polar coordinates.*

Sol. In the given integral, y varies from 0 to $\sqrt{2x - x^2}$ and x varies from 0 to 2.

$$y = \sqrt{2x - x^2} \;\Rightarrow\; y^2 = 2x - x^2 \;\Rightarrow\; x^2 + y^2 = 2x.$$

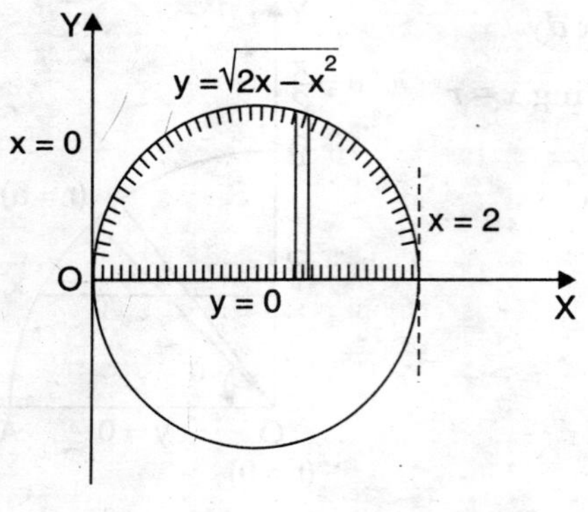

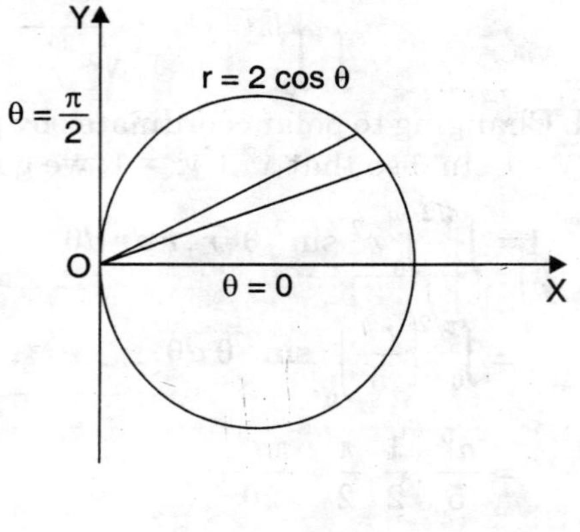

In polar co-ordinates, we have $r^2 = 2r \cos\theta$ or $r = 2\cos\theta$.

$\therefore$ For the region of integration, r varies from 0 to $2\cos\theta$ and θ varies from 0 to $\frac{\pi}{2}$.

In the given integral, replacing x by $r\cos\theta$, y by $r\sin\theta$, $dy\,dx$ by $r\,dr\,d\theta$, we have

$$I = \int_0^{\pi/2}\int_0^{2\cos\theta} \frac{r\cos\theta \,.\, r\,dr\,d\theta}{r} = \int_0^{\pi/2}\int_0^{2\cos\theta} r\cos\theta\,dr\,d\theta$$

$$= \int_0^{\pi/2} \cos\theta \left[\frac{r^2}{2}\right]_0^{2\cos\theta} d\theta = \int_0^{\pi/2} 2\cos^3\theta\,d\theta = 2\cdot\frac{2}{3} = \frac{4}{3}.$$

Example 4. *Evaluate* $\iint \sqrt{a^2 - x^2 - y^2}\,dx\,dy$ *over the semi-circle* $x^2 + y^2 = ax$ *in the positive quadrant.*

Sol. Changing to polar co-ordinates, $x^2 + y^2 = ax$ transforms into $r = a\cos\theta$. For the region of integration R, r varies from 0 to $a\cos\theta$ and θ varies from 0 to $\frac{\pi}{2}$.

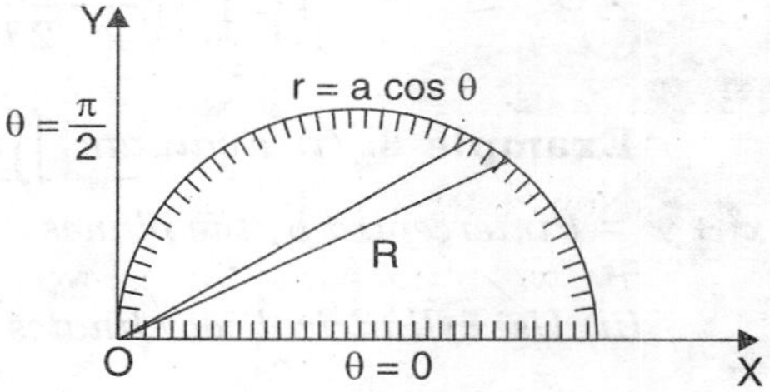

$$\therefore \iint_R \sqrt{a^2 - x^2 - y^2}\,dx\,dy$$

$$= \int_0^{\pi/2}\int_0^{a\cos\theta} \sqrt{a^2 - r^2}\,.\,r\,dr\,d\theta$$

$$= \int_0^{\pi/2}\int_0^{a\cos\theta} -\frac{1}{2}(a^2 - r^2)^{1/2}(-2r)\,dr\,d\theta$$

$$= \int_0^{\pi/2} -\frac{1}{2}\cdot\left[\frac{(a^2 - r^2)^{3/2}}{3/2}\right]_0^{a\cos\theta} d\theta$$

$$= -\frac{1}{3}\int_0^{\pi/2}(a^3\sin^3\theta - a^3)\,d\theta = -\frac{a^3}{3}\int_0^{\pi/2}(\sin^3\theta - 1)\,d\theta$$

$$= -\frac{a^3}{3}\left[\frac{2}{3} - \frac{\pi}{2}\right] = \frac{a^3}{3}\left(\frac{\pi}{2} - \frac{2}{3}\right).$$

Example 5. *Evaluate* $\iint \sqrt{\frac{1 - x^2 - y^2}{1 + x^2 + y^2}}\,dx\,dy$ *over the positive quadrant of the circle* $x^2 + y^2 = 1$.

Sol. Changing to polar coordinates by putting $x = r\cos\theta$, $y = r\sin\theta$; $x^2 + y^2 = 1$ transforms into $r = 1$. For the region of integration R, r varies from 0 to 1 and θ varies from 0 to $\frac{\pi}{2}$.

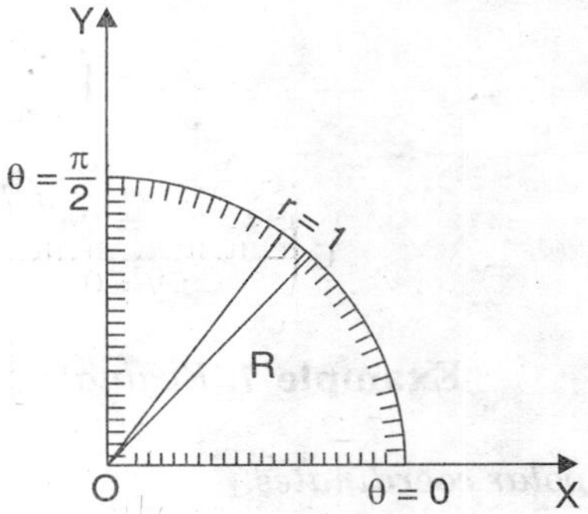

$$I = \iint_R \sqrt{\frac{1 - x^2 - y^2}{1 + x^2 + y^2}}\,dx\,dy$$

$$= \int_0^{\pi/2}\int_0^1 \sqrt{\frac{1 - r^2}{1 + r^2}}\,r\,dr\,d\theta = \int_0^{\pi/2}\int_0^1 \frac{r(1 - r^2)}{\sqrt{1 - r^4}}\,dr\,d\theta$$

Now, $$\int_0^1 \frac{r(1-r^2)}{\sqrt{1-r^4}}\,dr = \int_0^1 \left(\frac{r}{\sqrt{1-r^4}} - \frac{r^3}{\sqrt{1-r^4}} \right) dr$$

$$= \frac{1}{2}\int_0^1 \frac{2r}{\sqrt{1-r^4}}\,dr + \frac{1}{4}\int_0^1 -4r^3\,(1-r^4)^{-1/2}\,dr$$

$$= \frac{1}{2}\int_0^1 \frac{dt}{\sqrt{1-t^2}} + \frac{1}{4}\cdot\left[\frac{(1-r^4)^{1/2}}{1/2}\right]_0^1, \text{ where } t = r^2$$

$$= \frac{1}{2}\left[\sin^{-1} t\right]_0^1 + \frac{1}{2}\,(0-1) = \frac{1}{2}\left(\frac{\pi}{2}\right) - \frac{1}{2} = \frac{\pi}{4} - \frac{1}{2}$$

$$\therefore \qquad I = \int_0^{\pi/2} \left(\frac{\pi}{4} - \frac{1}{2}\right) d\theta = \left(\frac{\pi}{4} - \frac{1}{2}\right)\frac{\pi}{2} = \frac{\pi^2}{8} - \frac{\pi}{4}.$$

Example 6. *(i) Evaluate* $\iiint z\,(x^2+y^2)\,dx\,dy\,dz$ *over the volume of the cylinder* $x^2+y^2=1$ *intercepted by the planes* $z = 2$ *and* $z = 3$.

(ii) Use cylindrical coordinates to evaluate $\iiint_V (x^2+y^2)\,dx\,dy\,dz$ *taken over the region* V *bounded by the paraboloid* $z = 9 - x^2 - y^2$ *and the plane* $z = 0$.

Sol. *(i)* $$I = \iiint z\,(x^2+y^2)\,dx\,dy\,dz$$

$$= \int_{z=2}^{3}\int_{\theta=0}^{2\pi}\int_{r=0}^{1} z\,.\,r^2\,.\,r\,dr\,d\theta\,dz \qquad |\because\ |J| = r$$

$$= \int_2^3\int_0^{2\pi} z\left(\frac{r^4}{4}\right)_0^1 d\theta\,dz = \frac{1}{4}\,(2\pi)\left(\frac{z^2}{2}\right)_2^3 = \frac{\pi}{4}\,(9-4) = \frac{5\pi}{4}.$$

(ii) $$I = \iiint_V (x^2+y^2)\,dx\,dy\,dz = \int_0^{2\pi}\int_0^3\int_0^{9-r^2} r^2\,.\,r\,dz\,dr\,d\theta \qquad |\because\ |J| = r$$

$$= \int_0^{2\pi}\int_0^3 r^3\,(9-r^2)\,dr\,d\theta = \int_0^{2\pi}\left(\frac{9r^4}{4} - \frac{r^6}{6}\right)_0^3 d\theta$$

$$= 2\pi\left(\frac{729}{4} - \frac{729}{6}\right) = \frac{243\pi}{2}.$$

Example 7. *Evaluate* $\displaystyle\int_0^1\int_0^{\sqrt{1-x^2}}\int_0^{\sqrt{1-x^2-y^2}} \frac{dz\,dy\,dx}{\sqrt{1-x^2-y^2-z^2}}$, *by changing to spherical polar coordinates.*

Sol. Here the region of integration is bounded by

$z = 0, \qquad z = \sqrt{1 - x^2 - y^2} \qquad$ (*i.e.*, $x^2 + y^2 + z^2 = 1$)

$y = 0, \qquad y = \sqrt{1 - x^2} \qquad$ (*i.e.*, $x^2 + y^2 = 1$)

$x = 0, \qquad x = 1$

which is the volume of the sphere $x^2 + y^2 + z^2 = 1$ in the positive octant.

Changing to spherical polar co-ordinates by putting $x = r \sin \theta \cos \phi$, $y = r \sin \theta \sin \phi$, $z = r \cos \theta$ so that $x^2 + y^2 + z^2 = r^2$.

For the volume of sphere $x^2 + y^2 + z^2 = 1$ in the positive octant, r varies from 0 to 1, θ varies from 0 to $\frac{\pi}{2}$ and ϕ varies from 0 to $\frac{\pi}{2}$.

Replacing $dz\, dy\, dx$ by $r^2 \sin \theta\, dr\, d\theta\, d\phi$, we have

$$I = \int_0^{\pi/2} \int_0^{\pi/2} \int_0^1 \frac{r^2 \sin \theta\, dr\, d\theta\, d\phi}{\sqrt{1 - r^2}} = \int_0^{\pi/2} \int_0^{\pi/2} \int_0^1 \frac{1 - (1 - r^2)}{\sqrt{1 - r^2}} \sin \theta\, dr\, d\theta\, d\phi$$

$$= \int_0^{\pi/2} \int_0^{\pi/2} \int_0^1 \left(\frac{1}{\sqrt{1 - r^2}} - \sqrt{1 - r^2} \right) \sin \theta\, dr\, d\theta\, d\phi$$

$$= \int_0^{\pi/2} \int_0^{\pi/2} \sin \theta \left[\sin^{-1} r - \left(\frac{r\sqrt{1 - r^2}}{2} + \frac{1}{2} \sin^{-1} r \right) \right]_0^1 d\theta\, d\phi$$

$$= \int_0^{\pi/2} \int_0^{\pi/2} \sin \theta \left[\frac{\pi}{2} - \frac{1}{2} \cdot \frac{\pi}{2} \right] d\theta\, d\phi = \int_0^{\pi/2} \frac{\pi}{4} \Big[-\cos \theta \Big]_0^{\pi/2} d\phi = \int_0^{\pi/2} \frac{\pi}{4}\, d\phi$$

$$= \frac{\pi}{4} \cdot \frac{\pi}{2} = \frac{\pi^2}{8}.$$

Note. For the whole volume of the sphere $x^2 + y^2 + z^2 = a^2$, $0 \le r \le a$, $0 \le \theta \le \pi$, $0 \le \phi \le 2\pi$.

Example 8. (*i*) *Evaluate the following integrals through the volume of the sphere* $x^2 + y^2 + z^2 = 1$ *by changing into spherical polar coordinates:*

(*a*) $\iiint z^2\, dx\, dy\, dz$ (*b*) $\iiint (x^2 + y^2 + z^2)^m\, dx\, dy\, dz$, $(m > 0)$

(*ii*) *Evaluate* $\iiint xyz\, dx\, dy\, dz$ *over the positive octant of the sphere* $x^2 + y^2 + z^2 = b^2$ *by transforming to spherical polar coordinates.*

(*iii*) *Evaluate* $\iiint_V xyz\, (x^2 + y^2 + z^2)^{n/2}\, dx\, dy\, dz$ *taken through the positive octant of the sphere* $x^2 + y^2 + z^2 = b^2$ *provided* $n + 5 > 0$.

(*iv*) *Use spherical coordinates to evaluate the integral* $\iiint \frac{dx\, dy\, dz}{\sqrt{x^2 + y^2 + z^2}}$ *taken over the region V in the first octant bounded by the cones* $\theta = \frac{\pi}{4}$ *and* $\theta = \tan^{-1} 2$ *and the sphere* $x^2 + y^2 + z^2 = 6$.

Sol. (*i*) (*a*)
$$I = \int_0^{2\pi}\int_0^{\pi}\int_0^{1} r^2 \cos^2\theta \,.\, r^2 \sin\theta \, dr\, d\theta\, d\phi$$

$$= \int_0^{2\pi}\int_0^{\pi}\left(\frac{r^5}{5}\right)_0^1 \cos^2\theta \sin\theta \, d\theta\, d\phi$$

$$= \frac{2}{5}\int_0^{2\pi}\int_0^{\pi/2} \cos^2\theta \sin\theta\, d\theta\, d\phi = \frac{2}{5}\int_0^{2\pi}\int_0^{\pi/2} (\sin\theta - \sin^3\theta)\, d\theta\, d\phi$$

$$= \frac{2}{5}\int_0^{2\pi}\left[(-\cos\theta)_0^{\pi/2} - \frac{2}{3}\right] d\phi = \frac{2}{5}\cdot\frac{1}{3}\cdot 2\pi = \frac{4\pi}{15}.$$

(*b*)
$$I = \int_0^{2\pi}\int_0^{\pi}\int_0^{1} r^{2m} \,.\, r^2 \sin\theta\, dr\, d\theta\, d\phi = \frac{1}{2m+3}\int_0^{2\pi}\int_0^{\pi} \sin\theta\, d\theta\, d\phi$$

$$= \frac{1}{2m+3}\int_0^{2\pi} (-\cos\theta)_0^{\pi}\, d\phi = \frac{4\pi}{2m+3}.$$

(*ii*)
$$I = \int_0^{\pi/2}\int_0^{\pi/2}\int_0^{b} (r\sin\theta\cos\phi)(r\sin\theta\sin\phi)(r\cos\theta)\, r^2 \sin\theta\, dr\, d\theta\, d\phi$$

$$= \frac{b^6}{6}\int_0^{\pi/2}\int_0^{\pi/2} \sin^3\theta\cos\theta\cos\phi\sin\phi\, d\theta\, d\phi$$

$$= \frac{b^6}{6}\cdot\frac{\Gamma 1\,\Gamma 1}{2\,\Gamma 2}\cdot\frac{\Gamma 2\,\Gamma 1}{2\,\Gamma 3} = \frac{b^6}{48}.$$

(*iii*)
$$I = \int_0^{\pi/2}\int_0^{\pi/2}\int_0^{b} (r\sin\theta\cos\phi)(r\sin\theta\sin\phi)(r\cos\theta)\, r^n\, r^2 \sin\theta\, dr\, d\theta\, d\phi$$

$$= \frac{b^{n+6}}{n+6}\int_0^{\pi/2} \cos\phi\sin\phi\, d\phi \int_0^{\pi/2} \sin^3\theta\cos\theta\, d\theta$$

$$= \frac{b^{n+6}}{8(n+6)}.$$

(*iv*)
$$I = \int_0^{\pi/2}\int_{\pi/4}^{\tan^{-1}2}\int_0^{\sqrt{6}} \frac{1}{r}\,.\, r^2 \sin\theta\, dr\, d\theta\, d\phi = 3\int_0^{\pi/2} (-\cos\theta)_{\pi/4}^{\tan^{-1}2}\, d\phi$$

$$= -\frac{3\pi}{2}\cdot\left(\frac{1}{\sqrt{5}} - \frac{1}{\sqrt{2}}\right) = \frac{3\pi}{2}\left(\frac{1}{\sqrt{2}} - \frac{1}{\sqrt{5}}\right).$$

Example 9. *Evaluate* $\iiint (x^2y^2 + y^2z^2 + z^2x^2)\, dx\, dy\, dz$ over the volume of the sphere $x^2 + y^2 + z^2 = a^2$.

Sol. $x = r\sin\theta\cos\phi,\ y = r\sin\theta\sin\phi,\ z = r\cos\theta$

$$0 \le r \le a,\quad 0 \le \theta \le \pi,\quad 0 \le \phi \le 2\pi$$

$$\iiint (x^2y^2 + y^2z^2 + z^2x^2)\, dx\, dy\, dz$$

$$= \int_0^{2\pi}\int_0^{\pi}\int_0^{a} (r^2 \sin^2\theta \cos^2\phi \,.\, r^2 \sin^2\theta \sin^2\phi + r^2\sin^2\theta \sin^2\phi \,.\, r^2\cos^2\theta + r^2 \cos^2\theta \,.\, r^2\sin^2\theta\cos^2\phi) r^2\sin\theta\, dr\, d\theta\, d\phi$$

$$= \int_0^{2\pi}\int_0^{\pi}\int_0^{a} r^6(\sin^5\theta \cos^2\phi \sin^2\phi + \sin^3\theta \cos^2\theta \sin^2\phi + \sin^3\theta \cos^2\theta \cos^2\phi)\, dr\, d\theta\, d\phi$$

$$= \frac{a^7}{7} \times 8 \int_0^{\pi/2}\int_0^{\pi/2} (\sin^5\theta \cos^2\phi \sin^2\phi + \sin^3\theta \cos^2\theta) d\theta\, d\phi$$

$$= \frac{8a^7}{7}\left[\frac{4}{5}\cdot\frac{2}{3}\cdot\frac{\Gamma(3/2)\,\Gamma(3/2)}{2\,\Gamma(3)} + \frac{\Gamma(2)\,\Gamma(3/2)}{2\,\Gamma(7/2)}\cdot\frac{\pi}{2}\right]$$

$$= \frac{8a^7}{7}\left[\frac{\pi}{30} + \frac{\pi}{15}\right] = \frac{4}{35}\pi a^7$$

Example 10. *Evaluate* $\iint_R (x + y)^2\, dx\, dy$, *where R is the parallelogram in the xy-plane with vertices (1, 0), (3, 1), (2, 2), (0, 1), using the transformation* $u = x + y$ *and* $v = x - 2y$.

(A.K.T.U. 2020)

Sol. The vertices A(1, 0), B(3, 1), C(2, 2), D(0, 1) of the parallelogram ABCD in the xy-plane become A′(1, 1), B′(4, 1), C′(4, – 2) and D′(1, – 2) in the uv-plane under the given transformation.

The region R in the xy-plane becomes the region R′ in the uv-plane which is a rectangle bounded by the line $u = 1$, $u = 4$ and $v = -2$, $v = 1$. Solving the given equations for x and y, we have $x = \frac{1}{3}(2u + v)$, $y = \frac{1}{3}(u - v)$

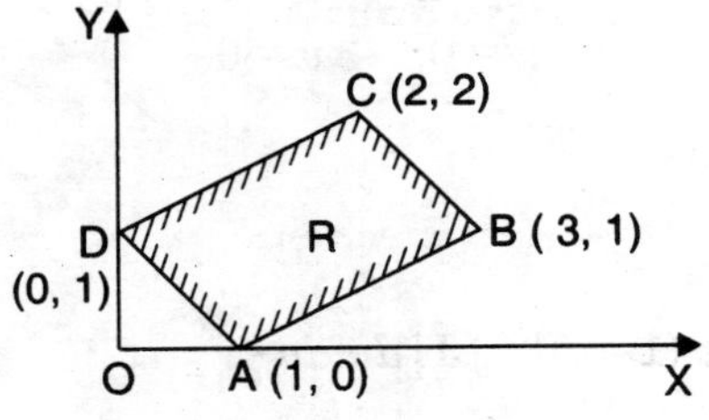

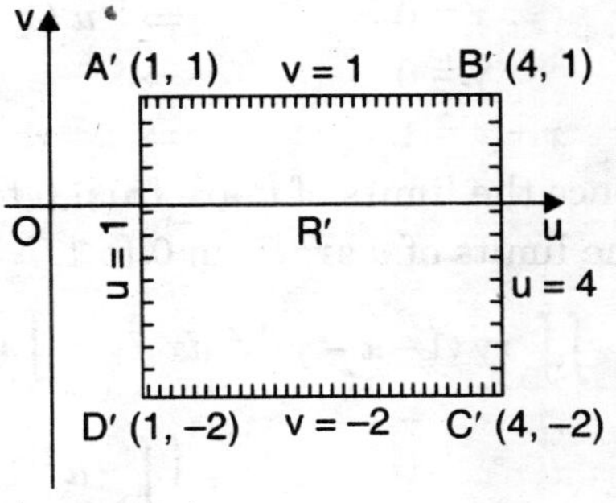

$$\therefore \quad J = \frac{\partial(x, y)}{\partial(u, v)} = \begin{vmatrix} \frac{\partial x}{\partial u} & \frac{\partial x}{\partial v} \\ \frac{\partial y}{\partial u} & \frac{\partial y}{\partial v} \end{vmatrix} = \begin{vmatrix} \frac{2}{3} & \frac{1}{3} \\ \frac{1}{3} & -\frac{1}{3} \end{vmatrix} = -\frac{1}{3}$$

$$\therefore \quad \iint_R (x + y)^2\, dx\, dy = \iint_{R'} u^2\, |J|\, du\, dv = \int_{-2}^{1}\int_1^4 u^2 \,.\, \frac{1}{3}\, du\, dv$$

$$= \int_{-2}^{1} \frac{1}{3}\left[\frac{u^3}{3}\right]_1^4 dv = \int_{-2}^{1} 7\, dv = 7 \times 3 = 21.$$

Example 11. (*i*) *Let D be the region in the first quadrant bounded by x = 0, y = 0 and x + y = 1. Change the variables x, y to u, v, where x + y = u, y = uv and evaluate*

$$\iint_D xy\,(1-x-y)^{1/2}\,dx\,dy.$$

(*ii*) *Using the transformation x + y = u, y = uv; show that*

$$\int_0^1\int_0^{1-x} e^{y/(x+y)}\,dy\,dx = \frac{1}{2}\,(e-1)$$

Sol. (*i*) Since $\quad x = u\,(1-v),\ \ y = uv$

$$\therefore \quad J = \frac{\partial(x, y)}{\partial(u, v)} = \begin{vmatrix} \dfrac{\partial x}{\partial u} & \dfrac{\partial x}{\partial v} \\ \dfrac{\partial y}{\partial u} & \dfrac{\partial y}{\partial v} \end{vmatrix} = \begin{vmatrix} 1-v & -u \\ v & u \end{vmatrix} = u - uv + uv = u$$

$$\therefore \quad dx\,dy = |\,J\,|\,du\,dv = u\,du\,dv \qquad \text{...(1)}$$

$$xy\,(1-x-y)^{1/2} = u\,(1-v)\,uv\,(1-u)^{1/2} \qquad \text{...(2)}$$

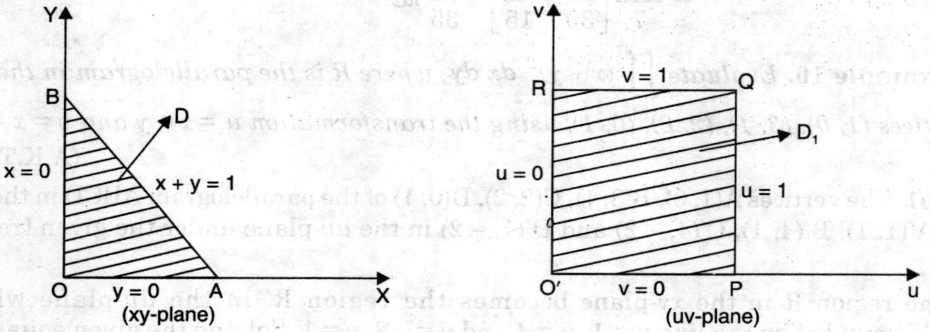

Also,

$$x = 0 \quad \Rightarrow \quad u\,(1-v) = 0 \quad \Rightarrow \quad u = 0, \quad v = 1$$

$$y = 0 \quad \Rightarrow \quad uv = 0 \quad \Rightarrow \quad u = 0, \quad v = 0$$

$$x + y = 1 \quad \Rightarrow \quad u = 1$$

Hence the limits of u are from 0 to 1

and the limits of v are from 0 to 1.

$$\therefore \quad \iint_D xy\,(1-x-y)^{1/2}\,dx\,dy = \iint_{D_1} u\,(1-v)\,uv\,(1-u)^{1/2}\,|\,J\,|\,du\,dv$$

$$= \iint_{D_1} u^2(1-u)^{1/2}\,v\,(1-v)\,u\,du\,dv$$

$$= \iint_{D_1} u^3(1-u)^{1/2}\,v\,(1-v)\,du\,dv$$

$$\therefore \quad \text{Required integral} = \int_0^1 u^3(1-u)^{1/2}\,du \int_0^1 v(1-v)\,dv$$

$$= \int_0^1 u^{4-1}\,(1-u)^{(3/2)-1}\,du \int_0^1 v^{2-1}\,(1-v)^{2-1}\,dv = \beta(4, 3/2)\,\beta(2, 2)$$

$$= \frac{\Gamma(4)\,\Gamma(\frac{3}{2})}{\Gamma(\frac{11}{2})} \cdot \frac{\Gamma(2)\,\Gamma(2)}{\Gamma(4)} = \frac{\Gamma(\frac{3}{2})\,\Gamma\,2\,\Gamma\,2}{\Gamma(\frac{11}{2})}$$

$$= \frac{\frac{1}{2}\sqrt{\pi}\,.\,1\,.\,1}{\frac{9}{2}\cdot\frac{7}{2}\cdot\frac{5}{2}\cdot\frac{3}{2}\cdot\frac{1}{2}\sqrt{\pi}} = \frac{16}{945}.$$

(*ii*) Since $x = u(1 - v),\ y = uv$

$$\therefore \quad J = \frac{\partial(x, y)}{\partial(u, v)} = \begin{vmatrix} \frac{\partial x}{\partial u} & \frac{\partial x}{\partial v} \\ \frac{\partial y}{\partial u} & \frac{\partial y}{\partial v} \end{vmatrix} = \begin{vmatrix} 1-v & -u \\ v & u \end{vmatrix} = u - uv + uv = u$$

$\therefore \quad dx\,dy = |\,J\,|\,du\,dv = u\,du\,dv$

Also,
$x = 0 \Rightarrow u(1 - v) = 0 \Rightarrow u = 0, v = 1$
$y = 0 \Rightarrow uv = 0 \Rightarrow u = 0, v = 0$
$x + y = 1 \Rightarrow u = 1$

Hence the limits of u are 0 to 1 and the limits of v are 0 to 1.

$$\therefore \quad \int_0^1 \int_0^{1-x} e^{y/(x+y)}\,dy\,dx = \int_0^1 \int_0^1 e^{uv/u}\,|\,J\,|\,du\,dv$$

$$= \int_0^1 \int_0^1 ue^v\,du\,dv = \left(\frac{u^2}{2}\right)_0^1 \left(e^v\right)_0^1 = \frac{1}{2}(e - 1).$$

Example 12. *Using the transformation* $x + y = u, y = v$, *evaluate* $\int_0^{\pi}\int_0^{\pi} |\cos(x + y)|\,dx\,dy$.

Sol. Let $\quad I = \int_0^{\pi}\int_0^{\pi} |\cos(x + y)|\,dx\,dy$ ...(1)

Consider transformation $\quad x + y = u, \quad y = v$

then, $\quad x = u - v, \quad y = v$

$$\text{Now,} \quad J = \frac{\partial(x, y)}{\partial(u, v)} = \begin{vmatrix} \frac{\partial x}{\partial u} & \frac{\partial x}{\partial v} \\ \frac{\partial y}{\partial u} & \frac{\partial y}{\partial v} \end{vmatrix} = \begin{vmatrix} 1 & -1 \\ 0 & 1 \end{vmatrix} = 1$$

$\therefore \quad dx\,dy = |\,J\,|\,du\,dv = du\,dv$ ...(2)

Also,
$y = 0 \Rightarrow v = 0$
$y = \pi \Rightarrow v = \pi$
$x = 0 \Rightarrow u = v$
$x = \pi \Rightarrow u = \pi + v$

Now, (1) becomes,

$$I = \int_0^{\pi} dv \int_v^{\pi + v} |\cos u|\,du \qquad \text{...(3)}$$

$$\int_v^{\pi+v} |\cos u|\,du = \int_v^{\pi/2} |\cos u|\,du + \int_{\pi/2}^{\pi} |\cos u|\,du + \int_{\pi}^{\pi+v} |\cos u|\,du$$

$$= \int_v^{\pi/2} \cos u\,du + \int_{\pi/2}^{\pi} -\cos u\,du + \int_{\pi}^{\pi+v} -\cos u\,du$$

$$= (\sin u)_v^{\pi/2} - (\sin u)_{\pi/2}^{\pi} - (\sin u)_{\pi}^{\pi+v}$$
$$= 1 - \sin v - (0 - 1) - \{\sin (\pi + v) - 0\}$$
$$= 1 - \sin v + 1 + \sin v = 2$$

$\therefore$ From (2), $\quad I = \int_0^{\pi} 2\, dv = 2\pi.$

Example 13. *Using the transformation $x - y = u$, $x + y = v$, show that*

$\iint_R \sin\left(\dfrac{x-y}{x+y}\right) dx\, dy = 0$, *where R is bounded by the co-ordinate axes and $x + y = 1$ in first quadrant.*

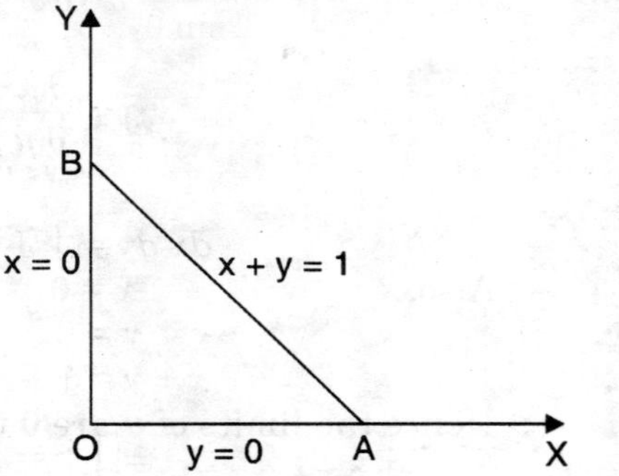

Sol. R is ΔOAB in xy-plane.

$$J = \frac{\partial(u, v)}{\partial(x, y)} = \begin{vmatrix} \dfrac{\partial u}{\partial x} & \dfrac{\partial u}{\partial y} \\ \dfrac{\partial v}{\partial x} & \dfrac{\partial v}{\partial y} \end{vmatrix} = \begin{vmatrix} 1 & -1 \\ 1 & 1 \end{vmatrix} = 2$$

$\therefore \quad du\, dv = 2\, dx\, dy \quad$ ...(1) $\mid \because \quad \mid J \mid = 2$

We have $\quad x - y = u, \quad x + y = v \quad$ ...(2)

By (2), $\quad 2x = u + v$ and $2y = v - u \quad$ ...(3)

If $x = 0$, $y = 0$ then (3) gives $\quad u = -v$, $u = v$

If $x + y = 1$ then (2) gives $\quad v = 1$

Corresponding domain in uv-plane is bounded by $u = -v$, $u = v$, $v = 1$.

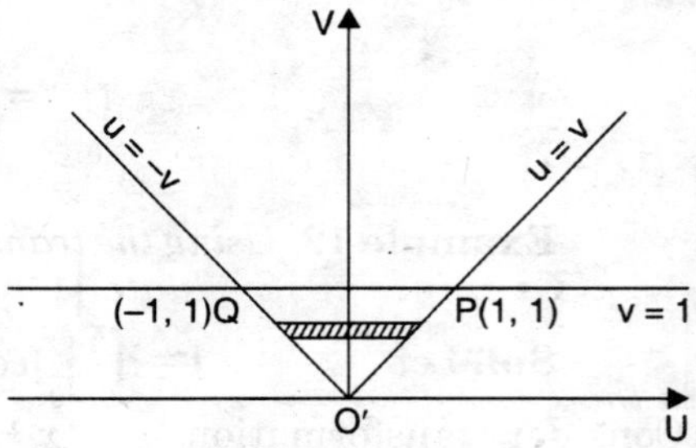

The domain in uv-plane is ΔPO′Q.

Take a strip parallel to u-axis. One end of this strip lies on $u = -v$ and the other end on $u = v$. For this strip, v varies from $v = 0$ to $v = 1$.

$$\therefore \quad I = \int_0^1 \int_{-v}^{v} \sin\left(\frac{u}{v}\right) \frac{1}{2}\, du\, dv = \frac{1}{2} \int_0^1 dv \left\{\frac{-\cos(u/v)}{(1/v)}\right\}_{-v}^{v}$$

$$= \frac{1}{2} \int_0^1 v\, \{-\cos 1 + \cos(-1)\}\, dv = 0.$$

Example 14. *Transform* $\int_0^{\pi/2} \int_0^{\pi/2} \sqrt{\dfrac{\sin \phi}{\sin \theta}}\, d\phi\, d\theta$ *by the substitution $x = \sin \phi \cos \theta$, $y = \sin \phi \sin \theta$ and show that its value is π.*

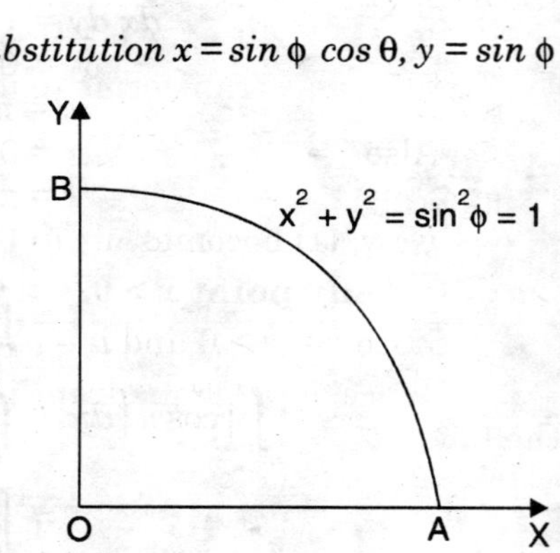

Sol. $x = \sin \phi \cos \theta$, $y = \sin \phi \sin \theta$.

Evidently, $\quad x^2 + y^2 = \sin^2 \phi \quad$ and $\quad \dfrac{y}{x} = \tan \theta$

θ and ϕ both varies from 0 to $\pi/2$.

This shows that the given integral is taken over the positive quadrant OABO of the circle $x^2 + y^2 = 1$. Limits of y are from 0 to 1. Limits of x in terms of y are from 0 to $\sqrt{1 - y^2}$.

$$J\left(\frac{x, y}{\theta, \phi}\right) = \begin{vmatrix} \frac{\partial x}{\partial \theta} & \frac{\partial x}{\partial \phi} \\ \frac{\partial y}{\partial \theta} & \frac{\partial y}{\partial \phi} \end{vmatrix}$$

$$= \begin{vmatrix} -\sin\phi \sin\theta & \cos\phi\cos\theta \\ \sin\phi\cos\theta & \cos\phi\sin\theta \end{vmatrix} = -\sin\phi\cos\phi$$

Now, $$\int_0^{\pi/2}\int_0^{\pi/2} \sqrt{\frac{\sin\phi}{\sin\theta}}\, d\phi\, d\theta = \int_0^1\int_0^{\sqrt{1-y^2}} \left\{\sqrt{\frac{\sin\phi}{\sin\theta}}\right\} \cdot \left\{\frac{1}{|J|}\right\} dy\, dx$$

$$= \int_0^1\int_0^{\sqrt{1-y^2}} \sqrt{\frac{\sin\phi}{\sin\theta}} \cdot \frac{dy\, dx}{\sin\phi\cos\phi} = \int_0^1\int_0^{\sqrt{1-y^2}} \frac{1}{\cos\phi} \cdot \frac{dy\, dx}{\sqrt{\sin\phi\sin\theta}}$$

$$= \int_0^1\int_0^{\sqrt{1-y^2}} \frac{1}{\sqrt{1-\sin^2\phi}} \cdot \frac{1}{\sqrt{y}}\, dy\, dx = \int_0^1\int_0^{\sqrt{1-y^2}} \frac{dy\, dx}{\sqrt{y}\sqrt{1-(x^2+y^2)}}$$

$$= \int_0^1 \frac{dy}{\sqrt{y}} \int_0^{\sqrt{1-y^2}} \frac{dx}{\sqrt{(1-y^2)-x^2}} = \int_0^1 \frac{dy}{\sqrt{y}} \cdot \left[\sin^{-1}\left(\frac{x}{\sqrt{1-y^2}}\right)\right]_0^{\sqrt{1-y^2}}$$

$$= \int_0^1 \frac{\pi}{2} \frac{dy}{\sqrt{y}} = \frac{\pi}{2}\left(2\sqrt{y}\right)_0^1 = \pi.$$

Example 15. *Evaluate* $\iint_R e^{-(x+y)} \sin\left(\frac{\pi y}{x+y}\right) dx\, dy$, *where R is the entire first quadrant in the xy-plane.*

Sol. Put $x + y = u$, $y = v$ so that $x = u - v$, $y = v$.

(*i*) **origin (0, 0) in xy-plane:**

$x = 0 \Rightarrow u - v = 0$

$y = 0 \Rightarrow v = 0$

$\therefore$ Origin (0, 0) in xy-plane corresponds to (0, 0) in uv-plane.

(*ii*) **x-axis:** $y = 0 \Rightarrow v = 0 \quad \therefore \quad x = u$

Since $x \ge 0 \quad \therefore \quad u \ge 0$

$\therefore$ x-axis transforms to u-axis in uv-plane.

(*iii*) **y-axis:** $x = 0 \Rightarrow u = v$

Since $y \ge 0 \Rightarrow v \ge 0$

$\therefore$ y-axis transforms to line $u = v$ in uv-plane.

(*iv*) **Any point** $x > 0, y > 0$**:**

Since $v = y > 0$ and $u - v = x > 0 \quad \therefore \quad u > v$

Hence entire I quadrant in xy-plane gets transformed to the region between u-axis and the line $u = v$.

$$\therefore \quad I = \int_0^\infty\int_0^u e^{-u} \sin\left(\frac{\pi v}{u}\right) dv\, du \qquad \left|\ \because\ J = \frac{\partial(x, y)}{\partial(u, v)} = 1\right.$$

$$= \int_0^\infty e^{-u}\left[\left(-\cos\frac{\pi v}{u}\right)\frac{u}{\pi}\right]_0^u du$$

$$= -\int_0^\infty \frac{u}{\pi} e^{-u}(-2)\,du = \frac{2}{\pi}\int_0^\infty ue^{-u}\,du$$

$$= \frac{2}{\pi}\left[(-ue^{-u})_0^\infty + \int_0^\infty 1.e^{-u}\,du\right] = \frac{2}{\pi}.$$

Example 16. *Prove that the area in the positive quadrant bounded by the curves $y^2 = 4ax$, $y^2 = 4bx$, $xy = c^2$ and $xy = d^2$ is $\frac{1}{3}(d^2 - c^2)\log\left(\frac{b}{a}\right)$; $d > c$, $b > a$.*

Sol. Let $\frac{y^2}{x} = u$ and $xy = v$ so that $y = u^{1/3}v^{1/3}$ and $x = u^{-1/3}v^{2/3}$ ...(1)

$$J = \frac{\partial(x, y)}{\partial(u, v)} = \begin{vmatrix} \frac{\partial x}{\partial u} & \frac{\partial x}{\partial v} \\ \frac{\partial y}{\partial u} & \frac{\partial y}{\partial v} \end{vmatrix} = \begin{vmatrix} -\frac{1}{3}u^{-4/3}v^{2/3} & \frac{2}{3}u^{-1/3}v^{-1/3} \\ \frac{1}{3}u^{-2/3}v^{1/3} & \frac{1}{3}u^{1/3}v^{-2/3} \end{vmatrix}$$

$$= -\frac{1}{9}u^{-1} - \frac{2}{9}u^{-1} = -\frac{1}{3u}$$

$$\therefore \quad |J| = \frac{1}{3u}$$

The bounding curves then become $u = 4a$, $u = 4b$ and $v = c^2$, $v = d^2$.

$$\text{Required area} = \iint_R dx\,dy = \iint_{R'} |J|\,du\,dv = \int_{c^2}^{d^2}\int_{4a}^{4b} \frac{1}{3u}\,du\,dv$$

$$= \frac{1}{3}\int_{c^2}^{d^2} (\log u)_{4a}^{4b}\,dv = \frac{1}{3}\log\left(\frac{b}{a}\right)\int_{c^2}^{d^2} dv$$

$$= \frac{1}{3}(d^2 - c^2)\log\left(\frac{b}{a}\right)$$

Example 17. *Transform the integral $\iiint (x + y + z)\,x^2y^2z^2\,dx\,dy\,dz$ taken over the volume bounded by $x = 0$, $y = 0$, $z = 0$, $x + y + z = 1$ substituting $u = x + y + z$, $x + y = uv$, $y = uvw$ and then evaluate it.*

Sol. We have, $u = x + y + z$, $x + y = uv$, $y = uvw$ so that
$x = uv(1 - w)$, $y = uvw$ and $z = u(1 - v)$

$$J = \frac{\partial(x, y, z)}{\partial(u, v, w)} = \begin{vmatrix} v(1-w) & u(1-w) & -uv \\ vw & uw & uv \\ 1-v & -u & 0 \end{vmatrix} = u^2v$$

$\therefore$ $dx\,dy\,dz = u^2v\,du\,dv\,dw$

Also, $x = 0 \Rightarrow uv(1 - w) = 0 \Rightarrow u = 0, v = 0, w = 1$

$y = 0 \Rightarrow uvw = 0 \Rightarrow u = 0, v = 0, w = 0$

$z = 0 \Rightarrow u(1 - v) = 0 \Rightarrow u = 0, v = 1$

$$\therefore \quad I = \int_0^1\int_0^1\int_0^1 u\,.\,u^2v^2(1-w)^2\,.\,u^2v^2w^2\,.\,u^2(1-v)^2\,.\,u^2v\,.\,du\,dv\,dw$$

$$= \int_0^1\int_0^1\int_0^1 u^9\,.\,v^5(1-v)^2\,.\,w^2(1-w)^2\,du\,dv\,dw$$

$$= \int_0^1 w^2(1-w)^2\, dw \,.\, \int_0^1 v^5(1-v)^2\, dv \,.\, \int_0^1 u^9\, du$$

$$= \beta(3, 3)\,.\,\beta(6, 3)\,.\,\frac{1}{10} = \frac{\Gamma 3\,\Gamma 3}{\Gamma 6}\,.\,\frac{\Gamma 6\,\Gamma 3}{\Gamma 9}\,.\,\frac{1}{10} = \frac{1}{50400}.$$

Example 18. *Evaluate by changing the variables, $\iint_R (x+y)^2\, dx\, dy$ where R is the region bounded by the parallelogram $x + y = 0$, $x + y = 2$, $3x - 2y = 0$ and $3x - 2y = 3$.*

(A.K.T.U. 2014, 2021)

Sol. Let $x + y = u$ and $3x - 2y = v$ so that

$$x = \frac{1}{5}(2u + v) \quad \text{and} \quad y = \frac{1}{5}(3u - v)$$

Now,
$$J = \frac{\partial(x, y)}{\partial(u, v)} = \begin{vmatrix} \frac{2}{5} & \frac{1}{5} \\ \frac{3}{5} & -\frac{1}{5} \end{vmatrix} = -\frac{1}{5} \quad \therefore \quad |\,J\,| = \frac{1}{5}$$

$$\therefore \qquad dx\, dy = |\,J\,|\, du\, dv = \frac{1}{5}\, du\, dv$$

The limits then become $u = 0$ to $u = 2$ and $v = 0$ to $v = 3$.

Given Integral
$$I = \iint_R (x+y)^2\, dx\, dy = \int_0^3 \int_0^2 u^2\,.\,\frac{1}{5}\, du\, dv$$

$$= \frac{1}{5}\int_0^3 \left(\frac{u^3}{3}\right)_0^2 dv = \frac{1}{5}\,.\,\frac{8}{3}\,.\,3 = \frac{8}{5}$$

TEST YOUR KNOWLEDGE

1. Evaluate $\iint \sin \pi (x^2 + y^2)\, dx\, dy$ over the region bounded by the circle $x^2 + y^2 = 1$ by changing to polar co-ordinates.

2. (*i*) Evaluate $\iint (a^2 - x^2 - y^2)\, dx\, dy$ over the semi-circle $x^2 + y^2 = ax$ in the positive quadrant by changing to polar co-ordinates.

   (*ii*) Evaluate $\int_0^{2a}\int_0^{\sqrt{2ax - x^2}} (x^2 + y^2)\, dx\, dy$ by changing into polar co-ordinates. (*U.P.T.U. 2014*)

3. Evaluate $\iint (x^2 + y^2)^{7/2}\, dx\, dy$ over the circle $x^2 + y^2 = 1$.

4. Evaluate $\iint xy\,(x^2 + y^2)^{3/2}\, dx\, dy$ over the positive quadrant of the circle $x^2 + y^2 = 1$.

5. Evaluate the following by changing into polar co-ordinates:

   (*i*) $\int_0^a \int_y^a \frac{x\, dx\, dy}{x^2 + y^2}$ (*ii*) $\int_{-a}^{a} \int_{-\sqrt{a^2 - x^2}}^{\sqrt{a^2 - x^2}} dy\, dx$ (*iii*) $\int_0^{4a}\int_{y^2/4a}^{y} \frac{x^2 - y^2}{x^2 + y^2}\, dx\, dy$

(iv) $\int_0^a \int_0^{\sqrt{a^2 - y^2}} (x^2 + y^2)\, dx\, dy$ (G.B.T.U. 2012) (v) $\int_0^a \int_y^a \frac{x^2\, dx\, dy}{\sqrt{x^2 + y^2}}$.

6. Evaluate $\iint_D e^{-(x^2+y^2)}\, dy\, dx$, where D is the region bounded by $x^2 + y^2 = a^2$.

7. Evaluate $\iint xy\,(x^2 + y^2)^{n/2}\, dx\, dy$ over the positive quadrant of $x^2 + y^2 = 4$, supposing $n + 3 > 0$.

8. Evaluate $\iint \sqrt{\dfrac{1 - \frac{x^2}{a^2} - \frac{y^2}{b^2}}{1 + \frac{x^2}{a^2} + \frac{y^2}{b^2}}}\, dx\, dy$ over the positive quadrant of the ellipse $\frac{x^2}{a^2} + \frac{y^2}{b^2} = 1$.

[**Hint.** Put $x = a\mathrm{X}$, $y = b\mathrm{Y}$ so that $\mathrm{I} = \iint \sqrt{\dfrac{1 - \mathrm{X}^2 - \mathrm{Y}^2}{1 + \mathrm{X}^2 + \mathrm{Y}^2}}\,.\, ab\, d\mathrm{X}\, d\mathrm{Y}$]

9. Transforming to cartesian form, Evaluate $\int_0^{\pi} \int_0^{a} r^3 \sin\theta \cos\theta\, dr\, d\theta$.

10. Evaluate the integral $\iint_R \sqrt{x^2 + y^2}\, dx\, dy$ by changing to polar coordinates where R is the region in the xy-plane bounded by the circles $x^2 + y^2 = 4$ and $x^2 + y^2 = 9$.

$\left[\textbf{Hint. } \mathrm{I} = \int_0^{2\pi} \int_2^3 r\,(r\, dr\, d\theta)\right]$

11. By transforming into cylindrical coordinates, evaluate the integral $\iiint (x^2 + y^2 + z^2)\, dx\, dy\, dz$ taken over the region $0 \le z \le x^2 + y^2 \le 1$.

$\left[\textbf{Hint. } \mathrm{I} = \int_0^1 \int_0^{2\pi} \int_0^1 (r^2 + z^2)\,.\, r\, dr\, d\theta\, dz\right]$

12. Evaluate $\iiint \dfrac{dx\, dy\, dz}{(x^2 + y^2 + z^2)}$ taken throughout the volume of the sphere $x^2 + y^2 + z^2 = a^2$.

$\left[\textbf{Hint. } \mathrm{I} = \int_0^{2\pi} \int_0^{\pi} \int_0^{a} \frac{1}{r^2}\,.\, r^2 \sin\theta\, dr\, d\theta\, d\phi\right]$

13. Using the transformation $x + y = u$, $y = uv$, prove that $\iint [xy(1 - x - y)]^{1/2}\, dx\, dy = \dfrac{2\pi}{105}$ taken over the area of triangle bounded by the lines $x = 0$, $y = 0$, $x + y = 1$.

14. Using the transformation $u = x - y$, $v = x + y$, prove that $\iint_R \cos \dfrac{x - y}{x + y}\, dx\, dy = \dfrac{1}{2} \sin 1$ where R is bounded by $x = 0$, $y = 0$, $x + y = 1$

$\left[\textbf{Hint. } x = \frac{1}{2}(u + v),\ y = \frac{1}{2}(v - u) \text{ so that } |\,\mathrm{J}\,| = \frac{1}{2}\right]$

15. Evaluate the integral $\iint_R (x - y)^2 \cos^2 (x + y)\, dx\, dy$ where R is the rhombus with vertices at $(\pi, 0)$, $(2\pi, \pi)$, $(\pi, 2\pi)$ and $(0, \pi)$.

$\left[\textbf{Hint. } \text{Put } y - x = u,\ y + x = v\,;\ |\,\mathrm{J}\,| = \frac{1}{2}\right]$

16. Evaluate $\iiint_w \exp(x^2 + y^2 + z^2)^{3/2}\, dx\, dy\, dz$ where w is a spherical ball of radius 1.

$\left[\textbf{Hint. } I = \int_0^1 \int_0^{\pi} \int_0^{2\pi} e^{r^3} . r^2 \sin\theta\, dr\, d\theta\, d\phi\right]$.

17. Evaluate $\iint_R (x^2 + y^2)\, dx\, dy$ where R is the region in the first quadrant bounded by $x^2 - y^2 = a$, $x^2 - y^2 = b$, $2xy = c$, $2xy = d$, $0 < a < b$, $0 < c < d$.

$\left[\textbf{Hint. } \text{Take } x^2 - y^2 = u, xy = v \text{ ; } J = \frac{1}{2(x^2 + y^2)}\right]$

18. Evaluate $\iint_D (y - x)\, dx\, dy$ where D is the region in xy-plane bounded by the straight lines $y = x + 1$, $y = x - 3$, $y = -\frac{1}{3}x + \frac{7}{3}$, $y = -\frac{1}{3}x + 5$. $\left[\textbf{Hint. } \text{Put } y - x = u, y + \frac{1}{3}x = v; \ |J| = \frac{3}{4}\right]$

19. Evaluate $\iint_R (x - y)^4\, e^{x+y}\, dx\, dy$ where R is the square with vertices at (1, 0), (2, 1), (1, 2) and (0, 1).

$\left[\textbf{Hint. } \text{Put } x + y = u, x - y = v; \ |J| = \frac{1}{2}\right]$ *(M.T.U. 2013)*

20. Evaluate $\iint_R xy\, dx\, dy$ where R is the region in the first quadrant bounded by the hyperbolas $x^2 - y^2 = a^2$ and $x^2 - y^2 = b^2$ and the circles $x^2 + y^2 = c^2$, $x^2 + y^2 = d^2$ with $0 < a < b < c < d$.
[**Hint.** $x^2 - y^2 = u$, $x^2 + y^2 = v$; $J = 8xy$]

21. The function $\phi(x) = e^{-x^2}$ is fundamental in probability

(*a*) Sketch the graph of $\phi(x)$. Explain why it is called a "bell curve".

(*b*) Evaluate $I = \int_{-\infty}^{\infty} e^{-x^2} dx$ using the following brilliant strategy of Gauss.

(*i*) Instead of computing I, compute $I^2 = \left(\int_{-\infty}^{\infty} e^{-x^2} dx\right)\left(\int_{-\infty}^{\infty} e^{-y^2} dy\right)$

(*ii*) Rewrite I^2 as in integral of the form $\iint_R \phi(x, y)\, dx\, dy$, where R is the entire cartesian plane.

(*iii*) Convert that integral to polar coordinates.

(*iv*) Evaluate to find I^2.

Answers

1. 2 **2.** (*i*) $\frac{5\pi a^4}{64}$ (*ii*) $\frac{3\pi a^4}{4}$ **3.** $\frac{2\pi}{9}$

4. $\frac{1}{14}$

5. (*i*) $\frac{\pi a}{4}$ (*ii*) πa^2 (*iii*) $8\left(\frac{\pi}{2} - \frac{5}{3}\right)a^2$ (*iv*) $\frac{\pi a^4}{8}$

(*v*) $\frac{a^3}{3} \log(\sqrt{2} + 1)$

6. $\pi(1 - e^{-a^2})$ **7.** $\frac{2^{n+3}}{n+4}$ **8.** $\frac{\pi ab}{8}(\pi - 2)$ **9.** 0

10. $\dfrac{38\pi}{3}$ 11. $\dfrac{5\pi}{6}$ 12. $4\pi a$ 15. $\dfrac{\pi^4}{3}$

16. $\dfrac{4}{3}\pi(e-1)$ 17. $\dfrac{(b-a)(d-c)}{4}$ 18. -8 19. $\dfrac{e(e^2-1)}{5}$

20. $\dfrac{1}{8}(b^2-a^2)(d^2-c^2)$. 21. (b) $\sqrt{\pi}$

4.7. AREA BY DOUBLE INTEGRATION

(a) **Cartesian Co-ordinates.** The area A of the region bounded by the curves $y = f_1(x)$, $y = f_2(x)$ and the lines $x = a$, $x = b$ is given by $A = \int_a^b \int_{f_1(x)}^{f_2(x)} dy\, dx$.

The area A of the region bounded by the curves $x = f_1(y)$, $x = f_2(y)$ and the lines $y = c$, $y = d$ is given by $A = \int_c^d \int_{f_1(y)}^{f_2(y)} dx\, dy$.

(b) **Polar Co-ordinates.** The area A of the region bounded by the curves $r = f_1(\theta)$, $r = f_2(\theta)$ and the lines $\theta = \alpha$, $\theta = \beta$ is given by $A = \int_\alpha^\beta \int_{f_1(\theta)}^{f_2(\theta)} r\, dr\, d\theta$.

4.7.1. Area of Curves Given by Parametric Equations

We know that the area between the curve $y = f(x)$, the axis of x and the ordinates at $x = a$ and $x = b$ is equal to $\int_a^b y\, dx$ or $\int_a^b f(x)\, dx$.

Also, the area between the curve $x = f(y)$, the y-axis and the abscissae $y = a$ and $y = b$ is given by $\int_a^b x\, dy$ or $\int_a^b f(y)\, dy$.

In case of parametric equations, the above formulae will be replaced by

$$\int_{t=\alpha}^{\beta} y\frac{dx}{dt}\, dt \quad \text{or} \quad \int_{t=\alpha}^{\beta} x\frac{dy}{dt}\, dt$$

where α and β are values of t corresponding to the values a and b of x or y.

4.8. VOLUME AS A DOUBLE INTEGRAL

(a) **Cartesian Co-ordinates.** Consider a surface

$$z = f(x, y) \quad ...(1)$$

Let the orthogonal projection of its region R′ on the xy-plane be the region R given by

$$\phi(x, y) = 0 \quad ...(2)$$

Now eqn. (2) represents a cylinder with generators parallel to z-axis and the guiding curve given by (2). Let V be the volume of this cylinder between R and R′.

Divide R into elementary rectangles of area $\delta x\, \delta y$ by drawing lines parallel to the axis of x and y. On each of these rectangles, erect prisms of lengths parallel to z-axis. Volume of this prism between R and R′ is $z\, \delta x\, \delta y$. The volume V is composed of such prisms.

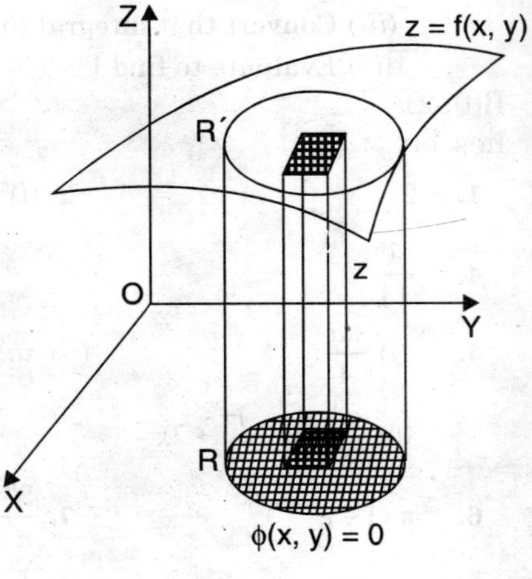

$$\therefore \quad V = \underset{\substack{\delta x \to 0 \\ \delta y \to 0}}{\text{Lt}} \sum z \, \delta x \, \delta y = \iint_R z \, dx \, dy.$$

(*b*) **Cylindrical Co-ordinates.** Let the equation of the surface be $z = f(r, \phi)$. Replacing $dx\,dy$ by $r\,dr\,d\phi$, we get $V = \iint_P zr\,dr\,d\phi$.

ILLUSTRATIVE EXAMPLES

Example 1. *Find the area lying between the parabola $y = 4x - x^2$ and above the line $y = x$.* (A.K.T.U. 2020)

Sol. Let two curves intersect at points whose abscissae are given by $4x - x^2 = x$

or $x^2 - 3x = 0$ *i.e.,* $x = 0, 3$.

Using vertical strips, the required area lies between $x = 0$, $x = 3$ and $y = x$, $y = 4x - x^2$.

$$\therefore \quad \text{Required area} = \int_0^3 \int_x^{4x - x^2} dy\,dx$$

$$= \int_0^3 \Big[y \Big]_x^{4x - x^2} dx$$

$$= \int_0^3 (4x - x^2 - x)\,dx = \int_0^3 (3x - x^2)\,dx$$

$$= \left[\frac{3x^2}{2} - \frac{x^3}{3} \right]_0^3 = \frac{27}{2} - 9 = 4.5.$$

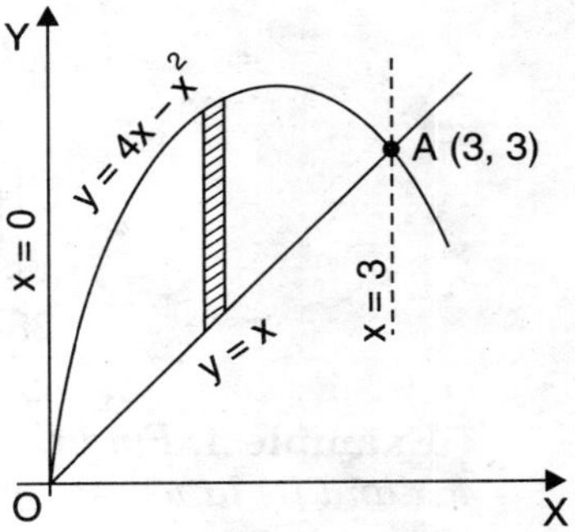

Example 2. *Find the smaller of the areas bounded by the ellipse $4x^2 + 9y^2 = 36$ and the straight line $2x + 3y = 6$.*

Sol. The equation of the ellipse is $\dfrac{x^2}{9} + \dfrac{y^2}{4} = 1$ ...(1)

and the line is $\dfrac{x}{3} + \dfrac{y}{2} = 1$ ...(2)

Both meet x-axis at A(3, 0) and y-axis at B(0, 2). Using horizontal strips, the required area lies between

$$x = \frac{3}{2}(2 - y),\ x = \frac{3}{2}\sqrt{4 - y^2} \ \text{ and } \ y = 0, y = 2.$$

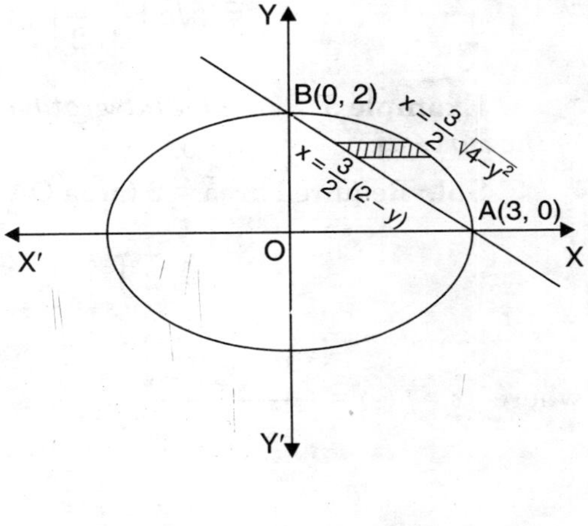

$$\therefore \quad \text{Required area} = \int_0^2 \int_{\frac{3(2-y)}{2}}^{\frac{3\sqrt{4-y^2}}{2}} dx\,dy$$

$$= \int_0^2 \frac{3}{2} [\sqrt{4 - y^2} - (2 - y)]\,dy$$

$$= \frac{3}{2}\left[\frac{y\sqrt{4-y^2}}{2} + \frac{4}{2}\sin^{-1}\frac{y}{2} - 2y + \frac{y^2}{2}\right]_0^2$$

$$= \frac{3}{2}[2\sin^{-1}(1) - 4 + 2] = \frac{3}{2}\left(2 \cdot \frac{\pi}{2} - 2\right) = \frac{3}{2}(\pi - 2).$$

Example 3. *Determine the area of region bounded by the curves $xy = 2$, $4y = x^2$, $y = 4$.* (U.P.T.U. 2014, 2015)

Sol. Required area of shaded region

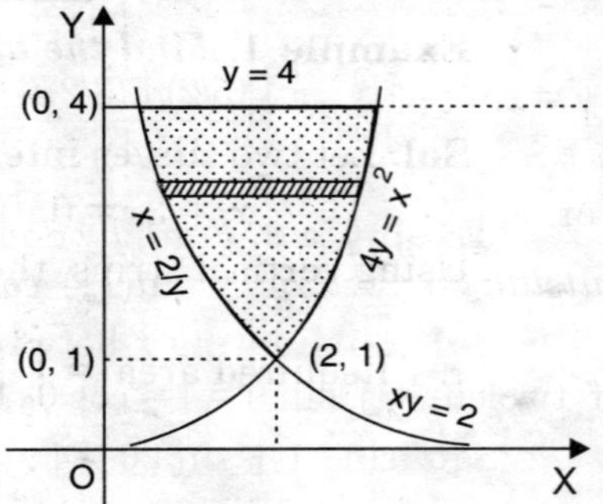

$$= \int_{y=1}^{4}\int_{x=\frac{2}{y}}^{2\sqrt{y}} dx\, dy$$

$$= \int_1^4 \left(2\sqrt{y} - \frac{2}{y}\right) dy$$

$$= 2\left(\frac{2}{3}y^{3/2} - \log y\right)_1^4$$

$$= 2\left[\left(\frac{16}{3} - 2\log 2\right) - \frac{2}{3}\right]$$

$$= \frac{28}{3} - 4\log 2$$

Example 4. *Find the area bounded by the parabola $y^2 = 4ax$ and its latus rectum.*

Sol. Required area

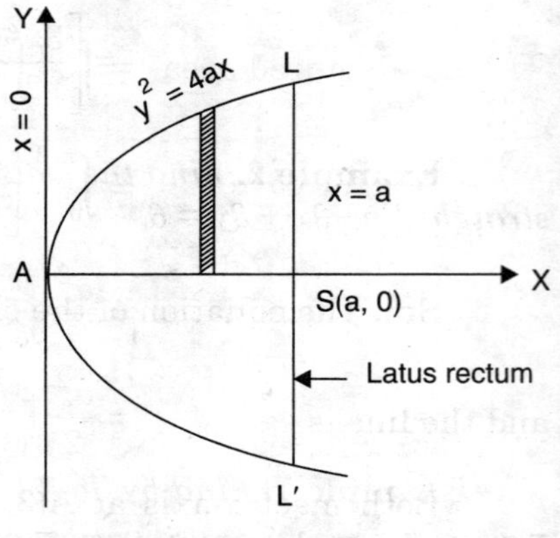

$$= 2(\text{area ASL})$$

$$= 2\int_0^a \int_0^{2\sqrt{ax}} dy\, dx$$

$$= 2\int_0^a 2\sqrt{ax}\, dx$$

$$= 4\sqrt{a}\left(\frac{x^{3/2}}{3/2}\right)_0^a = \frac{8a^2}{3}$$

Example 5. *By double integration, find the whole area of the curve $a^2x^2 = y^3(2a - y)$.*

Sol. Required area = 2 (area OAB)

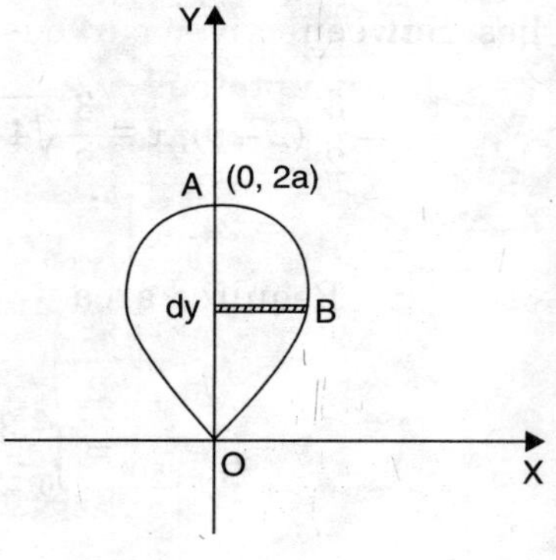

$$= 2\int_{y=0}^{2a}\int_{x=0}^{f(y)} dy\, dx \quad ...(1)$$

where $x = f(y) = \dfrac{y^{3/2}\sqrt{2a-y}}{a}$

From (1), Required area

$$I = 2\int_0^{2a} f(y)\,dy = \frac{2}{a}\int_0^{2a} y^{3/2}\sqrt{2a-y}\,dy$$

$$= \frac{2}{a}\int_0^{\pi/2}(2a\sin^2\theta)^{3/2}\sqrt{2a-2a\sin^2\theta}\,.\,4a\sin\theta\cos\theta\,d\theta$$

Put $y = 2a\sin^2\theta$

$\therefore\ dy = 4a\sin\theta\cos\theta\,d\theta$

$$= 32a^2\int_0^{\pi/2}\sin^4\theta\cos^2\theta\,d\theta$$

$$= 32a^2\,.\,\frac{\Gamma(5/2)\,\Gamma(3/2)}{2\,\Gamma(4)} = 32a^2\,.\,\frac{\frac{3}{2}\,.\,\frac{1}{2}\sqrt{\pi}\,.\,\frac{1}{2}\,.\,\sqrt{\pi}}{2\cdot3\cdot2\cdot1} = \pi a^2.$$

Example 6. *Find, by double integration, the area lying inside the circle $r = a\sin\theta$ and outside the cardioid $r = a(1 - \cos\theta)$.*

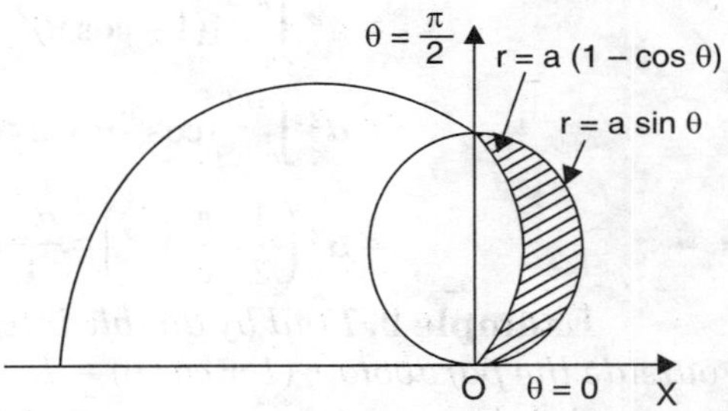

Sol. Eliminating r between the equations of two curves, $\sin\theta = 1 - \cos\theta$ or $\sin\theta + \cos\theta = 1$

Squaring $1 + \sin 2\theta = 1$ or $\sin 2\theta = 0$

$\therefore\quad 2\theta = 0$ or π *i.e.*, $\theta = 0$ or $\frac{\pi}{2}$

For the required area, r varies from $a(1-\cos\theta)$ to $a\sin\theta$ and θ varies from 0 to $\frac{\pi}{2}$.

$$\therefore\ \text{Required area} = \int_0^{\pi/2}\int_{a(1-\cos\theta)}^{a\sin\theta} r\,dr\,d\theta$$

$$= \int_0^{\pi/2}\left[\frac{r^2}{2}\right]_{a(1-\cos\theta)}^{a\sin\theta} d\theta = \frac{1}{2}\int_0^{\pi/2} a^2\,[\sin^2\theta - (1-\cos\theta)^2]\,d\theta$$

$$= \frac{a^2}{2}\int_0^{\pi/2}(\sin^2\theta - 1 - \cos^2\theta + 2\cos\theta)]\,d\theta$$

$$= \frac{a^2}{2}\int_0^{\pi/2}(-2\cos^2\theta + 2\cos\theta)\,d\theta = a^2\left(1 - \frac{\pi}{4}\right).$$

Example 7. *Find by double integration the area included between the curves $r = a(\sec\theta + \cos\theta)$ and its asymptotes.*

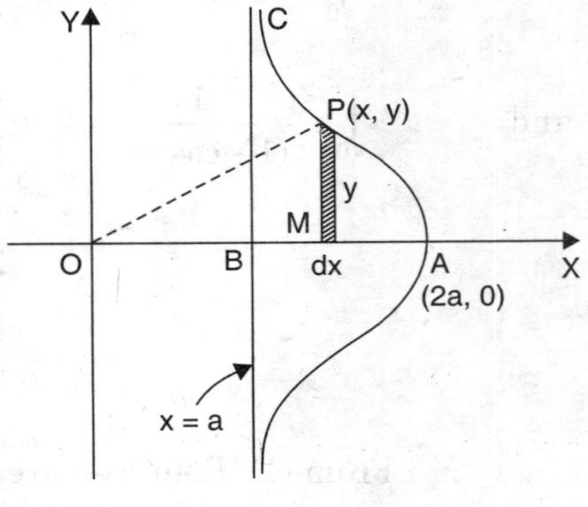

Sol. The given curve is $r = a(\sec\theta + \cos\theta)$

Its asymptote is $\qquad r = a\sec\theta$

$\therefore$ Required area

$$= 2\int_0^{\pi/2}\int_{a\sec\theta}^{a(\sec\theta+\cos\theta)} r\,dr\,d\theta$$

$$= 2\int_0^{\pi/2}\left(\frac{r^2}{2}\right)_{a\sec\theta}^{a(\sec\theta+\cos\theta)} d\theta$$

$$= a^2\int_0^{\pi/2}[(\sec\theta + \cos\theta)^2 - \sec^2\theta]\,d\theta$$

$$= a^2 \int_0^{\pi/2} (\cos^2 \theta + 2)\, d\theta$$

$$= a^2 \left(\frac{1}{2} \cdot \frac{\pi}{2} + 2 \cdot \frac{\pi}{2} \right) = a^2 \left(\frac{\pi}{4} + \pi \right) = \frac{5\pi}{4} a^2 .$$

Example 8. *Find, by double integration, the area lying inside the cardioid $r = a(1 + \cos\theta)$ and outside the circle $r = a$.*

Sol. Required area = area ABCDA = 2 (area ABDA)

$$= 2 \int_{\theta=0}^{\pi/2} \int_{r \text{ for circle}}^{\text{for cardioid}} r\, d\theta\, dr$$

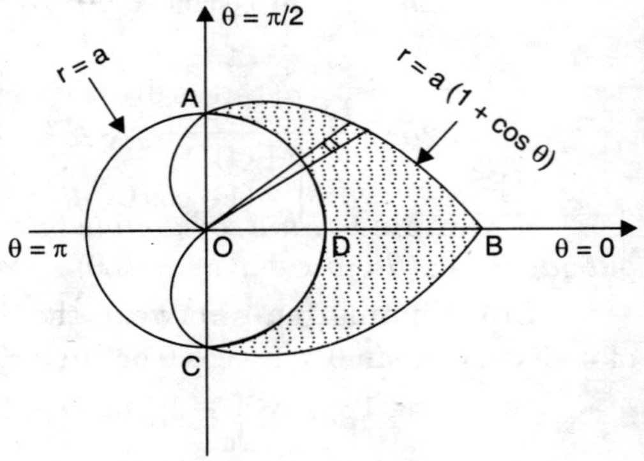

$$= 2 \int_0^{\pi/2} \int_a^{a(1+\cos\theta)} r\, d\theta\, dr$$

$$= 2 \int_0^{\pi/2} \left(\frac{r^2}{2} \right)_a^{a(1+\cos\theta)} d\theta$$

$$= a^2 \int_0^{\pi/2} [(1 + \cos\theta)^2 - 1]\, d\theta$$

$$= a^2 \int_0^{\pi/2} (\cos^2 \theta + 2\cos\theta)\, d\theta$$

$$= a^2 \left(\frac{1}{2} \cdot \frac{\pi}{2} + 2 \right) = \frac{a^2}{4} (\pi + 8) .$$

Example 9. *Find by double integration, the area lying inside a cardioid $r = 1 + \cos\theta$ and outside the parabola $r(1 + \cos\theta) = 1$.*

Sol. Required area = area CAFBEDC

= 2 (area DAFBED)

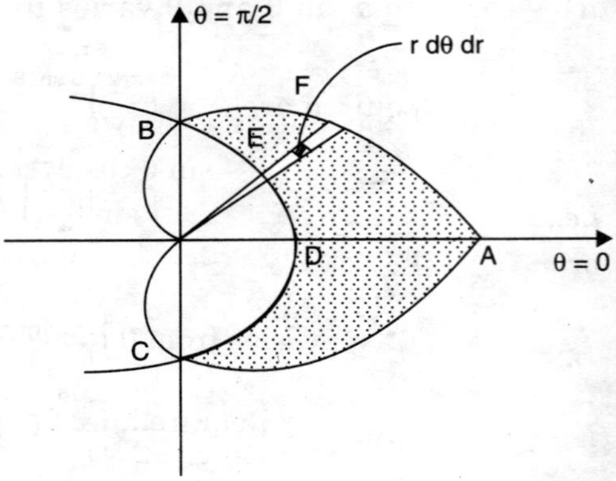

$$= 2 \int_0^{\pi/2} \int_{r \text{ for parabola}}^{\text{for cardioid}} r\, d\theta\, dr$$

$$= 2 \int_0^{\pi/2} \left(\frac{r^2}{2} \right)_{\frac{1}{1+\cos\theta}}^{1+\cos\theta} d\theta$$

$$= \int_0^{\pi/2} \left[(1 + \cos\theta)^2 - \frac{1}{(1 + \cos\theta)^2} \right] d\theta \quad \text{...(1)}$$

Now, $\int_0^{\pi/2} (1 + \cos\theta)^2\, d\theta = \int_0^{\pi/2} (1 + \cos^2\theta + 2\cos\theta)\, d\theta = \frac{\pi}{2} + \frac{1}{2} \cdot \frac{\pi}{2} + 2 = \frac{3\pi}{4} + 2$

and $\int_0^{\pi/2} \frac{1}{(1 + \cos\theta)^2}\, d\theta = \frac{1}{4} \int_0^{\pi/2} \sec^4 \frac{\theta}{2}\, d\theta = \frac{1}{4} \int_0^{\pi/4} \sec^4 \phi\, (2\, d\phi)$ | Put $\frac{\theta}{2} = \phi$ $\therefore \; d\theta = 2d\phi$

$$= \frac{1}{2} \int_0^{\pi/4} (1 + \tan^2 \phi) \sec^2 \phi\, d\phi = \frac{1}{2} \int_0^1 (1 + t^2)\, dt \text{, where } t = \tan\phi$$

$$= \frac{1}{2} \left(t + \frac{t^3}{3} \right)_0^1 = \frac{2}{3}$$

$\therefore$ From (1), Required area $= \frac{3\pi}{4} + 2 - \frac{2}{3} = \frac{3\pi}{4} + \frac{4}{3}.$

Example 10. *Find the area included between the curve* $x = a(\theta - \sin\theta)$, $y = a(1 - \cos\theta)$ *and its base.*

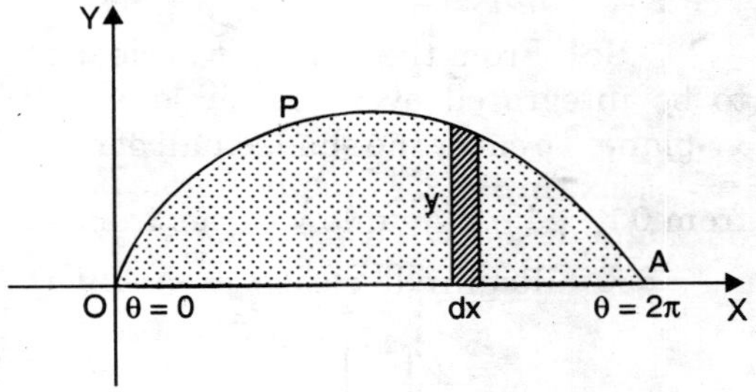

Sol. When $\theta = 0$, $x = 0$ and $y = 0$

When $\theta = 2\pi$, $x = 2a\pi$ and $y = 0$

$\therefore$ The limits for θ are from 0 to 2π.

$\therefore$ Required area = area OPAO

$$= \int_0^{2\pi} y \frac{dx}{d\theta} d\theta$$

$$= \int_0^{2\pi} a(1-\cos\theta)\, a(1-\cos\theta)\, d\theta$$

$$= a^2 \int_0^{2\pi} (1-\cos\theta)^2\, d\theta = 4a^2 \int_0^{2\pi} \sin^4 \theta/2\, d\theta$$

$$= 8a^2 \int_0^{\pi} \sin^4 \phi\, d\phi \qquad \left(\text{Put } \frac{\theta}{2} = \phi\right)$$

$$= 8a^2 . 2 \int_0^{\pi/2} \sin^4 \phi\, d\phi = 16a^2 . \frac{3}{4} . \frac{1}{2} \frac{\pi}{2} = 3\pi a^2.$$

Example 11. *Find the area of the loop of the curve* $x^3 + y^3 = 3axy$

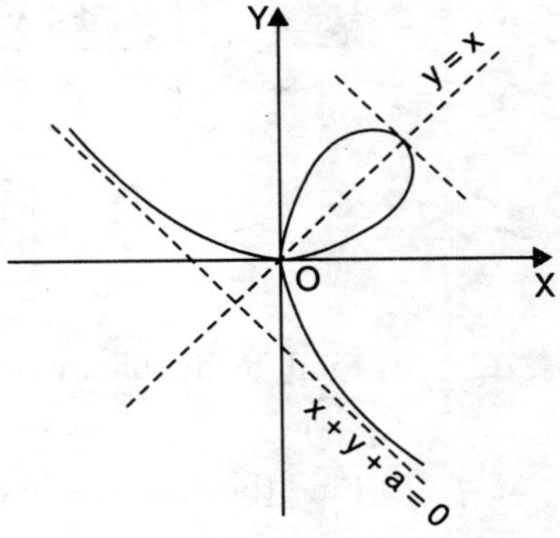

Sol. Changing to polar co-ordinates, we get

$$r^3(\cos^3\theta + \sin^3\theta) = 3ar^2 \sin\theta\cos\theta$$

$$\Rightarrow \qquad r = \frac{3a\sin\theta\cos\theta}{\sin^3\theta + \cos^3\theta} \qquad ...(1)$$

For a loop, puting $r = 0$, we get

$$\sin\theta\cos\theta = 0$$

i.e., $\sin\theta = 0$ and $\cos\theta = 0$

or $\theta = 0$ and $\theta = \pi/2$

(Two consecutive values)

Hence as θ varies from 0 to $\pi/2$, we get the loop.

$$\therefore \qquad \text{Required area} = \int_0^{\pi/2}\int_0^{r} r\, dr\, d\theta$$

$$= \int_0^{\pi/2} \left(\frac{r^2}{2}\right)_0^{\frac{3a\sin\theta\cos\theta}{\sin^3\theta+\cos^3\theta}} = \frac{9a^2}{2}\int_0^{\pi/2} \frac{\sin^2\theta\cos^2\theta}{(\sin^3\theta+\cos^3\theta)^2} d\theta$$

$$= \frac{9a^2}{2}\int_0^{\pi/2} \frac{\tan^2\theta\sec^2\theta}{(1+\tan^3\theta)^2} d\theta$$

$$\left(\begin{array}{r} \text{Put } \tan^3\theta = t \\ \therefore\ 3\tan^2\theta . \sec^2\theta\, d\theta = dt \end{array}\right)$$

$$= \frac{3a^2}{2}\int_0^{\infty} \frac{dt}{(1+t)^2} = \frac{3a^2}{2}\left(\frac{-1}{1+t}\right)_0^{\infty} = \frac{3}{2}a^2.$$

Example 12. *Find the volume bounded by the cylinder* $x^2 + y^2 = 4$ *and the planes* $y + z = 4$ *and* $z = 0$. (A.K.T.U. 2022)

Sol. From the figure, it is clear that $z = 4 - y$ is to be integrated over the circle $x^2 + y^2 = 4$ in the xy-plane. To cover the shaded half of this circle, x varies from 0 to $\sqrt{4-y^2}$ while y varies from -2 to 2.

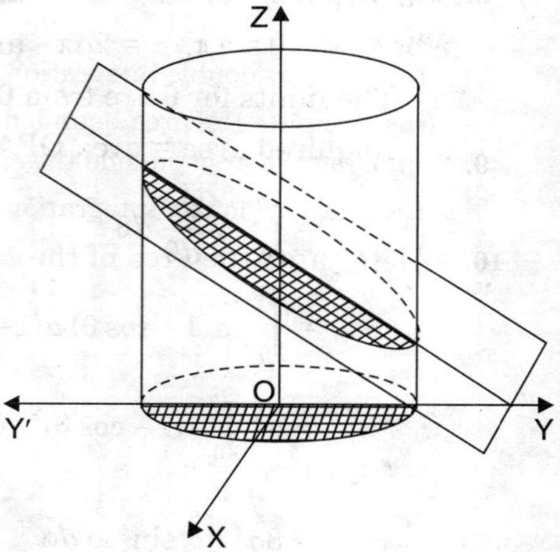

$\therefore$ Required volume

$$= 2\int_{-2}^{2}\int_{0}^{\sqrt{4-y^2}} z\,dx\,dy$$

$$= 2\int_{-2}^{2}\int_{0}^{\sqrt{4-y^2}} (4-y)\,dx\,dy$$

$$= 2\int_{-2}^{2} (4-y)\sqrt{4-y^2}\,dy$$

$$= 2\int_{-2}^{2} 4\sqrt{4-y^2}\,dy - 2\int_{-2}^{2} y\sqrt{4-y^2}\,dy$$

$$= 8\int_{-2}^{2} \sqrt{4-y^2}\,dy = 16\int_{0}^{2} \sqrt{4-y^2}\,dy$$

$$= 16\left[\frac{y\sqrt{4-y^2}}{2} + \frac{4}{2}\sin^{-1}\frac{y}{2}\right]_0^2 = 16\,[2\sin^{-1} 1] = 32 \times \frac{\pi}{2} = 16\pi.$$

TEST YOUR KNOWLEDGE

1. (*i*) Find, by double integration, the area enclosed by the ellipse $\dfrac{x^2}{a^2} + \dfrac{y^2}{b^2} = 1$.

   (*ii*) Find the area of a plate in the form of a quadrant of the ellipse $\dfrac{x^2}{a^2} + \dfrac{y^2}{b^2} = 1$.

2. (*i*) Find, by double integration, the smaller of the areas bounded by the circle $x^2 + y^2 = 9$ and the line $x + y = 3$.

   (*ii*) Find, by double integration, the area of the region enclosed by the curves $x^2 + y^2 = a^2$, $x + y = a$ in the first quadrant. [*U.P.T.U. 2014*]

3. Show, by double integration, that the area between the parabolas $y^2 = 4ax$ and $x^2 = 4ay$ is $\dfrac{16}{3}a^2$. (*U.P.T.U. 2015*)

4. (*i*) Find, by double integration, the whole area of the curve $a^2y^2 = x^3(2a - x)$.

   (*ii*) Find, by double integration, the volume of the sphere $x^2 + y^2 + z^2 = 9$.

5. (*i*) Show that the area between the curves $y(x^2 + 2) = 3x$ and $x^2 = 4y$ is $\left(\dfrac{3}{2}\log 3 - \dfrac{2}{3}\right)$.

   $$\left[\textbf{Hint. } A = \int_0^2\int_{x^2/4}^{3x/x^2+2} dy\,dx\right]$$

   (*ii*) Find the area of the region occupied by the curves $y^2 = x$ and $y^2 = 4 - x$. (*G.B.T.U. 2013*)

6. Find the whole area of the curve given by the equations $x = a \cos^3 t$, $y = b \sin^3 t$.
7. Find the area bounded by the circles $r = 2 \sin \theta$ and $r = 4 \sin \theta$.
8. (*i*) Find, by double integration, the area of one loop of the lemniscate $r^2 = a^2 \cos 2\theta$. (*G.B.T.U. 2011*)
   (*ii*) Find, by double integration, the area of the cardioid $r = a(1 + \cos \theta)$.
   (*iii*) Compute the area bounded by the lemniscate $r^2 = a^2 \cos 2\theta$. (*U.P.T.U. 2014*)
9. (*i*) Find the area outside the circle $r = 2a \cos \theta$ and inside the cardioid $r = a(1 + \cos \theta)$.
   (*ii*) Find, by double integration, the area bounded by the curve $y^2(2a - x) = x^3$ and its asymptote.
10. Determine the area of the region that lies inside $r = 3 + 2 \sin \theta$ and outside $r = 2$.

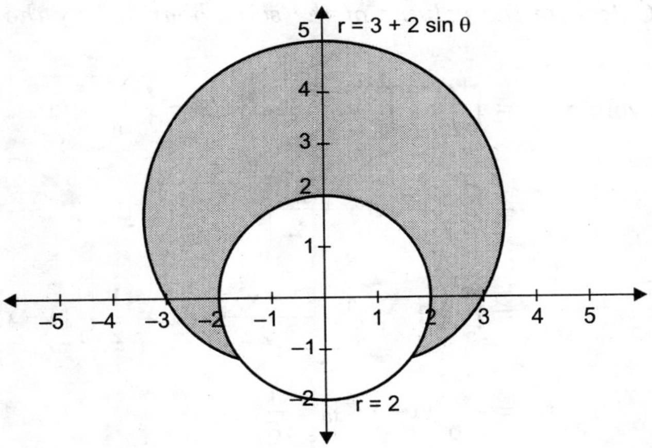

Answers

1. (*i*) πab (*ii*) $\frac{\pi ab}{4}$
2. (*i*) $\frac{9}{4}(\pi - 2)$ (*ii*) $\frac{a^2(\pi - 2)}{4}$
4. (*i*) πa^2 (*ii*) 36π
5. (*ii*) $\frac{16\sqrt{2}}{3}$
6. $\frac{3}{8}\pi ab$
7. 3π
8. (*i*) $\frac{a^2}{2}$ (*ii*) $\frac{3}{2}\pi a^2$ (*iii*) a^2
9. (*i*) $\frac{1}{2}\pi a^2$ (*ii*) $3\pi a^2$
10. $\frac{14\pi}{3} + \frac{11}{2}\sqrt{3}$

4.9. VOLUME AS A TRIPLE INTEGRAL

The volume V of a three-dimensional region is given by $V = \iiint_V dx\,dy\,dz$.

If the region is bounded by $z = f_1(x, y)$, $z = f_2(x, y)$; $y = \phi_1(x)$, $y = \phi_2(x)$ and $x = a$, $x = b$

$$V = \int_a^b \int_{\phi_1(x)}^{\phi_2(x)} \int_{f_1(x, y)}^{f_2(x, y)} dz\,dy\,dx$$

The order of integration may be changed with a suitable change in the limits of integration.

In cylindrical co-ordinates, we have $V = \iiint_V r \, dr \, d\phi \, dz$

In spherical polar co-ordinates, we have $V = \iiint_V r^2 \sin\theta \, dr \, d\theta \, d\phi$.

ILLUSTRATIVE EXAMPLES

Example 1. *Calculate the volume of the solid bounded by the surface $x = 0$, $y = 0$, $x + y + z = 1$ and $z = 0$.* (A.K.T.U. 2021, 2019)

Sol. Required volume $= \int_0^1 \int_0^{1-x} \int_0^{1-x-y} dz \, dy \, dx = \int_0^1 \int_0^{1-x} (1 - x - y) \, dy \, dx$

$$= \int_0^1 \left\{ (1-x)\, y - \frac{y^2}{2} \right\}_0^{1-x} dx$$

$$= \int_0^1 \left\{ (1-x)^2 - \frac{(1-x)^2}{2} \right\} dx = \frac{1}{2} \int_0^1 (1-x)^2 \, dx$$

$$= -\frac{1}{6} [(1-x)^3]_0^1 = \frac{1}{6}$$

Example 2. *A triangular prism is formed by planes whose equations are $ay = bx$, $y = 0$ and $x = a$. Find the volume of the prism between the planes $z = 0$ and surface $z = c + xy$.* (G.B.T.U. 2010)

Sol. Required volume $= \int_0^a \int_0^{\frac{bx}{a}} \int_0^{c + xy} dz \, dy \, dx$

$$= \int_0^a \int_0^{\frac{bx}{a}} (c + xy) \, dy \, dx = \int_0^a \left(cy + \frac{xy^2}{2} \right)_0^{\frac{bx}{a}} dx$$

$$= \int_0^a \left(\frac{cbx}{a} + \frac{b^2}{2a^2} x^3 \right) dx = \frac{bc}{a} \left(\frac{x^2}{2} \right)_0^a + \frac{b^2}{2a^2} \left(\frac{x^4}{4} \right)_0^a$$

$$= \frac{abc}{2} + \frac{b^2 a^2}{8} = \frac{ab}{8}(4c + ab).$$

Example 3. *Find the volume of the cylindrical column standing on the area common to the parabolas $x = y^2$, $y = x^2$ as base and cut-off by the surface $z = 12 + y - x^2$.* (U.P.T.U. 2014)

Sol. Required volume $= \int_0^1 dx \int_{x^2}^{\sqrt{x}} dy \int_0^{12 + y - x^2} dz$

$$= \int_0^1 dx \int_{x^2}^{\sqrt{x}} (12 + y - x^2) \, dy$$

$$= \int_0^1 dx \left(12y + \frac{y^2}{2} - x^2 y \right)_{x^2}^{\sqrt{x}}$$

$$= \int_0^1 \left(12\sqrt{x} + \frac{x}{2} - x^{5/2} - 12x^2 - \frac{x^4}{2} + x^4 \right) dx$$

$$= \left(8x^{3/2} + \frac{x^2}{4} - \frac{2}{7}x^{7/2} - 4x^3 - \frac{x^5}{10} + \frac{x^5}{5} \right)_0^1$$

$$= 8 + \frac{1}{4} - \frac{2}{7} - 4 - \frac{1}{10} + \frac{1}{5} = \frac{569}{140}.$$

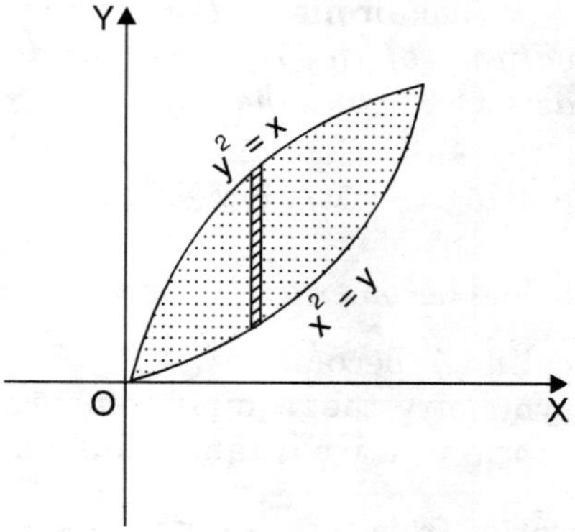

Example 4. *Find by triple integration, the volume of the paraboloid of revolution $x^2 + y^2 = 4z$ cut-off by the plane $z = 4$.*

Sol. By symmetry, the required volume is 4 times the volume in the positive octant. The volume in the positive octant is bounded on the sides by the zx and yz-planes; from above by the plane $z = 4$ and below by the curved surface $x^2 + y^2 = 4z$.

The section of the paraboloid by the plane $z = 4$ is the circle $x^2 + y^2 = 16$, $z = 4$ and its projection on the xy-plane is the circle $x^2 + y^2 = 16$, $z = 0$.

The volume in the positive octant is bounded by $z = \frac{x^2 + y^2}{4}$, $z = 4$, $y = 0$, $y = \sqrt{16 - x^2}$ and $x = 0$, $x = 4$.

$\therefore$ Required volume

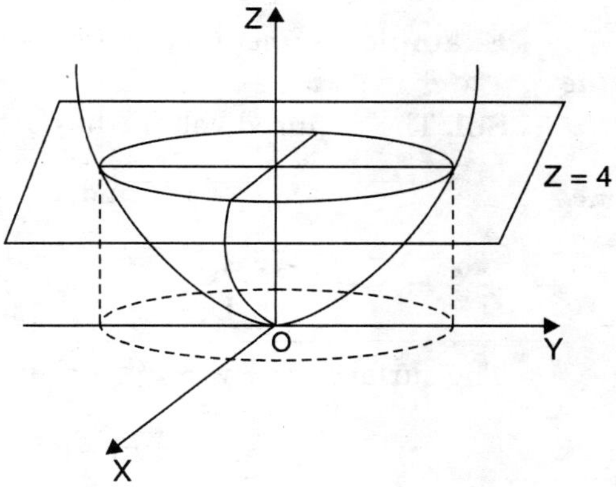

$$= 4 \int_0^4 \int_0^{\sqrt{16 - x^2}} \int_{(x^2 + y^2)/4}^4 dz\, dy\, dx$$

$$= 4 \int_0^4 \int_0^{\sqrt{16 - x^2}} \left(4 - \frac{x^2 + y^2}{4} \right) dy\, dx$$

$$= 4 \int_0^4 \left[\left(4 - \frac{x^2}{4} \right) y - \frac{1}{4} \cdot \frac{y^3}{3} \right]_0^{\sqrt{16 - x^2}} dx$$

$$= 4 \int_0^4 \left[\left(4 - \frac{x^2}{4} \right) \sqrt{16 - x^2} - \frac{1}{12} (16 - x^2)^{3/2} \right] dx$$

$$= 4 \int_0^4 \left[\frac{1}{4} (16 - x^2) \sqrt{16 - x^2} - \frac{1}{12} (16 - x^2)^{3/2} \right] dx$$

$$= \frac{2}{3} \int_0^4 (16 - x^2)^{3/2}\, dx = \frac{2}{3} \int_0^{\pi/2} (16)^{3/2} . \cos^3 \theta . 4 \cos \theta\, d\theta, \text{ where } x = 4 \sin \theta$$

$$= \frac{512}{3} \int_0^{\pi/2} \cos^4 \theta\, d\theta = \frac{512}{3} \cdot \frac{3 \cdot 1}{4 \cdot 2} \cdot \frac{\pi}{2} = 32\pi .$$

Example 5. *Find, by triple integration, the volume of the region bounded by the paraboloid $az = x^2 + y^2$ and the cylinder $x^2 + y^2 = R^2$.*

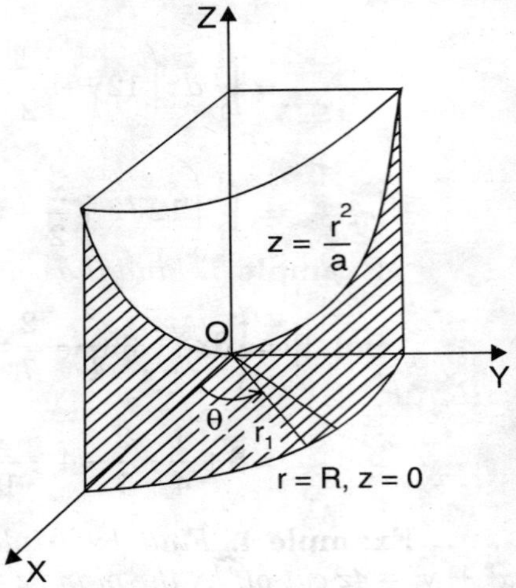

Sol. Changing to cylindrical co-ordinates, by putting $x = r\cos\theta$, $y = r\sin\theta$, the equation of the paraboloid becomes $az = r^2$ or $z = \frac{r^2}{a}$ and the equation of the cylinder becomes $r^2 = R^2$ or $r = R$. On account of symmetry, the required volume is four times the volume in the positive octant. Thus, in the common region, z varies from 0 to $\frac{r^2}{a}$, r varies from 0 to R and θ varies from 0 to $\frac{\pi}{2}$.

$$\therefore \quad \text{Required volume} = 4\int_0^{\pi/2}\int_0^{R}\int_0^{r^2/a} r\,dz\,dr\,d\theta$$

$$= 4\int_0^{\pi/2}\int_0^{R} r\cdot\frac{r^2}{a}\,dr\,d\theta = 4\int_0^{\pi/2}\left[\frac{r^4}{4a}\right]_0^{R} d\theta$$

$$= \frac{1}{a}\int_0^{\pi/2} R^4\,d\theta = \frac{R^4}{a}\cdot\frac{\pi}{2} = \frac{\pi R^4}{2a}.$$

Example 6. *Find the volume bounded above by the sphere $x^2 + y^2 + z^2 = a^2$ and below by the cone $x^2 + y^2 = z^2$.*

Sol. The required volume lies between $z^2 = x^2 + y^2$ and $x^2 + y^2 + z^2 = a^2$

i.e., $$z = \sqrt{x^2 + y^2} \quad \text{and} \quad z = \sqrt{a^2 - x^2 - y^2}$$

$$\therefore \quad V = \iint_A \int_{\sqrt{x^2+y^2}}^{\sqrt{a^2-x^2-y^2}} dz\,dx\,dy$$

The surfaces $x^2 + y^2 + z^2 = a^2$ and $x^2 + y^2 = z^2$ intersect on the cylinder

$$x^2 + y^2 + (x^2 + y^2) = a^2 \quad \Rightarrow \quad x^2 + y^2 = \frac{1}{2}a^2$$

So the base region A is the area inside the circle with the same equation. This area can be more easily covered by converting to polar co-ordinates and taking radial strips. Then A is enclosed by $r = 0$, $r = \frac{a}{\sqrt{2}}$, $\theta = 0$, $\theta = 2\pi$.

$$\therefore \quad \text{Volume} \quad V = \int_0^{2\pi}\int_0^{a/\sqrt{2}}\int_r^{\sqrt{a^2-r^2}} dz\,.\,(r\,dr\,d\theta)$$

$$= \int_0^{2\pi}\int_0^{a/\sqrt{2}} \Big(z\Big)_r^{\sqrt{a^2-r^2}} r\,dr\,d\theta = \int_0^{2\pi}\int_0^{a/\sqrt{2}} (\sqrt{a^2-r^2} - r)\,.\,r\,dr\,d\theta$$

$$= \int_0^{2\pi} \left[\frac{(a^2 - r^2)^{3/2}}{-2.(3/2)} - \frac{1}{3} r^3 \right]_0^{a/\sqrt{2}} d\theta = \frac{1}{3} \int_0^{2\pi} \left[-\left(\frac{a^2}{2} \right)^{3/2} + a^3 - \frac{a^3}{2\sqrt{2}} \right] d\theta$$

$$= \frac{1}{3} a^3 \left(1 - \frac{1}{\sqrt{2}} \right) \left(\theta \right)_0^{2\pi} = \frac{1}{3} a^3 \left(1 - \frac{1}{\sqrt{2}} \right) . 2\pi = (2 - \sqrt{2})\pi \frac{a^3}{3}.$$

Example 7. *Find the volume common to the cylinders $x^2 + y^2 = a^2$ and $x^2 + z^2 = a^2$.*

Sol. Required volume $= \int_{-a}^{a} \int_{-\sqrt{a^2 - x^2}}^{\sqrt{a^2 - x^2}} \int_{-\sqrt{a^2 - x^2}}^{\sqrt{a^2 - x^2}} dz\, dy\, dx$

$$= \int_{-a}^{a} \int_{-\sqrt{a^2 - x^2}}^{\sqrt{a^2 - x^2}} 2\sqrt{a^2 - x^2}\, dy\, dx = 4 \int_{-a}^{a} (a^2 - x^2)\, dx$$

$$= 8 \int_0^a (a^2 - x^2)\, dx = 8 \left(a^2 x - \frac{x^3}{3} \right)_0^a = \frac{16a^3}{3}.$$

Example 8. *The sphere $x^2 + y^2 + z^2 = a^2$ is pierced by the cylinder $(x^2 + y^2)^2 = a^2(x^2 - y^2)$. Prove that the volume of the sphere that lies inside the cylinder is* $\frac{8}{3}\left(\frac{\pi}{4} + \frac{5}{3} - \frac{4\sqrt{2}}{3} \right) a^3.$

Sol. Limits of z are from $-\sqrt{a^2 - x^2 - y^2}$ to $\sqrt{a^2 - x^2 - y^2}$

$$\therefore \quad V = \iint \int_{-\sqrt{a^2 - x^2 - y^2}}^{\sqrt{a^2 - x^2 - y^2}} dz\, dy\, dx = 2 \iint \sqrt{a^2 - x^2 - y^2}\, dy\, dx \quad \ldots(1)$$

Equation of cylinder is

$$(x^2 + y^2)^2 = a^2(x^2 - y^2) \quad \ldots(2)$$

Changing (2) to polar coordinates, we get

$$r^4 = a^2 r^2 \cos 2\theta \quad \Rightarrow \quad r^2 = a^2 \cos 2\theta$$

Also, $dx\, dy = r\, dr\, d\theta$ | Since | J | = r

Limits of r are from $-a\sqrt{\cos 2\theta}$ to $a\sqrt{\cos 2\theta}$.

Limits of θ are from $-\frac{\pi}{4}$ to $\frac{\pi}{4}$.

$$\therefore \quad \text{Required volume} = 2 \int_{-\pi/4}^{\pi/4} \int_{-a\sqrt{\cos 2\theta}}^{a\sqrt{\cos 2\theta}} \sqrt{a^2 - r^2}\, .\, r\, dr\, d\theta = 8 \int_0^{\pi/4} \left[\frac{(a^2 - r^2)^{3/2}}{(-3/2).2} \right]_0^{a\sqrt{\cos 2\theta}} d\theta$$

$$= \frac{8}{3} a^3 \int_0^{\pi/4} [1 - (1 - \cos 2\theta)^{3/2}]\, d\theta = \frac{8}{3} a^3 \int_0^{\pi/4} (1 - 2\sqrt{2} \sin^3 \theta)\, d\theta$$

$$= \frac{8}{3} a^3 \int_0^{\pi/4} \left[1 - 2\sqrt{2} \left(\frac{3}{4} \sin \theta - \frac{\sin 3\theta}{4} \right) \right] d\theta$$

$$= \frac{8}{3} a^3 \left[\frac{\pi}{4} - \frac{3}{\sqrt{2}} (-\cos \theta)_0^{\pi/4} + \frac{1}{\sqrt{2}} \left(\frac{-\cos 3\theta}{3} \right)_0^{\pi/4} \right]$$

$$= \frac{8}{3} a^3 \left[\frac{\pi}{4} - \frac{3}{\sqrt{2}} + \frac{3}{2} + \frac{1}{3\sqrt{2}} + \frac{1}{6} \right] = \frac{8}{3} a^3 \left(\frac{\pi}{4} + \frac{5}{3} - \frac{4\sqrt{2}}{3} \right).$$

TEST YOUR KNOWLEDGE

1. Find, by triple integration, the volume in the positive octant bounded by the co-ordinate planes and the plane $x + 2y + 3z = 4$.

2. Find the volume of the portion of the sphere $x^2 + y^2 + z^2 = a^2$ lying inside the cylinder $x^2 + y^2 = ax$.

$$\left[\textbf{Hint. } V = \int_0^{2\pi}\int_0^{a\cos\theta}\int_0^{\sqrt{a^2-r^2}} r\,dz\,dr\,d\theta\right]$$

3. Find the volume bounded above by the sphere $x^2 + y^2 + z^2 = 2a^2$ and below by the paraboloid $az = x^2 + y^2$.

$$\left[\textbf{Hint. } V = \int_0^{2\pi}\int_0^{a}\int_{r^2/a}^{\sqrt{2a^2-r^2}} dz\,(rdr\,d\theta)\right]$$

4. Find by triple integration, the volume of a solid bounded by the sphere $x^2 + y^2 + z^2 = 4$ and the paraboloid $x^2 + y^2 = 3z$.

$$\left[\textbf{Hint. } V = \int_0^{2\pi}\int_0^{\sqrt{3}}\int_{r^2/3}^{\sqrt{4-r^2}} r\,dz\,dr\,d\theta\right]$$

5. (*i*) Find by triple integration, the volume cut off from the cylinder $x^2 + y^2 = ax$ by the planes $z = mx$ and $z = nx$.

(*ii*) Find the volume of the region bounded by the surface $y = x^2$, $x = y^2$ and the planes $z = 0$, $z = 3$.

(*A.K.T.U. 2020*)

6. (*i*) Find the volume bounded by the *xy*-plane, the paraboloid $2z = x^2 + y^2$ and the cylinder $x^2 + y^2 = 4$.

(*ii*) Find the volume of the solid which is bounded by the surfaces $2z = x^2 + y^2$ and $z = x$.

(*M.T.U. 2012*)

7. Find the volume of the region bounded by $z = x^2 + y^2$, $z = 0$, $x = -a$, $x = a$ and $y = -a$, $y = a$.

8. Prove that the volume enclosed between the cylinders $x^2 + y^2 = 2ax$ and $z^2 = 2ax$ is $\dfrac{128a^3}{15}$.

9. Find the volume bounded by the *xy*-plane, the cylinder $x^2 + y^2 = 1$ and the plane $x + y + z = 3$.

10. Find the volume bounded by the cylinder $x^2 + y^2 = 4$ and the hyperboloid $x^2 + y^2 - z^2 = 1$.

11. Find the volume bounded by the paraboloid $x^2 + y^2 = az$, the cylinder $x^2 + y^2 = 2ay$ and the plane $z = 0$.

12. Find the volume of the solid under the surface $az = x^2 + y^2$ and whose base R is the circle $x^2 + y^2 = a^2$.

13. Find the volume of the cylinder $x^2 + y^2 - 2ax = 0$ intercepted between the paraboloid $x^2 + y^2 = 2az$ and the *xy*-plane.

14. (*i*) Find the volume enclosed between the two surfaces $z = 8 - x^2 - y^2$ and $z = x^2 + 3y^2$.

(*G.B.T.U. 2011*)

[**Hint.** Limits of z are from $x^2 + 3y^2$ to $8 - x^2 - y^2$.]

(*ii*) Find the volume bounded by the elliptic paraboloids $z = x^2 + 9y^2$ and $z = 18 - x^2 - 9y^2$.

(*G.B.T.U. 2012*)

$$\left[\textbf{Hint. } V = \int_{-3}^{3}\int_{-\frac{1}{3}\sqrt{9-x^2}}^{\frac{1}{3}\sqrt{9-x^2}}\int_{x^2+9y^2}^{18-x^2-9y^2} dz\,dy\,dx\right]$$

15. (*i*) Compute $\iiint_v x^2\, dx\, dy\, dz$ over volume of tetrahedron bounded by $x = 0$, $y = 0$, $z = 0$ and $\frac{x}{a} + \frac{y}{b} + \frac{z}{c} = 1$. *(A.K.T.U. 2018)*

(*ii*) Evaluate $\iiint x^2yz\, dx\, dy\, dz$ throughout the volume bounded by planes $x = 0$, $y = 0$, $z = 0$ and $\frac{x}{a} + \frac{y}{b} + \frac{z}{c} = 1$. *(A.K.T.U. 2017)*

Answers

1. $\dfrac{16}{9}$ **2.** $\dfrac{2a^3}{9}(3\pi - 4)$ **3.** $\left(\dfrac{4}{3}\sqrt{2} - \dfrac{7}{6}\right)\pi a^3$

4. $\dfrac{19\pi}{6}$ **5.** (*i*) $(n - m)\dfrac{\pi a^3}{8}$ (*ii*) 1

6. (*i*) 4π (*ii*) $\dfrac{\pi}{4}$ **7.** $\dfrac{8}{3}a^4$ **9.** $3\pi - \dfrac{4}{3}$

10. $4\sqrt{3}\pi$ **11.** $\dfrac{3\pi a^3}{2}$ **12.** $\dfrac{\pi a^3}{2}$

13. $3\pi a^3$ **14.** (*i*) $8\pi\sqrt{2}$ (*ii*) 27π.

15. (*i*) $\dfrac{a^3bc}{60}$ (*ii*) $\dfrac{a^3b^2c^2}{2520}$

4.10. CALCULATION OF MASS

(*a*) *For a plane lamina* of area A, if the density at a point P(x, y) be $\rho = f(x, y)$, then its total mass M is given by
$$M = \iint_A \rho\, dx\, dy = \iint_A f(x, y)\, dx\, dy$$

In polar co-ordinates, if $\rho = f(r, \theta)$ at a point P(r, θ), then
$$M = \iint_A \rho\, r\, dr\, d\theta = \iint_A f(r, \theta)\, r\, dr\, d\theta.$$

Also, if $(\bar{x}, \bar{y})$ is the required centre of gravity then,
$$\bar{x} = \frac{\iint \rho x\, dx\, dy}{\iint \rho\, dx\, dy}, \quad \bar{y} = \frac{\iint \rho y\, dx\, dy}{\iint \rho\, dx\, dy}$$

In polar coordinates,
$$\bar{x} = \frac{\iint \rho . r\cos\theta . r\, dr\, d\theta}{\iint \rho . r\, dr\, d\theta}, \quad \bar{y} = \frac{\iint \rho . r\sin\theta . r\, dr\, d\theta}{\iint \rho . r\, dr\, d\theta}$$

where ρ is the density.

(b) *For a solid* of volume V, if the density at a point P(x, y, z) be $\rho = f(x, y, z)$, then its total mass M is given by

$$M = \iiint_V \rho \, dx\, dy\, dz = \iiint_V f(x, y, z)\, dx\, dy\, dz.$$

ILLUSTRATIVE EXAMPLES

Example 1. *Find the mass of a plate in the shape of the curve* $\left(\frac{x}{a}\right)^{2/3} + \left(\frac{y}{b}\right)^{2/3} = 1$, *the density being given by* $\rho = \mu\, xy$.

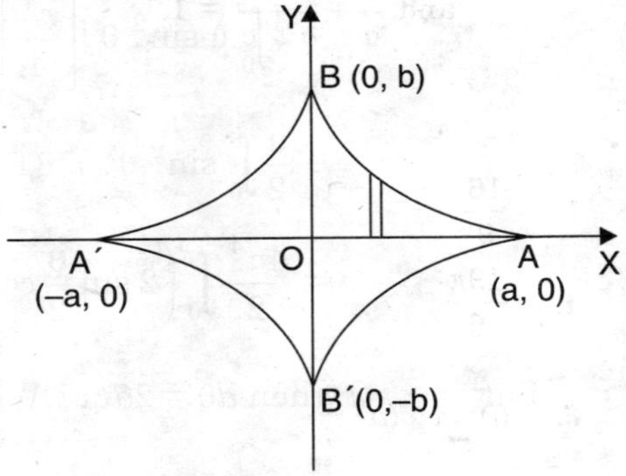

Sol. Let the required mass be M which is four times the mass in the first quadrant.

From the equation of the curve

$$\left(\frac{y}{b}\right)^{2/3} = 1 - \left(\frac{x}{a}\right)^{2/3}$$

or
$$y = b\left[1 - \left(\frac{x}{a}\right)^{2/3}\right]^{3/2} = y_1 \quad \text{(say)}$$

For the region OAB, x varies from 0 to a and y varies from 0 to y_1.

$$\therefore \quad M = 4\int_0^a \int_0^{y_1} \rho\, dy\, dx = 4\int_0^a \int_0^{y_1} \mu\, xy\, dy\, dx$$

$$= 4\int_0^a \mu x \cdot \left[\frac{y^2}{2}\right]_0^{y_1} dx = 2\mu \int_0^a x{y_1}^2\, dx$$

$$= 2\mu \int_0^a xb^2 \left[1 - \left(\frac{x}{a}\right)^{2/3}\right]^3 dx$$

Put $x = a \sin^3 \theta$, then $dx = 3a \sin^2 \theta \cos \theta\, d\theta$

When $x = 0, \theta = 0$; when $x = a, \theta = \frac{\pi}{2}$

$$\therefore \quad M = 2\mu b^2 \int_0^{\pi/2} a \sin^3 \theta\, (1 - \sin^2 \theta)^3 \,.\, 3a \sin^2 \theta \cos \theta\, d\theta$$

$$= 6\mu a^2 b^2 \int_0^{\pi/2} \sin^5 \theta \cos^7 \theta\, d\theta = 6\mu a^2 b^2 \,.\, \frac{4\cdot 2\cdot 6\cdot 4\cdot 2}{12\cdot 10\cdot 8\cdot 6\cdot 4\cdot 2} = \frac{\mu a^2 b^2}{20}.$$

Example 2. *Find the mass of a lamina in the form of the cardioid* $r = a(1 + \cos \theta)$ *whose density at any point varies as the square of its distance from the initial line.*

Sol. Let the required mass be M which is twice the mass above the initial line.

Since the distance of any point (r, θ) from the initial line is $r \sin \theta$, the density at (r, θ) is given by

$$\rho = \mu\, (r \sin \theta)^2 = \mu\, r^2 \sin^2 \theta.$$

For the region above the initial line, θ varies from 0 to π and r varies from 0 to $a(1 + \cos\theta)$.

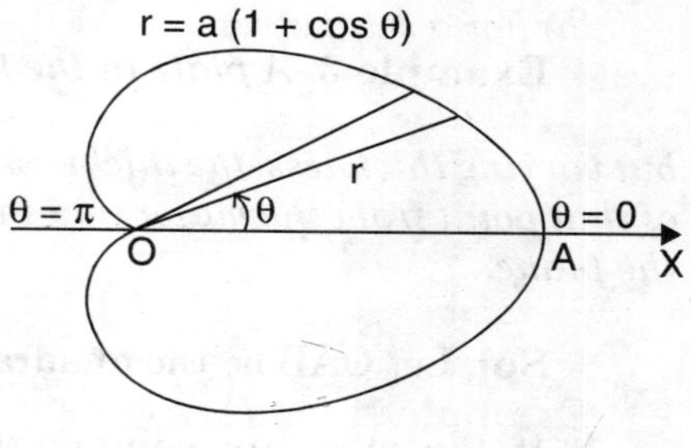

$$\therefore \quad M = 2\int_0^{\pi}\int_0^{a(1+\cos\theta)} \rho\, r\, dr\, d\theta$$

$$= 2\int_0^{\pi}\int_0^{a(1+\cos\theta)} \mu\, r^3 \sin^2\theta\, dr\, d\theta$$

$$= 2\int_0^{\pi} \mu \sin^2\theta \,.\left[\frac{r^4}{4}\right]_0^{a(1+\cos\theta)} d\theta$$

$$= \frac{\mu}{2}\int_0^{\pi} \sin^2\theta \,.\, a^4\,(1+\cos\theta)^4\, d\theta$$

$$= \frac{4a^4}{2}\int_0^{\pi}\left(2\sin\frac{\theta}{2}\cos\frac{\theta}{2}\right)^2 .\left(2\cos^2\frac{\theta}{2}\right)^4 d\theta = 32\mu a^4\int_0^{\pi}\sin^2\frac{\theta}{2}\cos^{10}\frac{\theta}{2}\, d\theta$$

Put $\frac{\theta}{2} = t$, then $d\theta = 2dt$; When $\theta = 0, t = 0$; when $\theta = \pi,\ t = \frac{\pi}{2}$

$$\therefore \quad M = 32\mu a^4\int_0^{\pi/2} 2\sin^2 t\cos^{10} t\, dt = 64\mu a^4 \,.\, \frac{1\cdot 9\cdot 7\cdot 5\cdot 3\cdot 1}{12\cdot 10\cdot 8\cdot 6\cdot 4\cdot 2}\cdot\frac{\pi}{2} = \frac{21}{32}\mu\pi a^4 .$$

4.11. CENTRE OF GRAVITY (C.G.)

(a) The C.G. $(\bar{x}, \bar{y})$ *of a plane lamina* occupying an area A in the xy plane and having density $\rho = f(x, y)$ is given by

$$\bar{x} = \frac{\iint_A x\,\rho\, dx\, dy}{\iint_A \rho\, dx\, dy}, \quad \bar{y} = \frac{\iint_A y\,\rho\, dx\, dy}{\iint_A \rho\, dx\, dy}$$

In polar co-ordinates, if $\rho = f(r, \theta)$ at a point $P(r, \theta)$, then the C.G. $(\bar{x}, \bar{y})$ is given by

$$\bar{x} = \frac{\iint_A r\cos\theta \,.\, \rho r\, dr\, d\theta}{\iint_A \rho r\, dr\, d\theta} = \frac{\iint_A r^2\cos\theta\, \rho d\, r\, d\theta}{\iint_A \rho r\, dr\, d\theta}$$

$$\bar{y} = \frac{\iint_A r\sin\theta\, \rho r\, dr\, d\theta}{\iint_A \rho r\, dr\, d\theta} = \frac{\iint_A r^2\sin\theta\, \rho\, dr\, d\theta}{\iint_A \rho r\, dr\, d\theta}$$

(b) The C.G. $(\bar{x}, \bar{y}, \bar{z})$ *of a solid* occupying a volume V and having density $\rho = f(x, y, z)$ is given by

$$\bar{x} = \frac{\iiint_V x\,\rho\, dx\, dy\, dz}{\iint_V \rho\, dx\, dy\, dz}, \quad \bar{y} = \frac{\iint_V y\,\rho\, dx\, dy\, dz}{\iint_V \rho\, dx\, dy\, dz}, \quad \bar{z} = \frac{\iiint_V z\,\rho\, dx\, dy\, dz}{\iiint_V \rho\, dx\, dy\, dz}.$$

Example 3. *A plate in the form of a quadrant of the ellipse* $\left(\frac{x}{a}\right)^2 + \left(\frac{y}{b}\right)^2 = 1$ *is of small but varying thickness, the thickness at any point being proportional to the product of the distances of that point from the major and the minor axes. Find the co-ordinates of the centre of gravity of the plane.*

Sol. Let OAB be the quadrant of the ellipse $\frac{x^2}{a^2} + \frac{y^2}{b^2} = 1$ with OA = a and OB = b.

If P(x, y) be any point on the plate, then density

$$\rho = kxy$$

For the area OAB, x varies from 0 to a and y varies from 0 to $b\sqrt{1 - \frac{x^2}{a^2}} = \frac{b}{a}\sqrt{a^2 - x^2}$

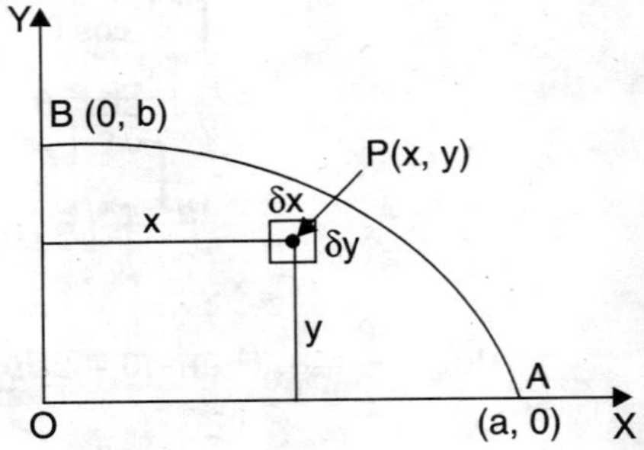

Let G($\bar{x}$, $\bar{y}$) be its C.G., then

$$\bar{x} = \frac{\int_0^a \int_0^{(b/a)\sqrt{a^2 - x^2}} x\,\rho dy\,dx}{\int_0^a \int_0^{(b/a)\sqrt{a^2 - x^2}} \rho dy\,dx}$$

$$= \frac{\int_0^a \int_0^{(b/a)\sqrt{a^2 - x^2}} x\,.\,kxy\,dy\,dx}{\int_0^a \int_0^{(b/a)\sqrt{a^2 - x^2}} kxy\,dy\,dx} = \frac{\int_0^a x^2 \left[\frac{y^2}{2}\right]_0^{(b/a)\sqrt{a^2 - x^2}} dx}{\int_0^a x \left[\frac{y^2}{2}\right]_0^{(b/a)\sqrt{a^2 - x^2}} dx}$$

$$= \frac{\int_0^a x^2 (a^2 - x^2)\,dx}{\int_0^a x(a^2 - x^2)\,dx} = \frac{\left[a^2 . \frac{x^3}{3} - \frac{x^5}{5}\right]_0^a}{\left[a^2 . \frac{x^2}{2} - \frac{x^4}{4}\right]_0^a} = \frac{\frac{2}{15}a^5}{\frac{1}{4}a^4} = \frac{8a}{15}$$

Similarly, $$\bar{y} = \frac{\int_0^a \int_0^{(b/a)\sqrt{a^2 - x^2}} y\,\rho dy\,dx}{\int_0^a \int_0^{(b/a)\sqrt{a^2 - x^2}} \rho dy\,dx} = \frac{8b}{15}$$

Hence $\quad G \rightarrow \left(\frac{8a}{15}, \frac{8b}{15}\right)$.

Example 4. *Find the centroid of a loop of the lemniscate* $r^2 = a^2 \cos 2\theta$.

Sol. The loop being symmetrical about the initial line, its C.G. lies on OX, so that $\bar{y} = 0$. Since the density is not given, let it be k, a constant.

For the loop, θ varies from $-\frac{\pi}{4}$ to $\frac{\pi}{4}$ and r varies from 0 to $a\sqrt{\cos 2\theta}$.

$$\therefore \quad \bar{x} = \frac{\int_{-\pi/4}^{\pi/4}\int_0^{a\sqrt{\cos 2\theta}} r\cos\theta \,.\, \rho r\, dr\, d\theta}{\int_{-\pi/4}^{\pi/4}\int_0^{a\sqrt{\cos 2\theta}} \rho\, r dr\, d\theta}$$

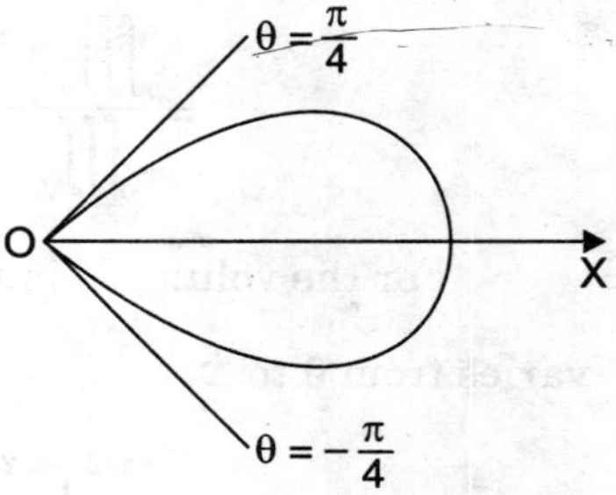

$$= \frac{\int_{-\pi/4}^{\pi/4}\int_0^{a\sqrt{\cos 2\theta}} k\, r^2 \cos\theta\, dr\, d\theta}{\int_{-\pi/4}^{\pi/4}\int_0^{a\sqrt{\cos 2\theta}} kr\, dr\, d\theta}$$

$$= \frac{\int_{-\pi/4}^{\pi/4} \cos\theta \left[\frac{r^3}{3}\right]_0^{a\sqrt{\cos 2\theta}} d\theta}{\int_{-\pi/4}^{\pi/4} \left[\frac{r^2}{2}\right]_0^{a\sqrt{\cos 2\theta}} d\theta} = \frac{\int_{-\pi/4}^{\pi/4} \frac{a^3}{3} \cos\theta\, (\cos 2\theta)^{3/2}\, d\theta}{\int_{-\pi/4}^{\pi/4} \frac{a^2}{2} \cos 2\theta\, d\theta}$$

$$= \frac{\frac{2a}{3}\int_0^{\pi/4} \cos\theta\, (1-\sin^2\theta)^{3/2}\, d\theta}{\left[\frac{\sin 2\theta}{2}\right]_0^{\pi/4}} = \frac{4a}{3}\int_0^{\pi/4} \cos\theta\, (1-2\sin^2\theta)^{3/2}\, d\theta$$

Put $\sqrt{2}\sin\theta = \sin t$, then $\sqrt{2}\cos\theta\, d\theta = \cos t\, dt$

When $\theta = 0,\ t = 0$; when $\theta = \frac{\pi}{4},\ t = \frac{\pi}{2}$

$$\therefore \quad \bar{x} = \frac{4a}{3}\int_0^{\pi/2} (1-\sin^2 t)^{3/2} \,.\, \frac{1}{\sqrt{2}} \cos t\, dt$$

$$= \frac{2\sqrt{2}a}{3}\int_0^{\pi/2} \cos^4 t\, dt = \frac{2\sqrt{2}a}{3} \cdot \frac{3\,.\,1}{4\,.\,2} \cdot \frac{\pi}{2} = \frac{\pi a\sqrt{2}}{8}$$

Hence $\quad G \to \left(\frac{\pi a\sqrt{2}}{8}, 0\right)$.

Example 5. *A solid is in the form of the positive octant of the sphere $x^2 + y^2 + z^2 = a^2$. The density ρ at any point (x, y, z) is given by $\rho = \mu\, xyz$, where μ is a constant. Find the co-ordinates of the centre of gravity of the solid.*

Sol. Let V be the volume of the solid in the positive octant. If $G(\bar{x}, \bar{y}, \bar{z})$ is the C.G. of the solid, then

$$\bar{x} = \frac{\iiint_V x\rho\, dx\, dy\, dz}{\iiint_V \rho\, dx\, dy\, dz} = \frac{\iiint_V \mu\, x^2 yz\, dx\, dy\, dz}{\iiint_V \mu\, xyz\, dx\, dy\, dz}$$

Changing to spherical polar co-ordinates by putting $x = r\sin\theta\cos\phi$, $y = r\sin\theta\sin\phi$, $z = r\cos\theta$ and $dx\, dy\, dz = r^2 \sin\theta\, dr\, d\theta\, d\phi$

$$\bar{x} = \frac{\iiint_V (r\sin\theta\cos\phi)^2\, (r\sin\theta\sin\phi)(r\cos\theta)\, r^2 \sin\theta\, dr\, d\theta\, d\phi}{\iiint_V (r\sin\theta\cos\phi)(r\sin\theta\sin\phi)(r\cos\theta)\, r^2\sin\theta\, dr\, d\theta\, d\phi}$$

$$= \frac{\iiint_V r^6 \sin^4\theta \cos\theta \sin\phi \cos^2\phi \, dr \, d\theta \, d\phi}{\iiint_V r^5 \sin^3\theta \cos\theta \sin\phi \cos\phi \, dr \, d\theta \, d\phi}$$

For the volume V in the positive octant, r varies from 0 to a, θ varies from 0 to $\frac{\pi}{2}$ and ϕ varies from 0 to $\frac{\pi}{2}$.

$$\therefore \quad \bar{x} = \frac{\int_0^{\pi/2}\int_0^{\pi/2}\int_0^a r^6 \sin^4\theta \cos\theta \sin\phi \cos^2\phi \, dr \, d\theta \, d\phi}{\int_0^{\pi/2}\int_0^{\pi/2}\int_0^a r^5 \sin^3\theta \cos\theta \sin\phi \cos\phi \, dr \, d\theta \, d\phi}$$

$$= \frac{\int_0^{\pi/2}\int_0^{\pi/2}\left[\frac{r^7}{7}\right]_0^a \sin^4\theta \cos\theta \sin\phi \cos^2\phi \, d\theta \, d\phi}{\int_0^{\pi/2}\int_0^{\pi/2}\left[\frac{r^6}{6}\right]_0^a \sin^3\theta \cos\theta \sin\phi \cos\phi \, d\theta \, d\phi}$$

$$= \frac{6a}{7} \cdot \frac{\int_0^{\pi/2}\int_0^{\pi/2} \sin^4\theta \, \cos\theta \sin\phi \cos^2\theta \, d\theta \, d\phi}{\int_0^{\pi/2}\int_0^{\pi/2} \sin^3\theta \cos\theta \sin\phi \cos\phi \, d\theta \, d\phi} = \frac{6a}{7} \cdot \frac{\int_0^{\pi/2} \frac{3 \cdot 1}{5 \cdot 3 \cdot 1} \sin\phi \cos^2\phi \, d\phi}{\int_0^{\pi/2} \frac{2}{4 \cdot 2} \sin\phi \cos\phi \, d\phi}$$

$$= \frac{24a}{35} \frac{\int_0^{\pi/2} \sin\phi \cos^2\phi \, d\phi}{\int_0^{\pi/2} \sin\phi \cos\phi \, d\phi} = \frac{24a}{35} \cdot \frac{\frac{1}{3 \cdot 1}}{\frac{1}{2}} = \frac{16a}{35}$$

By symmetry, $\bar{x} = \bar{y} = \bar{z} = \frac{16a}{35}$

Hence $\quad G \rightarrow \left(\frac{16a}{35}, \frac{16a}{35}, \frac{16a}{35}\right)$.

Example 6. *A triangular thin plate with vertices (0, 0), (2, 0) and (2, 4) has density* $\rho = 1 + x + y$. *Then find:*
(A.K.T.U. 2021)

(i) the mass of the plate

(ii) the position of its centre of gravity G.

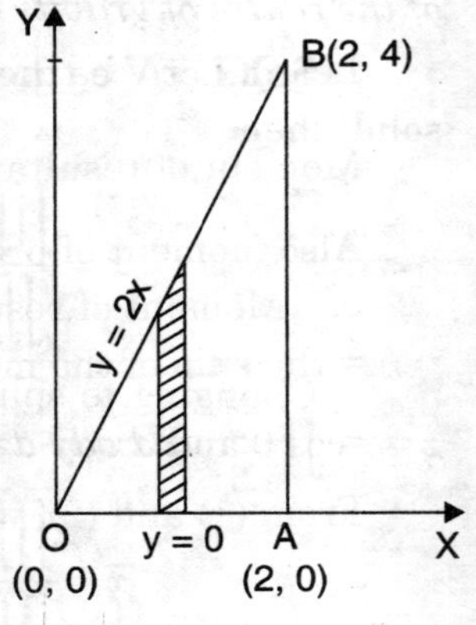

Sol. (*i*)
$$\text{Mass} = \iint \rho \, dx \, dy$$

$$= \int_0^2 \int_0^{2x} (1 + x + y) \, dy \, dx$$

$$= \int_0^2 \left[(1 + x) \, y + \frac{y^2}{2}\right]_0^{2x} dx$$

$$M = \int_0^2 (2x + 4x^2) \, dx$$

$$= \left(x^2 + \frac{4}{3} x^3\right)_0^2 = \frac{44}{3}$$

(ii)
$$\bar{x} = \frac{\iint \rho x \, dx \, dy}{M} = \frac{3}{44} \int_0^2 \int_0^{2x} (1 + x + y) \, x \, dy \, dx$$

$$= \frac{3}{44} \int_0^2 (2x + 4x^2) \, x \, dx = \frac{3}{44} \left(\frac{2}{3} x^3 + x^4 \right)_0^2 = \frac{16}{11}$$

$$\bar{y} = \frac{\iint \rho y \, dx \, dy}{M} = \frac{3}{44} \int_0^2 \int_0^{2x} (1 + x + y) \, y \, dy \, dx$$

$$= \frac{3}{44} \int_0^2 \left[(1 + x) \frac{y^2}{2} + \frac{y^3}{3} \right]_0^{2x} dx = \frac{3}{44} \int_0^2 \left[(1 + x) \, 2x^2 + \frac{8}{3} x^3 \right] dx$$

$$= \frac{3}{44} \left(\frac{2x^3}{3} + \frac{14}{3} \cdot \frac{x^4}{4} \right)_0^2 = \frac{18}{11}$$

$\therefore$ C.G. $\rightarrow \left(\frac{16}{11}, \frac{18}{11} \right)$

4.12. CENTRE OF PRESSURE

If a plane area is immersed in a homogeneous liquid, then the point in the plane area at which the resultant pressure of the liquid acts, is called the *Centre of Pressure* of the area.

Let a plane area A be immersed vertically in a homogeneous liquid. Take the line of intersection of the given plane with the free surface of the liquid as the x-axis and any line perpendicular to it downwards lying in this plane as the y-axis.

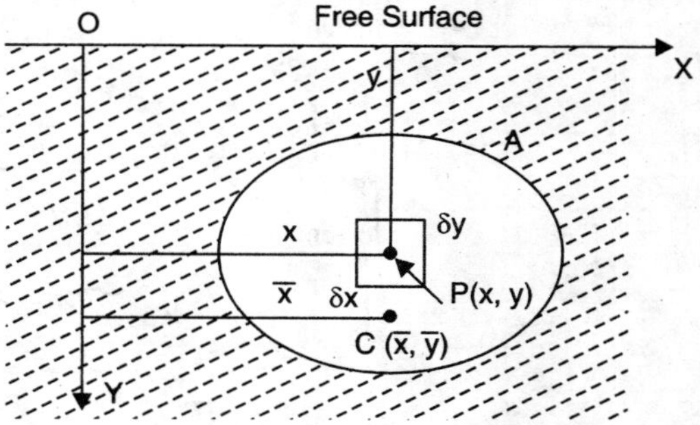

Let p be the pressure at a point P(x, y) of the plane area. Then the pressure on an elementary area $\delta x \, \delta y$ surrounding the point P is $p \, \delta x \, \delta y$ which is normal to the plane.

$\therefore$ The resultant pressure on A $= \iint_A p \, dx \, dy$

Let C($\bar{x}$, $\bar{y}$) be the centre of pressure where the resultant pressure acts.

Moment of resultant pressure about y-axis $= \bar{x} \iint_A p \, dx \, dy$ ...(1)

Also moment of pressure at P about y-axis $= x.p \, \delta x \, \delta y$

$\therefore$ Moment of resultant pressure about y-axis

= the sum of the moments of the individual pressures such as $p \, \delta x \, \delta y$ at P about y-axis.

$= \iint_A x \, . \, p \, dx \, dy$ ...(2)

From (1) and (2), we have

$$\bar{x} \iint_A p \, dx \, dy = \iint_A px \, dx \, dy \quad \Rightarrow \quad \bar{x} = \frac{\iint_A px \, dx \, dy}{\iint_A p \, dx \, dy}$$

Similarly, taking moments about the x-axis, we have $\bar{y} = \dfrac{\iint_A py\, dx\, dy}{\iint_A p\, dx\, dy}$

In polar co-ordinates, the above formulae become

$$\bar{x} = \frac{\iint_A pr^2 \cos\theta\, dr\, d\theta}{\iint_A pr\, dr\, d\theta}, \quad \bar{y} = \frac{\iint_A pr^2 \sin\theta\, dr\, d\theta}{\iint_A pr\, dr\, d\theta}$$

(by replacing x by $r\cos\theta$, y by $r\sin\theta$ and $dx\, dy$ by $r\, dr\, d\theta$).

Example 7. *A quadrant of a circle of radius a is just immersed vertically in a homogeneous liquid with one edge in the surface. Find the centre of pressure.*

Sol. Take the co-ordinate axes as shown in the figure, then the equation of the circle is $x^2 + y^2 = a^2$. For the quadrant OAB, x varies from 0 to a and y varies from 0 to $\sqrt{a^2 - x^2}$. If w is the weight of the liquid per unit volume, then the pressure p at the point P(x, y) is given by $p = wy$.

If C$(\bar{x}, \bar{y})$ be the centre of pressure, then

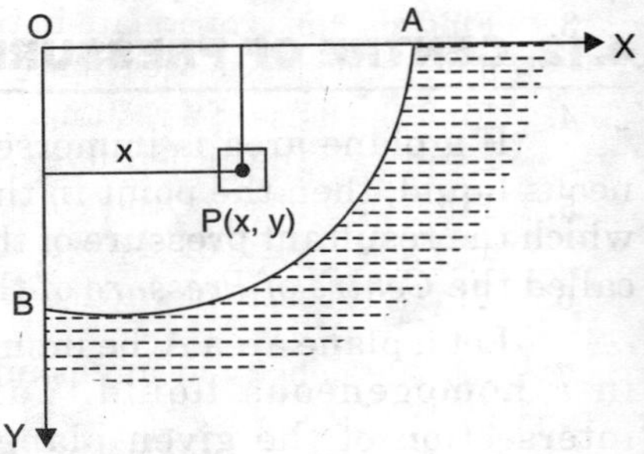

$$\bar{x} = \frac{\int_0^a \int_0^{\sqrt{a^2-x^2}} px\, dy\, dx}{\int_0^a \int_0^{\sqrt{a^2-x^2}} p\, dy\, dx} = \frac{\int_0^a \int_0^{\sqrt{a^2-x^2}} wyx\, dy\, dx}{\int_0^a \int_0^{\sqrt{a^2-x^2}} wy\, dy\, dx}$$

$$= \frac{\int_0^a x \cdot \left[\frac{y^2}{2}\right]_0^{\sqrt{a^2-x^2}} dx}{\int_0^a \left[\frac{y^2}{2}\right]_0^{\sqrt{a^2-x^2}} dx} = \frac{\int_0^a x(a^2 - x^2)\, dx}{\int_0^a (a^2 - x^2)\, dx} = \frac{\left[a^2 \frac{x^2}{2} - \frac{x^4}{2}\right]_0^a}{\left[a^2 x - \frac{x^3}{3}\right]_0^a} = \frac{\frac{a^4}{4}}{\frac{2a^3}{3}} = \frac{3a}{8}$$

$$\bar{y} = \frac{\int_0^a \int_0^{\sqrt{a^2-x^2}} py\, dy\, dx}{\int_0^a \int_0^{\sqrt{a^2-x^2}} p\, dy\, dx} = \frac{\int_0^a \int_0^{\sqrt{a^2-x^2}} wy \cdot y\, dy\, dx}{\int_0^a \int_0^{\sqrt{a^2-x^2}} wy\, dy\, dx}$$

$$= \frac{\int_0^a \left[\frac{y^3}{3}\right]_0^{\sqrt{a^2-x^2}} dx}{\int_0^a \left[\frac{y^2}{2}\right]_0^{\sqrt{a^2-x^2}} dx} = \frac{2}{3} \cdot \frac{\int_0^a (a^2 - x^2)^{3/2}\, dx}{\int_0^a (a^2 - x^2)\, dx}$$

$$= \frac{2}{3} \cdot \frac{\int_0^{\pi/2} (a^2 - a^2 \sin^2 \theta)^{3/2} \cdot a \cos \theta \, d\theta}{\left[a^2 x - \frac{x^3}{3} \right]_0^a} \quad \text{where } x = a \sin \theta$$

$$= \frac{2}{3} \cdot \frac{\int_0^{\pi/2} a^4 \cos^4 \theta \, d\theta}{\frac{2a^3}{3}} = a \cdot \frac{3 \cdot 1}{4 \cdot 2} \cdot \frac{\pi}{2} = \frac{3\pi a}{16}$$

Hence the centre of pressure is $C \to \left(\frac{3a}{8}, \frac{3\pi a}{16} \right)$.

TEST YOUR KNOWLEDGE

1. Find the mass of an elliptic plate $\frac{x^2}{a^2} + \frac{y^2}{b^2} = 1$, if the density at any point (x, y) on it is μxy.
2. Find the mass of the area bounded by the curves $y^2 = x$ and $y = x^3$, if $\rho = \mu(x^2 + y^2)$.
3. Find the mass of a plate in the form of one loop of the lemniscate $r^2 = a^2 \cos 2\theta$, if the density at a point varies as the square of its distance from the pole.
4. Find the mass of a rectangular parallelopiped bounded by the planes $x = 0, x = a$; $y = 0, y = b$ and $z = 0, z = c$, if the density ρ at a point $P(x, y, z)$ is given by $\rho = x + y + z$.
5. Find the mass of the solid bounded by the planes $y = 0, z = 0$ and $z = h$ $(h > 0)$ and the cylinder $x^2 + y^2 = a^2$ $(x > 0)$, if $\rho = \mu yz$.
6. Find the centre of gravity of the area of the circle $x^2 + y^2 = a^2$ lying in the first quadrant.
7. (*i*) The density at any point (x, y) of a lamina is $\frac{\sigma}{a}(x + y)$, where σ and a are constants. The lamina is bounded by the lines $x = 0$, $y = 0$, $x = a$, $y = b$. Find the position of its centre of gravity.
   (*ii*) Find the position of the C.G. of a semicircular lamina of radius a if its density varies as the square of the distance from the diameter. (*A.K.T.U. 2022*)
8. Find the centre of gravity of the area in the first quadrant lying between the curves $y^2 = x^3$ and $y = x$.
9. Find the centre of gravity of the area bounded by the parabola $y^2 = x$ and the line $x + y = 2$.
10. Find the centre of gravity of the area between $y = 6x - x^2$ and $y = x$.
11. Find the centre of gravity of the area of the cardioid $r = a(1 + \cos \theta)$.
12. Find the centroid of the tetrahedron bounded by the co-ordinate planes and the plane $x + y + z = 1$, the density at any point varying as its distance from the face $z = 0$.
13. Find the centroid of the region R bounded by the parabolic cylinder $z = 4 - x^2$, and the planes $x = 0, y = 0, y = 6, z = 0$ $(x > 0)$.
14. Find the centre of gravity of the volume cut off from the cylinder $x^2 + y^2 - 2ax = 0$ by the planes $z = mx$ and $z = nx$.
15. An ellipse $\left(\frac{x}{a}\right)^2 + \left(\frac{y}{b}\right)^2 = 1$ is immersed vertically in a fluid with its major axis horizontal. If its centre be at a depth h, find the depth of its centre of pressure.
16. Find the depth of the centre of pressure of a square lamina immersed in the liquid, with one vertex in the surface and the diagonal vertical.
17. A horizontal boiler has a flat bottom and its ends are plane and semi-circular. If it is just full of water, show that the depth of the centre of pressure of either end is $0.7 \times$ total depth approximately.

18. A semi-circular lamina of radius a is immersed vertically in a liquid with its highest point in the surface and the bounding diameter horizontal. Find the depth of its centre of pressure below the surface of the liquid.

Answers

1. $\dfrac{\mu^2 a^2 b^2}{2}$ 2. $\dfrac{23\mu}{105}$ 3. $\dfrac{\mu\pi a^4}{16}$

4. $\dfrac{abc}{2}(a+b+c)$ 5. $\dfrac{1}{3}\mu a^3 h^2$ 6. $\left(\dfrac{4}{3\pi}, \dfrac{4a}{3\pi}\right)$

7. (i) $\left[\dfrac{a(4a+3b)}{6(a+b)}, \dfrac{b(3a+4b)}{6(a+b)}\right]$ (ii) $\dfrac{32a}{15\pi}$ 8. $\left(\dfrac{10}{21}, \dfrac{5}{12}\right)$

9. $\left(\dfrac{8}{5}, -\dfrac{1}{2}\right)$ 10. $\left(\dfrac{5}{2}, 5\right)$ 11. $\left(\dfrac{5a}{6}, 0\right)$

12. $\left(\dfrac{1}{5}, \dfrac{1}{5}, \dfrac{2}{5}\right)$ 13. $\left(\dfrac{3}{4}, 3, \dfrac{8}{5}\right)$ 14. $\left[\dfrac{5a}{4}, 0, \dfrac{5a}{8}(m+n)\right]$

15. $h + \dfrac{b^2}{4h}$ 16. Divides the diagonal in the ratio 7 : 5

18. $\dfrac{15\pi - 32}{4(3\pi - 4)} a.$

4.13. GAMMA FUNCTION [A.K.T.U. 2018]

If n is positive, then the definite integral $\int_0^\infty e^{-x} x^{n-1}\, dx$, which is a function of n, is called the Gamma function (or Eulerian integral of second kind) and is denoted by $\Gamma(n)$. Thus

$$\boxed{\Gamma(n) = \int_0^\infty e^{-x} x^{n-1}\, dx,\ n > 0.}$$

In particular, $\Gamma(1) = \int_0^\infty e^{-x}\, dx = \Big| -e^{-x} \Big|_0^\infty = 1.$

REDUCTION FORMULA FOR $\Gamma(n)$

$$\Gamma(n+1) = \int_0^\infty e^{-x} x^n\, dx$$

Integrating by parts, we have

$$\Gamma(n+1) = \Big| -x^n e^{-x} \Big|_0^\infty + n\int_0^\infty x^{n-1} e^{-x}\, dx = n\int_0^\infty e^{-x} x^{n-1}\, dx \qquad \left[\because \lim_{x\to\infty} \frac{x^n}{e^x} = 0\right]$$

$$= n\Gamma(n)$$

$\therefore$ $\boxed{\Gamma(n+1) = n\Gamma(n)}$ which is the reduction or recurrence formula for $\Gamma(n)$.

Note 1. If n is a positive integer, then by repeated application of above formula, we get

$$\begin{aligned}\Gamma(n+1) &= n\Gamma(n)\\ &= n(n-1)\,\Gamma(n-1)\\ &= n(n-1)(n-2)\,\Gamma(n-2)\\ &\dots\dots\dots\\ &= n(n-1)(n-2)\dots\dots 3\cdot 2\cdot 1\,\Gamma(1)\\ &= n!,\quad \text{since } \Gamma(1)=1\end{aligned}$$

Hence $\boxed{\Gamma(n+1) = n!}$ when n is a positive integer.

Note 2. If n is a positive fraction, then by repeated application of above formula, we get

$$\Gamma(n) = (n-1)(n-2) \times \text{go on decreasing by 1 ......}$$

the series of factors being continued so long as the factors remain positive, multiplied by Γ (last factor).

Thus $\quad \Gamma\left(\frac{11}{4}\right) = \frac{7}{4}\Gamma\left(\frac{7}{4}\right) = \frac{7}{4}\cdot\frac{3}{4}\Gamma\left(\frac{3}{4}\right)$

The value of $\Gamma\left(\frac{3}{4}\right)$ can be obtained from the table of gamma functions.

Note 3. $\quad \Gamma(n+1) = n\,\Gamma(n)$

$$\Rightarrow \quad \Gamma(n) = \frac{\Gamma(n+1)}{n},\ n \neq 0$$

$$= \frac{(n+1)\,\Gamma(n+1)}{n(n+1)} = \frac{\Gamma(n+2)}{n(n+1)},\ n \neq 0, -1$$

$$= \frac{(n+2)\,\Gamma(n+2)}{n(n+1)(n+2)} = \frac{\Gamma(n+3)}{n(n+1)(n+2)},\ n \neq 0, -1, -2$$

$$\dots\dots\dots\dots$$

$$= \frac{\Gamma(n+k+1)}{n(n+1)(n+2)\dots(n+k)},\ n \neq 0, -1, -2\dots, -k$$

This result defines $\Gamma(n)$ for $n < 0$, k being the least positive integer such that $n + k + 1 > 0$.

For example, to evaluate $\Gamma(-3.4)$

$$n + k + 1 > 0 \ \Rightarrow\ -3.4 + k + 1 > 0 \ \Rightarrow\ k > 2.4$$

We choose $k = 3$

$$\therefore \quad \Gamma(-3.4) = \frac{\Gamma(-3.4+3+1)}{(-3.4)(-2.4)(-1.4)(-.4)} = \frac{\Gamma(.6)}{(3.4)(2.4)(1.4)(.4)}$$

The value of $\Gamma(.6)$ can be obtained from the table of gamma functions.

Also we observe that $\Gamma(n)$ is infinite when $n = 0$ or a negative integer.

4.14.1. Value of $\Gamma\left(\frac{1}{2}\right)$

$$\Gamma\left(\frac{1}{2}\right) = \int_0^\infty e^{-t}\, t^{-1/2}\, dt$$

Putting $t = x^2$ so that $dt = 2xdx$

$$\Gamma\left(\frac{1}{2}\right) = \int_0^\infty e^{-x^2}\cdot\frac{1}{x}\cdot 2xdx = 2\int_0^\infty e^{-x^2}\,dx \qquad \dots(1)$$

Writing y for x, we have $\Gamma\left(\frac{1}{2}\right) = 2\int_0^\infty e^{-y^2}\, dy$ ...(2)

Now we use the following result from double integrals:

If $f(x)$ and $g(y)$ are functions of x and y only, and the limits of integration are constants, then the double integral can be represented as a product of two integrals. Thus

$$\int_c^d \int_a^b f(x)\, g(y)\, dx\, dy = \int_a^b f(x)\, dx \cdot \int_c^d g(y)\, dy$$

$\therefore$ From (1) and (2), we have

$$\left[\Gamma\left(\frac{1}{2}\right)\right]^2 = 4\int_0^\infty e^{-x^2}\, dx. \int_0^\infty e^{-y^2}\, dy = 4\int_0^\infty \int_0^\infty e^{-(x^2+y^2)}\, dx\, dy$$

Changing to polar co-ordinates with $x = r\cos\theta$, $y = r\sin\theta$, $dx\, dy = r\, dr\, d\theta$; the region of integration in this integral is the complete positive quadrant, to cover which, r must vary from 0 to ∞ and θ from 0 to $\frac{\pi}{2}$.

$$\therefore \quad \left[\Gamma\left(\frac{1}{2}\right)\right]^2 = 4\int_0^{\pi/2}\int_0^\infty e^{-r^2}\, r\, dr\, d\theta = 4\int_0^{\pi/2}\left[-\frac{1}{2}e^{-r^2}\right]_0^\infty d\theta = 2\int_0^{\pi/2} d\theta = \pi$$

Hence, $\boxed{\Gamma\left(\frac{1}{2}\right) = \sqrt{\pi}.}$

4.15. TRANSFORMATIONS OF GAMMA FUNCTIONS

(1) $\boxed{\Gamma(n) = k^n \int_0^\infty e^{-kx}\, x^{n-1}\, dx}$ [A.K.T.U. 2016]

Proof. We have, $\Gamma(n) = \int_0^\infty e^{-x}\, x^{n-1}\, dx$

$$= \int_0^\infty e^{-ky}\, k^{n-1}\, y^{n-1}\, k\, dy = k^n \int_0^\infty e^{-ky}\, y^{n-1}\, dy \qquad \left|\begin{array}{l} \text{Put } x = ky \\ \therefore\ dx = k\, dy \end{array}\right.$$

$$\Rightarrow \quad \Gamma(n) = k^n \int_0^\infty e^{-kx}\, x^{n-1}\, dx$$

(2) $\boxed{\Gamma(n) = \int_0^1 \left[\log\left(\frac{1}{x}\right)\right]^{n-1} dx\ ; \quad n > 0}$

Proof. We have, $\Gamma(n) = \int_0^\infty e^{-x} \cdot x^{n-1}\, dx$

Put $e^{-x} = y$

$\therefore$ $-x = \log y \Rightarrow x = \log\left(\frac{1}{y}\right)$ and $dx = -\frac{dy}{y}$

$$\therefore \quad \Gamma(n) = -\int_1^0 y\left(\log\frac{1}{y}\right)^{n-1}\frac{dy}{y} = \int_0^1 \left(\log\frac{1}{y}\right)^{n-1} dy$$

Hence, $\Gamma(n) = \int_0^1 \left[\log\left(\frac{1}{x}\right)\right]^{n-1} dx.$

(3) $$\boxed{\Gamma(n) = \frac{1}{n}\int_0^\infty e^{-x^{1/n}}\, dx}$$

Proof. We have $\Gamma(n) = \int_0^\infty e^{-x} \cdot x^{n-1}\, dx$

Put $x^n = y \quad \Rightarrow \quad x = y^{1/n}$

$\therefore \quad dx = \frac{1}{n} y^{(1/n)-1} dy$

Now, $$\Gamma(n) = \int_0^\infty e^{-y^{1/n}}\, y^{\frac{n-1}{n}} \cdot \frac{1}{n} y^{\frac{1}{n}-1}\, dy = \frac{1}{n}\int_0^\infty e^{-y^{1/n}}\, dy$$

or $$\Gamma(n) = \frac{1}{n}\int_0^\infty e^{-x^{1/n}}\, dx.$$

4.15.1. Deduction. Put $n = \frac{1}{2}$ in (3), we get

$$\Gamma(1/2) = 2\int_0^\infty e^{-x^2} dx = 2\left(\frac{\sqrt{\pi}}{2}\right) = \sqrt{\pi}.$$

4.16. BETA FUNCTION [*A.K.T.U.2018*]

If m, n are positive, then the definite integral $\int_0^1 x^{m-1}(1-x)^{n-1}\, dx$, which is a function of m and n, is called the Beta Function (or Eulerian integral of first kind) and is denoted by $\beta(m, n)$. Thus,

$$\boxed{\beta(m, n) = \int_0^1 x^{m-1}(1-x)^{n-1} dx,\ m > 0,\ n > 0.}$$

4.17. SYMMETRY OF BETA FUNCTION i.e., $\beta(m, n) = \beta(n, m)$

$$\beta(m, n) = \int_0^1 x^{m-1}(1-x)^{n-1} dx,\ m > 0,\ n > 0$$

Since $\int_0^a f(x)\, dx = \int_0^a f(a-x)\, dx$

$$\therefore \quad \beta(m, n) = \int_0^1 (1-x)^{m-1}\,[1-(1-x)]^{n-1}\, dx = \int_0^1 x^{n-1}\,(1-x)^{m-1}\, dx = \beta(n, m)$$

Hence, $\boxed{\beta(m, n) = \beta(n, m).}$

4.18. TRANSFORMATIONS OF BETA FUNCTION

(1) $$\boxed{\beta(m, n) = \int_0^\infty \frac{x^{m-1}}{(1+x)^{m+n}}\, dx}$$

Proof. $\beta(m, n) = \int_0^1 x^{m-1} (1-x)^{n-1}\, dx$ | Put $x = \frac{1}{1+y}$ $\therefore$ $dx = -\frac{1}{(1+y)^2}\, dy$

$$= \int_\infty^0 \frac{1}{(1+y)^{m-1}} \left(1 - \frac{1}{1+y}\right)^{n-1} \left\{\frac{-1}{(1+y)^2}\right\} dy$$

$$= \int_0^\infty \frac{y^{n-1}}{(1+y)^{m-1} (1+y)^{n-1} (1+y)^2}\, dy = \int_0^\infty \frac{y^{n-1}}{(1+y)^{m+n}}\, dy$$

$$\beta(m, n) = \int_0^\infty \frac{x^{n-1}}{(1+x)^{m+n}}\, dx$$

$\therefore$ $$\beta(n, m) = \int_0^\infty \frac{x^{m-1}}{(1+x)^{m+n}}\, dx$$

But $\beta(m, n) = \beta(n, m)$

$\therefore$ $$\beta(m, n) = \int_0^\infty \frac{x^{m-1}}{(1+x)^{m+n}}\, dx$$

(2) $$\boxed{\beta(m, n) = 2 \int_0^{\pi/2} \sin^{2m-1}\theta \cos^{2n-1}\theta\, d\theta}$$

Proof. $\beta(m, n) = \int_0^1 x^{m-1} (1-x)^{n-1}\, dx$ | Put $x = \sin^2\theta$ $\therefore$ $dx = 2\sin\theta\cos\theta\, d\theta$

$$= \int_0^{\pi/2} \sin^{2m-2}\theta\, (1-\sin^2\theta)^{n-1} \cdot 2\sin\theta\cos\theta\, d\theta$$

$$= 2\int_0^{\pi/2} \sin^{2m-1}\theta \cos^{2n-1}\theta\, d\theta.$$

4.19. RELATION BETWEEN BETA *AND* GAMMA FUNCTIONS

$$\boldsymbol{\beta(m, n) = \frac{\Gamma(m)\Gamma(n)}{\Gamma(m+n)}}$$ *[A.K.T.U. 2018, 2019, 2022]*

Proof. From (1) of Art. 4.15, we have

$$\Gamma(n) = k^n \int_0^\infty e^{-kx} x^{n-1}\, dx$$

$$= z^n \int_0^\infty e^{-zx} x^{n-1}\, dx \qquad \text{(Replace } k \text{ by } z\text{)}$$

Multiplying both sides by $e^{-z} z^{m-1}$, we get

$$\Gamma(n) \cdot e^{-z} z^{m-1} = \int_0^\infty z^n \cdot e^{-zx} \cdot x^{n-1} \cdot e^{-z} \cdot z^{m-1}\, dx = \int_0^\infty z^{n+m-1} e^{-z(1+x)} x^{n-1}\, dx$$

Integrating both sides w.r.t. z from 0 to ∞, we get

$$\Gamma(n) \int_0^\infty e^{-z} z^{m-1}\, dz = \int_0^\infty x^{n-1} \left\{ \int_0^\infty e^{-z(1+x)}\, z^{m+n-1}\, dz \right\} dx$$

$$\Rightarrow \qquad \Gamma n\, \Gamma m = \int_0^\infty x^{n-1} \left\{ \int_0^\infty e^{-y} \cdot \frac{y^{m+n-1}}{(1+x)^{m+n-1}} \frac{dy}{(1+x)} \right\} dx$$

where $z(1 + x) = y$ so that $dz = \dfrac{dy}{1+x}$

$$= \int_0^\infty \frac{x^{n-1}}{(1+x)^{m+n}} \left\{ \int_0^\infty e^{-y}\, y^{m+n-1}\, dy \right\} dx$$

$$= \int_0^\infty \frac{x^{n-1}}{(1+x)^{m+n}}\, dx\, \Gamma(m+n) = \Gamma(m+n)\, \beta(m, n)$$

$$\therefore \qquad \boxed{\beta(m, n) = \frac{\Gamma(m)\Gamma(n)}{\Gamma(m+n)}}$$

Aliter:

We know that $\qquad \Gamma(m) = \int_0^\infty e^{-t}\, t^{m-1}\, dt$

Putting $t = x^2$ so that $dt = 2x\, dx$

$$\Gamma(m) = 2 \int_0^\infty e^{-x^2} x^{2m-1}\, dx \qquad ...(1)$$

Similarly, $\qquad \Gamma(n) = 2 \int_0^\infty e^{-y^2}\, y^{2n-1}\, dy$

$$\Gamma(m)\, \Gamma(n) = 4 \int_0^\infty e^{-x^2} x^{2m-1}\, dx \cdot \int_0^\infty e^{-y^2}\, y^{2n-1}\, dy$$

$$= 4 \int_0^\infty \int_0^\infty e^{-(x^2+y^2)} x^{2m-1}\, y^{2n-1}\, dx\, dy$$

Changing to polar co-ordinates, we have

$$\Gamma(m)\, \Gamma(n) = 4 \int_0^{\pi/2} \int_0^\infty e^{-r^2}\, r^{2(m+n)-1} \cos^{2m-1}\theta \sin^{2n-1}\theta\, dr\, d\theta$$

$$= 4 \int_0^\infty e^{-r^2}\, r^{2(m+n)-1}\, dr \cdot \int_0^{\pi/2} \cos^{2m-1}\theta \sin^{2n-1}\theta\, d\theta \qquad ...(2)$$

$$= \left[2 \int_0^\infty e^{-r^2}\, r^{2(m+n)-1}\, dr \right] \cdot \left[2 \int_0^{\pi/2} \cos^{2m-1}\theta \sin^{2n-1}\theta\, d\theta \right]$$

$$= \Gamma(m+n)\, \beta(m, n)$$ | Using (2) of 4.18

Hence, $\qquad \boxed{\beta(m, n) = \frac{\Gamma(m)\Gamma(n)}{\Gamma(m+n)}}$

4.20. TO EVALUATE $\int_0^{\pi/2} \sin^m\theta \cos^n\theta \, d\theta$ *(U.K.T.U. 2011)*

From (2) of Art. 4.18, we have

$$\int_0^{\pi/2} \sin^{2m-1}\theta \cos^{2n-1}\theta \, d\theta = \frac{1}{2}\beta(m, n) \quad \text{...(1)}$$

Using the relation $\beta(m, n) = \dfrac{\Gamma(m)\,\Gamma(n)}{\Gamma(m+n)}$, we get from (1)

$$\int_0^{\pi/2} \sin^{2m-1}\theta \cos^{2n-1}\theta \, d\theta = \frac{\Gamma(m)\,\Gamma(n)}{2\,\Gamma(m+n)} \quad \text{...(2)}$$

Replacing m by $\dfrac{m+1}{2}$ and n *by* $\dfrac{n+1}{2}$ in (2), we get

$$\boxed{\int_0^{\pi/2} \sin^m\theta \cos^n\theta \, d\theta = \frac{\Gamma\left(\frac{m+1}{2}\right)\Gamma\left(\frac{n+1}{2}\right)}{2\Gamma\left(\frac{m+n+2}{2}\right)}}$$

Cor. 1 Putting $m = n = 0$, we have

$$\frac{(\Gamma\frac{1}{2})^2}{2\,\Gamma(1)} = \int_0^{\pi/2} d\theta = \frac{\pi}{2} \quad \Rightarrow \quad \Gamma(\tfrac{1}{2}) = \sqrt{\pi} \qquad | \text{ Since } \Gamma 1 = 1$$

Cor. 2 Putting $n = 0$, we get

$$\int_0^{\pi/2} \sin^m\theta \, d\theta = \frac{\Gamma\left(\frac{m+1}{2}\right)}{\Gamma\left(\frac{m+2}{2}\right)} \cdot \frac{\sqrt{\pi}}{2} \qquad \Big| \because \Gamma(\tfrac{1}{2}) = \sqrt{\pi}$$

and similarly, putting $m = 0$, we get

$$\int_0^{\pi/2} \cos^n\theta \, d\theta = \frac{\Gamma\left(\frac{n+1}{2}\right)}{\Gamma\left(\frac{n+2}{2}\right)} \cdot \frac{\sqrt{\pi}}{2}$$

4.20.1. Deductions

Using $\displaystyle\int_0^\infty \frac{x^{n-1}}{1+x}\,dx = \frac{\pi}{\sin n\pi}$, where $0 < n < 1$, we can deduce the following important results.

(1) $$\boxed{\Gamma n \Gamma(1-n) = \frac{\pi}{\sin n\pi}}$$ *(G.B.T.U. 2010)*

Proof. We have,

$$\beta(m, n) = \frac{\Gamma(m)\,\Gamma(n)}{\Gamma(m+n)}$$

$$\Rightarrow \qquad \int_0^\infty \frac{x^{n-1}}{(1+x)^{m+n}}\,dx = \frac{\Gamma(m)\,\Gamma(n)}{\Gamma(m+n)}$$

Setting $m + n = 1$ so that $m = 1 - n$, we get

$$\int_0^\infty \frac{x^{n-1}}{1+x}\,dx = \frac{\Gamma(1-n)\,\Gamma(n)}{\Gamma(1)}$$

$$\Rightarrow \qquad \frac{\pi}{\sin n\pi} = \Gamma(n)\,\Gamma(1-n) \qquad | \because \Gamma(1) = 1$$

(2) $\boxed{\Gamma(\tfrac{1}{2}) = \sqrt{\pi}}$

Proof. We have

$$\Gamma(n)\,\Gamma(1-n) = \frac{\pi}{\sin n\pi}$$

As a particular case, put $n = \frac{1}{2}$, we get

$$\Gamma(\tfrac{1}{2})\,\Gamma(\tfrac{1}{2}) = \frac{\pi}{\sin\frac{\pi}{2}}$$

$$[\Gamma(\tfrac{1}{2})]^2 = \pi \qquad \Rightarrow \qquad \Gamma(\tfrac{1}{2}) = \sqrt{\pi}$$

(3) $\boxed{\Gamma(\tfrac{1}{4})\,\Gamma(\tfrac{3}{4}) = \pi\sqrt{2}}$

Proof. Putting $n = \frac{1}{4}$ in result (1), we obtain

$$\Gamma(\tfrac{1}{4})\,\Gamma(1-\tfrac{1}{4}) = \frac{\pi}{\sin\frac{\pi}{4}}$$

$$\Rightarrow \qquad \Gamma(\tfrac{1}{4})\,\Gamma(\tfrac{3}{4}) = \frac{\pi}{\left(\frac{1}{\sqrt{2}}\right)} \qquad \text{or} \qquad \Gamma(\tfrac{1}{4})\,\Gamma(\tfrac{3}{4}) = \pi\sqrt{2}$$

Note. Similarly $\Gamma(1/3)\,\Gamma(2/3) = \dfrac{\pi}{\sin \pi/3} = \dfrac{2\pi}{\sqrt{3}}$.

4.21. DUPLICATION FORMULA

$\Gamma(m)\,\Gamma\left(m + \dfrac{1}{2}\right) = \dfrac{\sqrt{\pi}}{(2)^{2m-1}}\,\Gamma(2m)$ **where *m* is positive.** *(A.K..T.U. 2022)*

We have already established

$$\int_0^{\pi/2} \sin^{2m-1}\theta\,\cos^{2n-1}\theta\,d\theta = \frac{\Gamma(m)\,\Gamma(n)}{2\,\Gamma(m+n)} \qquad \text{...(1)}$$

Putting $2n - 1 = 0$ or $n = \dfrac{1}{2}$ in (1), we obtain

$$\int_0^{\pi/2} \sin^{2m-1}\theta\,d\theta = \frac{\Gamma(m)\sqrt{\pi}}{2\,\Gamma\left(m+\frac{1}{2}\right)} \qquad \text{...(2)} \left[\because \Gamma(1/2) = \sqrt{\pi}\right]$$

Again putting $n = m$ in equation (1), we obtain

$$\int_0^{\pi/2} \sin^{2m-1}\theta\,\cos^{2m-1}\theta\,d\theta = \frac{(\Gamma m)^2}{2\,\Gamma(2m)}$$

i.e., $$\frac{1}{2^{2m-1}}\int_0^{\pi/2}(2\sin\theta\cos\theta)^{2m-1}\,d\theta=\frac{(\Gamma m)^2}{2\,\Gamma(2m)}$$

i.e., $$\frac{1}{2^{2m}}\int_0^{\pi/2}(\sin 2\theta)^{2m-1}\cdot 2d\theta=\frac{(\Gamma m)^2}{2\,\Gamma(2m)}$$

Putting $2\theta=\phi$ so that $2\,d\theta=d\phi$, this reduces to

$$\frac{1}{2^{2m}}\int_0^{\pi}\sin^{2m-1}\phi\,d\phi=\frac{(\Gamma\,m)^2}{2\,\Gamma\,(2m)}$$

i.e., $$\frac{2}{2^{2m}}\int_0^{\pi/2}\sin^{2m-1}\phi\,d\phi=\frac{(\Gamma m)^2}{2\,\Gamma(2m)}$$

Replacing ϕ by θ, we finally obtain

$$\int_0^{\pi/2}\sin^{2m-1}\theta\,d\theta=\frac{2^{2m-1}\,(\Gamma m)^2}{2\,\Gamma(2m)} \qquad ...(3)$$

From (2) and (3), we get

$$\frac{\Gamma(m)\sqrt{\pi}}{2\,\Gamma\left(m+\frac{1}{2}\right)}=\frac{2^{2m-1}\,(\Gamma m)^2}{2\,\Gamma(2m)}$$

$$\Rightarrow \qquad \boxed{\Gamma(m)\,\Gamma\left(m+\frac{1}{2}\right)=\frac{\sqrt{\pi}}{2^{2m-1}}\,\Gamma(2m)}$$

4.22. TO SHOW THAT

$$\Gamma\left(\frac{1}{n}\right)\Gamma\left(\frac{2}{n}\right)\Gamma\left(\frac{3}{n}\right)...\,\Gamma\left(\frac{n-1}{n}\right)=\frac{(2\pi)^{\left(\frac{n-1}{2}\right)}}{n^{1/2}}$$

where n is a positive integer greater than one.

Let $$P=\Gamma\left(\frac{1}{n}\right)\Gamma\left(\frac{2}{n}\right)\Gamma\left(\frac{3}{n}\right)...\,\Gamma\left(\frac{n-2}{n}\right)\Gamma\left(\frac{n-1}{n}\right)$$

$$=\Gamma\left(\frac{1}{n}\right)\Gamma\left(\frac{2}{n}\right)\Gamma\left(\frac{3}{n}\right)...\,\Gamma\left(1-\frac{2}{n}\right)\Gamma\left(1-\frac{1}{n}\right) \qquad ...(1)$$

Writing the value of P in the reverse order, we have

$$P=\Gamma\left(1-\frac{1}{n}\right)\Gamma\left(1-\frac{2}{n}\right)...\,\Gamma\left(\frac{3}{n}\right)\Gamma\left(\frac{2}{n}\right)\Gamma\left(\frac{1}{n}\right) \qquad ...(2)$$

Multiplying (1) and (2), we get

$$P^2=\left(\Gamma(1/n)\,\Gamma\left(1-\frac{1}{n}\right)\right)\left(\Gamma(2/n)\,\Gamma\left(1-\frac{2}{n}\right)\right)...\left(\Gamma\left(1-\frac{2}{n}\right)\Gamma\,(2/n)\right)\left(\Gamma\left(1-\frac{1}{n}\right)\Gamma\,(1/n)\right)$$

$$P^2=\frac{\pi}{\sin\left(\frac{\pi}{n}\right)}\cdot\frac{\pi}{\sin\left(\frac{2\pi}{n}\right)}\cdot\frac{\pi}{\sin\left(\frac{3\pi}{n}\right)}...\frac{\pi}{\sin\frac{(n-1)\,\pi}{n}} \qquad \because\ \Gamma(n)\,\Gamma(1-n)=\frac{\pi}{\sin n\pi}$$

$$\Rightarrow \qquad P^2=\frac{\pi^{n-1}}{\sin\left(\frac{\pi}{n}\right)\sin\left(\frac{2\pi}{n}\right)\sin\left(\frac{3\pi}{n}\right)...\sin\left\{\frac{(n-1)\,\pi}{n}\right\}} \qquad ...(3)$$

But from Trigonometry, we know that

$$\frac{\sin n\theta}{\sin\theta} = 2^{n-1}\sin\left(\theta+\frac{\pi}{n}\right)\sin\left(\theta+\frac{2\pi}{n}\right)\ldots\sin\left\{\theta+\frac{(n-1)\pi}{n}\right\} \quad \ldots(4)$$

Take Limit as $\theta \to 0$,

$$\operatorname*{Lt}_{\theta\to 0}\frac{\sin n\theta}{\sin\theta} = \operatorname*{Lt}_{\theta\to 0}\left(n\cdot\frac{\sin n\theta}{n\theta}\cdot\frac{\theta}{\sin\theta}\right) = n$$

$$\therefore \quad n = 2^{n-1}\sin\frac{\pi}{n}\sin\frac{2\pi}{n}\sin\frac{3\pi}{n}\ldots\sin\frac{(n-1)\pi}{n} \quad | \text{ From (4)}$$

Substituting this in equation (3), we obtain

$$P^2 = \frac{\pi^{n-1}}{\left(\dfrac{n}{2^{n-1}}\right)} = \frac{(2\pi)^{n-1}}{n}$$

$$\therefore \quad P = \frac{(2\pi)^{(n-1)/2}}{n^{1/2}}$$

or

$$\boxed{\Gamma\left(\frac{1}{n}\right)\Gamma\left(\frac{2}{n}\right)\Gamma\left(\frac{3}{n}\right)\ldots\Gamma\left(\frac{n-1}{n}\right) = \frac{(2\pi)^{(n-1)/2}}{n^{1/2}}}$$

4.23. TO SHOW THAT

(*i*) $\displaystyle\int_0^\infty e^{-ax}x^{n-1}\cos bx\,dx = \frac{\Gamma(n)\cos n\theta}{(a^2+b^2)^{n/2}}$

(*ii*) $\displaystyle\int_0^\infty e^{-ax}x^{n-1}\sin bx\,dx = \frac{\Gamma(n)\sin n\theta}{(a^2+b^2)^{n/2}}$, **where** $\theta = \tan^{-1}\left(\dfrac{b}{a}\right)$

We know that $\displaystyle\int_0^\infty e^{-ax}\cdot x^{n-1}\,dx = \frac{\Gamma(n)}{a^n}$, where a, n are (+)ve.

Put $ax = z$ so that $dx = \dfrac{dz}{a}$

$$\therefore \quad \int_0^\infty e^{-ax}x^{n-1}dx = \int_0^\infty e^{-z}\left(\frac{z}{a}\right)^{n-1}\cdot\frac{dz}{a} = \frac{1}{a^n}\int_0^\infty e^{-z}z^{n-1}\,dz = \frac{\Gamma(n)}{a^n}.$$

4.23.1. Deduction

Replacing a by $(a+ib)$, we have

$$\int_0^\infty e^{-(a+ib)x}x^{n-1}\,dx = \frac{\Gamma(n)}{(a+ib)^n} \quad \ldots(1)$$

Now $e^{-(a+ib)x} = e^{-ax}\cdot e^{-ibx} = e^{-ax}(\cos bx - i\sin bx)$

Putting $a = r\cos\theta$ and $b = r\sin\theta$ so that $r^2 = a^2+b^2$ and $\theta = \tan^{-1}\dfrac{b}{a}$

$$(a+ib)^n = (r\cos\theta + ir\sin\theta)^n = r^n(\cos\theta + i\sin\theta)^n$$
$$= r^n(\cos n\theta + i\sin n\theta) \quad \text{[De Moivre's Theorem]}$$

∴ From (1), we have

$$\int_0^\infty e^{-ax}(\cos bx - i\sin bx)\, x^{n-1}\, dx = \frac{\Gamma(n)}{r^n(\cos n\theta + i\sin n\theta)}$$

$$= \frac{\Gamma(n)}{r^n}(\cos n\theta + i\sin n\theta)^{-1} = \frac{\Gamma(n)}{r^n}(\cos n\theta - i\sin n\theta)$$

Now equating real and imaginary parts on the two sides, we get

$$\int_0^\infty e^{-ax}\, x^{n-1}\cos bx\, dx = \frac{\Gamma(n)}{r^n}\cos n\theta$$

and $$\int_0^\infty e^{-ax}\, x^{n-1}\sin bx\, dx = \frac{\Gamma(n)}{r^n}\sin n\theta$$

where $r^2 = a^2 + b^2$ and $\theta = \tan^{-1}\dfrac{b}{a}$.

ILLUSTRATIVE EXAMPLES

Example 1. *Using Beta and Gamma functions, evaluate:*

(*i*) $\int_0^\infty x^{1/4}\, e^{-\sqrt{x}}\, dx$ (*ii*) $\int_0^1 \left(\frac{x^3}{1-x^3}\right)^{1/2} dx$ (A.K.T.U. 2014, 2018)

(*iii*) $\int_0^1 x^5(1-x^3)^{10}\, dx$

Sol. (*i*) Let $$I = \int_0^\infty x^{1/4}\, e^{-\sqrt{x}}\, dx \qquad ...(1)$$

Put $\sqrt{x} = y \Rightarrow x = y^2$ so that $dx = 2y\, dy$ then equation (1) becomes

$$I = \int_0^\infty y^{1/2} e^{-y}\cdot 2y\, dy = 2\int_0^\infty e^{-y}\, y^{3/2}\, dy$$

$$= 2\int_0^\infty e^{-y}\, y^{(5/2)-1} dy = 2\,\Gamma(5/2) \qquad \text{| By definition}$$

$$= 2\cdot\frac{3}{2}\cdot\frac{1}{2}\sqrt{\pi} = \frac{3}{2}\sqrt{\pi} \qquad |\ \because\ \Gamma(n+1) = n\,\Gamma(n)$$

(*ii*) Let $$I = \int_0^1 x^{3/2}(1-x^3)^{-1/2}\, dx \qquad ...(1)$$

Put $x^3 = y \Rightarrow x = y^{1/3}$ so that $dx = \frac{1}{3}y^{-2/3}\, dy$ then equation (1) becomes

$$I = \int_0^1 y^{1/2}(1-y)^{-1/2}\cdot\frac{1}{3}y^{-2/3}\, dy = \frac{1}{3}\int_0^1 y^{-1/6}(1-y)^{-1/2} dy$$

$$= \frac{1}{3}\int_0^1 y^{\left(\frac{5}{6}\right)-1}(1-y)^{\left(\frac{1}{2}\right)-1} dy = \frac{1}{3}\beta\left(\frac{5}{6}, \frac{1}{2}\right)$$

$$= \frac{1}{3}\frac{\Gamma(5/6)\,\Gamma(1/2)}{\Gamma(4/3)} \qquad \left|\ \because\ \beta(m,n) = \frac{\Gamma(m)\,\Gamma(n)}{\Gamma(m+n)}\right.$$

$$= \frac{\sqrt{\pi}}{3} \cdot \frac{\Gamma(5/6)}{\frac{1}{3}\Gamma(1/3)} = \sqrt{\pi} \cdot \frac{\Gamma(5/6)\,\Gamma(1/6)\,\Gamma(2/3)}{\Gamma(1/6)\,\Gamma(1/3)\,\Gamma(2/3)} \qquad \left| \because \Gamma(n+1) = n\Gamma n \right.$$

$$= \sqrt{\pi} \cdot \frac{\Gamma(2/3)}{\Gamma(1/6)} \cdot \frac{\pi}{\sin\frac{\pi}{6}} \cdot \frac{\sin\frac{\pi}{3}}{\pi} \qquad \left| \because \Gamma n\, \Gamma(1-n) = \frac{\pi}{\sin n\pi} \right.$$

$$= \sqrt{3\pi}\, \frac{\Gamma(2/3)}{\Gamma(1/6)}$$

(iii) Let $\qquad I = \int_0^1 x^5 (1-x^3)^{10}\, dx \qquad$ ...(1)

Put $x^3 = y \Rightarrow x = y^{1/3}$ so that $dx = \frac{1}{3} y^{-2/3} dy$ then equation (1) becomes

$$I = \int_0^1 y^{5/3} (1-y)^{10} \cdot \frac{1}{3} y^{-2/3} dy$$

$$= \frac{1}{3}\int_0^1 y(1-y)^{10} dy = \frac{1}{3}\beta(2, 11) = \frac{1}{3}\frac{\Gamma 2\, \Gamma(11)}{\Gamma(13)} = \frac{1}{3} \cdot \frac{1}{12 \cdot 11} = \frac{1}{396}$$

Example 2. *Prove that:*

(i) $\beta(l, m) \cdot \beta(l+m, n) \cdot \beta(l+m+n, p) = \frac{\Gamma l\, \Gamma m\, \Gamma n\, \Gamma p}{\Gamma(l+m+n+p)}$

(ii) $\int_0^\infty \frac{x^c}{c^x} dx = \frac{\Gamma(c+1)}{(\log c)^{c+1}}\,; c > 1$

(iii) $\beta(m, m) = 2^{1-2m}\, \beta(m, 1/2)$

(iv) $\int_0^{\pi/2} \tan^n x\, dx = \frac{\pi}{2} \sec \frac{n\pi}{2}$

Sol. *(i)* $\qquad$ LHS $= \beta(l, m) \cdot \beta(l+m, n) \cdot \beta(l+m+n, p)$

$$= \frac{\Gamma l\, \Gamma m}{\Gamma(l+m)} \cdot \frac{\Gamma(l+m)\,.\,\Gamma n}{\Gamma(l+m+n)} \cdot \frac{\Gamma l+m+n\, \Gamma p}{\Gamma(l+m+n+p)}$$

$$= \frac{\Gamma l\, \Gamma m\, \Gamma n\, \Gamma p}{\Gamma(l+m+n+p)} = \text{RHS}$$

(ii) Let $\qquad I = \int_0^\infty \frac{x^c}{c^x} dx = \int_0^\infty e^{-x \log c}\, x^c\, dx \qquad$ ...(1)

Put $x \log c = y \Rightarrow x = \frac{y}{\log c}$ so that $dx = \frac{dy}{\log c}$ then equation (1) becomes

$$I = \int_0^\infty e^{-y} \left(\frac{y}{\log c}\right)^c \frac{dy}{\log c} = \frac{1}{(\log c)^{c+1}} \int_0^\infty e^{-y}\, y^c\, dy$$

$$= \frac{\Gamma(c+1)}{(\log c)^{c+1}}\,; c > 1$$

(iii) $$\beta(m, 1/2) = \frac{\Gamma m\, \Gamma(1/2)}{\Gamma(m+1/2)} \qquad \Big|\ \because\ \beta(m,n) = \frac{\Gamma(m)\,\Gamma(n)}{\Gamma(m+n)}$$

$$= \frac{(\Gamma m)^2 \sqrt{\pi}}{\Gamma m\, \Gamma\left(m+\frac{1}{2}\right)} = \frac{\Gamma(m)^2 \sqrt{\pi}}{\left(\frac{\sqrt{\pi}}{2^{2m-1}}\right)\Gamma 2m} \qquad \text{| By Duplication formula}$$

$$= 2^{2m-1} \frac{\Gamma m\, \Gamma m}{\Gamma(2m)} = 2^{2m-1}\, \beta(m, m)$$

$\Rightarrow$ $$\beta(m, m) = 2^{1-2m}\, \beta(m, 1/2)$$

(iv) Let $$I = \int_0^{\pi/2} \tan^n x\, dx = \int_0^{\pi/2} \sin^n x \cos^{-n} x\, dx$$

$$= \frac{\Gamma\left(\frac{n+1}{2}\right)\Gamma\left(\frac{-n+1}{2}\right)}{2\,\Gamma\left(\frac{n-n+2}{2}\right)} = \frac{1}{2}\Gamma\left(\frac{n+1}{2}\right)\Gamma\left(1-\left(\frac{n+1}{2}\right)\right)$$

$$= \frac{1}{2} \cdot \frac{\pi}{\sin\left(\frac{n+1}{2}\right)\pi} = \frac{\pi}{2\cos\frac{n\pi}{2}} = \frac{\pi}{2} \sec \frac{n\pi}{2}$$

Example 3. *Evaluate:*

(i) $\int_0^\infty \frac{x^8(1-x^6)}{(1+x)^{24}}\, dx$ (M.T.U. 2013) (ii) $\int_0^\infty \frac{x^4(1+x^5)}{(1+x)^{15}}\, dx$.

Sol. (i) $$I = \int_0^\infty \frac{x^8}{(1+x)^{24}}\, dx - \int_0^\infty \frac{x^{14}}{(1+x)^{24}}\, dx = \int_0^\infty \frac{x^{9-1}}{(1+x)^{9+15}}\, dx - \int_0^\infty \frac{x^{15-1}}{(1+x)^{15+9}}\, dx$$

$$= \beta(9, 15) - \beta(15, 9) = 0 \qquad \Big|\ \because\ \beta(m,n) = \beta(n,m)$$

(ii) $$I = \int_0^\infty \frac{x^4}{(1+x)^{15}}\, dx + \int_0^\infty \frac{x^9}{(1+x)^{15}}\, dx = \int_0^\infty \frac{x^{5-1}}{(1+x)^{5+10}}\, dx + \int_0^\infty \frac{x^{10-1}}{(1+x)^{10+5}}\, dx$$

$$= \beta(5, 10) + \beta(10, 5) = 2\beta(5, 10) = 2\,\frac{\Gamma(5)\,\Gamma(10)}{\Gamma(15)} \qquad \Big|\ \because\ \beta(m,n) = \frac{\Gamma(m)\,\Gamma(n)}{\Gamma(m+n)}$$

$$= \frac{1}{5005}. \qquad \text{| On simplification}$$

Example 4. *Evaluate:*

(i) $\int_0^2 x(8-x^3)^{1/3}\, dx$ (ii) $\int_0^\infty \frac{dx}{1+x^4}$ (iii) $\int_0^1 \frac{dx}{\sqrt{1+x^4}}$ (G.B.T.U. 2011)

(A.K.T.U. 2019)

Sol. (*i*) Putting $x^3 = 8y$ or $x = 2y^{1/3}$ so that $dx = \frac{2}{3}y^{-2/3}\,dy$, we get

$$I = \int_0^1 2y^{1/3}(8-8y)^{1/3} \cdot \frac{2}{3}y^{-2/3}\,dy$$

$$= \frac{8}{3}\int_0^1 y^{-1/3}(1-y)^{1/3}\,dy = \frac{8}{3}\beta\left(\frac{2}{3}, \frac{4}{3}\right)$$

$$= \frac{8}{3}\frac{\Gamma(\frac{2}{3})\,\Gamma(\frac{4}{3})}{\Gamma(2)} = \frac{8}{9}\Gamma(\tfrac{2}{3})\,\Gamma(\tfrac{1}{3}) = \frac{8}{9}\cdot\frac{\pi}{\sin\frac{\pi}{3}} = \frac{16\pi}{9\sqrt{3}}.$$

(*ii*) Putting $x^4 = y$ so that $x = y^{1/4}$ and $dx = \frac{1}{4}y^{-3/4}\,dy$, we get

$$I = \int_0^\infty \frac{\frac{1}{4}y^{-3/4}}{1+y}\,dy = \frac{1}{4}\int_0^\infty \frac{y^{\frac{1}{4}-1}}{1+y}\,dy = \frac{1}{4}\frac{\pi}{\sin\frac{\pi}{4}} = \frac{\pi\sqrt{2}}{4}. \qquad \because \int_0^\infty \frac{x^{n-1}}{1+x}\,dx = \frac{\pi}{\sin n\pi}$$

(*iii*) $I = \int_0^1 \frac{dx}{\sqrt{1+x^4}}$

$$\Rightarrow \quad I = \int_0^{\pi/4} \frac{1}{\sec\theta}\cdot\frac{\sec^2\theta}{2\sqrt{\tan\theta}}\,d\theta$$

Put $x^2 = \tan\theta \Rightarrow x = \sqrt{\tan\theta}$

$\therefore \quad dx = \frac{1}{2\sqrt{\tan\theta}}\sec^2\theta\,d\theta$

$$= \frac{1}{2}\int_0^{\pi/4}\frac{d\theta}{\sqrt{\sin\theta\cos\theta}} = \frac{1}{2}\int_0^{\pi/4}\frac{d\theta}{\sqrt{\sin 2\theta}}$$

$$\Rightarrow \quad I = \frac{1}{2\sqrt{2}}\int_0^{\pi/2}\sin^{-1/2}t\,dt$$

Put $2\theta = t \quad \therefore \quad d\theta = \frac{dt}{2}$

$$= \frac{1}{2\sqrt{2}}\frac{\Gamma\left(\frac{(-1/2)+1}{2}\right)\Gamma\left(\frac{0+1}{2}\right)}{2\Gamma\left(\frac{(-1/2)+0+2}{2}\right)} = \frac{1}{4\sqrt{2}}\cdot\frac{\Gamma(1/4)\,\Gamma(1/2)}{\Gamma(3/4)}$$

$$= \frac{\sqrt{\pi}}{4\sqrt{2}}\cdot\frac{\Gamma(1/4)^2}{\Gamma(1/4)\,\Gamma(3/4)} = \frac{\sqrt{\pi}}{4\sqrt{2}}\cdot\frac{\Gamma(1/4)^2}{\left(\frac{\pi}{\sin\pi/4}\right)} \qquad \because \Gamma(n)\,\Gamma(1-n) = \frac{\pi}{\sin n\pi};\ 0 < n < 1$$

$$= \frac{1}{8\sqrt{\pi}}\left(\Gamma\frac{1}{4}\right)^2.$$

Example 5. *Prove that:*

(*i*) $\int_0^1 y^{q-1}\left(\log\frac{1}{y}\right)^{p-1}dy = \frac{\Gamma(p)}{q^p}$, *where* $p > 0$, $q > 0$.

(*ii*) $\int_0^\infty \frac{x^{m-1}}{(a+bx)^{m+n}}\,dx = \frac{\beta(m,n)}{a^n b^m}$, *where* m, n, a, b *are positive.*

Sol. (*i*) Put $\log \frac{1}{y} = x$ so that $\frac{1}{y} = e^x$ or $y = e^{-x}$ and $dy = -e^{-x}\,dx$

$$\therefore \quad \int_0^1 y^{q-1}\left(\log \frac{1}{y}\right)^{p-1} dy = \int_\infty^0 e^{-(q-1)x} \cdot x^{p-1}(-e^{-x})\,dx = \int_0^\infty e^{-qx} x^{p-1}\,dx$$

$$= \int_0^\infty e^{-t} \cdot \left(\frac{t}{q}\right)^{p-1} \cdot \frac{dt}{q}, \text{ where } qx = t$$

$$= \frac{1}{q^p}\int_0^\infty e^{-t} t^{p-1}\,dt = \frac{\Gamma(p)}{q^p}.$$

(*ii*) Put $bx = at$ *i.e.* $x = \frac{at}{b}$ so that $dx = \frac{a}{b}dt$

$$\therefore \quad \int_0^\infty \frac{x^{m-1}}{(a+bx)^{m+n}}dx = \int_0^\infty \frac{\left(\frac{at}{b}\right)^{m-1}}{(a+at)^{m+n}} \cdot \frac{a}{b}\,dt = \frac{1}{a^n b^m}\int_0^\infty \frac{t^{m-1}}{(1+t)^{m+n}}dt = \frac{\beta(m,n)}{a^n b^m}.$$

Example 6. *Show that* $\beta(p, q) = \int_0^1 \frac{x^{p-1} + x^{q-1}}{(1+x)^{p+q}}\,dx$. (M.T.U. 2012)

Sol. $\beta(p, q) = \int_0^1 x^{p-1}(1-x)^{q-1}\,dx$

Putting $x = \frac{1}{1+y}$ so that $dx = -\frac{1}{(1+y)^2}dy$

$$\beta(p, q) = \int_\infty^0 \left(\frac{1}{1+y}\right)^{p-1}\left(\frac{y}{1+y}\right)^{q-1} \cdot \frac{-1}{(1+y)^2}dy = \int_0^\infty \frac{y^{q-1}}{(1+y)^{p+q}}dy$$

$$= \int_0^1 \frac{y^{q-1}}{(1+y)^{p+q}}dy + \int_1^\infty \frac{y^{q-1}}{(1+y)^{p+q}}dy \qquad \text{...(1)}$$

Now, putting $y = \frac{1}{z}$ in the second integral, we have

$$\int_1^\infty \frac{y^{q-1}}{(1+y)^{p+q}}dy = \int_1^0 \frac{\left(\frac{1}{z}\right)^{q-1}}{\left(1+\frac{1}{z}\right)^{p+q}}\left(-\frac{1}{z^2}\right)dz = \int_0^1 \frac{z^{p-1}}{(1+z)^{p+q}}dz$$

$\therefore$ From (1), we have

$$\beta(p, q) = \int_0^1 \frac{y^{q-1}}{(1+y)^{p+q}}dy + \int_0^1 \frac{z^{p-1}}{(1+z)^{p+q}}dz = \int_0^1 \frac{x^{p-1} + x^{q-1}}{(1+x)^{p+q}}dx.$$

Example 7. *Prove that* $\beta(m, n) = \beta(m + 1, n) + \beta(m, n + 1)$.

Sol. $\beta\,(m + 1, n) + \beta\,(m, n + 1)$

$$= \frac{\Gamma(m+1)\,\Gamma n}{\Gamma(m+n+1)} + \frac{\Gamma m\,\Gamma(n+1)}{\Gamma(m+n+1)} \qquad \because\ \beta(m, n) = \frac{\Gamma m\,\Gamma n}{\Gamma(m+n)}$$

$$= \frac{1}{\Gamma(m+n+1)}[m\,\Gamma m\,\Gamma n + \Gamma m \cdot n\,\Gamma(n)] \qquad \Big| \because \quad \Gamma(n+1) = n\,\Gamma(n)$$

$$= \frac{\Gamma m\,\Gamma n}{\Gamma(m+n)\,(m+n)}(m+n) = \frac{\Gamma m\,\Gamma n}{\Gamma(m+n)} = \beta(m, n).$$

Example 8. *Prove that* $\int_a^b (x-a)^m (b-x)^n\, dx = (b-a)^{m+n+1}\,\beta(m+1, n+1)$

Sol. Let $\quad I = \int_a^b (x-a)^m (b-x)^n\, dx \quad$ ...(1)

Put $x = a + (b-a)z \Rightarrow x - a = (b-a)z$ and $b - x = (b-a)(1-z)$ so that $dx = (b-a)\,dz$ then (1) becomes

$$I = \int_0^1 (b-a)^m z^m \cdot (b-a)^n (1-z)^n (b-a)\, dz$$

$$= (b-a)^{m+n+1} \int_0^1 z^m (1-z)^n\, dz$$

$$= (b-a)^{m+n+1}\,\beta(m+1, n+1)$$

Example 9. *Prove the following results:*

(*i*) $\int_0^{\pi/2} \sqrt{\tan\theta}\, d\theta = \int_0^{\pi/2} \sqrt{\cot\theta}\, d\theta = \frac{\pi}{\sqrt{2}}$ (A.K.T.U. 2019)

(*ii*) $\int_0^{\pi/2} \frac{d\theta}{\sqrt{\sin\theta}} \times \int_0^{\pi/2} \sqrt{\sin\theta}\, d\theta = \pi.$ (U.P.T.U. 2014)

Sol. (*i*) $\int_0^{\pi/2} \sqrt{\tan\theta}\, d\theta$

$$= \int_0^{\pi/2} \sqrt{\cot\theta}\, d\theta \quad \text{...(1)} \quad \Big| \because \int_0^a f(x)\,dx = \int_0^a f(a-x)\,dx$$

$$= \int_0^{\pi/2} \cos^{1/2}\theta \sin^{-1/2}\theta\, d\theta$$

$$= \frac{\Gamma\left(\frac{-\frac{1}{2}+1}{2}\right)\Gamma\left(\frac{\frac{1}{2}+1}{2}\right)}{2\,\Gamma\left(\frac{-\frac{1}{2}+\frac{1}{2}+2}{2}\right)} = \frac{\Gamma(\frac{1}{4})\,\Gamma(\frac{3}{4})}{2\,\Gamma(1)} = \frac{1}{2}\Gamma(\tfrac{1}{4})\,\Gamma(1-\tfrac{1}{4})$$

$$= \frac{1}{2}\,\frac{\pi}{\sin\frac{\pi}{4}} = \frac{\pi}{\sqrt{2}} \quad \text{...(2)} \quad \Big| \because \Gamma(n)\,\Gamma(1-n) = \frac{\pi}{\sin n\pi}$$

(*ii*) LHS

$$= \int_0^{\pi/2} \sin^{-1/2}\theta \cos^0\theta\, d\theta \times \int_0^{\pi/2} \sin^{1/2}\theta \cos^0\theta\, d\theta$$

$$= \frac{\Gamma\left(\frac{-\frac{1}{2}+1}{2}\right)\Gamma\left(\frac{0+1}{2}\right)}{2\,\Gamma\left(\frac{-\frac{1}{2}+0+2}{2}\right)} \times \frac{\Gamma\left(\frac{\frac{1}{2}+1}{2}\right)\Gamma\left(\frac{0+1}{2}\right)}{2\,\Gamma\left(\frac{\frac{1}{2}+0+2}{2}\right)} = \frac{\Gamma(\frac{1}{4})\,\Gamma(\frac{1}{2})}{2\,\Gamma(\frac{3}{4})} \times \frac{\Gamma(\frac{3}{4})\,\Gamma(\frac{1}{2})}{2\,\Gamma(\frac{5}{4})}$$

$$= \frac{\Gamma(\frac{1}{4})\sqrt{\pi}}{4} \times \frac{\sqrt{\pi}}{\frac{1}{4}\Gamma(\frac{1}{4})} = \pi = \text{RHS} \quad \Big| \because \Gamma(n+1) = n\,\Gamma(n) \text{ and } \Gamma(\tfrac{1}{2}) = \sqrt{\pi}$$

Example 10. *Evaluate:*

(*i*) $\int_0^\infty \cos x^2 \, dx$ (*ii*) $\int_{-\infty}^\infty \cos \frac{\pi}{2} x^2 \, dx$ (*iii*) $\int_0^1 \log \Gamma(x) \, dx$.

Sol. (*i*) We know that

$$\int_0^\infty e^{-ax} \cdot x^{n-1} \cos bx \, dx = \frac{\Gamma(n) \cos n\theta}{(a^2+b^2)^{n/2}} \text{ where } \theta = \tan^{-1}\left(\frac{b}{a}\right)$$

Put $a = 0$, $\int_0^\infty x^{n-1} \cos bx \, dx = \frac{\Gamma(n)}{b^n} \cos \frac{n\pi}{2}$

Put $x^n = z$ so that $x^{n-1} \, dx = \frac{dz}{n}$ and $x = z^{1/n}$

then, $$\int_0^\infty \cos bz^{1/n} \, dz = \frac{n \, \Gamma(n)}{b^n} \cos \frac{n\pi}{2}$$

or $$\int_0^\infty \cos (bx^{1/n}) \, dx = \frac{\Gamma(n+1)}{b^n} \cos \frac{n\pi}{2} \quad \text{...(1)}$$

Here $b = 1$, $n = \frac{1}{2}$

$$\therefore \quad \int_0^\infty \cos x^2 \, dx = \Gamma(\tfrac{3}{2}) \cos \frac{\pi}{4} = \frac{\sqrt{\pi}}{2} \cdot \frac{1}{\sqrt{2}} = \frac{1}{2}\sqrt{\frac{\pi}{2}}.$$

(*ii*) $$\text{I} = \int_{-\infty}^\infty \cos \frac{\pi x^2}{2} \, dx = 2 \int_0^\infty \cos \frac{\pi x^2}{2} \, dx \quad \text{...(2)}$$

Putting $b = \frac{\pi}{2}$ and $n = \frac{1}{2}$ in equation (1), we get

$$\int_0^\infty \cos\left(\frac{\pi}{2} x^2\right) dx = \frac{\Gamma\left(\frac{3}{2}\right)}{\left(\frac{\pi}{2}\right)^{1/2}} \cos \frac{\pi}{4}$$

$$\therefore \quad \text{From (2),} \quad \text{I} = 2 \frac{\Gamma\left(\frac{3}{2}\right)}{\left(\frac{\pi}{2}\right)^{1/2}} \cos \frac{\pi}{4} = 2 \cdot \frac{1}{2}\sqrt{\pi} \cdot \frac{\sqrt{2}}{\sqrt{\pi}} \cdot \frac{1}{\sqrt{2}} = 1.$$

(*iii*) Let $$\text{I} = \int_0^1 \log \Gamma(x) \, dx \quad \text{...(1)}$$

$$= \int_0^1 \log \Gamma(1-x) \, dx \quad \text{...(2)}$$

Adding (1) and (2),

$$2\text{I} = \int_0^1 (\log \Gamma(x) + \log \Gamma(1-x)) \, dx$$

$$= \int_0^1 \log (\Gamma(x) \, \Gamma(1-x)) \, dx = \int_0^1 \log\left(\frac{\pi}{\sin \pi x}\right) dx$$

$$= \int_0^1 (\log \pi - \log \sin \pi x)\, dx = \int_0^1 \log \pi\, dx - \int_0^1 \log \sin \pi x\, dx$$

$$= I_1 - I_2 \qquad \text{...(3)}$$

where $I_1 = \int_0^1 \log \pi\, dx = \log \pi$

$$I_2 = \int_0^1 \log \sin \pi x\, dx = \int_0^\pi \log \sin t \left(\frac{dt}{\pi}\right) \qquad \left|\ \text{Put } \pi x = t \Rightarrow dx = \frac{1}{\pi}\, dt\right.$$

$$= \frac{1}{\pi} \cdot 2 \int_0^{\pi/2} \log \sin t\, dt = \frac{2}{\pi}\left(-\frac{\pi}{2}\log 2\right) = -\log 2$$

From (3), $2I = \log \pi + \log 2 = \log 2\pi$

$$I = \frac{1}{2}\log 2\pi.$$

Example 11. *Prove that:*

(*a*) $\iint_D x^{l-1} y^{m-1} dxdy = \dfrac{\Gamma(l)\,\Gamma(m)}{\Gamma(l+m+1)} a^{l+m}$, *where D is the domain* $x \geq 0, y \geq 0$ *and* $x + y \leq a$.

(*b*) *Establish Dirichlet's integral:* $\iiint_V x^{l-1} y^{m-1} z^{n-1}\, dx\, dy\, dz = \dfrac{\Gamma(l)\,\Gamma(m)\,\Gamma(n)}{\Gamma(l+m+n+1)}$

where V is the region $x \geq 0$, $y \geq 0$, $z \geq 0$ *and* $x + y + z \leq 1$.

Sol. (*a*) Putting $x = a\text{X}$ and $y = a\text{Y}$, the given integral reduces to

$$I = \iint_{D'} (a\text{X})^{l-1} (a\text{Y})^{m-1} a^2\, d\text{X}d\text{Y}$$

where D′ is the domain $\text{X} \geq 0$, $\text{Y} \geq 0$ and $\text{X} + \text{Y} \leq 1$

$$I = a^{l+m} \int_0^1 \int_0^{1-\text{X}} \text{X}^{l-1}\, \text{Y}^{m-1}\, d\text{Y}d\text{X}$$

$$= a^{l+m} \int_0^1 \text{X}^{l-1} \left| \frac{\text{Y}^m}{m} \right|_0^{1-\text{X}} d\text{X} = \frac{a^{l+m}}{m} \int_0^1 \text{X}^{l-1} (1-\text{X})^m\, d\text{X}$$

$$= \frac{a^{l+m}}{m} \beta(l, m+1) = \frac{a^{l+m}}{m} \cdot \frac{\Gamma(l)\,\Gamma(m+1)}{\Gamma(l+m+1)} = a^{l+m} \frac{\Gamma(l)\,\Gamma(m)}{\Gamma(l+m+1)}$$

(*b*) Taking $y + z \leq 1 - x = a$ (say), the given integral

$$I = \int_0^1 \int_0^{1-x} \int_0^{1-x-y} x^{l-1} y^{m-1} z^{n-1}\, dzdydx$$

$$= \int_0^1 x^{l-1} \left[\int_0^a \int_0^{a-y} y^{m-1} z^{n-1}\, dz\, dy \right] dx$$

$$= \int_0^1 x^{l-1} \frac{\Gamma(m)\,\Gamma(n)}{\Gamma(m+n+1)} a^{m+n}\, dx \qquad \text{[by } (a)\text{]}$$

$$= \frac{\Gamma(m)\,\Gamma(n)}{\Gamma(m+n+1)} \int_0^1 x^{l-1} (1-x)^{m+n}\, dx, \text{ since } a = 1 - x$$

$$= \frac{\Gamma(m)\,\Gamma(n)}{\Gamma(m+n+1)} B(l, m+n+1) = \frac{\Gamma(m)\,\Gamma(n)}{\Gamma(m+n+1)} \cdot \frac{\Gamma(l)\,\Gamma(m+n+1)}{\Gamma(l+m+n+1)}$$

$$= \frac{\Gamma(l)\,\Gamma(m)\,\Gamma(n)}{\Gamma(l+m+n+1)}.$$

Example 12. *The plane* $\frac{x}{a}+\frac{y}{b}+\frac{z}{c}=1$ *meets the axes in A, B and C. Apply Dirichlet's integral to find the volume of the tetrahedron OABC. Also find its mass if the density at any point is kxyz.* (A.K.T.U. 2012, 2018)

Sol. Put $\frac{x}{a}=u, \frac{y}{b}=v, \frac{z}{c}=w$, then $u \ge 0,\ v \ge 0,\ w \ge 0$ and $u+v+w \le 1$.

Also, $dx = a\,du,\ dy = b\,dv,\ dz = c\,dw$.

$$\text{Volume OABC} = \iiint_D dx\,dy\,dz$$

$$= \iiint_{D'} abc\,du\,dv\,dw, \quad \text{where } u+v+w \le 1$$

$$= abc \iiint_{D'} u^{1-1}\,v^{1-1}\,w^{1-1}\,du\,dv\,dw$$

$$= abc\,\frac{\Gamma(1)\,\Gamma(1)\,\Gamma(1)}{\Gamma(1+1+1+1)} = \frac{abc}{3!} = \frac{abc}{6}$$

$$\text{Mass} = \iiint_D kxyz\,dx\,dy\,dz = \iiint_{D'} k(au)\,(bv)\,(cw)\,abc\,du\,dv\,dw$$

$$= ka^2b^2c^2 \iiint_{D'} u^{2-1}v^{2-1}w^{2-1}\,du\,dv\,dw$$

$$= ka^2b^2c^2\,\frac{\Gamma(2)\,\Gamma(2)\,\Gamma(2)}{\Gamma(2+2+2+1)} = ka^2b^2c^2\,\frac{1!\,1!\,1!}{6!} = \frac{ka^2b^2c^2}{720}.$$

Example 13. *Evaluate* $I = \iiint_V x^{\alpha-1}\,y^{\beta-1}\,z^{\gamma-1}\,dx\,dy\,dz$ *where v is the region in the first octant bounded by sphere* $x^2+y^2+z^2=1$ *and the coordinate planes.*

Sol. Let $x^2 = u \Rightarrow x = \sqrt{u} \quad \therefore \quad dx = \frac{1}{2\sqrt{u}}\,du$

$y^2 = v \Rightarrow y = \sqrt{v} \quad \therefore \quad dy = \frac{1}{2\sqrt{v}}\,dv$

$z^2 = w \Rightarrow z = \sqrt{w} \quad \therefore \quad dz = \frac{1}{2\sqrt{w}}\,dw$

Then, $u+v+w=1$. Also, $u \ge 0,\ v \ge 0,\ w \ge 0$.

$$I = \iiint_V (\sqrt{u})^{\alpha-1}\,(\sqrt{v})^{\beta-1}\,(\sqrt{w})^{\gamma-1}\,\frac{du}{2\sqrt{u}} \cdot \frac{dv}{2\sqrt{v}} \cdot \frac{dw}{2\sqrt{w}}$$

$$= \frac{1}{8} \iiint u^{(\alpha/2)-1}\,v^{(\beta/2)-1}\,w^{(\gamma/2)-1}\,du\,dv\,dw$$

$$= \frac{1}{8}\,\frac{\Gamma(\alpha/2)\,\Gamma(\beta/2)\,\Gamma(\gamma/2)}{\Gamma((\alpha/2)+(\beta/2)+(\gamma/2)+1)}.$$

Example 14. *Show that if l, m, n are all positive,*

$$\iiint x^{l-1}y^{m-1}z^{n-1}dx\,dy\,dz = \frac{a^l b^m c^n}{8} \cdot \frac{\Gamma(l/2)\,\Gamma(m/2)\Gamma(n/2)}{\Gamma(l/2+m/2+n/2+1)},$$ *where the triple integral is taken throughout the part of ellipsoid* $\frac{x^2}{a^2}+\frac{y^2}{b^2}+\frac{z^2}{c^2}=1$ *which lies in the positive octant.*

Sol. Put $\left(\frac{x}{a}\right)^2 = u, \left(\frac{y}{b}\right)^2 = v, \left(\frac{z}{c}\right)^2 = w$ so that $dx = \frac{a\,du}{2\sqrt{u}}$ etc.

$$\therefore \quad \iiint x^{l-1}y^{m-1}z^{n-1}dx\,dy\,dz$$

$$= \iiint a^{l-1}(\sqrt{u})^{l-1}b^{m-1}(\sqrt{v})^{m-1}c^{n-1}(\sqrt{w})^{n-1}\frac{abc\,du\,dv\,dw}{8\sqrt{u}\sqrt{v}\sqrt{w}}$$

$$= \frac{a^l b^m c^n}{8}\iiint u^{\frac{l}{2}-1}\,v^{\frac{m}{2}-1}\,w^{\frac{n}{2}-1}\,du\,dv\,dw \qquad \text{subject to } u+v+w=1$$

$$= \frac{a^l b^m c^n}{8}\,\frac{\Gamma(l/2)\,\Gamma(m/2)\,\Gamma(n/2)}{\Gamma\frac{l}{2}+\frac{m}{2}+\frac{n}{2}+1}.$$

Example 15. *Evaluate the integral* $\iiint x^{l-1}y^{m-1}z^{n-1}\,dx\,dy\,dz$ *where x, y, z are all positive but limited by the condition* $\left(\frac{x}{a}\right)^p+\left(\frac{y}{b}\right)^q+\left(\frac{z}{c}\right)^r \le 1.$

[G.B.T.U. 2011]

Sol. Let $\left(\frac{x}{a}\right)^p = u \quad \Rightarrow \quad x = au^{1/p} \quad \therefore \quad dx = \frac{a}{p}u^{\frac{1}{p}-1}\,du$

$\left(\frac{y}{b}\right)^q = v \quad \Rightarrow \quad y = bv^{1/q} \quad \therefore \quad dy = \frac{b}{q}v^{\frac{1}{q}-1}\,dv$

$\left(\frac{z}{c}\right)^r = w \quad \Rightarrow \quad z = cw^{1/r} \quad \therefore \quad dz = \frac{c}{r}w^{\frac{1}{r}-1}\,dw$

Then, $u+v+w \le 1$. Also $u \ge 0, v \ge 0, w \ge 0$ since x, y, z are all positive.

$$\text{I} = \iiint a^{l-1}u^{\frac{l-1}{p}}\,b^{m-1}v^{\frac{m-1}{q}}\,c^{n-1}w^{\frac{n-1}{r}}\,\frac{abc}{pqr}\,u^{\frac{1}{p}-1}\,v^{\frac{1}{q}-1}\,w^{\frac{1}{r}-1}\,du\,dv\,dw$$

$$= \frac{a^l b^m c^n}{pqr}\iiint u^{\frac{l}{p}-1}\,v^{\frac{m}{q}-1}\,w^{\frac{n}{r}-1}\,du\,dv\,dw$$

$$= \frac{a^l b^m c^n}{pqr}\,\frac{\Gamma\left(\frac{l}{p}\right)\Gamma\left(\frac{m}{q}\right)\Gamma\left(\frac{n}{r}\right)}{\Gamma\left(\frac{l}{p}+\frac{m}{q}+\frac{n}{r}+1\right)}$$

Example 16. *Find the volume of the solid bounded by the co-ordinate planes and the surface* $\sqrt{\dfrac{x}{a}}+\sqrt{\dfrac{y}{b}}+\sqrt{\dfrac{z}{c}}=1$.

Sol. Put $\sqrt{\dfrac{x}{a}}=u, \sqrt{\dfrac{y}{b}}=v, \sqrt{\dfrac{z}{c}}=w$ then $u \ge 0, v \ge 0, w \ge 0$ and $u+v+w=1$

Also, $dx = 2au\, du, \quad dy = 2bv\, dv, dz = 2cw\, dw$

$$\text{Required volume} = \iiint_D dx\, dy\, dz$$

$$= \iiint_{D'} 8\, abc\, uvw\, du\, dv\, dw, \qquad \text{where } u+v+w=1$$

$$= 8\, abc \iiint_{D'} u^{2-1}\, v^{2-1}\, w^{2-1}\, du\, dv\, dw$$

$$= 8\, abc\, \frac{\Gamma 2\, \Gamma 2\, \Gamma 2}{\Gamma(2+2+2+1)} = 8\, abc \cdot \frac{1 \cdot 1 \cdot 1}{\Gamma(7)} = \frac{abc}{90}.$$

Example 17. *Apply Dirichlet's integral to find the volume and the mass contained in the solid region in the first octant of the ellipsoid* $\dfrac{x^2}{a^2}+\dfrac{y^2}{b^2}+\dfrac{z^2}{c^2}=1$ *if the density at any point is* $\rho(x, y, z) = kxyz$. *(U.P.T.U. 2015)*

Sol. Put $\dfrac{x^2}{a^2}=u, \dfrac{y^2}{b^2}=v, \dfrac{z^2}{c^2}=w$ then $u \ge 0, v \ge 0, w \ge 0$ and $u+v+w=1$

Also, $dx = \dfrac{a}{2\sqrt{u}}\, du, dy = \dfrac{b}{2\sqrt{v}}\, dv, dz = \dfrac{c}{2\sqrt{w}}\, dw$

$$\text{Required Volume} = \iiint_D dx\, dy\, dz = \iiint_{D'} \frac{abc}{8\sqrt{u}\,\sqrt{v}\,\sqrt{w}}\, du\, dv\, dw$$

$$= \frac{abc}{8} \iiint_{D'} u^{\frac{1}{2}-1}\, v^{\frac{1}{2}-1}\, w^{\frac{1}{2}-1}\, du\, dv\, dw$$

$$= \frac{abc}{8}\, \frac{\Gamma\left(\frac{1}{2}\right)\Gamma\left(\frac{1}{2}\right)\Gamma\left(\frac{1}{2}\right)}{\Gamma\left(\frac{5}{2}\right)} = \frac{abc}{8} \cdot \frac{\pi\sqrt{\pi}}{\frac{3}{2}\cdot\frac{1}{2}\sqrt{\pi}}$$

$$= \frac{\pi\, abc}{6}$$

and $$\text{Required Mass} = \iiint_D k\, xyz\, dx\, dy\, dz$$

$$= \iiint_{D'} k \cdot a\sqrt{u} \cdot b\sqrt{v} \cdot c\sqrt{w} \cdot \frac{abc}{8\sqrt{u}\,\sqrt{v}\,\sqrt{w}}\, du\, dv\, dw$$

$$= k\frac{a^2b^2c^2}{8}\iiint_{D'} du\,dv\,dw = k\frac{a^2b^2c^2}{8}\iiint_{D'} u^{1-1}\,v^{1-1}\,w^{1-1}\,du\,dv\,dw$$

$$= k\frac{a^2b^2c^2}{8}\frac{\Gamma(1)\,\Gamma(1)\,\Gamma(1)}{\Gamma(4)} = k\frac{a^2b^2c^2}{48}.$$

Example 18. *Find the volume of the solid surrounded by the surface*

$$\left(\frac{x}{a}\right)^{2/3} + \left(\frac{y}{b}\right)^{2/3} + \left(\frac{z}{c}\right)^{2/3} = 1.$$

Sol. Let $\left(\frac{x}{a}\right)^{2/3} = u \Rightarrow x = au^{3/2} \quad \therefore \quad dx = \frac{3a}{2}u^{1/2}\,du$

$\left(\frac{y}{b}\right)^{2/3} = v \Rightarrow y = bv^{3/2} \quad \therefore \quad dy = \frac{3b}{2}v^{1/2}\,dv$

$\left(\frac{z}{c}\right)^{2/3} = w \Rightarrow z = cw^{3/2} \quad \therefore \quad dz = \frac{3c}{2}w^{1/2}\,dw$

For the positive octant,

$x \geq 0 \Rightarrow au^{3/2} \geq 0 \Rightarrow u \geq 0$

$y \geq 0 \Rightarrow bv^{3/2} \geq 0 \Rightarrow v \geq 0$

$z \geq 0 \Rightarrow cw^{3/2} \geq 0 \Rightarrow w \geq 0$

Then, we have $u + v + w = 1$, $u \geq 0$, $v \geq 0$, $w \geq 0$.

Required volume $= 8\iiint dx\,dy\,dz$

$$= 8\iiint \frac{3a}{2}u^{1/2}\cdot\frac{3b}{2}v^{1/2}\cdot\frac{3c}{2}w^{1/2}\,du\,dv\,dw$$

$$= 27\,abc\iiint u^{\frac{3}{2}-1}\,v^{\frac{3}{2}-1}\,w^{\frac{3}{2}-1}\,du\,dv\,dw$$

$$= 27\,abc\cdot\frac{\Gamma\left(\frac{3}{2}\right)\Gamma\left(\frac{3}{2}\right)\Gamma\left(\frac{3}{2}\right)}{\Gamma\left(\frac{11}{2}\right)} = \frac{4\pi abc}{35}$$

Example 19. *Show that the area bounded by the curve $x^n + y^n = a^n$ and the co-ordinate axes in the first quadrant is* $\dfrac{a^2\,\Gamma\left(\frac{1}{n}\right)^2}{2n\,\Gamma\left(\frac{2}{n}\right)}$.

Sol. Required area $A = \iint_D dx\,dy$ ...(1)

Let, $\left(\frac{x}{a}\right)^n = u$ so that $x = au^{1/n} \quad \therefore \quad dx = \frac{a}{n}u^{\frac{1}{n}-1}\,du$

and $\left(\frac{y}{a}\right)^n = v$ so that $y = av^{1/n} \quad \therefore \quad dy = \frac{a}{n}v^{\frac{1}{n}-1}\,dv$

Then, $u = 0, v = 0$ and $u + v = 1$

From (1),
$$A = \frac{a^2}{n^2} \iint_{D'} u^{\frac{1}{n}-1} v^{\frac{1}{n}-1} \, du\, dv = \frac{a^2}{n^2} \frac{\Gamma\left(\frac{1}{n}\right)\Gamma\left(\frac{1}{n}\right)}{\Gamma\left(\frac{2}{n}+1\right)}$$

$$\Rightarrow \quad A = \frac{a^2}{n^2} \cdot \frac{1}{\left(\frac{2}{n}\right)} \cdot \frac{\left[\Gamma\left(\frac{1}{n}\right)\right]^2}{\Gamma\left(\frac{2}{n}\right)} = \frac{a^2}{2n} \frac{\left[\Gamma\left(\frac{1}{n}\right)\right]^2}{\Gamma\left(\frac{2}{n}\right)}.$$

Example 20. *Find the area and the mass contained in the first quadrant enclosed by the curve* $\left(\frac{x}{a}\right)^{\alpha} + \left(\frac{y}{b}\right)^{\beta} = 1$ *where* $\alpha > 0, \beta > 0$ *given that density at any point* $\rho(x, y)$ *is* $k\sqrt{xy}$.

Sol. The area A of the plane region is $A = \iint_D dx\, dy$ ...(1)

Let $\left(\frac{x}{a}\right)^{\alpha} = u$ so that $x = au^{1/\alpha}$ $\quad \therefore \quad dx = \frac{a}{\alpha} u^{\frac{1}{\alpha}-1} du$

and $\left(\frac{y}{b}\right)^{\beta} = v$ so that $y = bv^{1/\beta}$ $\quad \therefore \quad dy = \frac{b}{\beta} v^{\frac{1}{\beta}-1} dv$

Then, $u > 0, v > 0$ and $u + v = 1$.

From (1),
$$A = \frac{ab}{\alpha\beta} \iint_{D'} u^{\frac{1}{\alpha}-1} v^{\frac{1}{\beta}-1} \, du\, dv = \frac{ab}{\alpha\beta} \frac{\Gamma\left(\frac{1}{\alpha}\right)\Gamma\left(\frac{1}{\beta}\right)}{\Gamma\left(\frac{1}{\alpha}+\frac{1}{\beta}+1\right)}$$

Now, the total mass M contained in the plane region A is
$$M = \iint_D \rho(x, y)\, dx\, dy = k \iint_D \sqrt{xy}\, dx\, dy$$

$$= k \iint_{D'} \sqrt{a}\, u^{\frac{1}{2\alpha}} \sqrt{b}\, v^{\frac{1}{2\beta}} \cdot \frac{a}{\alpha} u^{\frac{1}{\alpha}-1} \cdot \frac{b}{\beta} v^{\frac{1}{\beta}-1} \, du\, dv$$

$$= k \frac{(ab)^{3/2}}{\alpha\beta} \iint_{D'} u^{\frac{3}{2\alpha}-1} v^{\frac{3}{2\beta}-1} \, du\, dv$$

$$= k \frac{(ab)^{3/2}}{\alpha\beta} \cdot \frac{\Gamma(3/2\alpha)\,\Gamma(3/2\beta)}{\Gamma\left(\frac{3}{2\alpha}+\frac{3}{2\beta}+1\right)}$$

TEST YOUR KNOWLEDGE

Prove that (1–22):

1. (i) $\int_0^\infty x^2 e^{-x^2} dx = \frac{\sqrt{\pi}}{4}$. (ii) $\int_0^\infty \sqrt{x}\, e^{-x^3} dx = \frac{\sqrt{\pi}}{3}$. (iii) $\int_0^\infty e^{-x^2} x^{2n-1} dx = \frac{1}{2}\Gamma(n)$.

2. (i) $\int_0^1 x^5(1-x^3)^3 dx = \frac{1}{60}$ (ii) $\int_0^1 x^3(1-x)^{4/3} dx = \frac{243}{7280}$

3. $\int_0^2 (8-x^3)^{-1/3} dx = \frac{1}{3}\Gamma(\tfrac{1}{3})\Gamma(\tfrac{2}{3}) = \frac{2\pi}{3\sqrt{3}}$.

4. (i) $\int_0^\infty 4x^4 e^{-x^4} dx = \Gamma(\tfrac{5}{4})$ (ii) $\int_0^\infty x^6 e^{-2x} dx = \frac{45}{8}$ (iii) $\int_0^1 \sqrt{\frac{1-x}{x}}\, dx = \frac{\pi}{2}$

5. (i) $\int_0^\infty \frac{e^{-st}}{\sqrt{t}} dt = \sqrt{\frac{\pi}{s}}$, $s > 0$ (ii) $\int_0^1 \frac{dx}{\sqrt{-\log x}} = \sqrt{\pi}$. (*A.K.T.U. 2022*)

(iii) $\int_0^1 \left(\sqrt{\frac{x^3}{1-x^3}}\right)^{1/2} dx = \frac{\Gamma\frac{7}{12}\Gamma\frac{3}{4}}{\Gamma\frac{1}{3}}$

6. (i) $\Gamma(.1)\,\Gamma(.2)\,\Gamma(.3)\ldots\Gamma(.9) = \frac{(2\pi)^{9/2}}{\sqrt{10}}$

(ii) $\Gamma\left(\frac{3}{2}-p\right)\Gamma\left(\frac{3}{2}+p\right) = \left(\frac{1}{4}-p^2\right)\pi \sec p\pi,\ -1 < 2p < 1$

7. (i) $\int_0^1 \frac{dx}{\sqrt{1-x^n}} = \frac{\sqrt{\pi}}{n}\,\frac{\Gamma\left(\frac{1}{n}\right)}{\Gamma\left(\frac{1}{n}+\frac{1}{2}\right)}$ (ii) $\int_0^1 \frac{dx}{\sqrt{1-x^4}} = \frac{1}{4\sqrt{2\pi}}(\Gamma\tfrac{1}{4})^2$.

8. (i) $\int_0^\infty \frac{x\,dx}{1+x^6} = \frac{\pi}{3\sqrt{3}}$ (ii) $\int_0^\infty \frac{x^2\,dx}{1+x^4} = \frac{\pi}{2\sqrt{2}}$

(iii) $\int_0^2 \frac{x^2}{\sqrt{2-x}} dx = \frac{64\sqrt{2}}{15}$ (iv) $\int_0^3 \frac{dx}{\sqrt{3x-x^2}} = \pi$.

9. (i) $\int_0^\infty \sqrt{y}\, e^{-y^2} dy \times \int_0^\infty \frac{e^{-y^2}}{\sqrt{y}} dx = \frac{\pi}{2\sqrt{2}}$. (ii) $\int_0^\infty \frac{e^{-x^2}}{\sqrt{x}} dx \times \int_0^\infty x^2 e^{-x^4} dx = \frac{\pi}{4\sqrt{2}}$.

10. (i) $\int_0^\infty x^2 e^{-x^4} dx \cdot \int_0^\infty e^{-x^4} dx = \frac{\pi}{8\sqrt{2}}$ (ii) $\int_0^1 \frac{x^2 dx}{\sqrt{1-x^4}} \cdot \int_0^1 \frac{dx}{\sqrt{1+x^4}} = \frac{\pi}{4\sqrt{2}}$.

11. (i) $\int_0^{\pi/2} \sin^3 x \cos^{5/2} x\, dx = \frac{8}{77}$ (ii) $\int_0^a x^{n-1}(a-x)^{m-1} dx = a^{m+n-1}\beta(m,n)$

12. (i) $\int_0^1 \frac{dx}{\sqrt{1-x^6}} = \frac{\sqrt{3}}{2}\int_0^1 \frac{dx}{\sqrt{1-x^3}} = \frac{(\Gamma 1/3)^3}{(2)^{7/3}\pi}$ (ii) $\int_0^1 \frac{x\,dx}{\sqrt{1-x^5}} = \frac{1}{5}\beta\left(\frac{2}{5},\frac{1}{2}\right)$.

13. (i) $\frac{\Gamma(\frac{1}{3})\Gamma(\frac{5}{6})}{\Gamma(\frac{2}{3})} = (2)^{1/3}\sqrt{\pi}$ (*A.K.T.U. 2017*) (ii) $\Gamma(-\tfrac{3}{2}) = \frac{4}{3}\sqrt{\pi}$ (*G.B.T.U. 2013*)

(iii) $\Gamma(\tfrac{1}{6}) = 2^{-1/3}\pi^{-1/2}\sqrt{3}\,(\Gamma\tfrac{1}{3})^2$ (iv) $\Gamma(-\tfrac{1}{2}) = -2\sqrt{\pi}$.

14. $\int_0^1 x^m (\log x)^n \, dx = \dfrac{(-1)^n n!}{(m+1)^{n+1}}$, where n is a positive integer and $m > -1$.

15. $\int_0^\infty \int_0^\infty e^{-(ax^2+by^2)} x^{2m-1} y^{2n-1} \, dx \, dy = \dfrac{\Gamma(m)\Gamma(n)}{4a^m b^n}$, where a, b, m, n are positive.

16. (*i*) $\beta(m, m) \cdot \beta\left(m + \frac{1}{2}, m + \frac{1}{2}\right) = \dfrac{\pi m^{-1}}{(2)^{4m-1}}$ (*ii*) $\beta(m, n) = a^m b^n \int_0^\infty \dfrac{x^{m-1}}{(ax+b)^{m+n}} dx$

(*U.P.T.U. 2015*)

17. $\dfrac{B(p, q+1)}{q} = \dfrac{B(p+1, q)}{p} = \dfrac{B(p, q)}{p+q}$, $(p > 0, q > 0)$ (*U.P.T.U. 2015; U.K.T.U. 2012*)

18. $\dfrac{B(m+2, n-2)}{B(m, n)} = \dfrac{m(m+1)}{(n-1)(n-2)}$.

19. $2^n \Gamma\left(n + \frac{1}{2}\right) = 1 \cdot 3 \cdot 5 \,.....(2n-1)\sqrt{\pi}$

20. $\int_0^1 \dfrac{x^{m-1}(1-x)^{n-1}}{(a+bx)^{m+n}} dx = \dfrac{B(m, n)}{a^n (a+b)^m}$. $\left[\textbf{Hint.} \text{ Put } \dfrac{x}{a+bx} = \dfrac{z}{a+b \cdot 1}.\right]$

21. $\int_0^{\pi/2} \dfrac{d\theta}{(a\cos^4\theta + b\sin^4\theta)^{1/2}} = \dfrac{(\Gamma \frac{1}{4})^2}{4(ab)^{1/4}\sqrt{\pi}}$ [**Hint.** Put $\tan\theta = t$, then $bt^4 = az$.]

22. $\int_0^\infty x^n e^{-a^2x^2} dx = \dfrac{1}{2a^{n+1}} \Gamma\left(\dfrac{n+1}{2}\right)$, $n > 1$. Deduce that $\int_{-\infty}^\infty e^{-a^2x^2} dx = \dfrac{\sqrt{\pi}}{a}$.

23. Assuming $\Gamma n \, \Gamma(1-n) = \pi \operatorname{cosec} n\pi$, $0 < n < 1$, show that $\int_0^\infty \dfrac{x^{p-1}}{1+x} dx = \dfrac{\pi}{\sin p\pi}$; $0 < p < 1$.

(*A.K.T.U. 2022*)

24. Show that $\iint x^{m-1} y^{n-1} \, dx \, dy$ over the positive octant of the ellipse $\left(\dfrac{x}{a}\right)^2 + \left(\dfrac{y}{b}\right)^2 = 1$ is $\dfrac{a^m b^n}{2n} \beta\left(\dfrac{m}{2}, \dfrac{n}{2} + 1\right)$.

25. Find the volume of the solid bounded by co-ordinate planes and the surface

$$\left(\frac{x}{a}\right)^{2n} + \left(\frac{y}{b}\right)^{2n} + \left(\frac{z}{c}\right)^{2n} = 1, n \text{ being a positive integer.}$$

26. (*i*) Find the volume of the solid bounded by the ellipsoid $\dfrac{x^2}{a^2} + \dfrac{y^2}{b^2} + \dfrac{z^2}{c^2} = 1$.

(*ii*) Find the volume contained in the solid region in the first octant of the ellipsoid $\dfrac{x^2}{a^2} + \dfrac{y^2}{b^2} + \dfrac{z^2}{c^2} = 1$. (*A.K.T.U. 2014, 2022*)

(*iii*) Evaluate $\iiint xyz \, dx \, dy \, dz$ for all positive value of variables of ellipsoid $\dfrac{x^2}{a^2} + \dfrac{y^2}{b^2} + \dfrac{z^2}{c^2} = 1$.

27. (*i*) Find the mass of the region bounded by ellipsoid $\dfrac{x^2}{a^2} + \dfrac{y^2}{b^2} + \dfrac{z^2}{c^2} = 1$ if the density varies as the square of the distance from the centre. [**Hint.** $\rho = k(x^2 + y^2 + z^2)$] (*G.B.T.U. 2010*)

(*ii*) Find the mass of a solid $\left(\dfrac{x}{a}\right)^p + \left(\dfrac{y}{b}\right)^q + \left(\dfrac{z}{c}\right)^r = 1$, the density at any point being $\rho = kx^{l-1} y^{m-1} z^{n-1}$ where x, y, z are all positive. (*A.K.T.U. 2016*)

28. Evaluate $\iiint_V (ax^2 + by^2 + cz^2)\, dx\, dy\, dz$ where V is the region bounded by $x^2 + y^2 + z^2 \le 1$.
(*G.B.T.U. 2013*)

29. Compute $\iiint_v x^2\, dx\, dy\, dz$ over valume of tetraheron bounded by $x = 0,\ y = 0,\ z = 0$ and $\frac{x}{a} + \frac{y}{b} + \frac{z}{c} = 1.$
(*A.K.T.U. 2017*)

30. Evaluate $\iiint_v x^2 yz\, dx\, dy\, dz$ throughout the volume bounded by planes $x = 0,\ y = 0,\ z = 0$ and $\frac{x}{a} + \frac{y}{b} + \frac{z}{c} = 1.$
(*A.K.T.U. 2017*)

31. Use Beta and Gamma function to solve :

$$\int_0^\infty \frac{1}{1+x^4}\, dx \,.\, \int_0^{\pi/2} \sqrt{\cot\theta}\, d\theta.$$
(*A.K.T.U. 2019*)

32. Apply Dirichlet integral to find the volume of an octant of the sphere $x^2 + y^2 + z^2 = 25$.
(*A.K.T.U. 2019*)

Answers

25. $\dfrac{abc}{12n^2} \dfrac{\left(\Gamma \frac{1}{2n}\right)^3}{\Gamma\left(\frac{3}{2n}\right)}$ **26.** (*i*) $\frac{4}{3}\pi\, abc$ (*ii*) $\dfrac{\pi\, abc}{6}$ (*iii*) $\dfrac{a^2b^2c^2}{48}$

27. (*i*) $\dfrac{k\pi\, abc}{16}(a^2 + b^2 + c^2)$ (*ii*) $\dfrac{ka^l b^m c^n}{pqr} \dfrac{(l/p)!\,(m/q)!\,(n/r)!}{\left(\frac{l}{p} + \frac{m}{q} + \frac{n}{r} + 1\right)!}$

28. $\dfrac{\pi(a+b+c)}{30}$ **29.** $\dfrac{a^3bc}{60}$ **30.** $\dfrac{a^3b^2c^2}{2520}$

31. $\dfrac{\pi^2}{4}$ **32.** $\dfrac{125\pi}{6}$

4.24. LIOUVILLE'S EXTENSION OF DIRICHLET THEOREM

If the variables x, y, z are all positive such that $h_1 < (x + y + z) < h_2$ then

$$\iiint f(x+y+z)x^{l-1}y^{m-1}z^{n-1}dx\, dy\, dz = \frac{\Gamma(l)\,\Gamma(m)\,\Gamma(n)}{\Gamma(l+m+n)} \int_{h_1}^{h_2} f(u)\, u^{l+m+n-1}\, du.$$

Proof. Let, $I = \iiint x^{l-1}y^{m-1}z^{n-1}\, dx\, dy\, dz$ under the condition $x + y + z \le u$ then

$$I = \left(\frac{\Gamma(l)\,\Gamma(m)\,\Gamma(n)}{\Gamma(l+m+n+1)}\right) u^{l+m+n} \quad \text{...(1) | By Dirichlet's Theorem}$$

If $x + y + z \le u + \delta u$, then

$$I = (u+\delta u)^{l+m+n} \frac{\Gamma(l)\,\Gamma(m)\,\Gamma(n)}{\Gamma(l+m+n+1)} \quad \text{...(2)}$$

Now, if $u < x + y + z < (u + \delta u)$, then

$$\iiint x^{l-1}y^{m-1}z^{n-1}\, dx\, dy\, dz = \frac{\Gamma(l)\,\Gamma(m)\,\Gamma(n)}{\Gamma(l+m+n+1)}\left[(u+\delta u)^{l+m+n} - (u)^{l+m+n}\right]$$

$$= \frac{\Gamma(l)\,\Gamma(m)\,\Gamma(n)}{\Gamma(l+m+n+1)}\,u^{l+m+n}\left[1+\left(\frac{\delta u}{u}\right)^{l+m+n} - 1\right]$$

$$= \frac{\Gamma(l)\,\Gamma(m)\,\Gamma(n)}{\Gamma(l+m+n+1)}\,u^{l+m+n}\left[1+(l+m+n)\frac{\delta u}{u}+\ldots-1\right]$$

$$= \frac{\Gamma(l)\,\Gamma(m)\,\Gamma(n)}{\Gamma(l+m+n+1)}\,u^{l+m+n}(l+m+n)\frac{\delta u}{u}$$

$$= \frac{\Gamma(l)\,\Gamma(m)\,\Gamma(n)}{\Gamma(l+m+n)}\,u^{l+m+n-1}\,\delta u$$

Now, consider $\iiint f(x+y+z)x^{l-1}y^{m-1}z^{n-1}\, dx\, dy\, dz$ under the condition $h_1 \le (x+y+z) \le h_2$. When $x+y+z$ lies between u and $u+\delta u$, the value of $f(x+y+z)$ can only differ from $f(u)$ by a small quantity of the same order as δu. Hence,

$$\iiint f(x+y+z)x^{l-1}y^{m-1}z^{n-1}\, dx\, dy\, dz = \frac{\Gamma(l)\,\Gamma(m)\,\Gamma(n)}{\Gamma(l+m+n)}\int f(u)\,u^{l+m+n-1}\,\delta u$$

where $x+y+z$ lies between u and $u+\delta u$.

Therefore, $$\iiint f(x+y+z)x^{l-1}\, y^{m-1}\, z^{n-1}\, dx\, dy\, dz = \frac{\Gamma(l)\,\Gamma(m)\,\Gamma(n)}{\Gamma\,(l+m+n)}\int_{h_1}^{h_2} f(u)\,u^{l+m+n-1}\, du.$$

ILLUSTRATIVE EXAMPLES

Example 1. *Evaluate* $\iiint \log(x+y+z)\, dx\, dy\, dz$, *the integral extending over all positive and zero values of x, y, z subject to* $x+y+z<1$.

Sol. $0 \le x+y+z < 1$

$$\therefore \quad \iiint \log(x+y+z)\, dx\, dy\, dz = \iiint x^{1-1}y^{1-1}z^{1-1}\log(x+y+z)\, dx\, dy\, dz$$

$$= \frac{\Gamma(1)\,\Gamma(1)\,\Gamma(1)}{\Gamma(3)}\int_0^1 t^{1+1+1-1}\log t\, dt$$ | By Liouville's extension

$$= \frac{1}{2}\int_0^1 t^2\log t\, dt = \frac{1}{2}\left[\left(\frac{t^3}{3}\log t\right)_0^1 - \int_0^1 \frac{t^3}{3}\cdot\frac{1}{t}\, dt\right] = \frac{1}{2}\left[-\frac{1}{3}\left(\frac{t^3}{3}\right)_0^1\right] = -\frac{1}{18}.$$

Example 2. *Evaluate* $\iiint\ldots\int_n \dfrac{dx_1 dx_2 \ldots dx_n}{\sqrt{1-x_1^2-x_2^2\ldots-x_n^2}}$, *integral being extended to all positive values of the variables for which the expression is real.*

Sol. The expression will be real, if

$$1 - x_1^2 - x_2^2 - \ldots - x_n^2 > 0$$

or $$x_1^2 + x_2^2 + \ldots + x_n^2 < 1$$

Hence the given integral is extended for all positive value of the variables $x_1, x_2, \ldots, x_n$ such that $0 < x_1^2 + x_2^2 + \ldots x_n^2 < 1$.

Let us put $x_1^2 = u_1$ *i.e.*, $x_1 = \sqrt{u_1}$ so that, $dx_1 = \dfrac{1}{2\sqrt{u_1}} du_1$ etc.

Then the condition becomes, $0 < u_1 + u_2 + \ldots + u_n < 1$.

$\therefore$ Required integral $= \dfrac{1}{2^n} \iiint \ldots \int_n \dfrac{u_1^{-1/2} u_2^{-1/2} \ldots u_n^{-1/2}}{\sqrt{1 - u_1 - u_2 - \ldots - u_n}} du_1 \, du_2 \ldots du_n$

$$= \frac{1}{2^n} \iiint \ldots \int_n \frac{u_1^{1/2-1} u_2^{1/2-1} \ldots u_n^{1/2-1}}{\sqrt{1 - u_1 - u_2 - \ldots - u_n}} du_1 \, du_2 \ldots du_n$$

$$= \frac{1}{2^n} \frac{\Gamma(1/2)\,\Gamma(1/2) \ldots \Gamma(1/2)}{\Gamma(1/2 + 1/2 + \ldots + 1/2)} \int_0^1 \frac{1}{\sqrt{1-u}} u^{(1/2 + 1/2 + \ldots + 1/2) - 1} du$$

| By Liouville's Extension

$$= \frac{1}{2^n} \cdot \frac{[\Gamma(1/2)]^n}{\Gamma(n/2)} \int_0^1 \frac{1}{\sqrt{1-u}} u^{\frac{n}{2}-1} du$$

$$= \frac{1}{2^n} \frac{(\sqrt{\pi})^n}{\Gamma(n/2)} \int_0^{\pi/2} \frac{1}{\sqrt{1 - \sin^2 \theta}} (\sin^2 \theta)^{\frac{n}{2}-1} \cdot 2 \sin \theta \cos \theta \, d\theta$$

(Putting $u = \sin^2 \theta$)

$$= \frac{1}{2^{n-1}} \cdot \frac{(\sqrt{\pi})^n}{\Gamma(n/2)} \int_0^{\pi/2} \sin^{n-1} \theta \, d\theta = \frac{1}{2^{n-1}} \cdot \frac{(\pi)^{n/2}}{\Gamma(n/2)} \cdot \frac{\Gamma(n/2)\,\Gamma(1/2)}{2\,\Gamma\left(\frac{n+1}{2}\right)}$$

$$= \frac{1}{2^n} \cdot \frac{(\pi)^{\frac{n+1}{2}}}{\Gamma\left(\frac{n+1}{2}\right)}.$$

Example 3. *Show that* $\iiint \dfrac{dx\,dy\,dz}{(x+y+z+1)^3} = \dfrac{1}{2} \log 2 - \dfrac{5}{16}$, *the integral being taken throughout the volume bounded by planes* $x = 0$, $y = 0$, $z = 0$ *and* $x + y + z = 1$.

Sol. $0 \le x + y + z \le 1$

$$\therefore \quad \iiint \frac{dx\,dy\,dz}{(x+y+z+1)^3} = \iiint \frac{x^{1-1} y^{1-1} z^{1-1}}{(x+y+z+1)^3} dx\,dy\,dz$$

$$= \frac{\Gamma(1)\,\Gamma(1)\,\Gamma(1)}{\Gamma(1+1+1)} \int_0^1 \frac{1}{(u+1)^3} u^{1+1+1-1} du$$

$$= \frac{1}{2}\int_0^1 \frac{u^2}{(u+1)^3}\, du \qquad \text{| By Liouville's extension}$$

Put $u + 1 = t$ so that $du = dt$

$\therefore$ Required integral

$$= \frac{1}{2}\int_1^2 \frac{(t-1)^2}{t^3}\, dt = \frac{1}{2}\int_1^2 \left(\frac{t^2 - 2t + 1}{t^3}\right) dt$$

$$= \frac{1}{2}\int_1^2 \left(\frac{1}{t} - \frac{2}{t^2} + \frac{1}{t^3}\right) dt = \frac{1}{2}\left[\log t + \frac{2}{t} - \frac{1}{2t^2}\right]_1^2$$

$$= \frac{1}{2}\left[\log 2 + 1 - \frac{1}{8} - 2 + \frac{1}{2}\right] = \frac{1}{2}\left[\log 2 - \frac{5}{8}\right] = \frac{1}{2}\log 2 - \frac{5}{16}.$$

Example 4. *Evaluate* $\iiint \sqrt{\dfrac{1 - x^2 - y^2 - z^2}{1 + x^2 + y^2 + z^2}}\, dx\, dy\, dz$, *integral being taken over all positive values of x, y, z such that* $x^2 + y^2 + z^2 \le 1$.

Sol. Putting $x^2 = u$, $y^2 = v$, $z^2 = w$ so that $u + v + w \le 1$

Also, $x = \sqrt{u} \Rightarrow dx = \dfrac{1}{2\sqrt{u}}\, du$

$y = \sqrt{v} \Rightarrow dy = \dfrac{1}{2\sqrt{v}}\, dv$

$z = \sqrt{w} \Rightarrow dz = \dfrac{1}{2\sqrt{w}}\, dw$

$\therefore$ The given integral

$$= \iiint \sqrt{\frac{1 - (u+v+w)}{1 + (u+v+w)}} \cdot \frac{du\, dv\, dw}{8\sqrt{uvw}}$$

$$= \frac{1}{8}\iiint u^{1/2-1}\, v^{1/2-1}\, w^{1/2-1} \sqrt{\frac{1-(u+v+w)}{1+(u+v+w)}}\, du\, dv\, dw$$

$$= \frac{1}{8} \cdot \frac{\Gamma(\frac{1}{2})\,\Gamma(\frac{1}{2})\,\Gamma(\frac{1}{2})}{\Gamma(\frac{1}{2} + \frac{1}{2} + \frac{1}{2})} \int_0^1 \sqrt{\frac{1-u}{1+u}}\, u^{\frac{1}{2}+\frac{1}{2}+\frac{1}{2}-1}\, du \qquad \text{| Using Liouville's extension}$$

$$= \frac{1}{8}\frac{(\Gamma\frac{1}{2})^3}{\frac{1}{2}\Gamma(\frac{1}{2})}\int_0^1 \frac{(1-u)}{\sqrt{1-u^2}}\, u^{1/2} du = \frac{\pi}{4}\int_0^1 \frac{(1-\sqrt{t})}{\sqrt{1-t}}\, t^{1/4}\, \frac{dt}{2\sqrt{t}}, \quad \text{where } u^2 = t$$

$$= \frac{\pi}{8}\int_0^1 \frac{(1-\sqrt{t})\, t^{-1/4}}{\sqrt{1-t}}\, dt = \frac{\pi}{8}\left[\int_0^1 t^{\frac{3}{4}-1}(1-t)^{1/2-1}dt - \int_0^1 t^{5/4-1}(1-t)^{1/2-1}dt\right]$$

$$= \frac{\pi}{8}\left[\beta\left(\frac{3}{4}, \frac{1}{2}\right) - \beta\left(\frac{5}{4}, \frac{1}{2}\right)\right].$$

TEST YOUR KNOWLEDGE

1. Evaluate $\iiint e^{x+y+z} dx\, dy\, dz$, taken over positive octant such that $x + y + z \le 1$.

2. (*i*) Evaluate the integral $\iiint \frac{dx\, dy\, dz}{\sqrt{a^2 - x^2 - y^2 - z^2}}$ the integral being extended to all positive values of the variables for which the expression is real. [*U.K.T.U. 2010*]

   (*ii*) Prove that $\iiint \frac{dx\, dy\, dz}{\sqrt{1 - x^2 - y^2 - z^2}} = \frac{\pi^2}{8}$, the integral being extended to all positive values of the variables for which the expression is real. (*A.K.T.U. 2016; U.P.T.U. 2014*)

3. Apply Dirichlet's integral to find the moment of inertia about z-axis of an octant of the ellipsoid $\frac{x^2}{a^2} + \frac{y^2}{b^2} + \frac{z^2}{c^2} = 1$.

4. Show that $\iiint \frac{dx\, dy\, dz}{(x + y + z + 1)^2} = \frac{3}{4} - \log 2$, the integral being taken throughout the volume bounded by the planes $x = 0, y = 0, z = 0$ and $x + y + z = 1$.

5. Evaluate $\iiint (x + y + z)^9\, dx\, dy\, dz$ over the region defined by $x \ge 0, y \ge 0, z \ge 0$ and $x + y + z \le 1$.

6. Evaluate $\iiint \sqrt{\frac{1 - x - y - z}{xyz}}\, dx\, dy\, dz$, extended to all positive values of the variables subject to the condition $x + y + z < 1$.

7. Evaluate $\iint x^{l-1} y^{-l} e^{x+y} dx\, dy$, extended to all positive values subject to $x + y < h$.

8. Evaluate $\iiint xyz \sin (x + y + z)\, dx\, dy\, dz$, the integral being extended to all positive values of the variables subject to the condition $x + y + z \le \frac{\pi}{2}$. (*A.K.T.U. 2022*)

9. Evaluate $\iint \frac{\sqrt{1 - (x^2/a^2) - (y^2/b^2)}}{\sqrt{1 + (x^2/a^2) + (y^2/b^2)}}\, dx\, dy$, where $\frac{x^2}{a^2} + \frac{y^2}{b^2} \le 1$

10. Evaluate $\iiint \sqrt{a^2b^2c^2 - b^2c^2x^2 - c^2a^2y^2 - a^2b^2z^2}\, dx\, dy\, dz$ taken throughout the ellipsoid $\frac{x^2}{a^2} + \frac{y^2}{b^2} + \frac{z^2}{c^2} = 1$.

11. Evaluate $\iiint_V e^{-(x + y + z)}\, dx\, dy\, dz$, where the region of integration is bounded by planes $x = 0, y = 0, z = 0$ and $x + y + z = a, a > 0$. [*U.P.T.U. (SUM) 2008*]

12. For all values of the variables for which $x^2 + y^2 + z^2 + w^2$ is not less than a^2 and not greater than b^2, prove that:

$$\iiiint dx\, dy\, dz\, dw = \frac{\pi^2}{32}(b^4 - a^4).$$

Answers

1. $\frac{e}{2}-1$
2. (i) $\frac{\pi^2 a^2}{8}$
3. $\frac{\rho abc}{30}(a^2+b^2)\pi$
5. $\frac{1}{24}$
6. $\frac{1}{4}\pi^2$
7. $\frac{\pi}{\sin \pi l}(e^h-1)$
8. $\frac{1}{5!}\left[\frac{5}{16}\pi^4-15\pi^2+120\right]$
9. $\frac{\pi ab}{8}(\pi-2)$
10. $\frac{1}{4}\pi^2 a^2 b^2 c^2$
11. $1-e^{-a}\left[1+a+\frac{a^2}{2}\right]$.

ASSIGNMENT IV

1. Evaluate the area enclosed between the parabola $y = x^2$ and the straight line $y = x$. *(A.K.T.U. 2018)*
2. Evaluate $\int_0^1 \int_1^2 xy\, dy\, dx$.
3. Evaluate $\int_0^1 \int_1^2 \int_2^3 xyz\ dx\, dy\, dz$. *(G.B.T.U. 2012)*
4. Evaluate $\int_0^2 \int_1^3 \int_1^2 xy^2 z\ dz\, dy\, dx$.
5. Find the value of integral $\iiint_R dx\, dy\, dz$ where R: $-1 \le x \le 1, -2 \le y \le 2, -3 \le z \le 3$.
6. Find the volume of the solid under the surface $az = x^2 + y^2$ and whose base R is the circle $x^2 + y^2 = a^2$.
7. Evaluate: (i) $\int_1^{\log 8} \int_0^{\log y} e^{x+y} dx\, dy$ (ii) $\int_0^1 \int_0^{x^2} xe^y\ dy\, dx$ *(A.K.T.U. 2018)*
8. Evaluate $\iint_D (x^2 + y^2)\, dx\, dy$ where D is bounded by $y = x$ and $y^2 = 4x$.
9. Evaluate $\iint_D x^3 y\, dx\, dy$ where D is the region enclosed by the ellipse $\frac{x^2}{a^2}+\frac{y^2}{b^2} = 1$ in the first quadrant.
10. Evaluate $\int_3^4 \int_1^2 \frac{dy\, dx}{(x+y)^2}$.
11. Evaluate $\int_{-1/2}^{1} \int_{-x}^{1+x} (x^2 + y)\, dy\, dx$.
12. Find the area bounded by the curves $y^2 = x^3$ and $x^2 = y^3$.
13. Find the area bounded by the curves $\sqrt{x} + \sqrt{y} = \sqrt{a}$ and $x + y = a$.
14. Find the area bounded by the parabola $y^2 = 4ax$ and the straight line $x + y = 3a$.
15. Evaluate $\int_0^1 \int_0^x \int_0^{x+y} (x + y + z)\, dz\, dy\, dx$.
16. Find the total mass of the region in the case $0 \le x \le 1, 0 \le y \le 1, 0 \le z \le 1$ with density at any point being xyz.

17. Find the volume of the solid bounded by the plane $2x + 3y + 4z = 12$, xy-plane and the cylinder $x^2 + y^2 = 1$.

18. Evaluate $\iint_D x^p y^q \, dx\, dy$, where D is the region bounded by $x = 0$, $y = 0$ and $\left(\frac{x}{a}\right)^m + \left(\frac{y}{b}\right)^n = 1$.

19. Find the mass of the region in xy-plane bounded by $x = 0$, $y = 0$, $x + y = 1$ with density $k\sqrt{xy}$.

20. Evaluate $\int_0^\infty e^{-x^4} dx$.

21. Evaluate $\iint_R \sqrt{x^2 + y^2} \, dx\, dy$, where R is the region in xy-plane bounded by $x^2 + y^2 = 4$ and $x^2 + y^2 = 9$.

22. Change the order of integration:

$$\int_0^a \int_0^{2\sqrt{ay}} f(x, y)\, dx\, dy + \int_a^{3a} \int_0^{3a - y} f(x, y)\, dx\, dy.$$

23. Evaluate $\int_0^\infty e^{-x^2} dx$. *(A.K.T.U. 2011, 2018)*

24. Change the order of integration is $\int_0^\infty \int_x^\infty f(x, y)\, dy\, dx$

25. Evaluate $\int_0^a \int_0^x xy\, dy\, dx$. *(U.P.T.U. 2014)*

26. Changing the order of integration in the double integral $I = \int_0^8 \int_{\pi/4}^2 f(x, y)\, dx\, dy$ leads to $I = \int_r^s \int_p^q f(x, y)\, dy\, dx$, say. What is p? *(M.T.U. 2013)*

27. Determine the volume of the space below the paraboloid $x^2 + y^2 + z - 4 = 0$ and above the square in the xy-plane with vertices at (0, 0), (0, 1), (1, 0) and (1, 1).

28. If V is the solid region in the first octant bounded by the unit sphere $x^2 + y^2 + z^2 = 1$, find the mass of the ellipsoid $\left(\frac{x}{a}\right)^2 + \left(\frac{y}{b}\right)^2 + \left(\frac{z}{c}\right)^2 = 1$ with density $\rho(x, y, z) = (x + y + z)^4$.

29. What does $\int_0^1 \int_0^{1-x} dy\, dx$ represent?

30. If the volume of an object expressed in spherical coordinates as following:

$$V = \int_0^{2\pi} \int_0^{\pi} \int_0^1 r^2 \sin\phi \, dr\, d\phi\, d\theta. \text{ Evaluate V.}$$ *(A.K.T.U. 2017)*

31. Determine the area enclosed by the curve $\left(\frac{x}{a}\right)^{2m} + \left(\frac{y}{b}\right)^{2n} = 1$; m, n being positive integers.

32. Find the area enclosed by the curve $\left(\frac{x}{a}\right)^4 + \left(\frac{y}{b}\right)^{10} = 1$.

33. Find the volume of the region bounded by $z = 0$, $x^2 + y^2 = 1$, $x + y + z = 3$.

34. Find the area of the cardioid $r = a(1 + \cos\theta)$.

35. Evaluate $\iint_R \frac{x^2 y^2}{x^2 + y^2}\, dx\, dy$ where R is annular region between the circles $x^2 + y^2 = a^2$ and $x^2 + y^2 = b^2$ with $b > a$.

36. Find the volume and mass contained in a solid below the plane $z = 1 + y$, bounded by the co-ordinate planes and the vertical plane $2x + y = 2$ and having density $f(x, y, z) = x + z$.

37. Evaluate:

(*i*) $\int_0^2 \int_1^z \int_0^{yz} xyz\, dx\, dy\, dz$ (*ii*) $\int_1^2 y\, dy \int_y^{y^2} dx \int_0^{\log x} e^z\, dz$

(*iii*) $\int_0^{\pi/2} \int_x^{\pi/2} \int_0^{xy} \cos \frac{z}{x}\, dz\, dy\, dx$ (*iv*) $\int_0^1 \int_0^x \int_0^y xyz\, dz\, dy\, dx$

38. Evaluate:

(*i*) $\int_{1/2}^1 \int_0^{\pi/2} \rho\, d\theta\, d\rho$ (*ii*) $\int_0^{\pi} \int_{2\sin\theta}^{4\sin\theta} r^3\, dr\, d\theta$

(*iii*) $\int_0^{\pi} \int_0^{\cos\theta} \rho \sin\theta\, d\rho\, d\theta$ (*iv*) $\int_0^1 \int_0^{\sqrt{1-y^2}} (x^2 + y^2)\, dx\, dy$

39. Compute the mass of a sphere of radius r if the density varies inversely as the square of the distance from the centre.

40. The parabolic arc $y = \sqrt{x}$, $1 \le x \le 2$ is revolved around x-axis. Find the volume of solid of revolution. *(A.K.T.U. 2017)*

41. Evaluate: $\int_0^a \int_{\sqrt{ax - x^2}}^{\sqrt{a^2 - x^2}} \frac{dy\, dx}{\sqrt{a^2 - x^2 - y^2}}$.

42. Compute the volume bounded by $xy = z$, $z = 0$ and $(x - 1)^2 + (y - 1)^2 = 1$.

43. Compute the volume in the first octant bounded by the cylinder $x = 4 - y^2$ and the planes $z = y$, $x = 0$, $z = 0$.

44. Find the area bounded by curve $y^2 = x$ and $x^2 = y$. *(A.K.T.U. 2022)*

45. Find the value of $\int_0^1 \int_0^x \int_0^{x+y} dx\, dy\, dz$. *(A.K.T.U. 2022)*

46. Evaluate $\int_0^2 \int_0^1 (x^2 + 3y^2)\, dy\, dx$. *(A.K.T.U. 2020)*

47. Find the value of integral $\int_0^{\infty} e^{-ax} x^{n-1} dx$. *(A.K.T.U. 2016)*

48. Evaluate $\int_0^{\infty} \sqrt{x}\, e^{-x}\, dx$. *(U.P.T.U. 2014)*

49. Find the value of $\Gamma\left(-\frac{5}{2}\right)$. *(A.K.T.U. 2014, 2019)*

50. Evaluate $\Gamma(-3/2)$. *(G.B.T.U. 2013)*

51. Find the value of $\Gamma\left(-\frac{1}{2}\right)$. *(G.B.T.U. 2011)*

52. Evaluate : $\dfrac{\Gamma(8/3)}{\Gamma(2/3)}$

53. Show that $\int_{-1}^{1} \sqrt{\dfrac{1+t}{1-t}}\, dt = \pi$.

54. Evaluate $\beta(2, 1) + \beta(1, 2)$.

55. Evaluate $\iint_D x^3 y\, dx\, dy$ where D is the region enclosed by the ellipse $\dfrac{x^2}{a^2} + \dfrac{y^2}{b^2} = 1$ in the first quadrant.

56. Evaluate $\dfrac{\beta(m+1, n)}{\beta(m, n)}$. *(G.B.T.U. 2011; U.K.T.U. 2012)*

57. Evaluate $\Gamma\left(\frac{1}{4}\right)\Gamma\left(\frac{3}{4}\right)$. *(U.P.T.U. 2014)*

58. Evaluate $\int_0^{\pi/2} \sin^7 \theta \sqrt{\cos \theta}\, d\theta$ in terms of Beta function.

59. Evaluate:

(*i*) $\Gamma(3.5)$ (*ii*) $\beta\left(\frac{1}{2}, \frac{1}{2}\right)$

60. Evaluate $\int_0^{\infty} e^{-x^2}\, dx$. *(A.K.T.U. 2011, 2018)*

61. Evaluate $\int_0^{\infty} e^{-x^4}\, dx$.

62. Use Beta function to evaluate : $\int_0^{\infty} \dfrac{x^8(1-x^6)}{(1+x)^{24}}\, dx$. *(M.T.U. 2013)*

63. Find the total mass of the region in the case $0 \le x \le 1,\ 0 \le y \le 1,\ 0 \le z \le 1$ with density at any point being xyz.

64. Evaluate $\iint_D x^p y^q\, dx\, dy$, where D is the region bounded by $x = 0$, $y = 0$ and $\left(\frac{x}{a}\right)^m + \left(\frac{y}{b}\right)^n = 1$.

65. Find the mass of the region in xy-plane bounded by $x = 0$, $y = 0$, $x + y = 1$ with density $k\sqrt{xy}$.

66. Use Dirichlet's integral to evaluate $\iiint xyz\ dx\, dy\, dz$ throughout the volume bounded by $x = 0$, $y = 0$, $z = 0$ and $x + y + z = 1$. *(U.P.T.U. 2014)*

67. Evaluate (in terms of Beta and Gamma Functions)

(*i*) $\int_0^1 \sqrt[3]{x \log\left(\frac{1}{x}\right)}\, dx$ (*ii*) $\int_0^{\pi/2} \dfrac{\sqrt[3]{\sin^8 x}}{\sqrt{\cos x}}\, dx$

(*iii*) $\int_0^1 \left(\frac{x}{1-x^3}\right)^{1/2} dx$ (*U.P.T.U. 2014*) (*iv*) $\int_0^\infty \frac{x^2}{(1+x^4)^3}\, dx$

68. **Evaluate:** $\int_{-\infty}^{\infty} e^{-x^2}\, dx$

69. **Evaluate:** $\int_0^\infty \frac{x^4}{(x+1)^{10}}\, dx$

70. **Evaluate:** $\int_0^1 x^4 (1-x)^3\, dx$

71. **Evaluate:** $\int_0^1 \frac{dx}{\sqrt{-\ln x}}$. (*A.K.T.U. 2022*)

72. **Determine the area enclosed by the curve** $\left(\frac{x}{a}\right)^{2m} + \left(\frac{y}{b}\right)^{2n} = 1$ **;** m, n **being positive integers.**

73. **Find the area enclosed by the curve** $\left(\frac{x}{a}\right)^4 + \left(\frac{y}{b}\right)^{10} = 1$.

74. **Compute the mass of a sphere of radius** r **if the density varies inversely as the square of the distance from the centre.**

75. **If V is the solid region in the first octant bounded by the unit sphere** $x^2 + y^2 + z^2 = 1$**, find the mass of the ellipsoid** $\left(\frac{x}{a}\right)^2 + \left(\frac{y}{b}\right)^2 + \left(\frac{z}{c}\right)^2 = 1$ **with density** $\rho(x, y, z) = (x + y + z)^4$.

Answers

1. $\frac{1}{6}$ **2.** $\frac{3}{4}$ **3.** $\frac{15}{8}$ **4.** 8

5. 48 **6.** $\frac{\pi}{2}a^3$ **7.** (*i*) $8 \log 8 - 16 + e$ (*ii*) $\frac{e}{2} - 1$

8. $\frac{768}{35}$ **9.** $\frac{b^2 a^4}{24}$ **10.** $\ln\left(\frac{25}{24}\right)$ **11.** $\frac{63}{32}$

12. $\frac{1}{5}$ **13.** $\frac{a^2}{3}$ **14.** $\frac{10}{3}a^2$ **15.** $\frac{7}{8}$

16. $\frac{1}{8}$ **17.** 3π **18.** $\frac{a^{p+1} b^{q+1}}{mn} \frac{\Gamma\left(\frac{p+1}{m}\right)\Gamma\left(\frac{q+1}{n}\right)}{\Gamma\left(\frac{p+1}{m}+\frac{q+1}{n}+1\right)}$

19. $\frac{k\pi}{24}$ **20.** $\frac{1}{4}\Gamma\left(\frac{1}{4}\right)$ **21.** $\frac{38\pi}{3}$

22. $\int_0^{2a}\int_{\frac{x^2}{4a}}^{3a-x} f(x, y)\, dy\, dx$ **23.** $\frac{1}{2}\sqrt{\pi}$ **24.** $\int_0^\infty \int_0^y f(x, y)\, dx\, dy$

25. $\frac{1}{8}$ **26.** 0 **27.** $\frac{10}{3}$

28. $\frac{4\pi\, abc\,(a^2+b^2+c^2)}{35}$

29. Area of a triangle with vertices (0, 0), (1, 0) and (0, 1)

30. $\frac{4\pi}{3}$

31. $\frac{ab}{4mn}\cdot\frac{\Gamma\left(\frac{1}{2m}\right)\Gamma\left(\frac{1}{2n}\right)}{\Gamma\left(\frac{1}{2m}+\frac{1}{2n}+1\right)}$

32. $\frac{ab}{40}\cdot\frac{\Gamma\left(\frac{1}{4}\right)\Gamma\left(\frac{1}{10}\right)}{\Gamma\left(\frac{27}{20}\right)}$

33. 3π

34. $\frac{3}{2}\pi a^2$

35. $\frac{\pi}{16}(b^4-a^4)$

36. $\frac{5}{3}, 2$

37. (i) $\frac{7}{2}$ (ii) $\frac{47}{24}$ (iii) $\frac{\pi}{2}-1$ (iv) $\frac{1}{48}$

38. (i) $\frac{3\pi}{16}$ (ii) $\frac{45\pi}{2}$ (iii) $\frac{1}{3}$ (iv) $\frac{\pi}{8}$

39. $4k\pi r$

40. $\frac{3\pi}{2}$

41. a

42. π

43. 4

44. $\frac{1}{3}$

45. $\frac{1}{2}$

46. $\frac{14}{3}$

47. $\frac{n!}{a^n}$

48. $\frac{1}{2}\sqrt{\pi}$

49. $\frac{-8}{15}\sqrt{\pi}$

50. $\frac{4}{3}\sqrt{\pi}$

51. $-2\sqrt{\pi}$

52. $\frac{10}{9}$

54. 1

55. $\frac{b^2a^4}{24}$

56. $\frac{m}{m+n}$

57. $\pi\sqrt{2}$

58. $\frac{1}{2}\beta\left(4,\frac{3}{4}\right)$

59. (i) $\frac{15}{8}\sqrt{\pi}$ (ii) π

60. $\frac{1}{2}\sqrt{\pi}$

61. $\frac{1}{4}\Gamma\left(\frac{1}{4}\right)$

62. 0

63. $\frac{1}{8}$

64. $\frac{a^{p+1}b^{q+1}}{mn}\cdot\frac{\Gamma\left(\frac{p+1}{m}\right)\Gamma\left(\frac{q+1}{n}\right)}{\Gamma\left(\frac{p+1}{m}+\frac{q+1}{n}+1\right)}$

65. $\frac{k\pi}{24}$

66. $\frac{1}{720}$

67. (i) $\left(\frac{3}{4}\right)^{4/3}\Gamma\left(\frac{4}{3}\right)$ (ii) $\frac{1}{2}\beta\left(\frac{11}{6},\frac{1}{4}\right)$ (iii) $\frac{1}{3}\beta\left(\frac{5}{6},\frac{1}{2}\right)$ (iv) $\frac{1}{4}\beta\left(\frac{3}{4},\frac{9}{4}\right)$

68. $\sqrt{\pi}$

69. $\frac{1}{630}$

70. $\frac{1}{280}$

71. $2\sqrt{\pi}$

72. $\frac{ab}{4mn}\cdot\frac{\Gamma\left(\frac{1}{2m}\right)\Gamma\left(\frac{1}{2n}\right)}{\Gamma\left(\frac{1}{2m}+\frac{1}{2n}+1\right)}$

73. $\frac{ab}{40}\cdot\frac{\Gamma\left(\frac{1}{4}\right)\Gamma\left(\frac{1}{10}\right)}{\Gamma\left(\frac{27}{20}\right)}$

74. $4k\pi r$

75. $\frac{4\pi\, abc\,(a^2+b^2+c^2)}{35}$.

MODULE 5

Vector Calculus

5.1. SCALAR AND VECTOR FUNCTIONS

Scalar function $f(x, y, z)$ is a function defined at each point in a certain domain D in space. Its value is real and depends only on the point $P(x, y, z)$ in space but not on any particular coordinate system being used.

If to each value of a scalar variable t, there corresponds a value of a vector $\vec{r}$, then $\vec{r}$ is called a vector function of the scalar variable t and we write $\vec{r} = \vec{r}(t)$ or $\vec{r} = \vec{f}(t)$

For example, the position vector $\vec{r}$ of a particle moving along a curved path is a vector function of time t, a scalar.

Since every vector can be uniquely expressed as a linear combination of three fixed non-coplanar vectors, therefore we may write $\vec{f}(t) = f_1(t)\,\hat{i} + f_2(t)\,\hat{j} + f_3(t)\,\hat{k}$ where $\hat{i}, \hat{j}, \hat{k}$ denote unit vectors along the axis of x, y, z respectively and $f_1(t), f_2(t), f_3(t)$ are called the components of the vector $\vec{f}(t)$ along the coordinate axes.

5.2. DERIVATIVE OF A VECTOR FUNCTION WITH RESPECT TO A SCALAR

Let $\vec{r} = \vec{f}(t)$ be a vector function of the scalar variable t. Let δt be a small increment in t and $\delta\vec{r}$, the corresponding increment in $\vec{r}$. Then

$$\vec{r} + \delta\vec{r} = \vec{f}(t + \delta t) \quad \text{so that} \quad \delta\vec{r} = \vec{f}(t + \delta t) - \vec{f}(t) \quad \text{and} \quad \frac{\delta\vec{r}}{\delta t} = \frac{\vec{f}(t + \delta t) - \vec{f}(t)}{\delta t}$$

If $\underset{\delta t \to 0}{\text{Lt}} \dfrac{\delta\vec{r}}{\delta t} = \underset{\delta t \to 0}{\text{Lt}} \dfrac{\vec{f}(t + \delta t) - \vec{f}(t)}{\delta t}$ exists, then the value of this limit is denoted by $\dfrac{d\vec{r}}{dt}$ and is called the derivative of $\vec{r}$ with respect to t.

Since $\dfrac{d\vec{r}}{dt}$ is itself a vector function of t, its derivative is denoted by $\dfrac{d^2\vec{r}}{dt^2}$ and is called the second derivative of $\vec{r}$ with respect to t. Similarly, we can define higher order derivatives of $\vec{r}$.

5.3. GENERAL RULES FOR DIFFERENTIATION

If $\vec{a}$, $\vec{b}$ and $\vec{c}$ are vector functions of a scalar t and ϕ is a scalar function of t, then

(i) $\dfrac{d}{dt}(\vec{a} \pm \vec{b}) = \dfrac{d\vec{a}}{dt} \pm \dfrac{d\vec{b}}{dt}$

(ii) $\dfrac{d}{dt}(\vec{a}.\vec{b}) = \vec{a}.\dfrac{d\vec{b}}{dt} + \dfrac{d\vec{a}}{dt}.\vec{b}$

(iii) $\dfrac{d}{dt}(\vec{a} \times \vec{b}) = \vec{a} \times \dfrac{d\vec{b}}{dt} + \dfrac{d\vec{a}}{dt} \times \vec{b}$

(iv) $\dfrac{d}{dt}(\phi\vec{a}) = \phi\dfrac{d\vec{a}}{dt} + \dfrac{d\phi}{dt}\vec{a}$

(v) $\dfrac{d}{dt}[\vec{a}\ \vec{b}\ \vec{c}] = \left[\dfrac{d\vec{a}}{dt}\vec{b}\,\vec{c}\right] + \left[\vec{a}\dfrac{d\vec{b}}{dt}\vec{c}\right] + \left[\vec{a}\,\vec{b}\dfrac{d\vec{c}}{dt}\right]$

(vi) $\dfrac{d}{dt}\{\vec{a} \times (\vec{b} \times \vec{c})\} = \dfrac{d\vec{a}}{dt} \times (\vec{b} \times \vec{c}) + \vec{a} \times \left(\dfrac{d\vec{b}}{dt} \times \vec{c}\right) + \vec{a} \times \left(\vec{b} \times \dfrac{d\vec{c}}{dt}\right)$

5.4. DERIVATIVE OF A CONSTANT VECTOR

A vector is called constant if both its magnitude and direction are fixed. If either of these changes, the vector is not constant.

Let $\vec{r}$ be a constant vector function of the scalar variable t.

Let $\vec{r} = \vec{f}(t)$, then $\vec{r} = \vec{f}(t + \delta t)$ so that $\vec{f}(t + \delta t) - \vec{f}(t) = \vec{0}$

$$\therefore \quad \frac{d\vec{r}}{dt} = \underset{\delta t \to 0}{\text{Lt}} \frac{\vec{f}(t + \delta t) - \vec{f}(t)}{\delta t} = \underset{\delta t \to 0}{\text{Lt}}\ \vec{0} = \vec{0}$$

Thus, the derivative of a constant vector is equal to the null vector.

Note. $\hat{i}, \hat{j}, \hat{k}$ being fixed unit vectors are constant vectors.

$$\therefore \quad \frac{d\hat{i}}{dt} = \frac{d\hat{j}}{dt} = \frac{d\hat{k}}{dt} = \vec{0}.$$

5.5. DERIVATIVE OF A VECTOR FUNCTION IN TERMS OF ITS COMPONENTS

Let $\vec{r}$ be a vector function of the scalar variable t.

Let $\vec{r} = x\hat{i} + y\hat{j} + z\hat{k}$, where the components x, y, z are scalar functions of t.

$$\frac{d\vec{r}}{dt} = \frac{d}{dt}(x\hat{i}) + \frac{d}{dt}(y\hat{j}) + \frac{d}{dt}(z\hat{k})$$

$$= x\frac{d\hat{i}}{dt} + \frac{dx}{dt}\hat{i} + y\frac{d\hat{j}}{dt} + \frac{dy}{dt}\hat{j} + z\frac{d\hat{k}}{dt} + \frac{dz}{dt}\hat{k}$$

$$= \frac{dx}{dt}\hat{i} + \frac{dy}{dt}\hat{j} + \frac{dz}{dt}\hat{k}, \text{ since } \frac{d\hat{i}}{dt} = \frac{d\hat{j}}{dt} = \frac{d\hat{k}}{dt} = \vec{0}.$$

If $x = f_1(t)$, $y = f_2(t)$, $z = f_3(t)$; then $\vec{r} = f_1(t)\,\hat{i} + f_2(t)\,\hat{j} + f_3(t)\,\hat{k}$

$$\Rightarrow \quad \frac{d\vec{r}}{dt} = f_1'(t)\,\hat{i} + f_2'(t)\,\hat{j} + f_3'(t)\,\hat{k}$$

Therefore to differentiate a vector, differentiate its components.

5.6. IF $\vec{F}(t)$ HAS A CONSTANT MAGNITUDE, THEN $\vec{F} \cdot \frac{d\vec{F}}{dt} = 0$

$$|\vec{F}(t)| = \text{constant}$$

$$\therefore \quad \vec{F}(t) \cdot \vec{F}(t) = |\vec{F}(t)|^2 = \text{constant}$$

$$\Rightarrow \quad \frac{d}{dt}\{\vec{F}(t) \cdot \vec{F}(t)\} = 0$$

$$\Rightarrow \quad \vec{F} \cdot \frac{d\vec{F}}{dt} + \frac{d\vec{F}}{dt} \cdot \vec{F} = 0$$

$$\Rightarrow \quad 2\vec{F} \cdot \frac{d\vec{F}}{dt} = 0 \Rightarrow \vec{F} \cdot \frac{d\vec{F}}{dt} = 0$$

5.7. IF $\vec{F}(t)$ HAS A CONSTANT DIRECTION, THEN $\vec{F} \times \frac{d\vec{F}}{dt} = \vec{0}$

Let $|\vec{F}(t)| = f(t)$. Let $\hat{G}(t)$ be a unit vector in the direction of $\vec{F}(t)$ so that $\vec{F}(t) = f(t)\,\hat{G}(t)$

$$\therefore \quad \frac{d\vec{F}}{dt} = f\frac{d\hat{G}}{dt} + \frac{df}{dt}\hat{G} \qquad ...(1)$$

If $\vec{F}(t)$ has constant direction, so that $\hat{G}(t)$. Thus, $\hat{G}(t)$ is a constant vector and $\frac{d\hat{G}}{dt} = \vec{0}$

From (1), $$\frac{d\vec{F}}{dt} = \frac{df}{dt}\hat{G}$$

$$\therefore \quad \vec{F} \times \frac{d\vec{F}}{dt} = f\hat{G} \times \left(\frac{df}{dt}\hat{G}\right) = f\frac{df}{dt}\hat{G} \times \hat{G} = \vec{0}$$

5.8. GEOMETRICAL INTERPRETATION OF $\frac{d\vec{r}}{dt}$

Let O be the origin of reference. Let the position vector of a point P be given by $\vec{r} = \vec{f}(t)$. As t varies continuously, P traces out a curve C as shown in the figure. Thus, a vector function $\vec{f}(t)$ represents a curve in space.

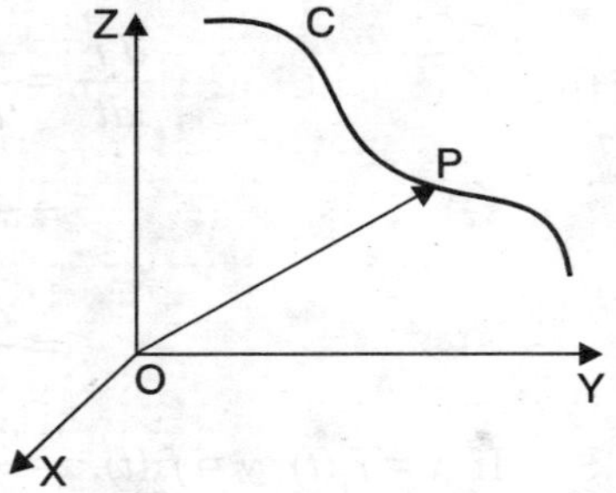

For example, the vector equation $\vec{r} = at^2\hat{i} + 2at\hat{j}$ represents the parabola $y^2 = 4ax$ in the xy-plane because its parametric equations are $x = at^2$, $y = 2at$.

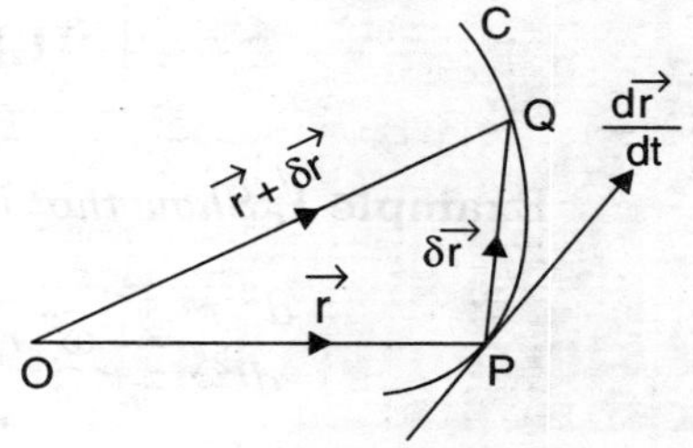

Now, let $\vec{r} = \vec{f}(t)$ be the vector equation of a curve C in space. Let $\vec{r}$ and $\vec{r} + \delta\vec{r}$ be the position vectors of two neighbouring points P and Q on this curve.

$$\vec{PQ} = \vec{OQ} - \vec{OP} = (\vec{r} + \delta\vec{r}) - \vec{r} = \delta\vec{r}$$

$\therefore$ $\dfrac{\delta\vec{r}}{\delta t}$ is directed along the chord PQ.

As $\delta t \to 0$, Q $\to$ P, chord PQ $\to$ tangent to the curve at P.

$\therefore$ $\displaystyle\operatorname*{Lt}_{\delta t \to 0} \frac{\delta\vec{r}}{\delta t} = \frac{d\vec{r}}{dt}$ *is a vector along the tangent to the curve at P.*

Suppose the scalar parameter t is replaced by s, where s denotes the arc length from any convenient point A on the curve upto P. Thus, arc AP = s and arc AQ = $s + \delta s$ so that δs = arc PQ. In this case $\dfrac{d\vec{r}}{ds}$ will be a vector along the tangent at P. Also

$$\left|\frac{d\vec{r}}{ds}\right| = \operatorname*{Lt}_{\delta s \to 0}\left|\frac{\delta\vec{r}}{\delta s}\right| = \operatorname*{Lt}_{Q \to P} \frac{\text{chord PQ}}{\text{arc PQ}} = 1.$$

Thus, $\dfrac{d\vec{r}}{ds}$ *is the unit vector $\hat{T}$ along the tangent at P.*

5.9. VELOCITY AND ACCELERATION

If the scalar variable t denotes the time and $\vec{r}$ is the position vector of a moving particle P, then $\delta\vec{r}$ is the displacement of the particle in time δt. The vector $\dfrac{\delta\vec{r}}{\delta t}$ is the average velocity of the particle during the interval δt. If $\vec{v}$ represents the velocity vector of the particle at P, then

$\displaystyle\vec{v} = \operatorname*{Lt}_{\delta t \to 0} \frac{\delta\vec{r}}{\delta t} = \frac{d\vec{r}}{dt}$ and its direction is along the tangent at P.

If $\delta\vec{v}$ be the change in velocity $\vec{v}$ during the time δt, then $\dfrac{\delta\vec{v}}{\delta t}$ is the average acceleration of the particle during the interval δt. If $\vec{a}$ represents the acceleration of the particle at P, then

$$\vec{a} = \operatorname*{Lt}_{\delta t \to 0} \frac{\delta\vec{v}}{\delta t} = \frac{d\vec{v}}{dt} = \frac{d}{dt}\left(\frac{d\vec{r}}{dt}\right) = \frac{d^2\vec{r}}{dt^2}.$$

ILLUSTRATIVE EXAMPLES

Example 1. *Show that if* $\vec{r} = \vec{a} \sin \omega t + \vec{b} \cos \omega t$, *where* $\vec{a}, \vec{b}$, ω *are constants, then*

$$\frac{d^2 \vec{r}}{dt^2} = -\omega^2 \vec{r} \quad \text{and} \quad \vec{r} \times \frac{d\vec{r}}{dt} = -\omega \vec{a} \times \vec{b}.$$

Sol.
$$\vec{r} = \vec{a} \sin \omega t + \vec{b} \cos \omega t$$

$$\Rightarrow \quad \frac{d\vec{r}}{dt} = \vec{a}\,\omega \cos \omega t - \vec{b}\,\omega \sin \omega t$$

$$\therefore \quad \frac{d^2 \vec{r}}{dt^2} = -\vec{a}\,\omega^2 \sin \omega t - \vec{b}\,\omega^2 \cos \omega t = -\omega^2 (\vec{a} \sin \omega t + \vec{b} \cos \omega t) = -\omega^2 \vec{r}$$

Also
$$\vec{r} \times \frac{d\vec{r}}{dt} = (\vec{a} \sin \omega t + \vec{b} \cos \omega t) \times (\vec{a}\,\omega \cos \omega t - \vec{b}\,\omega \sin \omega t)$$

$$= \omega(-\vec{a} \times \vec{b} \sin^2 \omega t + \vec{b} \times \vec{a} \cos^2 \omega t) \qquad [\because \vec{a} \times \vec{a} = \vec{0} = \vec{b} \times \vec{b}]$$

$$= \omega(-\vec{a} \times \vec{b} \sin^2 \omega t - \vec{a} \times \vec{b} \cos^2 \omega t) = -\omega\, \vec{a} \times \vec{b}.$$

Example 2. *If* $\vec{r} = a \cos t\hat{i} + a \sin t\hat{j} + at \tan \alpha\, \hat{k}$, *find* $\left| \frac{d\vec{r}}{dt} \times \frac{d^2\vec{r}}{dt^2} \right|$ *and* $\left[\frac{d\vec{r}}{dt}\ \frac{d^2\vec{r}}{dt^2}\ \frac{d^3\vec{r}}{dt^3} \right]$.

Sol.
$$\vec{r} = a \cos t\hat{i} + a \sin t\hat{j} + at \tan \alpha \hat{k}$$

$$\Rightarrow \quad \frac{d\vec{r}}{dt} = -a \sin t\hat{i} + a \cos t\hat{j} + a \tan \alpha \hat{k}$$

$$\frac{d^2\vec{r}}{dt^2} = -a \cos t\hat{i} - a \sin t\hat{j} \quad \text{and} \quad \frac{d^3\vec{r}}{dt^3} = a \sin t\hat{i} - a \cos t\hat{j}$$

$$\therefore \quad \frac{d\vec{r}}{dt} \times \frac{d^2\vec{r}}{dt^2} = \begin{vmatrix} \hat{i} & \hat{j} & \hat{k} \\ -a \sin t & a \cos t & a \tan \alpha \\ -a \cos t & -a \sin t & 0 \end{vmatrix}$$

$$= a^2 \sin t \tan \alpha \hat{i} - a^2 \cos t \tan \alpha \hat{j} + a^2 \hat{k}$$

$$\therefore \quad \left| \frac{d\vec{r}}{dt} \times \frac{d^2\vec{r}}{dt^2} \right| = \sqrt{a^4 \sin^2 t \tan^2 \alpha + a^4 \cos^2 t \tan^2 \alpha + a^4} = a^2 \sqrt{\tan^2 \alpha + 1} = a^2 \sec \alpha$$

Also
$$\left[\frac{d\vec{r}}{dt}\ \frac{d^2\vec{r}}{dt^2}\ \frac{d^3\vec{r}}{dt^3} \right] = \begin{vmatrix} -a \sin t & a \cos t & a \tan \alpha \\ -a \cos t & -a \sin t & 0 \\ a \sin t & -a \cos t & 0 \end{vmatrix} \qquad \text{(Expanding by third column)}$$

$$= a \tan \alpha (a^2 \cos^2 t + a^2 \sin^2 t) = a^3 \tan \alpha.$$

Example 3. *Find the unit tangent vector at any point on the curve $x = t^2 + 2$, $y = 4t - 5$, $z = 2t^2 - 6t$, where t is any variable. Also determine the unit tangent vector at the point $t = 2$.*

Sol. If $\vec{r}$ is the position vector of any point (x, y, z) on the given curve, then

$$\vec{r} = x\hat{i} + y\hat{j} + z\hat{k} = (t^2 + 2)\hat{i} + (4t - 5)\hat{j} + (2t^2 - 6t)\hat{k}$$

The vector $\frac{d\vec{r}}{dt}$ is along the tangent at the point (x, y, z) to the given curve.

Now $\frac{d\vec{r}}{dt} = 2t\hat{i} + 4\hat{j} + (4t - 6)\hat{k}$ and $\left|\frac{d\vec{r}}{dt}\right| = \sqrt{(2t)^2 + (4)^2 + (4t-6)^2} = 2\sqrt{5t^2 - 12t + 13}$

$\therefore$ The unit tangent vector $\hat{T} = \dfrac{\frac{d\vec{r}}{dt}}{\left|\frac{d\vec{r}}{dt}\right|} = \dfrac{t\hat{i} + 2\hat{j} + (2t - 3)\hat{k}}{\sqrt{5t^2 - 12t + 13}}$

Also the unit tangent vector at the point $t = 2$ is $\dfrac{2\hat{i} + 2\hat{j} + (2\times 2 - 3)\hat{k}}{\sqrt{5\times 4 - 12\times 2 + 13}} = \frac{1}{3}(2\hat{i} + 2\hat{j} + \hat{k})$.

Example 4. *Find the angle between the tangents to the curve $\vec{r} = t^2\hat{i} + 2t\hat{j} + t^3\hat{k}$ at the points $t = \pm 1$.*

Sol. $\frac{d\vec{r}}{dt} = 2t\hat{i} + 2\hat{j} + 3t^2\hat{k}$ is a vector along the tangent at any point 't'.

If $\vec{T_1}$ and $\vec{T_2}$ are the vectors along the tangents at $t = 1$ and $t = -1$ respectively, then

$$\vec{T_1} = 2\hat{i} + 2\hat{j} + 3\hat{k} \quad \text{and} \quad \vec{T_2} = -2\hat{i} + 2\hat{j} + 3\hat{k}$$

If θ is the angle between $\vec{T_1}$ and $\vec{T_2}$, then

$$\cos\theta = \frac{\vec{T_1}\cdot\vec{T_2}}{|\vec{T_1}||\vec{T_2}|} = \frac{2(-2) + 2(2) + 3(3)}{\sqrt{4+4+9}\,.\sqrt{4+4+9}} = \frac{9}{17}$$

$$\therefore \quad \theta = \cos^{-1}\left(\frac{9}{17}\right).$$

Example 5. *A particle moves on the curve $x = 2t^2$, $y = t^2 - 4t$, $z = 3t - 5$, where t is the time. Find the components of velocity and acceleration at time $t = 1$ in the direction $\hat{i} - 3\hat{j} + 2\hat{k}$.*

Sol. If $\vec{r}$ is the position vector of any point (x, y, z) on the given curve, then

$$\vec{r} = x\hat{i} + y\hat{j} + z\hat{k} = 2t^2\hat{i} + (t^2 - 4t)\hat{j} + (3t - 5)\hat{k}$$

Velocity $\quad \vec{v} = \frac{d\vec{r}}{dt} = 4t\hat{i} + (2t - 4)\hat{j} + 3\hat{k} = 4\hat{i} - 2\hat{j} + 3\hat{k}$ at $t = 1$

Acceleration $\quad \vec{a} = \frac{d^2\vec{r}}{dt^2} = 4\hat{i} + 2\hat{j}$ at $t = 1$

Now unit vector in the given direction $\hat{i} - 3\hat{j} + 2\hat{k} = \dfrac{\hat{i} - 3\hat{j} + 2\hat{k}}{\sqrt{14}} = \hat{n}$ (say)

$\therefore$ The component of velocity in the given direction

$$= \vec{v} \, . \, \hat{n} = (4\hat{i} - 2\hat{j} + 3\hat{k}) \, . \, \frac{\hat{i} - 3\hat{j} + 2\hat{k}}{\sqrt{14}}$$

$$= \frac{4(1) - 2(-3) + 3(2)}{\sqrt{14}} = \frac{16\sqrt{14}}{14} = \frac{8\sqrt{14}}{7}$$

and the component of acceleration in the given direction

$$= \vec{a} \, . \, \hat{n} = (4\hat{i} + 2\hat{j}) \, . \, \frac{\hat{i} - 3\hat{j} + 2\hat{k}}{\sqrt{14}} = \frac{-2}{\sqrt{14}} = -\frac{\sqrt{14}}{7} \, .$$

Example 6. *Show that the radial and transverse acceleration of a particle moving in a plane are:*

$$\frac{d^2 r}{dt^2} - r\left(\frac{d\theta}{dt}\right)^2 \text{ and } 2\,\frac{dr}{dt}\,\frac{d\theta}{dt} + r\,\frac{d^2\theta}{dt^2} \text{ respectively.}$$

Sol. Let i and j be two mutually perpendicular unit vectors in the plane of paper. Also let $\hat{e}_r$ and $\hat{e}_\theta$ be unit vectors along and perpendicular to radius vector OP.

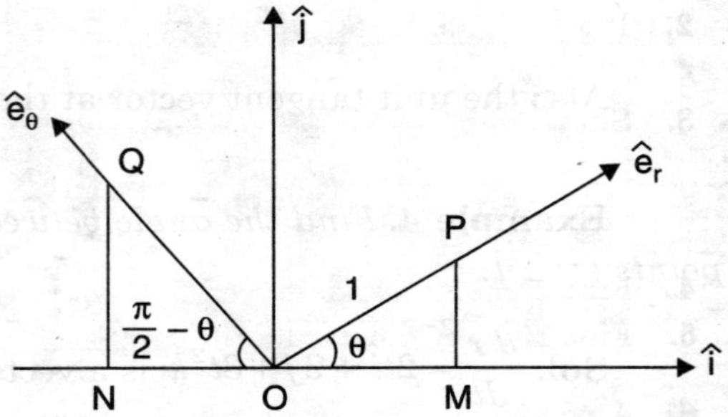

Then, $\hat{e}_r = \overline{\text{OP}} = \overline{\text{OM}} + \overline{\text{MP}}$

$$= (\cos\theta)\,\hat{i} + (\sin\theta)\,\hat{j} \quad ...(1)$$

$$\hat{e}_\theta = \cos(90° + \theta)\,\hat{i} + \sin(90° + \theta)\,\hat{j}$$

$$= (-\sin\theta)\,\hat{i} + (\cos\theta)\,\hat{j} \quad ...(2)$$

Also, $\vec{r} = r\hat{e}_r = r\cos\theta\,\hat{i} + r\sin\theta\,\hat{j}$ ...(3)

$$\frac{d\hat{e}_r}{dt} = (-\sin\theta\,\hat{i} + \cos\theta\,\hat{j})\,\frac{d\theta}{dt} = \hat{e}_\theta\,\frac{d\theta}{dt} = \dot{\theta}\hat{e}_\theta \quad ...(4)$$

$$\frac{d\hat{e}_\theta}{dt} = (-\cos\theta\,\hat{i} - \sin\theta\,\hat{j})\,\frac{d\theta}{dt} = -\,\hat{e}_r\,\frac{d\theta}{dt} = -\,\dot{\theta}\hat{e}_r \quad ...(5)$$

Now, $$\frac{d\vec{r}}{dt} = \frac{d}{dt}\,(r\cos\theta\,\hat{i} + r\sin\theta\,\hat{j})$$

$$= \left(\frac{dr}{dt}\cos\theta - r\sin\theta\,\frac{d\theta}{dt}\right)\hat{i} + \left(\frac{dr}{dt}\sin\theta + r\cos\theta\,\frac{d\theta}{dt}\right)\hat{j}$$

$$= \frac{dr}{dt}\,(\cos\theta\,\hat{i} + \sin\theta\,\hat{j}) + r\,\frac{d\theta}{dt}\,(-\sin\theta\,\hat{i} + \cos\theta\,\hat{j})$$

$$= \dot{r}\,\hat{e}_r + r\,\dot{\theta}\hat{e}_\theta$$

Acceleration $$= \frac{d^2\vec{r}}{dt} = \ddot{r}\,\hat{e}_r + \dot{r}\,\frac{d\hat{e}_r}{dt} + \dot{r}\,\dot{\theta}\,\hat{e}_\theta + r\,\ddot{\theta}\,\hat{e}_\theta + r\,\dot{\theta}\,\frac{d\hat{e}_\theta}{dt}$$

$$= \ddot{r}\,\hat{e}_r + 2\dot{r}\,\dot{\theta}\,\hat{e}_\theta + r\,\ddot{\theta}\,\hat{e}_\theta - r\dot{\theta}^2\,\hat{e}_r$$

$$= (\ddot{r} - r\,\dot{\theta}^2)\,\hat{e}_r + (2\,\dot{r}\,\dot{\theta} + r\,\ddot{\theta})\,\hat{e}_\theta$$

$$= \left\{\frac{d^2r}{dt^2} - r\left(\frac{d\theta}{dt}\right)^2\right\}\hat{e}_r + \frac{1}{r}\frac{d}{dt}\left(r^2\frac{d\theta}{dt}\right)\hat{e}_\theta$$

Hence, Radial acceleration = acceleration along $\hat{e}_r = \frac{d^2r}{dt^2} - r\left(\frac{d\theta}{dt}\right)^2$

and Transverse acceleration = acceleration along $\hat{e}_\theta = \frac{1}{r}\frac{d}{dt}\left(r^2\frac{d\theta}{dt}\right)$

TEST YOUR KNOWLEDGE

1. If $\vec{r} = \sin t\,\hat{i} + \cos t\,\hat{j} + t\,\hat{k}$, find $\left|\frac{d^2\vec{r}}{dt^2}\right|$.

2. If $\vec{r} = (\cos nt)\,\hat{i} + (\sin nt)\,\hat{j}$, where n is a constant and t varies, show that $\vec{r} \times \frac{d\vec{r}}{dt} = n\hat{k}$.

3. Show that $\vec{r} = \vec{a}\,e^{mt} + \vec{b}\,e^{nt}$ is the solution of the differential equation

$$\frac{d^2\vec{r}}{dt^2} - (m+n)\frac{d\vec{r}}{dt} + mn\,\vec{r} = 0.$$

4. Find the unit tangent vector at any point on the curve $x = 3\cos t$, $y = 3\sin t$, $z = 4t$.
5. Find the angle between the tangents to the curve $x = t$, $y = t^2$, $z = t^3$ at $t = \pm 1$.

6. A particle, whose position vector is $\vec{r}$, is moving in a circle with constant angular velocity w. Show, by vector methods, that the acceleration is equal to $-w^2\vec{r}$.

7. The position vector of a point at time t is given by $\vec{r} = e^t(\cos t\,\hat{i} + \sin t\,\hat{j})$. Show that

   (i) $\vec{a} = 2(\vec{v} - \vec{r})$ where $\vec{a}, \vec{v}$ are acceleration and velocity of the particle.
   (ii) The angle between the radius vector and the acceleration is constant.

8. A particle moves so that its position vector is given by $\vec{r} = \cos wt\,\hat{i} + \sin wt\,\hat{j}$. Show that the velocity $\vec{v}$ of the particle is perpendicular to $\vec{r}$ and $\vec{r} \times \vec{v}$ is a constant vector.
9. A particle moves along a plane curve such that its linear velocity is perpendicular to the radius vector. Show that the path of the particle is a circle. *(M.T.U. 2013)*

10. The position vector of a particle at time t is $\vec{r} = \cos(t-1)\,\hat{i} + \sinh(t-1)\,\hat{j} + \alpha t^3\,\hat{k}$. Find the condition imposed on α by requiring that at time $t = 1$, the acceleration is normal to the position vector.

Answers

1. 1   4. $\frac{1}{5}(-3\sin t\,\hat{i} + 3\cos t\,\hat{j} + 4\hat{k})$ 5. $\cos^{-1}(3/7)$ 10. $\alpha = \pm\frac{1}{\sqrt{6}}$.

5.10. SCALAR *AND* VECTOR FIELDS

A variable quantity whose value at any point in a region of space depends upon the position of the point, is called a *point function*. There are two types of point functions.

(*a*) **Scalar Point Function.** Let R be a region of space at each point of which a scalar $\phi = \phi(x, y, z)$ is given, then ϕ is called a *scalar function* and R is called a *scalar field.*

The temperature distribution in a medium, the distribution of atmospheric pressure in space are examples of scalar point functions.

(*b*) **Vector Point Function.** Let R be a region of space at each point of which a vector $\vec{v} = \vec{v}\,(x, y, z)$ is given, then $\vec{v}$ is called a *vector point function* and R is called a *vector field.* Each vector $\vec{v}$ of the field is regarded as a localised vector attached to the corresponding point (x, y, z).

The velocity of a moving fluid at any instant, the gravitational force are examples of vector point functions.

5.11. GRADIENT OF A SCALAR FIELD (*A.K.T.U. 2019*)

Let $\phi(x, y, z)$ be a function defining a scalar field, then the vector $\hat{i}\dfrac{\partial\phi}{\partial x} + \hat{j}\dfrac{\partial\phi}{\partial y} + \hat{k}\dfrac{\partial\phi}{\partial z}$ is called the **gradient** of the scalar field ϕ and is denoted by grad ϕ.

Thus, $$\boxed{\text{grad } \phi = \hat{i}\frac{\partial\phi}{\partial x} + \hat{j}\frac{\partial\phi}{\partial y} + \hat{k}\frac{\partial\phi}{\partial z}}$$

The gradient of scalar field ϕ is obtained by operating on ϕ by the vector operator

$$\hat{i}\frac{\partial}{\partial x} + \hat{j}\frac{\partial}{\partial y} + \hat{k}\frac{\partial}{\partial z}$$

This operator is denoted by the symbol ∇, read as **del** (also called nabla).

Thus, grad $\phi = \nabla\phi$.

5.12. GEOMETRICAL INTERPRETATION OF GRADIENT

If a surface $\phi\,(x, y, z) = c$ is drawn through any point P such that at each point on the surface, the function has the same value as at P, then such a surface is called a *level surface* through P. For example, if $\phi\,(x, y, z)$ represents potential at the point (x, y, z), the **equipotential surface** $\phi\,(x, y, z) = c$ is a level surface.

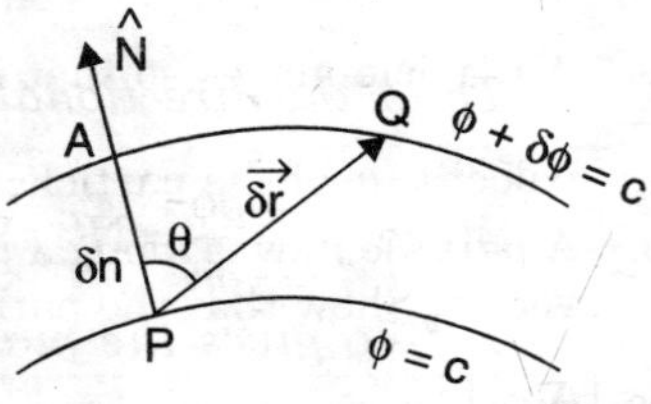

Through any point passes one and only one level surface. Moreover, no two level surfaces can intersect.

Consider the level surface through P at which the function has value ϕ and another level surface through a neighbouring point Q where the value is $\phi + \delta\phi$.

Let $\vec{r}$ and $\vec{r} + \delta\vec{r}$ be the position vectors of P and Q respectively, then $\vec{PQ} = \delta\vec{r}$. Now,

$$\nabla\phi \,.\, \delta\vec{r} = \left(\hat{i}\frac{\partial\phi}{\partial x} + \hat{j}\frac{\partial\phi}{\partial y} + \hat{k}\frac{\partial\phi}{\partial z}\right).(\hat{i}\,\delta x + \hat{j}\delta y + \hat{k}\delta z) = \frac{\partial\phi}{\partial x}\delta x + \frac{\partial\phi}{\partial y}\delta y + \frac{\partial\phi}{\partial z}\delta z = \delta\phi \qquad ...(1)$$

If Q lies on the same level surface as P, then $\delta\phi = 0$,

$\therefore$ (1) reduces to $\nabla\phi \cdot \delta\vec{r} = 0$.

Thus, $\nabla\phi$ is perpendicular to every $\delta\vec{r}$ lying in the surface.
Hence $\nabla\phi$ **is normal to the surface** $\phi(x, y, z) = c$.

Let $\nabla\phi = |\nabla\phi|\,\hat{N}$, where $\hat{N}$ is a unit vector normal to the surface. Let PA = δn be the perpendicular distance between the two level surfaces through P and Q. Then the rate of change of ϕ in the direction of normal to the surface through P is

$$\frac{\partial\phi}{\partial n} = \underset{\delta n \to 0}{\text{Lt}} \frac{\delta\phi}{\delta n} = \underset{\partial n \to 0}{\text{Lt}} \frac{\nabla\phi \cdot \delta\vec{r}}{\delta n} \qquad \text{[By (1)]}$$

$$= \underset{\delta n \to 0}{\text{Lt}} \frac{|\nabla\phi|\,\hat{N} \cdot \delta\vec{r}}{\delta n} = |\nabla\phi| \qquad (\because\ \hat{N} \cdot \delta\vec{r} = |\hat{N}|\,|\delta\vec{r}| \cos\theta = |\delta\vec{r}| \cos\theta = \delta n)$$

$$\therefore \quad |\nabla\phi| = \frac{\partial\phi}{\partial n}.$$

Hence the gradient of a scalar field ϕ is a vector normal to the surface $\phi = c$ and has a magnitude equal to the rate of change of ϕ along this normal.

5.13. DIRECTIONAL DERIVATIVE

Let PQ = δr, then $\underset{\delta r \to 0}{\text{Lt}} \frac{\delta\phi}{\delta r} = \frac{\partial\phi}{\partial r}$ is called the directional derivative of ϕ at P in the direction PQ.

Let $\hat{N}'$ be a unit vector in the direction PQ, then $\delta r = \frac{\delta n}{\cos\theta} = \frac{\delta n}{\hat{N} \cdot \hat{N}'}$

$$\therefore \quad \frac{\partial\phi}{\partial r} = \underset{\delta r \to 0}{\text{Lt}} \left[\hat{N} \cdot \hat{N}' \frac{\delta\phi}{\delta n}\right] = \hat{N} \cdot \hat{N}' \frac{\partial\phi}{\partial n}$$

$$= \hat{N}' \cdot \hat{N} \frac{\partial\phi}{\partial n} = \hat{N}' \cdot \hat{N}\,|\nabla\phi| = \hat{N}' \cdot \nabla\phi \qquad \left(\because\ |\nabla\phi| = \frac{\partial\phi}{\partial n} \text{ and } \hat{N}'|\nabla\phi| = \nabla\phi\right)$$

Thus, the directional derivative $\frac{\partial\phi}{\partial r}$ is the resolved part of $\nabla\phi$ in the direction $\hat{N}'$.

Since, $\quad \frac{\partial\phi}{\partial r} = \hat{N}' \cdot \nabla\phi = |\nabla\phi| \cos\theta \le |\nabla\phi|$.

$\therefore$ *$\nabla\phi$ gives the maximum rate of change of ϕ and the magnitude of this maximum is $|\nabla\phi|$.*

In other words, we can say that the directional derivative of a scalar field f at a point P(x, y, z) in the direction of unit vector $\hat{a}$ is given by

$$\boxed{\frac{df}{ds} = (\text{grad } f) \cdot \hat{a}}$$

5.14. PROPERTIES OF GRADIENT

(*a*) If ϕ is a constant scalar point function, then $\nabla\phi = \vec{0}$

(*b*) If ϕ_1 and ϕ_2 are two scalar point functions, then

(*i*) $\nabla(\phi_1 \pm \phi_2) = \nabla\phi_1 \pm \nabla\phi_2$

(*ii*) $\nabla(c_1\phi_1 + c_2\phi_2) = c_1\nabla\phi_1 + c_2\nabla\phi_2$, where c_1, c_2 are constant

(*iii*) $\nabla(\phi_1\phi_2) = \phi_1\nabla\phi_2 + \phi_2\nabla\phi_1$ (*iv*) $\nabla\left(\dfrac{\phi_1}{\phi_2}\right) = \dfrac{\phi_2\nabla\phi_1 - \phi_1\nabla\phi_2}{\phi_2^{\,2}}$, $\phi_2 \neq 0$.

ILLUSTRATIVE EXAMPLES

Example 1. *Find grad ϕ when ϕ is given by $\phi = 3x^2y - y^3z^2$ at the point (1, – 2, – 1).*

Sol. $$\text{grad } \phi = \nabla\phi = \left(\hat{i}\frac{\partial}{\partial x} + \hat{j}\frac{\partial}{\partial y} + \hat{k}\frac{\partial}{\partial z}\right)(3x^2y - y^3z^2)$$

$$= \hat{i}\frac{\partial}{\partial x}(3x^2y - y^3z^2) + \hat{j}\frac{\partial}{\partial y}(3x^2y - y^3z^2) + \hat{k}\frac{\partial}{\partial z}(3x^2y - y^3z^2)$$

$$= \hat{i}(6xy) + \hat{j}(3x^2 - 3y^2z^2) + \hat{k}(-2y^3z)$$

$$= -12\hat{i} - 9\hat{j} - 16\hat{k} \text{ at the point } (1, -2, -1).$$

Example 2. *Show that $\nabla r^n = nr^{n-2}\vec{r}$ and hence evaluate $\nabla\dfrac{1}{r}$ where $\vec{r} = x\hat{i} + y\hat{j} + z\hat{k}$.*

Sol. $$\nabla r^n = \left(\hat{i}\frac{\partial}{\partial x} + \hat{j}\frac{\partial}{\partial y} + \hat{k}\frac{\partial}{\partial z}\right)r^n$$

$$= \hat{i}\left(nr^{n-1}\frac{\partial r}{\partial x}\right) + \hat{j}\left(nr^{n-1}\frac{\partial r}{\partial y}\right) + \hat{k}\left(nr^{n-1}\frac{\partial r}{\partial z}\right) \quad \text{...(1)}$$

$$\vec{r} = x\hat{i} + y\hat{j} + z\hat{k}$$

$\therefore$ $$r = |\vec{r}| = \sqrt{x^2 + y^2 + z^2} \quad \text{or} \quad r^2 = x^2 + y^2 + z^2 \quad \text{...(2)}$$

Differentiating (2) partially w.r.t x, we get

$$2r\frac{\partial r}{\partial x} = 2x$$

$\Rightarrow$ $$\frac{\partial r}{\partial x} = \frac{x}{r}$$

Similarly, $$\frac{\partial r}{\partial y} = \frac{y}{r} \quad \text{and} \quad \frac{\partial r}{\partial z} = \frac{z}{r}$$

$\therefore$ From (1), $$\nabla r^n = \hat{i}\left(nr^{n-1}.\frac{x}{r}\right) + \hat{j}\left(nr^{n-1}.\frac{y}{r}\right) + \hat{k}\left(nr^{n-1}.\frac{z}{r}\right)$$

$$= nr^{n-2}(x\hat{i} + y\hat{j} + z\hat{k})$$

$$\Rightarrow \qquad \nabla r^n = n r^{n-2} \vec{r} \qquad ...(3)$$

Putting $n = -1$ in (3), we get

$$\nabla\left(\frac{1}{r}\right) = -\frac{\vec{r}}{r^3}.$$

Example 3. *(i) If $u = x + y + z$, $v = x^2 + y^2 + z^2$, $w = yz + zx + xy$, prove that grad u, grad v and grad w are coplanar vectors.* [U.P.T.U. 2015; U.K.T.U. 2010]

(ii) Show that $\nabla(\vec{a} \,.\, \vec{r}) = \vec{a}$ where $\vec{r} = x\hat{i} + y\hat{i} + z\hat{k}$ and $\vec{a}$ is a constant vector.

Sol. *(i)*

$$\text{grad } u = \left(\hat{i}\frac{\partial}{\partial x} + \hat{j}\frac{\partial}{\partial y} + \hat{k}\frac{\partial}{\partial z}\right)(x + y + z) = \hat{i} + \hat{j} + \hat{k}$$

$$\text{grad } v = \left(\hat{i}\frac{\partial}{\partial x} + \hat{y}\frac{\partial}{\partial y} + \hat{k}\frac{\partial}{\partial z}\right)(x^2 + y^2 + z^2) = 2x\hat{i} + 2y\hat{j} + 2z\hat{k}$$

$$\text{grad } w = \left(\hat{i}\frac{\partial}{\partial x} + \hat{j}\frac{\partial}{\partial y} + \hat{k}\frac{\partial}{\partial z}\right)(yz + zx + xy)$$

$$= \hat{i}(z + y) + \hat{j}(z + x) + \hat{k}(y + x)$$

$$\text{grad } u \,.\, (\text{grad } v \times \text{grad } w) = \begin{vmatrix} 1 & 1 & 1 \\ 2x & 2y & 2z \\ z+y & z+x & y+x \end{vmatrix} = 2\begin{vmatrix} 1 & 1 & 1 \\ x & y & z \\ z+y & z+x & y+x \end{vmatrix}$$

$$= 2\begin{vmatrix} 1 & 1 & 1 \\ x+z+y & y+z+x & z+y+z \\ z+y & z+x & y+x \end{vmatrix} \quad | \text{ Applying } R_2 \to R_2 + R_3$$

$$= 2(x + y + z)\begin{vmatrix} 1 & 1 & 1 \\ 1 & 1 & 1 \\ y+z & z+x & x+y \end{vmatrix} = 0$$

Hence grad u, grad v and grad w are coplanar vectors.

(ii) Let $\vec{a} = a_1\hat{i} + a_2\hat{j} + a_3\hat{k}$ where a_1, a_2, a_3 are constants.

$$\vec{a} \cdot \vec{r} = a_1 x + a_2 y + a_3 z$$

$$\therefore \quad \nabla(\vec{a} \cdot \vec{r}) = \left(\hat{i}\frac{\partial}{\partial x} + \hat{j}\frac{\partial}{\partial y} + \hat{k}\frac{\partial}{\partial z}\right)(a_1 x + a_2 y + a_3 z) = a_1\hat{i} + a_2\hat{j} + a_3\hat{k} = \vec{a}.$$

Example 4. *(i) Find a unit vector normal to the surface $x^3 + y^3 + 3xyz = 3$ at the point (1, 2, – 1).* (U.P.T.U. 2014)

(ii) Find a unit vector normal to the surface $x^2y + 2xz = 4$ at the point (2, – 2, 3). (U.P.T.U. 2014)

Sol. (*i*) Let $\phi = x^3 + y^3 + 3xyz - 3$, then $\dfrac{\partial \phi}{\partial x} = 3x^2 + 3yz$, $\dfrac{\partial \phi}{\partial y} = 3y^2 + 3xz$, $\dfrac{\partial \phi}{\partial z} = 3xy$

$$\therefore \quad \nabla\phi = \hat{i}\frac{\partial \phi}{\partial x} + \hat{j}\frac{\partial \phi}{\partial y} + \hat{k}\frac{\partial \phi}{\partial z} = (3x^2 + 3yz)\hat{i} + (3y^2 + 3xz)\hat{j} + (3xy)\hat{k}$$

At (1, 2, – 1), $\nabla\phi = -3\hat{i} + 9\hat{j} + 6\hat{k}$

which is a vector normal to the given surface at (1, 2, – 1).

Hence a unit vector normal to the given surface at (1, 2, – 1)

$$= \frac{-3\hat{i} + 9\hat{j} + 6\hat{k}}{\sqrt{[(-3)^2 + (9)^2 + (6)^2]}} = \frac{-3\hat{i} + 9\hat{j} + 6\hat{k}}{3\sqrt{14}} = \frac{1}{\sqrt{14}}(-\hat{i} + 3\hat{j} + 2\hat{k}).$$

(*ii*) Let $\phi = x^2y + 2xz - 4$

$$\therefore \quad \text{grad } \phi = \left(\hat{i}\frac{\partial}{\partial x} + \hat{j}\frac{\partial}{\partial y} + \hat{k}\frac{\partial}{\partial z}\right)(x^2y + 2xz - 4)$$

$$= (2xy + 2z)\,\hat{i} + x^2\,\hat{j} + 2x\hat{k} = -2\hat{i} + 4\hat{j} + 4\hat{k} \text{ at } (2, -2, 3)$$

$$|\text{ grad } \phi \,| = \sqrt{4 + 16 + 16} = 6$$

$$\therefore \quad \text{Unit normal vector} = \frac{\text{grad } \phi}{|\text{grad } \phi|} = \frac{-\hat{i} + 2\hat{j} + 2\hat{k}}{3}$$

Example 5. *What is greatest rate of increase of $u = xyz^2$ at the point (1, 0, 3)?*

Sol. $u = xyz^2$

$$\therefore \quad \text{grad } u = i\frac{\partial u}{\partial x} + j\frac{\partial u}{\partial y} + k\frac{\partial u}{\partial z} = yz^2 i + xz^2 j + 2xyz\, k = 9j \text{ at } (1, 0, 3)$$

∴ The greatest rate of increase of u at (1, 0, 3) = | grad u | = | $9j$ | = 9.

Example 6. *If $\nabla\phi = (y^2 - 2xyz^3)\, i + (3 + 2xy - x^2z^3)\, j + (6z^3 - 3x^2yz^2)\, k$, find ϕ.*

(U.P.T.U. 2015)

Sol. Let $\vec{F} = \nabla\phi$

$$\Rightarrow \quad \vec{F} \cdot d\vec{r} = \nabla\phi \cdot d\vec{r} = \left(\frac{\partial \phi}{\partial x}\hat{i} + \frac{\partial \phi}{\partial y}\hat{j} + \frac{\partial \phi}{\partial z}\hat{k}\right) \cdot (dx\hat{i} + dy\hat{j} + dz\hat{k})$$

$$= \frac{\partial \phi}{\partial x}dx + \frac{\partial \phi}{\partial y}dy + \frac{\partial \phi}{\partial z}dz = d\phi$$

$$\therefore \quad d\phi = \vec{F} \cdot d\vec{r}$$

$$= \{(y^2 - 2xyz^3)\,\hat{i} + (3 + 2xy - x^2z^3)\,\hat{j} + (6z^3 - 3x^2yz^2)\hat{k}\} \cdot (dx\hat{i} + dy\,\hat{j} + dz\hat{k})$$

$$= (y^2 - 2xyz^3)\, dx + (3 + 2xy - x^2z^3)\, dy + (6z^3 - 3x^2yz^2)\, dz$$

$$= (y^2\, dx + 2xy\, dy) - (2xyz^3\, dx + x^2z^3\, dy + 3x^2yz^2\, dz) + 3\, dy + 6z^3\, dz$$

$$= d(xy^2) - d(x^2\, yz^3) + d(3y) + d\left(\frac{3}{2}z^4\right)$$

$$\therefore \quad \phi = xy^2 - x^2yz^3 + 3y + \frac{3}{2}z^4 + c \qquad \text{(on integration)}$$

where c is a constant.

Example 7. *(i) Find the angle between the surfaces $x^2 + y^2 + z^2 = 9$ and $z = x^2 + y^2 - 3$ at the point (2, – 1, 2).*

(ii) If θ is the acute angle between the surfaces $xy^2z = 3x + z^2$ and $3x^2 - y^2 + 2z = 1$ at the point (1, – 2, 1), show that $\cos\theta = \dfrac{3}{7\sqrt{6}}$.

Sol. *(i)* Angle between two surfaces at a point is the angle between the normals to the surfaces at that point.

Let $\phi_1 \equiv x^2 + y^2 + z^2 - 9 = 0$ and $\phi_2 \equiv x^2 + y^2 - z - 3 = 0$

Then $\text{grad } \phi_1 = 2x\hat{i} + 2y\hat{j} + 2z\hat{k}$ and $\text{grad } \phi_2 = 2x\hat{i} + 2y\hat{j} - \hat{k}$

Let $\vec{n_1} = \text{grad } \phi_1$ at the point (2, – 1, 2) and $\vec{n_2} = \text{grad } \phi_2$ at the point (2, – 1, 2). Then

$$\vec{n_1} = 4\hat{i} - 2\hat{j} + 4\hat{k} \quad \text{and} \quad \vec{n_2} = 4\hat{i} - 2\hat{j} - \hat{k}$$

The vectors $\vec{n_1}$ and $\vec{n_2}$ are along normals to the two surfaces at the point (2, – 1, 2). If θ is the angle between these vectors, then

$$\cos\theta = \frac{\vec{n_1}\,.\,\vec{n_2}}{|\vec{n_1}|.|\vec{n_2}|} = \frac{4(4) - 2(-2) + 4(-1)}{\sqrt{16+4+16}\,.\sqrt{16+4+1}} = \frac{16}{6\sqrt{21}}$$

$$\therefore \quad \theta = \cos^{-1}\left(\frac{8}{3\sqrt{21}}\right)$$

(ii) Let $\phi_1 \equiv xy^2z - 3x - z^2 = 0$ and $\phi_2 \equiv 3x^2 - y^2 + 2z - 1 = 0$

Then,
$$\nabla\phi_1 = \left(\hat{i}\frac{\partial}{\partial x} + \hat{j}\frac{\partial}{\partial y} + \hat{k}\frac{\partial}{\partial z}\right)(xy^2 z - 3x - z^2)$$

$$= (y^2z - 3)\hat{i} + 2xyz\hat{j} + (xy^2 - 2z)\hat{k}$$

$$= \hat{i} - 4\hat{j} + 2\hat{k} \text{ at } (1, -2, 1)$$

and
$$\Delta\phi_2 = \left(\hat{i}\frac{\partial}{\partial x} + \hat{j}\frac{\partial}{\partial y} + \hat{k}\frac{\partial}{\partial z}\right)(3x^2 - y^2 + 2z - 1) = 6x\hat{i} - 2y\hat{j} + 2\hat{k}$$

$$= 6\hat{i} + 4\hat{j} + 2\hat{k} \text{ at } (1, -2, 1)$$

We know that $\nabla\phi_1$ and $\nabla\phi_2$ are along normals to the given surfaces at any point.
If ϕ is the angle between these vectors then,

$$\cos\phi = \frac{\nabla\phi_1 . \nabla\phi_2}{|\nabla\phi_1||\nabla\phi_2|}$$

$$= \frac{(1)(6) + (-4)(4) + (2)(2)}{\sqrt{1+16+4}\,\sqrt{36+16+4}} = \frac{-6}{\sqrt{21}\,\sqrt{56}} = \frac{-3}{7\sqrt{6}}$$

Let θ be the acute angle such that θ + ϕ = π, then,

$$\cos\theta = \frac{3}{7\sqrt{6}}.$$

Example 8. *Find the directional derivative of the function $f = x^2 - y^2 + 2z^2$ at the point P(1, 2, 3) in the direction of the line PQ where Q is the point (5, 0, 4).* [U.K.T.U. 2011]

In what direction it will be maximum? Find also the magnitude of this maximum.

Sol. We have $\nabla f = \hat{i}\frac{\partial f}{\partial x} + \hat{j}\frac{\partial f}{\partial y} + \hat{k}\frac{\partial f}{\partial z} = 2x\hat{i} - 2y\hat{j} + 4z\hat{k} = 2\hat{i} - 4\hat{j} + 12\hat{k}$ at P(1, 2, 3)

Also $\overrightarrow{PQ} = \overrightarrow{OQ} - \overrightarrow{OP} = (5\hat{i} + 4\hat{k}) - (\hat{i} + 2\hat{j} + 3\hat{k}) = 4\hat{i} - 2\hat{j} + \hat{k}$

If $\hat{n}$ is a unit vector in the direction $\overrightarrow{PQ}$, then $\hat{n} = \dfrac{4\hat{i} - 2\hat{j} + \hat{k}}{\sqrt{16+4+1}} = \dfrac{1}{\sqrt{21}}(4\hat{i} - 2\hat{j} + \hat{k})$

$\therefore$ Directional derivative of f in the direction $\overrightarrow{PQ} = (\nabla f) \cdot \hat{n}$

$$= (2\hat{i} - 4\hat{j} + 12\hat{k}) \cdot \frac{1}{\sqrt{21}}(4\hat{i} - 2\hat{j} + \hat{k}) = \frac{1}{\sqrt{21}}[2(4) - 4(-2) + 12(1)]$$

$$= \frac{28}{\sqrt{21}} = \frac{4}{3}\sqrt{21}$$

The directional derivative of f is maximum in the direction of the normal to the given surface *i.e.*, in the direction of $\nabla f = 2\hat{i} - 4\hat{j} + 12\hat{k}$

The maximum value of this directional derivative = $|\nabla f|$

$$= \sqrt{(2)^2 + (-4)^2 + (12)^2} = \sqrt{164} = 2\sqrt{41}.$$

Example 9. (*i*) *Suppose $D_b\,\phi(P) = 1$, $D_u\,\phi(P) = 3$, $\hat{b} = \dfrac{3\hat{i} + 4\hat{j}}{5}$, $\hat{u} = \dfrac{4\hat{i} - 3\hat{j}}{5}$. Find $\nabla\phi(P)$.*

(*ii*) *If $\phi(x, y) = x^2 - xy - y + y^2$, find all points where the directional derivative in the direction $\hat{b} = \dfrac{\hat{i} + \sqrt{3}\hat{j}}{2}$ is zero.*

Sol. (*i*) $\nabla\phi(P) = \hat{i}\dfrac{\partial\phi}{\partial x} + \hat{j}\dfrac{\partial\phi}{\partial y}$

We have, $\nabla\phi(P) \cdot \hat{b} = 1 \Rightarrow 3\dfrac{\partial\phi}{\partial x} + 4\dfrac{\partial\phi}{\partial y} = 5$ ...(1)

and $\nabla\phi(P) \cdot \hat{u} = 3 \Rightarrow 4\dfrac{\partial\phi}{\partial x} - 3\dfrac{\partial\phi}{\partial y} = 15$ ...(2)

Solving eqns. (1) and (2), we get

$$\frac{\partial\phi}{\partial x} = 3, \quad \frac{\partial\phi}{\partial y} = -1$$

Hence, $\nabla\phi(P) = 3\hat{i} - \hat{j}$

(ii) $$\nabla\phi = (2x - y)\hat{i} + (-x - 1 + 2y)\hat{j}$$

We have, $$\nabla\phi \cdot \hat{b} = 0$$

$$2x - y + \sqrt{3}\,(-x - 1 + 2y) = 0$$

$$(2 - \sqrt{3})x + (2\sqrt{3} - 1)\,y = \sqrt{3}$$

Hence all required points lie on the above line.

Example 10. *Find the directional derivative of* $\phi = (x^2 + y^2 + z^2)^{-1/2}$ *at the point P(3, 1, 2) in the direction of the vector* $yz\hat{i} + zx\hat{j} + xy\hat{k}$. (U.P.T.U. 2014)

Sol. $$\phi = (x^2 + y^2 + z^2)^{-1/2}$$

$$\therefore \quad \text{grad}\,\phi = \hat{i}\frac{\partial}{\partial x}(x^2 + y^2 + z^2)^{-1/2} + \hat{j}\frac{\partial}{\partial y}(x^2 + y^2 + z^2)^{-1/2} + \hat{k}\frac{\partial}{\partial z}(x^2 + y^2 + z^2)^{-1/2}$$

$$= \hat{i}\left\{-\frac{1}{2}(x^2 + y^2 + z^2)^{-3/2}(2x)\right\} + \hat{j}\left\{-\frac{1}{2}(x^2 + y^2 + z^2)^{-3/2}(2y)\right\} + \hat{k}\left\{-\frac{1}{2}(x^2 + y^2 + z^2)^{-3/2}(2z)\right\}$$

$$= -\frac{(x\hat{i} + y\hat{j} + z\hat{k})}{(x^2 + y^2 + z^2)^{3/2}} = -\frac{3\hat{i} + \hat{j} + 2\hat{k}}{14\sqrt{14}} \text{ at } (3, 1, 2)$$

Let $\hat{a}$ be the unit vector in the given direction, then

$$\hat{a} = \frac{yz\hat{i} + zx\hat{j} + xy\hat{k}}{\sqrt{y^2z^2 + z^2x^2 + x^2y^2}} = \frac{2\hat{i} + 6\hat{j} + 3\hat{k}}{7} \text{ at } (3, 1, 2)$$

$$\therefore \quad \frac{d\phi}{ds} = \hat{a} \cdot \text{grad}\,\phi = \frac{2i + 6j + 3k}{7} \cdot \left(-\frac{3i + j + 2k}{14\sqrt{14}}\right) = -\frac{(2)(3) + (6)(1) + (3)(2)}{7 \cdot 14\sqrt{14}} = -\frac{9}{49\sqrt{14}}.$$

Example 11. *For the function* $\phi = \dfrac{y}{x^2 + y^2}$, *find the magnitude of the directional derivative making an angle 30° with the positive x-axis at the point (0, 1).*

Sol. Here, $$\phi = \frac{y}{x^2 + y^2}$$

$$\therefore \quad \frac{\partial\phi}{\partial x} = -\frac{2xy}{(x^2 + y^2)^2}, \quad \frac{\partial\phi}{\partial y} = \frac{x^2 - y^2}{(x^2 + y^2)^2} \quad \text{and} \quad \frac{\partial\phi}{\partial z} = 0$$

$$\therefore \quad \text{grad}\,\phi = \hat{i}\frac{\partial\phi}{\partial x} + \hat{j}\frac{\partial\phi}{\partial y} + \hat{k}\frac{\partial\phi}{\partial z}$$

$$\Rightarrow \quad \nabla\phi = -\frac{2xy}{(x^2 + y^2)^2}\hat{i} + \frac{(x^2 - y^2)}{(x^2 + y^2)^2}\hat{j} = -\hat{j} \text{ at } (0, 1)$$

Let $\hat{a}$ be the unit vector along the line making an angle 30° with the positive x-axis at the point (0, 1), then

$$\hat{a} = \cos 30°\,\hat{i} + \sin 30°\,\hat{j}$$

Hence the directional derivative is given by

$$\frac{d\phi}{ds} = \hat{a} \cdot \nabla\phi = (\cos 30°\,\hat{i} + \sin 30°\,\hat{j}) \cdot (-\hat{j}) = -\sin 30° = -\frac{1}{2}.$$

Example 12. *If the directional derivative of $\phi = ax^2y + by^2z + cz^2x$ at the point (1, 1, 1) has maximum magnitude 15 in the direction parallel to the line $\frac{x-1}{2} = \frac{y-3}{-2} = \frac{z}{1}$, find the values of a, b and c.*

Sol.
$$\phi = ax^2y + by^2z + cz^2x$$

$$\therefore \quad \text{grad } \phi = \hat{i}\frac{\partial \phi}{\partial x} + \hat{j}\frac{\partial \phi}{\partial y} + \hat{k}\frac{\partial \phi}{\partial z}$$

$$= (2axy + cz^2)\hat{i} + (ax^2 + 2byz)\hat{j} + (by^2 + 2czx)\hat{k}$$

$$= (2a + c)\hat{i} + (a + 2b)\hat{j} + (b + 2c)\hat{k} \text{ at } (1, 1, 1)$$

Now, the directional derivative is maximum along the normal to the surface *i.e.*, along grad ϕ.

$$|\text{ grad } \phi \text{ }| = \sqrt{(2a+c)^2 + (a+2b)^2 + (b+2c)^2}$$

$$\Rightarrow \quad 15 = \sqrt{(2a+c)^2 + (a+2b)^2 + (b+2c)^2}$$

or,
$$(2a + c)^2 + (a + 2b)^2 + (b + 2c)^2 = 225 \quad \text{...(1)}$$

But we are given that the directional derivative is maximum in the direction parallel to the line $\frac{x-1}{2} = \frac{y-3}{-2} = \frac{z}{1}$ *i.e.*, parallel to the vector $2\hat{i} - 2\hat{j} + \hat{k}$. Hence,

$$\frac{2a+c}{2} = \frac{a+2b}{-2} = \frac{2c+b}{1}$$

$$\Rightarrow \quad 2a + c = -a - 2b \quad \text{or} \quad 3a + 2b + c = 0 \quad \text{...(2)}$$

and
$$2b + a = -4c - 2b \quad \text{or} \quad a + 4b + 4c = 0 \quad \text{...(3)}$$

By cross-multiplication, we get

$$\frac{a}{4} = \frac{b}{-11} = \frac{c}{10} = \lambda \text{ (say)}$$

$$\Rightarrow \quad a = 4\lambda,\ b = -11\lambda,\ c = 10\lambda \quad \text{...(4)}$$

From (1), $(8\lambda + 10\lambda)^2 + (4\lambda - 22\lambda)^2 + (-11\lambda + 20\lambda)^2 = 225$

$$\Rightarrow \quad \lambda = \pm\frac{5}{9}$$

$$\therefore \quad a = \pm\frac{20}{9},\ b = \mp\frac{55}{9},\ c = \pm\frac{50}{9} \quad \text{| From (4)}$$

Example 13. *Find the directional derivative of $\phi = 5x^2y - 5y^2z + \frac{5}{2}z^2x$ at the point P(1, 1, 1) in the direction of the line $\frac{x-1}{2} = \frac{y-3}{-2} = \frac{z}{1}$.* (A.K.T.U. 2019)

Sol.
$$\phi = 5x^2y - 5y^2z + \frac{5}{2}z^2x$$

$$\therefore \quad \text{grad } \phi = \hat{i}\frac{\partial \phi}{\partial x} + \hat{j}\frac{\partial \phi}{\partial y} + \hat{k}\frac{\partial \phi}{\partial z}$$

$$= \left(10xy + \frac{5}{2}z^2\right)\hat{i} + (5x^2 - 10yz)\hat{j} + (-5y^2 + 5zx)\hat{k}$$

$$= \frac{25}{2}\hat{i} - 5\hat{j} \text{ at } (1, 1, 1)$$

Here $\hat{a} = \dfrac{2\hat{i} - 2\hat{j} + \hat{k}}{3}$

$\therefore$ Directional derivative $= (\text{grad } \phi) \,.\, \hat{a}$

$$= \left(\frac{25}{2}\hat{i} - 5\hat{j}\right) . \left(\frac{2}{3}\hat{i} - \frac{2}{3}\hat{j} + \frac{1}{3}\hat{k}\right) = \frac{25}{3} + \frac{10}{3} = \frac{35}{3}.$$

Example 14. *Find the values of λ and μ so that surfaces $\lambda x^2 - \mu yz = (\lambda + 2)x$ and $4x^2y + z^3 = 4$ intersect orthogonally at the point (1, – 1, 2).*

Sol. Let $\phi_1 \equiv \lambda x^2 - \mu yz - (\lambda + 2)x = 0$ ...(1)

$\phi_2 \equiv 4x^2y + z^3 - 4 = 0$ ...(2)

Then, $\nabla\phi_1 = (2\lambda x - \lambda - 2)\hat{i} - \mu z\hat{j} - \mu y\hat{k}$

$= (\lambda - 2)\hat{i} - 2\mu\hat{j} + \mu\hat{k}$ at (1, – 1, 2)

$\nabla\phi_2 = 8xy\hat{i} + 4x^2\hat{j} + 3z^2\hat{k} = -8\hat{i} + 4\hat{j} + 12\hat{k}$ at (1, – 1, 2)

Condition for surfaces (1) and (2) to cut orthogonally is

$\nabla\phi_1 \,.\, \nabla\phi_2 = 0$

$\Rightarrow \quad -8(\lambda - 2) - 8\mu + 12\mu = 0$

$\Rightarrow \quad \mu - 2\lambda + 4 = 0$ ...(3)

Also since the point (1, – 1, 2) lies on both surfaces; hence, from (1),

$\lambda + 2\mu - \lambda - 2 = 0 \quad \Rightarrow \quad \mu = 1$

$\therefore$ From (3), $\lambda = 5/2$

Hence reqd. values are $\lambda = \dfrac{5}{2}$, $\mu = 1$.

Example 15. *Find the values of constants a, b, c so that the maximum value of the directional derivative of $\phi = axy^2 + byz + cz^2x^3$ at (1, 2, – 1) has a magnitude 64 in the direction parallel to z-axis.*

Sol. $\phi = axy^2 + byz + cz^2x^3$

$\therefore$ $\text{grad } \phi = (ay^2 + 3cz^2x^2)\hat{i} + (2ayx + bz)\hat{j} + (by + 2czx^3)\hat{k}$

$= (4a + 3c)\hat{i} + (4a - b)\hat{j} + (2b - 2c)\hat{k}$ at (1, 2, – 1).

Now, the directional derivative is maximum along the normal to the surface *i.e.*, along grad ϕ. But, we are given that directional derivative is maximum in the direction parallel to z-axis *i.e.*, parallel to $\hat{k}$.

Hence coefficients of $\hat{i}$ and $\hat{j}$ in grad ϕ should be zero and the coefficient of k positive.

Thus, $4a + 3c = 0$ ...(1)

$4a - b = 0$ ...(2)

and $2b - 2c > 0 \quad \Rightarrow \quad b > c$ ...(3)

Then, $\text{grad } \phi = 2(b - c)\hat{k}$

Also, Max. value of directional derivative $= |\text{ grad } \phi \,|$

$$\therefore \quad 64 = 2(b - c)$$

$$\Rightarrow \quad b - c = 32 \qquad \ldots(4)$$

Solving (1), (2) and (4), we get $a = 6$, $b = 24$ and $c = -8$

Example 16. *Find the directional derivative of $f(x, y, z) = e^{2x} \cos yz$ at (0, 0, 0) in the direction of the tangent to the curve $x = a \sin t$, $y = a \cos t$, $z = at$ at $t = \pi/4$.*

Sol. $f = e^{2x} \cos yz$

$$\therefore \quad \nabla f = \hat{i}\frac{\partial f}{\partial x} + \hat{j}\frac{\partial f}{\partial y} + \hat{k}\frac{\partial f}{\partial z}$$

$$= (2e^{2x} \cos yz)\,\hat{i} - (ze^{2x} \sin yz)\,\hat{j} - (ye^{2x} \sin yz)\,\hat{k} = 2\hat{i} \text{ at } (0, 0, 0)$$

Tangent to the given curve is

$$\frac{d\vec{r}}{dt} = \hat{i}\frac{dx}{dt} + \hat{j}\frac{dy}{dt} + \hat{k}\frac{dz}{dt} = a\,(\cos t\hat{i} - \sin t\hat{j} + \hat{k})$$

$$= a\left(\frac{\hat{i}}{\sqrt{2}} - \frac{\hat{j}}{\sqrt{2}} + \hat{k}\right) \quad \text{at } t = \frac{\pi}{4}$$

$$\hat{n} = \frac{(d\bar{r}/dt)}{|d\bar{r}/dt|} = \frac{\dfrac{a}{\sqrt{2}}\hat{i} - \dfrac{a}{\sqrt{2}}\hat{j} + a\hat{k}}{\sqrt{\dfrac{a^2}{2} + \dfrac{a^2}{2} + a^2}} = \frac{\hat{i}}{2} - \frac{\hat{j}}{2} + \frac{\hat{k}}{\sqrt{2}}$$

$$\therefore \quad \text{Directional derivative} = \nabla f \,.\, \hat{n} = 2\hat{i} \,.\left(\frac{\hat{i}}{2} - \frac{\hat{j}}{2} + \frac{\hat{k}}{\sqrt{2}}\right) = 2\left(\frac{1}{2}\right) = 1.$$

TEST YOUR KNOWLEDGE

1. Find grad ϕ when ϕ is given by

   (*i*) $\phi = x^2 + yz$ (*ii*) $\phi = x^3 + y^3 + 3xyz$ (*iii*) $\phi = \log (x^2 + y^2 + z^2)$.

2. (*i*) Find a unit vector normal to the surface $xy^3z^2 = 4$ at the point $(-1, -1, 2)$

   (*ii*) Find a unit normal vector $\hat{n}$ of the cone of revolution $z^2 = 4(x^2 + y^2)$ at the point P(1, 0, 2).

3. What is the greatest rate of increase of $u = x^2 + yz^2$ at the point $(1, -1, 3)$?

4. Calculate the angle between the normals to the surface $xy = z^2$ at the points (4, 1, 2) and $(3, 3, -3)$. (*U.K.T.U. 2012*)

5. Find the directional derivative of the function

   (*i*) $f(x, y, z) = xy^2 + yz^3$ at the point $(2, -1, 1)$ in the direction of the vector $\hat{i} + 2\hat{j} + 2\hat{k}$.

   (*ii*) $f(x, y, z) = 2xy + z^2$ at the point $(1, -1, 3)$ in the direction of the vector $\hat{i} + 2\hat{j} + 2\hat{k}$.

   (*iii*) $\phi = 4xz^3 - 3x^2yz^2$ at $(2, -1, 2)$ along z-axis.

(*iv*) $\phi = xy^2 + yz^3$ at the point (2, – 1, 1) in the direction of the normal to the surface
$x \log z - y^2 + 4 = 0$ at (2, – 1, 1). (*G.B.T.U. 2011*)

(*v*) $f(x, y, z) = 2x^2 + 3y^2 + z^2$ at the point P (2, 1, 3) in the direction of the vector $\vec{a} = \hat{i} - 2\hat{k}$.

(*vi*) $\psi(x, y, z) = 4e^{x + 5y - 13z}$ at the point (1, 2, 3) in the direction towards the point (– 3, 5, 7).

6. In what direction from (3, 1, – 2) is the directional derivative of $\phi = x^2y^2z^4$ maximum and what is its magnitude?

7. The temperature at a point (x, y, z) in space is given by $T(x, y, z) = x^2 + y^2 - z$. A mosquito located at (1, 1, 2) desires to fly in such a direction that it will get warm as soon as possible. In what direction should it fly?

8. Find the directional derivative of $\phi(x, y, z) = x^2 - 2y^2 + 4z^2$ at the point (1, 1, – 1) in the direction of $2i + j - k$. In what direction will the directional derivative be maximum and what is its magnitude?

9. Find the directional derivative of scalar function $f(x, y, z) = xyz$ at point P(1, 1, 3) in the direction of the outward drawn normal of the surface $x^2 + y^2 + z^2 = 11$ through the point P.

10. (*i*) Find the directional derivative of $\vec{V}^2$ where $\vec{V} = xy^2\hat{i} + zy^2\hat{j} + xz^2\hat{k}$ at the point (2, 0, 3) in the direction of the outward normal to the sphere $x^2 + y^2 + z^2 = 14$ at the point (3, 2, 1). (*M.T.U. 2013*)

(*ii*) Find the directional derivative of $\phi(x, y, z) = x^2yz + 4xz^2$ at (1, –2, 1) in the direction of $2\hat{i} - \hat{j} - 2\hat{k}$. Find also the greatest rate of increase of ϕ. (*A.K.T.U. 2020*)

Answers

1. (*i*) $2x\hat{i} + z\hat{j} + y\hat{k}$ (*ii*) $3(x^2 + yz)\,\hat{i} + 3(y^2 + xz)\,\hat{j} + 3xy\,\hat{k}$ (*iii*) $\dfrac{2(x\hat{i} + y\hat{j} + z\hat{k})}{x^2 + y^2 + z^2}$

2. (*i*) $-\dfrac{1}{\sqrt{11}}(\hat{i} + 3\hat{j} - \hat{k})$ (*ii*) $\dfrac{2\hat{i} - \hat{k}}{\sqrt{5}}$ **3.** 11 **4.** $\cos^{-1}\left(\dfrac{-1}{\sqrt{22}}\right)$

5. (*i*) $-\dfrac{11}{3}$ (*ii*) $\dfrac{14}{3}$ (*iii*) 144 (*iv*) $-3\sqrt{2}$

(*v*) $-\dfrac{4}{\sqrt{5}}$ (*vi*) $-4\sqrt{41}\,e^{-28}$

6. $96(\hat{i} + 3\hat{j} - 3\hat{k}), 96\sqrt{19}$ **7.** $\dfrac{1}{3}(2\hat{i} + 2\hat{j} - \hat{k})$ **8.** $\dfrac{8}{\sqrt{6}}$; $2i - 4j - 8k$; $2\sqrt{21}$

9. $\dfrac{9}{\sqrt{11}}$ **10.** (*i*) $\dfrac{1404}{\sqrt{14}}$ (*ii*) $\dfrac{-13}{3}$; $\sqrt{37}$.

5.15. GRADIENT IN POLAR COORDINATES

Let P be a point whose position vector is $\vec{r}$. Let $\hat{e}_r$ and $\hat{e}_\theta$ be the unit vectors along and perpendicular to the position vector $\vec{r}$. If s be the distance in the direction of $\vec{r}$, then

$$ds = dr$$

Also directional derivative along $\hat{e}_r$ is $\hat{e}_r \,.\, \nabla\phi$.

Thus, $$\frac{\partial \phi}{\partial r} = \hat{e}_r \,.\, \nabla \phi \qquad ...(1)$$

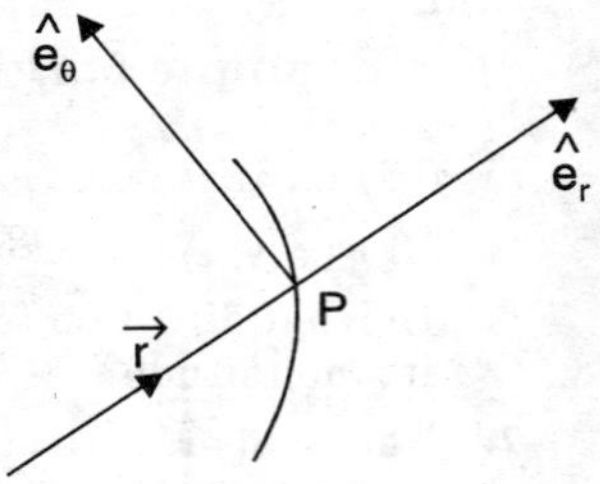

Now, if we take distance s in the direction perpendicular to $\vec{r}$, then $ds = r d\theta$. Also directional derivative along $\hat{e}_\theta$ is $\hat{e}_\theta \cdot \nabla \phi$.

Thus, $$\frac{\partial \phi}{r \partial \theta} = \hat{e}_\theta \cdot \nabla \phi \quad \text{...(2)}$$

Again, components of $\nabla\phi$ along and perpendicular to $\vec{r}$ are $\hat{e}_r \cdot \nabla\phi$ and $\hat{e}_\theta \cdot \nabla\phi$ respectively.

Hence $$\nabla\phi = (\hat{e}_r \cdot \nabla\phi)\, \hat{e}_r + (\hat{e}_\theta \cdot \nabla\phi)\, \hat{e}_\theta$$

$$\Rightarrow \boxed{\nabla\phi = \frac{\partial \phi}{\partial r}\, \hat{e}_r + \frac{1}{r}\frac{\partial \phi}{\partial \theta}\, \hat{e}_\theta} \quad \text{...(3)} \quad | \text{ From (1) and (2)}$$

Eqn. (3) represents gradient in polar coordinates.

ILLUSTRATIVE EXAMPLES

Example 1. *If* $\vec{r} = x\hat{i} + y\hat{j} + z\hat{k}$, *then show that*

(*i*) *grad* $r = \frac{\vec{r}}{r}$ (*ii*) *grad* $\frac{1}{r} = -\frac{\vec{r}}{r^3}$

(*iii*) *grad* $r^n = nr^{n-2}\, \vec{r}$ *where* $r = |\vec{r}|$

Sol. (*i*) $$\text{grad } r = \frac{\partial}{\partial r}(r)\, \hat{r} = \hat{r} = \frac{\vec{r}}{r}$$

(*ii*) $$\text{grad } \frac{1}{r} = \frac{\partial}{\partial r}\left(\frac{1}{r}\right) \hat{r} = -\frac{1}{r^2}\, \hat{r} = -\frac{\vec{r}}{r^3}$$

(*iii*) $$\text{grad } r^n = \frac{\partial}{\partial r}(r^n)\, \hat{r} = nr^{n-1}\, \hat{r} = nr^{n-2}\, \vec{r}$$

where $\hat{r}$ is the unit vector in the direction of $\vec{r}$.

Example 2. *Find* $\nabla\, |\vec{r}|^2$.

Sol. $$\nabla\, |\vec{r}|^2 = \nabla\, r^2 = \frac{\partial}{\partial r}(r^2)\, \hat{r} = 2r\hat{r} = 2\vec{r}.$$

Example 3. *Evaluate grad* e^{r^2}.

Sol. $$\nabla\, e^{r^2} = \frac{\partial}{\partial r}(e^{r^2})\, \hat{r} = e^{r^2}\, 2r\hat{r} = 2e^{r^2}\, \vec{r}. \quad (\because\ r\hat{r} = \vec{r})$$

Example 4. *Show that grad* $f(r) \times \vec{r} = \vec{0}$.

Sol. $$\text{grad } f(r) = \frac{\partial}{\partial r}\{f(r)\}\, \hat{r} = f'(r)\, \hat{r} = f'(r)\, \frac{\vec{r}}{r}$$

$$\therefore \quad \text{grad } f(r) \times \vec{r} = \frac{f'(r)}{r}(\vec{r} \times \vec{r}) = \vec{0}. \quad (\because\ \vec{r} \times \vec{r} = \vec{0})$$

Example 5. *If* $V(x, y) = \frac{1}{2} \log (x^2 + y^2)$, *prove that*

$$\text{grad } V = \frac{\vec{r} - \vec{k}(\vec{k} \cdot \vec{r})}{\{\vec{r} - \vec{k}(\vec{k} \cdot \vec{r})\} \cdot \{\vec{r} - \vec{k}(\vec{k} \cdot \vec{r})\}}.$$

Sol.
$$V = \frac{1}{2} \log (x^2 + y^2) = \log \sqrt{x^2 + y^2} = \log r$$

$$\therefore \quad \text{grad } V = \frac{\partial}{\partial r} (\log r)\, \hat{r} = \frac{1}{r} \hat{r} = \frac{\vec{r}}{r^2}$$

Now, $\vec{k} \cdot \vec{r} = \vec{k} \cdot (x\hat{i} + y\hat{j}) = 0 \quad \therefore \quad \text{RHS} = \frac{\vec{r}}{\vec{r} \cdot \vec{r}} = \frac{\vec{r}}{r^2}$

Hence the result.

Example 6. *If* $f = r \cos \theta + \tan \theta$, *find grad f in polar coordinates.*

Sol.
$$\text{grad } f = \frac{\partial f}{\partial r} \hat{e}_r + \frac{1}{r} \frac{\partial f}{\partial \theta} \hat{e}_\theta$$

$$= \frac{\partial}{\partial r} (r \cos \theta + \tan \theta)\, \hat{e}_r + \frac{1}{r} \frac{\partial}{\partial \theta} (r \cos \theta + \tan \theta)\, \hat{e}_\theta$$

$$= \cos \theta\, \hat{e}_r + \frac{1}{r} (- r \sin \theta + \sec^2 \theta)\, \hat{e}_\theta.$$

Example 7. *Find the directional derivative of* $\frac{1}{r}$ *in the direction of* $\vec{r}$ *where* $\vec{r} = x\hat{i} + y\hat{j} + z\hat{k}$. (U.K.T.U. 2011)

Sol. Here,
$$\phi = \frac{1}{r}$$

$$\therefore \quad \nabla\phi = \nabla \frac{1}{r} = -\frac{1}{r^2} \hat{r} = -\frac{\vec{r}}{r^3}$$

Let $\hat{a}$ be the unit vector in the direction of $\vec{r}$ then, $\hat{a} = \frac{\vec{r}}{|\vec{r}|} = \frac{\vec{r}}{r}$.

$$\therefore \quad \text{directional derivative } = \nabla\phi \cdot \hat{a} = \left(-\frac{\vec{r}}{r^3}\right) \cdot \frac{\vec{r}}{r} = -\frac{1}{r^4} (\vec{r} \cdot \vec{r}) = -\frac{r^2}{r^4} = -\frac{1}{r^2}$$

Example 8. (*i*) *Find the directional derivative of* $\frac{1}{r^2}$ *in the direction of* $\vec{r}$ *where* $\vec{r} = x\hat{i} + y\hat{j} + z\hat{k}$. (A.K.T.U. 2017)

(*ii*) *Find the directional derivative of* $\frac{1}{r^n}$ *in the direction of* $\vec{r}$ *where* $\vec{r} = x\hat{i} + y\hat{j} + z\hat{k}$.

Sol. (*i*)
$$\nabla \left(\frac{1}{r^2}\right) = -\frac{2}{r^3} \hat{r} = -\frac{2}{r^4} \vec{r}$$

Let $\hat{a}$ be the unit vector in the direction of $\vec{r}$ then $\hat{a} = \hat{r} = \dfrac{\vec{r}}{r}$

$$\therefore \quad \text{Directional derivative} = -\frac{2}{r^4}\vec{r} \cdot \frac{\vec{r}}{r} = -\frac{2}{r^5}(r^2) = -\frac{2}{r^3}$$

$$(ii) \qquad \nabla\left(\frac{1}{r^n}\right) = \frac{-n}{r^{n+1}}\hat{r} = \frac{-n}{r^{n+2}}\vec{r} \quad ; \quad \hat{a} = \hat{r} = \frac{\vec{r}}{r}$$

$$\therefore \quad \text{Directional derivative} = \nabla\left(\frac{1}{r^n}\right) \cdot \hat{a} = \frac{-n}{r^{n+2}}\vec{r} \cdot \frac{\vec{r}}{r} = -\frac{n}{r^{n+3}}(r^2) = -\frac{n}{r^{n+1}}$$

TEST YOUR KNOWLEDGE

1. Prove that $(\vec{a} \cdot \nabla)\vec{r} = \vec{a}$.
2. Prove that $\nabla \log r = \dfrac{\vec{r}}{r^2}$.
3. (*i*) Find $\nabla\left(\dfrac{e^{\mu r}}{r}\right)$ (*ii*) Prove that $\nabla f(r) = f'(r)\nabla r$ (*iii*) Find grad $f'(r) \times \vec{r}$
4. Find (*i*) $\nabla |\vec{r}|^3$ (*ii*) ∇r^{-3} (*iii*) $\nabla \log r^n$.
5. If $\vec{a}, \vec{b}$ are constant vectors, prove that $\nabla(\vec{r}\ \vec{a}\ \vec{b}) = \vec{a} \times \vec{b}$.
6. If $\phi = r(1 + \cos\theta)$, find ∇f in polar coordinates.
7. If $f = \sin\theta + r\cos\theta$, find ∇f in polar coordinates.

Answers

3. (*i*) $\dfrac{e^{\mu r}(\mu r - 1)\vec{r}}{r^3}$ (*iii*) $\vec{0}$

4. (*i*) $3r\vec{r}$ (*ii*) $-3r^{-5}\vec{r}$ (*iii*) $\dfrac{n\vec{r}}{r^2}$

6. $(1 + \cos\theta)\hat{e}_r - \sin\theta\,\hat{e}_\theta$ 7. $(\cos\theta)\hat{e}_r + \dfrac{1}{r}(\cos\theta - r\sin\theta)\hat{e}_\theta$

5.16. DIVERGENCE OF A VECTOR POINT FUNCTION

The divergence of a differentiable vector point function $\vec{V}$ is denoted by div $\vec{V}$ and is defined as

$$\text{div}\,\vec{V} = \nabla \cdot \vec{V} = \left(\hat{i}\frac{\partial}{\partial x} + \hat{j}\frac{\partial}{\partial y} + \hat{k}\frac{\partial}{\partial z}\right) \cdot \vec{V} = \hat{i} \cdot \frac{\partial \vec{V}}{\partial x} + \hat{j} \cdot \frac{\partial \vec{V}}{\partial y} + \hat{k} \cdot \frac{\partial \vec{V}}{\partial z}.$$

Obviously, the divergence of a vector point function is a scalar point function.

If $\qquad \vec{V} = V_1\hat{i} + V_2\hat{j} + V_3\hat{k}$, then

$$\text{div } \vec{V} = \nabla \cdot \vec{V} = \left(\hat{i} \frac{\partial}{\partial x} + \hat{j} \frac{\partial}{\partial y} + \hat{k} \frac{\partial}{\partial z} \right) \cdot [V_1 \hat{i} + V_2 \hat{j} + V_3 \hat{k}] = \frac{\partial V_1}{\partial x} + \frac{\partial V_2}{\partial y} + \frac{\partial V_3}{\partial z}.$$

since $\hat{i} \cdot \hat{i} = \hat{j} \cdot \hat{j} = \hat{k} \cdot \hat{k} = 1$ and $\hat{i} \cdot \hat{j} = \hat{j} \cdot \hat{k} = \hat{k} \cdot \hat{i} = 0.$

5.17. CURL OF A VECTOR POINT FUNCTION [G.B.T.U. 2012]

The curl (or rotation) of a differentiable vector point function $\vec{V}$ is denoted by curl $\vec{V}$ and is defined as

$$\text{curl } \vec{V} = \nabla \times \vec{V} = \left(\hat{i} \frac{\partial}{\partial x} + \hat{j} \frac{\partial}{\partial y} + \hat{k} \frac{\partial}{\partial z} \right) \times \vec{V} = \hat{i} \times \frac{\partial \vec{V}}{\partial x} + \hat{j} \times \frac{\partial \vec{V}}{\partial y} + \hat{k} \times \frac{\partial \vec{V}}{\partial z}.$$

Obviously, the curl of a vector point function is a vector point function.

If $\vec{V} = V_1 \hat{i} + V_2 \hat{j} + V_3 \hat{k}$

then
$$\text{curl } \vec{V} = \nabla \times \vec{V} = \left(\hat{i} \frac{\partial}{\partial x} + \hat{j} \frac{\partial}{\partial y} + \hat{k} \frac{\partial}{\partial z} \right) \times [V_1 \hat{i} + V_2 \hat{j} + V_3 \hat{k}]$$

$$= \begin{vmatrix} \hat{i} & \hat{j} & \hat{k} \\ \frac{\partial}{\partial x} & \frac{\partial}{\partial y} & \frac{\partial}{\partial z} \\ V_1 & V_2 & V_3 \end{vmatrix} = \hat{i} \left(\frac{\partial V_3}{\partial y} - \frac{\partial V_2}{\partial z} \right) + \hat{j} \left(\frac{\partial V_1}{\partial z} - \frac{\partial V_3}{\partial x} \right) + \hat{k} \left(\frac{\partial V_2}{\partial x} - \frac{\partial V_1}{\partial y} \right).$$

ILLUSTRATIVE EXAMPLES

Example 1. *If* $\vec{r} = x\hat{i} + y\hat{j} + z\hat{k}$, *show that*

(*i*) *div* $\vec{r} = 3$ (U.P.T.U. 2014) (*ii*) *curl* $\vec{r} = \vec{0}$. (A.K.T.U. 2014, 2018)

Sol. (*i*) $\text{div } \vec{r} = \nabla \cdot \vec{r} = \frac{\partial}{\partial x}(x) + \frac{\partial}{\partial y}(y) + \frac{\partial}{\partial z}(z) = 1 + 1 + 1 = 3.$

(*ii*)
$$\text{curl } \vec{r} = \nabla \times \vec{r} = \begin{vmatrix} \hat{i} & \hat{j} & \hat{k} \\ \frac{\partial}{\partial x} & \frac{\partial}{\partial y} & \frac{\partial}{\partial z} \\ x & y & z \end{vmatrix}$$

$$= \hat{i} \left[\frac{\partial}{\partial y}(z) - \frac{\partial}{\partial z}(y) \right] + \hat{j} \left[\frac{\partial}{\partial z}(x) - \frac{\partial}{\partial x}(z) \right] + \hat{k} \left[\frac{\partial}{\partial x}(y) - \frac{\partial}{\partial y}(x) \right]$$

$$= \hat{i}(0) + \hat{j}(0) + \hat{k}(0) = \vec{0}.$$

Example 2. *Find the divergence and curl of the vector* $\vec{V} = (xyz)\hat{i} + (3x^2y)\hat{j} + (xz^2 - y^2z)\hat{k}$ *at the point (2, – 1, 1).*

Sol.
$$\text{div } \vec{V} = \frac{\partial}{\partial x}(xyz) + \frac{\partial}{\partial y}(3x^2y) + \frac{\partial}{\partial z}(xz^2 - y^2z)$$
$$= yz + 3x^2 + 2xz - y^2 = -1 + 12 + 4 - 1 = 14 \text{ at } (2, -1, 1)$$
$$\text{curl } \vec{V} = \begin{vmatrix} \hat{i} & \hat{j} & \hat{k} \\ \frac{\partial}{\partial x} & \frac{\partial}{\partial y} & \frac{\partial}{\partial z} \\ xyz & 3x^2y & xz^2 - y^2z \end{vmatrix} = \hat{i}(-2yz - 0) + \hat{j}(xy - z^2) + \hat{k}(6xy - xz)$$
$$= 2\hat{i} - 3\hat{j} - 14\hat{k} \text{ at } (2, -1, 1).$$

Example 3. *Find div* $\vec{F}$ *and curl* $\vec{F}$ *where* $\vec{F} = \text{grad}(x^3 + y^3 + z^3 - 3xyz)$.

Sol. Let $\phi = x^3 + y^3 + z^3 - 3xyz$, then
$$\vec{F} = \text{grad } \phi = \left(\hat{i}\frac{\partial}{\partial x} + \hat{j}\frac{\partial}{\partial y} + \hat{k}\frac{\partial}{\partial z}\right)(x^3 + y^3 + z^3 - 3xyz)$$
$$= (3x^2 - 3yz)\hat{i} + (3y^2 - 3zx)\hat{j} + (3z^2 - 3xy)\hat{k}$$
$$\therefore \quad \text{div } \vec{F} = \frac{\partial}{\partial x}(3x^2 - 3yz) + \frac{\partial}{\partial y}(3y^2 - 3zx) + \frac{\partial}{\partial z}(3z^2 - 3xy)$$
$$= 6x + 6y + 6z = 6(x + y + z)$$
and
$$\text{curl } \vec{F} = \begin{vmatrix} \hat{i} & \hat{j} & \hat{k} \\ \frac{\partial}{\partial x} & \frac{\partial}{\partial y} & \frac{\partial}{\partial z} \\ 3x^2 - 3yz & 3y^2 - 3zx & 3z^2 - 3xy \end{vmatrix}$$
$$= \hat{i}(-3x + 3x) + \hat{j}(-3y + 3y) + \hat{k}(-3z + 3z) = \vec{0}.$$

Example 4. (*i*) *If* $\vec{F}(x, y, z) = xz^3\hat{i} - 2x^2yz\hat{j} + 2yz^4\hat{k}$, *find divergence and curl of* $\vec{F}(x, y, z)$.

(*ii*) *Find the divergence and curl of the vector field* $\vec{V} = x^2y^2\hat{i} + 2xy\,\hat{j} + (y^2 - xy)\,\hat{k}$.

(*iii*) *Find the divergence and curl of the vector function* $\vec{F} = e^{xyz}(xy^2\hat{i} + yz^2\hat{j} + zx^2\hat{k})$ *at the point (1, 2, 3).*

Sol. (*i*)
$$\text{div } \vec{F} = \frac{\partial}{\partial x}(xz^3) - \frac{\partial}{\partial y}(2x^2yz) + \frac{\partial}{\partial z}(2yz^4) = z^3 - 2x^2z + 8yz^3$$
$$\text{curl } \vec{F} = \begin{vmatrix} \hat{i} & \hat{j} & \hat{k} \\ \partial/\partial x & \partial/\partial y & \partial/\partial z \\ xz^3 & -2x^2yz & 2yz^4 \end{vmatrix}$$
$$= \hat{i}[2z^4 + 2x^2y] - \hat{j}[-3xz^2] + \hat{k}[-4xyz]$$
$$= 2(x^2y + z^4)\,\hat{i} + 3xz^2\,\hat{j} - 4xyz\hat{k}$$

(ii) $\text{div } \vec{V} = \frac{\partial}{\partial x}(x^2y^2) + \frac{\partial}{\partial y}(2xy) + \frac{\partial}{\partial z}(y^2 - xy) = 2xy^2 + 2x$

$$\text{curl } \vec{V} = \begin{vmatrix} \hat{i} & \hat{j} & \hat{k} \\ \partial/\partial x & \partial/\partial y & \partial/\partial z \\ x^2y^2 & 2xy & y^2 - xy \end{vmatrix}$$

$$= \hat{i}\,[2y - x] - \hat{j}\,[-y] + \hat{k}\,(2y - 2x^2y) = (2y - x)\,\hat{i} + y\,\hat{j} + 2y(1 - x^2)\,\hat{k}$$

(iii) $\text{div } \vec{F} = \frac{\partial}{\partial x}(xy^2e^{xyz}) + \frac{\partial}{\partial y}(yz^2\,e^{xyz}) + \frac{\partial}{\partial z}(zx^2\,e^{xyz})$

$$= y^2\,[e^{xyz} + xyz\,e^{xyz}] + z^2(e^{xyz} + yxz\,e^{xyz}) + x^2\,(e^{xyz} + zxy\,e^{xyz})$$
$$= (1 + xyz)\,(x^2 + y^2 + z^2)\,e^{xyz} = 98e^6 \text{ at } (1, 2, 3)$$

$$\text{curl } \vec{F} = \begin{vmatrix} \hat{i} & \hat{j} & \hat{k} \\ \partial/\partial x & \partial/\partial y & \partial/\partial z \\ xy^2\,e^{xyz} & yz^2\,e^{xyz} & zx^2e^{xyz} \end{vmatrix}$$

$$= \hat{i}\,[x^3z^2e^{xyz} - y\,\{2ze^{xyz} + xyz^2\,e^{xyz}\}]$$
$$- \hat{j}\left[z\{2xe^{xyz} + x^2yz\,e^{xyz}\} - x^2y^3\,e^{xyz}\right] + \hat{k}\left[y^2z^3\,e^{xyz} - x\{2y\,e^{xyz} + y^2zx\,e^{xyz}\}\right]$$

$$= (x^3z^2 - xy^2z^2 - 2yz)\,e^{xyz}\,\hat{i} - (2xz + x^2yz^2 - x^2z^3)\,e^{xyz}\,\hat{j} + (y^2z^3 - 2xy - x^2y^2z)\,e^{xyz}\,\hat{k}$$

$$= -39\,e^6\,\hat{i} + 3e^6\,\hat{j} + 92\,e^6\,\hat{k} \qquad \text{at } (1, 2, 3)$$

Example 5. *Prove that* $\vec{a} \times (\nabla \times \vec{r}) = \nabla\,(\vec{a}\,.\,\vec{r}) - (\vec{a}\,.\nabla)\,\vec{r}$ *where* $\vec{a}$ *is a constant vector and* $\vec{r} = x\hat{i} + y\hat{j} + z\hat{k}$.

Sol. $\text{LHS} = \vec{a} \times (\nabla \times \vec{r}) = \vec{a} \times \vec{0} = \vec{0}$ $\qquad | \because \ \text{curl } \vec{r} = \vec{0}$

$$\text{RHS} = \nabla(\vec{a}\,.\,\vec{r}) - (\vec{a}\,.\,\nabla)\,\vec{r}$$

$$= \nabla\,(a_1x + a_2y + a_3z) - \left(a_1\frac{\partial}{\partial x} + a_2\frac{\partial}{\partial y} + a_3\frac{\partial}{\partial z}\right)\vec{r}$$

$$= \hat{i}(a_1) + \hat{j}(a_2) + \hat{k}(a_3) - (a_1\hat{i} + a_2\,\hat{j} + a_3\,\hat{k}) = \vec{0} = \text{LHS}$$

5.18. PHYSICAL INTERPRETATION OF DIVERGENCE (*G.B.T.U. 2010*)

Consider a fluid having density $\rho = \rho\,(x, y, z, t)$ and velocity $\vec{v} = \vec{v}\,(x, y, z, t)$ at a point (x, y, z) at time t. Let $\vec{V} = \rho\,\vec{v}$, then $\vec{V}$ is a vector having the same direction as $\vec{v}$ and magnitude $\rho\,|\,\vec{v}\,|$. It is known as **flux.** Its direction gives the direction of the fluid flow, and its magnitude gives the mass of the fluid crossing per unit time a unit area placed perpendicular to the direction of flow.

Consider the motion of the fluid having velocity $\vec{V} = V_x\hat{i} + V_y\hat{j} + V_z\hat{k}$ at a point $P(x, y, z)$. Consider a small parallelopiped with edges $\delta x, \delta y, \delta z$ parallel to the axes with one of its corners at P.

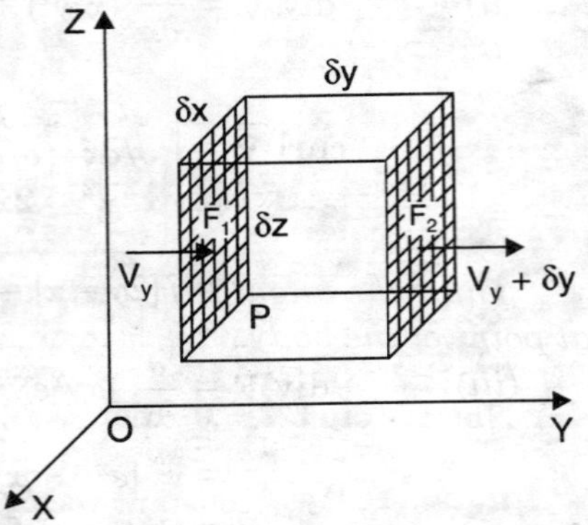

The mass of the fluid entering through the face F_1 per unit time is $V_y\, \delta x\, \delta z$ and that flowing out through the opposite face F_2 is $V_{y+\delta y}\, \delta x\, \delta z = \left(V_y + \frac{\partial V_y}{\partial y}\delta y\right)\delta x\, \delta z$ by using Taylor's series.

$\therefore$ The net decrease in the mass of fluid flowing across these two faces

$$= \left(V_y + \frac{\partial V_y}{\partial y}\delta y\right)\delta x\, \delta z - V_y\, \delta x\, \delta z = \frac{\partial V_y}{\partial y}\delta x\, \delta y\, \delta z$$

Similarly, considering the other two pairs of faces, we get the total decrease in the mass of fluid inside the parallelopiped per unit time $= \left(\frac{\partial V_x}{\partial x} + \frac{\partial V_y}{\partial y} + \frac{\partial V_z}{\partial z}\right)\delta x\, \delta y\, \delta z.$

Dividing this by the volume $\delta x\, \delta y\, \delta z$ of the parallelopiped, we have the rate of loss of fluid per unit time $= \frac{\partial V_x}{\partial x} + \frac{\partial V_y}{\partial y} + \frac{\partial V_z}{\partial z} = \text{div}\, \vec{V}$

Hence div $\vec{V}$ gives the rate of outflow per unit volume at a point of the fluid.

Note. If **div** $\vec{V} = \mathbf{0}$ everywhere in some region R of space, then $\vec{V}$ is called **solenoidal vector point function.**

5.19. PHYSICAL INTERPRETATION OF CURL *(G.B.T.U. 2010)*

Consider a rigid body rotating about a fixed axis through O with uniform angular velocity

$$\vec{\omega} = \omega_1\hat{i} + \omega_2\hat{j} + \omega_3\hat{k}$$

The velocity $\vec{V}$ of any point $P(x, y, z)$ on the body is given by $\vec{V} = \vec{\omega} \times \vec{r}$, where $\vec{r} = x\hat{i} + y\hat{j} + z\hat{k}$ is the position vector of P.

$$\therefore \quad \vec{V} = \vec{\omega} \times \vec{r} = \begin{vmatrix} \hat{i} & \hat{j} & \hat{k} \\ \omega_1 & \omega_2 & \omega_3 \\ x & y & z \end{vmatrix}$$

$$= (\omega_2 z - \omega_3 y)\,\hat{i} + (\omega_3 x - \omega_1 z)\,\hat{j} + (\omega_1 y - \omega_2 x)\,\hat{k}$$

$$\text{curl}\, \vec{V} = \nabla \times \vec{V} = \begin{vmatrix} \hat{i} & \hat{j} & \hat{k} \\ \frac{\partial}{\partial x} & \frac{\partial}{\partial y} & \frac{\partial}{\partial z} \\ \omega_2 z - \omega_3 y & \omega_3 x - \omega_1 z & \omega_1 y - \omega_2 x \end{vmatrix}$$

$$= (\omega_1 + \omega_1)\,\hat{i} + (\omega_2 + \omega_2)\,\hat{j} + (\omega_3 + \omega_3)\,\hat{k}$$

[$\because$ $\omega_1, \omega_2, \omega_3$ are constants]

$$= 2\,(\omega_1\hat{i} + \omega_2\hat{j} + \omega_3\hat{k}) = 2\vec{\omega}$$

$$\Rightarrow \quad \boxed{\vec{\omega} = \tfrac{1}{2}\,\text{curl}\,\vec{V}}$$ (*G.B.T.U. 2010*)

Thus, the angular velocity at any point is equal to half the curl of the linear velocity at that point of the body.

Note. If **curl** $\vec{\mathbf{V}} = \vec{\mathbf{0}}$, then $\vec{V}$ is said to be an **irrotational vector.**

Remark. For a constant vector $\vec{a}$, div $\vec{a} = 0$, curl $\vec{a} = \vec{0}$

5.20. VECTOR IDENTITIES

1. $$\boxed{\textbf{grad}\ (\vec{a}\,.\,\vec{b}) = \vec{a}\times \textbf{curl}\ \vec{b} + \vec{b}\times\textbf{curl}\ \vec{a} + (\vec{a}\,.\,\nabla)\,\vec{b} + (\vec{b}\,.\,\nabla)\,\vec{a}}$$

Proof. $$\text{grad}\,(\vec{a}\,.\,\vec{b}) = \sum \hat{i}\,\frac{\partial}{\partial x}(\vec{a}\,.\,\vec{b}) = \sum \hat{i}\left(\vec{a}\,.\,\frac{\partial \vec{b}}{\partial x} + \frac{\partial \vec{a}}{\partial x}\,.\,\vec{b}\right)$$

$$= \sum\left(\vec{a}\,.\,\frac{\partial \vec{b}}{\partial x}\right)\hat{i} + \sum\left(\vec{b}\,.\,\frac{\partial \vec{a}}{\partial x}\right)\hat{i} \qquad \text{...(1)}$$

We know that,

$$\vec{a}\times(\vec{b}\times\vec{c}) = (\vec{a}\,.\,\vec{c})\vec{b} - (\vec{a}\,.\,\vec{b})\vec{c}$$

$$\Rightarrow \quad (\vec{a}\,.\,\vec{b})\,\vec{c} = (\vec{a}\,.\,\vec{c})\vec{b} - \vec{a}\times(\vec{b}\times\vec{c})$$

$$\therefore \quad \left(\vec{a}\,.\,\frac{\partial \vec{b}}{\partial x}\right)\hat{i} = (\vec{a}\,.\,\hat{i})\,\frac{\partial \vec{b}}{\partial x} - \vec{a}\times\left(\frac{\partial \vec{b}}{\partial x}\times\hat{i}\right) = (\vec{a}\,.\,\hat{i})\,\frac{\partial \vec{b}}{\partial x} + \vec{a}\times\left(\hat{i}\times\frac{\partial \vec{b}}{\partial x}\right)$$

$$\therefore \quad \sum\left(\vec{a}\,.\,\frac{\partial \vec{b}}{\partial x}\right)\hat{i} = \left(\vec{a}\,.\sum\hat{i}\,\frac{\partial}{\partial x}\right)\vec{b} + \vec{a}\times\sum\left(\hat{i}\times\frac{\partial \vec{b}}{\partial x}\right)$$

$$= (\vec{a}\,.\,\nabla)\,\vec{b} + \vec{a}\times\text{curl}\,\vec{b} \qquad \text{...(2)}$$

Interchanging $\vec{a}$ and $\vec{b}$, we get

$$\sum\left(\vec{b}\,.\,\frac{\partial \vec{a}}{\partial x}\right)\hat{i} = (\vec{b}\,.\,\nabla)\,\vec{a} + \vec{b}\times\text{curl}\,\vec{a} \qquad \text{...(3)}$$

Substituting values from (2) and (3) in (1), we get

$$\text{grad}\,(\vec{a}\,.\,\vec{b}) = \vec{a}\times\text{curl}\,\vec{b} + \vec{b}\times\text{curl}\,\vec{a} + (\vec{a}\,.\,\nabla)\,\vec{b} + (\vec{b}\,.\,\nabla)\,\vec{a}$$

2. If $\vec{a}$ is a vector function and u is a scalar function then,

$$\boxed{\text{div}\,(u\vec{a}) = u\,\text{div}\,\vec{a} + (\text{grad}\,u)\,.\,\vec{a}}$$

Proof. $\text{div}\,(u\vec{a}) = \sum \hat{i}\,.\,\frac{\partial}{\partial x}(u\vec{a})$

$$= \sum \hat{i}\,.\left(u\frac{\partial \vec{a}}{\partial x} + \frac{\partial u}{\partial x}\vec{a}\right) = u\sum \hat{i}\,.\,\frac{\partial \vec{a}}{\partial x} + \left(\sum \hat{i}\,\frac{\partial u}{\partial x}\right).\,\vec{a}$$

$$= u\,\text{div}\,\vec{a} + (\text{grad}\,u)\,.\,\vec{a}$$

3. $\boxed{\text{div}\,(\vec{a}\times\vec{b}) = \vec{b}\,.\,\text{curl}\,\vec{a}\ \text{Đ}\vec{a}\,.\,\text{curl}\,\vec{b}}$ *(G.B.T.U. 2012)*

Proof. $\text{div}\,(\vec{a}\times\vec{b}) = \sum \hat{i}\,.\,\frac{\partial}{\partial x}(\vec{a}\times\vec{b}) = \sum \hat{i}\,.\left(\frac{\partial \vec{a}}{\partial x}\times\vec{b} + \vec{a}\times\frac{\partial \vec{b}}{\partial x}\right)$

$$= \sum \hat{i}\,.\left(\frac{\partial \vec{a}}{\partial x}\times\vec{b}\right) - \sum \hat{i}\,.\left(\frac{\partial \vec{b}}{\partial x}\times\vec{a}\right)$$

$$= \sum\left(\hat{i}\times\frac{\partial \vec{a}}{\partial x}\right).\,\vec{b} - \sum\left(\hat{i}\times\frac{\partial \vec{b}}{\partial x}\right).\,\vec{a} \qquad |\ \text{Since } \vec{a}\,.\,(\vec{b}\times\vec{c}) = (\vec{a}\times\vec{b})\,.\,\vec{c}$$

$$= (\text{curl}\,\vec{a})\,.\,\vec{b} - (\text{curl}\,\vec{b})\,.\,\vec{a} = \vec{b}\,.\,\text{curl}\,\vec{a} - \vec{a}\,.\,\text{curl}\,\vec{b}$$

4. If $\vec{a}$ is a vector function and u is a scalar function then,

$$\boxed{\text{curl}\,(u\vec{a}) = u\,\text{curl}\,\vec{a} + (\text{grad}\,u)\times\vec{a}}$$

Proof. $\text{curl}\,(u\vec{a}) = \sum \hat{i}\times\frac{\partial}{\partial x}(u\vec{a}) = \sum \hat{i}\times\left(u\frac{\partial \vec{a}}{\partial x} + \frac{\partial u}{\partial x}\vec{a}\right)$

$$= u\left(\sum \hat{i}\times\frac{\partial \vec{a}}{\partial x}\right) + \sum \hat{i}\,\frac{\partial u}{\partial x}\times\vec{a} = u\,\text{curl}\,\vec{a} + (\text{grad}\,u)\times\vec{a}$$

5. $\boxed{\text{curl}\,(\vec{a}\times\vec{b}) = \vec{a}\,\text{div}\,\vec{b}\ \text{Đ}\vec{b}\,\text{div}\,\vec{a} + (\vec{b}\,.\,\nabla)\,\vec{a}\ \text{Đ}(\vec{a}\,.\,\nabla)\,\vec{b}}$ *(G.B.T.U. 2013)*

Proof. $\text{curl}\,(\vec{a}\times\vec{b}) = \sum \hat{i}\times\frac{\partial}{\partial x}(\vec{a}\times\vec{b})$

$$= \sum \hat{i}\times\left(\frac{\partial \vec{a}}{\partial x}\times\vec{b} + \vec{a}\times\frac{\partial \vec{b}}{\partial x}\right) = \sum \hat{i}\times\left(\frac{\partial \vec{a}}{\partial x}\times\vec{b}\right) + \sum \hat{i}\times\left(\vec{a}\times\frac{\partial \vec{b}}{\partial x}\right)$$

$$= \sum \left[(\hat{i} \,.\, \vec{b}) \frac{\partial \vec{a}}{\partial x} - \left(\hat{i} \,.\, \frac{\partial \vec{a}}{\partial x} \right) \vec{b} \right] + \sum \left[\left(\hat{i} \,.\, \frac{\partial \vec{b}}{\partial x} \right) \vec{a} - (\hat{i} \,.\, \vec{a}) \frac{\partial \vec{b}}{\partial x} \right]$$

| Since $\vec{a} \times (\vec{b} \times \vec{c}) = (\vec{a} \,.\, \vec{c})\, \vec{b} - (\vec{a} \,.\, \vec{b})\, \vec{c}$

$$= \sum (\vec{b} \,.\, \hat{i}) \frac{\partial \vec{a}}{\partial x} - \left(\sum \hat{i} \,.\, \frac{\partial \vec{a}}{\partial x} \right) \vec{b} + \sum \left(\hat{i} \,.\, \frac{\partial \vec{b}}{\partial x} \right) \vec{a} - \sum (\vec{a} \,.\, \hat{i}) \frac{\partial \vec{b}}{\partial x}$$

$$= \left(\vec{b} \,.\, \sum i \frac{\partial}{\partial x} \right) \vec{a} - \vec{b} \text{ div } \vec{a} + \vec{a} \text{ div } \vec{b} - \left(\vec{a} \,.\, \sum i \frac{\partial}{\partial x} \right) \vec{b}$$

$$= (\vec{b} \,.\, \nabla)\, \vec{a} - \vec{b} \text{ div } \vec{a} + \vec{a} \text{ div } \vec{b} - (\vec{a} \,.\, \nabla)\, \vec{b}$$

$$= \vec{a} \text{ div } \vec{b} - \vec{b} \text{ div } \vec{a} + (\vec{b} \,.\, \nabla)\, \vec{a} - (\vec{a} \,.\, \nabla)\, \vec{b}$$

6. div (grad ϕ) = $\nabla^2 \phi$

Proof. div (grad ϕ) $= \nabla \,.\, (\nabla \phi) = \left(\hat{i} \frac{\partial}{\partial x} + \hat{j} \frac{\partial}{\partial y} + \hat{k} \frac{\partial}{\partial z} \right) . \left(\hat{i} \frac{\partial \phi}{\partial x} + \hat{j} \frac{\partial \phi}{\partial y} + \hat{k} \frac{\partial \phi}{\partial z} \right)$

$$= \frac{\partial}{\partial x} \left(\frac{\partial \phi}{\partial x} \right) + \frac{\partial}{\partial y} \left(\frac{\partial \phi}{\partial y} \right) + \frac{\partial}{\partial z} \left(\frac{\partial \phi}{\partial z} \right)$$

$$= \frac{\partial^2 \phi}{\partial x^2} + \frac{\partial^2 \phi}{\partial y^2} + \frac{\partial^2 \phi}{\partial z^2} = \left(\frac{\partial^2}{\partial x^2} + \frac{\partial^2}{\partial y^2} + \frac{\partial^2}{\partial z^2} \right) \phi$$

$$= \nabla^2 \phi \quad \text{where } \nabla^2 \equiv \frac{\partial^2}{\partial x^2} + \frac{\partial^2}{\partial y^2} + \frac{\partial^2}{\partial z^2}$$

$\nabla^2 \phi$ is called the Laplacian operator and $\nabla^2 \phi = 0$ is called Laplace's equation.

7. curl (grad ϕ) = $\nabla \times \nabla \phi = \vec{0}$ *(A.K.T.U. 2011, 2018)*

Proof. curl (grad ϕ) $= \nabla \times \nabla \phi = \nabla \times \left(\hat{i} \frac{\partial \phi}{\partial x} + \hat{j} \frac{\partial \phi}{\partial y} + \hat{k} \frac{\partial \phi}{\partial z} \right)$

$$= \begin{vmatrix} \hat{i} & \hat{j} & \hat{k} \\ \frac{\partial}{\partial x} & \frac{\partial}{\partial y} & \frac{\partial}{\partial z} \\ \frac{\partial \phi}{\partial x} & \frac{\partial \phi}{\partial y} & \frac{\partial \phi}{\partial z} \end{vmatrix} = \sum \hat{i} \left(\frac{\partial^2 \phi}{\partial y \partial z} - \frac{\partial^2 \phi}{\partial z \partial y} \right) = \vec{0}$$

8. div (curl $\vec{V}$) = $\nabla \,.\, (\nabla \times \vec{V}) = 0$ *(A.K.T.U. 2011, 2016, 2018)*

Proof. Let $\vec{V} = V_1 \hat{i} + V_2 \hat{j} + V_3 \hat{k}$ then,

$$\text{curl } \vec{V} = \begin{vmatrix} \hat{i} & \hat{j} & \hat{k} \\ \frac{\partial}{\partial x} & \frac{\partial}{\partial y} & \frac{\partial}{\partial z} \\ V_1 & V_2 & V_3 \end{vmatrix} = \hat{i} \left(\frac{\partial V_3}{\partial y} - \frac{\partial V_2}{\partial z} \right) + \hat{j} \left(\frac{\partial V_1}{\partial z} - \frac{\partial V_3}{\partial x} \right) + \hat{k} \left(\frac{\partial V_2}{\partial x} - \frac{\partial V_1}{\partial y} \right)$$

$\therefore \quad \text{div}(\text{curl}\ \vec{V}) = \nabla . (\nabla \times \vec{V})$

$$= \frac{\partial}{\partial x}\left(\frac{\partial V_3}{\partial y} - \frac{\partial V_2}{\partial z}\right) + \frac{\partial}{\partial y}\left(\frac{\partial V_1}{\partial z} - \frac{\partial V_3}{\partial x}\right) + \frac{\partial}{\partial z}\left(\frac{\partial V_2}{\partial x} - \frac{\partial V_1}{\partial y}\right)$$

$$= \left(\frac{\partial^2 V_3}{\partial x \partial y} - \frac{\partial^2 V_2}{\partial x \partial z}\right) + \left(\frac{\partial^2 V_1}{\partial y \partial z} - \frac{\partial^2 V_3}{\partial y \partial x}\right) + \left(\frac{\partial^2 V_2}{\partial z \partial x} - \frac{\partial^2 V_1}{\partial z \partial y}\right) = 0.$$

9. $\boxed{\textbf{curl (curl } \vec{V}) = \textbf{grad (div } \vec{V}) - \nabla^2 \vec{V}}$

Proof. Let $\vec{V} = V_1\hat{i} + V_2\hat{j} + V_3\hat{k}$

then $\text{curl}\ \vec{V} = \begin{vmatrix} \hat{i} & \hat{j} & \hat{k} \\ \frac{\partial}{\partial x} & \frac{\partial}{\partial y} & \frac{\partial}{\partial z} \\ V_1 & V_2 & V_3 \end{vmatrix} = \hat{i}\left(\frac{\partial V_3}{\partial y} - \frac{\partial V_2}{\partial z}\right) + \hat{j}\left(\frac{\partial V_1}{\partial z} - \frac{\partial V_3}{\partial x}\right) + \hat{k}\left(\frac{\partial V_2}{\partial x} - \frac{\partial V_1}{\partial y}\right)$

$$\therefore \quad \text{curl}(\text{curl}\ \vec{V}) = \begin{vmatrix} \hat{i} & \hat{j} & \hat{k} \\ \frac{\partial}{\partial x} & \frac{\partial}{\partial y} & \frac{\partial}{\partial z} \\ \frac{\partial V_3}{\partial y} - \frac{\partial V_2}{\partial z} & \frac{\partial V_1}{\partial z} - \frac{\partial V_3}{\partial x} & \frac{\partial V_2}{\partial x} - \frac{\partial V_1}{\partial y} \end{vmatrix}$$

$$= \Sigma\hat{i}\left\{\frac{\partial}{\partial y}\left(\frac{\partial V_2}{\partial x} - \frac{\partial V_1}{\partial y}\right) - \frac{\partial}{\partial z}\left(\frac{\partial V_1}{\partial z} - \frac{\partial V_3}{\partial x}\right)\right\}$$

$$= \Sigma\hat{i}\left\{\frac{\partial}{\partial x}\left(\frac{\partial V_2}{\partial y} + \frac{\partial V_3}{\partial z}\right) - \left(\frac{\partial^2 V_1}{\partial y^2} + \frac{\partial^2 V_1}{\partial z^2}\right)\right\}$$

$$= \Sigma\hat{i}\left\{\frac{\partial}{\partial x}\left(\frac{\partial V_1}{\partial x} + \frac{\partial V_2}{\partial y} + \frac{\partial V_3}{\partial z}\right) - \left(\frac{\partial^2 V_1}{\partial x^2} + \frac{\partial^2 V_1}{\partial y^2} + \frac{\partial^2 V_1}{\partial z^2}\right)\right\}$$

$$= \Sigma\hat{i}\left\{\frac{\partial}{\partial x}(\nabla . \vec{V}) - (\nabla^2 V_1)\right\} = \Sigma\hat{i}\,\frac{\partial}{\partial x}(\nabla . \vec{V}) - \nabla^2\,\Sigma\hat{i}\,V_1$$

$$= \nabla(\nabla . \vec{V}) - \nabla^2\vec{V} = \text{grad}(\text{div}\ \vec{V}) - \nabla^2\,\vec{V}.$$

Example 6. *A vector field is given by* $\vec{A} = (x^2 + xy^2)\,\hat{i} + (y^2 + x^2y)\,\hat{j}$. *Show that the field is irrotational and find the scalar potential.*

Sol. Field $\vec{A}$ is irrotational if curl $\vec{A} = \vec{0}$

Now $$\text{curl } \vec{A} = \begin{vmatrix} \hat{i} & \hat{j} & \hat{k} \\ \dfrac{\partial}{\partial x} & \dfrac{\partial}{\partial y} & \dfrac{\partial}{\partial z} \\ x^2 + xy^2 & y^2 + x^2 y & 0 \end{vmatrix} = \hat{i}\,(0 - 0) - \hat{j}(0 - 0) + \hat{k}\,(2xy - 2xy) = \vec{0}.$$

$\therefore$ Field $\vec{A}$ is irrotational.

If ϕ is the scalar potential then,

$$\vec{A} = \text{grad } \phi$$

$$\Rightarrow \quad \vec{A} \cdot d\vec{r} = (\text{grad } \phi) \cdot d\vec{r} = \left(\hat{i}\,\frac{\partial \phi}{\partial x} + \hat{j}\,\frac{\partial \phi}{\partial y} + \hat{k}\,\frac{\partial \phi}{\partial z}\right) \cdot (dx\,\hat{i} + dy\,\hat{j} + dz\,\hat{k})$$

$$= \frac{\partial \phi}{\partial x}\,dx + \frac{\partial \phi}{\partial y}\,dy + \frac{\partial \phi}{\partial z}\,dz = d\phi$$

$$\therefore \quad d\phi = \vec{A} \cdot d\vec{r}$$

$$= (x^2 + xy^2)\,dx + (y^2 + x^2y)\,dy = x^2\,dx + y^2\,dy + xy\,(y\,dx + x\,dy)$$

$$= d\left(\frac{x^3}{3}\right) + d\left(\frac{y^3}{3}\right) + xy\,d(xy) = d\left(\frac{x^3}{3}\right) + d\left(\frac{y^3}{3}\right) + d\left(\frac{x^2y^2}{2}\right)$$

Integration yields,

$$\phi = \frac{x^3}{3} + \frac{y^3}{3} + \frac{x^2y^2}{2} + c$$ where c is a constant.

Example 7. *A fluid motion is given by* $\vec{V} = (y + z)\,\hat{i} + (z + x)\,\hat{j} + (x + y)\,\hat{k}$

(*i*) *Is this motion irrotational? If so, find the velocity potential.* (A.K.T.U. 2016)

(*ii*) *Is the motion possible for an incompressible fluid?*

Sol. We have $\vec{V} = (y + z)\hat{i} + (z + x)\,\hat{j} + (x + y)\,\hat{k}$.

(*i*) The motion is irrotational if curl $\vec{V} = \vec{0}$

$$\therefore \quad \text{curl } \vec{V} = \begin{vmatrix} \hat{i} & \hat{j} & \hat{k} \\ \dfrac{\partial}{\partial x} & \dfrac{\partial}{\partial y} & \dfrac{\partial}{\partial z} \\ y + z & z + x & x + y \end{vmatrix}$$

$$= \hat{i}\left\{\frac{\partial}{\partial y}(x + y) - \frac{\partial}{\partial z}(z + x)\right\} - \hat{j}\left\{\frac{\partial}{\partial x}(x + y) - \frac{\partial}{\partial z}(y + z)\right\}$$

$$+ \hat{k}\left\{\frac{\partial}{\partial x}(z + x) - \frac{\partial}{\partial y}(y + z)\right\} = \hat{i}(1 - 1) - \hat{j}(1 - 1) + \hat{k}(1 - 1)$$

$$= \hat{i}(0) - \hat{j}(0) + \hat{k}(0) = \vec{0}$$

Hence the motion is irrotational.

Now, let $\vec{V}$ = grad ϕ, where ϕ is velocity potential

$$\Rightarrow \quad \vec{V}\cdot d\vec{r} = \nabla\phi \cdot d\vec{r} = \left(\frac{\partial\phi}{\partial x}\hat{i} + \frac{\partial\phi}{\partial y}\hat{j} + \frac{\partial\phi}{\partial z}\hat{k}\right)\cdot(dx\,\hat{i} + dy\,\hat{j} + dz\,\hat{k})$$

$$= \frac{\partial\phi}{\partial x}dx + \frac{\partial\phi}{\partial y}dy + \frac{\partial\phi}{\partial z}dz = d\phi$$

$$\therefore \quad d\phi = \vec{V}\cdot d\vec{r}$$

$$= (y+z)\hat{i} + (z+x)\hat{j} + (x+y)\hat{k}\cdot(dx\,\hat{i} + dy\,\hat{j} + dz\,\hat{k})$$
$$= (y+z)\,dx + (z+x)\,dy + (x+y)\,dz$$
$$= (y\,dx + x\,dy) + (z\,dx + x\,dz) + (z\,dy + y\,dz)$$
$$= d(xy) + d(zx) + d(zy)$$

Integrating, we get $\phi = xy + yz + zx + c$.

(*ii*) For an incompressible fluid,

$$\text{div}\,\vec{V} = 0$$

Now, $$\text{div}\,\vec{V} = \left(\hat{i}\frac{\partial}{\partial x} + \hat{j}\frac{\partial}{\partial y} + \hat{k}\frac{\partial}{\partial z}\right)\cdot\{(y+z)\,\hat{i} + (z+x)\,\hat{j} + (x+y)\,\hat{k}\}$$

$$= \frac{\partial}{\partial x}(y+z) + \frac{\partial}{\partial y}(z+x) + \frac{\partial}{\partial z}(x+y) = 0 + 0 + 0 = 0$$

Hence the motion is possible for an incompressible fluid.

Example 8. *If $\vec{a}$ is a constant vector, evaluate div $(\vec{r}\times\vec{a})$ and curl $(\vec{r}\times\vec{a})$ where $\vec{r} = x\hat{i} + y\hat{j} + z\hat{k}$.*

Sol. We know that

$$\text{div}\,(\vec{a}\times\vec{b}) = \vec{b}\cdot\text{curl}\,\vec{a} - \vec{a}\cdot\text{curl}\,\vec{b}$$

$$\therefore \quad \text{div}\,(\vec{r}\times\vec{a}) = \vec{a}\cdot\text{curl}\,\vec{r} - \vec{r}\cdot\text{curl}\,\vec{a}$$

$$= \vec{a}\cdot\vec{0} - \vec{r}\cdot\vec{0} = \vec{0} - \vec{0} = \vec{0}$$

$| \because$ curl $\vec{r} = \vec{0}$ and curl of a constant vector is $\vec{0}$

Also, we know that

$$\text{curl}\,(\vec{a}\times\vec{b}) = \vec{a}\,\text{div}\,\vec{b} - \vec{b}\,\text{div}\,\vec{a} + (\vec{b}\cdot\nabla)\,\vec{a} - (\vec{a}\cdot\nabla)\,\vec{b}$$

$$\therefore \quad \text{curl}\,(\vec{r}\times\vec{a}) = \vec{r}\,\text{div}\,\vec{a} - \vec{a}\,\text{div}\,\vec{r} + (\vec{a}\cdot\nabla)\,\vec{r} - (\vec{r}\cdot\nabla)\,\vec{a}$$

$$= \vec{0} - 3\vec{a} + \vec{a} - \vec{0} \quad | \because \ \text{div}\,\vec{a} = 0,\ (\vec{r}\cdot\nabla)\,\vec{a} = \vec{0} \text{ and } (\vec{a}\cdot\nabla)\,\vec{r} = \vec{a}$$

$$= -2\vec{a}.$$

Example 9. *If* $\vec{r} = x\hat{i} + y\hat{j} + z\hat{k}$ *and* $r = |\vec{r}|$, *show that*

(i) $div\,(\vec{r}\,\phi) = 3\phi + \vec{r}\,.\,grad\,\phi$ (ii) $div\,(\hat{r}) = \dfrac{2}{r}$.

Sol. We know that $\text{div}\,(u\vec{a}) = u\,\text{div}\,\vec{a} + \vec{a}\,.\,\text{grad}\,u$

$$\therefore \quad \text{div}\,(\phi\vec{r}) = \phi\,\text{div}\,\vec{r} + \vec{r}\,.\,\text{grad}\,\phi = 3\phi + \vec{r}\,.\,\text{grad}\,\phi \qquad |\because \text{div}\,\vec{r} = 3$$

and
$$\text{div}\,(\hat{r}) = \text{div}\left(\frac{1}{r}\vec{r}\right) \qquad \because \hat{r} = \frac{\vec{r}}{r}$$

$$= \frac{1}{r}\,\text{div}\,\vec{r} + \vec{r}\,.\,\text{grad}\,\frac{1}{r} = \frac{3}{r} + \vec{r}\,.\left(-\frac{1}{r^2}\hat{r}\right)$$

$$= \frac{3}{r} + \vec{r}\,.\left(-\frac{\vec{r}}{r^3}\right) = \frac{3}{r} - \frac{r^2}{r^3} \qquad |\because \vec{r}\,.\,\vec{r} = r^2$$

$$= \frac{3}{r} - \frac{1}{r} = \frac{2}{r}$$

Example 10. *Prove that div (grad* $r^n) = \nabla^2(r^n) = n(n+1)r^{n-2}$ *where* $\vec{r} = x\hat{i} + y\hat{j} + z\hat{k}$. *Hence show that* $\nabla^2\left(\dfrac{1}{r}\right) = 0$. *Hence or otherwise evaluate* $\nabla \times \left(\dfrac{\vec{r}}{r^2}\right)$. (U.P.T.U. 2014; U.K.T.U. 2012)

Sol. Here, $r^2 = x^2 + y^2 + z^2$ so that $\dfrac{\partial r}{\partial x} = \dfrac{x}{r}, \dfrac{\partial r}{\partial y} = \dfrac{y}{r}, \dfrac{\partial r}{\partial z} = \dfrac{z}{r}$

$$\text{grad}\,r^n = nr^{n-1}\,\hat{r} = nr^{n-1}\,\frac{\vec{r}}{r} = nr^{n-2}\,\vec{r}$$

$$\therefore \quad \text{div}\,(\text{grad}\,r^n) = \text{div}\,(nr^{n-2}\,\vec{r}) = nr^{n-2}\,\text{div}\,\vec{r} + \text{grad}\,(nr^{n-2})\,.\,\vec{r}$$

$$= 3nr^{n-2} + n(n-2)r^{n-3}\,\hat{r}\,.\,\vec{r} \qquad |\because \text{div}\,\vec{r} = 3$$

$$= 3nr^{n-2} + n(n-2)r^{n-4}\,(\vec{r}\,.\,\vec{r}) = 3nr^{n-2} + n(n-2)\,r^{n-4}\,(r^2)$$

$$= 3nr^{n-2} + n(n-2)r^{n-2}$$

$$\Rightarrow \quad \nabla^2(r^n) = n(n+1)r^{n-2}$$

Putting $n = -1$, we get

$$\nabla^2\left(\frac{1}{r}\right) = 0$$

We know that,

$$\text{curl}\,(u\vec{a}) = u\,\text{curl}\,\vec{a} + (\text{grad}\,u) \times \vec{a}$$

$$\therefore \quad \text{curl}\left(\frac{\vec{r}}{r^2}\right) = \frac{1}{r^2}\text{curl } \vec{r} + \left(\text{grad } \frac{1}{r^2}\right) \times \vec{r} = \vec{0} - \left(\frac{2}{r^3}\hat{r}\right) \times \vec{r} \qquad | \because \text{curl } \vec{r} = \vec{0}$$

$$= -\frac{2}{r^4}(\vec{r} \times \vec{r}) = \vec{0}.$$

Example 11. (*i*) *Prove that vector* $f(r)\ \vec{r}$ *is irrotational.*

(*ii*) *Prove that* $\nabla^2 f(r) = f''(r) + \dfrac{2}{r} f'(r)$. *Hence evaluate* $\nabla^2 (\log r)$ *if* $r = (x^2 + y^2 + z^2)^{1/2}$

(U.K.T.U. 2010; G.B.T.U. 2012)

Sol. (*i*) $\text{curl } \{f(r)\vec{r}\} = f(r) \text{ curl } \vec{r} + \{\text{grad } f(r)\} \times \vec{r}$ | Using vector identity

$$= \vec{0} + f'(r)\,\hat{r} \times \vec{r} \qquad | \text{ curl } \vec{r} = \vec{0}$$

$$= \frac{f'(r)}{r}(\vec{r} \times \vec{r}) = \vec{0} \qquad | \ \vec{r} \times \vec{r} = \vec{0}$$

Hence $f(r)\ \vec{r}$ is irrotational.

(*ii*) $\quad \text{grad } f(r) = f'(r)\hat{r} = \dfrac{1}{r} f'(r)\ \vec{r}$

$\text{div } \{\text{grad } f(r)\} = \nabla^2 f(r)$

$$= \text{div}\left\{\frac{f'(r)}{r}\vec{r}\right\} = \frac{f'(r)}{r}\text{div } \vec{r} + \text{grad}\left\{\frac{f'(r)}{r}\right\}.\vec{r}$$

$$= \frac{3}{r}f'(r) + \left\{\frac{rf''(r) - f'(r)}{r^2}\right\}\hat{r}.\vec{r} = \frac{3}{r}f'(r) + \left\{\frac{rf''(r) - f'(r)}{r^3}\right\}(\vec{r}.\vec{r})$$

$$= \frac{3}{r}f'(r) + \left\{\frac{rf''(r) - f'(r)}{r}\right\} \qquad | \because \vec{r}.\vec{r} = r^2$$

$$\Rightarrow \quad \nabla^2 f(r) = f''(r) + \frac{2}{r}f'(r)$$

Now, $\quad \nabla^2 \log r = -\dfrac{1}{r^2} + \dfrac{2}{r}\left(\dfrac{1}{r}\right) = \dfrac{1}{r^2} = \dfrac{1}{x^2 + y^2 + z^2}$

Example 12. *If* $u\vec{F} = \nabla v$ *where u, v are scalar fields and* $\vec{F}$ *is a vector field show that* $\vec{F}.\text{curl } \vec{F} = 0$.

Sol. $\quad \text{curl } \vec{F} = \text{curl}\left(\dfrac{1}{u}\nabla v\right) = \dfrac{1}{u}\text{curl }(\nabla v) + \left(\text{grad } \dfrac{1}{u}\right) \times \nabla v$

$$= \vec{0} + \left(\nabla\frac{1}{u}\right) \times \nabla v = \left(\nabla\frac{1}{u}\right) \times \nabla v \qquad | \because \nabla \times (\nabla v) = \vec{0}$$

$$\therefore \qquad \vec{F} \cdot \text{curl } \vec{F} = \frac{1}{u}(\nabla v) \cdot \left\{\left(\nabla \frac{1}{u}\right) \times (\nabla v)\right\} = 0$$

being the scalar triple product in which two factors are equal.

Example 13. (*i*) *Show that,*

$$\textit{curl}\left(\hat{k} \times \textit{grad}\,\frac{1}{r}\right) + \textit{grad}\left(\hat{k} \cdot \textit{grad}\,\frac{1}{r}\right) = \vec{0}$$ *where r is the distance of a point (x, y, z) from the origin and* $\hat{k}$ *is a unit vector in the direction of OZ.*

(*ii*) *If f and g are two scalar point functions prove that*

$$div\,(f\nabla g) = f\nabla^2 g + \nabla f \cdot \nabla g$$

Sol. (*i*) We know that

$$\text{curl }(\vec{a} \times \vec{b}) = \vec{a} \text{ div } \vec{b} - \vec{b} \text{ div } \vec{a} + (\vec{b} \cdot \nabla)\,\vec{a} - (\vec{a} \cdot \nabla)\,\vec{b}$$

$$\therefore \qquad \text{curl}\left(\hat{k} \times \text{grad}\,\frac{1}{r}\right) = \hat{k}\,\text{div}\left(\text{grad}\,\frac{1}{r}\right) - \left(\text{grad}\,\frac{1}{r}\right)\text{div}\,\hat{k}$$

$$+ \left(\text{grad}\,\frac{1}{r} \cdot \nabla\right)\hat{k} - (\hat{k} \cdot \nabla)\,\text{grad}\,\frac{1}{r}$$

$$= 0 - 0 + \left(\text{grad}\,\frac{1}{r} \cdot \nabla\right)\hat{k} - (\hat{k} \cdot \nabla)\,\text{grad}\,\frac{1}{r}$$

$$\Big| \because \nabla^2\left(\frac{1}{r}\right) = 0 \text{ and div } \hat{k} = 0$$

$$= \left(\text{grad}\,\frac{1}{r} \cdot \nabla\right)\hat{k} - (\hat{k} \cdot \nabla)\,\text{grad}\,\frac{1}{r} \qquad \text{...(1)}$$

We also know that,

$$\text{grad }(\vec{a} \cdot \vec{b}) = \vec{a} \times \text{curl } \vec{b} + \vec{b} \times \text{curl } \vec{a} + (\vec{a} \cdot \nabla)\,\vec{b} + (\vec{b} \cdot \nabla)\,\vec{a}$$

$$\therefore \qquad \text{grad}\left(\hat{k} \cdot \text{grad}\,\frac{1}{r}\right) = \hat{k} \times \text{curl}\left(\text{grad}\,\frac{1}{r}\right) + \left(\text{grad}\,\frac{1}{r}\right) \times \text{curl}\,\hat{k}$$

$$+ (\hat{k} \cdot \nabla)\,\text{grad}\,\frac{1}{r} + \left(\text{grad}\,\frac{1}{r} \cdot \nabla\right)\hat{k}$$

$$= \vec{0} + \vec{0} + (\hat{k} \cdot \nabla)\,\text{grad}\,\frac{1}{r} + \left(\text{grad}\,\frac{1}{r} \cdot \nabla\right)\hat{k} \qquad \Bigg| \because \text{curl (grad } \phi) = \vec{0} \text{ and curl } \hat{k} = \vec{0}$$

$$= (\hat{k} \cdot \nabla)\,\text{grad}\,\frac{1}{r} + \left(\text{grad}\,\frac{1}{r} \cdot \nabla\right)\hat{k} \qquad \text{...(2)}$$

Adding (1) and (2), we get

$$\text{curl}\left(\hat{k} \times \text{grad}\,\frac{1}{r}\right) + \text{grad}\left(\hat{k} \cdot \text{grad}\,\frac{1}{r}\right) = 2\left(\text{grad}\,\frac{1}{r} \cdot \nabla\right)\hat{k} = 2\left(-\frac{1}{r^2}\,\hat{r} \cdot \nabla\right)\hat{k} = -\frac{2}{r^3}(\vec{r} \cdot \nabla)\,\hat{k}$$

$$= -\frac{2}{r^3}\left(x\frac{\partial}{\partial x} + y\frac{\partial}{\partial y} + z\frac{\partial}{\partial z}\right)\hat{k} = \vec{0}$$

(ii) $\text{div}\,(f\nabla g) = f\,\text{div}\,(\nabla g) + (\text{grad}\,f)\,.\,\nabla g$ | Using vector identity

$$= f\nabla\,.(\nabla g) + (\text{grad}\,f)\,.\,\nabla g = f\nabla^2 g + \nabla f\,.\,\nabla g.$$

Example 14. *Show that the vector field* $\vec{F} = \frac{\vec{r}}{r^3}$ *is irrotational as well as solenoidal. Find the scalar potential.* (A.K.T.U. 2018)

Sol. For the vector field $\vec{F}$ to be irrotational, curl $\vec{F} = \vec{0}$

We know that, $\text{curl}\,(u\vec{a}) = u\,\text{curl}\,\vec{a} + (\text{grad}\,u) \times \vec{a}$

$$\therefore \quad \text{curl}\left(\frac{1}{r^3}\vec{r}\right) = \frac{1}{r^3}\,\text{curl}\,\vec{r} + \left(\text{grad}\,\frac{1}{r^3}\right) \times \vec{r}$$

$$= \frac{1}{r^3}(\vec{0}) + \left(-\frac{3}{r^4}\hat{r}\right) \times \vec{r} \qquad | \because \ \text{curl}\,\vec{r} = \vec{0}$$

$$= \vec{0} - \frac{3}{r^5}(\vec{r} \times \vec{r}) = \vec{0} - \vec{0} = \vec{0}$$

Hence vector field $\vec{F}$ is irrotational.

Again, for the vector field $\vec{F}$ to be solenoidal, div $\vec{F} = 0$

We know that $\text{div}\,(u\vec{a}) = u\,\text{div}\,\vec{a} + \vec{a}\,.\,\text{grad}\,u$

$$\therefore \quad \text{div}\left(\frac{\vec{r}}{r^3}\right) = \frac{1}{r^3}\,\text{div}\,\vec{r} + \vec{r}\,.\,\text{grad}\left(\frac{1}{r^3}\right)$$

$$= \frac{3}{r^3} + \vec{r}\,.\left(-\frac{3}{r^4}\frac{\vec{r}}{r}\right) \qquad | \because \ \text{div}\,\vec{r} = 3$$

$$= \frac{3}{r^3} - \frac{3}{r^5}r^2 = \frac{3}{r^3} - \frac{3}{r^3} = 0 \qquad | \because \ \vec{r}\,.\,\vec{r} = r^2$$

Hence vector field $\vec{F}$ is solenoidal.

Now, Let $\vec{F} = \nabla\phi$ where ϕ is scalar potential.

$$\therefore \quad \vec{F}\,.\,d\vec{r} = \nabla\phi\,.\,d\vec{r}$$

$$\vec{F}\,.\,d\vec{r} = d\phi \qquad | \because \ \nabla\phi\,.\,d\vec{r} = d\phi$$

$$\therefore \quad d\phi = \frac{(xi + yj + zk)}{(x^2 + y^2 + z^2)^{3/2}}\,.\,(dx\,i + dy\,j + dz\,k) \qquad | \because \ \vec{r} = xi + yj + zk$$

$$= \frac{xdx + ydy + zdz}{(x^2 + y^2 + z^2)^{3/2}} = d\,\{-(x^2 + y^2 + z^2)^{-1/2}\}$$

Integrating, we get $\phi = -\dfrac{1}{\sqrt{x^2+y^2+z^2}} + c$

or $\phi = -\dfrac{1}{r} + c.$

Example 15. *Prove that div* $(\nabla\phi \times \nabla\psi) = 0$.

Sol. We know that

$$\text{div}\,(\vec{a} \times \vec{b}) = \vec{b}\,.\,\text{curl}\,\vec{a} - \vec{a}\,.\,\text{curl}\,\vec{b}$$

$$\therefore \quad \text{div}\,(\nabla\phi \times \nabla\psi) = \nabla\psi\,.\,\text{curl}\,(\nabla\phi) - \nabla\phi\,.\,\text{curl}\,(\nabla\psi)$$

$$= \nabla\psi\,.\,\vec{0} - \nabla\phi\,.\,\vec{0} = 0 \qquad \Big|\because \ \text{curl}\,(\text{grad}\,f) = \vec{0}$$

Example 16. *Show that* $\nabla^2\left(\dfrac{x}{r^3}\right) = 0$ *where r is magnitude of position vector* $\vec{r} = x\hat{i} + y\hat{j} + z\hat{k}$.

Sol. $$\frac{\partial}{\partial x}\left(\frac{x}{r^3}\right) = \frac{1}{r^3} - \frac{3x}{r^4}\frac{\partial r}{\partial x} = \frac{1}{r^3} - \frac{3x^2}{r^5} \qquad \Big|\because \ \frac{\partial r}{\partial x} = \frac{x}{r}$$

$$\therefore \quad \frac{\partial^2}{\partial x^2}\left(\frac{x}{r^3}\right) = \frac{\partial}{\partial x}\left(\frac{1}{r^3} - \frac{3x^2}{r^5}\right) = -\frac{3}{r^4}\frac{\partial r}{\partial x} - \frac{6x}{r^5} + \frac{15x^2}{r^6}\frac{\partial r}{\partial x}$$

$$= -\frac{3x}{r^5} - \frac{6x}{r^5} + \frac{15x^3}{r^7} = -\frac{9x}{r^5} + \frac{15x^3}{r^7} \qquad \text{...(1)}$$

$$\frac{\partial}{\partial y}\left(\frac{x}{r^3}\right) = -\frac{3x}{r^4}\frac{\partial r}{\partial y} = -\frac{3xy}{r^5}$$

$$\therefore \quad \frac{\partial^2}{\partial y^2}\left(\frac{x}{r^3}\right) = -3x\frac{\partial}{\partial y}\left(\frac{y}{r^5}\right) = -\frac{3x}{r^5} + \frac{15xy^2}{r^7} \qquad \text{...(2)}$$

Similarly, $$\frac{\partial^2}{\partial z^2}\left(\frac{x}{r^3}\right) = -\frac{3x}{r^5} + \frac{15xz^2}{r^7} \qquad \text{...(3)}$$

Adding (1), (2) and (3), we get

$$\nabla^2\left(\frac{x}{r^3}\right) = -\frac{15x}{r^5} + \frac{15x}{r^7}(x^2+y^2+z^2)$$

$$= -\frac{15x}{r^5} + \frac{15x}{r^5} \qquad \Big|\because \ x^2+y^2+z^2 = r^2$$

$$= 0$$

TEST YOUR KNOWLEDGE

1. If $\vec{V} = \dfrac{x\hat{i} + y\hat{j} + z\hat{k}}{\sqrt{x^2+y^2+z^2}}$, show that $\nabla\,.\,\vec{V} = \dfrac{2}{\sqrt{x^2+y^2+z^2}}$ and $\nabla \times \vec{V} = \vec{0}$.
2. Find the divergence and curl of the vector $\vec{R} = (x^2+yz)\,\hat{i} + (y^2+zx)\,\hat{j} + (z^2+xy)\,\hat{k}$. *(U.P.T.U. 2015)*

3. (*i*) If $\vec{F} = (x + y + 1)\,\hat{i} + \hat{j} - (x + y)\,\hat{k}$, show that $\vec{F}\,.\,\text{curl}\,\vec{F} = 0$.

(*ii*) If $\vec{A} = (3xz^2)\,\hat{i} - (yz)\,\hat{j} + (x + 2z)\,\hat{k}$, find curl (curl $\vec{A}$).

4. (*i*) Show that vector $\vec{V} = (x + 3y)\,\hat{i} + (y - 3z)\,\hat{j} + (x - 2z)\,\hat{k}$ is solenoidal. (*A.K.T.U. 2020, 2021*)

(*ii*) Show that the vector field $\vec{V} = (\sin y + z)\,\hat{i} + (x \cos y - z)\,\hat{j} + (x - y)\,\hat{k}$ is irrotational.

(*iii*) Find most general $f(r)$ such that $f(r)\,\vec{r}$ is solenoidal. (*G.B.T.U. 2013*)

5. If $\vec{F} = (\vec{a}\,.\,\vec{r})\,\vec{r}$ where $\vec{a}$ is a constant vector, find curl $\vec{F}$ and prove that it is perpendicular to $\vec{a}$. (*M.T.U. 2012*)

6. (*i*) Prove that div $(r^n\,\vec{r}) = (n + 3)\,r^n$, further show that $r^n\,\vec{r}$ is solenoidal only if $n = -3$.

(*ii*) If $\vec{E}$ and $\vec{H}$ are irrotational, prove that $\vec{E} \times \vec{H}$ is solenoidal. [*G.B.T.U. (SUM) 2010*]

7. If $u = x^2 + y^2 + z^2$ and $\vec{V} = x\hat{i} + y\hat{j} + z\hat{k}$, show that div $(u\,\vec{V}) = 5u$.

8. A fluid motion is given by $\vec{v} = (y \sin z - \sin x)\,\hat{i} + (x \sin z + 2yz)\,\hat{j} + (xy \cos z + y^2)\,\hat{k}$. Is the motion irrotational? If so, find the velocity potential. (*A.K.T.U. 2021*)

9. (*i*) Show that the vector field $\vec{A}$, where $\vec{A} = (x^2 - y^2 + x)\,\hat{i} - (2xy + y)\,\hat{j}$ is irrotational, and find a scalar ϕ such that $\vec{A}$ = grad ϕ. Hence evaluate the line integral from (1, 2) to (2, 1). (*U.P.T.U. 2014*)

(*ii*) Show that the vector field defined by $\vec{F} = 2xyz^3\,\hat{i} + x^2z^3\,\hat{j} + 3x^2yz^2\,\hat{k}$ is irrotational. Find the scalar potential u such that $\vec{F}$ = grad u.

(*iii*) Show that $\vec{A} = (6xy + z^3)\,\hat{i} + (3x^2 - z)\,\hat{j} + (3xz^2 - y)\,\hat{k}$ is irrotational. Find the velocity potential ϕ such that $\vec{A} = \nabla\phi$. (*U.P.T.U. 2014; U.K.T.U. 2011*)

(*iv*) Prove that $\vec{F} = (y^2 - z^2 + 3yz - 2x)\,\hat{i} + (3xz + 2xy)\,\hat{j} + (3xy - 2xz + 2z)\,\hat{k}$ is both solenoidal and irrotational. Also find the scalar potential ϕ such that $\vec{F}$ = grad ϕ. (*A.K.T.U. 2019*)

(*v*) Find the constants a, b, c so that $\vec{F} = (x + 2y + az)\,\hat{i} + (bx - 3y - z)\,\hat{j} + (4x + cy + 2z)\,\hat{k}$ is irrotational. If $\vec{F}$ = grad ϕ, show that $\phi = \dfrac{x^2}{2} - \dfrac{3y^2}{2} + z^2 + 2xy + 4xz - yz$. (*A.K.T.U. 2018, 2014, 2022*)

10. If $\vec{A} = xz^2\,\hat{i} + 2y\,\hat{j} - 3xz\,\hat{k}$ and $\vec{B} = 3xz\,\hat{i} + 2yz\,\hat{j} - z^2\hat{k}$, find the value of $\vec{A} \times (\nabla \times \vec{B})$ and $(\vec{A} \times \nabla) \times \vec{B}$. (*A.K.T.U. 2017*)

11. Find the directional derivative of $\nabla.\,(\nabla\phi)$ at the point (1, – 2, 1) in the direction of the normal to the surface $xy^2z = 3x + z^2$, where $\phi = 2x^3y^2z^4$. (*A.K.T.U. 2014, 2022*)

12. If $\vec{r} = x\hat{i} + y\hat{j} + z\hat{k}$, prove that

(i) curl $(r^n \vec{r}) = \vec{0}$ (ii) $\nabla^2 (r^n \vec{r}) = n(n+3)r^{n-2}\vec{r}$

(iii) $\nabla . \left\{ r\nabla\left(\frac{1}{r^3}\right)\right\} = \frac{3}{r^4}$ (iv) $\nabla^2 \left[\nabla . \left(\frac{\vec{r}}{r^2}\right)\right] = 2r^{-4}$.

13. (i) Prove that curl $\{(\vec{r} \times \vec{a}) \times \vec{b}\} = \vec{b} \times \vec{a}$ where $\vec{a}$ and $\vec{b}$ are constant vectors.

(ii) Prove that div $\{(\vec{r} \times \vec{a}) \times \vec{b}\} = -2\vec{b} . \vec{a}$, where $\vec{a}$ and $\vec{b}$ are constant vectors.

14. If $\vec{V_1}$ and $\vec{V_2}$ be the vectors joining the fixed points (x_1, y_1, z_1) and (x_2, y_2, z_2) respectively to a variable point (x, y, z), prove that

(i) div $(\vec{V_1} \times \vec{V_2}) = 0$ (ii) grad $(\vec{V_1} . \vec{V_2}) = \vec{V_1} + \vec{V_2}$

(iii) curl $(\vec{V_1} \times \vec{V_2}) = 2(\vec{V_1} - \vec{V_2})$.

15. Prove that

(i) curl $\{(\vec{a} \times \vec{r})\, r^n\} = (n+2)\, r^n \vec{a} - nr^{n-2} (\vec{a} . \vec{r})\, \vec{r}$ (ii) curl $(\hat{r}) = \vec{0}$.

16. Prove that curl $\left(\frac{\vec{a} \times \vec{r}}{r^3}\right) = -\frac{\vec{a}}{r^3} + \frac{3\vec{r}\,(\vec{a} . \vec{r})}{r^5}$.

17. Show that the vector field $\vec{V} = \frac{a(x\hat{i} + y\hat{j})}{x^2 + y^2}$ is a "source" or "sink" field according as $a > 0$ or $a < 0$. **[Hint.** If div $\vec{V} > 0$ then $\vec{V}$ is a source field and if div $\vec{V} < 0$ then $\vec{V}$ is a sink field.**]**

Answers

2. $2(x + y + z)$; $\vec{0}$ **3.** (ii) $-6x\hat{i} + (6z - 1)\hat{k}$

4. (iii) $f(r) = \frac{c}{r^3}$. **5.** $\vec{a} \times \vec{r}$ **8.** Yes ; $\phi = xy \sin z + \cos x + y^2 z + c$

9. (i) $\frac{1}{3} x^3 - xy^2 + \frac{1}{2} (x^2 - y^2) + c$, 22/3 (ii) $x^2yz^3 + c$ (iii) $\phi = 3x^2y + xz^3 - zy + c$

(iv) $xy^2 - xz^2 + z^2 - x^2 + 3xyz + c$ (v) $a = 4$, $b = 2$, $c = -1$

10. $9x^2z\hat{i} + 6xyz\,\hat{j} + (3x^2z^2 + 4y^2)\,\hat{k}$; $-2yz\hat{i} + (4y^2 + 15\,xz^2 + 3x^2z^2)\,\hat{k}$

11. $\frac{1724}{\sqrt{21}}$

5.21. INTEGRATION OF VECTOR FUNCTIONS

Let $\vec{f}(t)$ and $\vec{F}(t)$ be two vector functions of a scalar variable t such that $\frac{d}{dt}\vec{F}(t) = \vec{f}(t)$, then $\vec{F}(t)$ is called an integral of $\vec{f}(t)$ with respect to t and we write $\int \vec{f}(t)\, dt = \vec{F}(t)$

If $\vec{c}$ is any *arbitrary constant vector* independent of t, then $\dfrac{d}{dt}\{\vec{F}(t) + \vec{c}\} = \vec{f}(t)$

This is equivalent to $\int \vec{f}(t)\, dt = \vec{F}(t) + \vec{c}$

$\vec{F}(t)$ is called the *indefinite integral* of $\vec{f}(t)$. The constant vector $\vec{c}$ is called the *constant of integration* and can be determined if some initial conditions are given.

The *definite integral* of $\vec{f}(t)$ between the limits $t = a$ and $t = b$ is written as

$$\int_a^b \vec{f}(t)\, dt = \left[\vec{F}(t)\right]_a^b = \vec{F}(b) - \vec{F}(a).$$

Note 1. If $\vec{f}(t) = f_1(t)\,\hat{i} + f_2(t)\hat{j} + f_3(t)\hat{k}$, then $\int \vec{f}(t)dt = \hat{i}\int f_1(t)dt + \hat{j}\int f_2(t)dt + \hat{k}\int f_3(t)dt$

Thus in order to integrate a vector function, integrate its components.

Note 2. We can obtain some standard results for integration of vector functions by considering the derivatives of suitable vector functions. For example,

(*i*) $\dfrac{d}{dt}(\vec{r}\,.\,\vec{s}) = \dfrac{d\vec{r}}{dt}\,.\,\vec{s} + \vec{r}\,.\,\dfrac{d\vec{s}}{dt} \quad \Rightarrow \quad \int\left(\dfrac{d\vec{r}}{dt}\,.\,\vec{s} + \vec{r}\,.\,\dfrac{d\vec{s}}{dt}\right)dt = \vec{r}\,.\,\vec{s} + c$

Hence c is a scalar quantity since the integrand is a scalar.

(*ii*) $\dfrac{d}{dt}(\vec{r}^{\,2}) = 2\,\vec{r}\,.\,\dfrac{d\vec{r}}{dt} \quad \Rightarrow \quad \int\left(2\,\vec{r}\,.\,\dfrac{d\vec{r}}{dt}\right)dt = \vec{r}^{\,2} + c$ where c is a scalar quantity.

(*iii*) $\dfrac{d}{dt}\left(\vec{r}\times\dfrac{d\vec{r}}{dt}\right) = \dfrac{d\vec{r}}{dt}\times\dfrac{d\vec{r}}{dt} + \vec{r}\times\dfrac{d^2\vec{r}}{dt^2} = \vec{r}\times\dfrac{d^2\vec{r}}{dt^2} \quad \Rightarrow \quad \int\left(\vec{r}\times\dfrac{d^2\vec{r}}{dt^2}\right)dt = \vec{r}\times\dfrac{d\vec{r}}{dt} + \vec{c}$

Here $\vec{c}$ is a vector quantity since the integrand is a vector.

(*iv*) If $\vec{a}$ is a constant vector, then

$$\frac{d}{dt}(\vec{a}\times\vec{r}) = \frac{d\vec{a}}{dt}\times\vec{r} + \vec{a}\times\frac{d\vec{r}}{dt} = \vec{a}\times\frac{d\vec{r}}{dt} \quad \Rightarrow \quad \int\left(\vec{a}\times\frac{d\vec{r}}{dt}\right)dt = \vec{a}\times\vec{r} + \vec{c}$$

where $\vec{c}$ is a vector quantity.

ILLUSTRATIVE EXAMPLES

Example 1. *If* $\vec{r}(t) = 5t^2\hat{i} + t\hat{j} - t^3\hat{k}$, *prove that* $\displaystyle\int_1^2\left(\vec{r}\times\frac{d^2\vec{r}}{dt^2}\right)dt = -\,14\hat{i} + 75\hat{j} - 15\hat{k}$.

Sol. Since $\qquad \dfrac{d}{dt}\left(\vec{r}\times\dfrac{d\vec{r}}{dt}\right) = \dfrac{d\vec{r}}{dt}\times\dfrac{d\vec{r}}{dt} + \vec{r}\times\dfrac{d^2\vec{r}}{dt^2} = \vec{r}\times\dfrac{d^2\vec{r}}{dt^2}$

$$\therefore \quad \int_1^2 \left(\vec{r} \times \frac{d^2\vec{r}}{dt^2} \right) dt = \left[\vec{r} \times \frac{d\vec{r}}{dt} \right]_1^2 \quad ...(1)$$

Now,
$$\vec{r} \times \frac{d\vec{r}}{dt} = \begin{vmatrix} \hat{i} & \hat{j} & \hat{k} \\ 5t^2 & t & -t^3 \\ 10t & 1 & -3t^2 \end{vmatrix} = -2t^3\hat{i} + 5t^4\hat{j} - 5t^2\hat{k}$$

$\therefore$ From (1), we have

$$\int_1^2 \left(\vec{r} \times \frac{d^2\vec{r}}{dt^2} \right) dt = \left[-2t^3\hat{i} + 5t^4\hat{j} - 5t^2\hat{k} \right]_1^2 = -14\hat{i} + 75\hat{j} - 15\hat{k}$$

Example 2. *Given that* $\vec{r}(t) = \begin{cases} 2\hat{i} - \hat{j} + 2\hat{k} & when\ t = 2 \\ 4\hat{i} - 2\hat{j} + 3\hat{k} & when\ t = 3 \end{cases}$

show that $\displaystyle\int_2^3 \left(\vec{r} \cdot \frac{d\vec{r}}{dt} \right) dt = 10.$

Sol. Since $\dfrac{d}{dt}(\vec{r}^{\,2}) = 2\vec{r} \cdot \dfrac{d\vec{r}}{dt}$ $\quad\therefore\quad \displaystyle\int_2^3 \left(\vec{r} \cdot \frac{d\vec{r}}{dt} \right) dt = \frac{1}{2}\left[\vec{r}^{\,2} \right]_2^3$ ...(1)

When $t = 3$, $\quad \vec{r}^{\,2} = (4\hat{i} - 2\hat{j} + 3\hat{k}) \cdot (4\hat{i} - 2\hat{j} + 3\hat{k}) = 16 + 4 + 9 = 29$

When $t = 2$, $\quad \vec{r}^{\,2} = (2\hat{i} - \hat{j} + 2\hat{k}) \cdot (2\hat{i} - \hat{j} + 2\hat{k}) = 4 + 1 + 4 = 9$

$\therefore$ From (1), we have $\displaystyle\int_2^3 \left(\vec{r} \cdot \frac{d\vec{r}}{dt} \right) dt = \frac{1}{2}[29 - 9] = 10.$

5.22. LINE INTEGRALS

Any integral which is to be evaluated along a curve is called a line integral.

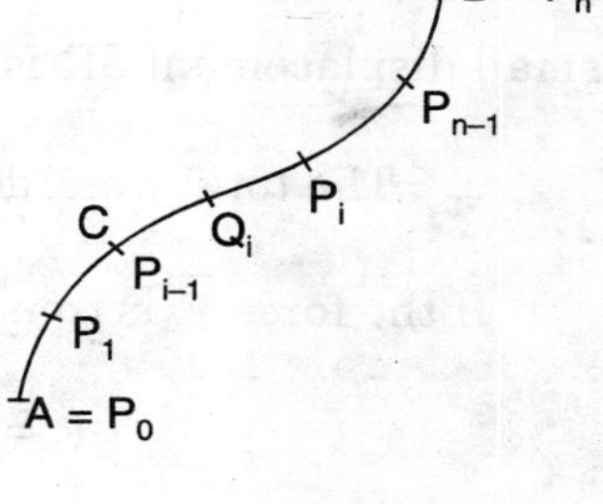

Let $\vec{F}(P)$ be a continuous vector point function defined at every point of a curve C in space. Divide the curve C into n parts by the points $A = P_0, P_1, P_2, ..., P_n = B$

and let $\vec{R}_0, \vec{R}_1, \vec{R}_2, ..., \vec{R}_n$ be the position vectors of these points. Let Q_i be any point on the arc $P_{i-1} P_i$. Then the limit of the sum

$$\sum_{i=1}^{n} \vec{F}(Q_i) \cdot \delta\vec{R}_i \quad \text{where} \quad \delta\vec{R}_i = \vec{R}_i - \vec{R}_{i-1} \quad ...(1)$$

as $n \to \infty$ and every $|\delta\vec{R}_i| \to 0$, if it exists, is called a line integral of $\vec{F}$ along C and is denoted by

$$\int_C \vec{F} \cdot d\vec{R} \qquad \text{or} \qquad \int_C \vec{F} \cdot \frac{d\vec{R}}{dt} dt.$$

Clearly, it is a scalar. It is called the *tangential line integral* of $\vec{F}$ along C.

If the scalar products in (1) are replaced by vector products, then the corresponding line integral is defined as $\int_C \vec{F} \times d\vec{R}$ which is a vector.

If the vector function $\vec{F}$ is replaced by a scalar function ϕ, then the corresponding line integral is defined as $\int_C \phi d\vec{R}$, which is a vector.

If $\vec{F}(x, y, z) = f_1\hat{i} - f_2\hat{j} + f_3\hat{k}$ and $\vec{R} = x\hat{i} - y\hat{j} + z\hat{k}$ then $d\vec{R} = \hat{i}dx - \hat{j}dy + \hat{k}dz$

$$\therefore \qquad \int_C \vec{F} . d\vec{R} = \int_C (f_1 dx + f_2 dy + f_3 dz)$$

If the parametric equations of the curve C are $x = x(t)$, $y = y(t)$, $z = z(t)$ and $t = t_1$ at A, $t = t_2$ at B, then

$$\int_C \vec{F} \cdot d\vec{R} = \int_{t_1}^{t_2} \left(f_1 \frac{dx}{dt} + f_2 \frac{dy}{dt} + f_3 \frac{dz}{dt} \right) dt$$

If C is a closed curve, then the integral sign $\int_C$ is replaced by $\oint_C$.

5.23. CIRCULATION

(G.B.T.U. 2010)

In Fluid Dynamics, if $\vec{V}$ represents the velocity of a fluid particle and C is a closed curve, then the integral $\oint_C \vec{V} \cdot d\vec{R}$ is called the *circulation* of $\vec{V}$ round the curve C.

If the circulation of $\vec{V}$ round every closed curve in a region D vanishes, then $\vec{V}$ is said to be *irrotational* in D.

5.24. WORK DONE BY A FORCE

Let $\vec{F}$ represent the force acting on a particle moving along an arc AB. The work done during a small displacement $\delta\vec{R}$ is $\vec{F} \cdot \delta\vec{R}$.

$\therefore$ The total work done by $\vec{F}$ during displacement from A to B is given by $\int_A^B \vec{F} \cdot d\vec{R}$

If the force $\vec{F}$ is conservative, then there exists a scalar function ϕ such that

$$\vec{F} = \nabla\phi = \hat{i}\frac{\partial \phi}{\partial x} + \hat{j}\frac{\partial \phi}{\partial y} + \hat{k}\frac{\partial \phi}{\partial z}$$

$\therefore$ The work done by $\vec{F}$ during displacement from A to B $= \int_A^B \vec{F} \cdot d\vec{R}$

$$= \int_A^B \left(\hat{i}\frac{\partial \phi}{\partial x} + \hat{j}\frac{\partial \phi}{\partial y} + \hat{k}\frac{\partial \phi}{\partial z} \right) . (\hat{i}\ dx + \hat{j}\ dy + \hat{k}\ dz)$$

$$= \int_A^B \left(\frac{\partial \phi}{\partial x} dx + \frac{\partial \phi}{\partial y} dy + \frac{\partial \phi}{\partial z} dz \right) = \int_A^B d\phi = \Big[\phi \Big]_A^B = \phi_B - \phi_A$$

Thus, in a conservative field, the work done depends on the value of ϕ at the end points A and B, and not on the path joining A and B.

Example 3. *Suppose $\vec{F}(x, y, z) = x^3\hat{i} + y\hat{j} + z\hat{k}$ is the force field. Find the work done by $\vec{F}$ along the line from (1, 2, 3) to (3, 5, 7).*

Sol. Equations of the line are

$$\frac{x-1}{3-1} = \frac{y-2}{5-2} = \frac{z-3}{7-3} = t(\text{say})$$

$$\Rightarrow \qquad x = 2t + 1,\ y = 3t + 2,\ z = 4t + 3$$

Now, $\vec{r} = x\hat{i} + y\hat{j} + z\hat{k} = (2t + 1)\ \hat{i} + (3t + 2)\ \hat{j} + (4t + 3)\ \hat{k}$

$\therefore \qquad d\vec{r} = (2\hat{i} + 3\hat{j} + 4\hat{k})\ dt$

At (1, 2, 3), $t = 0$ and At (3, 5, 7), $t = 1$

$$\text{Work done} = \int_C \vec{F} \cdot d\vec{r} = \int_0^1 [(2t+1)^3\ \hat{i} + (3t+2)\ \hat{j} + (4t+3)\ \hat{k}] \cdot (2\hat{i} + 3\hat{j} + 4\hat{k})\ dt$$

$$= \int_0^1 [2(2t+1)^3 + 3(3t+2) + 4(4t+3)]\ dt$$

$$= \int_0^1 (16t^3 + 24t^2 + 37t + 20)\ dt = \left(4t^4 + 8t^3 + \frac{37}{2}t^2 + 20t\right)_0^1$$

$$= 4 + 8 + \frac{37}{2} + 20 = \frac{101}{2}.$$

Example 4. *Find the total work done by a force $\vec{F} = (x^2 + y^2)\ \hat{i} - 2xy\hat{j}$ in moving a point from (0, 0) to (a, b) along the rectangle bounded by the lines $x = 0$, $x = a$, $y = 0$ and $y = b$.*

(U.P.T.U. 2014)

Sol. $\vec{F} = (x^2 + y^2)\ \hat{i} - 2xy\hat{j}$

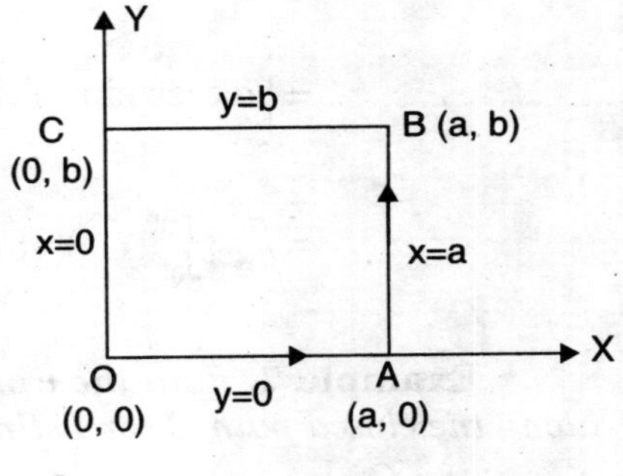

$d\vec{r} = dx\hat{i} + dy\hat{j}$

$\therefore \qquad \vec{F} \cdot d\vec{r} = (x^2 + y^2)\ dx - 2xy\ dy$

$$\text{Total work done} = \int_{OA} \vec{F} \cdot d\vec{r} + \int_{AB} \vec{F} \cdot d\vec{r}$$

$$= I_1 + I_2 \qquad ...(1)$$

where, $I_1 = \int_{OA} \vec{F} \cdot d\vec{r}$

$$= \int_0^a x^2\ dx = \frac{a^3}{3} \qquad | \text{ Along OA, } y = 0 \ \therefore\ dy = 0$$

$$I_2 = \int_{AB} \vec{F} \cdot d\vec{r} = \int_0^b -2ay\ dy \qquad | \text{ Along AB, } x = a \ \therefore\ dx = 0$$

$$= -ab^2$$

From (1), Total work done $= \dfrac{a^3}{3} - ab^2$

Example 5. *If $\vec{F} = 3xy\hat{i} - y^2\hat{j}$, evaluate $\int_C \vec{F} \cdot d\vec{r}$, where C is the arc of the parabola $y = 2x^2$ from (0, 0) to (1, 2).*

Sol. Let $x = t$, then the parametric equations of the parabola $y = 2x^2$ are $x = t$, $y = 2t^2$.

At the point (0, 0), $x = 0$ and so $t = 0$

At the point (1, 2), $x = 1$ and so $t = 1$

If $\vec{r}$ is the position vector of any point (x, y) in C, then $\vec{r} = x\hat{i} + y\hat{j} = t\hat{i} + 2t^2\hat{j}$

Also in terms of t, $\vec{F} = 6t^3\hat{i} - 4t^4\hat{j}$

$$\therefore \quad \int_C \vec{F} \cdot d\vec{r} = \int_0^1 (6t^3\hat{i} - 4t^4\hat{j}) \cdot (\hat{i} + 4t\hat{j})\, dt$$

$$= \int_0^1 (6t^3 - 16t^5)\, dt = \left[\frac{6t^4}{4} - \frac{16t^6}{6}\right]_0^1 = \frac{3}{2} - \frac{8}{3} = -\frac{7}{6}.$$

Example 6. *A vector field is given by $\vec{F} = (\sin y)\,\hat{i} + x(1 + \cos y)\,\hat{j}$. Evaluate the line integral over the circular path given by $x^2 + y^2 = a^2$, $z = 0$.*

Sol. The parametric equations of the circular path are $x = a \cos t$, $y = a \sin t$, $z = 0$ where t varies from 0 to 2π.

Since the particle moves in the xy-plane ($z = 0$), we can take $\vec{r} = x\hat{i} + y\hat{j}$ so that $d\vec{r} = dx\hat{i} + dy\hat{j}$

$$\therefore \quad \oint_C \vec{F} \cdot d\vec{r} = \oint_C [\sin y\hat{i} + x(1 + \cos y)\hat{j}] \cdot (dx\hat{i} + dy\hat{j})$$

$$= \oint_C [\sin y\, dx + x\,(1 + \cos y)\, dy] = \oint_C [(\sin y\, dx + x \cos y\, dy) + x\, dy]$$

$$= \oint_C d(x \sin y) + \oint_C x\, dy = \int_0^{2\pi} d[a \cos t \sin (a \sin t)] + \int_0^{2\pi} a \cos t \cdot a \cos t\, dt$$

$$= \Big[a \cos t \sin (a \sin t)\Big]_0^{2\pi} + a^2 \int_0^{2\pi} \cos^2 t\, dt$$

$$= \frac{a^2}{2} \int_0^{2\pi} (1 + \cos 2t)\, dt = \frac{a^2}{2}\left[t + \frac{\sin 2t}{2}\right]_0^{2\pi} = \frac{a^2}{2}(2\pi) = \pi a^2.$$

Example 7. *Find the work done by the force $\vec{F} = x\hat{i} - z\hat{j} + 2y\hat{k}$ in the displacement along the closed path C consisting of the segments C_1, C_2 and C_3 where*

on C_1,	$0 \le x \le 1$,	$y = x$,	$z = 0$
on C_2,	$0 \le z \le 1$,	$x = 1$,	$y = 1$
on C_3,	$1 \ge x \ge 0$,	$y = z = x$.	

Sol. Total work done $= \oint_C \vec{F} \cdot d\vec{r}$

$$= \oint_C (x\hat{i} - z\hat{j} + 2y\hat{k}) \cdot (dx\hat{i} + dy\hat{j} + dz\hat{k})$$

$$= \oint_C (xdx - zdy + 2ydz)$$

$$= \oint_{C_1} (xdx - zdy + 2ydz) + \oint_{C_2} (xdx - zdy + 2ydz) + \oint_{C_3} (xdx - zdy + 2ydz)$$

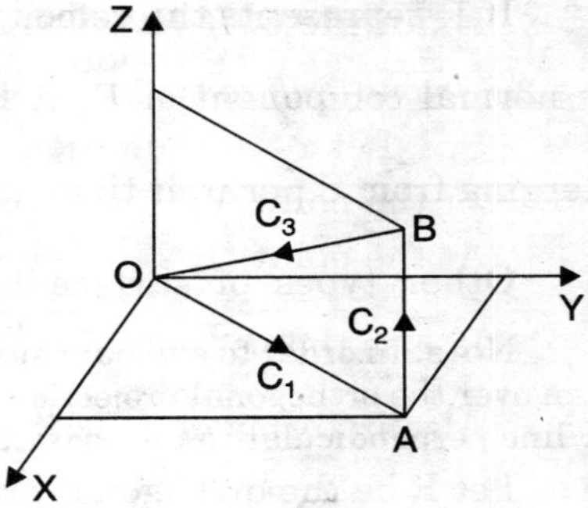

$$= W_1 + W_2 + W_3$$

where W_1, W_2, W_3 denote the work done in displacement along C_1, C_2 and C_3 respectively.

On C_1, $0 \le x \le 1$, $y = x, z = 0, dy = dx, \quad dz = 0$

$$\therefore \quad W_1 = \oint_{C_1} x\,dx = \int_0^1 x\,dx = \tfrac{1}{2}$$

On C_2, $0 \le z \le 1$, $\quad x = 1, \quad y = 1, \quad dx = 0, \quad dy = 0$

$$\therefore \quad W_2 = \oint_{C_2} 2dz = 2\int_0^1 dz = 2$$

On C_3, $1 \ge x \ge 0$, $\quad y = z = x, \quad dy = dz = dx$

$$\therefore \quad W_3 = \oint_{C_3} (x\,dx - x\,dx + 2\,xdx) = 2\int_1^0 x\,dx = -1.$$

$$\text{Total work done} = W_1 + W_2 + W_3 = \frac{1}{2} + 2 - 1 = \frac{3}{2}.$$

5.25. SURFACE INTEGRALS

Any integral which is to be evaluated over a surface is called a surface integral.

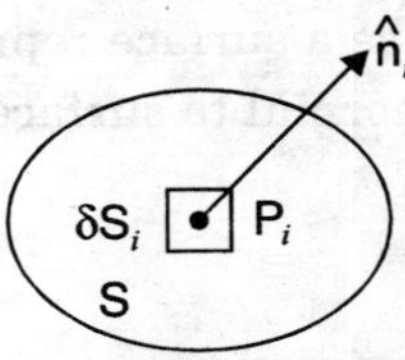

Let $\vec{F}(P)$ be a continuous vector point function and S a two sided surface. Divide S into a finite number of sub-surfaces. $\delta S_1, \delta S_2 \ldots, \delta S_k$. Let P_i be any point in δS_i and $\hat{n}_i$ be the unit vector at P in the direction of outward drawn normal to the surface at P_i. Then the limit of the sum $\sum_{i=1}^{k} \vec{F}(P_i) \cdot \hat{n}_i \, \delta S_i$, as $k \to \infty$ and each $\delta S_i \to 0$ is called the *normal surface integral* of $\vec{F}(P)$ over S and is denoted by $\iint_S \vec{F} \cdot \hat{n}\, dS$.

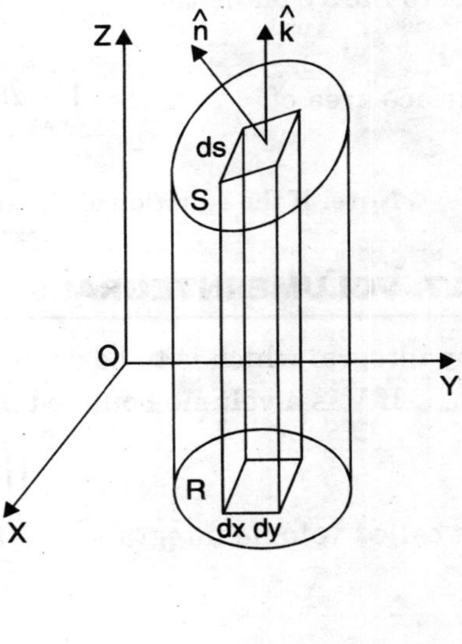

The surface element $\delta\vec{S}$ surrounding any point P can be regarded as a vector whose magnitude is area δS and the direction that of the outward drawn normal $\hat{n}$ *i.e.*, $\delta\vec{S} = \hat{n}\,\delta S$. The surface integral may alternatively be written as $\iint_S \vec{F} \cdot d\vec{S}$.

If $\vec{F}$ represents the velocity of a fluid at any point P on a closed surface S, then $\vec{F} \cdot \hat{n}$ is the normal component of $\vec{F}$ at P and $\iint_S \vec{F} \cdot \hat{n} \, dS = \iint_S \vec{F} \cdot d\vec{S}$ is a measure of volume emerging from S per unit time, *i.e.*, it measures the *flux of* $\vec{F}$ *over S.*

Other types of surface integrals are $\iint_S \vec{F} \times d\vec{S}$, $\iint_S \phi d\vec{S}$.

Note. In order to evaluate surface integrals, it is convenient to express them as double integrals taken over the orthogonal projection of S on one of the coordinate planes. But this is possible only when any line perpendicular to the coordinate plane chosen meets the surface S in not more than one point.

Let R be the orthogonal projection of S on the *xy*-plane.

Let $\hat{n} = \cos\alpha \hat{i} + \cos\beta \hat{j} + \cos\gamma \hat{k}$ where cos α, cos β, cos γ are the direction cosines of $\hat{n}$.

Now, $dxdy$ = projection of dS on the *xy*-plane = $dS \cos\gamma \quad \Rightarrow \quad dS = \dfrac{dx\,dy}{\cos\gamma}$

Also $\quad |\hat{k} \cdot \hat{n}| = \cos\gamma \quad \therefore \quad dS = \dfrac{dxdy}{|\hat{k} \cdot \hat{n}|}$

Hence $\iint_R \vec{F} \cdot \hat{n} \, dS = \iint_R \vec{F} \cdot \hat{n} \dfrac{dxdy}{|\hat{k} \cdot \hat{n}|}$.

5.26. SURFACE AREA OF A CURVED SURFACE

Let S be a surface represented by $F(x, y, z) = 0$ ...(1)

Unit normal to surface S is

$$\hat{n} = \frac{\nabla F}{|\nabla F|} = \frac{F_x \hat{i} + F_y \hat{j} + F_z \hat{k}}{\sqrt{F_x^2 + F_y^2 + F_z^2}}$$

where F_x, F_y, F_z are partial derivatives of F w.r.t. *x*, *y*, *z* respectively. Let D be the projection of S onto the *xy*-plane then,

Surface area of $\quad S = \iint_S dS = \iint_D \dfrac{dx\,dy}{|\hat{n} \cdot \hat{k}|} = \iint_D \dfrac{\sqrt{F_x^2 + F_y^2 + F_z^2}}{|F_z|} dx\,dy$

Note. If the equation of the surface S is $z = f(x, y)$, then surface area $= \iint \sqrt{1 + z_x^2 + z_y^2} \, dx\,dy$

5.27. VOLUME INTEGRALS

Any integral which is to be evaluated over a volume is called a volume integral.

If V is a volume bounded by a surface S, then the triple integrals

$$\iiint_V \phi \, dV \quad \text{and} \quad \iiint_V \vec{F} \, dV$$

are called volume integrals. The first of these is a scalar and the second is a vector.

If we sub-divide the volume V into small cuboids by drawing planes parallel to the coordinate planes, then $dV = dxdydz$.

$$\therefore \quad \iiint_V \phi \, dV = \iiint_V \phi\,(x, y, z)\, dxdydz$$

If $\vec{F} = F_1\hat{i} + F_2\hat{j} + F_3\hat{k}$, then

$$\iiint_V \vec{F}\, dV = \hat{i}\iiint_V F_1\,(x, y, z)\, dxdydz + \hat{j}\iiint_V F_2\,(x, y, z)\, dxdydz + \hat{k}\iiint_V F_3\,(x, y, z)\, dxdydz.$$

Example 8. *Evaluate* $\iint_S \vec{A}\,.\,\hat{n}\, dS$, *where* $\vec{A} = (x + y^2)\,\hat{i} - 2x\hat{j} + 2yz\hat{k}$ *and S is the surface of the plane* $2x + y + 2z = 6$ *in the first octant.*

Sol. A vector normal to the surface S is given by

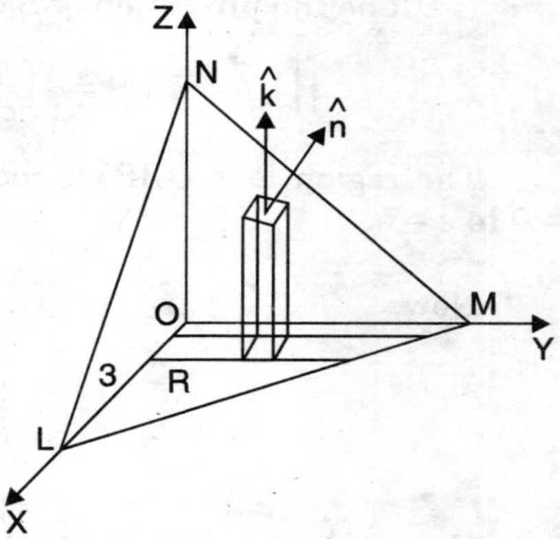

$$\nabla\,(2x + y + 2z) = 2\hat{i} + \hat{j} + 2\hat{k}$$

$\therefore \quad \hat{n}$ = a unit vector normal to surface S

$$= \frac{2\hat{i} + \hat{j} + 2\hat{k}}{\sqrt{(2)^2 + (1)^2 + (2)^2}} = \frac{2}{3}\hat{i} + \frac{1}{3}\hat{j} + \frac{2}{3}\hat{k}$$

$$\hat{k}\,.\,\hat{n} = \hat{k}\,.\left(\frac{2}{3}\hat{i} + \frac{1}{3}\hat{j} + \frac{2}{3}\hat{k}\right) = \frac{2}{3}$$

$$\therefore \quad \iint_S \vec{A}\,.\,\hat{n}\, dS = \iint_R \vec{A}\,.\,\hat{n}\, \frac{dxdy}{|\hat{k}\,.\,\hat{n}|},$$

where R is the projection of S, *i.e.*, triangle LMN on the *xy*-plane. The region R, *i.e.*, triangle OLM is bounded by *x*-axis, *y*-axis and the line $2x + y = 6$, $z = 0$.

$$\text{Now } \vec{A}\,.\,\hat{n} = [(x + y^2)\,\hat{i} - 2x\,\hat{j} + 2yz\hat{k}]\,.\left(\frac{2}{3}\hat{i} + \frac{1}{3}\hat{j} + \frac{2}{3}\hat{k}\right)$$

$$= \frac{2}{3}(x + y^2) - \frac{2}{3}x + \frac{4}{3}yz = \frac{2}{3}y^2 + \frac{4}{3}yz$$

$$= \frac{2}{3}y^2 + \frac{4}{3}y\left(\frac{6 - 2x - y}{2}\right) \qquad \Bigg|\ \because \text{ On the plane } 2x + y + 2z = 6,\ z = \frac{6 - 2x - y}{2}$$

$$= \frac{2}{3}y(y + 6 - 2x - y) = \frac{4}{3}y(3 - x)$$

$$\text{Hence } \iint_S \vec{A}\,.\,\hat{n}\, dS = \iint_R \vec{A}\,.\,\hat{n}\, \frac{dxdy}{|\hat{k}\,.\,\hat{n}|}$$

$$= \iint_R \frac{4}{3}y(3 - x)\,.\,\frac{2}{3}\, dxdy = \int_0^3 \int_0^{6-2x} \frac{8}{9}y(3 - x)\, dydx$$

$$= \int_0^3 \frac{8}{9}(3 - x)\,.\left[\frac{y^2}{2}\right]_0^{6-2x} dx = \frac{4}{9}\int_0^3 (3 - x)\,(6 - 2x)^2\, dx$$

$$= \frac{16}{9}\int_0^3 (3 - x)^3\, dx = \frac{16}{9}\,.\left[\frac{(3 - x)^4}{4(-1)}\right]_0^3 = \frac{4}{9}(81) = 36.$$

Example 9. *Evaluate* $\iint_S \vec{A} \cdot \hat{n}\, dS$, *where* $\vec{A} = z\hat{i} + x\hat{j} - 3y^2z\hat{k}$ *and S is the surface of the cylinder* $x^2 + y^2 = 16$ *included in the first octant between* $z = 0$ *and* $z = 5$.

Sol. A vector normal to the surface S is given by $\nabla(x^2 + y^2) = 2x\hat{i} + 2y\hat{j}$

$\therefore$ $\hat{n}$ = a unit vector normal to surface S

$$= \frac{2x\hat{i} + 2y\hat{j}}{\sqrt{(2x)^2 + (2y)^2}} = \frac{x\hat{i} + y\hat{j}}{\sqrt{x^2 + y^2}} = \frac{x\hat{i} + y\hat{j}}{4}$$

| $\because$ On the surface of cylinder, $x^2 + y^2 = 16$

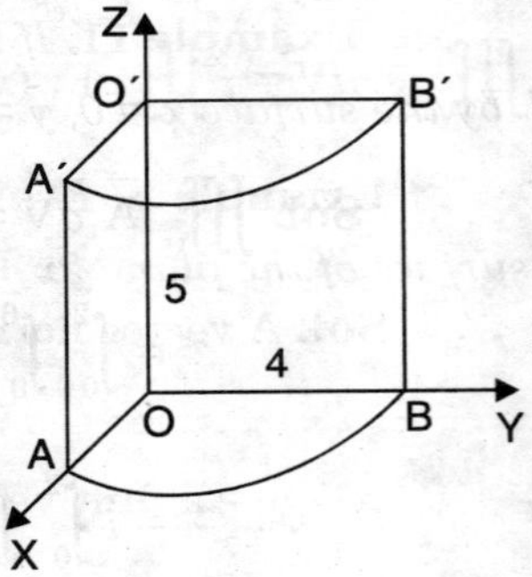

Let R be the projection of S on *yz*-plane, then

$$\iint_S \vec{A} \cdot \hat{n}\, dS = \iint_R \vec{A} \cdot \hat{n} \frac{dydz}{|\hat{i} \cdot \hat{n}|}$$

The region R is OBB′O′ enclosed by $y = 0$ to $y = 4$ and $z = 0$ to $z = 5$.

Now,
$$\hat{i} \cdot \hat{n} = \hat{i} \cdot \left(\frac{1}{4}x\hat{i} + \frac{1}{4}y\hat{j}\right) = \frac{1}{4}x$$

$$\vec{A} \cdot \hat{n} = (z\hat{i} + x\hat{j} - 3y^2z\hat{k}) \cdot \left(\frac{1}{4}x\hat{i} + \frac{1}{4}y\hat{j}\right) = \frac{1}{4}x(y + z).$$

Hence
$$\iint_R \vec{A} \cdot \hat{n}\, dS = \iint_R \vec{A} \cdot \hat{n} \frac{dydz}{|\hat{i} \cdot \hat{n}|} = \iint_R \frac{1}{4}x(y+z)\frac{dydz}{\frac{1}{4}x} = \int_0^5 \int_0^4 (y + z)\, dy\, dz$$

$$= \int_0^5 \left[\frac{y^2}{2} + zy\right]_0^4 dz = \int_0^5 (8 + 4z)\, dz = \left[8z + 2z^2\right]_0^5 = 40 + 50 = 90.$$

Example 10. *If* $\vec{F} = (2x^2 - 3z)\hat{i} - 2xy\hat{j} - 4x\hat{k}$, *then evaluate* $\iiint_V \nabla \cdot \vec{F}\, dV$, *where V is bounded by the planes* $x = 0$, $y = 0$, $z = 0$ *and* $2x + 2y + z = 4$.

Sol. $\nabla \cdot \vec{F} = \frac{\partial}{\partial x}(2x^2 - 3z) + \frac{\partial}{\partial y}(-2xy) + \frac{\partial}{\partial z}(-4x) = 4x - 2x = 2x$

$$\therefore \iiint_V \nabla \cdot \vec{F}\, dV = \iiint_V 2x\, dx\, dy\, dz$$

$$= \int_0^2 \int_0^{2-x} \int_0^{4-2x-2y} 2x\, dz\, dy\, dx = \int_0^2 \int_0^{2-x} 2x \left[z\right]_0^{4-2x-2y} dy\, dx$$

$$= \int_0^2 \int_0^{2-x} 2x(4 - 2x - 2y)\, dy\, dx$$

$$= \int_0^2 \left[4x(2-x)y - 2xy^2\right]_0^{2-x} dx = \int_0^2 [4x(2-x)^2 - 2x(2-x)^2]\, dx$$

$$= \int_0^2 2x(2-x)^2\,dx = 2\int_0^2 (4x - 4x^2 + x^3)\,dx$$

$$= 2\left[2x^2 - 4\frac{x^3}{3} + \frac{x^4}{4}\right]_0^2 = 2\left(8 - \frac{32}{3} + 4\right) = \frac{8}{3}.$$

Example 11. *If $\vec{A} = 2xz\hat{i} - x\hat{j} + y^2\hat{k}$, evaluate $\iiint_V \vec{A}\,dV$, where V is the region bounded by the surface $x = 0$, $y = 0$, $x = 2$, $y = 6$, $z = x^2$, $z = 4$.*

Sol. $\iiint_V \vec{A}\,dV = \iiint_V (2xz\hat{i} - x\hat{j} + y^2\hat{k})\,dx\,dy\,dz$

$$= \int_0^2\int_0^6\int_{x^2}^4 (2xz\hat{i} - x\hat{j} + y^2\hat{k})\,dz\,dy\,dx$$

$$= \hat{i}\int_0^2\int_0^6 \left[xz^2\right]_{x^2}^4 dy\,dx - \hat{j}\int_0^2\int_0^6 \left[xz\right]_{x^2}^4 dy\,dx + \hat{k}\int_0^2\int_0^6 \left[y^2 z\right]_{x^2}^4 dy\,dx$$

$$= \hat{i}\int_0^2\int_0^6 (16x - x^5)\,dy\,dx - \hat{j}\int_0^2\int_0^6 (4x - x^3)\,dy\,dx + \hat{k}\int_0^2\int_0^6 y^2(4 - x^2)\,dy\,dx$$

$$= \hat{i}\int_0^2 (16x - x^5)\left[y\right]_0^6 dx - \hat{j}\int_0^2 (4x - x^3)\left[y\right]_0^6 dx + \hat{k}\int_0^2 (4 - x^2)\left[\frac{y^3}{3}\right]_0^6 dx$$

$$= 6\hat{i}\left[8x^2 - \frac{x^6}{6}\right]_0^2 - 6\hat{j}\left[2x^2 - \frac{x^4}{4}\right]_0^2 + 72\hat{k}\left[4x - \frac{x^3}{3}\right]_0^2 = 128\hat{i} - 24\hat{j} + 384\hat{k}.$$

Example 12. *Evaluate $\iint_S (yz\hat{i} + zx\hat{j} + xy\hat{k})\cdot d\vec{S}$ where S is the surface of the sphere $x^2 + y^2 + z^2 = a^2$ in the first octant.*

Sol. $\vec{F} = yz\hat{i} + zx\hat{j} + xy\hat{k}$

$\hat{n}$ = a unit vector normal to surface S $= \dfrac{x\hat{i} + y\hat{j} + z\hat{k}}{\sqrt{x^2 + y^2 + z^2}} = \dfrac{x\hat{i} + y\hat{j} + z\hat{k}}{a}$

$$\iint_S (yz\hat{i} + zx\hat{j} + xy\hat{k})\cdot d\vec{S} = \iint_S \vec{F}\cdot\hat{n}\,d\vec{S}$$

$$= \iint_R \frac{3xyz}{a}\,\frac{dx\,dy}{|\hat{n}\cdot\hat{k}|}$$ | R is projection of S on the *xy*-plane

$$= 3\iint_R xy\,dy\,dx$$ $\left|\ \because\ \hat{n}\cdot\hat{k} = \dfrac{z}{a}\right.$

$$= 3\int_0^a\int_0^{\sqrt{a^2 - x^2}} xy\,dy\,dx$$

$$= \frac{3}{2}\int_0^a x(a^2 - x^2)\,dx = \frac{3}{8}a^4.$$

TEST YOUR KNOWLEDGE

1. Evaluate the line integrals $\int_C [(x^2 + xy)dx + (x^2 + y^2)dy]$, where C is the square formed by the lines $y = \pm 1$ and $x = \pm 1$.

2. Find the circulation of $\vec{F}$ round the curve C, where $\vec{F} = e^x \sin y\hat{i} + e^x \cos y\hat{j}$ and C is the rectangle whose vertices are (0, 0), (1, 0), $\left(1, \frac{\pi}{2}\right)$ and $\left(0, \frac{\pi}{2}\right)$.

3. (*i*) Show that $\int_C \vec{F} \cdot d\vec{r} = 3\pi$, given that $\vec{F} = z\hat{i} + x\hat{j} + y\hat{k}$ and C being the arc of the curve $\vec{r} = \cos t\hat{i} + \sin t\hat{j} + t\hat{k}$ from $t = 0$ to $t = 2\pi$.

   (*ii*) If $\vec{A} = (x - y)\hat{i} + (x + y)\hat{j}$, evaluate $\oint_C \vec{A} \cdot d\vec{r}$ around the curve C consisting of $y = x^2$ and $y^2 = x$. (*A.K.T.U. 2014, 2018*)

4. If $\vec{F} = (3x^2 + 6y)\hat{i} - 14yz\hat{j} + 20xz^2\hat{k}$, evaluate the line integral $\int_C \vec{F} \cdot d\vec{r}$ from (0, 0, 0) to (1, 1, 1) along the following paths C:

   (*i*) $x = t, y = t^2, z = t^3$

   (*ii*) the straight line joining (0, 0, 0) to (1, 1, 1)

   (*iii*) the straight line from (0, 0, 0) to (1, 0, 0), then to (1, 1, 0) and then to (1, 1, 1).

5. (*i*) Evaluate $\int_C [(2x^2y + y + z^2)\hat{i} + 2(1 + yz^3)\hat{j} + (2z + 3y^2z^2)\hat{k}] \cdot d\vec{r}$ along the curve C: $y^2 + z^2 = a^2, x = 0$.

   (*ii*) Find the circulation of $\vec{F}$ round the curve C, where $\vec{F} = y\hat{i} + z\hat{j} + x\hat{k}$ and C is the circle $x^2 + y^2 = 1, z = 0$.

   (*iii*) Evaluate $\int_C \vec{F} \cdot d\vec{r}$ along the curve $x^2 + y^2 = 1, z = 1$ in the positive direction from (0, 1, 1) to (1, 0, 1) where : $\vec{F} = (yz + 2x)\hat{i} + xz\hat{j} + (xy + 2z)\hat{k}$. (*M.T.U. 2013*)

6. Find the work done when a force $\vec{F} = (x^2 - y^2 + x)\hat{i} - (2xy + y)\hat{j}$ moves a particle in the xy-plane from (0, 0) to (1, 1) along the parabola $y^2 = x$. Is the work done different when the path is the straight line $y = x$?

7. (*i*) Find the work done in moving a particle once round the circle $x^2 + y^2 = 9$ in the xy-plane if the field of force is $\vec{F} = (2x - y - z)\hat{i} + (x + y - z^2)\hat{j} + (3x - 2y + 4z)\hat{k}$.

   (*ii*) Show that the vector field $\vec{F} = yz\hat{i} + (zx + 1)\hat{j} + xy\hat{k}$ is conservative. Find its scalar potential. Also find the work done by $\vec{F}$ in moving a particle from (1, 0, 0) to (2, 1, 4). (*G.B.T.U. 2013*)

8. (*i*) Compute the work done by the force $\vec{F} = (2y + 3)\hat{i} + xz\hat{j} + (yz - x)\hat{k}$ when it moves a particle from the point (0, 0, 0) to the point (2, 1, 1) along the curve $x = 2t^2$, $y = t$, $z = t^3$. (*G.B.T.U. 2011*)

(*ii*) Find the work done in moving a particle in the force field $\vec{F} = 3x^2\hat{i} + (2xz - y)\hat{j} + z\hat{k}$ along the curve defined by $x^2 = 4y$, $3x^3 = 8z$ from $x = 0$ to $x = 2$. (*M.T.U. 2012; U.P.T.U. 2014*)

9. (*i*) Evaluate $\iint_S \vec{F}\cdot\hat{n}\,dS$, where $\vec{F} = 18z\hat{i} - 12\hat{j} + 3y\hat{k}$ and S is the surface of the plane $2x + 3y + 6z = 12$ in the first octant. (*M.T.U. 2012*)

(*ii*) Evaluate $\iint_S \vec{A}\cdot\hat{n}\,dS$, where $\vec{A} = 12x^2y\hat{i} - 3yz\hat{j} + 2y\hat{k}$ and S is the portion of the plane $x + y + z = 1$ included in the first octant.

10. (*i*) Show that $\iint_S \vec{F}\cdot\hat{n}\,dS = \frac{3}{2}$ where $\vec{F} = 4xz\hat{i} - y^2\hat{j} + yz\hat{k}$ and S is the surface of the cube bounded by the planes $x = 0, x = 1, y = 0, y = 1, z = 0, z = 1$.

(*ii*) Evaluate $\iint_S \vec{F}\cdot\hat{n}\,dS$, where $\vec{F} = 4x\hat{i} - 2y^2\hat{j} + z^2\hat{k}$ and S is the surface bounding the region

$$x^2 + y^2 = 4,\ z = 0,\ z = 3.$$

11. Evaluate $\iiint_V \phi\,dV$, where $\phi = 45x^2y$ and V is the closed region bounded by the planes

$$4x + 2y + z = 8,\ x = 0,\ y = 0,\ z = 0.$$

12. (*i*) Evaluate $\oint_c \vec{F}\cdot d\vec{r}$ where $\vec{F}(x, y, z) = e^{xyz}(yz\hat{i} + zx\hat{j} + xy\hat{k})$ and $\vec{r} = x\hat{i} + y\hat{j} + z\hat{k}$ and c is the boundary of $0 \le x \le 1$, $0 \le y \le 1$ and $z = 1$ clockwise.

(*ii*) Find the surface area of the plane $x + 2y + 2z = 12$ cut off by $x = 0$, $y = 0$ and $x^2 + y^2 = 16$. (*G.B.T.U. 2012*)

$$\left[\textbf{Hint. } A = \iint \frac{\sqrt{F_x^2 + F_y^2 + F_z^2}}{|F_z|}\,dx\,dy\right]$$

Answers

1. 0 **2.** 0 **3.** (*ii*) $\frac{2}{3}$

4. (*i*) 5 (*ii*) $\frac{13}{3}$ (*iii*) $\frac{23}{3}$

5. (*i*) 0 (*ii*) $-\pi$ (*iii*) 1 **6.** $-\frac{2}{3}$; No, it is same

7. (*i*) 18π (*ii*) $\phi = xyz + y + c$, 9

8. (*i*) $8\frac{8}{35}$ (*ii*) 16 **9.** (*i*) 24 (*ii*) $\frac{49}{120}$

10. (*ii*) 84π **11.** 128. **12.** (*i*) 0 (*ii*) 6π.

5.28. GAUSS DIVERGENCE THEOREM *(Relation between surface and volume integrals)*

[*G.B.T.U. 2012*]

If $\vec{F}$ is a vector point function having continuous first order partial derivatives in the region V bounded by a closed surface S, then

$$\boxed{\iint_S \vec{F} \cdot \hat{n} \, dS = \iiint_V div \, \vec{F} \, dV}$$

where $\hat{n}$ is the outward drawn unit normal vector to the surface S.

[*i.e.*, the volume integral of the divergence of a vector point function $\vec{F}$ taken over the volume V enclosed by a surface S, is equal to the surface integral of the normal component of $\vec{F}$ taken over the closed surface S].

Proof. Let $\vec{F} = F_1\hat{i} + F_2\hat{j} + F_3\hat{k}$, then

$$\nabla \cdot \vec{F} = \text{div } \vec{F} = \frac{\partial F_1}{\partial x} + \frac{\partial F_2}{\partial y} + \frac{\partial F_3}{\partial z}.$$

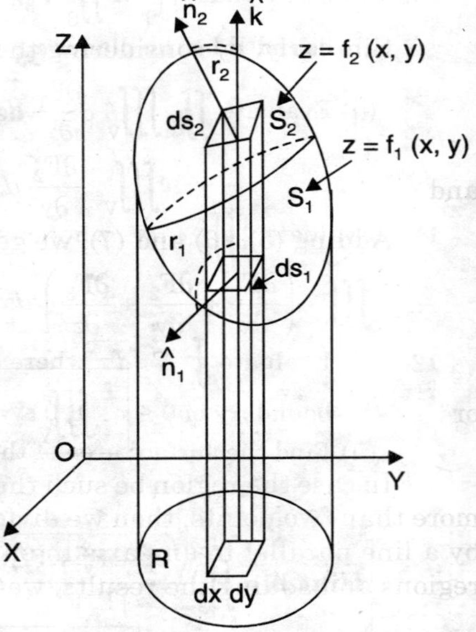

Let α, β, γ be the angles which the outwards drawn unit normal vector $\hat{n}$ makes with the positive directions of x, y, z-axes respectively, then cos α, cos β, cos γ are the direction cosines of $\hat{n}$ and

$$\hat{n} = \cos \alpha \hat{i} + \cos \beta \hat{j} + \cos \gamma \hat{k}$$

$$\therefore \vec{F} \cdot \hat{n} = (F_1\hat{i} + F_2\hat{j} + F_3\hat{k}) \cdot (\cos \alpha \hat{i} + \cos \beta \hat{j} + \cos \gamma \hat{k})$$

$$= F_1 \cos \alpha + F_2 \cos \beta + F_3 \cos \gamma$$

Therefore, the cartesian equivalent of divergence theorem is

$$\iiint_V \left(\frac{\partial F_1}{\partial x} + \frac{\partial F_2}{\partial y} + \frac{\partial F_3}{\partial z} \right) dx \, dy \, dz$$

$$= \iint_S (F_1 \cos \alpha + F_2 \cos \beta + F_3 \cos \gamma) \, dS \quad ...(1)$$

Suppose that S is such a closed surface that a line parallel to the co-ordinate axes meets it in two points only. Let S_1 and S_2 denote the lower and upper portions of S with equations $z = f_1(x, y)$ and $z = f_2(x, y)$ respectively.

Let R be the projection of S on the xy-plane, then

$$\iiint_V \frac{\partial F_3}{\partial z} dx \, dy \, dz = \iint_R \left[\int_{z = f_1(x, y)}^{z = f_2(x, y)} \frac{\partial F_3}{\partial z} dz \right] dx \, dy$$

$$= \iint_R \Big[F_3(x, y, z) \Big]_{f_1}^{f_2} dx \, dy = \iint_R [F_3(x, y, f_2) - F_3(x, y, f_1)] \, dx \, dy$$

$$= \iint_R F_3(x, y, f_2) \, dx \, dy - \iint_R F_3(x, y, f_1) \, dx \, dy \quad ...(2)$$

Now for the upper portion S_2 of S, the normal $\hat{n}_2$ to S_2 makes an acute angle γ_2 with $\hat{k}$.

$$\therefore \quad dx\,dy = \cos\gamma_2\, dS_2 = \hat{k}\,.\,\hat{n}_2\, dS_2$$

For the lower portion S_1 of S, the normal $\hat{n}_1$ to S_1 makes an obtuse angle γ_1 with $\hat{k}$.

$$\therefore \quad dx\,dy = -\cos\gamma_1\, dS_1 = -\,\hat{k}\,.\,\hat{n}_1\, dS_1$$

$$\therefore \quad \iint_R F_3(x, y, f_2)\, dx\,dy = \iint_{S_2} F_3\, \hat{k}\,.\,\hat{n}_2\, dS_2 \qquad ...(3)$$

and

$$\iint_R F_3(x, y, f_1)\, dx\,dy = \iint_{S_1} F_3\, \hat{k}\,.\,\hat{n}_1\, dS_1 \qquad ...(4)$$

Using (3) and (4), (2) becomes

$$\iiint_V \frac{\partial F_3}{\partial z}\, dx\,dy\,dz = \iint_{S_2} F_3\, \hat{k}\,.\,\hat{n}_2\, dS_2 + \iint_{S_1} F_3\, \hat{k}\,.\,\hat{n}_1\, dS_1$$

$$= \iint_S F_3 \hat{k}\,.\,\hat{n}\, dS = \iint_R F_3 \cos\gamma\, dS \qquad ...(5)$$

Similarly, by considering the projection of S on *yz* and *zx*-planes, we have

$$\iiint_V \frac{\partial F_1}{\partial x}\, dx\,dy\,dz = \iint_S F_1 \hat{i}\,.\,\hat{n}\, dS = \iint_S F_1 \cos\alpha\, dS \qquad ...(6)$$

and

$$\iiint_V \frac{\partial F_2}{\partial y}\, dx\,dy\,dz = \iint_S F_2\, \hat{j}\,.\,\hat{n}\, dS = \iint_S F_2 \cos\beta\, dS \qquad ...(7)$$

Adding (5), (6) and (7), we get (1) *i.e.*,

$$\iiint_V \left(\frac{\partial F_1}{\partial x} + \frac{\partial F_2}{\partial y} + \frac{\partial F_3}{\partial z}\right) dx\,dy\,dz = \iint_S (F_1 \cos\alpha + F_2 \cos\beta + F_3 \cos\gamma)\, dS$$

or

$$\boxed{\iiint_V \nabla\,.\,\vec{F}\, dV = \iint_S \vec{F}\,.\,\hat{n}\, dS}$$

In case the region be such that the lines drawn parallel to the coordinate axes meet it in more than two points, then we divide the region into various sub-regions each of which is met by a line parallel to any axis in only two points. Applying the theorem to each of these sub-regions and adding the results, we get the volume integral over the whole region.

ILLUSTRATIVE EXAMPLES

Example 1. *For any closed surface S, prove that* $\iint_S \text{curl}\, \vec{F}\,.\,\hat{n}\, dS = 0.$

Sol. By the divergence theorem, we have $\iint_S \text{curl}\, \vec{F}\,.\,\hat{n}\, dS = \iiint_V (\text{div curl}\, \vec{F})\, dV = 0,$

where V is the volume enclosed by S = 0, since div curl $\vec{F} = 0$.

Example 2. *Evaluate* $\iint_S \vec{r}\,.\,\hat{n}\, dS$, *where S is a closed surface and* $\vec{r} = x\hat{i} + y\hat{j} + z\hat{k}$.

Sol. By Gauss' divergence theorem,

$$\iint_S \vec{r}\,.\,\hat{n}\, dS = \iiint_V \text{div}\, \vec{r}\, dV = \iiint_V 3\, dV = 3V \qquad |\because \text{div}\, \vec{r} = 3$$

where V is the volume enclosed by S.

Example 3. *Use divergence theorem to show that* $\oint_S \nabla r^2 . \vec{dS} = 6V$*, where S is any closed surface enclosing a volume V.*

Sol. By the divergence theorem, we have

$$\oint_S \nabla r^2 . \vec{dS} = \int_V \text{div}\,(\nabla r^2)\, dV = \int_V \nabla . (\nabla r^2)\, dV = \int_V \nabla^2 r^2\, dV$$

$$= \int_V \left(\frac{\partial^2}{\partial x^2} + \frac{\partial^2}{\partial y^2} + \frac{\partial^2}{\partial z^2}\right)(x^2 + y^2 + z^2)\, dV = \int_V 6dV = 6V.$$

Example 4. *Find* $\iint_S \vec{F} . \hat{n}\, dS$*, where* $\vec{F} = (2x + 3z)\,\hat{i} - (xz + y)\,\hat{j} + (y^2 + 2z)\,\hat{k}$ *and S is the surface of the sphere having centre at (3, – 1, 2) and radius 3.*

Sol. Let V be the volume enclosed by the surface S. Then by Gauss divergence theorem, we have

$$\iint_S \vec{F} . \hat{n}\, dS = \iiint_V \text{div}\,\vec{F}\, dV$$

$$= \iiint_V \left[\frac{\partial}{\partial x}(2x + 3z) + \frac{\partial}{\partial y}(-xz - y) + \frac{\partial}{\partial z}(y^2 + 2z)\right] dV$$

$$= \iiint_V (2 - 1 + 2)\, dV = 3 \iiint_V dV = 3V$$

But V is the volume of a sphere of radius 3.

$$\therefore \qquad V = \frac{4}{3}\pi(3)^3 = 36\pi.$$

Hence $\iint_S \vec{F} . \hat{n}\, dS = 3 \times 36\pi = 108\pi.$

Example 5. *Evaluate* $\iint_S \vec{F} . \hat{n}\, dS$ *over the entire surface of the region above the xy-plane bounded by the cone* $z^2 = x^2 + y^2$ *and the plane* $z = 4$*, if* $\vec{F} = 4xz\hat{i} + xyz^2\hat{j} + 3z\hat{k}$.

Sol. If V is the volume enclosed by S, then V is bounded by the surfaces

$$z = 0,\; z = 4,\; z^2 = x^2 + y^2.$$

By divergence theorem, we have

$$\iint_S \vec{F} . \hat{n}\, dS = \iiint_V \text{div}\,\vec{F}\, dV$$

$$= \iiint_V \left[\frac{\partial}{\partial x}(4xz) + \frac{\partial}{\partial y}(xyz^2) + \frac{\partial}{\partial z}(3z)\right] dV = \iiint_V (4z + xz^2 + 3)\, dV$$

$$= \int_0^4 \int_{-z}^{z} \int_{-\sqrt{z^2 - y^2}}^{\sqrt{z^2 - y^2}} (4z + xz^2 + 3)\, dx\, dy\, dz$$

$$= 2 \int_0^4 \int_{-z}^{z} \int_0^{\sqrt{z^2 - y^2}} (4z + 3)\, dx\, dy\, dz \qquad \left|\; \text{Since } \int_{-\sqrt{z^2 - y^2}}^{\sqrt{z^2 - y^2}} x\, dx = 0\right.$$

$$= 2 \int_0^4 \int_{-z}^{z} (4z + 3)\sqrt{z^2 - y^2}\, dy\, dz = 4 \int_0^4 \int_0^z (4z + 3)\sqrt{z^2 - y^2}\, dy\, dz$$

$$= 4\int_0^4 (4z+3)\left[\frac{y\sqrt{z^2-y^2}}{2} + \frac{z^2}{2}\sin^{-1}\frac{y}{z}\right]_0^z dz$$

$$= 4\int_0^4 (4z+3)\left[\frac{z^2}{2}\sin^{-1} 1\right] dz = 4 \times \frac{\pi}{4}\int_0^4 (4z^3+3z^2)\, dz$$

$$= \pi\left[z^4 + z^3\right]_0^4 = \pi(256+64) = 320\pi.$$

Example 6. *The vector field* $\vec{F} = x^2\hat{i} + z\hat{j} + yz\hat{k}$ *is defined over the volume of the cuboid given by* $0 \le x \le a,\ 0 \le y \le b,\ 0 \le z \le c$ *enclosing the surface S, evaluate* $\iint_S \vec{F}\,.\,d\vec{S}$.

Sol. By Gauss Divergence theorem

$$\iint_S \vec{F}\,.\,d\vec{S} = \iiint_V \text{div } \vec{F}\, dV$$

$$= \iiint_V \left\{\frac{\partial}{\partial x}(x^2) + \frac{\partial}{\partial y}(z) + \frac{\partial}{\partial z}(yz)\right\} dV = \iiint_V (2x+y)\, dV$$

$$= \int_0^a\int_0^b\int_0^c (2x+y)\, dz\, dy\, dx$$

$$= \int_0^a\int_0^b (2x+y)\left(z\right)_0^c dy\, dx = c\int_0^a \left(2xy + \frac{y^2}{2}\right)_0^b dx$$

$$= bc\int_0^a \left(2x + \frac{b}{2}\right) dx = bc\left(x^2 + \frac{b}{2}x\right)_0^a = abc\left(a + \frac{b}{2}\right)$$

Example 7. *Evaluate* $\iint_S (\nabla \times \vec{A})\,.\,\hat{n}\, dS$, *where S is the surface of the cone* $z = 2 - \sqrt{x^2+y^2}$ *above the xy-plane and* $\vec{A} = (x-z)\,\hat{i} + (x^3+yz)\,\hat{j} - 3xy^2\hat{k}$.

Sol. Here S is not a closed surface. The surface $z = 2 - \sqrt{x^2+y^2}$ meets the xy-plane in a circle C given by $x^2 + y^2 = 4$, $z = 0$. Let S_1 be the plane region bounded by the circle C. If S′ is the surface consisting of the surfaces S and S_1, then S′ *is a closed surface.* By divergence theorem, we have

$$\iint_{S'} \text{curl } \vec{A}\,.\,\hat{n}\, dS = 0 \qquad \text{[See Example 1]}$$

$$\Rightarrow \quad \iint_S \text{curl } \vec{A}\,.\,\hat{n}\, dS + \iint_{S_1} \text{curl } \vec{A}\,.\,\hat{n}\, dS = 0 \qquad [\because \ S' \text{ consists of S and } S_1]$$

$$\Rightarrow \quad \int_S \text{curl } \vec{A}\,.\,\hat{n}\, dS - \iint_{S_1} \text{curl } \vec{A}\,.\,\hat{k}\, dS = 0 \qquad [\because \ \text{on } S_1,\ \hat{n} = \hat{k}]$$

$$\Rightarrow \quad \int_S \text{curl } \vec{A}\,.\,\hat{n}\, dS = \iint_{S_1} \text{curl } \vec{A}\,.\,\hat{k}\, dS$$

Now $\quad \text{curl } \vec{A} = \begin{vmatrix} \hat{i} & \hat{j} & \hat{k} \\ \dfrac{\partial}{\partial x} & \dfrac{\partial}{\partial y} & \dfrac{\partial}{\partial z} \\ x - z & x^3 + yz & -3xy^2 \end{vmatrix} = \hat{i}(-6xy - y) + \hat{j}(-1 + 3y^2) + \hat{k}(3x^2 - 0)$

$\therefore \quad \text{curl } \vec{A} \cdot \hat{k} = 3x^2$

$$\therefore \quad \iint_{S_1} \text{curl } \vec{A} \cdot \hat{n}\, dS = \iint_{S_1} 3x^2\, dS = \int_0^{2\pi}\int_0^2 3(r^2 \cos^2\theta)\, r\, dr\, d\theta$$

$$= 3\int_0^{2\pi}\int_0^2 r^3 \cos^2\theta\, dr\, d\theta = 3\int_0^{2\pi}\left[\frac{r^4}{4}\right]_0^2 \cos^2\theta\, d\theta = 12\int_0^{2\pi} \cos^2\theta\, d\theta = 12\pi.$$

Example 8. *Evaluate* $\iint_S (y^2z^2\hat{i} + z^2x^2\hat{j} + z^2y^2\hat{k}) \cdot \hat{n}\, dS$, *where S is the part of the sphere* $x^2 + y^2 + z^2 = 1$ *above the xy-plane and bounded by this plane.* (U.P.T.U. 2014)

Sol. Let V be the volume enclosed by the surface S. Then by divergence theorem, we have

$$\iint_S (y^2z^2\hat{i} + z^2x^2\hat{j} + z^2y^2\hat{k}) \cdot \hat{n}\, dS = \iiint_V \text{div}\,(y^2z^2\hat{i} + z^2x^2\hat{j} + z^2y^2\hat{k})\, dV$$

$$= \iiint_V \left[\frac{\partial}{\partial x}(y^2z^2) + \frac{\partial}{\partial y}(z^2x^2) + \frac{\partial}{\partial z}(z^2y^2)\right] dV$$

$$= \iiint_V 2zy^2\, dV = 2\iiint_V zy^2\, dV$$

Changing to spherical polar coordinates by putting

$$x = r\sin\theta\cos\phi,\ y = r\sin\theta\sin\phi,\ z = r\cos\theta$$

$$dV = r^2 \sin\theta\, dr\, d\theta\, d\phi$$

To cover V, the limits of r will be 0 to 1, those of θ will be 0 to $\dfrac{\pi}{2}$ and those of ϕ will be 0 to 2π.

$$\therefore \quad 2\iiint_V zy^2\, dV = 2\int_0^{2\pi}\int_0^{\pi/2}\int_0^1 (r\cos\theta)(r^2\sin^2\theta\sin^2\phi)\, r^2\sin\theta\, dr\, d\theta\, d\phi$$

$$= 2\int_0^{2\pi}\int_0^{\pi/2}\int_0^1 r^5 \sin^3\theta\cos\theta\sin^2\phi\, dr\, d\theta\, d\phi$$

$$= 2\int_0^{2\pi}\int_0^{\pi/2} \sin^3\theta\cos\theta\sin^2\phi\left[\frac{r^6}{6}\right]_0^1 d\theta\, d\phi$$

$$= \frac{2}{6}\int_0^{2\pi} \sin^2\phi \cdot \frac{2}{4.2}\, d\phi = \frac{1}{12}\int_0^{2\pi} \sin^2\phi\, d\phi = \frac{\pi}{12}.$$

Example 9. *Use divergence theorem to evaluate the surface integral* $\iint_S (x\, dy\, dz + y\, dz\, dx + z\, dx\, dy)$ *where S is the portion of the plane* $x + 2y + 3z = 6$ *which lies in the first octant.* (A.K.T.U. 2021)

Sol. By Gauss' Divergence theorem,

$$\iint_S \vec{F} \cdot \hat{n}\, dS = \iiint_V \text{div } \vec{F}\, dV \qquad \ldots(1)$$

Here, $\vec{F} = x\hat{i} + y\hat{j} + z\hat{j}, \hat{n}\, ds = dy\, dz\, \hat{i} + dz\, dx\, \hat{j} + dx\, dy\, \hat{k}$

Now, $$\text{div}\ \vec{F} = \left(\hat{i}\frac{\partial}{\partial x} + \hat{j}\frac{\partial}{\partial y} + \hat{k}\frac{\partial}{\partial z}\right) . (x\hat{i} + y\hat{j} + z\hat{k})$$
$$= \frac{\partial}{\partial x}(x) + \frac{\partial}{\partial y}(y) + \frac{\partial}{\partial z}(z) = 1 + 1 + 1 = 3.$$

From (1),
$$\iint_S (x\, dy\, dz + y\, dz\, dx + z\, dx\, dy) = \iiint_V 3\, dV$$
$$= 3\int_{x=0}^{6}\int_{y=0}^{\frac{6-x}{2}}\int_{z=0}^{\frac{6-x-2y}{3}} dz\, dy\, dx$$
$$= 3\int_{x=0}^{6}\int_{y=0}^{\frac{6-x}{2}}\left(\frac{6-x-2y}{3}\right) dy\, dx$$
$$= \int_{x=0}^{6}\left\{(6-x)y - y^2\right\}_0^{\frac{6-x}{2}} dx$$
$$= \int_{x=0}^{6}\left\{\frac{(6-x)^2}{2} - \frac{(6-x)^2}{4}\right\} dx = \int_{x=0}^{6}\frac{(6-x)^2}{4} dx$$
$$= \frac{1}{4}\left[\frac{(6-x)^3}{(-3)}\right]_0^6 = \frac{1}{12}(216) = 18.$$

Example 10. *Apply divergence theorem to evaluate* $\iint_S \overline{F} . \hat{n}\, dS$ *where* $\overline{F} = 4x^3\,\hat{i} - x^2y\,\hat{j} + x^2z\hat{k}$ *and S is the surface of the cylinder* $x^2 + y^2 = a^2$ *bounded by the planes* $z = 0$ *and* $z = b$.

Sol. $$\overline{F} = 4x^3\hat{i} - x^2y\,\hat{j} + x^2z\hat{k}$$

$\therefore$ $$\text{div}\ \overline{F} = \left(\hat{i}\frac{\partial}{\partial x} + \hat{j}\frac{\partial}{\partial y} + \hat{k}\frac{\partial}{\partial z}\right) . (4x^3\hat{i} - x^2y\,\hat{j} + x^2z\hat{k})$$
$$= \frac{\partial}{\partial x}(4x^3) + \frac{\partial}{\partial y}(-x^2y) + \frac{\partial}{\partial z}(x^2z) = 12x^2 - x^2 + x^2 = 12x^2$$

Now,
$$\iiint_V \text{div}\ \overline{F}\, dV = 12\int_{x=-a}^{a}\int_{y=-\sqrt{a^2-x^2}}^{\sqrt{a^2-x^2}}\int_{z=0}^{b} x^2\, dz\, dy\, dx$$
$$= 12\int_{x=-a}^{a}\int_{y=-\sqrt{a^2-x^2}}^{\sqrt{a^2-x^2}} x^2\left(z\right)_0^b dy\, dx$$
$$= 12b\int_{-a}^{a} x^2(y)_{-\sqrt{a^2-x^2}}^{\sqrt{a^2-x^2}} dx = 12b\int_{-a}^{a} x^2 . 2\sqrt{a^2-x^2}\, dx$$
$$= 24b\int_{-a}^{a} x^2\sqrt{a^2-x^2}\, dx = 48b\int_0^a x^2\sqrt{a^2-x^2}\, dx$$
$$= 48b\int_0^{\pi/2} a^2\sin^2\theta . a\cos\theta . a\cos\theta\, d\theta$$

| Put $x = a\sin\theta$ $\therefore$ $dx = a\cos\theta\, d\theta$

$$= 48ba^4 \int_0^{\pi/2} \sin^2\theta \,.\, \cos^2\theta \, d\theta = 48ba^4 \,.\, \frac{\Gamma(3/2)\Gamma(3/2)}{2\Gamma 3}$$

$$= 48ba^4 \,.\, \frac{\frac{1}{2}\sqrt{\pi} \,.\, \frac{1}{2}\sqrt{\pi}}{2\,.\,2} = 3ba^4\pi.$$

Example 11. *Evaluate* $\iint_S (a^2x^2 + b^2y^2 + c^2z^2)^{1/2} \, dS$ *where S is the surface of the ellipsoid* $ax^2 + by^2 + cz^2 = 1$.

Sol. Let $\phi = ax^2 + by^2 + cz^2 - 1$ ...(1)

$$\therefore \quad \text{grad } \phi = \left(\hat{i}\frac{\partial}{\partial x} + \hat{j}\frac{\partial}{\partial y} + \hat{k}\frac{\partial}{\partial z}\right)(ax^2 + by^2 + cz^2 - 1)$$

$$= 2ax\,\hat{i} + 2by\,\hat{j} + 2cz\,\hat{k} = 2(ax\,\hat{i} + by\,\hat{j} + cz\,\hat{k})$$

and $|\text{ grad } \phi \,| = 2\sqrt{a^2x^2 + b^2y^2 + c^2z^2}$

$$\therefore \quad \hat{n} = \frac{\text{grad } \phi}{|\text{grad } \phi|} = \frac{2(ax\,\hat{i} + by\,\hat{j} + cz\,\hat{k})}{2\sqrt{a^2x^2 + b^2y^2 + c^2z^2}} = \frac{ax\,\hat{i} + by\,\hat{j} + cz\,\hat{k}}{\sqrt{a^2x^2 + b^2y^2 + c^2z^2}} \quad ...(2)$$

According to the question,

$$\overline{\mathbf{F}} \,.\, \hat{n} = \sqrt{a^2x^2 + b^2y^2 + c^2z^2} \quad ...(3)$$

Consequently, $\overline{\mathbf{F}} = ax\,\hat{i} + by\,\hat{j} + cz\,\hat{k}$

$$\therefore \quad \text{div } \overline{\mathbf{F}} = \left(\hat{i}\frac{\partial}{\partial x} + \hat{j}\frac{\partial}{\partial y} + \hat{k}\frac{\partial}{\partial z}\right) . (ax\,\hat{i} + bx\,\hat{j} + cz\,\hat{k}) = a + b + c$$

Now, $$\iiint_V \text{div } \overline{\mathbf{F}} \, dV = (a + b + c) \iiint_V dV \quad ...(4)$$

Volume of ellipsoid $\left(\frac{x^2}{a^2} + \frac{y^2}{b^2} + \frac{z^2}{c^2} = 1\right)$ is equal to $\frac{4}{3}\pi \, abc$

$\therefore$ Volume of ellipsoid $(ax^2 + by^2 + cz^2 = 1)$ is equal to $\frac{4}{3}\pi \frac{1}{\sqrt{a}} \frac{1}{\sqrt{b}} \frac{1}{\sqrt{c}} = \frac{4\pi}{3\sqrt{abc}}$

$\therefore$ From eqn. (4), $$\iiint_V \text{div } \overline{\mathbf{F}} \, dV = (a + b + c) \frac{4\pi}{3\sqrt{abc}}$$

Hence, $$\iint_S (a^2x^2 + b^2y^2 + c^2z^2)^{1/2} \, dS = \frac{4\pi(a + b + c)}{3\sqrt{abc}}.$$

Example 12. *Verify divergence theorem for* $\overline{\mathbf{F}} = 4xz\,\hat{i} - y^2\,\hat{j} + yz\,\hat{k}$ *taken over the cube bounded by the planes* $x = 0$, $x = 1$, $y = 0$, $y = 1$, $z = 0$, $z = 1$. [A.K.T.U. 2020]

Sol. $\overline{\mathbf{F}} = 4xz\,\hat{i} - y^2\,\hat{j} + yz\,\hat{k}$ ...(1) | given

$$\therefore \quad \text{div } \vec{\mathbf{F}} = \left(\hat{i}\frac{\partial}{\partial x} + \hat{j}\frac{\partial}{\partial y} + \hat{k}\frac{\partial}{\partial z}\right) . (4xz\,\hat{i} - y^2\,\hat{j} + yz\,\hat{k})$$

$$= \frac{\partial}{\partial x}(4xz) + \frac{\partial}{\partial y}(-y^2) + \frac{\partial}{\partial z}(yz) = 4z - 2y + y = 4z - y$$

$$\iiint_V \text{div}\,\overline{\mathbf{F}}\,dV = \int_{x=0}^{1}\int_{y=0}^{1}\int_{z=0}^{1}(4z - y)\,dz\,dy\,dx$$

$$= \int_{x=0}^{1}\int_{y=0}^{1}(2z^2 - yz)_0^1\,dy\,dx = \int_{x=0}^{1}\int_{y=0}^{1}(2-y)\,dy\,dx$$

$$= \int_{x=0}^{1}\left(2y - \frac{y^2}{2}\right)_0^1 dx = \int_0^1 \frac{3}{2}\,dx = \frac{3}{2} \qquad ...(2)$$

To evaluate $\iint_S \overline{\mathbf{F}}\,.\,\hat{n}\,d\mathbf{S}$.

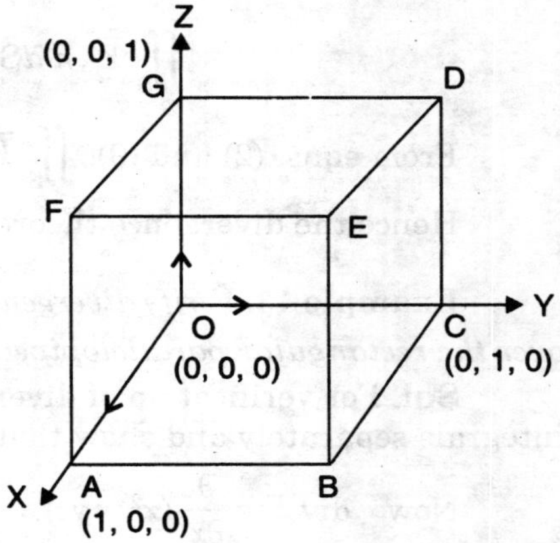

Here S is the surface of the cube bounded by the 6 plane surfaces.

Over the face OABC,

$$z = 0,\ dz = 0,\ \hat{n} = -\hat{k},\ dS = dx\,dy$$

$$\therefore \quad \iint \overline{\mathbf{F}}\,.\,\hat{n}\,dS = \int_0^1\int_0^1(-y^2\hat{j})\,.\,(-\hat{k})\,dx\,dy = 0 \qquad ...(3)$$

Over the face BCDE,

$$y = 1,\ dy = 0,\ \hat{n} = \hat{j},\ dS = dx\,dz$$

$$\therefore \quad \iint_D \overline{\mathbf{F}}\,.\,\hat{n}\,dS = \int_0^1\int_0^1(4xz\,\hat{i} - \hat{j} + z\hat{k})\,.\,\hat{j}\,dx\,dz$$

$$= -\int_0^1\int_0^1 dx\,dz = -(x)_0^1\,(z)_0^1 = -1 \qquad ...(4)$$

Over the face DEFG,

$$z = 1,\ dz = 0,\ \hat{n} = \hat{k},\ dS = dx\,dy$$

$$\therefore \quad \iint \overline{\mathbf{F}}.\hat{n}\,dS = \int_0^1\int_0^1(4x\,\hat{i} - y^2\hat{j} + y\hat{k}).\,\hat{k}\,dx\,dy$$

$$\int_0^1\int_0^1 y\,dx\,dy = \int_0^1 dx\int_0^1 y\,dy = (x)_0^1\left(\frac{y^2}{2}\right)_0^1 = \frac{1}{2} \qquad ...(5)$$

Over the face AOGF,

$$y = 0,\ dy = 0,\ \hat{n} = -\hat{j},\ dS = dx\,dz$$

$$\therefore \quad \iint \overline{\mathbf{F}}\,.\,\hat{n}\,dS = \int_0^1\int_0^1(4xz\,\hat{i})\,.\,(-\hat{j})\,dx\,dz = 0 \qquad ...(6)$$

Over the face OCDG,

$$\therefore \quad x = 0,\ dx = 0,\ \hat{n} = -\hat{i},\ dS = dy\,dz$$

$$\therefore \quad \iint \overline{\mathbf{F}}\,.\,\hat{n}\,dS = \int_0^1\int_0^1(-y^2\hat{j} + yz\,\hat{k})\,.\,(-\hat{i})\,dy\,dz = 0 \qquad ...(7)$$

Over the face ABEF,

$$x = 1,\ dx = 0,\ \hat{n} = \hat{i},\ dS = dy\,dz$$

$$\therefore \quad \iint \overline{F} \cdot \hat{n}\, dS = \int_0^1\int_0^1 (4z\,\hat{i} - y^2\hat{j} + yz\hat{k}) \cdot \hat{i}\, dy\, dz$$

$$= \int_0^1\int_0^1 4z\, dy\, dz = \int_0^1 dy \int_0^1 4z\, dz = (y)_0^1\,(2z^2)_0^1 = 2 \qquad \text{...(8)}$$

Adding eqns. (3), (4), (5), (6), (7) and (8), we get over the whole surface S,

$$\iint_S \overline{F} \cdot \hat{n}\, dS = 0 - 1 + \frac{1}{2} + 0 + 0 + 2 = 3/2 \qquad \text{...(9)}$$

From eqns. (2) and (9), $\iint_S \overline{F} \cdot \hat{n}\, dS = \iiint_V \text{div}\, \overline{F}\, dV$

Hence the divergence theorem is verified.

Example 13. *Verify divergence theorem for* $\vec{F} = (x^2 - yz)\,\hat{i} + (y^2 - zx)\,\hat{j} + (z^2 - xy)\,\hat{k}$ *taken over the rectangular parallelopiped* $0 \le x \le a,\ 0 \le y \le b,\ 0 \le z \le c.$ (M.T.U. 2013)

Sol. For verification of divergence theorem, we shall evaluate the volume and surface integrals separately and show that they are equal.

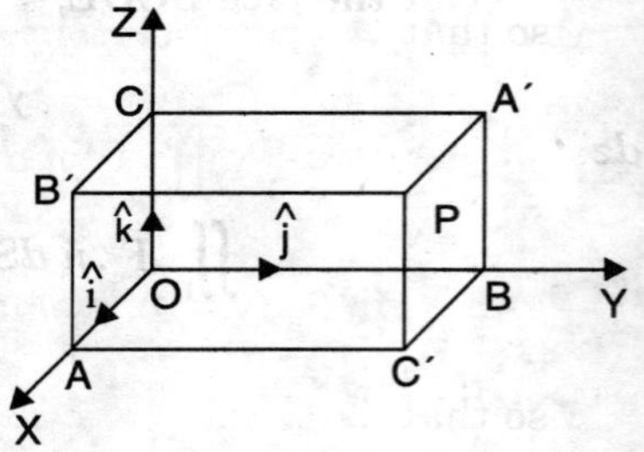

$$\text{Now}\quad \text{div}\,\vec{F} = \frac{\partial}{\partial x}(x^2 - yz) + \frac{\partial}{\partial y}(y^2 - zx) + \frac{\partial}{\partial z}(z^2 - xy)$$

$$= 2(x + y + z)$$

$$\therefore \quad \iiint_V \text{div}\,\vec{F}\, dV = \int_0^c\int_0^b\int_0^a 2(x + y + z)\, dx\, dy\, dz$$

$$= \int_0^c\int_0^b 2\left[\left(\frac{x^2}{2} + yx + zx\right)\right]_0^a dy\, dz$$

$$= \int_0^c\int_0^b 2\left(\frac{a^2}{2} + ya + za\right) dy\, dz = \int_0^c 2\left[\left(\frac{a^2}{2}y + \frac{ay^2}{2} + azy\right)\right]_0^b dz$$

$$= 2\int_0^c \left(\frac{a^2b}{2} + \frac{ab^2}{2} + abz\right) dz = 2\left[\frac{a^2b}{2}z + \frac{ab^2}{2}z + ab\,\frac{z^2}{2}\right]_0^c$$

$$= a^2bc + ab^2c + abc^2 = abc\,(a + b + c) \qquad \text{...(1)}$$

To evaluate the surface integral, divide the closed surface S of the rectangular parallelopiped into 6 parts.

S_1 : the face OAC′B, S_2 : the face CB′PA′, S_3 : the face OBA′C,
S_4 : the face AC′PB′, S_5 : the face OCB′A, S_6 : the face BA′PC′.

$$\text{Also} \iint_S \vec{F} \cdot \hat{n}\, dS = \iint_{S_1} \vec{F} \cdot \hat{n}\, dS + \iint_{S_2} \vec{F} \cdot \hat{n}\, dS + \iint_{S_3} \vec{F} \cdot \hat{n}\, dS$$

$$+ \iint_{S_4} \vec{F} \cdot \hat{n}\, dS + \iint_{S_5} \vec{F} \cdot \hat{n}\, dS + \iint_{S_6} \vec{F} \cdot \hat{n}\, dS$$

On S_1 $(z = 0)$, we have $\hat{n} = -\hat{k},\ \vec{F} = x^2\hat{i} + y^2\hat{j} - xy\hat{k}$

so that $\qquad \vec{F} \cdot \hat{n} = (x^2\hat{i} + y^2\hat{j} - xy\hat{k}) \cdot (-\hat{k}) = xy$

$$\therefore \quad \iint_S \vec{F} \cdot \hat{n}\, dS = \int_0^b \int_0^a xy\, dx\, dy = \int_0^b \left[y \frac{x^2}{2} \right]_0^a dy = \frac{a^2}{2} \int_0^b y\, dy = \frac{a^2 b^2}{4}$$

On S_2 ($z = c$), we have $\hat{n} = \hat{k}, \vec{F} = (x^2 - cy)\, \hat{i} + (y^2 - cx)\, \hat{j} + (c^2 - xy)\, \hat{k}$

so that $\quad \vec{F} \cdot \hat{n} = [(x^2 - cy)\, \hat{i} + (y^2 - cx)\, \hat{j} + (c^2 - xy)\, \hat{k}] \cdot \hat{k} = c^2 - xy$

$$\iint_{S_2} \vec{F} \cdot \hat{n}\, dS = \int_0^b \int_0^a (c^2 - xy)\, dx\, dy = \int_0^b \left(c^2 a - \frac{a^2}{2} y \right) dy = abc^2 - \frac{a^2 b^2}{4}$$

On S_3 ($x = 0$), we have $\quad \hat{n} = -\hat{i}, \vec{F} = -yz\hat{i} + y^2\hat{j} + z^2\hat{k}$

so that $\quad \vec{F} \cdot \hat{n} = (-yz\hat{i} + y^2\hat{j} + z^2\hat{k}) \cdot (-\hat{i}) = yz$

$$\therefore \quad \iint_{S_3} \vec{F} \cdot \hat{n}\, dS = \int_0^c \int_0^b yz\, dy\, dz = \int_0^c \frac{b^2}{2} x\, dz = \frac{b^2 c^2}{4}$$

On S_4 ($x = a$), we have $\quad \hat{n} = \hat{i}, \vec{F} = (a^2 - yz)\, \hat{i} + (y^2 - az)\, \hat{j} + (z^2 - ay)\, \hat{k}$

so that $\quad \vec{F} \cdot \hat{n} = [(a^2 - yz)\, \hat{i} + (y^2 - az)\, \hat{j} + (z^2 - ay)\, \hat{k}] \cdot \hat{i} = a^2 - yz$

$$\therefore \quad \iint_{S_4} \vec{F} \cdot \hat{n}\, dS = \int_0^c \int_0^b (a^2 - yz)\, dy\, dz = \int_0^c \left(a^2 b - \frac{b^2}{2} z \right) dz = a^2 bc - \frac{b^2 c^2}{4}$$

On S_5 ($y = 0$), we have $\hat{n} = -\hat{j}, \vec{F} = x^2\hat{i} - zx\hat{j} + z^2\hat{k}$

so that $\quad \vec{F} \cdot \hat{n} = (x^2\hat{i} - zx\hat{j} + z^2\hat{k}) \cdot (-\hat{j}) = zx$

$$\therefore \quad \iint_{S_5} \vec{F} \cdot \hat{n}\, dS = \int_0^a \int_0^c zx\, dz\, dx = \int_0^a \frac{c^2}{2} x\, dx = \frac{a^2 c^2}{4}$$

On S_6 ($y = b$), we have $\hat{n} = \hat{j}, \vec{F} = (x^2 - bz)\, \hat{i} + (b^2 - zx)\, \hat{j} + (z^2 - bx)\, \hat{k}$

so that $\quad \vec{F} \cdot \hat{n} = [(x^2 - bz)\, \hat{i} + (b^2 - zx)\, \hat{j} + (z^2 - bx)\, \hat{k}] \cdot \hat{j} = b^2 - zx$

$$\therefore \quad \iint_{S_6} \vec{F} \cdot \hat{n}\, dS = \int_0^a \int_0^c (b^2 - zx)\, dz\, dx = \int_0^a \left(b^2 c - \frac{c^2}{2} x \right) dx = ab^2 c - \frac{a^2 c^2}{4}$$

$$\therefore \quad \iint_S \vec{F} \cdot \hat{n}\, dS = \frac{a^2 b^2}{4} + abc^2 - \frac{a^2 b^2}{4} + \frac{b^2 c^2}{4} + a^2 bc - \frac{b^2 c^2}{4} + \frac{a^2 c^2}{4} + ab^2 c - \frac{a^2 c^2}{4}$$

$$= abc\,(a + b + c) \qquad \ldots(2)$$

The equality of (1) and (2) verifies divergence theorem.

Example 14. *Verify divergence theorem for* $\vec{F} = 4x\hat{i} - 2y^2\hat{j} + z^2\hat{k}$ *taken over the region bounded by the cylinder* $x^2 + y^2 = 4$, $z = 0$, $z = 3$.

Sol. Since div $\vec{F} = \frac{\partial}{\partial x}(4x) + \frac{\partial}{\partial y}(-2y^2) + \frac{\partial}{\partial z}(z^2) = 4 - 4y + 2z$

$$\therefore \iiint_V \text{div}\,\vec{F}\,dV = \iiint_V (4 - 4y + 2z)\,dx\,dy\,dz = \int_{-2}^{2}\int_{-\sqrt{4-x^2}}^{\sqrt{4-x^2}}\int_0^3 (4 - 4y + 2z)\,dz\,dy\,dx$$

$$= \int_{-2}^{2}\int_{-\sqrt{4-x^2}}^{\sqrt{4-x^2}} \Big[4z - 4yz + z^2\Big]_0^3 dy\,dx$$

$$= \int_{-2}^{2}\int_{-\sqrt{4-x^2}}^{\sqrt{4-x^2}} (12 - 12y + 9)\,dy\,dx$$

$$= \int_{-2}^{2} \Big[21y - 6y^2\Big]_{-\sqrt{4-x^2}}^{\sqrt{4-x^2}} dx$$

$$= \int_{-2}^{2} 42\sqrt{4 - x^2}\,dx = 84\int_0^2 \sqrt{4 - x^2}\,dx$$

$$= 84\left[\frac{x\sqrt{4-x^2}}{2} + \frac{4}{2}\sin^{-1}\frac{x}{2}\right]_0^2$$

$$= 84[2\sin^{-1} 1] = 84\left[2 \times \frac{\pi}{2}\right] = 84\pi \qquad \text{...(1)}$$

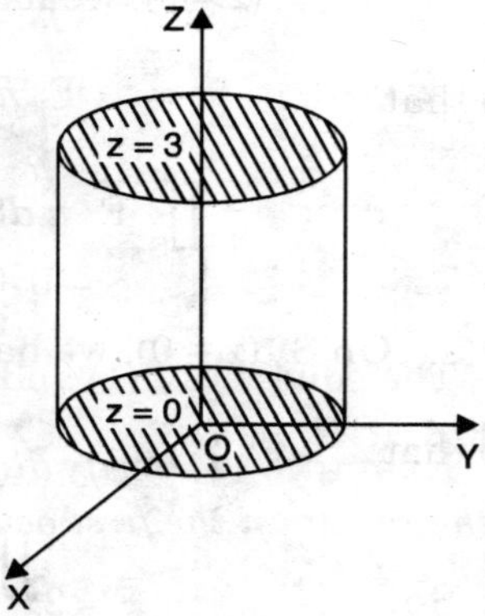

To evaluate the surface integral, divide the closed surface S of the cylinder into 3 parts.

S_1 : the circular base in the plane $z = 0$

S_2 : the circular top in the plane $z = 3$

S_3 : the curved surface of the cylinder, given by the equation $x^2 + y^2 = 4$.

Also $\iint_S \vec{F}\cdot\hat{n}\,dS = \iint_{S_1} \vec{F}\cdot\hat{n}\,dS + \iint_{S_2} \vec{F}\cdot\hat{n}\,dS + \iint_{S_3} \vec{F}\cdot\hat{n}\,dS$

On S_1 ($z = 0$), we have $\hat{n} = -\hat{k}$, $\vec{F} = 4x\hat{i} - 2y^2\hat{j}$

so that $\vec{F}\cdot\hat{n} = (4x\hat{i} - 2y^2\hat{j})\cdot(-\hat{k}) = 0$

$\therefore \iint_{S_1} \vec{F}\cdot\hat{n}\,dS = 0$

On S_2 ($z = 3$), we have $\hat{n} = \hat{k}$, $\vec{F} = 4x\hat{i} - 2y^2\hat{j} + 9\hat{k}$

so that $\vec{F}\cdot\hat{n} = (4x\hat{i} - 2y^2\hat{j} + 9\hat{k})\cdot\hat{k} = 9$

$$\therefore \iint_{S_2} \vec{F}\cdot\hat{n}\,dS = \iint_{S_2} 9\,dx\,dy = 9\iint_S dx\,dy$$

$$= 9 \times \text{area of surface } S_2 = 9\,(\pi\,.\,2^2) = 36\pi$$

On S_3, $x^2 + y^2 = 4$

$$\therefore \quad \hat{n} = \text{a unit vector normal to surface } S_3 = \frac{2x\hat{i} + 2y\hat{j}}{\sqrt{4x^2 + 4y^2}} = \frac{x\hat{i} + y\hat{j}}{2}$$

$$\vec{F}\cdot\hat{n} = (4x\hat{i} - 2y^2\hat{j} + z^2\hat{k})\cdot\left(\frac{x\hat{i} + y\hat{j}}{2}\right) = 2x^2 - y^3$$

Also, on S_3, *i.e.* $x^2 + y^2 = 4$, $x = 2\cos\theta$, $y = 2\sin\theta$ and $dS = 2d\theta\, dz$.

To cover the whole surface S_3, z varies from 0 to 3 and θ varies from 0 to 2π.

$$\therefore \quad \iint_{S_3} \vec{F}\cdot\hat{n}\, dS = \int_0^{2\pi}\int_0^3 [2(2\cos\theta)^2 - (2\sin\theta)^3]\, 2dz\, d\theta$$

$$= \int_0^{2\pi} 16(\cos^2\theta - \sin^3\theta)\times 3\, d\theta = 48\int_0^{2\pi}(\cos^2\theta - \sin^3\theta)\, d\theta = 48\pi$$

$$\therefore \quad \iint_S \vec{F}\cdot\hat{n}\, dS = 0 + 36\pi + 48\pi = 84\pi \qquad ...(2)$$

The equality of (1) and (2) verifies divergence theorem.

Example 15. *Verify divergence theorem for the function* $\overline{F} = 2x^2y\,\hat{i} - y^2\,\hat{j} + 4xz^2\,\hat{k}$ *taken over the region in the first octant bounded by* $y^2 + z^2 = 9$ *and* $x = 2$.

Sol. $$\overline{F} = 2x^2y\,\hat{i} - y^2\,\hat{j} + 4xz^2\,\hat{k} \qquad ...(1) \mid \text{given}$$

$$\text{div}\,\overline{F} = \left(\hat{i}\frac{\partial}{\partial x} + \hat{j}\frac{\partial}{\partial y} + \hat{k}\frac{\partial}{\partial z}\right)\cdot(2x^2y\hat{i} - y^2\hat{j} + 4xz^2\hat{k}) = 4xy - 2y + 8xz$$

$$\therefore \quad \iiint_V \text{div}\,\overline{F}\, dV = \int_{x=0}^{2}\int_{y=0}^{3}\int_{z=0}^{\sqrt{9-y^2}} (4xy - 2y + 8xz)\, dz\, dy\, dx$$

$$= \int_{x=0}^{2}\int_{y=0}^{3} [4xy\sqrt{9-y^2} - 2y\sqrt{9-y^2} + 4x(9-y^2)]\, dy\, dx$$

$$= \int_{x=0}^{2}\left[\frac{-4x}{2}\cdot\frac{2}{3}(9-y^2)^{3/2} + \frac{2}{3}(9-y^2)^{3/2} + 36xy - \frac{4}{3}xy^3\right]_0^3 dx$$

$$= \int_{x=0}^{2}(108x - 18x)\, dx = (54x^2 - 18x)_0^2$$

$$= 216 - 36 = 180 \qquad ...(2)$$

To evaluate $\iint_S \mathrm{F}\cdot\hat{n}\, dS$

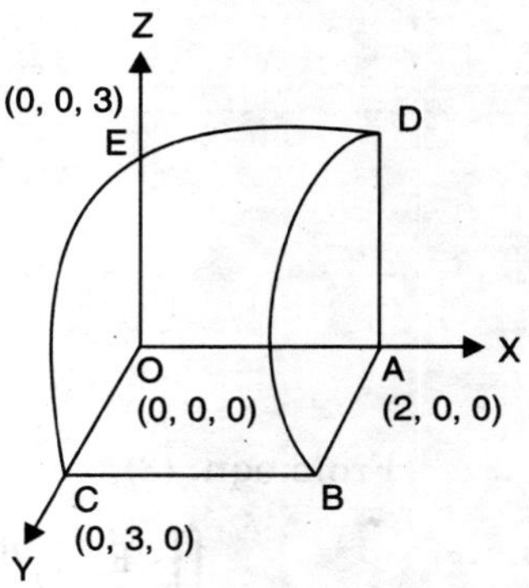

The surface S consists of five parts.

$$\iint_S \overline{F}\cdot\hat{n}\, dS = \iint_{OABC} \overline{F}\cdot\hat{n}\, dS + \iint_{OCE} \overline{F}\cdot\hat{n}\, dS + \iint_{OADE} \overline{F}\cdot\hat{n}\, dS$$

$$+ \iint_{ABD} \overline{F}\cdot\hat{n}\, dS + \iint_{BDEC} \overline{F}\cdot\hat{n}\, dS \qquad ...(3)$$

Over the surface OABC, $z = 0$, $dz = 0$, $\hat{n} = -\hat{k}$, $dS = dx\, dy$

$$\iint_{OABC} \overline{F}\cdot\hat{n}\, dS = \iint (2x^2y\,\hat{i} - y^2\hat{j})(-\hat{k})\, dx\, dy = 0 \qquad ...(4)$$

Over the surface OCE, $x = 0$, $dx = 0$, $\hat{n} = -\hat{i}$, $dS = dy\,dz$

$$\iint_{OCE} \overline{F}.\hat{n}\,dS = \iint (-y^2\hat{j}).(-\hat{i})\,dy\,dz = 0 \qquad ...(5)$$

Over the surface OADE, $y = 0$, $dy = 0$, $\hat{n} = -\hat{j}$, $dS = dz\,dx$

$$\iint_{OADE} \overline{F}.\hat{n}\,dS = \iint (4xz^2\,\hat{k}).(-\hat{j})\,dz\,dx = 0 \qquad ...(6)$$

Over the surface ABD, $x = 2$, $dx = 0$, $\hat{n} = \hat{i}$, $dS = dy\,dz$

$$\iint_{ABC} \overline{F}.\hat{n}\,dS = \iint (8y\,\hat{i} - y^2\hat{j} + 8z^2\,\hat{k}).\hat{i}\,dy\,dz$$

$$= \int_{z=0}^{3}\int_{y=0}^{\sqrt{9-z^2}} 8y\,dy\,dz = \int_{z=0}^{3} (4y^2)_0^{\sqrt{9-z^2}}\,dz$$

$$= 4\int_{z=0}^{3} (9 - z^2)\,dz = 4\left(9z - \frac{z^3}{3}\right)_0^3 = 4(27 - 9) = 72 \qquad ...(7)$$

Over the surface BDEC,

$$\hat{n} = \frac{\text{grad}\,\phi}{|\text{grad}\,\phi|} \quad \text{where } \phi \equiv y^2 + z^2 - 9 = 0$$

$$\therefore \quad \hat{n} = \frac{2(y\hat{j} + z\hat{k})}{2\sqrt{y^2 + z^2}} = \frac{y\hat{j} + z\hat{k}}{3}$$

$$\iint_{BDEC} \overline{F}.\hat{n}\,dS = \iint_{BDEC} (2x^2y\hat{i} - y^2\hat{j} + 4xz^2\hat{k}).\left(\frac{y\hat{j} + z\hat{k}}{3}\right)\frac{dx\,dy}{(\hat{n}.\hat{k})}$$

$$= \iint_{BDEC} \frac{(-y^3 + 4xz^3)}{3}\frac{dx\,dy}{\left(\frac{z}{3}\right)} = \int_{x=0}^{2}\int_{y=0}^{3}\left(-\frac{y^3}{z} + 4xz^2\right)dx\,dy$$

$$= \int_{x=0}^{2}\int_{y=0}^{\pi/2}\left[\frac{-27\sin^3\theta}{3\cos\theta} + 4x(9\cos^2\theta)\right] 3\cos\theta\,d\theta\,dx$$

| Putting $y = 3\sin\theta$ and $z = 3\cos\theta$

$$= \int_{x=0}^{2}\int_{y=0}^{\pi/2} (-27\sin^3\theta + 108x\cos^3\theta)\,d\theta\,dx$$

$$= \int_{x=0}^{2}\left(\frac{-54}{3} + 108x \times \frac{2}{3}\right)dx = \int_0^2 (-18 + 72x)\,dx$$

$$= (-18x + 36x^2)_0^2 = 108 \qquad ...(8)$$

From eqn. (3),

$$\iint_S \vec{F}.\hat{n}\,dS = 0 + 0 + 0 + 72 + 108 = 180 \qquad ...(9)$$

From results (2) and (8), it is clear that

$$\iint_S \vec{F}.\hat{n}\,dS = \iiint_V \text{div}\,\vec{F}\,dV$$

Hence, Gauss' divergence theorem is verified.

Example 16. *Evaluate* $\iint_S (a^2x^2 + b^2y^2 + c^2z^2)^{-1/2}\, d\overline{S}$ *where S is the surface of the ellipsoid* $ax^2 + by^2 + cz^2 = 1$. (M.T.U. 2012)

Sol. $\overline{F} \cdot \hat{n} = (a^2x^2 + b^2y^2 + c^2z^2)^{-1/2}$ ...(1) | given

Let $\phi \equiv ax^2 + by^2 + cz^2 - 1 = 0$

$$\text{grad } \phi = 2ax\,\hat{i} + 2by\,\hat{j} + 2cz\,\hat{k}$$

$$|\text{grad } \phi| = 2\sqrt{a^2x^2 + b^2y^2 + c^2z^2}$$

$$\therefore \quad \hat{n} = \frac{\text{grad } \phi}{|\text{grad } \phi|} = \frac{ax\,\hat{i} + by\,\hat{j} + cz\,\hat{k}}{\sqrt{a^2x^2 + b^2y^2 + c^2z^2}} \qquad ...(2)$$

From (1) and (2), it is clear that

$$\overline{F} = x\,\hat{i} + y\,\hat{j} + z\,\hat{k} \qquad | \text{ Since } ax^2 + by^2 + cz^2 = 1$$

$$\therefore \quad \text{div } \overline{F} = \frac{\partial}{\partial x}(x) + \frac{\partial}{\partial y}(y) + \frac{\partial}{\partial z}(z) = 1 + 1 + 1 = 3$$

By Divergence theorem,

$$\iint_S \overline{F} \cdot d\overline{S} = \iiint_V \text{div } \overline{F}\, dV = 3 \iiint_V dV = 3V$$

$$= 3\,\frac{4\pi}{3}\,\frac{1}{\sqrt{a}} \cdot \frac{1}{\sqrt{b}} \cdot \frac{1}{\sqrt{c}} \qquad \text{Volume of } (ax^2 + by^2 + cz^2 = 1) \text{ is } = \frac{4\pi}{3} \cdot \frac{1}{\sqrt{abc}}$$

$$= \frac{4\pi}{\sqrt{abc}}.$$

TEST YOUR KNOWLEDGE

1. (*i*) Using the divergence theorem, evaluate the surface integral $\iint_S (yz\, dy\, dz + zx\, dz\, dx + xy\, dx\, dy)$ where S : $x^2 + y^2 + z^2 = 4$.

   (*ii*) Evaluate $\int_S (yz\hat{i} + zx\hat{j} + xy\hat{k}) \cdot d\vec{S}$ where S is the surface of the sphere $x^2 + y^2 + z^2 = a^2$ in the first octant. (*U.P.T.U. 2015*)

2. If S is any closed surface enclosing a volume V and $\vec{F} = x\hat{i} + 2y\hat{j} + 3z\hat{k}$, prove that $\iint_S \vec{F} \cdot \hat{n}\, dS = 6V$.

3. If $\vec{F} = ax\hat{i} + by\hat{j} + cz\hat{k}$, *a*, *b*, *c* are constants, show that $\iint_S \vec{F} \cdot d\vec{S} = \frac{4}{3}\pi\,(a + b + c)$, where S is the surface of a unit sphere.

4. Use divergence theorem to evaluate $\iint_S \vec{F} \cdot d\vec{S}$, where $\vec{F} = x^3\hat{i} + y^3\hat{j} + z^3\hat{k}$ and S is the surface of the sphere $x^2 + y^2 + z^2 = a^2$.

5. Use divergence theorem to evaluate $\iint_S \vec{F} \cdot \hat{n}\, d\vec{S}$, where $\vec{F} = x^3\hat{i} + x^2y\hat{j} + x^2z\hat{k}$ and S is the surface bounding the region $x^2 + y^2 = a^2$, $z = 0$, $z = b$.

6. Evaluate $\iint_S (x\hat{i} + y\hat{j} + z^2\hat{k}) \cdot \hat{n}\; dS$, where S is the closed surface bounded by the cone $x^2 + y^2 = z^2$ and the plane $z = 1$.

7. Use divergence theorem to evaluate $\iint_S \vec{r} \cdot \hat{n}\; dS$ where S is the surface of the sphere $x^2 + y^2 + z^2 = 9$.

8. Verify divergence theorem for $\vec{F} = x^2\hat{i} + z\hat{j} + yz\hat{k}$ taken over the cube bounded by $x = 0$, $x = 1$; $y = 0$, $y = 1$; $z = 0$, $z = 1$. *(G.B.T.U. 2013)*

9. Verify divergence theorem for $\vec{F} = (x^3 - yz)\hat{i} - 2x^2y\hat{j} + 2\hat{k}$ taken over the cube bounded by the planes $x = 0$, $x = a$, $y = 0$, $y = a$, $z = 0$, $z = a$. *(A.K.T.U. 2017)*

10. Verify divergence theorem for $\vec{F} = y\hat{i} + x\hat{j} + z^2\hat{k}$ over the cylindrical region bounded by $x^2 + y^2 = a^2$, $z = 0$ and $z = h$.

11. Verify Gauss' divergence theorem for $\vec{F} = (x^3 - yz)\,\hat{i} + (y^3 - zx)\,\hat{j} + (z^3 - xy)\,\hat{k}$ taken over the cube bounded by the planes $x = 0$, $x = 1$, $y = 0$, $y = 1$ and $z = 0$, $z = 1$. *(A.K.T.U. 2019)*

12. Verify the divergence theorem for $\vec{F} = (2x - z)\,\hat{i} + x^2y\,\hat{j} - xz^2\hat{k}$ taken over the region bounded by $x = 0$, $x = 1$, $y = 0$, $y = 1$, $z = 0$, $z = 1$.

13. Evaluate $\iint_S xz^2\, dy\, dz + (x^2y - z^3)\, dz\, dx + (2xy + y^2z)\, dx\, dy$ where S is the surface of hemispherical region bounded by $z = \sqrt{a^2 - x^2 - y^2}$ and $z = 0$.

14. Verify divergence theorem for $\vec{F} = x^3\hat{i} - y^3\hat{j} + z^3\hat{k}$ taken over the surface of sphere $x^2 + y^2 + z^2 = a^2$. *(U.P.T.U. 2015)*

Answers

1. (*i*) 0	(*ii*) 0	**4.** $\frac{12}{5}\pi a^5$	**5.** $\frac{5}{4}\pi a^4 b$
6. 3π	**7.** 108π	**13.** $\frac{2}{5}\pi a^5$	

5.29. STOKE'S THEOREM (*Relation between line and surface integrals*) (U.K.T.U. 2012)

If S is an open surface bounded by a closed curve C and $\vec{F} = F_1\hat{i} + F_2\hat{j} + F_3\hat{k}$ is any vector point function having continuous first order partial derivatives, then

$$\oint_C \vec{F} \cdot d\vec{r} = \iint_S \text{curl } \vec{F} \cdot \hat{n}\; dS$$

where $\hat{n}$ is a unit normal vector at any point of S drawn in the sense in which a right handed screw would advance when rotated in the sense of description of C.

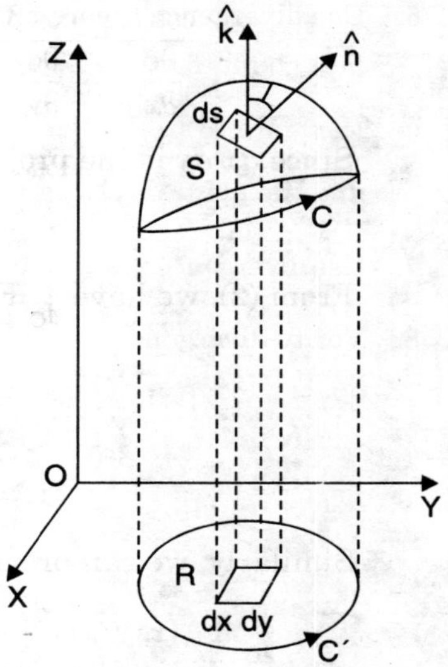

Proof. Let $\hat{n}$ make angles α, β, γ with positive directions of *x*, *y*, *z* axes, then

$$\hat{n} = \cos\alpha\hat{i} + \cos\beta\hat{j} + \cos\gamma\hat{k}.$$

Also $\qquad \vec{r} = x\hat{i} + y\hat{j} + z\hat{k}$

so that $\qquad d\vec{r} = \hat{i}dx + \hat{j}dy + \hat{k}\,dz$

$$\text{curl } \vec{F} = \nabla \times \vec{F} = \begin{vmatrix} \hat{i} & \hat{j} & \hat{k} \\ \frac{\partial}{\partial x} & \frac{\partial}{\partial y} & \frac{\partial}{\partial z} \\ F_1 & F_2 & F_3 \end{vmatrix}$$

$$= \left(\frac{\partial F_3}{\partial y} - \frac{\partial F_2}{\partial z}\right)\hat{i} + \left(\frac{\partial F_1}{\partial z} - \frac{\partial F_3}{\partial x}\right)\hat{j} + \left(\frac{\partial F_2}{\partial x} - \frac{\partial F_1}{\partial y}\right)\hat{k}$$

$$\therefore \quad \text{curl } \vec{F}\,.\,\hat{n} = \left(\frac{\partial F_3}{\partial y} - \frac{\partial F_2}{\partial z}\right)\cos\alpha + \left(\frac{\partial F_1}{\partial z} - \frac{\partial F_3}{\partial x}\right)\cos\beta + \left(\frac{\partial F_2}{\partial x} - \frac{\partial F_1}{\partial y}\right)\cos\gamma$$

Also $\quad \vec{F}\,.\,d\vec{r} = (F_1\hat{i} + F_2\hat{j} + F_3\hat{k})\,.\,(\hat{i}dx + \hat{j}dy + \hat{k}dz) = F_1dx + F_2dy + F_3dz$

∴ Stoke's theorem can be written as

$$\oint_C (F_1\,dx + F_2\,dy + F_3\,dz)$$

$$= \iint_S \left[\left(\frac{\partial F_3}{\partial y} - \frac{\partial F_2}{\partial z}\right)\cos\alpha + \left(\frac{\partial F_1}{\partial z} - \frac{\partial F_3}{\partial x}\right)\cos\beta + \left(\frac{\partial F_2}{\partial x} - \frac{\partial F_1}{\partial y}\right)\cos\gamma\right] dS \quad \text{...(1)}$$

Let $z = \phi(x, y)$ be the equation of the surface S whose projection on the *xy*-plane is R. Then the projection of C on the *xy*-plane is the curve C′ which bounds the region R.

$$\therefore \quad \oint_C F_1(x, y, z)\,dx = \oint_{C'} F_1(x, y, \phi)\,dx = \oint_{C'} [F_1(x, y, \phi)\,dx + 0\,dy]$$

$$= -\iint_R \frac{\partial}{\partial y} F_1(x, y, \phi)\,dx\,dy \qquad \text{by Green's theorem in plane for the region R}$$

$$= -\int_R \left[\frac{\partial F_1}{\partial y} + \frac{\partial F_1}{\partial z}\frac{\partial \phi}{\partial y}\right] dx\,dy \quad \text{...(2)}$$

Now the direction ratios of the normal $\hat{n}$ to the surface $z = \phi(x, y)$ are $\frac{\partial \phi}{\partial x}, \frac{\partial \phi}{\partial y}, -1$

$$\Rightarrow \quad \frac{\cos\alpha}{\frac{\partial\phi}{\partial x}} = \frac{\cos\beta}{\frac{\partial\phi}{\partial y}} = \frac{\cos\gamma}{-1} \quad \Rightarrow \quad \frac{\partial\phi}{\partial y} = \frac{\cos\beta}{\cos\gamma}$$

Since $dx\, dy$ is the projection of dS on the xy-plane

$$\therefore \quad dx\, dy = \cos\gamma\, d\text{S}$$

From (2), we have $\oint_C \text{F}_1(x, y, z)\, dx = -\iint_S \left[\frac{\partial \text{F}_1}{\partial y} - \frac{\partial \text{F}_1}{\partial z} \cdot \frac{\cos\beta}{\cos\gamma}\right] \cos\gamma\, d\text{S}$

$$= -\iint_S \left[\frac{\partial \text{F}_1}{\partial y}\cos\gamma - \frac{\partial \text{F}_1}{\partial z}\cos\beta\right] d\text{S}$$

$$= \iint_S \left[\frac{\partial \text{F}_1}{\partial z}\cos\beta - \frac{\partial \text{F}_1}{\partial y}\cos\gamma\right] d\text{S} \quad \ldots(3)$$

Similarly, we can prove that

$$\oint_C \text{F}_2(x, y, z)\, dy = \iint_S \left[\frac{\partial \text{F}_2}{\partial x}\cos\gamma - \frac{\partial \text{F}_2}{\partial z}\cos\alpha\right] d\text{S} \quad \ldots(4)$$

and
$$\oint_C \text{F}_3(x, y, z)\, dz = \iint_S \left[\frac{\partial \text{F}_3}{\partial y}\cos\alpha - \frac{\partial \text{F}_3}{\partial x}\cos\beta\right] d\text{S} \quad \ldots(5)$$

Adding (3), (4) and (5), we get (1). Hence the theorem is proved. In words, Stoke's theorem states that "The line integral of the tangential component of a vector point function $\vec{F}$ taken around a closed curve C is equal to the surface integral of the normal component of $\vec{F}$ taken over any surface S having C as its boundary".

ILLUSTRATIVE EXAMPLES

Example 1. *Verify Stoke's theorem for $\vec{F} = (x^2 + y^2)\,\hat{i} - 2xy\hat{j}$ taken round the rectangle bounded by the lines $x = \pm a$, $y = 0$, $y = b$.* (A.K.T.U. 2015, 2018)

Sol. Let C denote the boundary of the rectangle ABED, then

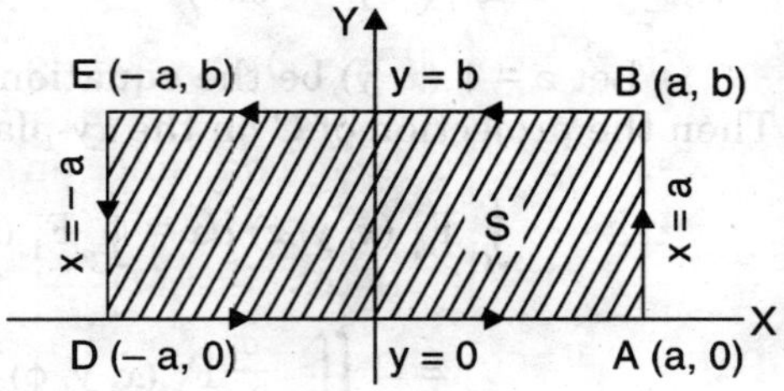

$$\oint_C \vec{F} \cdot d\vec{r} = \oint_C [(x^2 + y^2)\,\hat{i} - 2xy\hat{j}]\,(\hat{i}\,dx + \hat{j}\,dy)$$

$$= \oint_C [(x^2 + y^2)\, dx - 2xy\, dy]$$

The curve C consists of four lines AB, BE, ED and DA.

Along AB, $x = a$, $dx = 0$ and y varies from 0 to b.

$$\therefore \quad \int_{AB} [(x^2 + y^2)\, dx - 2xy\, dy] = \int_0^b -2ay\, dy = -a\left[y^2\right]_0^b = -ab^2 \quad \ldots(1)$$

Along BE, $y = b$, $dy = 0$ and x varies from a to $-a$.

$$\therefore \quad \int_{BE} [(x^2 + y^2)\, dx - 2xy\, dy] = \int_a^{-a} (x^2 + b^2)\, dx = \left[\frac{x^3}{3} + b^2 x\right]_a^{-a} = -\frac{2a^3}{3} - 2ab^2 \quad \ldots(2)$$

Along ED, $x = -a$, $dx = 0$ and y varies from b to 0.

$$\therefore \quad \int_{ED} [(x^2 + y^2)\, dx - 2xy\, dy] = \int_b^0 2ay\, dy = a\left[y^2\right]_b^0 = -ab^2 \qquad ...(3)$$

Along DA, $y = 0$, $dy = 0$ and x varies from $-a$ to a.

$$\therefore \quad \int_{DA} [(x^2 + y^2)\, dx - 2dx\, dy] = \int_{-a}^{a} x^2\, dx = \frac{2a^3}{3} \qquad ...(4)$$

Adding (1), (2), (3) and (4), we get

$$\oint_C \vec{F}\,.\,d\vec{r} = -ab^2 - \frac{2a^3}{3} - 2ab^2 - ab^2 + \frac{2a^3}{3} = -4ab^2 \qquad ...(5)$$

Now
$$\text{curl } \vec{F} = \begin{vmatrix} \hat{i} & \hat{j} & \hat{k} \\ \frac{\partial}{\partial x} & \frac{\partial}{\partial y} & \frac{\partial}{\partial z} \\ x^2 + y^2 & -2xy & 0 \end{vmatrix} = (-2y - 2y)\,\hat{k} = -4y\hat{k}$$

For the surface S, $\hat{n} = \hat{k}$

$$\text{curl } \vec{F}\,.\,\hat{n} = -4y\hat{k}\,.\,\hat{k} = -4y$$

$$\therefore \quad \iint_S \text{curl } \vec{F}\,.\,\hat{n}\, dS = \int_0^b \int_{-a}^{a} -4y\, dx\, dy = \int_0^b -4y\,[x]_{-a}^{a}\, dy$$

$$= -8a\int_0^b y\, dy = -8a\left[\frac{y^2}{2}\right]_0^b = -4ab^2 \qquad ...(6)$$

The equality of (5) and (6) verifies Stoke's theorem.

Example 2. *Verify Stoke's theorem for the vector field* $\vec{F} = (2x - y)\,\hat{i} - yz^2\,\hat{j} - y^2z\hat{k}$ *over the upper half surface of* $x^2 + y^2 + z^2 = 1$*, bounded by its projection on the xy-plane.*

Sol. Let S be the upper half surface of the sphere $x^2 + y^2 + z^2 = 1$. The boundary C of S is a circle in the xy-plane of radius unity and centre O. The equations of C are $x^2 + y^2 = 1$, $z = 0$ whose parametric form is $x = \cos t$, $y = \sin t$, $z = 0$, $0 \le t < 2\pi$.

$$\oint_C \vec{F}\,.\,d\vec{r} = \oint_C [2x - y)\,\hat{i} - yz^2\,\hat{j} - y^2z\hat{k}]\,.\,(\hat{i}\,dx + \hat{j}dy + \hat{k}dz)$$

$$= \oint_C [(2x - y)\, dx - yz^2\, dy - y^2z\, dz]$$

$$= \oint_C (2x - y)\, dx, \text{ since on C, } z = 0,\ dz = 0$$

$$= \int_0^{2\pi} (2\cos t - \sin t)\frac{dx}{dt}\, dt$$

$$= \int_0^{2\pi} (2\cos t - \sin t)(-\sin t)\, dt$$

$$= \int_0^{2\pi} (-\sin 2t + \sin^2 t)\, dt$$

$$= \int_0^{2\pi} \left(-\sin 2t + \frac{1 - \cos 2t}{2}\right) dt$$

$$= \left[\frac{\cos 2t}{2} + \frac{t}{2} - \frac{\sin 2t}{4}\right]_0^{2\pi} = \frac{1}{2} + \pi - \frac{1}{2} = \pi$$

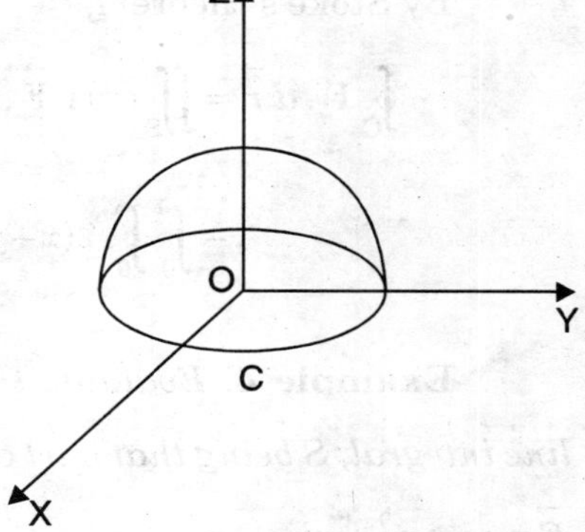

Also $\text{curl } \vec{F} = \begin{vmatrix} \hat{i} & \hat{j} & \hat{k} \\ \frac{\partial}{\partial x} & \frac{\partial}{\partial y} & \frac{\partial}{\partial z} \\ 2x - y & -yz^2 & -y^2 z \end{vmatrix} = (-2yz + 2yz)\hat{i} + (0 - 0)\hat{j} + (0 + 1)\hat{k} = \hat{k}$

$\text{curl } \vec{F} \cdot \hat{n} = \hat{k} \cdot \hat{n} = \hat{n} \cdot \hat{k}$

$$\iint_S \text{curl} \vec{F} \cdot \hat{n} \, dS = \iint_S \hat{n} \cdot \hat{k} \, dS = \iint_R \hat{n} \cdot \hat{k} \frac{dx \, dy}{|\hat{n} \cdot \hat{k}|}$$

where R is the projection of S on xy-plane.

$$= \iint_R dx \, dy = \int_{-1}^{1} \int_{-\sqrt{1-x^2}}^{\sqrt{1-x^2}} dy \, dx = \int_{-1}^{1} 2\sqrt{1-x^2} \, dx = 4 \int_0^1 \sqrt{1-x^2} \, dx$$

$$= 4 \left[\frac{x\sqrt{1-x^2}}{2} + \frac{1}{2} \sin^{-1} x \right]_0^1 = 4 \left[\frac{1}{2} \cdot \frac{\pi}{2} \right] = \pi$$

Since $\oint_C \vec{F} \cdot d\vec{r} = \iint_S \text{curl } \vec{F} \cdot \hat{n} \, dS$, Stoke's theorem is verified.

Example 3. *Evaluate* $\oint_C \vec{F} \cdot d\vec{r}$ *by Stoke's theorem, where* $\vec{F} = y^2 \hat{i} + x^2 \hat{j} - (x + z)\hat{k}$ *and C is the boundary of the triangle with vertices at (0, 0, 0), (1, 0, 0) and (1, 1, 0).*

(U.P.T.U. 2014; U.K.T.U. 2012)

Sol. Since z-coordinates of each vertex of the triangle is zero, therefore, the triangle lies in the xy-plane and $\hat{n} = \hat{k}$

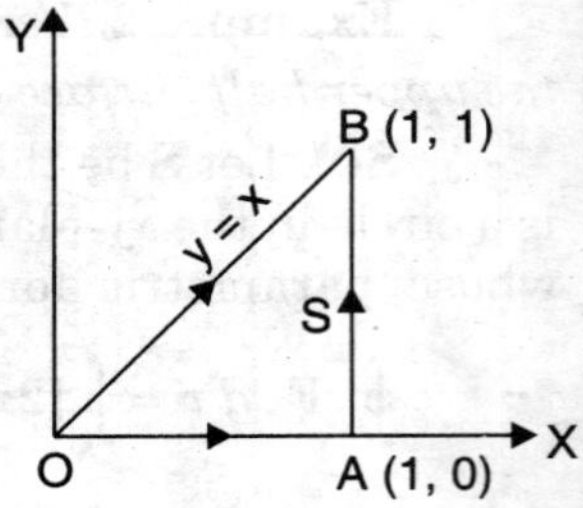

$$\text{curl } \vec{F} = \begin{vmatrix} \hat{i} & \hat{j} & \hat{k} \\ \frac{\partial}{\partial x} & \frac{\partial}{\partial y} & \frac{\partial}{\partial z} \\ y^2 & x^2 & -(x+z) \end{vmatrix} = \hat{j} + 2(x - y)\hat{k}$$

$\therefore \quad \text{curl } \vec{F} \cdot \hat{n} = [\hat{j} + 2(x - y)\hat{k}] \cdot \hat{k} = 2(x - y)$

The equation of line OB is $y = x$.

By Stoke's theorem,

$$\oint_C \vec{F} \cdot d\vec{r} = \iint_S \text{curl } \vec{F} \cdot \hat{n} \, dS$$

$$= \int_0^1 \int_0^x 2(x-y) \, dy \, dx = \int_0^1 2 \left[xy - \frac{y^2}{2} \right]_0^x dx = 2 \int_0^1 \left(x^2 - \frac{x^2}{2} \right) dx = \int_0^1 x^2 \, dx = \frac{1}{3}.$$

Example 4. *Evaluate the surface integral* $\iint_S \text{curl } \vec{F} \cdot \hat{n} \, dS$ *by transforming it into a line integral, S being that part of the surface of the paraboloid* $z = 1 - x^2 - y^2$*, for which* $z \geq 0$ *and* $\vec{F} = y\hat{i} + z\hat{j} + x\hat{k}$.

Sol. The boundary C of the surface S is the circle $x^2 + y^2 = 1$, $z = 0$ whose parametric equations are $x = \cos t$, $y = \sin t$, $z = 0$, $0 \leq t < 2\pi$.

By Stoke's theorem, we have

$$\iint_S \text{curl } \vec{F} \cdot \hat{n}\, dS = \oint_C \vec{F} \cdot d\vec{r} = \oint_C (y\hat{i} + z\hat{j} + x\hat{k})\,(\hat{i}\,dx + \hat{j}dy + \hat{k}dz)$$

$$= \oint_C (y\,dx + z\,dy + x\,dz) = \oint_C y\,dx, \text{ since on C, } z = 0,\ dz = 0$$

$$= \oint_C y\frac{dx}{dt}\,dt = \int_0^{2\pi} \sin t\,(-\sin t)\,dt = -\int_0^{2\pi} \sin^2 t\,dt$$

$$= -4\int_0^{\pi/2} \sin^2 t\,dt = -4 \times \frac{1}{2} \times \frac{\pi}{2} = -\pi.$$

Example 5. *Evaluate* $\iint_S \text{curl } \vec{F} \cdot d\vec{S}$, *where* $\vec{F} = (y^2 + z^2 - x^2)\,\hat{i} + (z^2 + x^2 - y^2)\,\hat{j} + (x^2 + y^2 - z^2)\,\hat{k}$ *and S is the portion of the surface* $x^2 + y^2 - 2ax + az = 0$ *above that plane* $z = 0$.

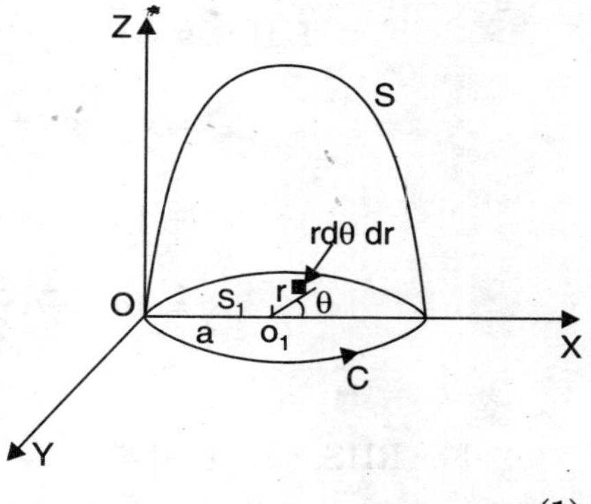

Sol. By Stoke's theorem, the surface integral over the surface S of the paraboloid is the same as the surface integral over S_1, the plane region S of the circles as both S and S_1 have the same boundary C.

The equation of given surface can be written as

$$(x - a)^2 + y^2 = -a(z - a) \qquad \text{...(1)}$$

This shows that the given surface is a paraboloid of revolution with its vertes at $(a, 0, a)$. It meets the plane $z = 0$ in the circle C given by

$$x^2 + y^2 - 2ax = 0, \quad z = 0$$

$$\Rightarrow \qquad (x - a)^2 + y^2 = a^2,\ z = 0 \qquad \text{...(2)}$$

Now,

$$\text{curl } \vec{F} = \begin{vmatrix} \hat{i} & \hat{j} & \hat{k} \\ \dfrac{\partial}{\partial x} & \dfrac{\partial}{\partial y} & \dfrac{\partial}{\partial z} \\ y^2 + z^2 - x^2 & z^2 + x^2 - y^2 & x^2 + y^2 - z^2 \end{vmatrix}$$

$$= 2(y - z)\hat{i} + 2(z - x)\hat{j} + 2(x - y)\,\hat{k}$$

Also, $\hat{n}$ (for the circle S_1) = k

$\therefore$ The surface integral over S = $\iint_S \text{curl } \vec{F} \cdot \hat{n}\, dS_1 = \iint_{S_1} 2(x - y)\, dS_1$

Let $dS_1 = r\,d\theta\,dr$ be an element of this circle (1) where r, θ are the polar coordinates of the element w.r.t. the origin O_1, the centre of the circle then,

$$x = a + r\cos\theta \quad \text{and} \quad y = r\sin\theta$$

$$\therefore \qquad \iint_{S_1} 2(x - y)\, dS_1 = 2\int_{r=0}^{a}\int_{\theta=0}^{2\pi} (a + r\cos\theta - r\sin\theta).\, r\,d\theta\,dr$$

$$= 2\int_0^2 \{a\theta + r(\sin\theta + \cos\theta)\}_0^{2\pi}\,.\,r\,dr$$

$$= 2\,.\,2a\pi\int_0^a r\,dr = 4\,a\pi\left(\frac{r^2}{2}\right)_0^a = 2\pi a^3.$$

Example 6. *Verify Stoke's theorem for the vector $\vec{F} = z\hat{i} + x\hat{j} + y\hat{k}$ taken over the half of the sphere $x^2 + y^2 + z^2 = a^2$ lying above xy-plane.*

Sol. Let S be the upper half surface of sphere $x^2 + y^2 + z^2 = a^2$. The boundary C of S is a circle in *xy*-plane of radius a and centre O. The equations of C are $x^2 + y^2 = a^2$, $z = 0$ whose parametric form is $x = a \cos t$, $y = a \sin t$, $z = 0$; $0 \le t < 2\pi$.

Stoke's theorem states that

$$\oint_C \vec{F} \cdot d\vec{r} = \iint_S \text{curl } \vec{F} \cdot \hat{n}\, dS \quad \ldots(1)$$

$$\text{L.H.S.} = \oint_C \vec{F} \cdot d\vec{r} = \oint_C (z\hat{i} + x\hat{j} + y\hat{k}) \cdot (dx\,\hat{i} + dy\,\hat{j} + dz\,\hat{k})$$

$$= \oint_C (z\,dx + x\,dy + y\,dz) = \oint_C x\,dy \qquad | \because z = 0 \Rightarrow dz = 0 \text{ above } xy\text{-plane}$$

$$= \int_0^{2\pi} a \cos t \cdot a \cos t\, dt = a^2 \int_0^{2\pi} \cos^2 t\, dt = 4a^2 \int_0^{\pi/2} \cos^2 t\, dt$$

$$= 4a^2 \cdot \frac{1}{2} \cdot \frac{\pi}{2} = \pi a^2 \quad \ldots(2)$$

$$\text{For RHS, curl } \vec{F} = \begin{vmatrix} \hat{i} & \hat{j} & \hat{k} \\ \frac{\partial}{\partial x} & \frac{\partial}{\partial y} & \frac{\partial}{\partial z} \\ z & x & y \end{vmatrix} = \hat{i} + \hat{j} + \hat{k}$$

$$\nabla\phi = \left(\hat{i}\frac{\partial}{\partial x} + \hat{j}\frac{\partial}{\partial y} + \hat{k}\frac{\partial}{\partial z}\right)(x^2 + y^2 + z^2 - a^2) = 2x\,\hat{i} + 2y\,\hat{j} + 2z\hat{k}$$

$$\hat{n} = \frac{\nabla\phi}{|\nabla\phi|} = \frac{2x\hat{i} + 2y\hat{j} + 2z\hat{k}}{\sqrt{4x^2 + 4y^2 + 4z^2}} = \frac{x\hat{i} + y\hat{j} + z\hat{k}}{a}$$

$$\text{curl } \vec{F} \cdot \hat{n} = (\hat{i} + \hat{j} + \hat{k}) \cdot \left(\frac{x\hat{i} + y\hat{j} + z\hat{k}}{a}\right) = \frac{x + y + z}{a}$$

$$= \frac{a \sin\theta \cos\phi + a \sin\theta \sin\phi + a \cos\theta}{a} = \sin\theta\cos\phi + \sin\theta\sin\phi + \cos\theta$$

$$\therefore \qquad \text{RHS} = \iint_S \text{curl } \vec{F} \cdot \hat{n}\, dS$$

$$= \iint_S (\sin\theta\cos\phi + \sin\theta\sin\phi + \cos\theta)\, dS \qquad \text{where } dS = a^2 \sin\theta\, d\theta\, d\phi$$

$$= a^2 \int_{\theta=0}^{\pi/2} \int_{\phi=0}^{2\pi} (\sin\theta\cos\phi + \sin\theta\sin\phi + \cos\theta) \sin\theta\, d\phi\, d\theta$$

$$= a^2 \int_{\theta=0}^{\pi/2} \int_{\phi=0}^{2\pi} \{\sin^2\theta(\cos\phi + \sin\phi) + \sin\theta\cos\theta\}\, d\phi\, d\theta$$

$$= a^2 \int_{\theta=0}^{\pi/2} \left\{\sin^2\theta \Big(\sin\phi - \cos\phi\Big)_0^{2\pi} + \frac{\sin 2\theta}{2}\Big(\phi\Big)_0^{2\pi}\right\} d\theta$$

$$= \frac{a^2}{2} \int_{\theta=0}^{\pi/2} \sin 2\theta \, (2\pi) \, d\theta = \pi a^2 \left(\frac{-\cos 2\theta}{2} \right)_0^{\pi/2} = \pi a^2 \qquad ...(3)$$

Hence from (2) and (3),

$$\oint_c \vec{F} . d\vec{r} = \iint_S \text{curl } \vec{F} . \hat{n} \, ds$$

Hence Stoke's theorem is verified.

Example 7. *Evaluate* $\iint_S (\overline{\nabla} \times \overline{F}) . \hat{n} \, dS$ *where S is the surface* $x^2 + y^2 + z^2 = 4$ *above the xy-plane and* $\overline{F} = (x^2 + y - 4) \, i + 3xy \, j + (2xz + z^2) \, k$.

Sol. Stoke's theorem is

$$\int_C \overline{F} . d\overline{r} = \iint_S \text{curl } \overline{F} . \hat{n} \, dS$$

where C is the circle, $x^2 + y^2 = 4, z = 0$

Also, $\overline{F} . d\overline{r} = [(x^2 + y - 4) \, \hat{i} + 3xy \, \hat{j} + (2xz + z^2) \, \hat{k}] . (dx \, \hat{i} + dy \, \hat{j} + dz \, \hat{k})$

$= (x^2 + y - 4)dx + 3xy \, dy + (2xz + z^2) \, dz$

On the circle C, $z = 0, \quad \therefore \quad dz = 0$

Let $x = 2 \cos \phi, \quad \therefore \quad dx = -2 \sin \phi \, d\phi$

and $y = 2 \sin \phi, \quad \therefore \quad dy = 2 \cos \phi \, d\phi$

$$\therefore \quad \int_C \overline{F} . d\overline{r} = \int_C [x^2 + y - 4) \, dx + 3xy \, dy]$$

$$= \int_0^{2\pi} (4 \cos^2 \phi + 2 \sin \phi - 4) (- 2 \sin \phi \, d\phi) + 3(2 \cos \phi) (2 \sin \phi) (2 \cos \phi \, d\phi)]$$

$$= \int_0^{2\pi} (- 8 \cos^2 \phi \sin \phi - 4 \sin^2 \phi + 8 \sin^2 \phi + 24 \cos^2 \phi \sin \phi) \, d\phi$$

$$= \int_0^{2\pi} (16 \cos^2 \phi \sin \phi - 4 \sin^2 \phi + 8 \sin \phi) \, d\phi$$

$$= -4 \int_0^{2\pi} \sin^2 \phi \, d\phi = -4 \times 4 \int_0^{\pi/2} \sin^2 \phi \, d\phi = -16 \left(\frac{1}{2} \frac{\pi}{2} \right) = -4\pi$$

Example 8. *Apply Stoke's theorem to evaluate* $\int_C [(x + y) \, dx + (2x - z) \, dy + (y + z) \, dz]$, *where C is the boundary of the triangle with vertices (2, 0, 0); (0, 3, 0) and (0, 0, 6).*

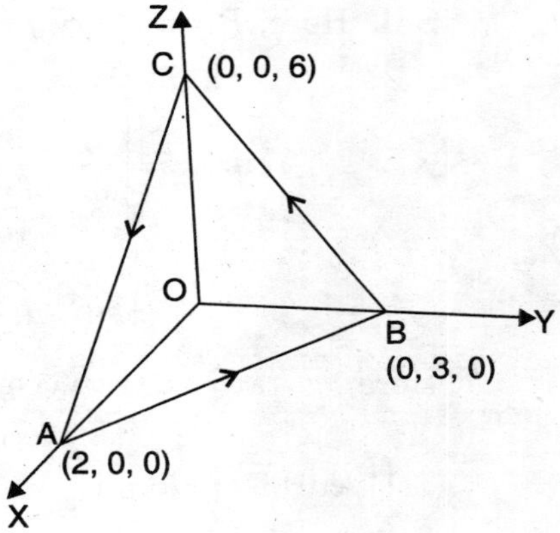

Sol. Let S be the plane surface of triangle ABC bounded by C. Let $\hat{n}$ be the unit normal vector to surface S. Then by Stoke's Theorem, we have

$$\oint_C \overline{F} . d\overline{r} = \iint_S \text{curl } \overline{F} . \hat{n} \, dS \qquad ...(1)$$

Here, $\overline{F} = (x + y)\hat{i} + (2x - z)\hat{j} + (y + z)\hat{k}$

$$\text{curl } \overline{\mathrm{F}} = \begin{vmatrix} \hat{i} & \hat{j} & \hat{k} \\ \dfrac{\partial}{\partial x} & \dfrac{\partial}{\partial y} & \dfrac{\partial}{\partial z} \\ x+y & 2x-z & y+z \end{vmatrix}$$

$$= \hat{i}\,(1+1) - \hat{j}(0-0) + \hat{k}\,(2-1) = 2\hat{i} + \hat{k}$$

Eqn. of the plane of ABC is $\dfrac{x}{2}+\dfrac{y}{3}+\dfrac{z}{6} = 1$

Let $\phi = \dfrac{x}{2}+\dfrac{y}{3}+\dfrac{z}{6} - 1 = 0$

Normal to the plane ABC is

$$\nabla\phi = \left(\hat{i}\,\frac{\partial}{\partial x} + \hat{j}\,\frac{\partial}{\partial y} + \hat{k}\,\frac{\partial}{\partial z}\right)\left(\frac{x}{2}+\frac{y}{3}+\frac{z}{6}-1\right) = \frac{\hat{i}}{2}+\frac{\hat{j}}{3}+\frac{\hat{k}}{6}$$

$$\text{Unit normal vector } (\hat{n}) = \frac{\dfrac{\hat{i}}{2}+\dfrac{\hat{j}}{3}+\dfrac{\hat{k}}{6}}{\sqrt{\dfrac{1}{4}+\dfrac{1}{9}+\dfrac{1}{36}}} = \frac{1}{\sqrt{14}}(3\hat{i}+2\hat{j}+\hat{k})$$

$$\therefore \quad \iint_{\mathrm{S}} \text{curl } \overline{\mathrm{F}} \,.\, \hat{n}\, d\mathrm{S} = \iint_{\mathrm{S}} (2\hat{i}+\hat{k}) \,.\, \frac{1}{\sqrt{14}}(3\hat{i}+2\hat{j}+\hat{k})\,\frac{dx\,dy}{(\hat{n}\,.\,\hat{k})}$$

$$= \iint_{\mathrm{S}} \frac{7}{\sqrt{14}}\,\frac{dx\,dy}{\left(\dfrac{1}{\sqrt{14}}\right)} = 7\iint_{\mathrm{S}} dx\,dy$$

$$= 7\,(\text{Area of } \Delta\, \text{AOB}) = 7\left(\frac{1}{2}\times 2\times 3\right) = 21 \qquad \text{...(2)}$$

$\therefore$ From (1) and (2),

$$\int_{\mathrm{C}} [(x+y)\,dx + (2x-z)\,dy + (y+z)\,dz] = 21.$$

Example 9. *Verify Stoke's theorem for the function $\overline{F} = x^2\hat{i} + xy\,\hat{j}$ integrated round the square whose sides are $x = 0$, $y = 0$, $x = a$, $y = a$ in the plane $z = 0$.* (A.K.T.U. 2021)

Sol. Here, $\overline{\mathrm{F}} = x^2\hat{i} + xy\,\hat{j}$

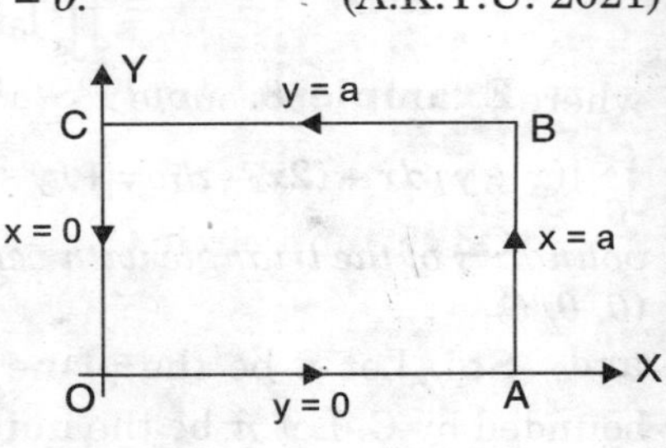

$$\therefore \quad \text{curl } \overline{\mathrm{F}} = \begin{vmatrix} \hat{i} & \hat{j} & \hat{k} \\ \dfrac{\partial}{\partial x} & \dfrac{\partial}{\partial y} & \dfrac{\partial}{\partial z} \\ x^2 & xy & 0 \end{vmatrix}$$

$$= \hat{i}\,(0-0) - \hat{j}(0-0) + \hat{k}\,(y) = y\hat{k}$$

Here, $\hat{n} = k$($\hat{n}$ is $\perp^r$ to xy-plane)

$$\iint_S \text{curl } \overline{\mathrm{F}} \,.\, \hat{n}\, d\mathrm{S} = \iint_{\mathrm{S}} yk \,.\, k\,dx\,dy = \int_0^a dx \int_0^a y\,dy = a\left(\frac{y^2}{2}\right)_0^a = \frac{a^3}{2} \qquad \text{...(1)}$$

Now, $\int_C \overline{F} \cdot d\overline{r} = \int_C (x^2\hat{i} + xy\,\hat{j}) \cdot (dx\,\hat{i} + dy\,\hat{j}) = \int_C (x^2\,dx + xy\,dy)$ ...(2)

where C is the path OABCO as shown above.

$$\int_{OABCO} \overline{F} \cdot d\overline{r} = \int_{OA} \overline{F} \cdot d\overline{r} + \int_{AB} \overline{F} \cdot d\overline{r} + \int_{BC} \overline{F} \cdot d\overline{r} + \int_{CO} \overline{F} \cdot d\overline{r} \quad ...(3)$$

Along OA, $y = 0$, $dy = 0$

$$\int_{OA} \overline{F} \cdot d\overline{r} = \int_0^a x^2\,dx = \left(\frac{x^3}{3}\right)_0^a = \frac{a^3}{3} \quad ...(4)$$

Along AB, $x = a$, $dx = 0$

$$\int_{AB} \overline{F} \cdot d\overline{r} = \int_0^a ay\,dy = a\left(\frac{y^2}{2}\right)_0^a = \frac{a}{2}(a^2) = \frac{a^3}{2} \quad ...(5)$$

Along BC, $y = a$, $dy = 0$

$$\int_{BC} \overline{F} \cdot d\overline{r} = \int_{BC} x^2\,dx = \int_a^0 x^2\,dx = \left(\frac{x^3}{3}\right)_a^0 = -\frac{a^3}{3} \quad ...(6)$$

Along CO, $x = 0$, $dx = 0$

$$\int_{CO} \overline{F} \cdot d\overline{r} = \int_{CO} x^2\,dx + xy\,dy = 0 \quad ...(7)$$

From (3), $\int_C \overline{F} \cdot d\overline{r} = \frac{a^3}{3} + \frac{a^3}{2} - \frac{a^3}{3} + 0 = \frac{a^3}{2}$ ...(8) | Using (4), (5), (6) and (7)

From eqns. (1) and (8), we have

$$\int_C \overline{F} \cdot d\overline{r} = \iint_S \text{curl}\,\overline{F} \cdot \hat{n}\,dS$$

Hence Stoke's theorem is verified.

Example 10. *Verify Stoke's theorem for $\overline{F} = y\hat{i} + z\hat{j} + x\hat{k}$ and surface S is the portion of the sphere $x^2 + y^2 + z^2 = 1$ above the xy-plane.* (U.P.T.U. 2015)

Sol. Stoke's theorem is

$$\oint_C \overline{F} \cdot d\overline{r} = \iint_S \text{curl}\,\overline{F} \cdot \hat{n}\,dS$$

where C is the unit circle $x^2 + y^2 = 1$, $z = 0$.

Also, $\overline{F} \cdot d\overline{r} = (y\hat{i} + z\hat{j} + x\hat{k}) \cdot (dx\,\hat{i} + dy\,\hat{j} + dz\,\hat{k}) = y\,dx + z\,dy + x\,dz$ ...(1)

On the unit circle C, $z = 0$ $\therefore$ $dz = 0$

Let $x = \cos\phi$ so that $dx = -\sin\phi\,d\phi$

and $y = \sin\phi$ so that $dy = \cos\phi\,d\phi$

$$\therefore \quad \oint_C \overline{F} \cdot d\overline{r} = \int_C y\,dx = \int_0^{2\pi} \sin\phi\,(-\sin\phi\,d\phi) = -\int_0^{2\pi} \sin^2\phi\,d\phi = -\pi \quad ...(2)$$

$$\text{Again,} \quad \text{curl}\,\overline{F} = \begin{vmatrix} \hat{i} & \hat{j} & \hat{k} \\ \dfrac{\partial}{\partial x} & \dfrac{\partial}{\partial y} & \dfrac{\partial}{\partial z} \\ y & z & x \end{vmatrix} = -\hat{i} - \hat{j} - \hat{k}$$

Using spherical polar coordinates,

$$\hat{n} = \sin\theta\cos\phi\,\hat{i} + \sin\theta\sin\phi\,\hat{j} + \cos\theta\,\hat{k}$$

$\therefore$ curl $\overline{\text{F}}\,.\,\hat{n} = -\sin\theta\cos\phi - \sin\theta\sin\phi - \cos\theta$

Hence,

$$\iint_S \text{curl}\,\overline{\text{F}}\,.\,\hat{n}\,dS = -\int_{\phi=0}^{\pi/2}\int_{\phi=0}^{2\pi} (\sin\theta\cos\phi + \sin\theta\sin\phi + \cos\theta)\sin\theta\,d\theta\,d\phi$$

$$= -\int_{\theta=0}^{\pi/2}\int_{\phi=0}^{\pi/2} [\sin\theta\sin\phi - \sin\theta\cos\phi + \phi\cos\theta]_0^{2\pi}\sin\theta\,d\theta$$

$$= -2\pi\int_{\theta=0}^{\pi/2}\sin\theta\cos\theta\,d\theta = -\pi\int_{\theta=0}^{\pi/2}\sin 2\theta\,d\theta$$

$$= \frac{\pi}{2}\left(\cos 2\theta\right)_0^{\pi/2} = -\pi \qquad \text{...(3)}$$

From (1) and (2), Stokes theorem is verified.

Example 11. *If* $\vec{F} = (x-z)\hat{i} + (x^3 + yz)\hat{j} - 3xy^2\,\hat{k}$ *and S is the surface of the curve* $z = a - \sqrt{x^2 + y^2}$ *above the xy-plane, show that* $\iint_S \textit{curl}\,\overline{F}\,.\,\overline{dS} = 3\pi a^4/4$.

Sol. $\vec{\text{F}} = (x-z)\hat{i} + (x^3 + yz)\hat{j} - 3xy^2\,\hat{k}$ ...(1) | given

By Stoke's theorem, we have

$$\iint_S \text{curl}\,\overline{\text{F}}\,.\,\vec{dS} = \oint_C \overline{\text{F}}\,.\,d\bar{r} \qquad \text{...(2)}$$

where S is the surface $x^2 + y^2 = a^2$, $z = 0$ above the xy-plane.

$$\overline{\text{F}}\,.\,d\bar{r} = \{(x-z)\hat{i} + (x^3 + yz)\hat{j} - 3xy^2\,\hat{k}\}\,.\,(dx\,\hat{i} + dy\,\hat{j} + dz\,\hat{k})$$

$$= (x-z)\,dx + (x^3 + yz)\,dy - 3xy^2\,dz \qquad \text{...(3)}$$

Let $x = a\cos\theta$ so that $dx = -a\sin\theta\,d\theta$

and $y = a\sin\theta$ so that $dy = a\cos\theta\,d\theta$

$$\oint_C \overline{\text{F}}\,.\,d\bar{r} = \int_0^{2\pi} (a\cos\theta)(-a\sin\theta\,d\theta) + a^3\cos^3\theta\,.\,a\cos\theta\,d\theta \qquad | \because z = 0 \quad \therefore \quad dz = 0$$

$$= -a^2\int_0^{2\pi}\sin\theta\cos\theta\,d\theta + a^4\int_0^{2\pi}\cos^4\theta\,d\theta$$

$$= a^4\int_0^{2\pi}\cos^4\theta\,d\theta = 4a^4\int_0^{\pi/2}\cos^4\theta\,d\theta = 4a^4\cdot\frac{3}{4}\cdot\frac{1}{2}\frac{\pi}{2} = \frac{3}{4}\pi a^4$$

$$\Rightarrow \iint_S \text{curl}\,\overline{\text{F}}\cdot\overline{dS} = \frac{3}{4}\pi a^4. \qquad | \text{ Using (2)}$$

Example 12. *If* $\overline{F} = 3y\,\hat{i} - xz\,\hat{j} + yz^2\,\hat{k}$ *and S is the surface of the paraboloid* $2z = x^2 + y^2$ *bounded by* $z = 2$*, show by using Stoke's theorem that* $\iint_S (\overline{\nabla}\times\overline{F})\,.\,\overline{dS} = -20\pi$.

Sol. $\overline{F} = 3y\,\hat{i} - xz\,\hat{j} + yz^2\,\hat{k}$ ...(1) | Given

$\therefore$ $\overline{F} \cdot d\overline{r} = 3y\,dx - xz\,dy + yz^2\,dz$ ...(2)

By Stoke's theorem, we know that

$$\iint_S \text{curl } \overline{F} \cdot \overline{dS} = \oint_C \overline{F} \cdot d\overline{r}$$

where S is the surface of the paraboloid $2z = x^2 + y^2$ bounded by $z = 2$.

Hence $S \equiv x^2 + y^2 = 4,\ z = 2$

$$\oint_C \overline{F} \cdot d\overline{r} = \oint_C (3y\,dx - xz\,dy + yz^2\,dz) \quad \text{...(3) | Using (2)}$$

Let $x = 2\cos\theta$ so that $dx = -2\sin\theta\,d\theta$

and $y = 2\sin\theta$ so that $dy = 2\cos\theta\,d\theta$

then from eqn. (3),

$$\oint_c \vec{F} \cdot d\vec{r} = \int_0^{2\pi} [3\,(2\sin\theta)(-2\sin\theta\,d\theta) - 2(2\cos\theta)(2\cos\theta\,d\theta)]$$

$| \because z = 2 \quad \therefore \quad dz = 0$

$$= \int_0^{2\pi} (-12\sin^2\theta - 8\cos^2\theta)d\theta$$

$$= -4\int_0^{\pi/2} (12\sin^2\theta + 8\cos^2\theta)d\theta$$

$$= -4\left[12 \cdot \frac{1}{2}\frac{\pi}{2} + 8 \cdot \frac{1}{2} \cdot \frac{\pi}{2}\right] = -4(3\pi + 2\pi) = -20\pi$$

TEST YOUR KNOWLEDGE

1. If $\vec{F} = (x^2 + y - 4)\hat{i} + 3xyz + (2xz + z^2)\hat{k}$, evaluate $\iint_S (\nabla + \vec{F}) \cdot \hat{n}\,dS$, where S is the surface of the sphere $x^2 + y^2 + z^2 = 16$ above the xy-plane.

2. Use Stoke's theorem to evaluate $\int_C [(x + 2y)\,dx + (x - z)dy + (y - z)dz]$ where c is the boundary of the triangle with vertices (2, 0, 0), (0, 3, 0) and (0, 0, 6) oriented in the anti-clockwise direction.

3. Evaluate by Stoke's theorem $\oint_C (yz\,dx + zx\,dy + xy\,dz)$, where C is the curve $x^2 + y^2 = 1,\ z = y^2$.

4. Use Stoke's theorem to evaluate $\iint_S (\nabla + \vec{F}) \cdot \hat{n}\,dS$, where $\vec{F} = y\hat{i} + (x - 2xz)\hat{j} - xy\hat{k}$ and S is the surface of the sphere $x^2 + y^2 + z^2 = a^2$ above the xy-plane.

5. Evaluate by Stoke's theorem $\oint_C (\sin z\,dx - \cos x\,dy + \sin y\,dz)$, where C is the boundary of the rectangle $0 \le x \le \pi,\ 0 \le y \le 1,\ z = 3$.

6. Evaluate $\int_C \vec{F} . d\vec{r}$ where $\vec{F} = y\hat{i} + xz^3\hat{j} - zy^3\hat{k}$, C is the circle $x^2 + y^2 = 4$, $z = 1.5$.

7. Apply Stoke's theorem to evaluate $\int_C (ydx + zdy + xdz)$, where C is the curve of intersection of $x^2 + y^2 + z^2 = a^2$ and $x + z = a$.

8. Prove by using Stoke's theorem that $\int_C (e^x\, dx + 2y\, dy - dz) = 0$, where C is the curve $x^2 + y^2 = 4$, $z = 2$.

9. Apply Stoke's theorem to prove that $\int_C (y\, dx + z\, dy + x\, dz) = -2\sqrt{2}\,\pi a^2$, where C is the curve given by $x^2 + y^2 + z^2 - 2ax - 2ay = 0$, $x + y = 2a$ and begins at the point $(2a, 0, 0)$ and goes at first below the xy-plane.

10. Verfiy Stoke's theorem for $\vec{F} = (y - z + 2)\hat{i} + (yz + 4)\hat{j} - xz\hat{k}$ over the surface of cube $x = 0$, $y = 0$, $z = 0$, $x = 2$, $y = 2$, $z = 2$ above the XOY plane (open the bottom). *(U.P.T.U. 2014)*

11. Verfiy Stoke's theorem for function $\vec{F} = z\hat{i} + x\hat{j} - y\hat{k}$, where C is a unit circle in xy-plane bounding the hemisphere $z = \sqrt{1 - x^2 - y^2}$.

12. Verfiy Stoke's theorem for function $\vec{F} = (x - y - z)\,\hat{i} + (y - z - x)\,\hat{j} + (z - x - y)\,\hat{k}$ over the unclosed surface of cylinder $\frac{x^2}{a^2} + \frac{y^2}{b^2} = 1$ bounded by the plane $z = h$ and open at the end $z = 0$.

13. Verify Stoke's theorem for $\vec{F}\,(x, y, z) = xz\hat{i} - y\hat{j} + x^2y\hat{k}$, where the surface S is the surface of the region bounded by $x = 0$, $y = 0$, $z = 0$, $2x + y + 2z = 8$ which is not included on xz-plane.

14. Verify Stoke's theorem by evaluating the line integral $\oint_C xe^z dx + ye^x\, dy + ze^y\, dz$ along the boundary of the triangle with vertices (0, 0, 1); (1, 0, 1); (1, 1, 1).

15. Verify Stoke's theorem for the vector field $\vec{F} = (x^2 - y^2)\,\hat{i} + 2xy\hat{j}$ integrated round the rectangle in the plane $z = 0$ and bounded by the lines $x = 0$, $y = 0$, $x = a$, $y = b$. *(A.K.T.U. 2020)*

16. Verify Stoke's theorem for $\vec{F} = xy^2\hat{i} + y\,\hat{j} + z^2x\hat{k}$ for the surface of a rectangular lamina bounded by $x = 0$, $y = 0$, $x = 1$, $y = 2$, $z = 0$.

17. Verify Stoke's theorem first by evaluating the $\oint_C (3x^2 - 6yz)\, dx + (2y + 3zx)\, dy + (1 - 4yz^2)\, dz$ from (0, 0, 0) to (1, 1, 1) along the path C given by the straight line from (1, 0, 0) to (1, 1, 0) then from (1, 1, 0) to (1, 1, 1) then from (1, 1, 1) to (1, 0, 0) in counter clockwise and secondly evaluating the integral as surface integral, the surface being enclosed by path c and the coordinate plane.

18. Verify Stoke's theorem for $\vec{F} = (2y + z, x - z, y - x)$ taken over the triangle ABC cut from the plane $x + y + z = 1$ by the coordinate planes. *(A.K.T.U. 2017)*

Answers

1. -16π **2.** 15 **3.** 0 **4.** 0

5. 2 **6.** $\dfrac{19\pi}{2}$ **7.** $-\dfrac{\pi a^2}{\sqrt{2}}$.

5.30. GREEN'S THEOREM

Statement. *If ϕ and ψ are two scalar point functions such that they and their derivatives in any direction are uniform and continuous in a certain region V bounded by a closed surface S, then*

$$\int_V (\phi\, \nabla^2 \psi - \psi\, \nabla^2\phi)\, dV = \int_S (\phi\, \nabla\, \psi - \psi\, \nabla\phi) \cdot \hat{n}\, dS$$

Proof. By Gauss' Divergence theorem, we have

$$\int_V \text{div}\, \vec{F}\, dV = \int_S \vec{F} \cdot \hat{n}\, dS$$

Put $\vec{F} = \phi\, \nabla \psi$, we have

$$\int_V \text{div}\, (\phi\, \nabla\, \psi)\, dV = \int_S \phi\, \nabla\, \psi \cdot \hat{n}\, dS$$

$$\Rightarrow \int_V \{\phi\, \text{div}\, \nabla\, \psi + (\text{grad}\, \phi) \cdot \nabla\, \psi\}\, dV = \int_S \phi\, \nabla\, \psi \cdot \hat{n}\, dS$$

$$\Rightarrow \int_V \{\phi\, (\nabla \cdot \nabla\psi) + \nabla\phi \cdot \nabla\, \psi\}\, dV = \int_S \phi\, \nabla\, \psi \cdot \hat{n}\, dS$$

$$\Rightarrow \int_V \{\phi\, \nabla^2\, \psi + \nabla\phi \cdot \nabla\psi\}\, dV = \int_S \phi\, \nabla\, \psi \cdot \hat{n}\, dS$$

$$\Rightarrow \int_V \phi\, \nabla^2\, \psi\, dV + \int_V \nabla\phi\, .\, \nabla\psi\, dV = \int_S \phi\, \nabla\, \psi\, .\, \hat{n}\, dS \quad ...(1)$$

Interchanging ϕ and ψ, we get

$$\int_V \psi\, \nabla^2\, \phi\, dV + \int_V \nabla\psi\, .\, \nabla\phi\, dV = \int_S \psi\, \nabla\, \phi\, .\, \hat{n}\, dS \quad ...(2)$$

(1) – (2) gives,

$$\boxed{\int_V (\phi\, \nabla^2\, \psi - \psi\, \nabla^2\, \phi)\, dV = \int_S (\phi\, \nabla\, \psi - \psi\, \nabla\, \phi)\, .\hat{n}\, dS}$$

Hence the theorem is proved.

5.31. GREEN'S THEOREM IN THE PLANE (*M.T.U. 2012*)

If C is a regular closed curve in the xy-plane and R be the region bounded by C, then

$$\int_C (M\, dx + N\, dy) = \iint_R \left(\frac{\partial N}{\partial x} - \frac{\partial M}{\partial y}\right) dx\, dy$$

where M (x, y) and N(x, y) are continuously differentiable functions inside and on C.

Proof. From Stoke's theorem, we have

$$\int_C \vec{F}.d\vec{r} = \iint_R \text{curl}\,\vec{F}.\hat{n}\,dS \quad ...(1)$$

Let $\vec{F} = M\hat{i} + N\hat{j}$ then

$$\text{curl}\,\vec{F} = \begin{vmatrix} \hat{i} & \hat{j} & \hat{k} \\ \frac{\partial}{\partial x} & \frac{\partial}{\partial y} & \frac{\partial}{\partial z} \\ M & N & 0 \end{vmatrix} = -\hat{i}\frac{\partial N}{\partial z} + \hat{j}\frac{\partial M}{\partial z} + \hat{k}\left(\frac{\partial N}{\partial x} - \frac{\partial M}{\partial y}\right),$$

Also as C is a closed curve in *xy*-plane

$$\therefore \quad \hat{n} = \hat{k}$$

$$\therefore \quad \text{curl}\,\vec{F}.\hat{n} = \frac{\partial N}{\partial x} - \frac{\partial M}{\partial y}$$

Also on *xy*-plane, we have $dS = dx\,dy$

$$\therefore \quad \iint_S \text{curl}\,\vec{F}.\hat{n}\,ds = \iint_R \left(\frac{\partial N}{\partial x} - \frac{\partial M}{\partial y}\right) dx\,dy \quad ...(2)$$

Also, $$\int_C \vec{F}.d\vec{r} = \int_C (M\hat{i} + N\hat{j}).(dx\hat{i} + dy\hat{j}) = \int_C M\,dx + N\,dy \quad ...(3)$$

Hence from (1), (2) and (3), we get

$$\boxed{\int_C M\,dx + N\,dy = \iint_R \left(\frac{\partial N}{\partial x} - \frac{\partial M}{\partial y}\right) dx\,dy.}$$

ILLUSTRATIVE EXAMPLES

Example 1. *By the use of Green's theorem, show that area bounded by a simple closed curve C is given by* $\frac{1}{2}\int (x\,dy - y\,dx)$. *Hence find the area of an ellipse.*

Sol. By Green's theorem in planes, we have

$$\int_C (M\,dx + N\,dy) = \iint_R \left(\frac{\partial N}{\partial x} - \frac{\partial M}{\partial y}\right) dx\,dy \quad ...(1)$$

Put $M = -y$ and $N = x$ so that

$$\frac{\partial M}{\partial y} = -1 \text{ and } \frac{\partial N}{\partial x} = 1$$

Hence from (1),

$$\int_C (-y\,dx + x\,dy) = \iint_R (1+1)\,dx\,dy = 2\iint_R dx\,dy = 2A$$

where A is the required area

$$\therefore \qquad A = \frac{1}{2}\int_C (x\,dy - y\,dx)$$

Any point (x, y) on the ellipse is given by

$x = a\cos\phi,\quad y = b\sin\phi$ where ϕ is the parameter.

$$\text{Hence area of ellipse } = \frac{1}{2}\int_0^{2\pi} (a\cos\phi)(b\cos\phi\,d\phi) - (b\sin\phi)(-a\sin\phi)\,d\phi$$

$$= \frac{1}{2}ab\int_0^{2\pi} (\cos^2\phi + \sin^2\phi)\,d\phi = \frac{1}{2}ab\,(2\pi) = \pi ab.$$

Example 2. *If C is a simple closed curve in the xy-plane not containing the origin, evaluate* $\int_C \vec{F}\cdot d\vec{r}$ *where* $\vec{F} = \dfrac{-iy + jx}{x^2+y^2}$.

Sol. Let $\qquad \vec{F} = M\hat{i} + N\hat{j}$ then

$$\vec{F}\cdot d\vec{r} = (M\hat{i} + N\hat{j})\cdot(dx\hat{i} + dy\hat{j}) = M dx + N dy$$

For given $\vec{F}$, we have

$$M = \frac{-y}{x^2+y^2},\ N = \frac{x}{x^2+y^2}$$

$$\therefore \qquad \frac{\partial M}{\partial y} = \frac{(x^2+y^2)(-1) - (-y)(2y)}{(x^2+y^2)^2} = \frac{y^2-x^2}{(x^2+y^2)^2}$$

and

$$\frac{\partial N}{\partial x} = \frac{(x^2+y^2).1 - x\,.\,2x}{(x^2+y^2)^2} = \frac{y^2-x^2}{(x^2+y^2)^2}$$

$$\Rightarrow \qquad \frac{\partial N}{\partial x} - \frac{\partial M}{\partial y} = 0.$$

Hence from Green's theorem, we have

$$\int_C \vec{F}\cdot d\vec{r} = \int_C (M\,dx + N\,dy) = \iint_R \left(\frac{\partial N}{\partial x} - \frac{\partial M}{\partial y}\right) dx\,dy = 0.$$

Example 3. *A vector field* $\vec{F}$ *is given by* $\vec{F} = \sin y\,\hat{i} + x(1+\cos y)\,\hat{j}$

Evaluate the integral $\int_C \vec{F}\cdot d\vec{r}$ *where C is the circular path given by* $x^2 + y^2 = a^2$.

Sol.

$$\int_C \vec{F}\cdot d\vec{r} = \int_C \{\sin y\hat{i} + x(1+\cos y)\hat{j}\}\cdot(dx\,\hat{i} + dy\,\hat{j})$$

$$= \int_C \{\sin y\,dx + x(1+\cos y)\,dy\}$$

$$= \iint_R \left[\frac{\partial}{\partial x}\{x(1+\cos y)\} - \frac{\partial}{\partial y}(\sin y)\right] dx\,dy$$

| Using Green's theorem

$$= \iint_R (1 + \cos y - \cos y)\, dx\, dy = \iint_R dx\, dy$$

$$= A = \text{area of circle} = \pi\, (\text{radius})^2 = \pi a^2.$$

Example 4. *Apply Green's theorem to evaluate $\int_C [(2x^2 - y^2)\, dx + (x^2 + y^2)\, dy]$ where C is the boundary of the area enclosed by the x-axis and the upper half of circle $x^2 + y^2 = a^2$.*

(A.K.T.U. 2022)

Sol. Let $\vec{F} = M\hat{i} + N\hat{j}$ then

$$\vec{F} . d\vec{r} = (M\hat{i} + N\hat{j}) . (dx\,\hat{i} + dy\,\hat{j}) = M\, dx + N\, dy$$

For given $\vec{F}$, we have

$$M = 2x^2 - y^2 \quad \text{and} \quad N = x^2 + y^2$$

$$\therefore \quad \frac{\partial M}{\partial y} = -2y, \quad \frac{\partial N}{\partial x} = 2x$$

$$\therefore \quad \int_C \vec{F} \cdot d\vec{r} = \int_C (2x^2 - y^2)\, dx + (x^2 + y^2)\, dy$$

$$= \iint_R (2x + 2y)\, dx\, dy \qquad | \text{ Using Green's theorem}$$

$$= 2\int_{x=-a}^{a} \int_{y=0}^{\sqrt{a^2 - x^2}} (x + y)\, dx\, dy = 2\int_{-a}^{a} \left(xy + \frac{y^2}{2} \right)_0^{\sqrt{a^2 - x^2}} dx$$

$$= 2\int_{-a}^{a} \left(x\sqrt{a^2 - x^2} + \frac{a^2 - x^2}{2} \right) dx = 0 + 2\int_0^a (a^2 - x^2)\, dx$$

$$= 2\left(a^2 x - \frac{x^3}{3} \right)_0^a = 2\left(a^3 - \frac{a^3}{3} \right) = \frac{4}{3} a^3.$$

Example 5. *Use Green's theorem to evaluate $\int_C (x^2 + xy)\, dx + (x^2 + y^2)\, dy$ where C is the square formed by the lines $y = \pm 1$, $x = \pm 1$.*

[A.K.T.U. 2015, 2018]

Sol. $\int_C (x^2 + xy)\, dx + (x^2 + y^2)\, dy$

$$= \iint_R \left[\frac{\partial}{\partial x}(x^2 + y^2) - \frac{\partial}{\partial y}(x^2 + xy) \right] dx\, dy = \iint_R (2x - x) dx\, dy$$

$$= \int_{x=-1}^{1} \int_{y=-1}^{1} x\, dx\, dy = \int_{-1}^{1} x \left(y \right)_{-1}^{1} dx$$

$$= \int_{-1}^{1} 2x\, dx = 0. \qquad \text{(Since integrand is odd)}$$

Example 6. *Using Green's theorem, evaluate $\int_C (x^2 y\, dx + x^2 dy)$ where C is the boundary described counter clockwise of the triangle with vertices (0, 0); (1, 0); (1, 1).*

(U.K.T.U. 2010)

Sol. $\int_C (x^2 y\, dx + x^2 dy)$

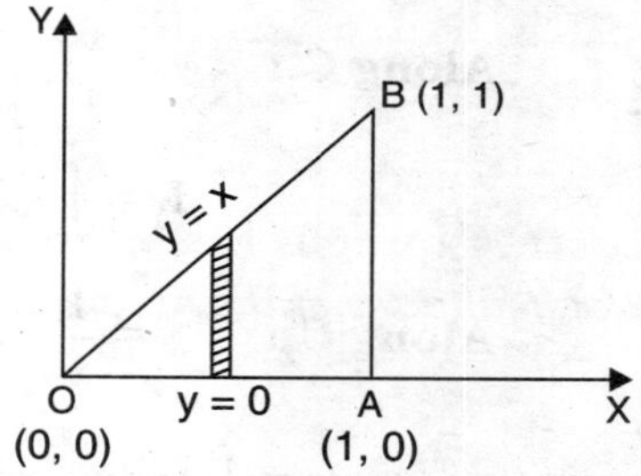

$$= \iint_R \left[\frac{\partial}{\partial x}(x^2) - \frac{\partial}{\partial y}(x^2 y) \right] dx\, dy$$

| Using Green's theorem

$$= \int_{x=0}^{1} \int_{y=0}^{x} (2x - x^2)\, dx\, dy$$

$$= \int_0^1 (2x^2 - x^3) dx = \left(\frac{2}{3} x^3 - \frac{x^4}{4} \right)_0^1$$

$$= \frac{2}{3} - \frac{1}{4} = \frac{5}{12}.$$

Example 7. *Evaluate* $\int_C [(y - \sin x)\, dx + \cos x\, dy]$ *where C is the triangle formed by* $y = 0,\ x = \frac{\pi}{2},\ y = \frac{2}{\pi} x.$

Sol. Here $\qquad M = y - \sin x,\ N = \cos x$

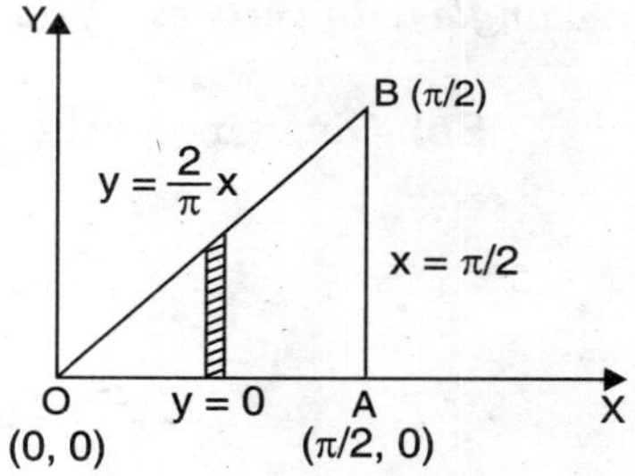

$$\therefore \quad \frac{\partial M}{\partial y} = 1, \qquad \frac{\partial N}{\partial x} = -\sin x$$

$$\therefore \quad \text{Given integral} = \iint_R \left(\frac{\partial N}{\partial x} - \frac{\partial M}{\partial y} \right) dx\, dy$$

$$= -\int_{x=0}^{\pi/2} \int_{y=0}^{2x/\pi} (\sin x + 1)\, dx\, dy$$

$$= -\frac{2}{\pi} \int_0^{\pi/2} (x + x \sin x)\, dx$$

$$= -\frac{2}{\pi} \left[\frac{x^2}{2} + x(-\cos x) + \sin x \right]_0^{\pi/2} = -\frac{2}{\pi}\left(\frac{\pi^2}{8} + 1 \right) = -\left(\frac{\pi}{4} + \frac{2}{\pi} \right).$$

Example 8. *Using Green's theorem, find the area of the region in the first quadrant bounded by the curves* $y = x,\ \ y = \frac{1}{x},\ \ y = \frac{x}{4}.$

Sol. By Green's theorem, area A of the region bounded by a closed curve c is given by

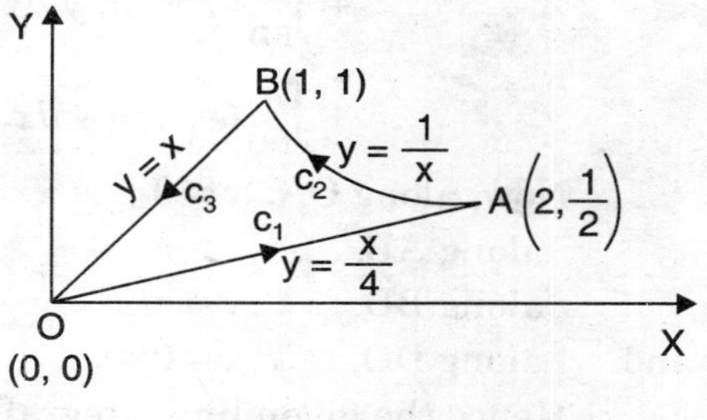

$$A = \frac{1}{2} \oint_c (x dy - y dx)$$

Here c consists of the curves

$$c_1 : y = \frac{x}{4},\ c_2 : y = \frac{1}{x},\ c_3 : y = x$$

$$\therefore \qquad A = \frac{1}{2}\left[\int_{c_1} (x dy - y dx) + \int_{c_2} (x dy - y dx) + \int_{c_3} (x dy - y dx) \right] = \frac{1}{2}(I_1 + I_2 + I_3)$$

Along C_1: $y = \dfrac{x}{4}$ $\therefore$ $dy = \dfrac{1}{4}\,dx$; $x : 0$ to 2

$$I_1 = \int_{c_1} (xdy - ydx) = \int_{c_1} \left(\frac{x}{4}\,dx - \frac{x}{4}\,dx \right) = 0$$

Along C_2: $y = \dfrac{1}{x}$ $\therefore$ $dy = -\dfrac{1}{x^2}\,dx$; $x : 2$ to 1

$$I_2 = \int_{c_2} (xdy - ydx) = \int_2^1 \left[x\left(\frac{-1}{x^2}\right) dx - \frac{1}{x}\,dx \right] = -2\int_2^1 \frac{1}{x}\,dx = -2\,(\log x)_2^1 = 2\log 2$$

Along C_3: $y = x$ $\therefore$ $dy = dx$; $x : 1$ to 0

$$I_3 = \int_{c_3} (xdy - ydx) = \int (xdx - xdx) = 0$$

$$\therefore \qquad A = \frac{1}{2}\,(I_1 + I_2 + I_3) = \log 2.$$

Example 9. *Evaluate by Green's theorem* $\int_C (e^{-x} \sin y\, dx + e^{-x} \cos y\, dy)$ *where C is the rectangle with vertices (0, 0), (π, 0), (π, π/2), (0, π/2) and hence verify Green's theorem.*

Sol. Given integral $= \iint_R \left[\dfrac{\partial}{\partial x}(e^{-x} \cos y) - \dfrac{\partial}{\partial y}(e^{-x} \sin y) \right] dx\, dy$ | by Green's theorem

$$= \int_{x=0}^{\pi} \int_{y=0}^{\pi/2} (-e^{-x} \cos y - e^{-x} \cos y)\, dx\, dy$$

$$= -2\int_0^{\pi} e^{-x}\, dx \int_0^{\pi/2} \cos y\, dy = -2\,(-e^{-x})_0^{\pi}\,(\sin y)_0^{\pi/2} = 2\,(e^{-\pi} - 1).$$

Verification of Theorem

For this purpose, let us evaluate the given line integral directly.

$$\int_C (e^{-x} \sin y\, dx + e^{-x} \cos y\, dy)$$

$$= \int_{OA} (e^{-x} \sin y\, dx + e^{-x} \cos y\, dy)$$

$$+ \int_{AB} (e^{-x} \sin y\, dx + e^{-x} \cos y\, dy)$$

$$+ \int_{BD} (e^{-x} \sin y\, dx + e^{-x} \cos y\, dy)$$

$$+ \int_{DO} (e^{-x} \sin y\, dx + e^{-x} \cos y\, dy)$$

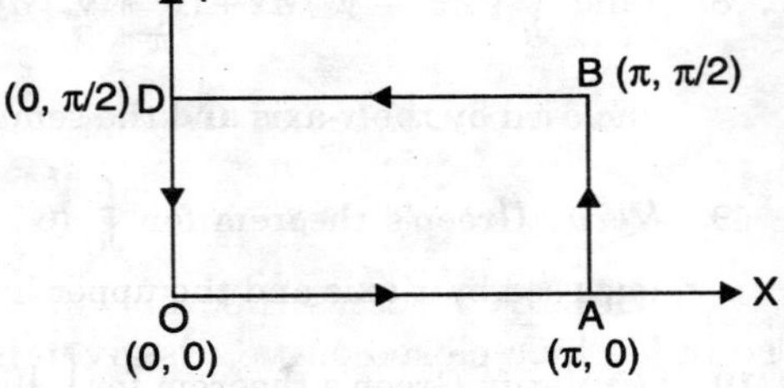

Now, along OA, $y = 0$ $\Rightarrow$ $dy = 0$

along AB, $x = \pi$ $\Rightarrow$ $dx = 0$

along BD, $y = \pi/2$ $\Rightarrow$ $dy = 0$

and along DO, $x = 0$ $\Rightarrow$ $dx = 0$

Hence the given line integral

$$= 0 + \int_0^{\pi/2} e^{-\pi} \cos y\, dy + \int_{\pi}^{0} e^{-x}\, dx + \int_{\pi/2}^{0} \cos y\, dy$$

$$= e^{-\pi} (\sin y)_0^{\pi/2} + (-e^{-x})_\pi^0 + (\sin y)_{\pi/2}^0$$
$$= e^{-\pi} - (1 - e^{-\pi}) + (-1) = 2(e^{-\pi} - 1)$$

Hence Green's theorem is verified.

TEST YOUR KNOWLEDGE

1. Evaluate by Green's theorem $\int_C (x^2 - \cosh y)\, dx + (y + \sin x)\, dy$ where C is the rectangle with vertices (0, 0); (π, 0); (π, 1); (0, 1).

2. Evaluate by Green's theorem $\int_C (\cos x \sin y - xy)\, dx + \sin x \cos y\, dy$ where C is the circle $x^2 + y^2 = 1$. *(U.P.T.U. 2015)*

3. Verify Green's theorem by evaluating $\int_C [(x^3 - xy^3)\, dx + (y^2 - 2xy)\, dy]$ where C is the square having the vertices at the points (0, 0), (2, 0), (2, 2) and (0, 2)

4. Verify Green's theorem in the plane for $\int_C [(xy + y^2)\, dx + x^2\, dy]$ where C is closed curve of the region bounded by $y = x$ and $y = x^2$.

5. Verify Green's theroem in the plane for $\int_C \{(3x^2 - 8y^2)\, dx + (4x - 6xy)\, dy\}$ where C is the region bounded by the parabolas $y = \sqrt{x}$ and $y = x^2$. *(U.K.T.U. 2011)*

6. Verify Green's theorem in plane for $\int_C (x^2 + 2xy)\, dx + (y^2 + x^3 y)\, dy$ where C is a square with vertices A(0, 0); B(1, 0); D(1, 1) and E(0, 1).

7. Using Green's theorem to evaluate $\int_C (2y^2\, dx + 3x\, dy)$ where C is the boundary of the closed region bounded by $y = x$ and $y = x^2$. *(U.K.T.U. 2011; A.K.T.U. 2018, 2016)*

8. Find $\int_C [(2x^2 - y^2)\, dx + (x^2 + y^2)\, dy]$ where C is the boundary of the surface in the xy-plane enclosed by the x-axis and the semi circle $y = \sqrt{1 - x^2}$.

9. Verify Green's theorem for $\oint_C (6 + x^2)\, dx + (1 - 2xy) dy$ where C is the boundary of the area enclosed by x-axis and the upper half of circle $x^2 + y^2 = 1$.

10. (*i*) Verify Green's theorem for $\int_C [(3x^2 - 8y^2)\, dx + (4y - 6xy) dy]$ where C is the boundary of the region enclosed by the lines $x = 0, y = 0, x + y = 1$.

    (*ii*) Verify Green's theorem for $\int_C [(2x^2 - y^2)\, dx + (x^2 + y^2)\, dy]$ where C is the boundary of the area enclosed by the X-axis and the upper half of the circle $x^2 + y^2 = a^2$.

11. Verify Green's theorem in plane for $\oint_C (x^2 - 2xy)\, dx + (x^2 y + 3)\, dy$ where C is the boundary of the region defined by $y^2 = 8x$ and $x = 2$. *(U.P.T.U. 2014)*

12. Let $\vec{F} = ax^2\,\hat{i} + zy^2\,\hat{j} + by^2\,\hat{k}$ and let C_1 be the straight line joining O(0, 0, 0) to the point D(*c*, *c*, *c*) and let C_2 be the curve composed of the three segments of straight lines.

C_3 : $x = 0, \quad y = 0, 0 < z < c$

C_4 : $x = 0, 0 < y < c, z = c$

C_5 : $0 < x < c, y = c, z = c$

Find $\int_C \vec{F} \cdot d\vec{r}$ over both the curves.

Explain why despite the fact that the beginning and end points of curves are the same, the two integrals are not equal.

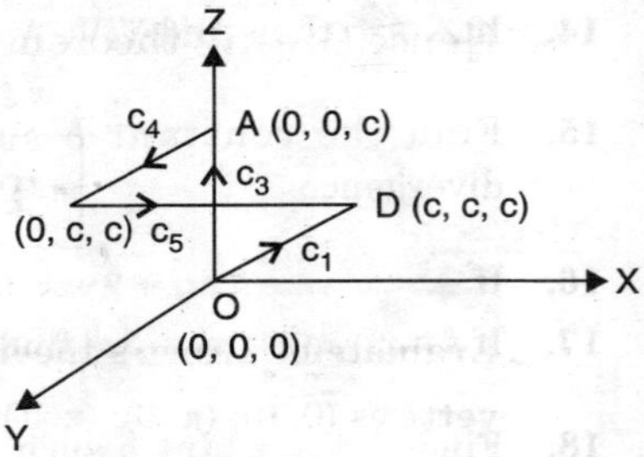

Answers

1. $\pi(\cosh 1 - 1)$ **2.** 0 **7.** 7/30. **8.** $\frac{4}{3}$

12. Along C_1, $I = \frac{1}{12}c^3(4a + 3c + 4b)$

Along C_2, $I = \frac{1}{3}c^3(c + a)$; $\vec{F}$ is not conservative.

ASSIGNMENT V

1. Prove that if $\vec{u}$ and $\vec{v}$ are irrotational then $\vec{u} \times \vec{v}$ will be solenoidal.
2. Find the directional derivative of $\phi = 2x^2y + 3xy^3$ in the direction of $3\hat{i} + 4\hat{j}$ at the point (1, 1).
3. Find the magnitude of the gradient of the function $f = xyz^3$ at (1, 0, 2). *(G.B.T.U. 2012)*
4. (*i*) Write the statement of divergence theorem for a given vector field $\vec{F}$. *(A.K.T.U. 2012, 2018)*
   (*ii*) State Stoke's theorem. *(A.K.T.U. 2016, 2022)*
5. Show that the vector $\vec{V} = 3y^4z^2\hat{i} + 4x^3z^2\hat{j} - 3x^2y^2\hat{k}$ is solenoidal. *(M.T.U. 2012)*
6. State Green's theorem for a plane region. *(A.K.T.U. 2021)*
7. Find $\nabla(\log r)$ if $\vec{r} = x\hat{i} + y\hat{j} + z\hat{k}$ is position vector.
8. Find $\vec{A} . \nabla f$ at (1, – 1, 1) if $\vec{A} = 2x^2\hat{i} - 3yz\hat{j} + xz^2\hat{k}$ and $f = 2z - x^3y$.
9. (*i*) Find the directional derivative of $f(x, y, z) = 4e^{2x-y+z}$ at the point (1, 1, – 1) in the direction towards the point (– 3, 5, 6).
   (*ii*) Find the directional derivative of $f(x, y, z) = xy + yz + zx$ in the direction of vector $\hat{i} + 2\hat{j} + 2\hat{k}$ at the point (1, 2, 0).
10. Determine the expression for $f(r)$ such that $\nabla f = \frac{\vec{r}}{r^5}$ and $f(1) = 0$.
11. Find the unit outward drawn normal to the surface $(x-1)^2 + y^2 + (z+2)^2 = 9$ at the point (3, 1, – 4).
12. A particle moves along the curve $\vec{r} = (t^3 - 4t)\hat{i} + (t^2 + 4t)\hat{j} + (8t^2 - 3t^3)\hat{k}$ where *t* is time, find the velocity at *t* = 2. *(U.P.T.U. 2014)*

13. Find p such that $\vec{A} = (px + 4y^2z)\hat{i} + (x^3 \sin z - 3y)\hat{j} - (e^x + 4\cos x^2y)\hat{k}$ is solenoidal.

14. If $\vec{A} = \hat{r}$ then find $\nabla(\nabla . \vec{A})$.

15. Find the constant b such that $\vec{A} = (bx^2y + yz)\hat{i} + (xy^2 - xz^2)\hat{j} + (2xyz - 2x^2y^2)\hat{k}$ has zero divergence.

16. If $\vec{A} = x^2y\hat{i} - 2xz\hat{j} + 2yz\hat{k}$ then find curl curl $\vec{A}$ at the point (1, 0, 2).

17. If $f = x^2yz$, $g = xy - 3z^2$ find $\nabla.(\nabla f \times \nabla g)$.

18. Find the constant b such that $\vec{A} = (bxy - z^3)\hat{i} + (b-2)x^2\hat{j} + (1-b)xz^2\hat{k}$ has its curl identically equal to zreo.

19. (*i*) Find the value of m if $\vec{F} = mx\hat{i} - 5y\hat{j} + 2z\hat{k}$ is a solenoidal vector. (*A.K.T.U. 2018*)

(*ii*) Find 'b' for a solenoidal vector $(bx + 4y^2z)\hat{i} + (x^3 \sin z - 3y)\hat{j} - (e^x + 4\cos x^2y)\hat{k}$.

20. If $f = \dfrac{xy}{z}$, find $\nabla^2 f$ at (2, 3, 1).

21. Find $\oint_C (x^3dy - y^3dx)$ where c is the circle $x^2 + y^2 = 1$.

22. If $\vec{F}$ is the velocity of a fluid particle, what does $\oint_C \vec{F}.d\vec{r}$ represent? (*G.B.T.U. 2010*)

23. (*i*) If $\vec{F} = xz\hat{i} - y^2\hat{j} + 2x^2y\hat{k}$, then find curl $\vec{F}$.

(*ii*) Find the curl of $\vec{F} = xy\hat{i} + y^2\hat{j} + xz\hat{k}$ at (–2, 4, 1). (*A.K.T.U. 2016*)

24. Find the constants a and b such that the curl of a vector $\vec{A} = (2xy + 3yz)\hat{i} + (x^2 + axz - 4z^2)\hat{j} + (3xy + 2byz)\hat{k}$ is zero.

25. If $\vec{F} = (x^2 + y^2)\hat{i} - 2xy\hat{j}$, then evaluate $\oint \vec{F}.d\vec{r}$. (*A.K.T.U. 2017*)

26. Find the divergence of $(2x^2z\hat{i} - xy^2z\hat{j} + 3yz^2\hat{k})$ at the point (1, 1, 1).

27. Find the directional derivative of $\nabla.\vec{U}$ at the point (4, 4, 2) in the direction of the corresponding outer normal of the sphere $x^2 + y^2 + z^2 = 36$ where $\vec{U} = x^4\hat{i} + y^4\hat{j} + z^4\hat{k}$.

28. Find div $(r^3\vec{r})$ if $\vec{r} = x\hat{i} + y\hat{j} + z\hat{k}$.

29. Find curl $(yz\hat{i} + 3zx\hat{j} + z\hat{k})$ at (2, 3, 4).

30. If $f = 3x^2z - y^2z^3 + 4x^3y + 2x - 3y - 5$, then find $\nabla^2 f$ at (1, 1, 0).

31. What is the condition for ∇f to be both solenoidal and irrotational?

32. If $\vec{F} = 2xy\hat{i} + (xyz^2 - \sin yz)\hat{j} + ze^{x+y}\hat{k})$, then find div (curl $\vec{F}$).

33. Find $\int_c \hat{t}.d\vec{r}$ where $\hat{t}$ is the unit tangent vector and c is the unit circle in the xy-plane about the origin.

34. Find the unit normal at the surface $z = x^2 + y^2$ at the point (1, 2, 5). (*G.B.T.U. 2013*)

35. For the scalar field $u = -\frac{x^2}{2} + \frac{y^2}{3}$, find the magnitude of gradient at the point (1, 3). *(A.K.T.U. 2017)*

36. If $\vec{F} = \frac{\vec{r}}{r^3}$, find curl $\vec{F}$. *(M.T.U. 2013)*

37. Using Green's theorem, evaluate the integral : $\oint_c (xy\,dy - y^2\,dx)$, where c is the square cut from the first quadrant by the lines $x = 1$, $y = 1$. *(M.T.U. 2013)*

38. Define del (∇) operator and gradient. *(A.K.T.U. 2019)*

39. If $\phi = 3x^2y - y^3z^2$, find grad ϕ at point (2, 0, –2). *(A.K.T.U. 2019)*

40. Find grad ϕ at the point (2, 1, 3) where $\phi = x^2 + yz$. *(A.K.T.U. 2020)*

41. Find a unit normal vector to the surface $z^2 = x^2 + y^2$ at the point (1, 0, –1). *(A.K.T.U. 2022)*

Answers

2. 13	**3.** 8	**7.** $\frac{\vec{r}}{r^2}$	**8.** 5
9. (*i*) $\frac{-20}{9}$	(*ii*) $\frac{10}{3}$	**10.** $\frac{1}{3} - \frac{1}{3}r^{-3}$	**11.** $\frac{2\hat{i} + \hat{j} - 2\hat{k}}{3}$
12. $8\hat{i} + 8\hat{j} - 4\hat{k}$	**13.** 3	**14.** $-\frac{2\vec{r}}{r^3}$	**15.** – 2
16. $4\hat{j}$	**17.** 0	**18.** 4	
19. (*i*) 3	(*ii*) 3	**20.** 12	**21.** $\frac{3\pi}{2}$
22. circulation	**23.** (*i*) $2x^2\hat{i} + (x - 4xy)\hat{j}$		(*ii*) $-\hat{j} + 2\hat{k}$
24. $a = 3, b = -4$	**25.** $\oint (x^2 + y^2)\,dx - 2xy\,dy$		**26.** 8
27. 272	**28.** $6r^3$	**29.** $-6\hat{i} + 3\hat{j} + 8\hat{k}$	**30.** 24
31. $\nabla^2 f = 0$	**32.** 0	**33.** 2π	**34.** $\frac{-2\hat{i} - 4\hat{j} + \hat{k}}{\sqrt{21}}$
35. $\sqrt{5}$	**36.** 0	**37.** 3/2	**39.** $12\hat{j}$
40. $4\hat{i} + 3\hat{j} + \hat{k}$	**41.** $\frac{\hat{i} + \hat{k}}{\sqrt{2}}$.		

EXAMINATION PAPERS

A.K.T.U., LUCKNOW

BAS 103

B. Tech. [SEMESTER-I]

THEORY EXAMINATION 2024-25

ENGINEERING MATHEMATICS-I

Time: 3 Hours *Max. Marks: 70*

Note: Attempt all Sections. In case of any missing data; choose suitably.

Section-A

1. Attempt *all* questions in brief. **(2 × 07 = 14)** **CO**

(*a*) Find the eigen values of the matrix $\begin{bmatrix} \cos\theta & -\sin\theta \\ -\sin\theta & -\cos\theta \end{bmatrix}$. (1)

(*b*) If $u = \dfrac{x^2 + y^2}{x + y}$, find the value of $x\dfrac{\partial u}{\partial x} + y\dfrac{\partial u}{\partial y}$. (2)

(*c*) What is the difference between total derivatives and partial derivatives? (2)

(*d*) What are the applications of Jacobians. (3)

(*e*) Write the statement of Liouville's Theorem. (4)

(*f*) Evaluate $\int_1^2\int_1^3 x^2y^2\ dxdy$. (4)

(*g*) Prove that $curl\ \vec{r} = 0$. (5)

Section-B

2. Attempt any *three* of the following: **(07 × 3 = 21)** **CO**

(*a*) Find two non-singular matrices P and Q such that PAQ is in normal form, (1)

where $A = \begin{bmatrix} 1 & 3 & 6 & -1 \\ 1 & 4 & 5 & 1 \\ 1 & 5 & 4 & 3 \end{bmatrix}$.

(*b*) Find the nth derivative of $\tan^{-1}\left(\dfrac{x}{a}\right)$. (2)

(*c*) Find the volume of the largest rectangular parallelepiped that can be inscribed in the ellipsoid $\dfrac{x^2}{a^2} + \dfrac{y^2}{b^2} + \dfrac{z^2}{c^2} = 1$. (3)

(*d*) Apply Dirichlet's theorem to evaluate $\iiint xyz\ dxdxdz$ taken throughout the ellipsoid $\dfrac{x^2}{a^2} + \dfrac{y^2}{b^2} + \dfrac{z^2}{c^2} \le 1$. (4)

(*e*) Show that the vector $f(r)\vec{r}$ is irrotational, where $\vec{r} = x\hat{i} + y\hat{j} + z\hat{k}$. (5)

Section-C

3. Attempt any *one* part of the following: **(07 × 1 = 07)** **CO**

(*a*) Find the eigen values and eigen vectors of the following matrices:

$$A = \begin{bmatrix} 2 & -1 & 1 \\ -1 & 2 & -1 \\ 1 & -1 & 2 \end{bmatrix}.$$ (1)

(*b*) Discuss for all values of k for the system of equations $x + y + 4z = 6$, $x + 2y - 2z = 6, kx + y + z = 6$ as regards existence and nature of solution. (1)

4. Attempt any *one* part of the following: **(07 × 1 = 07)** **CO**

(*a*) Trace the curve $y^2 (a + x) = x^2 (3a - x)$. (2)

(*b*) If $u = f(r)$, where $r^2 = x^2 + y^2$, prove that $\dfrac{\partial^2 u}{\partial x^2} + \dfrac{\partial^2 u}{\partial y^2} = f''(r) + \dfrac{1}{r} f'(r)$. (2)

5. Attempt any *one* part of the following: **(07 × 1 = 07)** **CO**

(*a*) If $u = xyz$, $v = x^2 + y^2 + z^2$ and $w = x + y + z$. *Find the jacobian* $\dfrac{\partial(x, y, z)}{\partial(u, v, w)}$. (3)

(*b*) Find the maxima and minima of the function $\sin x + \sin y + \sin (x + y)$. (3)

6. Attempt any *one* part of the following: **(07 × 1 = 07)** **CO**

(*a*) Find the area inside the circle $r = 2a \cos \theta$ and outside the circle $r = a$. (4)

(*b*) Change the order of integration and then evaluate

$$\int_0^{2a} \int_{\frac{x^4}{4a}}^{3a - x} (x^2 + y^2) dy dx$$ (4)

7. Attempt any *one* part of the following: **(07 × 1 = 07)** **CO**

(*a*) Show that div (grad r^n) $= n(n + 1)r^{n-2}$, where $\vec{r} = x\hat{i} + y\hat{j} + z\hat{k}$ (5)

(*b*) Verify Stokes theorem for $\vec{F} = (x^2 + y^2)\hat{i} - 2xy\hat{j}$ taken round the rectangle bounded by the lines $x = 0$, $x = a$, $y = 0$, $y = b$. (5)

B. Tech. [SEMESTER-I]

THEORY EXAMINATION 2024-25

KAS 103T

ENGINEERING MATHEMATICS-I

Time: 3 Hours | *Max. Marks: 100*

Note: Attempt all Sections. In case of any missing data; choose suitably.

Section-A

1. Attempt *all* questions in brief. **(2 × 10 = 20)** **CO**

(*a*) If the Matrix $A = \begin{bmatrix} 2 & 3-4i \\ 3+4i & 2 \end{bmatrix}$, then show that the Matrix A is Hermitian. (1)

(*b*) Describe Rank-Nullity theorem. (1)

(*c*) State Lagrange's Mean value theorem. (2)

(*d*) Find the envelop of the family of straight lines $m^2x - my + a = 0$ where m is a parameter. (2)

(*e*) What is the degree of homogeneous function $u(x, y) = \left(\sqrt{x} + \sqrt{y}\right)(x^n + y^n)$. (3)

(*f*) If $\dfrac{\partial(x, y, z)}{\partial(u, v, w)} = 5$, then find the value of $\dfrac{\partial(u, v, w)}{\partial(x, y, z)}$. (3)

(*g*) Evaluate: $\int_0^1 \int_0^{x^2} xe^y \, dydx$ (4)

(*h*) Change the order of integration in $\int_0^\infty \int_x^\infty f(x, y) \, dydx$. (4)

(*i*) For the scalar field $u = yz + zx + xy$, find the gradient of u at the point (1, 2, 3). (5)

(*j*) State Green's Theorem. (5)

Section-B

2. Attempt any *three* of the following: **(10 × 3 = 30)** **CO**

(*a*) Interpret the values of λ and μ such that the system (1)

$x + y + z = 6$, $x + 2y + 5z = 10$, $2x + 3y + \lambda z = \mu$ has

I. No solution

II. A unique solution

III. Infinite number of solutions.

(*b*) Verify Rolle's theorem for the function $f(x) = -x^2 - 6x$. In the interval [2, 4]. (2)

(*c*) Express the function $f(x, y) = e^x \sin y$ as Taylor's series expansion in powers of x and y. (3)

(*d*) Using double integration find the area of the region bounded by the curves $xy = 2$, $4y^2$, $yx = 4$. (4)

(*e*) Find the directional derivative of $\phi = (x^2 + y^2 + z^2)^{-1/2}$ at the point P (3, 1, 2) in the direction of the vector $yz\hat{i} + zx\hat{j} + xyk$ (5)

Section-C

3. Attempt any *one* part of the following: **(10 × 1 = 10)** **CO**

(*a*) Determine Eigen values and corresponding Eigen vectors for the Matrix (1)

$$A = \begin{bmatrix} 1 & 1 & 3 \\ 1 & 5 & 1 \\ 3 & 1 & 1 \end{bmatrix}$$

(*b*) Verify Cayley-Hamilton theorem for the Matrix $A = \begin{bmatrix} 1 & 2 & 3 \\ 2 & 4 & 5 \\ 3 & 5 & 6 \end{bmatrix}$ and hence compute A^{-1}. (1)

4. Attempt any *one* part of the following: **(10 × 1 = 10)** **CO**

(*a*) Use Leibnitz theorem to construct the following equation: (2)

$$(1 - x^2)y_{n+2} - (2n + 1)xy_{n+1} - (n^2 - a^2)y_n = 0,$$

if $y = \sin (a \sin^{-1} x)$.

Also calculate the value of $(y_n)_0$.

(*b*) Construct the curve: $y^2 (a + x) = x^2(3a - x)$. (2)

5. Attempt any *one* part of the following: **(10 × 1 = 10)** **CO**

(*a*) If $u^3 + v^3 + w^3 = x + y + z$, $u^2 + v^2 + w^2 = x^3 + y^3 + z^3$ and $u + v + w = x^2 + y^2 + z^2$, then find the value of $\dfrac{\partial (u, v, w)}{\partial (x, y, z)}$. (3)

(*b*) A rectangular box open at the top have capacity 32c.c.. Find the dimensions of the box requiring least material for its construction. (3)

6. Attempt any *one* part of the following: **(10 × 1 = 10)** **CO**

(*a*) Using the transformation $u = x + y$, $v = x - 2y$, find the value of $\iint_R (x + y)^2 \, dxdy$ where R is the parallelogram in the *xy*-plane with vertices (1, 0), (3, 1), (2, 2), (0, 1) (4)

(*b*) Compute $\iiint_V x^2 \, dxdydz$ over volume of tetrahedron bounded by $x = 0$, $y = 0$, $z = 0$ and $\dfrac{x}{a} + \dfrac{y}{b} + \dfrac{z}{c} = 1$. (4)

7. Attempt any *one* part of the following: **(10 × 1 = 10)** **CO**

(*a*) Verify Stoke's theorem of $\vec{F} = y^2\hat{i} + x^2\hat{j} - (x + z)\hat{k}$ and C is the boundary of the triangle with vertices at (0, 0, 0), (1, 0, 0) and (1, 1, 0). (5)

(*b*) Verify divergence theorem for $F = (x^2 - yz)\hat{i} + (y^2 - zx)\hat{j} + (z^2 - xy)\hat{k}$ taken over the rectangular parallelopiped $0 \le x \le a$, $0 \le y \le b$, $0 \le z \le c$. (5)

B. Tech. [SEMESTER-I]

THEORY EXAMINATION 2023-24 BAS–103

ENGINEERING MATHEMATICS-I

Time: 3 Hours *Max. Marks: 70*

Note: Attempt all Sections. If require any missing data, then choose suitably.

Section-A

1. Attempt *all* questions in brief. (2 × 7 = 14) CO

(*a*) Find the product and sum of the eigen values for A = $\begin{bmatrix} 8 & -4 \\ 2 & 2 \end{bmatrix}$. (2)

(*b*) Find all symmetry in the curve $y^2(a^2 + x^2) = x^2(a^2 - x^2)$. (2)

(*c*) Calculate the error in R if E = RI and possible errors in E and I are 30% and 20% respectively. (2)

(*d*) Determine the value of $\Gamma\frac{1}{4}\,\Gamma\frac{3}{4}$. (2)

(*e*) Prove that $B(p, q) = B(p + 1, q) + B(p, q + 1)$. (2)

(*f*) Prove that $\vec{A} = (6xy + z^3)\hat{i} + (3x^2 - z)\hat{j} + (3xz^2 - y)\hat{k}$ is irrotational. (2)

(*g*) Find a unit normal vector to the surface $xy^3z^2 = 4$ at the point (–1, –1, 2). (2)

Section-B

2. Attempt any *three* of the following: (7 × 3 = 21) CO

(*a*) Solve the system of homogenous equations: (7)

$$x_1 + x_2 + x_3 + x_4 = 0,\ x_1 + 3x_2 + 2x_3 + 4x_4 = 0,$$
$$2x_1 + x_3 - x_4 = 0$$

(*b*) If $u = y^2e^{y/x} + x^2\tan^{-1}\left(\frac{x}{y}\right)$, show that (7)

(*i*) $x\frac{\partial u}{\partial x} + y\frac{\partial u}{\partial y} = 2u.$

(*ii*) $x^2\frac{\partial^2 u}{\partial x^2} + 2xy\frac{\partial^2 u}{\partial x\partial y} + y^2\frac{\partial^2 u}{\partial y^2} = 2u.$

(*c*) Expand $f(x, y) = e^x \cos y$ about the point $\left(1, \frac{\pi}{4}\right)$ by Taylor's series. (7)

(*d*) Evaluate the integral $\iint_D (y - x)\,dxdy$; by changing the variables, D: (7)

Region in xy-plane bounded by the lines

$$y - x = -3,\ y - x = 1,\ y + \frac{1}{3}x = \frac{7}{3},\ y + \frac{1}{3}x = 5.$$

(*e*) Find the directional derivative of $f(x, y, z) = e^{2x} \cos yz$ at (0, 0, 0) in the direction of the tangent to the curve $x = a \sin \theta$, $y = a \cos \theta$, $z = a\theta$ at $\theta = \frac{\pi}{4}$. (7)

Section-C

3. Attempt any *one* part of the following: **(7 × 1 = 7) CO**

(*a*) Determine eigen vectors for the matrix $A = \begin{bmatrix} 1 & 2 & 2 \\ 0 & 2 & 1 \\ -1 & 2 & 2 \end{bmatrix}$. (7)

(*b*) Determine A^{-1}, A^{-2} and A^{-3} if $A = \begin{bmatrix} 4 & 6 & 6 \\ 1 & 3 & 2 \\ -1 & -4 & -3 \end{bmatrix}$ using Cayley-Hamilton theorem. (7)

4. Attempt any *one* part of the following: **(7 × 1 = 7) CO**

(*a*) If $y \cos(m \sin^{-1} x)$ then prove that

$(1 - x^2)y_{n+2} - (2n + 1)xy_{n+1} + (m^2 - n^2)y_n = 0$. Also find $(y_n)_0$ (7)

(*b*) If $z = f(x, y)$, $x = e^u + e^{-v}$, $y = e^{-u} - e^v$, then show that

$$\frac{\partial z}{\partial u} - \frac{\partial z}{\partial v} = x\frac{\partial z}{\partial x} - y\frac{\partial z}{\partial y}.$$ (7)

5. Attempt any *one* part of the following: **(7 × 1 = 7) CO**

(*a*) If $u^3 + v^3 = x + y$, $u^2 + v^2 = x^3 + y^3$, then show $\frac{\partial(u, v)}{\partial(x, y)} = \frac{y^2 - x^2}{2uv(u - v)}$. (7)

(*b*) The pressure P at any point (x, y, z) in space is $P = 400\, xyz^2$. Find the highest pressure at the surface of a unit sphere $x^2 + y^2 + z^2 = 1$ using Lagrange's method. (7)

6. Attempt any *one* part of the following: **(7 × 1 = 7) CO**

(*a*) Find the volume of the solid bounded by the coordinate planes and the surface

$$\left(\frac{x}{a}\right)^{1/2} + \left(\frac{y}{b}\right)^{1/2} + \left(\frac{z}{c}\right)^{1/2} = 1.$$ (7)

(*b*) Prove that $B(m, n) = \frac{\Gamma m \Gamma n}{\Gamma m + n}$. (7)

7. Attempt any *one* part of the following: **(7 × 1 = 7) CO**

(*a*) Applying Gauss Divergence theorem, evaluate

$\iint_S [e^x dydz - ye^x dzdx + 3zdxdy]$, where S is the surface of the cylinder $x^2 + y^2 = c^2$, $0 \le z \le h$. (7)

(*b*) Prove that $\nabla^2 r^n = n(n + 1)r^{n-2}$, where $\vec{r} = x\hat{i} + y\hat{j} + z\hat{k}$ and hence show that $\nabla^2\left(\frac{1}{r}\right) = 0$. (7)

Paper Id: *233418* **Sub. Code:** ***BAS–103***

B. Tech. (SEM–I)

A.K.T.U., LUCKNOW

THEORY EXAMINATION 2022–2023
ENGINEERING MATHEMATICS–I

Time: 3 Hours ***Total Marks: 70***

Note. *Attempt* ***all*** *Sections. If require any missing data; then choose suitably.*

Section–A

1. Attempt **all** questions in brief: **(2 × 7 = 14)**

(*a*) If A is a Hermitian matrix, then show that iA is Skew-Hermitian matrix.

(*b*) Find the eigen value of the matrix $A = \begin{bmatrix} 4 & 2 \\ 2 & 4 \end{bmatrix}$ corresponding to the eigen vector $\begin{bmatrix} 51 \\ 51 \end{bmatrix}$.

(*c*) If $y = \cos^{-1} x$, prove that $(1 - x^2)y_2 - xy_1 = 0$.

(*d*) If $u = \sin^{-1}\left(\dfrac{x^3 + y^3}{\sqrt{x} + \sqrt{y}}\right)$, then show that $x\dfrac{\partial u}{\partial x} + y\dfrac{\partial u}{\partial y} = \dfrac{5}{2}\tan u$.

(*e*) Find the percentage error in measuring the volume of a rectangular box when the error of 1% is made in measuring each side.

(*f*) Evaluate $\iint y\,dx\,dy$ over the part of the plane bounded by the line $y = x$ and the parabola $y = 4x - x^2$.

(*g*) Find curl of a vector field given by $\vec{F} = (x^2 + xy^2)\hat{i} + (y^2 + x^2y)\hat{j}$.

Section–B

2. Attempt any **three** of the following: **(7 × 3 = 21)**

(*a*) Verify Cayley-Hamilton theorem for the matrix $A = \begin{bmatrix} 2 & -1 & 1 \\ -1 & 2 & -1 \\ 1 & -1 & 2 \end{bmatrix}$ and hence find its inverse.

(*b*) If $y\sqrt{x^2 - 1} = \log_e(x + \sqrt{x^2 - 1})$, prove that $(x^2 - 1)y_{n+1} + (2n + 1)xy_n + n^2 y_{n-1} = 0$.

(*c*) Expand $f(x, y) = y^x$ about (1, 1) up to second degree terms and hence evaluate $(1.02)^{1/n}$.

(*d*) Evaluate the double integral $\int_0^a \int_{\sqrt{ax}}^a \dfrac{y^2}{\sqrt{(y^4 - a^2x^2)}}\,dx\,dy$ by changing the order of integration.

(*e*) Find the directional derivative of scalar function $f(x, y, z) = xyz$ at point P(1, 1, 3) in the direction of the outward drawn normal to the sphere $x^2 + y^2 + z^2 = 11$ through the point P.

Section–C

3. Attempt any **one** part of the following: **(7 × 1 = 7)**

(*a*) Test the consistency for the following system of equations and if system is consistent, solve them:

$$x + y + z = 6,$$
$$x + 2y + 3z = 14,$$
$$x + 4y + 7z = 30.$$

(*b*) Find the eigen values and corresponding eigen vectors of the matrix A.

$$A = \begin{bmatrix} 2 & 1 & 1 \\ 2 & 3 & 4 \\ -1 & -1 & -2 \end{bmatrix}$$

4. Attempt any **one** part of the following: **(7 × 1 = 7)**

(*a*) Trace the curve $x^2y^2 = (a^2 + y^2)(a^2 - y^2)$ in *xy*-plane, where *a* is constant.

(*b*) If $u = \dfrac{x^2y^2}{x^2 + y^2} + \cos\left(\dfrac{xy}{x^2 + y^2}\right)$, prove that

$$x^2\frac{\partial^2 u}{\partial x^2} + 2xy\frac{\partial^2 u}{\partial x \partial y} + y^2\frac{\partial^2 u}{\partial y^2} = 2\frac{x^2y^2}{x^2 + y^2}.$$

5. Attempt any **one** part of the following: **(7 × 1 = 7)**

(*a*) Find the Jacobian of the functions $y_1 = (x_1 - x_2)(x_2 + x_3)$, $y_2 = (x_1 + x_2)(x_2 - x_3)$, $y_3 = x_2(x_1 - x_3)$, hence show that the functions are not independent. Find the relation between them.

(*b*) A rectangular box, which is open at the top, has a capacity of 32 cubic feet. Determine, using Lagrange's method of multipliers, the dimensions of the box such that the least material is required for the construction of the box.

6. Attempt any **one** part of the following: **(7 × 1 = 7)**

(*a*) Evaluate $\iiint_y (x - 2y + z)\, dzdydx$, where R is the region determined by $0 \le x \le 1$, $0 \le y \le x^2$, $0 \le z \le x + y$.

(*b*) Use Dirichlet's integral to evaluate $\iiint xyz\, dx\, dy\, dz$ throughout the volume bounded by $x = 0, y = 0, z = 0$ and $x + y + z = 1$.

7. Attempt any **one** part of the following: **(7 × 1 = 7)**

(*a*) Apply Gauss divergence theorem to evaluate $\iiint_S \vec{F}.\hat{n}ds$, where $\vec{F} = 4x\hat{i} - 2y^2\hat{j} + z^2\hat{k}$ and S is the surface of the region bounded by the cylinder $x^2 + y^2 = 4$, $z = 0$, $z = 3$.

(*b*) Evaluate $\oint \vec{F}.\vec{dr}$ by Stoke's theorem, where $\vec{F} = y^2\hat{i} + x^2\hat{j} - (x + z)\hat{k}$ and C is the boundary of the triangle with vertices at (0, 0, 0), (1, 0, 0) and (1, 1, 0).

EXAMINATION PAPERS

B. Tech. [SEMESTER-I] **KAS103T**

THEORY EXAMINATION, 2021–2022

ENGINEERING MATHEMATICS–I

(PAPER ID: 411939)

Time: 3 Hours *Total Marks: 100*

Notes:

- **Attempt all sections and assume any missing data.**
- **Appropriate marks are allotted to each question, answer accordingly.**

SECTION–A

1. Attempt all of the following questions in brief. **(10 × 2 = 20)** **CO**

(*a*) If the matrix $A = \begin{bmatrix} -1 & 2 & 3 \\ 0 & 3 & 5 \\ 0 & 0 & -2 \end{bmatrix}$, then find the eigen value of $A^3 + 5A + 8I$. 1

(*b*) Reduce the matrix $\begin{bmatrix} 1 & 1 & 1 \\ 3 & 1 & 1 \end{bmatrix}$ into normal form and find its rank. 1

(*c*) Find the envelope of the family of straight line $y = mx + \dfrac{a}{m}$, where m is a parameter. 2

(*d*) Can mean value theorem be applied to $f(x) = \tan x$ in $[0, \pi]$. 2

(*e*) State Euler's Theorem on homogeneous function. 3

(*f*) Find the critical points of the function $f(x, y) = x^3 + y^3 - 3axy$. 3

(*g*) Find the area bounded by curve $y^2 = x$ and $x^2 = y$. 4

(*h*) Find the value of $\int_0^1 \int_0^x \int_0^{x+y} dx\, dy\, dz$. 4

(*i*) Find a unit normal vector to the surface $z^2 = x^2 + y^2$ at the point $(1, 0, -1)$. 5

(*j*) State Stoke's Theorem. 5

SECTION–B

2. Attempt **any three** of the following questions: **(3 × 10 = 30)** **CO**

(*a*) **Find the characteristic equation of the matrix** $A = \begin{bmatrix} 2 & 1 & 1 \\ 0 & 1 & 0 \\ 1 & 1 & 2 \end{bmatrix}$**, compute** A^{-1} **and** 1

prove that $A^8 - 5A^7 + 7A^6 - 3A^5 + A^4 - 5A^3 + 8A^2 - 2A + I = \begin{bmatrix} 8 & 5 & 5 \\ 0 & 3 & 0 \\ 5 & 5 & 8 \end{bmatrix}$

(*b*) **State Rolle's theorem and verify Rolle's theorem for the function** $f(x) = \dfrac{\sin x}{e^x}$ **in** $[0, \pi]$**.** 2

(*c*) **If** u, v **and** w **are the roots of** $(\lambda - x)^3 + (\lambda - y)^3 + (\lambda - z)^3 = 0$**, cubic in** λ**, find** $\dfrac{\partial(u, v, w)}{\partial(x, y, z)}$**.** 3

(*d*) **Find the volume bounded by the cylinder** $x^2 + y^2 = 4$ **and the plane** $y + z = 4$ **and** $z = 0$**.** 4

(*e*) **Apply Green's theorem to evaluate** $\int_C [(2x^2 - y^2)dx + (x^2 + y^2)dy]$**, where C is the boundary of the area enclosed by the** x**-axis and the upper half of the circle** $x^2 + y^2 = a^2$**.** 5

SECTION–C

3. Attempt **any one** of the following questions: **(1 × 10 = 10)** **CO**

(*a*) **Find the value of** k **for which the system of equations** $(3k - 8)x + 3y + 3z = 0$**,** $3x + (3k - 8)y + 3z = 0$**,** $3x + 3y + (3k - 8)z = 0$ **has a non-trivial solution.** 1

(*b*) **Find the eigen values and eigen vectors of matrix** $A = \begin{bmatrix} 2 & 1 & 1 \\ 2 & 3 & 2 \\ 3 & 3 & 4 \end{bmatrix}$**.** 1

4. Attempt **any one** of the following questions: **(1 × 10 = 10)** **CO**

(*a*) **If** $f(x) = \dfrac{x}{1 + e^{\frac{1}{x}}}$**;** $x \neq 0$ **and** $f(0) = 0$**, then show that the function is continuous but not differentiable at** $x = 0$**.** 2

(*b*) **If** $y = (x + \sqrt{1 + x^2})^m$**, find** $y_n(0)$**.** 2

5. Attempt **any one** of the following questions: **(1 × 10 = 10)** **CO**

(*a*) **Expand** x^y **in powers of** $(x - 1)$ **and** $(y - 1)$ **up to the third-degree terms and hence evaluate** $(1.1)^{1.02}$**.** 3

(*b*) A rectangular box which is open at the top having capacity 32 c.c. Find the dimension of the box such that the least material is required for its constructions. 3

6. Attempt **any one** of the following questions: **(1 × 10 = 10)** **CO**

(*a*) Change the order of integration in $I = \int_0^1 \int_{x^2}^{2-x} xy\,dy\,dx$ and hence evaluate the same. 4

(*b*) Find the position of the C.G. of a semicircular lamina of radius a, if its density varies as the square of the distance from the diameter. 4

7. Attempt **any one** of the following questions: **(1 × 10 = 10)** **CO**

(*a*) Find the directional derivative of $\nabla(\nabla f)$ at the point (1, –2, 1) in the direction of the normal to the surface $xy^2z = 3x + z^2$ where $f = 2x^3y^2z^4$. 5

(*b*) Find the constants a, b, c so that 5

$\vec{F} = (x + 2y + az)\hat{i} + (bx - 3y - z)\hat{j} + (4x + cy + 2z)\hat{k}$ is irrotational and hence find function ϕ such that $\vec{F} = \Delta\phi$.

B. Tech. [SEMESTER-I] **KAS103T**

THEORY EXAMINATION, 2020–2021

ENGINEERING MATHEMATICS –I

(*PAPER ID: 311288*)

Time: 3 Hours *Total Marks: 100*

Note: Attempt all Sections. If require any missing data, then choose suitably.

SECTION–A

1. Attempt **all** questions in brief. **(2 × 10 = 20)** **CO**

(*a*) Prove that the matrix $\frac{1}{\sqrt{3}}\begin{bmatrix} 1 & 1+i \\ 1-i & -1 \end{bmatrix}$ is unitary. 1

(*b*) State Rank-Nullity Theorem. 1

(*c*) State Rolle's Theorem. 2

(*d*) Discuss all the symmetry of the curve $x^2y^2 = x^2 - a^2$. 2

(*e*) If $u = f(y - z, z - x, x - y)$, prove that $\frac{\partial u}{\partial x} + \frac{\partial u}{\partial y} + \frac{\partial u}{\partial z} = 0$. 3

(*f*) If $x = e^v \sec u$, $y = e^v \tan u$, then evaluate $\frac{\partial(x, y)}{\partial(u, v)}$. 3

(*g*) Evaluate $\int_0^1 \int_0^{x^2} e^{y/x} dydx$. 4

(*h*) Calculate the volume of the solid bounded by the surface $x = 0, y = 0, x + y + z = 1$ and $z = 0$. 4

(*i*) Show that the vector $\vec{V} = (x + 3y)\hat{i} + (y - 3z)\hat{j} + (x - 2z)\hat{k}$ is solenoidal. 5

(*j*) State Green's theorem. 5

SECTION–B

2. Attempt **any three** of the following: **(10 × 3 = 30)** **CO**

(*a*) Find the inverse of the matrix $A = \begin{bmatrix} 2 & 3 & 4 \\ 4 & 3 & 1 \\ 1 & 2 & 4 \end{bmatrix}$. 1

(*b*) If $y = e^{\tan^{-1}x}$, prove that $(1 + x^2)\, y_{n+2} + [(2n + 2)\, x - 1]\, y_{n+1} + n\,(n + 1)\, y_n = 0$. 2

(*c*) If 3

$$u^3 + v + w = x + y^2 + z^2,$$
$$u + v^3 + w = x^2 + y + z^2,$$
$$u + v + w^3 = x^2 + y^2 + z$$

Show that:

$$\frac{\partial(u, v, w)}{\partial(x, y, z)} = \frac{1 - 4xy(xy + yz + zx) + 16xyz}{2 - 3(u^2 + v^2 + w^2) + 27u^2v^2w^2}.$$

(*d*) Evaluate by changing the variables, $\iint_R (x + y)^2 dx\, dy$ where R is the region bounded by the parallelogram $x + y = 0$, $x + y = 2$, $3x - 2y = 0$ and $3x - 2y = 3$. 4

(*e*) Use divergence theorem to evaluate the surface integral $\iint_S (xdydz + ydzdx + zdxdy)$ where S is the portion of the plane $x + 2y + 3z = 6$ which lies in the first octant. 5

SECTION–C

3. Attempt **any one** part of the following: **(10 × 1 = 10)** **CO**

(*a*) Find non-singular matrices P and Q such that PAQ is normal form. 1

$$\begin{bmatrix} 1 & 1 & 2 \\ 1 & 2 & 3 \\ 0 & 1 & 1 \end{bmatrix}$$

(*b*) Find the eigen values and the corresponding eigen vectors of the following matrix: 1

$$A = \begin{bmatrix} 2 & 0 & 1 \\ 0 & 3 & 0 \\ 1 & 0 & 2 \end{bmatrix}.$$

4. Attempt **any one** part of the following: **(10 × 1 = 10)** **CO**

(*a*) Find the envelope of the family of lines $\frac{x}{a} + \frac{y}{b} = 1$, where a and b are connected by the relation $a^n + b^n = c^n$ 2

(*b*) If $y = \sin(m \sin^{-1} x)$, find the value of y_n at $x = 0$. 2

5. Attempt **any one** part of the following: **(10 × 1 = 10)** **CO**

(*a*) Divide 24 into three parts such that continued product of first, square of second and cube of third is a maximum. 3

(*b*) If $u = \sec^{-1}\left(\frac{x^3 - y^3}{x + y}\right)$, prove that $x\frac{\partial u}{\partial x} + y\frac{\partial u}{\partial y} = 2 \cot u$. 3

Also evaluate $x^2\frac{\partial^2 u}{\partial x^2} + 2xy\frac{\partial^2 u}{\partial x \partial y} + y^2\frac{\partial^2 u}{\partial y^2}$.

6. Attempt **any one** part of the following: **(10 × 1 = 10)** **CO**

(*a*) Evaluate the following integral by changing the order of integration **4**

$$\int_0^\infty \int_0^\infty \frac{e^{-y}}{y}\, dydx.$$

(*b*) A triangular thin plate with vertices (0, 0), (2, 0) and (2, 4) has density $\rho = 1 + x + y$. Then find: **4**

(*i*) The mass of the plate.

(*ii*) The position of its centre of gravity G.

7. Attempt **any one** part of the following: **(10 × 1 = 10)** **CO**

(*a*) A fluid motion is given by $\vec{v} = (y \sin z - \sin x)\hat{i} + (x \sin z + 2yz)\hat{j} + (xy \cos z + y^2)\hat{k}$. Is the motion irrotational? If so, find the velocity potential. **5**

(*b*) Verify Stoke's theorem for the function $\vec{F} = x^2\hat{i} + xy\hat{j}$ integrated round the square whose sides are $x = 0, y = 0, x = a, y = a$ in the plane $z = 0$. **5**

B. Tech. [SEMESTER-I] KAS103

THEORY EXAMINATION, 2019–2020

ENGINEERING MATHEMATICS–I

(*PAPER ID: 199103*)

Time: 3 Hours *Total Marks: 100*

Note: Attempt all Sections. If require any missing data, then choose suitably.

SECTION–A

1. Attempt **all** questions. **(10 × 2 = 20)** **CO**

(*a*) Show that vectors (1, 6, 4), (0, 2, 3) and (0, 1, 2) are linearly independent. 1

(*b*) Define Lagrange's mean value theorem. 2

(*c*) If $u = x(1-y)$, $v = xy$, find $\dfrac{\partial(u, v)}{\partial(x, y)}$. 3

(*d*) Show that vector $\vec{V} = (x+3y)\hat{i} + (y-3z)\hat{j} + (x-2z)\hat{k}$ is solenoidal. 5

(*e*) Find the value of '*b*' so that rank of $A = \begin{bmatrix} 2 & 4 & 2 \\ 3 & 1 & 2 \\ 1 & 0 & b \end{bmatrix}$ is 2. 1

(*f*) Evaluate $\int_0^2 \int_0^1 (x^2 + 3y^2)\,dy\,dx$. 4

(*g*) Find grad ϕ at the point (2, 1, 3) where $\phi = x^2 + yz$. 5

(*h*) If $u = \cos^{-1}\left(\dfrac{x+y}{\sqrt{x}+\sqrt{y}}\right)$, then find the value of $x\dfrac{\partial u}{\partial x} + y\dfrac{\partial u}{\partial y}$. 3

(*i*) Find $\dfrac{du}{dt}$ if $u = x^3 + y^3$, $x = a\cos t$, $y = b\sin t$. 3

(*j*) Find the area lying between the parabola $y = 4x - x^2$ and above the line $y = x$. 4

SECTION–B

2. Attempt **any three** of the following: **(10 × 3 = 30)** **CO**

(*a*) Verify Cayley-Hamilton theorem for the matrix $A = \begin{bmatrix} 4 & 0 & 1 \\ 0 & 1 & 2 \\ 1 & 0 & 1 \end{bmatrix}$ and hence find A^{-1}. 1

(*b*) If $y = e^{m \cos^{-1} x}$, prove that $(1 - x^2)y_{n+2} - (2n + 1)xy_{n+1} - (n^2 + m^2)y_n = 0$. Hence find y_n when $x = 0$. 2

(*c*) If $u^3 + v^3 + w^3 = x + y + z$, $u^2 + v^2 + w^2 = x^3 + y^3 + z^3$ and $u + v + w = x^2 + y^2 + z^2$,

then show that $\dfrac{\partial(u, v, w)}{\partial(x, y, z)} = \dfrac{(x - y)(y - z)(z - x)}{(u - v)(v - w)(w - u)}$. 3

(*d*) Evaluate the integral by changing the order of integration: $I = \int_0^1 \int_{x^2}^{2-x} xy \, dy \, dx$. 4

(*e*) Verify Stoke's theorem for the vector field $\vec{F} = (x^2 - y^2)\hat{i} + 2xy\hat{j}$ integrated round the rectangle in the plane $z = 0$ and bounded by the lines $x = 0, y = 0, x = a, y = b$. 5

SECTION–C

3. Attempt **any one** part of the following: **(10 × 1 = 10)** **CO**

(*a*) For what values of λ and μ the system of linear equations: 1

$$x + y + z = 6$$
$$x + 2y + 5z = 10$$
$$2x + 3y + \lambda z = \mu$$

has (*i*) a unique solution (*ii*) no solution (*iii*) infinite solution.
Also find the solution for $\lambda = 2$ and $\mu = 8$.

(*b*) Find the rank of the matrix $A = \begin{bmatrix} 1 & 3 & 4 & 2 \\ 2 & -1 & 3 & 2 \\ 3 & -5 & 2 & 2 \\ 6 & -3 & 8 & 6 \end{bmatrix}$ by reducing it to normal form. 1

4. Attempt **any one** part of the following: **(10 × 1 = 10)** **CO**

(*a*) Verify the Cauchy's mean value theorem for the function e^x and e^{-x} in the interval $[a, b]$. Also show that 'c' is the arithmetic mean between a and b. 2

(*b*) Trace the curve $r^2 = a^2 \cos 2\theta$. 2

5. Attempt **any one** part of the following: **(10 × 1 = 10)** **CO**

(*a*) If $u = f(2x - 3y, 3y - 4z, 4z - 2x)$, prove that $\dfrac{1}{2}\dfrac{\partial u}{\partial x} + \dfrac{1}{3}\dfrac{\partial u}{\partial y} + \dfrac{1}{4}\dfrac{\partial u}{\partial z} = 0$. 3

(*b*) Find the volume of the largest rectangular parallelepiped that can be inscribed in the ellipsoid $\dfrac{x^2}{a^2} + \dfrac{y^2}{b^2} + \dfrac{z^2}{c^2} = 1$. 3

6.	Attempt **any one** part of the following:	**(10 × 1 = 10)**	**CO**
	(*a*) Evaluate $\iint (x+y)^2\, dx\, dy$, where R is the parallelogram in the *xy*-plane with vertices (1, 0), (3, 1), (2, 2), (0, 1) using the transformation $u = x + y$, $v = x - 2y$.		4
	(*b*) Find the volume of the region bounded by the surface $y = x^2$, $x = y^2$ and the planes $z = 0$, $z = 3$.		4
7.	Attempt **any one** part of the following:	**(10 × 1 = 10)**	**CO**
	(*a*) Verify the divergence theorem for $\vec{F} = 4xz\hat{i} - y^2\hat{j} + yz\hat{k}$ taken over the rectangular parallelepiped $0 \le x \le 1, 0 \le y \le 1, 0 \le z \le 1$.		5
	(*b*) Find the directional derivative of $\phi(x, y, z) = x^2yz + 4xz^2$ at (1, –2, 1) in the direction of $2\hat{i} - \hat{j} - 2\hat{k}$. Find also the greatest rate of increase of ϕ.		5